Encyclopedia of Mathematics and its Applications

Founding Editor G. C. Rota

All the titles listed below can be obtained from good booksellers or from Cambridge University Press. For a complete series listing visit http://publishing.cambridge.org/stm/mathematics/eom/

88. Teo Mora *Solving Polynomial Equation Systems, I*
89. Klaus Bichteler *Stochastic Integration with Jumps*
90. M. Lothaire *Algebraic Combinatorics on Words*
91. A. A. Ivanov & S. V. Shpectorov *Geometry of Sporadic Groups, 2*
92. Peter McMullen & Egon Schulte *Abstract Regular Polytopes*
93. G. Gierz *et al. Continuous Lattices and Domains*
94. Steven R. Finch *Mathematical Constants*
95. Youssef Jabri *The Mountain Pass Theorem*
96. George Gasper & Mizan Rahman *Basic Hypergeometric Series, 2nd ed.*
97. Maria Cristina Pedicchio & Walter Tholen *Categorical Foundations*
100. Enzo Olivieri & Maria Eulalia Vares *Large Deviations and Metastability*
102. R. J Wilson & L. Beineke *Topics in Algebraic Graph Theory*

Dynamic Data Assimilation
A Least Squares Approach

JOHN M. LEWIS
National Severe Storms Laboratory
and
Desert Research Institute

S. LAKSHMIVARAHAN
University of Oklahoma

SUDARSHAN DHALL
University of Oklahoma

CAMBRIDGE
UNIVERSITY PRESS

CAMBRIDGE UNIVERSITY PRESS
Cambridge, New York, Melbourne, Madrid, Cape Town, Singapore, São Paulo

Cambridge University Press
The Edinburgh Building, Cambridge CB2 2RU, UK

Published in the United States of America by Cambridge University Press, New York

www.cambridge.org
Information on this title: www.cambridge.org/9780521851558

First published 2006

Printed in the United Kingdom at the University Press, Cambridge

A catalog record for this publication is available from the British Library

ISBN-13 978-0-521-85155-8 hardback
ISBN-10 0-521-85155-6 hardback

To our teachers

They knew the place where we should be headed. As so eloquently expressed by American poet Don Lee Petersen:

They alone knew a certain address
and the map leading to it
and the place was a necessary stop
on the journey to truth (verity) and understanding
(Petersen, D. (1990). *Mentors*)

and

to our students

They demanded much from us and gave much to us. In the spirit of Hesse's master-pupil studies, '*when the student is ready the teacher will come*'. The complementarity between teacher and pupil is at the heart of all great advances in learning.

Contents

* The Appendices are available in the electronic version of this book or can be downloaded from the book's website.

Preface

What is dynamic data assimilation?

'Assimilate' is a word that conjures up a variety of meanings ranging from its use in the biological to the social to the physical sciences. In all of its uses and meanings, the word embraces the concept of incorporation. "Incorporation of what?" is central to the definition. In our case, we expand its usage by appending the words dynamic and data – where dynamic implies the use of a law or equation or set of equations, typically physical laws. Now we have dynamic law, we have data and we assimilate. It is this melding of data to law or matching of data and law or, in the spirit of the dictionary definition, incorporating data into the law that captures the meaning of our title, dynamic data assimilation.

Its modern-day usage stems from the efforts of meteorologists to estimate the 3-D state of the global atmosphere. The genesis of this effort began in the mid- to late-1960s when atmospheric general circulation models (now known as global or climate prediction models) came into prominence and the weather satellites began to collect data on a global scale. The major question surfaced: Is it possible or feasible to make long-term weather predictions, predictions on the order of weeks instead of days? Certainly a first step in such an endeavor is to estimate the atmospheric state in the global domain so that the deterministic model can use this state as initial condition and march out into the future via numerical integration of the governing equations. Over the populated continents of the world, Europe and the North American Continent, conventional data are generally sufficient to define a meaningful state. Over the less populated regions of the world, including the poles and the oceanic regions, reliance must be placed on the remotely sensed data from the satellite. Yet the data are generally incomplete, i.e., not all variables are measured. Furthermore, each data source has different error characteristics. And certainly the model has imperfections, especially in regard to processes such as turbulence and rainfall. The question looms: how best can we use information from the model (for example, a global forecast from an earlier time) and the various sources of data to produce an estimate of the atmospheric state that is better than the model or data alone. This was the question that faced meteorologists in the 1960s, and even though major advances in methods to obtain the estimate have accrued in the past several decades, the problem remains and continues to present new challenges in the face of more diverse observations and advances in computation and modeling.

In its broadest sense, dynamic data assimilation has its roots in orbital dynamics, the calculation of the orbits of the heavenly bodies. This effort began in earnest in the late seventeenth century, where the greatest stride was accomplished in 1801 by Gauss. The international competition to determine the future position of the planetoid Ceres, which had disappeared behind the sun, was the stimulus that drove Gauss to develop the method of least squares and to correctly predict Ceres's position and time of reappearance. Through the intervening years, approximately two centuries, the ideas and concepts of dynamic data assimilation have been advanced and refined and applied to essentially every discipline that relies on governing equations and data – even fields like econometrics where the laws are empirical/statistical.

In its most challenging arena, the modern-day operational weather prediction center, dynamic data assimilation demands expertise in both the mathematical tools (applied mathematics linked to numerical weather prediction) and a feeling or sense for the underlying physical processes. Researchers in the data assimilation groups at these centers generally exhibit strength in both the mathematical tools and synoptic meteorology – that component of meteorology that strives to understand the mechanism of weather systems including the system's motion, the cloud and precipitation process, and air/sea interaction. It is far too demanding to expect the reader of this textbook to master the skills necessary to operate in these most-demanding scientific environments. Nevertheless, our intention is to introduce the student to a series of simplified problems that exhibit some of the dynamics of the more complete system. We follow in the spirit of Professors Johann Burgers (University of Delft), George Platzman (University of Chicago), and Edward Lorenz (MIT), scientists who investigated prediction in the context of simplified, yet far from trivial, fluid dynamical models. Platzman laid a strong foundation for the study of truncated spectral models by thoroughly examining Burgers' equation (Platzman 1964) while Lorenz (1960, 1963, 1965) explored limits of predictability with a variety of low-order systems. Our philosophy and belief is that when the student understands the principles of dynamic data assimilation applied to the simpler yet nontrivial systems, applications to the larger dimensional system follow – not that the more complete systems don't present problems of their own, but that the fundamental components of the assimilation system are unchanged. Metaphorically, as youthful aspirants of the piano, we were admonished to refrain from trying to play a Brahms' concerto until we had exhibited proficiency on a succession of musical pieces such as "Twinkle Twinkle Little Star", "Go tell Aunt Rhody", "Maple Leaf Rag (Scott Joplin)", ...

The aim of this first year graduate level book is to distill out of the well-tested and time-honored principles and techniques for data assimilation, problems of interest in a variety of disciplines. Our goal and hope is that the student who undertakes this study will develop dexterity with the tools, but more importantly, develop the facility to identify and properly pose problems that will yield to these approaches and thereby incrementally advance the knowledge of his/her field.

Prerequisites

This book grew out of the lecture notes for a graduate class we taught four times over the past ten years (1996–2005). Over these years, this course has attracted a diverse group of graduate students drawn from Meteorology, Physics, Industrial Engineering, Petroleum and Geological Engineering and Computer Science. There was some unity in this diversity, however. These students, as part of their course work in their respective undergraduate curricula, were exposed to a similar set of courses in mathematics – two years of calculus, at least one course in differential equations, probability theory and statistics, numerical analysis and in some cases a first course in linear algebra. While everyone had some experience in computer programming (using FORTRAN or C), the depth of experience was not uniform.

While this is a respectable background, it became abundantly clear that there was a gap we needed to bridge. This gap is related to some of the advanced (graduate level) mathematical tools that are necessary to formulate and solve data assimilation problems. These tools include (but not limited to) vector spaces, matrix theory, multivariate calculus, statistical estimation theory, and theory and algorithms for optimization of multivariate functions. Since typical graduate level applied mathematics courses available to first year graduate students do not cover all of these topics in a single course, it became necessary for us to introduce these mathematical tools along with the principles and practices of data assimilation. Accordingly, we have strived to make the book self-contained by carefully developing the necessary tools side by side with their application to data assimilation.

An overview of the contents

The book has been designed in a modular fashion to accommodate a wide-spectrum audience. It consists of thirty-two chapters divided into eight parts with seven appendices. Equations, figures and tables are numbered serially within each chapter. Thus equation (I.J.K) refers to the Kth equation in Section J of Chapter I. A similar numbering scheme is used for tables and figures. Each chapter ends with two supplemental sections: Notes and References that provides pointers to the literature, and Exercises with ample hints. Many of the exercises form an integral part of the development and the reader is encouraged to spend considerable time on these exercises.

We encourage the use of computer projects as a part of classroom instruction and we recommend the use of MATLAB. Our choice of MATLAB is dictated by the relative ease of using the software and the availability of ready-to-use packages for solving matrix and optimization problems. It has an excellent graphical user interface to draw graphs, contours, and other 2-d and 3-d plots. Another important advantage is its PC/laptop base and its universal availability.

Here is a snapshot of the book's contents.

Part I Genesis of Data Assimilation (Chapters 1 through 4). In Chapter 1 we discuss the impetus for data assimilation and briefly view the components of the system. At this early stage, we offer our view of the various approaches to data assimilation – a view that hinges on the nature of governing equations (also known as constraints). Nomenclature and notation associated with this study follow. We pay particular attention to the coupling (and associated notation) between the observations and the models. Chapters 2 and 3 acquaint the student with the philosophy of data assimilation – the fundamental underpinning of the subject from both mathematical and statistical view and the nature of the problem to be solved. The applications introduced in Chapter 3 provide a test bed for various approaches and some of the applications appear as exercises in the book. Finally, Chapter 4 pays homage to pioneers of data assimilation and presents a simplified version of Gauss's problem.

Part II Data Assimilation: Deterministic/Static Models (Chapters 5 through 8). Chapter 5 develops the normal equation approach to the classical least squares problem. A geometric view of the least squares solution using projections (orthogonal and oblique) in finite dimensional vector spaces and the invariance of the least squares solution to linear transformations in both the model and observation spaces (which includes scaling as a special case) are covered in Chapter 6. A first look at the challenges of the nonlinear least squares problem using the first-order and second-order approximations is developed in Chapter 7. While these chapters deal with the off-line approach (all the data available before the estimation process begins), principles of on-line/recursive/sequential estimation (estimate updated as new data arrives) is covered in Chapter 8. Part II draws heavily upon the information in Appendices A, B, C, and D.

Part III Computational Techniques (Chapters 9 through 12). The solution to a least squares problem leads to solving a linear system with a symmetric positive definite (SPD) matrix as in the normal equation approach (Chapter 5), or to an iterative minimization problem as in the case of nonlinear least squares problem (Chapter 7). Accordingly, in Chapter 9 we provide an overview of three matrix methods – Cholesky, QR decomposition, singular value decomposition (SVD) for solving a linear system with a SPD matrix. Chapters 10 through 12 develop the three classes of optimization algorithms – steepest descent (also known as the gradient) method in Chapter 10, the classical conjugate gradient methods in Chapter 11 and Newton's and quasi-Newton family of algorithms in Chapter 12. This part draws from the information in Appendices A, B, C, and D. We encourage the use of the ubiquitous computing environ provided by MATLAB.

Part IV Statistical Estimation (Chapters 13 through 17). Since the core of data assimilation deals with estimation, it is our view that every student in this field must have an appreciation of the time-honored concepts from this theory. These are covered in Chapters 13 through 17 in Part IV. A classification of the statistical methods – least squares method of Gauss, maximum likelihood method of Fisher, Bayesian approach, and the (sequential) linear least squares estimation leading to Kalman filtering – is described in Chapter 13. Principles of statistical least squares method (which is a statistical analog of the deterministic least squares covered in Chapter 5) and the Gauss–Markov theorem are developed in Chapter 14. The principle of the maximum likelihood method is covered in Chapter 15 and Bayesian approach is developed in Chapter 16. Linear minimum variance estimation and a first look at Kalman filtering are covered in Chapter 17. This part draws from information in Appendices A, B, D, and F.

Part V Data Assimilation in Static/Stochastic Models (Chapters 18 through 21). The opening Chapter 18 is devoted to discussion of basic concepts leading to the formulation of this important class of problems of interest in geophysical sciences. Many of the known classical algorithms – including the polynomial approximation, successive correction, optimal interpolation, etc. – are described in Chapter 19. The modern view is to recast this problem as an estimation problem within the Bayesian framework and this view is pursued in Chapter 20. This framework provides a global solution. Using this framework we bring out the inherent duality between the model space and observation space approaches. Recent years have witnessed growth of interest in the use of digital filters in two related directions: first to smooth the observations over the grid using recursive versions of these filters, and second is to model the background error covariance using matrix models for implicit filter equations. An introduction to these two approaches is contained in Chapter 21. This part uses several facts from Appendices F and G.

Part VI Data Assimilation: Deterministic/Dynamic Models (Chapters 22 through 26). In Chapter 22, we introduce the "adjoint method" based on Lagrange's principle of undetermined multipliers. The computational details are first demonstrated by using simple dynamics embodied in the straight line problem. Chapter 23 develops the theory of the first-order adjoint method when the system is governed by linear dynamics. Similar developments for the nonlinear dynamics are covered in Chapter 24. Additionally, this chapter also develops the theory of first-order sensitivity analysis. Chapter 25 develops the theory of the second-order adjoint methods for computing the gradient and Hessian-vector product along with the treatment of adjoint approach to second-order sensitivity analysis. Finally, Chapter 26 develops the theory of sequential or recursive estimation techniques for estimating the state of a deterministic dynamical system and this brings out the similarities between the off-line approach described in Chapters 22 to 25 and the on-line sequential approach covered in Part VII. Part VI uses facts from Appendices B, C, and D.

Part VII Data Assimilation: Stochastic/Dynamic Models (Chapters 27 through 30). The sequential minimum variance method for estimating the state of a linear stochastic dynamical system leading to the Kalman filter algorithm is covered in Chapter 27. This is an extension of the method covered in Chapter 17 in Part IV. Various properties relating to sensitivity, divergence, stability, etc. of the Kalman filters are analyzed in Chapter 28. Chapter 29 develops the theory of nonlinear filters and methods for deriving various families of approximate filters. Chapter 30 develops the theory of computationally efficient reduced rank filters including ensemble filters. This part draws heavily upon the information in Appendices A through F.

Part VIII Predictability (Chapters 31 and 32). Predictability is a subject that assumed prominence in the late 19th century with the work of Poincaré. When dealing with dynamic data assimilation, it is crucially important for the student/researcher to understand the limits of predictability of the governing equations. Consequently, these chapters acquaint the student with the fundamental underpinning of this field of investigation. In particular, a stochastic view of predictability elaborating on the discrete counterparts of the Liouville and Kolmogorov forward equations are discussed in Chapter 31.This chapter also develops the theory of predictability based on approximate moment dynamics along with the classical Monte Carlo methods. Chapter 32 provides an overview of the deterministic view of predictability championed by Lorenz. This is largely based on the classical Lyapunov stability theory. This chapter concludes with the discussion of the deterministic ensemble approach to predictability. Part VIII uses results from Appendices A and B.

The accompanying website contains seven Appendices, A through G. An introduction to finite-dimensional vector spaces is given in Appendix A. Topics that are usually covered in a second course in matrix theory are covered in Appendix B. Concepts from multivariate calculus are reviewed in Appendix C. Characterization of the properties of optima of functions of several variables with and without constraints is given in Appendix D. An overview of the concepts relating to the definition of sensitivity of functions is given in Appendix E. Relevant concepts from Probability theory are reviewed in Appendix F. Finally, Appendix G provides a resume of concepts from the theory of Fourier transforms.

Relation to earlier work

Prior to the mid-twentieth century a student/investigator interested in dynamical data assimilation was forced to revisit classical papers by stalwarts such as Gauss, Poincaré, Wiener, and Kolmogorov. In most cases the journal articles or treatises required a solid background in applied mathematics. In the last half of the twentieth century, with the benefit of computational power there came a valuable collection of pedagogical books – Gandin (1963), Bengtsson, Ghil, and Kallen (1981), Menke (1984), Tarontola (1987), Thiébaux and Pedder (1987), Daley (1991), Parker (1994), Bennett (1992 & 2002), Wunch (1996), Enting (2002), Segers (2002), Kalnay (2003). In most cases these books are focused on a particular application, and are noteworthy for the depth of development along the lines of investigation. Our book has followed a different line of attack, dictated in part by the diverse set of students we have taught. In particular, we have encouraged the student to gain proficiency in a variety of data assimilation strategies. Further, as stated above, we have chosen to use simplified dynamical constraints in our examples and exercises – constraints that capture features of the more-realistic/real-world dynamics but are more manageable through reduced dimensionality and idealized structure. Certainly, the student will benefit from both lines of attack.

On the use of the book

This book could be used in several different ways to suit the demands of the varied groups of students interested in data assimilation. One possibility is use for a two semester course. The first course titled Mathematics for Data Assimilation, covering Parts II, III and IV along with Appendices A through D and F. The follow up course titled Methods for Data Assimilation covering Parts I, V through VIII. When such luxury is not possible, a one semester course on Dynamic Data Assimilation could focus on sections of Part I, V through VIII with occasional reference to other parts and Appendices dictated only by the mathematical maturity and preparedness of the students. We have also written the book with researchers in mind. That is, we hope it will serve as a resource book for practitioners.

A plea to the reader

We have strived to catch all the errors – both conceptual and typographical, but for such an ambitious venture, we realize it is not "bug free". We welcome your comments and identification of errors.

Salient features of the book

- A comprehensive review of the **mathematical tools** needed in data assimilation
- A self-contained introduction to **statistical estimation theory** – a basis for data assimilation
- A view of data assimilation based on model structure – **static/dynamic, deterministic/stochastic and linear/nonlinear** models
- An expansive view that includes side by side treatment of **first-order** and **second-order methods** for nonlinear problems
- A comprehensive coverage of both classical and Bayesian approach to the 3D-VAR problems
- A succinct introduction to **digital spatial filters** and their use in modeling background covariance and in smoothing spatial fields
- A comprehensive overview of the **first- and second-order adjoint** methods for 4DVAR data assimilation and sensitivity analysis
- An in-depth coverage of **Kalman filter**, **nonlinear filters**, **reduced rank** and **ensemble** filters

- A comprehensive review of methods for assessing **predictability** of dynamical models
- Discussion that promotes an appreciation for the **interaction** between theory, applications and computational issues
- Wide spectrum view of data assimilation that includes problems from **atmospheric chemistry**, **oceanography**, **astronomy**, **fluid dynamics, and meteorology**
- **Problems** of varied complexity at the end of each chapter
- Historical view of data assimilation

Acknowledgements

The impetus for this book came in fall 1995 when we began teaching a first year graduate level course on Data Assimilation at the School of Meteorology, University of Oklahoma (OU). All we had at that time was a set of handwritten notes (nearly fifty pages long) by John Lewis entitled "Adjoint Methods" that he used for a short course offered at the National Center for Atmospheric Research (NCAR) in 1990. Further development of this material, including a more expansive view of data assimilation, has resulted in the present book. A project of this magnitude spread over a decade by authors who live and work in different cities and in different time zones has been a challenge. We gladly extend our gratitude to several people who have selflessly contributed to this endeavor.

We have used portions of this book as a basis for three courses – METR 5803 "Data Assimilation" (which is a first year graduate level course) in the School of Meteorology and CS 4743/5743 Computational Sciences I (a senior/first year graduate level course) and CS 5753 Computational Sciences II (which is a Special Topics graduate level course) at the School of Computer Science at OU. We wish to express our gratitude to the School of Meteorology, especially to chairperson Fred Carr, Professors Kelvin Droegemier (who co-taught this course with us when it was offered in the School of Meteorology) and Eugenia Kalnay, and to the School of Computer Science, especially to Professor John Antonio, for the opportunity and encouragement to develop this book and teach these courses in the academic setting.

John Lewis thanks his mentor, Professor Yoshi Sasaki, who introduced him to this exciting world of variational methods in data assimilation. Furthermore, support for this effort came from staff of the National Severe Storms Laboratory and Storm Prediction Center including the following: Edwin Kessler, Robert Maddox, James Kimpel, Kevin Kelleher, David Rust, David Stensrud, and Steven Weiss. The Desert Research Institute strongly encouraged the effort, especially the Division of Atmospheric Sciences (directors Peter Barber and Kent Hoekman).

A subset of the dedicated students who took our courses provided valuable step-by-step criticism of the course material that led to an improved manuscript. Among these students are: Mark Askelson, Michael Baldwin, Li Bi, Chris Calvert, Yuh-Rong Chen, Daniel Dawson, Ren Diandong, Jili Dong, Yannong Dong, Scott Ellis,

Robert Fritchie, Sylvain Guinepain, Yaqing Gu, Mostafa El Hamly, Issac Hartley, Rafal Jabrzemski, Jeffrey Kilpatrick, Kristin Kuhlman, Tim Kwiatowski, Carrie Langston, Haixia Liu, Ning Liu, John Mewes, David Montroy, Ernani De Lima Nascimanto, Eelco Nederkoorn, Chris Porter, and Nusrat Yussouf.

The assiduous and painstaking effort that went into the review of the entire draft manuscript by Jim Purser and Andrew Lorenc has been extraordinary. The extended list of suggested revisions, which we followed faithfully, has led to an improved textbook. We also commend Tomi Vukicevic, Martin Ehrendorfer, Tom Schlatter, and Deszo Devenyi, for thorough reviews of large sections of the book. Other colleagues who supported the effort and offered valuable input are the following (listed alphabetically): John Derber, Tony Hollingsworth, Francois Le Dimet, Richard Menard, Tim Palmer, William Stockwell, Olivier Talagrand, David Wang, Yuenheng Wang, Luther White, Ming Xue, Qin Xu, Dusanka Zupenski, and the two anonymous Press-appointed reviewers – we are indebted to each of you and we shall not forget your unselfish contributions to this work.

Ms. Joan O'Bannon of the National Severe Storms Laboratory deserves a special mention for her exquisite care in drafting the figures in Chapters 2, 3, and 4 – done to professional quality. Our thanks are due to Mr. Tao Zheng and Mr. Reji Zacharia who worked tirelessly in transforming several versions of the handwritten manuscript into this present form using LaTeX.

The team effort by the editorial staff of Cambridge University Press has been noteworthy. From the earliest encouragement we received from Sally Thomas to the editorial management of the text by Ken Blake, Wendy Phillips, and David Tranah, to the exceptional copy editing by Jon Billam, the daunting task of book publication has been made bearable, and even uplifting on occasion, by this team's skill and coordination. Like the great umpires in the history of baseball, they don't interfere with the flow of the game yet they "call" the game flawlessly.

We thank members of our immediate families for their encouragement, patience, and understanding during this extended period of textbook development.

Last but not least, we credit that unique set of students with whom we have worked at doctoral and post doctoral levels. These individuals have positively impacted us and the net result is embodied in this book. Their names follow: Michael Baldwin, Jian-Wen Bao, Tony Barnston, Steve Bloom, Dave Chen, Wanglung Chung, John Derber, Nolan Doeskin, Rachel Fiedler, Lou Gidel, Tom Grayson, Peter Hildebrand, Yuki Honda, Jung Sing Jwo, Hartmut Kapitza, Dongsoo Kim, Roger Langland, Yong Li, Si-Shin Lo, Dong Lee, William Martin, Graham Mills, Lee Panetta, Seon Ki Park, Jim Purser, Chang Geun Song, Andy van Tyl, and Carl Youngblut. We thank all of you for your contributions especially to Rachel Fiedler whom we warmly remember with these words:

Our protégé Rachel Fielder (1960–2002), a joyful and exceedingly talented young woman who, in the presence of overwhelmingly unfavorable odds, fought through kidney failure to obtain her doctorate (OU 1997) and to give birth to two beautiful daughters (Lisa and Greta), before succumbing to complications from the disease.

PART I

Genesis of data assimilation

1
Synopsis

This opening chapter begins with a discussion of the **role** of dynamic data assimilation in applied sciences. After a brief review of the **models** and **observations**, we describe **four** basic forms – based on the model characteristics: **static** vs. **dynamic** and **deterministic** vs. **stochastic**. We then describe two related problems – analysis of **sensitivity** and **predictability** of the models. In the process, two basic goals are achieved: (*a*) we introduce the mathematical notation and concepts and (*b*) we provide a **top-down view** of data assimilation with pointers to various parts of the book where the basic forms and methodology for solving the problems are found.

1.1 Forecast: justification for data assimilation

It is the desire to forecast, to predict with accuracy, that demands a strategy to meld observations with model, a coupling that we call data assimilation. At the fountainhead of data assimilation is Carl Friedrich Gauss and his prediction of the reappearance of Ceres, a planetoid that disappeared behind the Sun in 1801, only to reappear a year later. And in a *tour de force*, the likes of which have rarely been seen in science, Gauss told astronomers where to point their telescopes to locate the wanderer. In the process, he introduced the method of least squares, the foundation of data assimilation. We briefly explore these historical aspects of data assimilation in Chapter 4.

Prediction came into prominence in the early seventeenth century with Johann Kepler's establishment of the three laws of planetary motion. These laws were put on a dynamical framework by Newton in the late seventeenth century, and determinism, prediction of the future state of a system dependent only on the initial state, or more precisely, on the control elements at an epoch to use the phraseology of astronomers, became the standard. Laplace became the champion of determinism or the mechanistic view of the universe. And it is this reliance of prediction on dynamical principles or dynamical laws that leads us to append the word "dynamical" to "data assimilation".

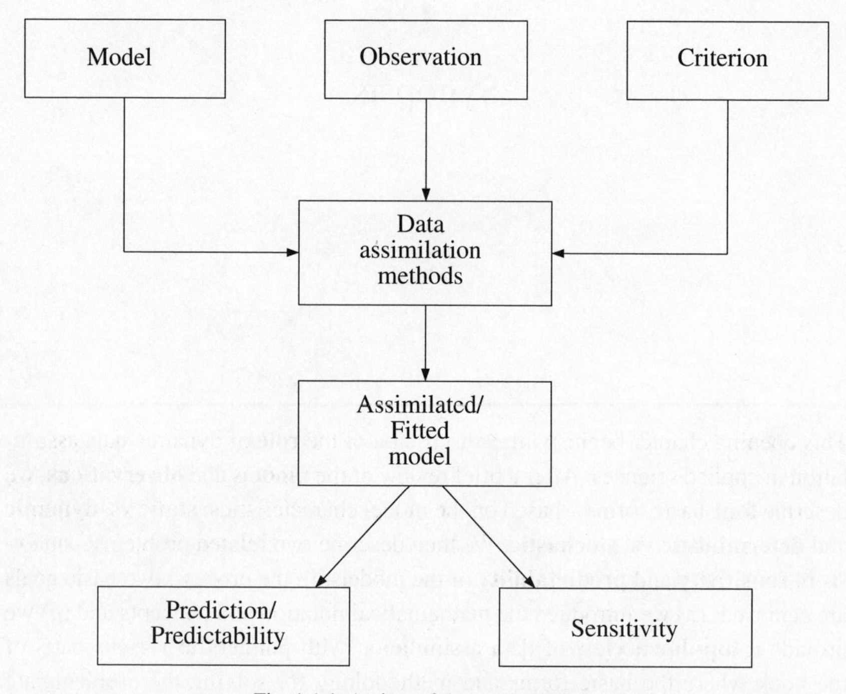

Fig. 1.1.1 A view of data assimilation.

Generally speaking, deterministic models are imperfect. The imperfection stems from an incompleteness, an inability to account for all relevant processes. What are the consequences of this incompleteness? As one might expect, the consequences can be severe or nearly inconsequential. In the case of prediction of the planetary motions in the solar system, the dynamics of two-body gravitational attraction yields excellent results. Furthermore, slight errors in the control vector (angular measurements of the planet's position at an epoch) are tolerated, i.e., the motion under the inverse-square force generally yields accuracy despite these initial inaccuracies.

Nevertheless, the incompleteness of the two-body dynamics was the source of one of astronomy's great discoveries, the discovery of the planet Neptune. In 1842, Uranus was found to considerably deviate from its expected orbit and a young Cambridge mathematician John C. Adams made calculations that led him to believe that another planet, yet unknown, was exerting an attractive force on Uranus that could account for the deviations in orbital motion. Indeed, despite initial disbelief by England's Astronomer Royal, George Airy, Adams' conjecture turned out to be correct and the massive planet Neptune, beyond Uranus, was sighted in 1846. Hoyle's book (see references) plays out the drama of discovery exquisitely, where the French mathematician LeVerrier and the German astronomer J. G. Galle are also principal participants. In this case, two-body dynamics was insufficient to explain the observed path.

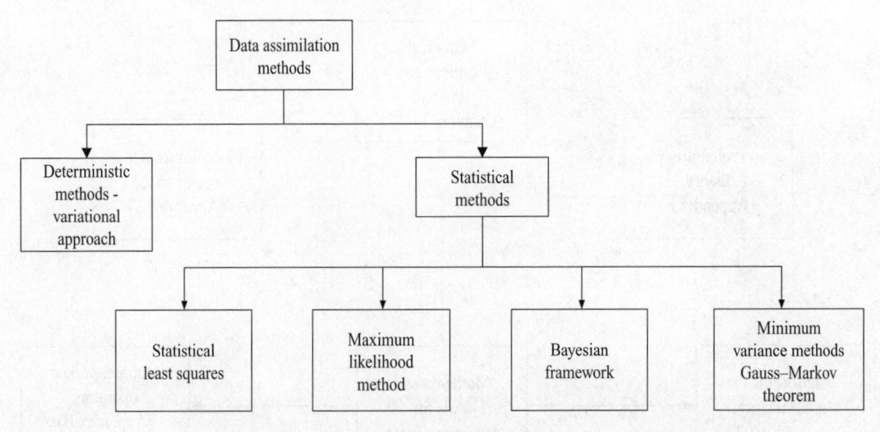

Fig. 1.1.2 A classification of data assimilation methods.

Although the two-body problem of gravitational attraction leads to accurate forecasts despite the inevitable error in the control vector, there are other dynamical systems where the small errors grow and eventually destroy the value of the forecast. The atmosphere, with its governing laws based on Newtonian dynamics and associated thermodynamic laws, is an unforgiving system, i.e., a system where the errors in the control vector/initial conditions grow with a doubling time of 2–3 days dependent on the scale of the phenomenon. These systems are labelled unstable, and in the case of the atmosphere, this instability leads to nonperiodicity – in complete opposition to the motion of the planets in our solar system.

It becomes immediately clear that when observations are used to improve forecasting, by either the specification of accurate initial conditions in the case of deterministic models, or by a process of updating the model evolution, i.e., altering the forecast state by accounting for observations of that state, the nature of the physical system must be kept utmost in mind. That is, is the model stable or unstable? If it is unstable, at what rate do the errors grow? Then, with knowledge of this growth rate, what is the prudent strategy for coupling imperfect observations with imperfect forecast? It thus becomes clear that knowledge of predictability is a fundamental component of data assimilation.

When we view data assimilation in its broadest perspective, we include three primary components as shown in Figure 1.1.1 – model, observations, and criterion. A classification of the data assimilation methods is given in Figure 1.1.2 and the mathematical tools germane to these methods are given in Figure 1.1.3. Prediction is generally the primary goal, a prediction that makes use of the optimal estimate. The dependence of the model output on the elements of the state vector (initial condition, boundary condition, and parameters), i.e., sensitivity of output to these elements, completes the macroscopic view of data assimilation.

In the remainder of Chapter 1, we elaborate on the structure of models that will be used in this course and the general relationship between model variables

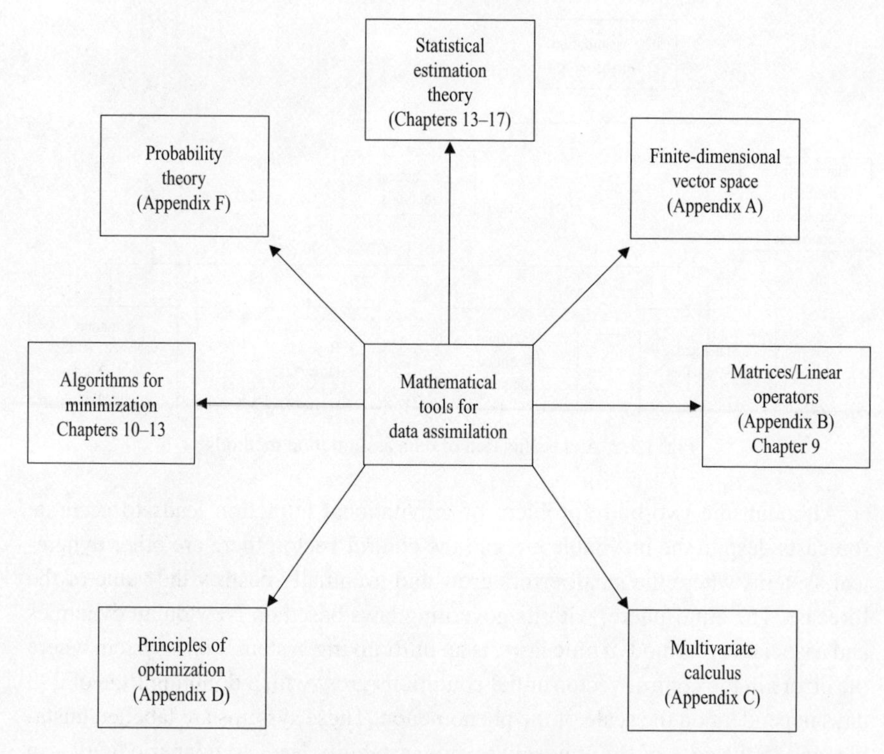

Fig. 1.1.3 Tools for data assimilation.

and observations. We are then in a position to view our primary stratification of models and associated observations based on two sets of structural criteria: (Set 1) Deterministic or Stochastic, and (Set 2) Dynamic or Static. Thus, there are four categories that are discussed separately. To complete the chapter, we argue for the inclusion of sensitivity and predictability as important components of data assimilation.

1.2 Models

Let $\mathbb{R}^n$, the ***n*-dimensional Euclidean space**, denote the **state space** of a **dynamic system** or the **model** under consideration, where we use the terms model and system interchangeably. The state space is also known as the **model space** or the **grid space** in meteorology and **phase space** in dynamic system theory and statistical mechanics. Let $x_k \in \mathbb{R}^n$ be the n-vector denoting the **state** of the system at **discrete time** $k \in \{0, 1, 2, \ldots\}$. Let $\mathbf{M} : \mathbb{R}^n \rightarrow \mathbb{R}^n$ denote a mapping of the state space into itself. That is, $\mathbf{M}(\mathbf{x}) = (M_1(\mathbf{x}), M_2(\mathbf{x}), \ldots, M_n(\mathbf{x}))^{\mathrm{T}}$ is a vector function of the vector $\mathbf{x}$ and T denotes the **transpose** (For a review of vectors and matrices refer to Appendices A and B, respectively). It is assumed that the state of the dynamic

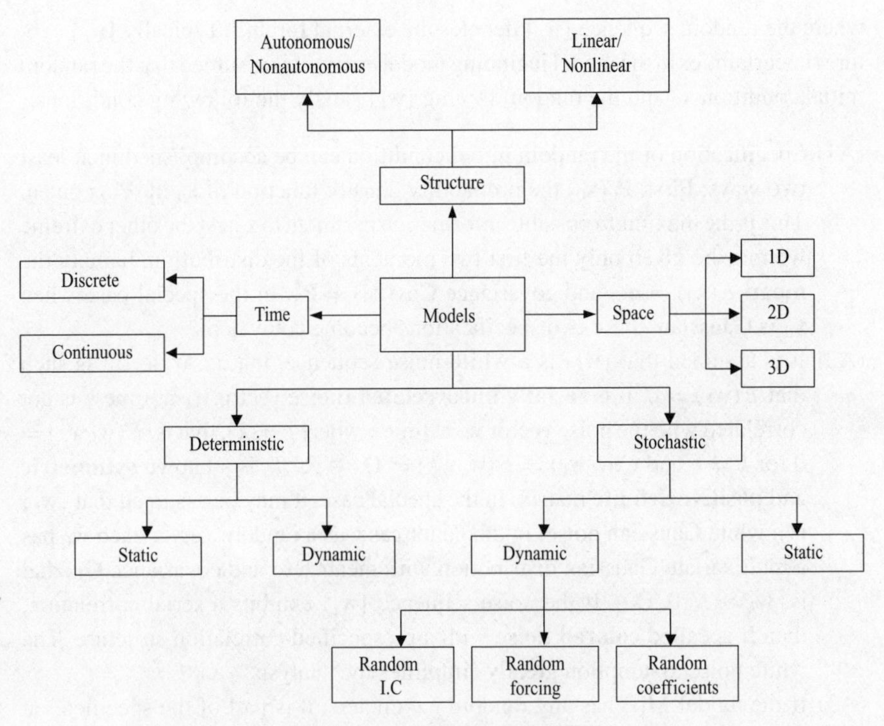

Fig. 1.2.1 A classification of models.

system evolves according to the **first-order nonlinear difference** equation

$$\mathbf{x}_{k+1} = \mathbf{M}(\mathbf{x}_k). \tag{1.2.1}$$

If the mapping $\mathbf{M}(\cdot)$ does not depend on the time index k, then (1.2.1) is called a **time-invariant** or **autonomous** system. If $\mathbf{M}(\cdot)$ also varies with time, that is $\mathbf{x}_{k+1} = \mathbf{M}_k(\mathbf{x}_k)$, then it is called a **time-varying** system. If $\mathbf{x}_{k+1} = \mathbf{M}\mathbf{x}_k$ for some $n \times n$ nonsingular matrix (that is, $\mathbf{M} \in \mathbb{R}^{n\times n}$), then (1.2.1) is called a **time-invariant linear** system. If the matrix $\mathbf{M}$ varies with time, that is, $\mathbf{x}_{k+1} = \mathbf{M}_k\mathbf{x}_k$, then it is called a **time-varying linear** or **non-autonomous** system. In the special case when $\mathbf{M}(\cdot)$ is an **identity map**, that is, $\mathbf{M}(\mathbf{x}) = \mathbf{x}$, then (1.2.1) is called a **static** system. Refer to Figure 1.2.1.

In the **deterministic** case, given $\mathbf{M}(\cdot)$ and the **initial condition** $\mathbf{x}_0$, equation (1.2.1) uniquely specifies the trajectory $\{\mathbf{x}_0, \mathbf{x}_1, \mathbf{x}_2, \ldots\}$ of the system. An immediate consequence of the uniqueness of the solution of the deterministic system is that the trajectories of the model equation (1.2.1), starting from different initial conditions, cannot intersect. **Randomness** in a model can enter in three ways: (i) random initial conditions, (ii) random forcing and (iii) random coefficients. A random or a **stochastic** model is given by

$$\mathbf{x}_{k+1} = \mathbf{M}(\mathbf{x}_k) + \mathbf{w}_{k+1} \tag{1.2.2}$$

where the random sequence $\{\mathbf{w}_k\}$ denotes the external forcing. Typically $\{\mathbf{w}_k\}$ captures uncertainties in the model including model errors. It is assumed that the random initial condition $\mathbf{x}_0$ and the random forcing $\{\mathbf{w}_k\}$ satisfy the following conditions:

(A1) Specification of the random initial condition can be accomplished in at least two ways: First, $\mathbf{P}_0(\mathbf{x}_0)$ the probability density function of $\mathbf{x}_0$ in $\mathbb{R}^n$ is given. This is the maximum possible information pertinent to $\mathbf{x}_0$. At the other extreme we may be given only the first two moments of the distribution, namely the **mean** $E(\mathbf{x}_0) = \mathbf{m}_0$ and covariance $\text{Cov}(\mathbf{x}_0) = \mathbf{P}_0$. In the special case when $\mathbf{x}_0$ is Gaussian, these two specifications become equivalent.

(A2) It is assumed that $\{\mathbf{w}_k\}$ is a **white** noise sequence, that is, $\mathbf{w}_k \in \mathbb{R}^n$ is such that $E(\mathbf{w}_k) = 0$. It is **serially uncorrelated** (noise vector $\mathbf{w}_k$ at time k is not correlated with the noise vector $\mathbf{w}_r$ at time r where $r \neq k$), that is, $E(\mathbf{w}_k \mathbf{w}_r^{\text{T}}) = 0$ for $k \neq r$ and $\text{Cov}(\mathbf{w}_k) = E(\mathbf{w}_k \mathbf{w}_k^{\text{T}}) = \mathbf{Q}_k \in \mathbb{R}^{n \times n}$, is a known **symmetric** and **positive definite** matrix. In the special case, it may be assumed that $\{\mathbf{w}_k\}$ is a white Gaussian noise. In this latter case, as a random vector each $\mathbf{w}_k$ has a multivariate Gaussian distribution with mean zero and covariance $\mathbf{Q}_k$, that is, $\mathbf{w}_k \sim N(0, \mathbf{Q}_k)$. If the noise sequence $\{\mathbf{w}_k\}$ exhibits a serial correlation, then it is called **colored noise** with a prespecified correlation structure. The white noise assumption greatly simplifies the analysis.

(A3) If the model $\mathbf{M}(\cdot)$ has any random parameters, it is part of the specification of the mapping $\mathbf{M}(\cdot)$. In the following it is assumed that $\mathbf{M}(\cdot)$ does **not** have any random parameters but it may have some fixed but unknown parameters. In this latter case, the goal of data assimilation includes estimation of these unknown parameters.

(A4) It is also assumed that the random initial condition $\mathbf{x}_0$, and the random forcing sequence $\{\mathbf{w}_k\}$ are **uncorrelated**.

Accordingly, for our purposes we arrive at four classes of models of interest as shown in Figure 1.2.2.

A number of observations concerning these classifications are in order.

(1) **Discrete vs. continuous time formulation** Model equations are usually derived by applying the laws of physics – the conservation laws of mass, energy, momentum, etc., Newton's laws of motion, laws of thermodynamics, laws of electromagnetics, and other generative and dissipative forces including absorption, emission, radiation, conduction, convection, evaporation, condensation, and turbulence. These equations by their very nature are continuous functions of time and are expressed as a system of ordinary or partial differential equations involving space and time variables. Based on these equations, one can directly formulate the dynamic data assimilation problem in continuous time. But such an approach would involve a good working knowledge of optimization in infinite dimensional space, functional analysis and calculus of variations. Training

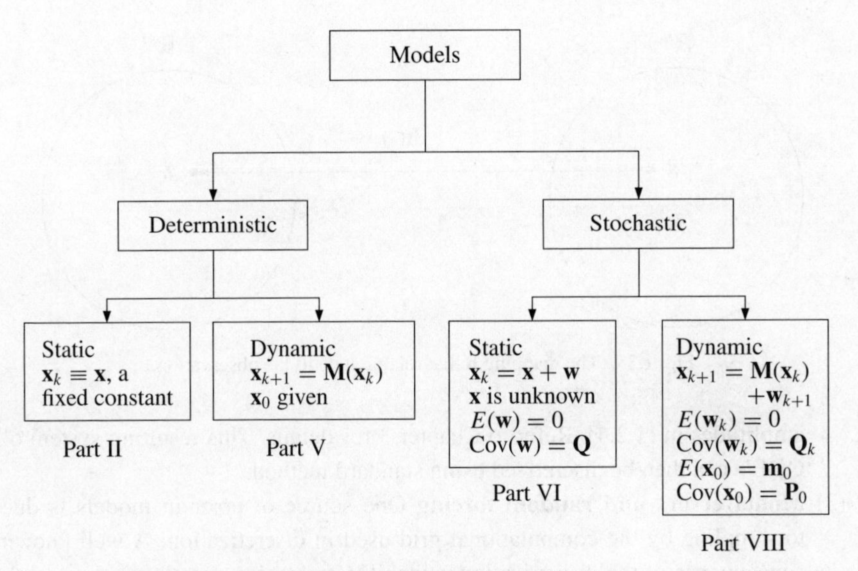

Fig. 1.2.2 Classes of models used in data assimilation including linkage to parts of the book.

in this theoretical domain is outside standard graduate curriculum in applied sciences. Discretization in time, however, gives us the luxury of converting the infinite dimensional problems to their finite dimensional counterparts. As such, finite dimensional data assimilation problems can be approached by invoking the theory of optimization in finite dimensional vector spaces and reliance on multivariate calculus. Besides, from a computational point of view, not withstanding the initial formulation, discretization is generally required to achieve the solution. Hence, in this book we adopt the discrete time formulation.

We will assume that the reader is familiar with the standard methods for converting a continuous time problem to its discrete time counterpart.

(2) **Spectral vs. (space–time) grid models** Any model described by a partial differential equation (PDE) such as the Burgers' equation

$$\frac{\partial u}{\partial t} + u\frac{\partial u}{\partial x} = 0 \tag{1.2.3}$$

where $u = u(t, x)$ can be discretized by embedding a grid in space and time leading to a gridded model. Alternatively, one may want to express the spatial variation of $u(t, x)$ in a Fourier or spectral expansion as

$$u(t, x) = \sum_{n=1}^{\infty} a_n \cos nx + \sum_{n=1}^{\infty} b_n \sin nx. \tag{1.2.4}$$

By substituting (1.2.4) into (1.2.3), we can convert the latter into a system of ordinary differential equations (ODE) that govern the evolution of the Fourier

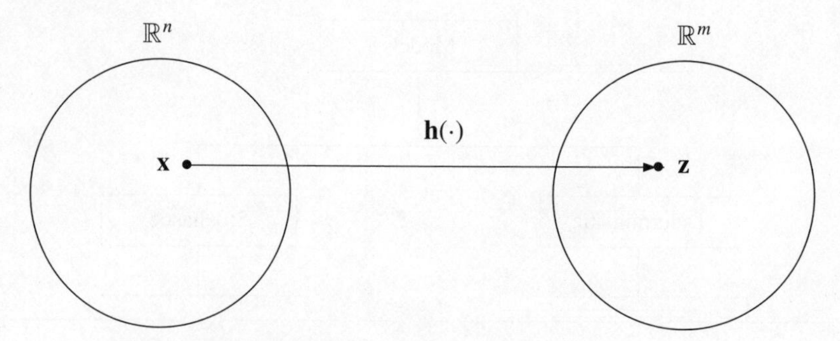

Fig. 1.3.1 The mapping **h** that relates state to the observation.

amplitudes in (1.2.4). Refer to Chapter 3 for details. This resulting system of ODE's can then be discretized using standard methods.

(3) **Model errors and random forcing** One source of error in models is due to sampling by the computational grid used in discretization. A well known consequence of the discretization is inability to resolve signals of wavelength smaller than $2\Delta x$ (The Nyquist criterion for a grid mesh of length Δx). These subgrid scale signals of smaller wavelength (or higher frequency) are often modeled by high frequency noise sequence $\{\mathbf{w}_k\}$ in (1.2.2). If these neglected subgrid scale signals do not have any temporal correlations, then we can require $\{\mathbf{w}_k\}$ to be a **white noise** sequence. Otherwise $\{\mathbf{w}_k\}$ is modeled as a **colored noise** (also called **red noise**) with a prespecified correlation structure.

Other sources of model error, generally more severe than sampling error, are the inexact specification of physical processes such as turbulence and cloud in the dynamical laws that govern atmospheric motion. In the absence of exact terms governing these processes, paramterizations of these processes are generally expressed in terms of the large-scale and better-known elements of the state vector, e.g., the large-scale gradients of wind and temperature. These parameterizations, far from perfect, often lead to systematic errors or "climate drift".

1.3 Observations

Let the m-dimensional Euclidean space, $\mathbb{R}^m$, denote the **observation space**. Let $\mathbf{z} \in \mathbb{R}^m$ denote the m-vector of observations. Let $\mathbf{h} : \mathbb{R}^n \to \mathbb{R}^m$ be a mapping from the model space, $\mathbb{R}^n$ to the observation space, $\mathbb{R}^m$, where $\mathbf{h}(\mathbf{x}) = (h_1(\mathbf{x}), h_2(\mathbf{x}), \ldots, h_m(\mathbf{x}))^{\mathrm{T}}$. Then

$$\mathbf{z} = \mathbf{h}(\mathbf{x}) \tag{1.3.1}$$

defines in general, a nonlinear relationship between the observations **z** and the state **x**. Refer to Figure 1.3.1. Typical examples of the mapping $\mathbf{h}(\cdot)$ are given in Table 1.3.

Table 1.3.1 *Examples of* $\mathbf{h}(\cdot)$ *function*

State $\mathbf{x}$	Observation $\mathbf{z}$	Function $\mathbf{h}(\cdot)$
Temperature T	Earth/atmosphere radiation measured by a satellite	Planck's law of black body radiation or Stefan's law
Rate of rainfall	Reflected energy as measured by a radar	Empirical relation between the radius of the raindrops and the reflectivity
Speed	Voltage (in cruise control)	Faraday's law

This mapping $\mathbf{h}(\cdot)$ is often derived using the physical or empirical laws governing the **sensors** used in observations. These sensors include voltmeters, pressure gauges, anemometers, antenna on radars, radiation sensors aboard satellites, to name a few. In Table 1.3.1, the first entry relates to the observations, radiation from a gas in the atmosphere at various wavelengths, and the model counterparts, weighted integrals of temperature over the depth of the atmosphere.

If $\mathbf{h}(\mathbf{x}_k) = \mathbf{H}_k\mathbf{x}_k$ for some matrix $\mathbf{H}_k \in \mathbb{R}^{m\times n}$ then (1.3.1) represents a **time-varying linear** observation system. If $\mathbf{h}(\mathbf{x}_k) = \mathbf{H}\mathbf{x}_k$ for some $\mathbf{H} \in \mathbb{R}^{m\times n}$, then (1.3.1) is a **time-invariant linear** observation system. In the geophysical literature, $\mathbf{h}(\cdot)$ is also known as the **forward operator**.

It is often the case that observations include additive errors which are often modeled by a random sequence. In such a case, (1.3.1) is modified as follows:

$$\mathbf{z}_k = \mathbf{h}(\mathbf{x}_k) + \mathbf{v}_k, \tag{1.3.2}$$

where $\mathbf{v}_k \in \mathbb{R}^m$ is a white noise sequence with

$$E(\mathbf{v}_k) = 0 \quad \text{and} \quad \mathrm{Cov}(\mathbf{v}_k) = \mathbf{R}_k \in \mathbb{R}^{m\times m}$$

and $\mathbf{R}_k$ is a real symmetric and positive definite matrix. Clearly, $\mathbf{R}_k$ relates to the quality of the sensors used in making the measurements. If we use the same set of sensors over time, then $\mathbf{R}_k \equiv \mathbf{R}$. Further, if the error in different sensors is uncorrelated, it is reasonable to assume that $\mathbf{R}$ is an $m \times m$ diagonal matrix where the non-zero diagonal entries denote the variance of the sensors.

Several comments are in order.

(1) **A condition for a well-designed observation system** First consider a linear time-invariant system defined by $\mathbf{h}(\mathbf{x}) = \mathbf{H}\mathbf{x}$ where $\mathbf{H} \in \mathbb{R}^{mm}$. Recall from Appendix B that the

$$\mathrm{Rank}(\mathbf{H}) \leq \min(m, n)\,. \tag{1.3.3}$$

If the $\mathrm{Rank}(\mathbf{H}) = \min(m, n)$, then $\mathbf{H}$ is said to be of **maximum rank**, otherwise it is called **rank deficient**. Rank deficiency indicates that the columns and/or the rows of $\mathbf{H}$ are not linearly independent. In meteorology, the recovery of atmospheric temperature from observed radiance is especially challenging because the rows/columns of the corresponding **H** matrix exhibit a lack of "strong" independence (See Section 3.8). When the rows/columns of **H** lack independence, the measurement system is **not** well conceived and is defective. In the following analysis, we assume without loss of generality that **H** is well conceived and hence is of full rank. When $\mathbf{h}(\mathbf{x})$ is nonlinear, an analogous condition requires that the Jacobian (refer to Appendix C) $\mathbf{D}_h(\mathbf{x}) \in \mathbb{R}^{m\times n}$ of $\mathbf{h}(\mathbf{x})$ is of maximum rank for all $\mathbf{x}$ along the trajectory of the model.

(2) **Representative errors** Beyond the random errors in observation, we must contend with a class of errors known as **representative** errors. This class of errors arises due to insufficient density of observations to give us an accurate portrayal of the field – its detailed variations or gradients. In meteorology, for example it is **not** unusual to have a dense set of observations on a land mass but only a sparse set of observations over the adjoining water mass (ocean or lake). In such cases, it is difficult to achieve continuity in analysis in the region that straddles the coastline. We classify such errors as "errors of representativeness".

(3) **Interpolation errors** The observation network, for example, consisting of m fixed sensors, is often fixed in time. There is typically an incompatibility between the network of grid points and the location of the observations. We do not expect coincidence between observation sites and grid points; further, the sites are generally nonuniform in distribution. This structure dictates some form of interpolation. No matter the care and sophistication that enters into the interpolation from observations to grid points or vice versa, error is introduced by this process.

1.4 Categorization of models used in data assimilation

In this section we describe segregation of data assimilation problems into several categories. Our classification is based on the model being **dynamic** or **static** and **deterministic** or **stochastic**. Accordingly, there are four types of problems that we address and we describe them below.

1.4.1 Deterministic/Static models

Let $\mathbf{x}_k \equiv \mathbf{x} \in \mathbb{R}^n$ be an **unknown** vector. Let $\mathbf{z} \in \mathbb{R}^m$ be a set of observations related to $\mathbf{x}$, where

$$\mathbf{z} = \mathbf{h}(\mathbf{x}). \tag{1.4.1}$$

Given $\mathbf{z}$ and the functional form of $\mathbf{h}(\cdot)$, our goal is to find $\mathbf{x}$ that satisfies a prespecified criterion. For simplicity in exposition, we consider two cases.

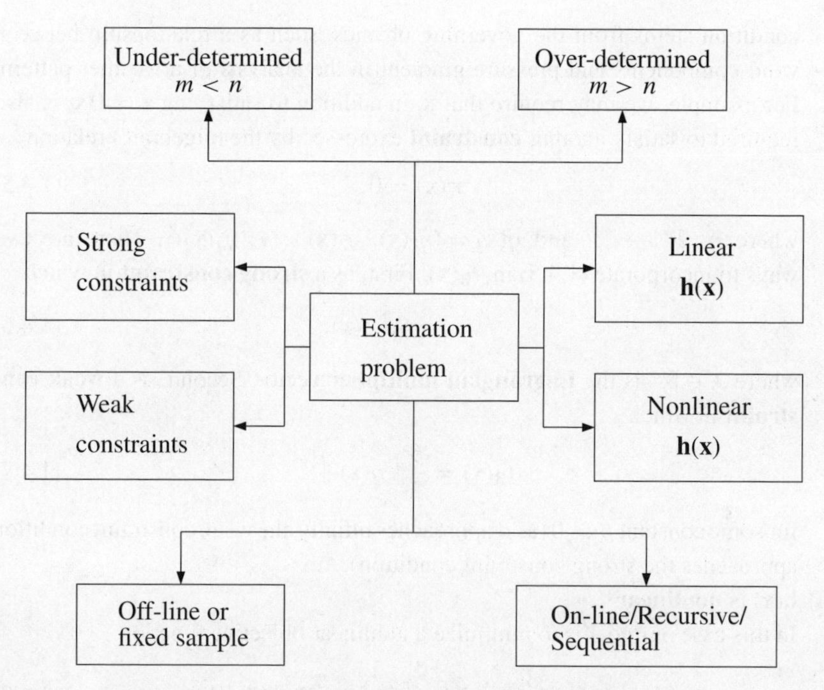

Fig. 1.4.1 A classification of the estimation problem.

(1) **$\mathbf{h}(\cdot)$ is linear**
In this case there is a matrix $\mathbf{H} \in \mathbb{R}^{m \times n}$ such that $\mathbf{h}(\mathbf{x}) = \mathbf{H}\mathbf{x}$. Depending on whether $m > n$ or $m < n$ we get an **over-determined** or an **under-determined** system, respectively. Refer to Figure 1.4.1. In the over-determined case, there is no solution to $\mathbf{z} = \mathbf{H}\mathbf{x}$ in the usual sense, and in the under-determined case there are infinitely many solutions to $\mathbf{z} = \mathbf{H}\mathbf{x}$. In the absence of a unique solution under these circumstances, the problem is reformulated by introducing a minimization condition. A functional $f : \mathbb{R}^n \to \mathbb{R}$ is introduced as follows:

$$f(\mathbf{x}) = f_r(\mathbf{x}) + f_{\mathrm{R}}(\mathbf{x}) + f_{\mathrm{B}}(\mathbf{x}) \tag{1.4.2}$$

where

$$2 f_r(\mathbf{x}) = (\mathbf{z} - \mathbf{H}\mathbf{x})^{\mathrm{T}}(\mathbf{z} - \mathbf{H}\mathbf{x}) = \| \mathbf{z} - \mathbf{H}\mathbf{x} \|^2 \tag{1.4.3}$$

is the square of the 2-**norm of the residual** $(\mathbf{z} - \mathbf{H}\mathbf{x})$ and is quadratic in $\mathbf{x}$. The term $f_{\mathrm{R}}(\mathbf{x})$ is called the **regularity condition** and typically takes the form

$$f_{\mathrm{R}}(\mathbf{x}) = \frac{\alpha}{2}\, \mathbf{x}^{\mathrm{T}}\mathbf{x} = \frac{\alpha}{2} \| \mathbf{x} \|^2 \tag{1.4.4}$$

for some real constant $\alpha \geq 0$. The addition of this term (with $\alpha > 0$) helps to provide a unified treatment of both the over-determined and under-determined cases. The last term $f_{\mathrm{B}}(\mathbf{x})$ denotes the **balance condition**. This balance condition stems from the governing physics, such as a relationship between wind components and pressure gradient in the analysis of a weather pattern. For example, we may require that $\mathbf{x}$, in addition to satisfying $\mathbf{z} = \mathbf{H}\mathbf{x}$, is also required to satisfy another **constraint** expressed by the (algebraic) relation

$$\boldsymbol{\eta}(\mathbf{x}) = 0 \tag{1.4.5}$$

where $\boldsymbol{\eta} : \mathbb{R}^n \to \mathbb{R}^q$ and $\boldsymbol{\eta}(\mathbf{x}) = (\eta_1(\mathbf{x}), \eta_2(\mathbf{x}), \ldots, \eta_q(\mathbf{x}))^{\mathrm{T}}$. There are two ways to incorporate (1.4.5) in $f_{\mathrm{B}}(\mathbf{x})$. First, as a **strong constraint** in which

$$f_{\mathrm{B}}(\mathbf{x}) = \boldsymbol{\lambda}^{\mathrm{T}}\boldsymbol{\eta}(\mathbf{x}) \tag{1.4.6}$$

where $\boldsymbol{\lambda} \in \mathbb{R}^q$ is the **Lagrangian multiplier** vector. Second, as a **weak constraint** in which

$$f_{\mathrm{B}}(\mathbf{x}) = \frac{\beta}{2} \| \boldsymbol{\eta}(\mathbf{x}) \|^2 \tag{1.4.7}$$

for some constant $\beta > 0$ (as β approaches infinity, the weak constraint condition approaches the strong constraint condition).

(2) **h(x) is nonlinear**
In this case our goal is to minimize a nonlinear objective function

$$f(\mathbf{x}) = \frac{1}{2}(\mathbf{z} - \mathbf{h}(\mathbf{x}))^{\mathrm{T}}(\mathbf{z} - \mathbf{h}(\mathbf{x})) + f_{\mathrm{B}}(\mathbf{x}) \tag{1.4.8}$$

where $f_{\mathrm{B}}(\mathbf{x})$ denotes the term arising from the balance condition.

Off-line vs. online problem

If observations spread over space and time are known a priori, then we can approach our data assimilation problem "off-line" – in short, we have a historical set of data that we treat in a collective fashion. In an "online" or sequential operation we wish to compute a "new" estimate of the unknown $\mathbf{x}$ as a function of the most recent estimate and the current observation. In this fashion, past information is accumulated as we step forward and thus the need to continually view the data set as a record spanning history is obviated. Online formulation is most useful in real-time applications.

Several observations are in order.

(1) **Direct vs. Inverse Problem** Evaluating the function $\mathbf{h}(\cdot)$ at the point $\mathbf{x}$ to compute $\mathbf{z}$ in (1.4.8) is called the **direct** or the **forward** problem. However, the problem of finding $\mathbf{x}$ given $\mathbf{z}$ is called the **inverse** problem. In the literature, terms 'data assimilation problem' and 'inverse problem' are used interchangeably.
(2) Analysis of data assimilation in the deterministic and static model context is covered in Part II. Off-line methods are covered in Chapters 5–7 and on-line or the sequential methods in Chapter 8.
(3) In the special case when $f(\mathbf{x})$ in (1.4.2) is quadratic in $\mathbf{x}$, minimization of $f(\mathbf{x})$ reduces to solving a special class of linear system with symmetric

and positive definite matrix. Methods for solving this special class of linear system is covered in Chapter 9. Otherwise, $f(\mathbf{x})$ is minimized by iterative minimization techniques. Iterative techniques rely on local approximations – **linear** or **first-order** and **quadratic** or **second-order** approximations to the function being minimized. The general iterative techniques include gradient, conjugate gradient, and quasi-Newton methods which are covered in Chapters 10–12, respectively.

1.4.2 Stochastic/Static models

Let $\mathbf{x}_k \equiv \mathbf{x} \in \mathbb{R}^n$ be an **unknown random** vector which is to be estimated based on a vector $\mathbf{z} \in \mathbb{R}^m$ of noisy observations related to $\mathbf{x}$ via

$$\mathbf{z} = \mathbf{h}(\mathbf{x}) + \mathbf{v} \tag{1.4.9}$$

where $\mathbf{h}(\mathbf{x})$ is a known function and $\mathbf{v}$ is the additive random noise vector with

$$E(\mathbf{v}) = 0 \quad \text{and} \quad \text{Cov}(\mathbf{v}) = R,$$

a known symmetric and positive definite matrix. This class of data assimilation problem relies heavily on **statistical estimation theory** (covered in Chapters 13–17). We consider two cases, one where there is no **prior** ("before") information, and one where we have the prior information. The terms **prior** and **posterior** respectively refer to availability of information **before** and **after** the estimation process. Typically, **a posteriori** probabilities are expressed in terms of **a priori** probabilities.

(1) **No prior information about x** In the absence of a priori information about $\mathbf{x}$, the problem is formulated as minimization of

$$f(\mathbf{x}) = \frac{1}{2}(\mathbf{z} - \mathbf{h}(\mathbf{x}))^{\mathrm{T}}\mathbf{R}^{-1}(\mathbf{z} - \mathbf{h}(\mathbf{x})). \tag{1.4.10}$$

(2) **Prior information is available** There are at least two different ways in which prior information about $\mathbf{x}$ can be incorporated in the estimation process: (*i*) probability density function $P(\mathbf{x})$ is known and (*ii*) only the first two moments of $\mathbf{x}$ are known.

(i) **Prior probability density is known**

Let $P(\mathbf{x}, \mathbf{z})$ denote the joint density of $\mathbf{x}$ and $\mathbf{z}$. Then, using Bayes' rule (Appendix F) we get

$$P(\mathbf{x} \mid \mathbf{z}) = \frac{P(\mathbf{z} \mid \mathbf{x})P(\mathbf{x})}{P(\mathbf{z})} = \frac{P(\mathbf{z} \mid \mathbf{x})P(\mathbf{x})}{\int_{\mathbb{R}^n} P(\mathbf{z} \mid \mathbf{x})P(\mathbf{x})\,\mathrm{d}\mathbf{x}} \tag{1.4.11}$$

where $P(\mathbf{x} \mid \mathbf{z})$ is called the **posterior** density of $\mathbf{x}$ given $\mathbf{z}$ and $P(\mathbf{x})$ is the given **prior** density of $\mathbf{x}$. It follows from (1.4.9) that $P(\mathbf{z} \mid \mathbf{x})$ depends on the distribution of the observation noise vector $\mathbf{v}$. Thus, $P(\mathbf{x} \mid \mathbf{z})$ combines the prior information in $P(\mathbf{x})$ and the information in the observation given by $P(\mathbf{z} \mid \mathbf{x})$ in a natural way. Using this Bayesian framework we can formulate a wide variety of criteria such as **minimum variance**, **maximizing a posteriori probability density**, etc. For details refer to Chapter 16.

(ii) **The first two moments of x are known**
Let $\mathbf{x_B}$ and $\mathbf{B} \in \mathbb{R}^{n\times n}$ be the mean and the covariance of $\mathbf{x}$. In meteorology $\mathbf{x_B}$ is called the **background** information and $\mathbf{B}$ is its covariance. In this case we often consider a combined objective function

$$f(\mathbf{x}) = f_b(\mathbf{x}) + f_0(\mathbf{x}) \tag{1.4.12}$$

where

$$f_b(\mathbf{x}) = \frac{1}{2}\,(\mathbf{x} - \mathbf{x_B})^{\mathrm{T}}\,\mathbf{B}^{-1}\,(\mathbf{x} - \mathbf{x_B}) \tag{1.4.13}$$

and

$$f_0(\mathbf{x}) = \frac{1}{2}(\mathbf{z} - \mathbf{h}(\mathbf{x}))^{\mathrm{T}}\,\mathbf{R}^{-1}(\mathbf{z} - \mathbf{h}(\mathbf{x})). \tag{1.4.14}$$

Clearly, $f_b(\mathbf{x})$ is the measure of the departure of the desired estimate $\mathbf{x}$ from $\mathbf{x_B}$ and $f_0(\mathbf{x})$ is the measure of the residual, the difference between the observations and the model counterparts of these observations.

The following comments are in order.

(1) These two approaches are not unrelated. When the underlying probability densities $P(\mathbf{x})$ and $P(\mathbf{z} \mid \mathbf{x})$ are Gaussian, then maximizing the a posteriori probability density (1.4.11) reduces to minimizing (1.4.12).
(2) It is often easy to obtain $\mathbf{x_B}$, the background information about $\mathbf{x}$. This can be the forecast from a previous time or climatology (the mean state based on a historical set of observations) or a combination of the two. But obtaining the background error covariance matrix $\mathbf{B}$ is often the most difficult part. In meteorology, we never know the true state of the atmosphere and this, of course, makes it difficult to precisely determine $\mathbf{B}$. Nevertheless, there are approximations to $\mathbf{B}$ obtained by a number of strategies (see references).
(3) Data assimilation for stochastic/static models is covered in Chapters 18–20 in Part IV. This type of data assimilation problem has come to be known as the **3-dimensional variational** (3DVAR) analysis in meteorology.
(4) Principles of **statistical estimation** techniques are covered in Chapters 13–17 in Part III.

1.4.3 Deterministic/Dynamic models

Let

$$\mathbf{x}_{k+1} = \mathbf{M}(\mathbf{x}_k) \tag{1.4.15}$$

be the given deterministic, dynamic model, where the initial condition $\mathbf{x}_0$ is **not** known exactly. By iteratively applying the operator **M**, we find

$$\mathbf{x}_k = \mathbf{M}^{(k)}(\mathbf{x}_0) \tag{1.4.16}$$

where the k-fold iterate $\mathbf{M}^{(k)}$ of **M** is defined by

$$\left.\begin{aligned} &\mathbf{M}^{(1)}(\mathbf{x}) = \mathbf{M}(\mathbf{x}) \\ \text{and} \quad & \\ &\mathbf{M}^{(k)}(\mathbf{x}) = \mathbf{M}^{(k-1)}(\mathbf{M}(\mathbf{x})) \end{aligned}\right\} \tag{1.4.17}$$

We are given a set of noisy observations

$$\mathbf{z}_k = \mathbf{h}(\mathbf{x}_k) + \mathbf{v}_k \tag{1.4.18}$$

where $\{\mathbf{v}_k\}$ is a white noise sequence with

$$E(\mathbf{v}_k) = 0 \quad \text{and} \quad \text{Cov}(\mathbf{v}_k) = \mathbf{R}_k, \tag{1.4.19}$$

a known real symmetric matrix for each k. Our goal is given $\mathbf{M}(\cdot)$, $\mathbf{h}(\cdot)$, $\{\mathbf{z}_k \mid k = 0, 1, \ldots, N\}$ and $\{\mathbf{R}_k \mid k = 0, 1, 2, \ldots, N\}$, find $\mathbf{x}_0$ that minimizes

$$J(\mathbf{x}_0) = \frac{1}{2}\sum_{k=0}^{N}(\mathbf{z}_k - \mathbf{h}(\mathbf{x}_k))^{\mathrm{T}}\mathbf{R}_k^{-1}(\mathbf{z}_k - \mathbf{h}(\mathbf{x}_k)) \tag{1.4.20}$$

where the states $\mathbf{x}_k$ are **constrained** to evolve according to (1.4.15). Substituting (1.4.16) into (1.4.20), we see that

$$J(\mathbf{x}_0) = \frac{1}{2}\sum_{k=0}^{N}[\mathbf{z}_k - \mathbf{h}(\mathbf{M}^{(k)}(\mathbf{x}_0))]^{\mathrm{T}}\mathbf{R}_k^{-1}[\mathbf{z}_k - \mathbf{h}(\mathbf{M}^{(k)}(\mathbf{x}_0))]. \tag{1.4.21}$$

The following comments are in order.

(1) **Observability** A dynamic system is said to be **observable** if its past state can be recovered based on future observations. For example, let $\mathbf{x}_k = \mathbf{x}_0 + \mathbf{v}k$ be the position of a particle at time k starting from the initial position $\mathbf{x}_0$ and travelling at a constant velocity **v**. Intuitively, if we only observe the velocity, $\mathbf{z} = \mathbf{v}$, it will be impossible to recover the location $\mathbf{x}_0$. But if we observe the position, $\mathbf{z}_k = \mathbf{x}_k$, then we can recover both $\mathbf{x}_0$ and **v** from two or more such measurements. Alternatively stated, observability requires that the observations contain sufficient information about the unknown such that it can be recovered.

(2) **Off-line adjoint method for the 4DVAR problem** In the literature on meteorology, this class of problems has come to be known as the 4-dimensional

variational (4DVAR) problem. Determination of the gradient of the cost function (gradient with respect to the elements of the control vector) is generally difficult when the governing dynamics is nonlinear. By finding the adjoint of the operator associated with the dynamical law, the gradient can be found in a most efficient manner. The **first-order adjoint** method is a recursive procedure for numerically computing the gradient of $J(\mathbf{x}_0)$. The **second-order adjoint method**, in addition to calculation of the gradient, also computes information about the Hessian of $J(\mathbf{x}_0)$ in the form of a Hessian-vector product. This gradient and/or Hessian information is then used in conjunction with the iterative minimization methods in Chapters 10–12 to obtain the minimizing $\mathbf{x}_0$. Adjoint method is covered in Chapters 22–25 in Part VI.

(3) **On-line or recursive least squares** On-line or recursive versions of the method for minimizing $J(\mathbf{x}_0)$ in (1.4.21) are derived in Chapter 26 of Part VI. This recursive least squares algorithm is the forerunner of the Kalman filtering algorithms covered in Part VII.

1.4.4 Stochastic/Dynamic models

Let

$$\mathbf{x}_{k+1} = \mathbf{M}(\mathbf{x}_k) + \mathbf{w}_{k+1} \tag{1.4.22}$$

be the stochastic dynamic model and let

$$\mathbf{z}_k = \mathbf{h}(\mathbf{x}_k) + \mathbf{v}_k \tag{1.4.23}$$

be the sequence of noisy observations related to the state $\mathbf{x}_k$ of the model in (1.4.22). Let $\mathbf{x}_0$ be the random initial condition. This problem is usually solved in a **sequential** or **on-line** or **recursive** framework and lies at the heart of **nonlinear filtering** theory. The general solution rests on the computation of the evolution of the conditional probability density function $\mathbf{P}_k(\mathbf{x}_k|\mathbf{z}_1, \mathbf{z}_2, \ldots, \mathbf{z}_k)$ called the **filter** density, and $\mathbf{P}_{k+1}(\mathbf{x}_{k+1}|\mathbf{z}_1, \mathbf{z}_2, \ldots, \mathbf{z}_k)$ called the **predictor** density. At the other extreme we may compute the **evolution** of the conditional mean $\widehat{\mathbf{x}}_k = E[\mathbf{x}_k|\mathbf{z}_1, \mathbf{z}_2, \ldots, \mathbf{z}_k]$ and its covariance $\widehat{\mathbf{P}}_k$. When the dynamics $\mathbf{M}(\cdot)$ is linear, $\mathbf{h}(\cdot)$ is linear and the random components $\{\mathbf{w}_k\}$, $\{\mathbf{v}_k\}$ and $\mathbf{x}_0$ are (uncorrelated) Gaussian random vectors, the solution is given by the classical **Kalman filter** equation. In the case when $\mathbf{M}(\cdot)$ and/or $\mathbf{h}(\cdot)$ is nonlinear, one can obtain a family of approximations to the dynamics of $\widehat{\mathbf{x}}_k$ and $\widehat{\mathbf{P}}_k$.

The following observations are in order.

(1) Linear and nonlinear filtering theory in discrete time is covered in Chapters 27–30 in Part VII.

(2) **Nonlinear filtering in continuous time** The theory of nonlinear filtering in continuous time is one of the most beautiful and well-understood parts of

data assimilation problems related to stochastic dynamic systems. The dynamical equation that governs the evolution of the conditional density of $\mathbf{x}_k$ given $\mathbf{z}_1, \mathbf{z}_2, \ldots, \mathbf{z}_k$ is called the **Kushner–Stratnovich** equation which is a **parabolic type partial differential equation** with stochastic forcing term related to observation. This equation is the generalization of the **Fokker–Planck** or the **Kolmogorov forward** equation that describes the forward evolution of the probability density of the states of a continuous time **Markov process**. A non-normalized version of the filter equation was derived by Zakai and is known as the **Zakai equation**. Derivation of the Kushner–Stratanovich or Zakai equations requires a good working knowledge of the **stochastic calculus** developed by K. Ito and **probability analysis over function spaces**. In the references, we refer the curious and ambitious reader to several excellent treatises on this topic.

1.5 Sensitivity analysis

Sensitivity is a pervasive term that is captured ever so eloquently in one of Blaise Pascal's Pensées, his short phrases and thoughts that were kept on loose scraps of paper:

> *A mere trifle consoles us, for a mere trifle distresses us.*
>
> *Pascal, Pensée no. 136 (Pascal 1932)*

Here he speaks of the relative ease with which the human spirit can be lifted or lowered – this spirit is changed significantly by a "trifle".

In science generally and data assimilation specifically, it is extremely valuable to know the sensitivity of a model's output to small changes in the elements of the control vector. To this end, let us describe a mathematical framework for quantifying sensitivity. Let $\Phi : \mathbb{R} \to \mathbb{R}$ be a scalar-valued function, typically the output of interest from our model. Let $\Delta\Phi(\mathbf{x}) = \Phi(\mathbf{x} + \Delta\mathbf{x}) - \Phi(\mathbf{x})$ be the induced change in $\Phi(\mathbf{x})$ resulting from a change $\Delta\mathbf{x}$ in $\mathbf{x}$. Then, the ratio $(\Delta\Phi(\mathbf{x})/\Phi(\mathbf{x}))$ is known as the **relative change** in $\Phi(\mathbf{x})$ resulting from the **relative change** $(\Delta\mathbf{x}/\mathbf{x})$ in $\mathbf{x}$. The ratio

$$S_\Phi(\mathbf{x}) = \frac{(\Delta\Phi(\mathbf{x})/\Phi(\mathbf{x}))}{(\Delta\mathbf{x}/\mathbf{x})} \tag{1.5.1}$$

is called the **first-order sensitivity coefficient** of $\Phi(\mathbf{x})$. Since $\Delta\Phi(\mathbf{x}) \approx (\mathrm{d}\Phi/\mathrm{d}\mathbf{x})\,\Delta\mathbf{x}$, to a first-order approximation, we get

$$S_\Phi(\mathbf{x}) \approx \left(\frac{\mathrm{d}\Phi}{\mathrm{d}\mathbf{x}}\right)\left(\frac{\mathbf{x}}{\Phi(\mathbf{x})}\right). \tag{1.5.2}$$

This relation is the basis for the usual claim that the derivative of a function is a measure of the first-order sensitivity. Extension of this idea to functionals and the related notion of second-order sensitivity coefficients along with illustrative examples are given in Appendix E.

In general, there are two ways to compute the sensitivity.

1.5.1 Direct method

This method is applicable when a quantity, say $\mathbf{x}^*$, is known **explicitly** as a function of the parameters with respect to which the sensitivity of $\mathbf{x}^*$ is to be computed.

For definiteness consider the data assimilation problem related to the stochastic/static model, in particular the problem of minimizing $f(\mathbf{x})$ in (1.4.12) when $\mathbf{h}(\mathbf{x})$ is linear, that is, $\mathbf{h}(\mathbf{x}) = \mathbf{Hx}$. Then,

$$f(\mathbf{x}) = \frac{1}{2}(\mathbf{z} - \mathbf{Hx})^{\mathrm{T}}\mathbf{R}^{-1}(\mathbf{z} - \mathbf{Hx}) + \frac{1}{2}(\mathbf{x} - \mathbf{x_B})^{\mathrm{T}}\mathbf{B}^{-1}(\mathbf{x} - \mathbf{x_B}). \qquad (1.5.3)$$

It can be verified (Appendix C and D) that the minimum of $f(\mathbf{x})$ in (1.5.3) is given by the solution of a linear system

$$(\mathbf{B}^{-1} + \mathbf{H}^{\mathrm{T}}\mathbf{R}^{-1}\mathbf{H})\mathbf{x}^* = \mathbf{H}^{\mathrm{T}}\mathbf{R}^{-1}\mathbf{z}$$

or

$$\mathbf{x}^* = (\mathbf{B}^{-1} + \mathbf{H}^{\mathrm{T}}\mathbf{R}^{-1}\mathbf{H})^{-1}\mathbf{H}^{\mathrm{T}}\mathbf{R}^{-1}\mathbf{z}. \qquad (1.5.4)$$

Clearly, the explicit dependence of $\mathbf{x}^*$ on $\mathbf{B}^{-1}$, $\mathbf{R}^{-1}$, $\mathbf{H}$ and $\mathbf{z}$ is known. It is most often the case that these input quantities are associated with errors stemming either from observational or measurement errors or resulting from the finite precision of the computer storage, etc. Sensitivity analysis relates to computing the induced change $\delta\mathbf{x}^*$ in $\mathbf{x}^*$ resulting from small changes in $\mathbf{B}^{-1}$, $\mathbf{R}^{-1}$, $\mathbf{H}$, and $\mathbf{z}$.

1.5.2 Adjoint method

When a quantity is only known **implicitly** as a function of the parameters with respect to which the sensitivity is desired, then adjoint method provides a mathematically elegant and efficient algorithmic framework for quantifying sensitivity.

As an example, consider the data assimilation problem in the context of deterministic/dynamic models described in Section 1.4.3. Then function $J(\mathbf{x}_0)$ in (1.4.21) is an implicit function of $\mathbf{x}_0$. In this case the adjoint method is used to compute the gradient of $J(\mathbf{x}_0)$ with respect to $\mathbf{x}_0$, i.e., $\nabla J(\mathbf{x}_0)$.

In this book direct methods for sensitivity analysis are covered in Chapters 6, 14 and 25 and adjoint methods are covered in Chapters 22–25.

1.6 Predictability

The desire to accurately predict as an impetus for data assimilation – the process of fitting models to data, was presented in the opening section of Chapter 1. In this section we provide an overview of the all-pervasive and intellectually challenging concept of **predictability** – the ability to predict and quantification of its goodness.

An old saying "the future will resemble the past and the unknown is similar to the known" provides the conceptual framework for making prediction. Accordingly, every prediction is contingent on the **information set**, $\mathcal{F}$, consisting of all the past experience and all the known facts about the phenomenon being predicted. The goodness of a prediction which is often measured by the magnitude of the difference between the predicted value and its actual realization, is a direct consequence of the quality of the information in $\mathcal{F}$. Thus, a natural starting point for making "good" prediction is to build a "good" information set.

For concreteness, we partition our discussion of predictability in accord with the basic form of the model – **deterministic** or **stochastic**.

1.6.1 Prediction using deterministic models

The work of Kepler, Newton, Gauss and others naturally led to the theory of deterministic dynamic system that was explored further by Poincaré and Birkhoff. As clearly presented by David Bohm (1957):

> The very precision of Newton's laws led, however, to new problems of a philosophical order. For, as these laws were found to be verified in wider and wider domains, the idea tended to grow that they have *universal* validity. Laplace, during the eighteenth century, was one of the first scientists to draw the full logical consequences of such an assumption. Laplace supposed that the *entire universe* consisted of nothing but bodies undergoing motions through space, motions which obeyed Newton's laws. While the forces acting between these bodies were not yet completely and accurately known in all cases, he also supposed that eventually these forces could be known with the aid of suitable experiments. This meant that once the positions and velocities of all the bodies were given at any instant of time, the future behavior of everything in the whole universe would be determined for all time.

The basic tenet of this theory is that given the error-free present state $\mathbf{x}_k$ of a dynamical system, the future states $\mathbf{x}_{k+T}$ for $T \geq 1$ are uniquely defined. In other words, a deterministic system is **perfectly predictable** under these conditions. It was further shown that a deterministic system can exhibit only one of the three modes of behavior depending on the initial condition: (*i*) **stable** behavior where the trajectories converge to one of the stable equilibria if any, (*ii*) **unstable** behavior where the trajectories diverge to infinity or (*iii*) **periodic** behavior where the trajectories converged to a well-defined limit cycle. The motion of the planets in our Solar System is an example of a deterministic system with periodic behavior, leading to accurate predictions of eclipses of Moon and Sun. Even slight errors in the "initial" or epohal conditions are forgiven.

The prevailing notion that a deterministic system is perfectly predictable was shattered in the 1960s when Edward Lorenz discovered that a deterministic system can also exhibit a new and a fourth type of **non-periodic** behavior that has come to be called **deterministic chaos**. One useful and intuitive way to understand chaos is to think of the movements of an energetic tiger trapped in a cage. Chaos is the result of unstable dynamics where the trajectories are prevented or disallowed from going to infinity. Consequently the trajectory folds back in a finite subdomain much like the agitated tiger confined to the cage. However, unlike the tiger whose gyrations likely lead to a criss-crossed path, the trajectory of a dynamical system cannot cross itself; thus, the folding trajectory quickly fills the space leading to the butterfly-like structures in the now famous Lorenz's system. The part of the phase space filled by the folding trajectory is called the **strange attractor** with **fractal** structure and **non-integer dimension**.

An immediate import of this seminal discovery is that a non-chaotic deterministic system is perfectly predictable but a chaotic deterministic system is **not**. Conditions under which deterministic systems exhibit this chaotic behavior are now well understood and this has led to a rich body of knowledge and phraseology that now inundates our culture.

Thus, while in principle we now have mathematical tools to discriminate between chaotic and non-chaotic deterministic systems, except in simple, lower-dimensional cases it is often difficult to check and verify these conditions, especially for large, complex systems of interest in geophysical sciences. Consequently, one quickly settles for **local** analysis for understanding and establishing the predictability of complex systems. In the following we describe such an idea that is routinely used in meteorological literature.

Let $\widehat{\mathbf{x}}_k$ be the optimal estimate of the state of a deterministic-dynamic system arising from data assimilation at time k. Let $\widehat{\mathbf{x}}_{k+T}$ be the unique forecast obtained from $\widehat{\mathbf{x}}_k$ using

$$\widehat{\mathbf{x}}_{k+T} = \mathbf{M}^{(T)}(\widehat{\mathbf{x}}_k) \tag{1.6.1}$$

where $\mathbf{M}^{(T)}$ is the T-fold iterate of $\mathbf{M}$ for some $T \geq 1$. Let $\mathbf{z}_{k+T}$ be the actual realization of the observation at time $(k + T)$. The forecast error is then given by

$$\mathbf{e}_{k+T} = \mathbf{z}_{k+T} - \mathbf{h}(\widehat{\mathbf{x}}_{k+T}). \tag{1.6.2}$$

If the magnitude of this forecast error is close to zero for all $T \geq 1$, then the model exhibits a faithfulness to the observations and has great value as a predictive tool. On the other hand, if the magnitude of the prediction error $\mathbf{e}_{k+T}$ grows as a function of T, then the predictive power of the model is limited in which case it is natural to ask: what is the **predictability limit** of the model?

To this end, if the prediction error is far from zero, then this error could be due to (*i*) the error in the model, (*ii*) error in $\mathbf{h}(\cdot)$ or (*iii*) due to **magnification** of the error $\mathbf{e}_k$ by the model $\mathbf{M}(\cdot)$ which is otherwise error-free. Assuming for the moment that the model $\mathbf{M}(\cdot)$ and the function $\mathbf{h}(\cdot)$ are both error-free, we now turn our attention to the way the model $\mathbf{M}(\cdot)$ processes the input error $\mathbf{e}_k$.

If the model $\mathbf{M}(\cdot)$ is **asymptotically stable** in the sense of **Lyapunov**, then any perturbation $\mathbf{e}_k$ of the input $\widehat{\mathbf{x}}_k$ will decrease in magnitude at an exponential rate, that is $\| \mathbf{e}_{k+T} \| \longrightarrow 0$ as $T \longrightarrow \infty$, for any $\mathbf{e}_k$. Proving stability of complex models is generally difficult. Thus, if $\mathbf{e}_{k+T}$ is large, then it indicates that the model may **not** be stable. In the following we describe a framework for assessing the **local stability** of models using first-order approximation.

Let $\widehat{\mathbf{x}}_{k+j}$ for $j > 0$ denote the trajectory of the model $\mathbf{M}$ starting from $\widehat{\mathbf{x}}_k$. If $\mathbf{e}_k$ is the error in $\widehat{\mathbf{x}}_k$, then the dynamics of the evolution of this error to a first-order approximation is given by the so-called **tangent linear system** (TLS)

$$\mathbf{e}_{k+1} = \mathbf{D_M}(k)\mathbf{e}_k \tag{1.6.3}$$

where $\mathbf{D_M}(k) = DM(\widehat{\mathbf{x}}_k)$ is the Jacobian of $\mathbf{M}(\cdot)$ at $\widehat{\mathbf{x}}_k$. Iterating this, we obtain

$$\mathbf{e}_{k+T} = \mathbf{D_M}(k + T - 1 : k)\mathbf{e}_k \tag{1.6.4}$$

where

$$\mathbf{D_M}(k + T - 1 : k) = \mathbf{D_M}(k + T - 1) \cdots \mathbf{D_M}(k + 1)\, \mathbf{D_M}(k) \tag{1.6.5}$$

is the product of the Jacobians along the trajectory. As a means of assessing the rate of growth of error in the time interval $[k, k + T]$, define the **Rayleigh** coefficient

$$r(k + T : k) = \frac{\mathbf{e}_{k+T}^{\mathrm{T}}\, \mathbf{e}_{k+T}}{\mathbf{e}_k^{\mathrm{T}} \mathbf{e}_k} \tag{1.6.6}$$

which is the ratio of the energy (as measured by the 2-norm) in the error at time $(k + T)$ to that at time k. Substituting (1.6.4) on the r.h.s.of (1.6.6), we get

$$r(k + T : k) = \frac{\mathbf{e}_k^{\mathrm{T}}\, [\mathbf{D}_{\mathbf{M}}^{\mathrm{T}}(k + T - 1 : k)\, \mathbf{D_M}(k + T - 1 : k)]\, \mathbf{e}_k}{\mathbf{e}_k^{\mathrm{T}}\, \mathbf{e}_k}. \tag{1.6.7}$$

Clearly, the value of this ratio is uniquely determined by the eigenvalues of the Grammian $\mathbf{A} = \mathbf{D}_{\mathbf{M}}^{\mathrm{T}}(k + T : k)\, \mathbf{D_M}(k + T : k)$. Let $(\lambda_i, \mathbf{v}_i)$ for $i = 1, 2, \ldots, n$ be the eigenvalue-eigenvector pair of this Grammian matrix $\mathbf{A}$, where, without loss of generality, let

$$\lambda_1 \geq \lambda_2 \geq \cdots \geq \lambda_r \geq \lambda_{r+1} \geq \cdots \geq \lambda_n. \tag{1.6.8}$$

It can be shown (Appendix B) that

$$\lambda_n \leq r(k + T : k) \leq \lambda_1. \tag{1.6.9}$$

Recall (Appendix B) that the system $\{\mathbf{v}_i \,|\, i = 1 \,\text{to}\, n\}$ of eigenvectors forms an orthonormal basis for the model space $\mathbb{R}^n$. Hence we can express the initial error

$\mathbf{e}_k$ as a linear combination of $\mathbf{v}_i$'s, that is,

$$\mathbf{e}_k = \sum_{i=1}^{n} a_i \, \mathbf{v}_i. \tag{1.6.10}$$

If $\mathbf{e}_{k+T}$ is much larger than $\mathbf{e}_k$ in magnitude, then it implies that there exists an index r such that

$$\lambda_r > 1 > \lambda_{r+1} \tag{1.6.11}$$

and the first r eigenvectors $\{\mathbf{v}_1, \mathbf{v}_2, \ldots, \mathbf{v}_r\}$ define the **unstable manifold**, local to the trajectory $\{\widehat{\mathbf{x}}_{k+j} \mid j \geq 0\}$ of the model starting at $\widehat{\mathbf{x}}_k$. Hence, if any one or more of the a_i's for $i = 1$ to r is non-zero in (1.6.9), $\widehat{\mathbf{e}}_{k+j}$ will grow with j at an exponential rate leading to the observed error $\mathbf{e}_{k+T}$ at time $(k+T)$. Refer to Part VIII for details.

The above analysis leads to a working definition of **predictability limit**. Given two initial conditions $\bar{\mathbf{x}}_0$ and $\mathbf{x}_0$ such that $\| \bar{\mathbf{x}}_0 - \mathbf{x}_0 \| \leq \varepsilon$ and if $k > 0$ is the **first time** at which

$$\| \bar{\mathbf{x}}_k - \mathbf{x}_k \| \geq \text{ a prespecified target} \tag{1.6.12}$$

then k is the **predictability limit of the model**.

A number of observations are in order.

(1) **Concept of analogs** Two initial conditions $\widehat{\mathbf{x}}_0$ and $\mathbf{x}_0$ separated by a fixed but a small distance ε are called **analogous** states. Condition (1.6.12) defines predictability limit as the first time when trajectories starting from two analogous states cease to be analogous.

(2) **Choice of the target in (1.6.12)** Depending on the nature and the type of applications, there is a wide latitude for choosing the appropriate target. A simple and a straightforward way to define predictability limit is to choose the target to be 2ε. Such a choice would imply that the predictability limit corresponds to the **doubling time** for the initial error. Another possibility stems from the knowledge of the covariance $\mathbf{P}_k$ of the error $\mathbf{e}_k$ in $\widehat{\mathbf{x}}_k$. In this case we can compute $\widehat{\mathbf{P}}_{k+T}$, the covariance of the forecast error $\mathbf{e}_{k+T}$ (to a first-order accuracy) by iterating the recurrence (Refer to Chapter 29)

$$\mathbf{P}_{k+1} = \mathbf{D}_{\mathbf{M}}(k)\, \mathbf{P}_k\, \mathbf{D}_{\mathbf{M}}^{\mathrm{T}}(k)$$

where $\mathbf{D}_{\mathbf{M}}(k) = \mathbf{D}_{\mathbf{M}}(\widehat{\mathbf{x}}_k)$ the Jacobian of $\mathbf{M}$ calculated along the trajectory $\widehat{\mathbf{x}}_k$ starting from $\widehat{\mathbf{x}}_0$, that is,

$$\mathbf{P}_{k+T} = \mathbf{D}_{\mathbf{M}}(k+T-1:k)\, \mathbf{P}_k\, \mathbf{D}_{\mathbf{M}}^{\mathrm{T}}(k+T-1:k). \tag{1.6.13}$$

Let $(s_1, s_2, \ldots, s_n)$ be the standard deviations which are the square roots of the diagonal elements of the matrix $\mathbf{P}_{k+T}$. Then, we can pick the threshold to be $\max_i \{s_i\}$ or the average $\overline{s}$ of the standard deviations s_1 to s_n.

1.6.2 Predictability in stochastic systems

Let

$$\mathbf{x}_{k+1} = \mathbf{M}(\mathbf{x}_k) + \mathbf{w}_{k+1} \tag{1.6.14}$$

be the stochastic-dynamic system with $\mathbf{x}_0$ being the random initial condition. Let $\mathbf{P}_0(\mathbf{x}_0)$ and $\mathbf{P}_{\mathbf{w}_k}(\mathbf{w}_k)$ be the probability density functions of the initial state and forcing noise term $\mathbf{w}_k$ in (1.6.14). Predictability in this case consists of computing the evolution of the probability density function $\mathbf{P}_k(\mathbf{x}_k)$ of the state $\mathbf{x}_k$ as a function of k. Once $\mathbf{P}_k(\mathbf{x}_k)$ is known, we can address questions such as: given an arbitrary subset $\mathbf{S}$ of the state space $\mathbb{R}^n$, what is the probability that the state $\mathbf{x}_k$ at time k will be a member of the set $\mathbf{S}$? This can be represented as

$$\text{Prob}[\mathbf{x}_k \in \mathbf{S}] = \int_{\mathbf{S}} \mathbf{P}_k(\mathbf{x}_k)\, d\mathbf{x}_k. \tag{1.6.15}$$

Several comments are in order.

(1) **Predictability in continuous time** For stochastic dynamic systems with random initial condition in continuous time (but without any data or observation after the initial time), the evolution of the probability density function of the state is given by the **Fokker–Planck** or **Kolmogorov's forward** equation. If there is no random forcing and the only randomness is through the random initial condition, then the evolution of the probability density function of the state is given by **Liouville's equation** which is a special case of Kolmogorov's forward equation.

(2) **Predictability in discrete time** Equations relating to the evolution of the probability density of the states of (1.6.14) in discrete time are developed in Chapter 29.

Notes and references

The student is encouraged to read some of the insightful treatises on the various components of data assimilation. In addition to the standard references on the subject mentioned in the front pages of this book, the following authors have stimulated our thought: Henri Poincaré (1952), Edward Lorenz (1993), Cornelius Lanczos (1970). Although Lanczos' book is general, his Chapter 2 on the foundations of variational mechanics is one of the most concise yet powerful introductions to the underpinnings of minimization principles under constraint. He is a gifted writer whose work takes on the aura of mentorship, much in the manner that Camille Jordan's *Cours d'analyse* (Jordan 1893-1896) was the impetus for G. H. Hardy to enter the field of pure mathematics (Hardy 1967).

Section 1.1 Fred Hoyle's book *Astronomy* (Hoyle 1962) gives a brief yet stimulating account of the issues and events that led to the discovery of Uranus.
Section 1.2 The fundamentals of model construction in science is a pervasive subject and the associated books tend to be restricted to a particular discipline. In atmospheric science, the book by Jacobson (2005) is highly recommended for its didactic discussion of models that include both meteorological and chemical processes. Principles of discretization of equations that govern a particular model are found in Anderson et al. (1984), Richtmyer (1957) and (1963), Richtmyer and Morton (1957), Issacson and Keller (1966), Gear (1971). Also refer to Arakawa (1966) and Richardson (1922).
Section 1.3 Observations: In addition to Poincaré's book mentioned above, the survey book by Beers (1957) gives a solid background on issues related to observations and associated analysis of errors as they enter into calculations in physics. Another source that identifies the errors of observation common to meteorology is the book on upper-air observations published by the British Meteorological Office (1961).
Section 1.4 References to estimating the background error covariance are found in Chapter 20. Nonlinear filtering techniques are covered extensively in Bucy and Joseph (1968), Jazwinski (1970), Kallianpur (1980), Krishnan (1984), Lipster and Shiryaev (1977)(1978), and Maybeck (1981)(1982).
Section 1.5–1.6 Sensitivity and predictability: An especially engaging article on these subjects, written in the style of a popular science lecture akin to those we have all heard at museums of science and industry, is Lorenz's *Atmospheric Predictability* (Lorenz 1966). Also, David Bohm's treatise (Bohm 1957) is a must read for those interested in the development of ideas in physics. In a more abbreviated fashion, Kenneth Ford (Ford 1963, Chapter 3) discusses the coupling of probability and dynamics. Finally, Einstein and Infeld (1938) present clear examples exhibiting the overlap and separateness of classic dynamics and quantum mechanics.

2
Pathways into data assimilation: illustrative examples

This chapter complements Chapter 1 by providing a **bottom-up** view of data assimilation through illustrative examples – one for each of the four classes of problems introduced there. We also include a discussion of problems associated with sensitivity and predictability. Using the standard least squares formulation, we provide a natural and intuitive interpretation of the solutions to these problems.

2.1 Least squares

The central criterion used in data assimilation is least squares. As stated earlier, it arose 200 years ago and history has bestowed simultaneity of discovery on both Gauss and Legendre. It assumes a variety of forms, but its fundamental tenet in data assimilation is minimization of the squared departure between the desired estimate and observations and/or other "background" information (typically a forecast). It was built on the foundation of variational calculus, the branch of mathematics that explores minimization of integrals – for example, integrals that express the path of quickest descent (the **brachistichrone** problem), path of least time (refraction of light), and the principle of least action. As such, there is a rich heritage of applied mathematical methods that can be brought to bear on these minimization problems.

2.2 Deterministic/Static problem

In its simplest form, the solution of a data assimilation problem underpinned by least squares reduces to averaging the observations. It is no more or no less than the "carpenter's rule of thumb": the best estimate of a length measured more than once with the same instrument is the average of the measurements. Let's put this adage in the context of a dynamical law where we choose the nonlinear advection constraint of Burgers (see Chapter 3). The governing equation for $u(x, t)$ is

$$\frac{\partial u}{\partial t} + u\frac{\partial u}{\partial x} = 0, \tag{2.2.1}$$

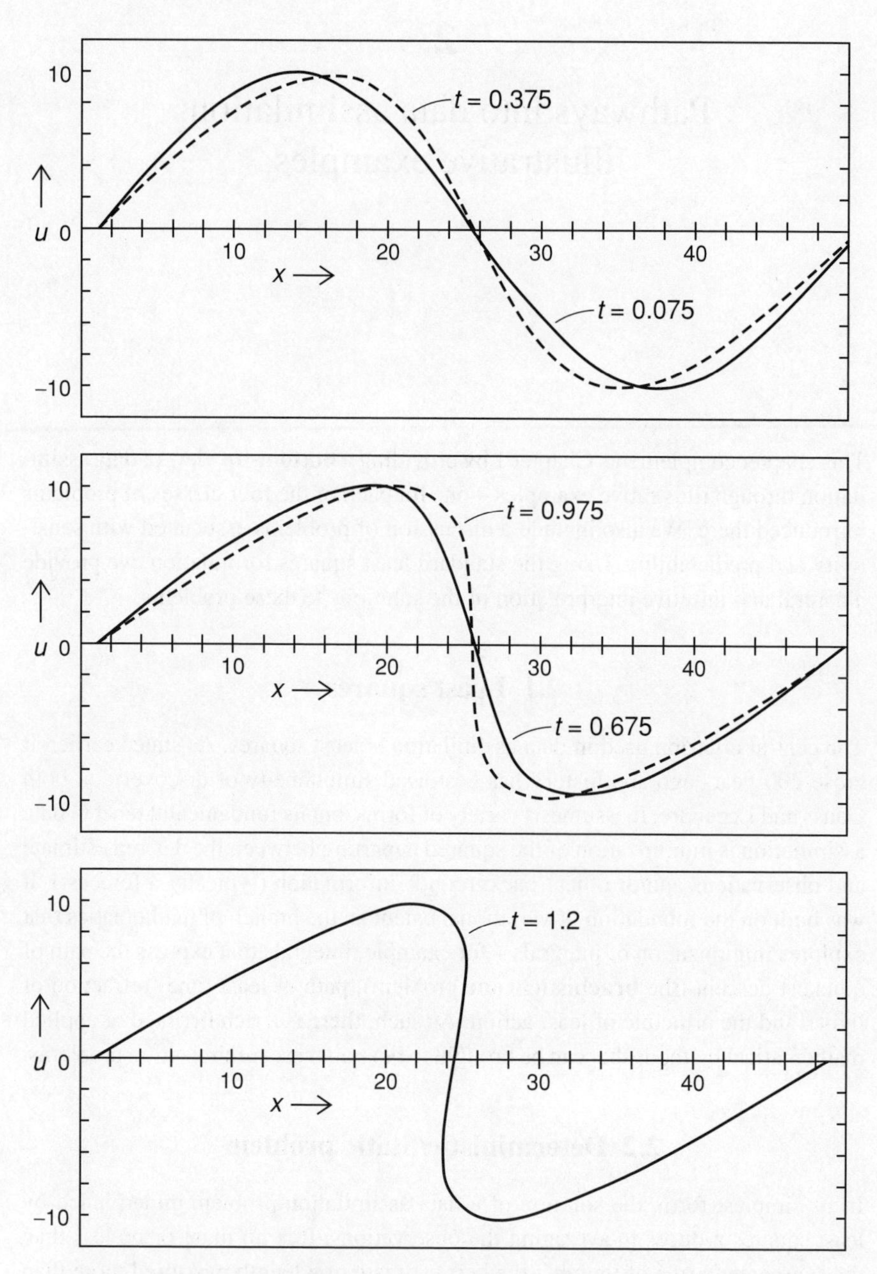

Fig. 2.2.1 Profile of the breaking wave at successive times where $u = \sin x$ at $t = 0$. The wave exhibits a multivalued nature (overlapping structure reminiscent of a wave ready to "break") at time $t > 1$.

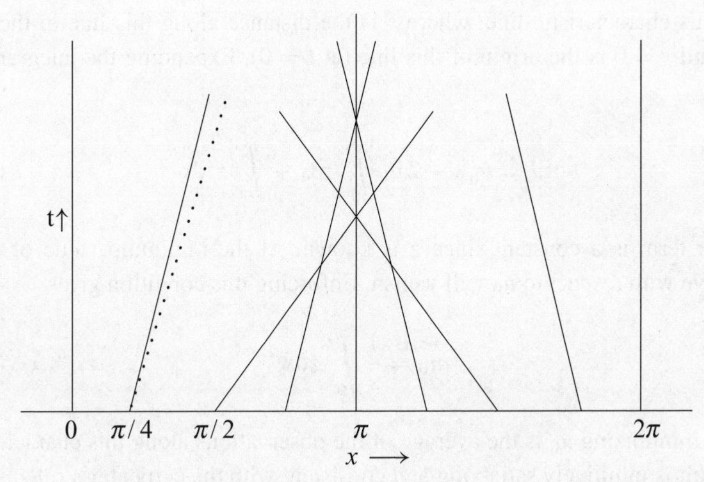

Fig. 2.2.2 A sample of characteristics for the equation (2.2.1).

where we choose $u(x, 0) = \sin x$, $0 \leq x \leq 2\pi$, and we further assume periodicity in space, $u(x \pm 2\pi, t) = u(x, t)$. The evolution of u represents a breaking wave as sketched in Figure 2.2.1.

The solution can be interpreted with the help of the characteristics associated with this equation. The characteristics are straight lines emanating from $t = 0$, where their slopes, $\mathrm{d}t/\mathrm{d}x$, are equal to $1/u(x, 0)$. In this case, the characteristics are shown by the solid lines in Figure 2.2.2.

The variable $u(x, t)$ is conserved along a characteristic line. Thus, the dynamical problem is reduced to a static problem. The solution becomes multivalued after the time of first crossing of a pair of characteristic lines – for physical meaning, we limit the time range such that the function is single valued ($t < 1$).

Assume we have an estimate of the initial condition, $u(x, 0)$, generally inexact (contaminated by noise in observations). These initial values determine the characteristics. For the sake of argument, assume that we examine the characteristic that emanates from $x = \pi/4$, and since we assume some error in the initial state, the characteristic is not the exact one, and we represent it by the dashed line in Figure 2.2.2.

Now assume we have observations of u along this characteristic, denoted by z. We would like to use these observations to get a better estimate of the initial condition. So let us pose the data assimilation problem as follows: Determine the initial value of u at $(\pi/4, 0)$. The mathematical expression for the data assimilation problem is

$$\text{minimize} \quad J(u_0) = \int_0^s (u_0 - z)^2 \mathrm{d}s \tag{2.2.2}$$

along this characteristic line, where s is the distance along this line in the (x, t) space and $s = 0$ is the origin of this line (at $t = 0$). Expanding the integrand, we get

$$J = u_0^2 s - 2 u_0 \int_0^s z \mathrm{d}s + \int_0^s z^2 \mathrm{d}s. \tag{2.2.3}$$

The last term is a constant since z is known. At the minimum value of J, the derivative with respect to u_0 will vanish. Enforcing this condition gives

$$u_0 = \frac{1}{s} \int_0^s z \mathrm{d}s, \tag{2.2.4}$$

and the minimizing u_0 is the average of the observations along this characteristic. The result is intuitively satisfying and consistent with the carpenter's rule.

The solution to this problem is not found in one step. As stated above, the values of u on the initial line contain error. And as depicted, the characteristic line emanating from a particular point on the initial line is subject to error; yet, this line is used to determine u_0. Once u_0 is determined, a new characteristic is defined and certainly the observations are not the same as those used to find u_0. It is clear that some form of iteration toward an optimal value is dictated.

2.3 Deterministic/Linear dynamics

2.3.1 Forecast errors do not grow

An interesting view of data assimilation is afforded by the following situation: you wish to find an estimate of the system state at $t = 0$ (the initial time for your forecast) by making use of observations at $t = 0$ and at earlier times denoted by $t = -T, -2T, -3T, \ldots, (-N + 1)T$, where observations have been collected at these increments of time. The system state is governed by a model that can advance the information forward, and in the case of reversible processes, the information can be moved backward in time by the dynamical law. Several interesting questions arise in this problem formulation, not the least of which is the propagation of error in the model evolution. Let us view the problem pictorially as shown in Figure 2.3.1.

We would like to combine the historical information with the analysis at $t = 0$ to get the best estimate of system state at $t = 0$. Let us denote the analysis at $t = 0$ by X_0, and the forecasts from earlier times by $X^{\mathrm{f}}_{-1}, X^{\mathrm{f}}_{-2}, \ldots, X^{\mathrm{f}}_{-N+1}$, respectively, for the forecasts from $t = -T, -2T, \ldots, (-N + 1)T$. We construct our estimate $\widehat{X}$ as follows:

$$\widehat{X} = W_0 X_0 + W_{-1} X^{\mathrm{f}}_{-1} + W_{-2} X^{\mathrm{f}}_{-2} + \cdots + W_{-N+1} X^{\mathrm{f}}_{-N+1} \tag{2.3.1}$$

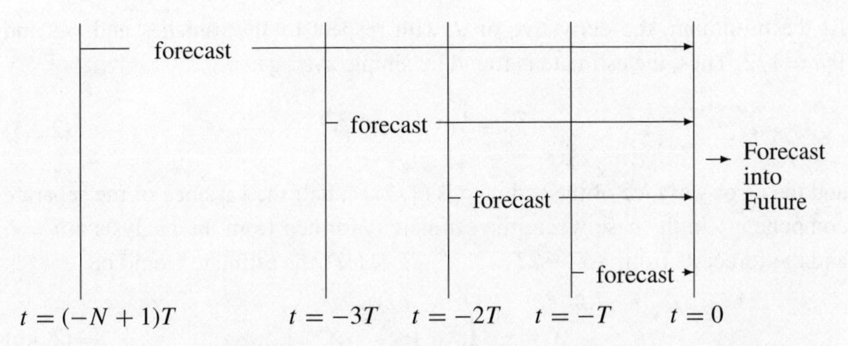

Fig. 2.3.1 A pictorial view of the information.

where the weights $W_0, W_{-1}, \ldots, W_{-N+1}$, are to be determined subject to minimization of an ensemble least squares criterion

$$J = \overline{(\widehat{X} - X_T)^2} \tag{2.3.2}$$

where $\overline{(\quad)}$ indicates average over many realizations or samples, X_T is the true state vector at $t = 0$, and the weights are normalized (to ensure unbiasedness) such that

$$W_0 + W_{-1} + W_{-2} + \cdots + W_{-N+1} = 1\,. \tag{2.3.3}$$

Let us look at the case where $N = 2$. Then

$$\widehat{X} = W_0 X_0 + W_{-1} X^{\mathrm{f}}_{-1}$$

and

$$J = \overline{(W_0 X_0 + (1 - W_0) X^{\mathrm{f}}_{-1} - X_T)^2}$$

or

$$J = \overline{[W_0(X_0 - X_T) + (1 - W_0)(X^{\mathrm{f}}_{-1} - X_T)]^2}.$$

The errors are:

$$\varepsilon_0 = X_0 - X_T$$
$$\varepsilon^{\mathrm{f}}_{-1} = X^{\mathrm{f}}_{-1} - X_T\,.$$

In terms of these errors, J becomes

$$J = \overline{(W_0\,\varepsilon_0 + (1 - W_0)\varepsilon^{\mathrm{f}}_{-1})^2}\,.$$

If the errors are uncorrelated and have zero mean, we get

$$J = W_0^2 \sigma_0^2 + (1 - W_0)^2 \sigma^2_{-1} \tag{2.3.4}$$

where σ_0^2 and σ^2_{-1} are the variances of ε_0 and $\varepsilon^{\mathrm{f}}_{-1}$, respectively. Assuming $\sigma_0^2 = \sigma^2_{-1} = \sigma^2$, we have

$$J = \sigma^2(2W_0^2 - 2W_0 + 1).$$

At the minimum, the derivative of J with respect to W_0 vanishes and we find $W_0 = 1/2$. Thus, the estimate is found by simple average,

$$\widehat{X} = \frac{1}{2}(X_0 + X^{\mathrm{f}}_{-1}), \tag{2.3.5}$$

and the error variance of the estimate is $(1/2)\,\varepsilon^2$, half the variance of the separate components. In the case where the estimate is formed from the analysis at $t = 0$ and the forecasts from $-T, -2T, \ldots, (-N+1)T$, the estimate would be

$$\widehat{X} = \frac{1}{N}\left(X_0 + \sum_{i=1}^{-N+1} X^{\mathrm{f}}_{-i}\right) \tag{2.3.6}$$

and the variance of the estimate would be ε^2/N, or an rms error of $\varepsilon/\sqrt{N}$, the well-known result from statistics; namely the error of the mean is reduced by a factor of square root of the number of members in the sample ("law of large numbers").

What has tacitly been assumed in this formulation is that the forecast error is the same no matter the duration of forecast. That is, the forecast error does not grow with time – the error at the initial time, an error related to the observational error, is propagated forward but does not grow. In this case, the least squares data assimilation again reduces to the simple averaging process.

2.3.2 Forecast errors grow/dampen

Let us now consider a case where the governing dynamics exhibits either exponential growth or decay. Most simply,

$$X(-k+1) = aX(-k) \tag{2.3.7}$$

where $X(-k)$ is the state at $t = -kT$ and where "a" is a constant, $a > 1$ signifying growth and $a < 1$ associated with decay. Assume we start with the estimate or analysis at $t = -k$, i.e., X_{-k}, and advance forward to $t = 0$, then the state at $t = 0$ is

$$X_{-k+1} = X_{-k}\, a^k. \tag{2.3.8}$$

Now, if error exists in the estimate of the state X_{-k}, i.e.,

$$X_{-k} = (X_T)_{-k} + \varepsilon_{-k}, \tag{2.3.9}$$

where $(X_T)_{-k}$ is the true state at $t = -kT$, then this error grows with time if $a > 1$ and damps with time if $a < 1$. It becomes clear that the weighting as expressed in (2.3.1) should reflect this growth/decay with time. An exercise further developing these ideas is found at the end of the chapter (Exercise 2.3). In these cases, equal weighting of the analysis and forecasts is no longer associated with optimal estimates under the least squares criterion.

2.4 Stochastic/Static problem

Let us return to the analysis of Burgers' equation that was discussed earlier (Section 2.2). We now assume that the constraint has some uncertainty represented by w, i.e.,

$$\frac{\partial u}{\partial t} + u\frac{\partial u}{\partial x} = w\,, \tag{2.4.1}$$

where we assume that this dynamical error has zero mean but possesses some error variance. In this case, u is not conserved along a characteristic line. Let us also assume that the observations along the characteristic have zero mean error and an error variance. Thus, the data assimilation problem becomes: under the assumption of **conservation** of u along the characteristic, but in the presence of assumed known error in this dynamical law, and assumed known error in the observations, find the initial value of u such that the sum of the squared departure between forecast and observation is minimized. We will weight the terms in the functional in accordance with inverse error variances. Thus, we have:

$$J(u_0) = \int_0^s \left\{ (u_0 - z)^2 \frac{1}{\sigma_z^2} + (u_0 - u^{\mathrm{f}})^2 \frac{1}{\sigma_{\mathrm{f}}^2} \right\} \mathrm{d}s \tag{2.4.2}$$

where z is the observation (with variance σ_z^2) and u^{f} is the forecast (with variance σ_{f}^2). We wish to find u_0 that will minimize $J(u_0)$ when we know z, σ_z^2, u^{f}, and σ_{f}^2. Enforcing the condition that the derivative of $J(u_0)$ with respect to the unknown u_0 vanish (at the minimum), we find

$$u_0 = \left(\frac{\sigma_z^2}{\sigma_{\mathrm{f}}^2 + \sigma_z^2} \right) u_f + \left(\frac{\sigma_{\mathrm{f}}^2}{\sigma_{\mathrm{f}}^2 + \sigma_z^2} \right) \frac{1}{s} \int_0^s z \mathrm{d}s. \tag{2.4.3}$$

If $\sigma_{\mathrm{f}}^2 \gg \sigma_z^2$, i.e., forecast error variance much greater than observation error variance, then the optimal state reduces to the average of the observations. And, if $\sigma_z^2 \gg \sigma_{\mathrm{f}}^2$, then the forecast prevails. Generally, both the forecast and observations contribute to the optimal state.

If the true state of the dynamics is given by

$$\frac{\partial u}{\partial t} + u\frac{\partial u}{\partial x} = \frac{1}{Re}\frac{\partial^2 u}{\partial x^2}\,, \tag{2.4.4}$$

then there would be a systematic error in the forecast under the constraint of strict conservation. In essence, the wave amplitude decreases with time and the multi-valued nature disappears. Even in the case when Re (Reynolds number) is large such that this term on the right-hand side is small in comparison to the terms on the left, there would still be a systematic decrease in the amplitude of u along a characteristic line. In order to address this situation, the systematic error would have to be determined and this would hinge on sufficiently good observations to

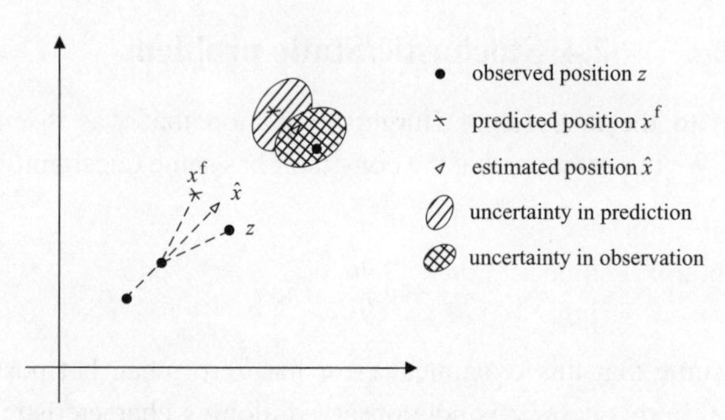

Fig. 2.5.1 Dead-reckoning: an illustration.

detect this error. Once detected and quantified, the law could be empirically altered to remove this bias. Only then would J, as formulated above, be appropriate.

2.5 Stochastic/Dynamic problem

Instead of performing data assimilation **off-line**, i.e., with previously collected data or a historical data set, it can be performed **online** or sequentially. This approach is especially appropriate for problems of in-flight correction of rocketry or tracking in real time. If the dynamical law used in the assimilation has uncertainties that can be quantified, we treat the problem stochastically – determinism coupled with probability. Ideally, there is an information base at the latest instant of time: error variance of the dynamical law, including error covariances between the elements of the state vector, and error variance of the observations. The observational error structure is generally given a priori. Modeling errors are more difficult to specify a priori, yet an estimate of these errors is required to begin the process. If during the assimilation it is found that the model prediction differs substantially from the a priori estimates, then a mechanism to make suitable adjustments sequentially is a great advantage.

To clarify the issues at hand, let us consider the radar tracking of a flying object, and for simplification, we will assume it travels in a horizontal 2-d frame. Further, let us assume that our prediction is based on the “dead-reckoning” principle – an extrapolation of an object’s position by reliance on its previous speed and direction (originally used by ship’s personnel when obscured skies or faulty instruments did not allow them to view the stars). The schematic shown in Figure 2.5.1 pictorially presents the problem.

In this diagram we are assuming that the radar’s positioning (observation) is available at every sweep of the beam (a constant time interval) and the forecast is made at these same intervals of time. As shown, the uncertainties in position through observation and forecast overlap, a Venn diagram of sorts. It is these uncertainties, generally expressed as error variances, that determine the estimate.

We demonstrate with the following strategy. Assume the estimate of the state vector X is a linear combination of the forecast and observation, i.e.,

$$\widehat{X} = w_1 X^{\mathrm{f}} + w_2 z$$

where the error variance of the forecast is σ_{f}^2 and the error variance of the observation is σ_z^2. We will also assume that the weights are normalized, i.e., $w_1 + w_2 = 1$, or restated in terms of weights W and $(1 - W)$,

$$\widehat{X} = W X^{\mathrm{f}} + (1 - W) z.$$

If X_T is the true position at the next time of observation/forecast, we can rewrite the expression as

$$\widehat{X} - X_T = W(X^{\mathrm{f}} - X_T) + (1 - W)(z - X_T)$$

or

$$\widehat{\varepsilon} = W \varepsilon^{\mathrm{f}} + (1 - W)\varepsilon_z$$

where $\widehat{\varepsilon}$, ε^{f}, and ε_z are errors associated with the estimate, the forecast, and the observation respectively, where

$$\overline{(\widehat{\varepsilon})^2} = \widehat{\sigma}^2, \quad \overline{(\varepsilon^{\mathrm{f}})^2} = \sigma_{\mathrm{f}}^2, \quad \overline{(\varepsilon_z)^2} = \sigma_z^2$$

the overbar indicating ensemble variance or variance over many trials. To find the optimal weight, we require that the error variance of the estimate be minimized, i.e., minimize

$$J = \overline{(\widehat{\varepsilon})^2} = \overline{(W \varepsilon^{\mathrm{f}} + (1 - W)\,\varepsilon_z)^2}\,.$$

If the forecast and observation errors are uncorrelated, then

$$\widehat{\sigma}^2 = W^2 \sigma_{\mathrm{f}}^2 + (1 - W)^2 \sigma_z^2\,.$$

The minimum is found at that value of W where the derivative of the error variance of the estimate vanishes

$$\frac{\partial \widehat{\sigma}^2}{\partial W} = 0 = 2W\sigma_{\mathrm{f}}^2 + 2(1 - W)(-\sigma_z^2)$$

or

$$W = \frac{\sigma_z^2}{\sigma_z^2 + \sigma_{\mathrm{f}}^2}\,.$$

Accordingly,

$$\widehat{X} = \frac{\sigma_z^2}{\sigma_z^2 + \sigma_{\mathrm{f}}^2} X^f + \frac{\sigma_{\mathrm{f}}^2}{\sigma_z^2 + \sigma_{\mathrm{f}}^2}\, z\,,$$

a result consistent with the example studied in Section 2.4. Again, if $\sigma_{\mathrm{f}}^2 \gg \sigma_z^2$, $\widehat{X}$ reduces to the observation, and if $\sigma_z^2 \gg \sigma_{\mathrm{f}}^2$, $\widehat{X}$ reduces to the forecast.

2.6 An intuitive view of least squares adjustment

Assume we have a model governing the movement of mid-latitude transitory weather systems – one such model is the conservation of vorticity. We further assume that hemispheric observations of this vorticity are available at two times (typically 12 hours apart). Let us take the "strong constraint" approach where the governing law is assumed to be perfect, but where the observations are assumed to contain error. The data assimilation problem is stated as follows:

> Under the exact constraint of vorticity conservation, obtain estimates of the vorticity at each time satisfying the constraint while minimizing the squared difference between this state and the observations.

Before we intuitively discuss the solution , let us view the constraint pictorially. Figure 2.6.1 (top) shows the circulation of air (the horizontal motion) around the northern hemisphere for a period in late winter (March 1988). This is called the "steering current" and it is a large-scale flow that is free from short wavelength features (often achieved by averaging the flow in both space and time). The vorticity conservation constraint is typically applied at a mid-tropospheric level such as the 500 mb level ($\sim$ 5.5 km above sea level). This is the case for the flow that is shown. The speed of flow is inversely proportional to the spacing of the heavy contour lines – as if the air were flowing in channels or conduits where the speed increases as the channel narrows. This steering current moves the vorticity pattern along the streamline.

Again, in the top figure, you will find a sequence of dots running from the northwestern USA down through the southern states and tracking to the east coast. These dots identify the successive positions of the center of one disturbance (a positive vorticity center); the positions of the center are shown at 12 h intervals – 10 intervals representing the movement over a 5-d period (from west to east). The thin dotted contours represent the vorticity distribution associated with the disturbance at the initial time ($t = 0$) and 3 days later. We assume these distributions (observations) are given at each time, (each 12 h increment).

The vorticity is a measure of the circulation associated with the disturbances. The vorticity is not directly observed, rather it is found from the pressure field (on constant height surfaces) or from the height field (or constant pressure surfaces). This form of vorticity is called the geostrophic vorticity which is a good approximation in mid-latitudes. The horizontal wind field exhibits this circulation most fundamentally as shown in the panel in the lower-left corner of Figure 2.6.1. Here we display the observed wind at 1–2 km above sea level at the time when the center of the disturbance is at point 7 (7th dot in the sequence). This circulation is usually in evidence at all levels in the troposphere for a given disturbance. The wind observations are found from instrumented balloons and the wind symbol (a horizontal wind) is directed along the line from tail ("feathers")

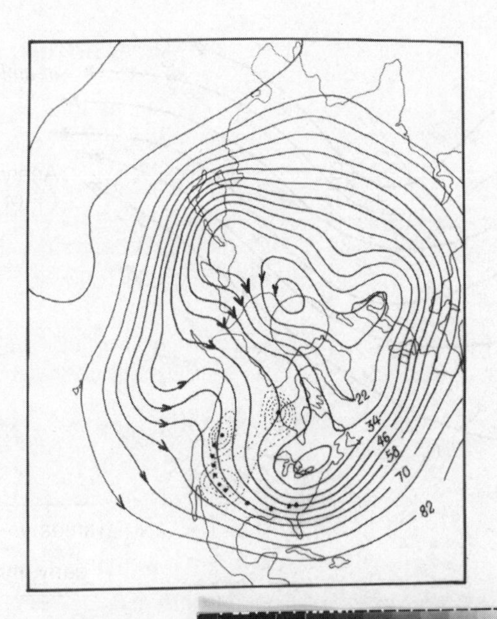

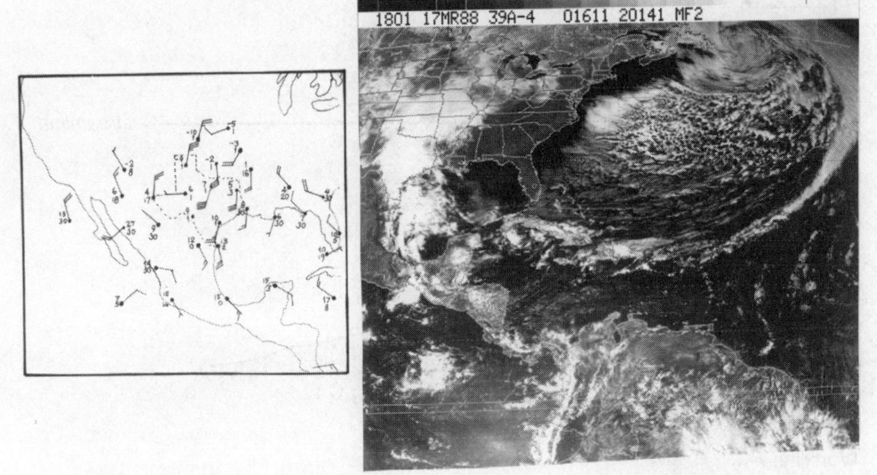

Fig. 2.6.1 Top panel shows average streamlines (geopotential lines) at the 500 mb level for the period 14 Mar 1988–19 Mar 1988. The numerals on the isolines represent the height of the surface in meters where the leading "5" and trailing "0" have been deleted, i.e., 82 represents 5820 m. The panel on lower left displays upper-air wind and temperature observations at the 850 mb level ($\sim$ 1500 m) on 17 March (1200UTC) where the wind direction is from "tail" (feathers) to "head" and a full barb represents 10 knots. The temperature (°C) is upper left on the station model and dew point depression (°C) is below it. The panel on the lower right is the visible satellite imagery from 17 March at 1800UTC.

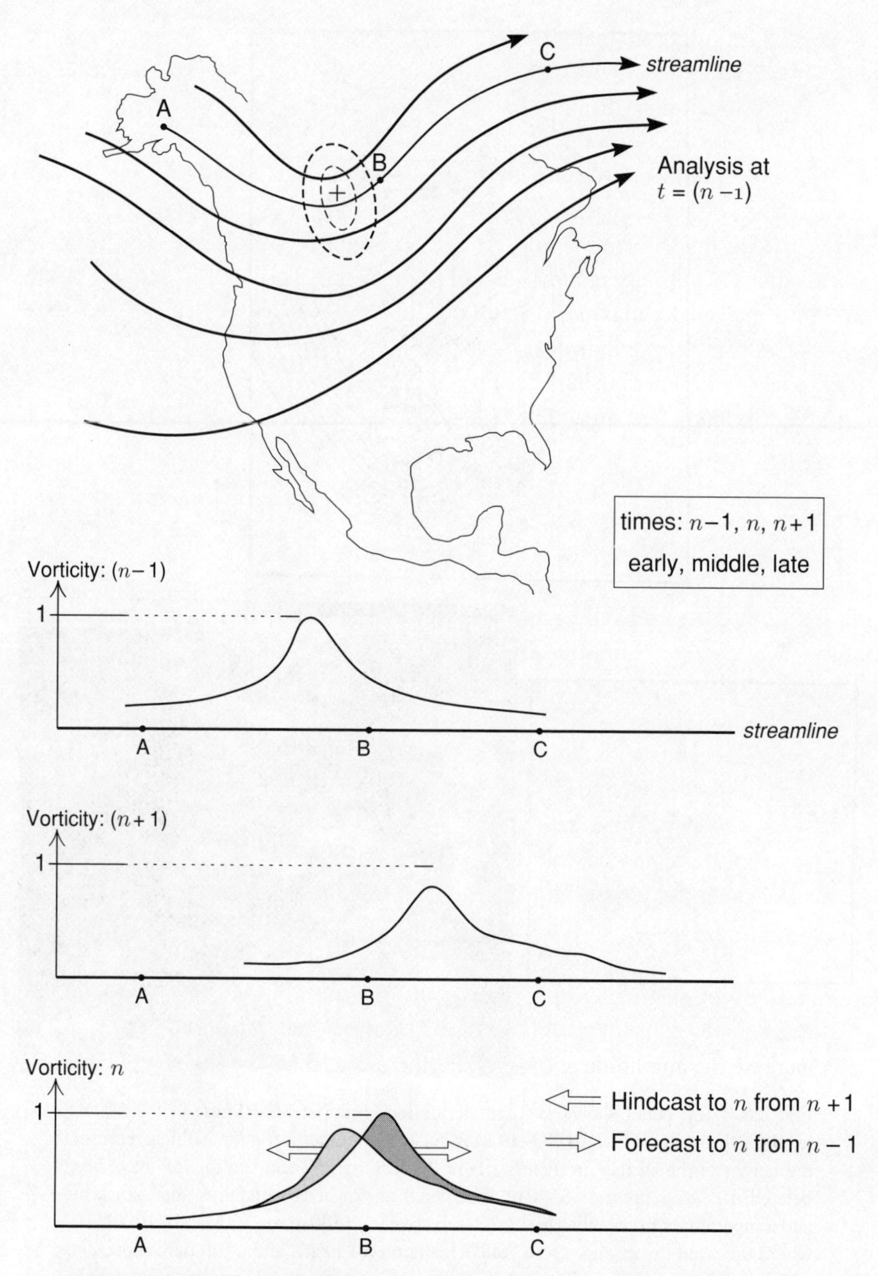

Fig. 2.6.2 Top panel shows idealized streamlines at a mid-tropospheric level where the dashed elliptical lines represent a perturbation or disturbance in the large-scale flow. The lower three panels exhibit the vorticity pattern at various times along the streamline labelled ABC.

to head. Each full barb (feather) represents a speed of 10 knots. We note a distinct cyclonic (counter-clockwise) circulation over Texas–Oklahoma at this level. The associated cloud/rain is displayed in the panel in the lower right. Air is being drawn northward from the Gulf of Mexico into the midwestern states.

The constraint can be applied to each streamline separately when we use this form of the vorticity equation. Thus, in Figure 2.6.2, we focus on the streamline $\overline{ABC}$. Along this streamline at times $(n-1)$ and $(n+1)$ [12 h apart], we have observations of vorticity plotted schematically on the 2nd and 3rd tier of the figure. We have indicated a maximum amplitude of 1 at $(n-1)$ and somewhat lower at $(n+1)$. Also, we have indicated that the pattern moves from left to right (west to east) over the 12 h period. The mathematical equation describing the advection of vorticity (ζ) along this streamline is

$$\frac{\partial \zeta}{\partial t} + V \frac{\partial \zeta}{\partial s} = 0$$

where t is time, s is curvilinear distance along the streamline and V, a function of s, is the speed of movement of the pattern (the steering current). If V were constant, the pattern would remain unchanged while it moved downstream, but since V is variable along the streamline, the spatial structure of the pattern can change – exhibiting either a compression or expansion.

One way to test the constraint is to compare the forecast of vorticity from $(n-1)$ to n [a 6 h forecast] with the hindcast from $(n+1)$ to n [a 6 h hindcast]. This is another form of the equation for the constraint. The forecast and hindcast to time-level n are shown on the lowest (4th) tier of the figure.

If the constraint and observations were without error, the forecasted and hindcasted vorticity patterns at n would match or coincide. As can be seen in our schematic, there is a mismatch where the maximum amplitudes are displaced relative to one another indicating that the separation of the features is inconsistent with the speed of the steering current. Intuitively, to minimize the degree of mismatch, the least squares adjustment will decrease the amplitude of the vorticity at $(n-1)$ and increase the amplitude at $(n+1)$. Furthermore, since the steering current cannot be adjusted (it is assumed to be given), the phase mismatch will be ameliorated by shifting the patterns further upstream and downstream at $(n-1)$ and $(n+1)$, respectively.

2.7 Sensitivity

The issue of sensitivity can enter data assimilation in a variety of ways. Prior to performing data assimilation, it is often advisable to determine the sensitivity of model output to the elements of the control vector (initial conditions, boundary

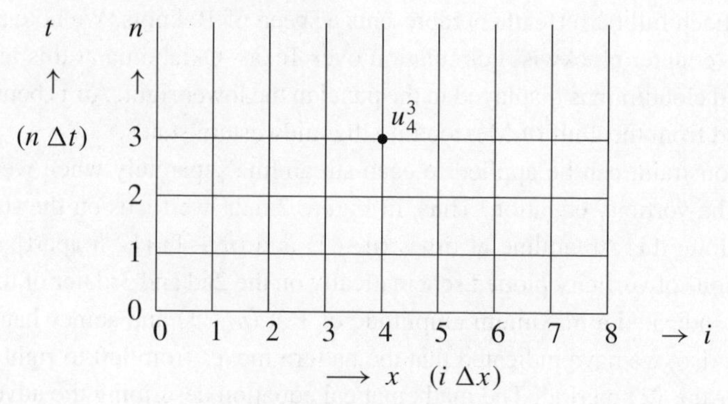

Fig. 2.7.1 The 2-d grid.

conditions, physical parameters, a priori estimates of observational errors, interpolation algorithm, ...). Typically, the scientist is interested in a particular output of the model, e.g., the forecast of rainfall (possibly the accumulated rainfall) over a particular region and over a particular time period. And if data assimilation is one of the goals of the research – e.g., estimates of the system state – then it is instructive to determine the sensitivity of the output of interest to the various elements of the control vector. This knowledge allows one to speculate, at least make educated guesses, on the relative importance of the variables on the forecast aspect. For example, it may be found that the water vapor field below 1 km and south of the rain area is extremely important to the accumulated rain forecast. This targeted area and targeted variable would dictate an assimilation strategy that aimed at precise analysis of vapor in that region.

The principal idea of sensitivity analysis can be presented with the following example. Advection of a property $u(x, t) = u(i\,\Delta x, n\,\Delta t) \equiv u_i^n$ is governed by the following difference equation:

$$u_i^{n+1} = u_i^n - \frac{1}{2}(u_{i+1}^n - u_{i-1}^n) \tag{2.7.1}$$

where we have conveniently chosen the grid spacing (Δx and Δt) and speed of propagation (c) such that $c\Delta t/\Delta x = 1$. Let us consider the grid in Figure 2.7.1. We ask the question: What is the sensitivity of the forecast at $(i, n) = (4, 3)$, i.e., u_4^3, to the initial conditions (u_i^0)? There are several ways to find the answer to this question, but we follow a path that is related to "backward forecasting", i.e., "hindcasting". As will be seen in the body of this text, the strategy of working backwards to find sensitivity (derivatives of the output with respect to elements of the control vector) is the basis of adjoint method. Referring to Figure 2.7.2, we can assign the rational numbers ("weights") to the grid as shown. The meaning is clear – multiplication of the variables at the particular grid points at time $n = 2$ by the "weights" yields the forecast of u_4^3 (at time $n = 3$). Working further backward, i.e., from the variables at $n = 2$ that influence u_4^3, we obtain the weights shown in Figure 2.7.3. That is,

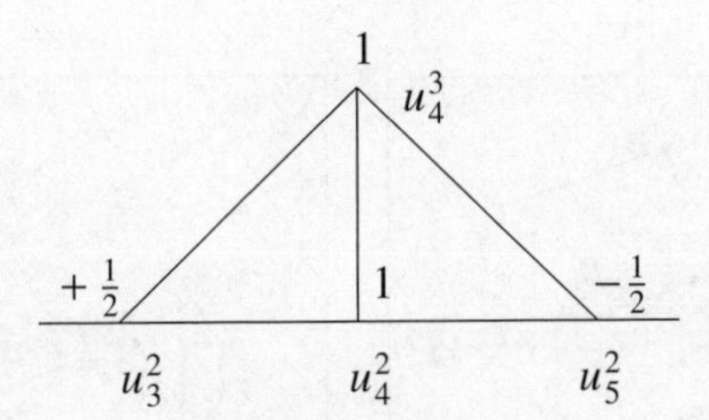

Fig. 2.7.2 The stencil defined by (2.7.1).

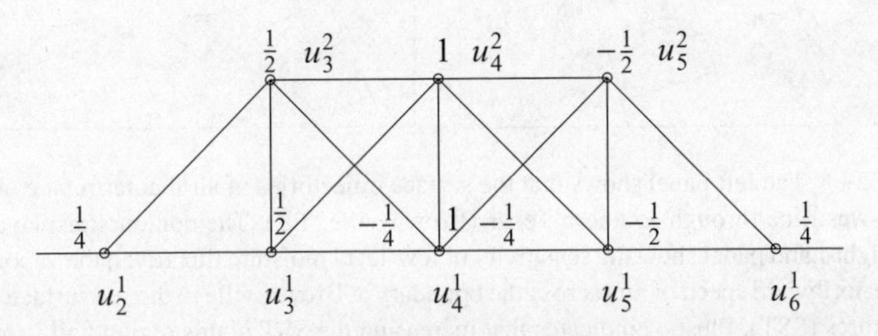

Fig. 2.7.3 An illustration of the backward analysis.

a forecast of $\frac{1}{2}u_3^2$ is given by $\frac{1}{4}u_2^1 + \frac{1}{2}u_3^1 - \frac{1}{4}u_4^1$. And, as can now be seen, working from these grid points at $n = 1$ to the points on the initial line ($n = 0$), the resulting rational numbers will give the sensitivity of the forecast u_4^3 to the initial values of u_i^0, $i = 1, 2, 3, \ldots, 7$. These rational numbers on the initial line are $\partial u_4^3/\partial u_i^0$, $i = 1, 2, 3, \ldots, 7$ – the measure of sensitivity.

Although this example is based on a simple linear difference equation, the same concept is used on complicated models. The backward integration can be performed in a manner analogous to the forward integration and the methodology works equally well with nonlinear systems.

An example of one sensitivity study with a realistic atmospheric prediction model is the following.

The moisture flux (horizontal transport of water vapor) into the northern coastal plain of the Gulf of Mexico was the output of interest. It was found that the 48-h forecast of this flux into Texas was crucially dependent on the sea surface temperatures (SSTs) along the paths of wind at low levels. These low-level trajectories and the associated sensitivity of the flux to the SSTs is shown in the Figure 2.7.4. As might be expected intuitively, the forecast of this flux was virtually insensitive to the SSTs in the eastern Gulf – far from the inflow along the coast of Texas. The implication is that an accurate forecast of moisture flux into Texas will depend on a swath of SSTs (boundary conditions) over the Gulf. The data assimilation strategy should aim at obtaining very good estimates of these SSTs.

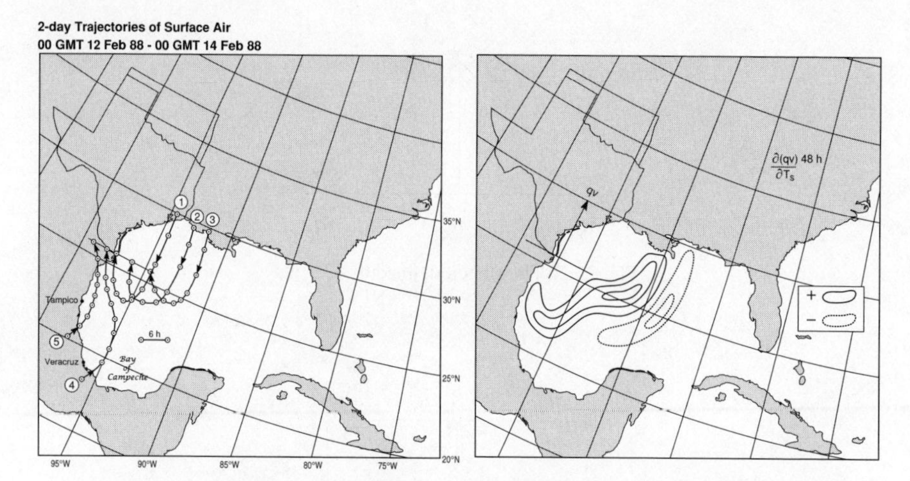

Fig. 2.7.4 The left panel shows that the surface trajectories of air that terminate on the east–west line through southern Texas (Brownsville, TX). The contours displayed in the right-hand panel show the sensitivity of low-level moisture flux (qv: q the vapor and v the northward speed of air) across the boundary at Brownsville to the sea surface temperatures (SST). Plus(+) indicates that increasing the SST in this region will increase the northward moisture flux across the boundary, whereas a minus(−) indicates that decreasing the SST in this region will lead to a decrease in the flux.

2.8 Predictability

In Section 1.1 we have made it clear that knowledge of error growth in models is critically important to data assimilation. Although predictability includes the analysis of error growth in models, its conceptual foundation is broader and rests on the tenets of dynamical systems, a subject pioneered by Poincaré, Lyapunov and Birkhoff among others. The stability of the system is central to understanding predictability. A physical system is said to be asymptotically stable if after it is perturbed, it will return to the undisturbed state. On the other hand, if the perturbation grows and the system does not return to the unperturbed state, it is classified as unstable. One of the simplest yet profound early studies in this direction was conducted by Lewis Richardson and Henry Stommel (Richardson and Stommel (1948)). Their paper begins with the unusual and eye-catching phrase:

> We have observed the relative motion of two floating pieces of parsnip, and have repeated the observation for many such pairs of different initial separations.

Parsnips (about 2 cm in diameter) were used because they were easily visible, and because they were almost completely immersed and thus free from the wind's influence. An optical device was used to track the "tracers" and the sea in which they floated was about 2 m deep (Blairmore Pier, Loch Long, Scotland). Their research was aimed at quantifying the diffusivity for turbulent flow, but indeed they were addressing the larger question of predictability in turbulent fluid motion.

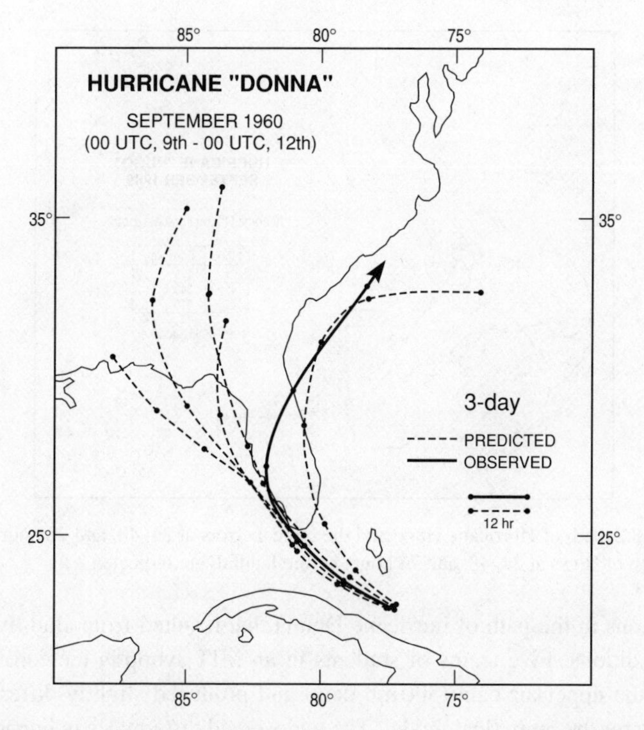

Fig. 2.8.1 Predicted and observed path of Hurricane Donna, where the predictions were based on slight differences in the initial conditions.

It is fruitful to imagine two states of a system, only slightly different, that we label as analogues. For example, these analogues could be the state of the global atmosphere at a mid-tropospheric level such as 500 mb ($\sim$ 5.5 km above sea level). By some measure such as standard deviation of the geopotential heights of this surface on 5° latitude $\times$ 5° longitude grids, let us say that February 15, 1953, is an analogue to March 2, 1967. We can then follow the evolution of these fields to see how long they remain analogues. What we have found for the atmosphere is that these states begin to diverge, where the doubling time (e.g., time where the standard deviation measure has increased by a factor of 2) is the order of several days. Typically within a week, the similarity between the two states is no closer than the difference between two arbitrarily chosen states for the season. In short, analogue forecasting generally fails rather quickly (beyond a few days typically). This fact supports the contention that the atmosphere is an unstable physical system and a consequence of this instability is that there is a limit to predictability.

From a deterministic forecasting viewpoint, we note that slight changes in the initial conditions often lead to drastically different forecasts. A case in point is the forecast of the track of a hurricane by a "steering" model, a model that moves the hurricane as if it were a permeable object in a stream (a vorticity conservation forecast of the storm as discussed earlier in Section 2.6). Figure 2.8.1 exhibits

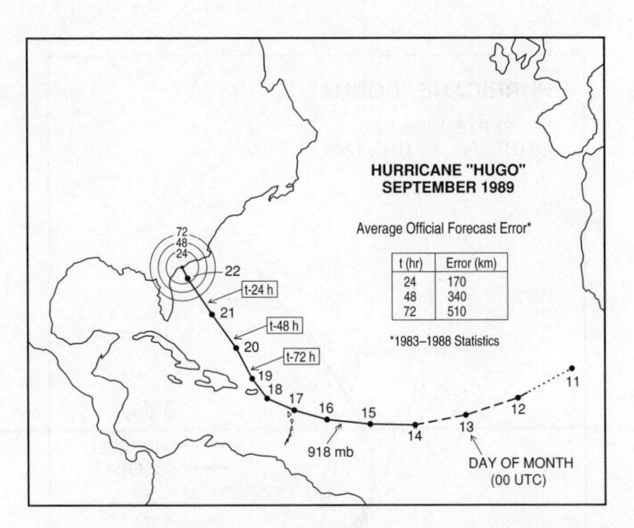

Fig. 2.8.2 Path of Hurricane Hugo and the typical errors at 24, 48, and 72 hours. The location of Hugo at 24, 48, and 72 hours before landfall are indicated.

the variations in the path of hurricane Donna that resulted from slightly different initial conditions. Five teams of students in an MIT synoptic meteorology class analyzed the upper-air data (500 mb data) and produced slightly different initial conditions for the numerical model. The widespread differences in hurricane track astounded the professor and his doctoral student [Fredrick Sanders and Robert Burpee, respectively. See Sanders and Burpee (1968)].

The errors associated with operational hurricane track forecasting are shown on the inset of Figure 2.8.2. In this figure, the estimated errors at landfall for hurricane Hugo are schematically represented by the concentric circles centered on the coast of South Carolina.

Insofar as predictability relates to data assimilation, it is instructive to revisit the results obtained by Lorenz in the early 1980s. He made use of the archive of analyses and 10-day forecasts generated by the state-of-the-art numerical weather prediction model at the European Centre for Medium-Range Weather Forecasts (ECMWF). He created analogues of the global 500 mb geopotential field by pairing the 1-day forecasts with the analysis of this field at verification time. From the archive, he was able to create 100 analogues over a winter season. He then measured the divergence of these analogues, and he could do this for 9 days since forecasts extended 10 days into the future. We sketch the results of this study. Referring to Figure 2.8.3, the solid line indicates the model error as a function of day (forecast). The dashed line shows the divergence of the analogues. Assuming the true atmospheric states would diverge at the same rate as these analogues, Lorenz speculated that two-week forecasts would be likely. This is based on the extrapolation of the dashed curve – it would appear that it flattens at about two weeks. After two weeks, the analogues exhibit a difference similar to the difference between two arbitrarily chosen states.

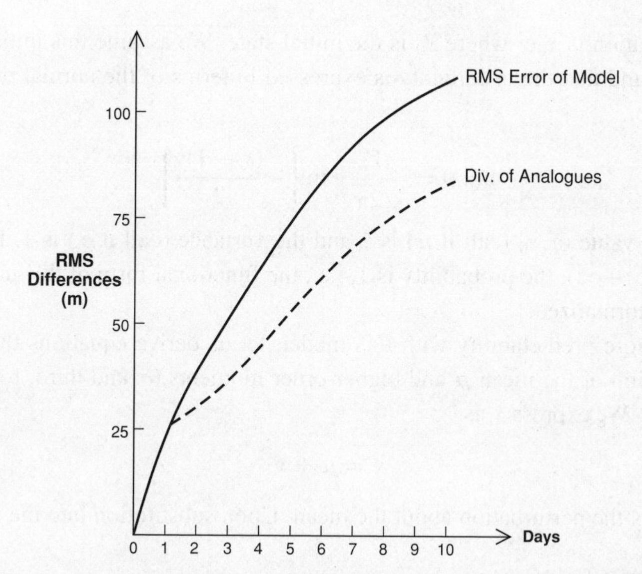

Fig. 2.8.3 Root-mean-square errors in the 500 mb forecast from the European Centre's (ECMWF) model in 1981 are shown by the solid line. The one-day forecast and analysis at that time are close analogues ($\sim$ 25 m rms difference on the average). The divergence of solutions starting from these analogues is displayed by the dashed curve.

Now, in order to extend the predictability limit, either the model must be improved, in which case the rms error (solid line) would approach the dashed (analogue divergence line), or the analysis error must be reduced. In this example, an improved analysis at $t = 0$ would lead to an improved 1-day forecast. That is, the dashed curve would follow a path below the one shown and likely lead to an extension of valuable forecast (beyond two weeks).

From a data assimilation viewpoint, predictability diagrams such as this give guidance regarding the goodness of analysis as a function of the limit of predictability. Further, the rate of divergence as a function of the size of the differences in analogues is a valuable piece of information (Results in Lorenz (1982) but not shown here). Lorenz found that the rate of divergence decreased with increase in analogue difference. This type of information offers valuable guidance on assimilation strategies, especially on the frequency of updating models and the predictability consequences of given error.

2.9 Stochastic/Dynamic prediction

In the spirit of this "pathways" chapter, let us exhibit elements of predictability with an example. Suppose we are given the dynamical law in the form

$$\frac{\mathrm{d}x}{\mathrm{d}t} = x,$$

whose solution is $x_0 e^t$ where x_0 is the initial state. We assume this initial state is uncertain and that the uncertainty is expressed in terms of the normal probability density, i.e.,

$$p(x_0) = \frac{1}{\sqrt{2\pi}} \exp\left\{-\frac{(x-1)^2}{2}\right\}.$$

The mean value of x_0 (call it μ) is 1 and the variance (call it σ) is 1. Integrated from $(-\infty, +\infty)$, the probability is 1, i.e., the functional form of this uncertainty has been normalized.

To explore predictability with this model, let us derive equations that govern the evolution of the mean μ and higher-order moments (σ and third, fourth, ..., moments). We express x as

$$x = \mu + x',$$

where x' is the perturbation about the mean. Upon substitution into the model we get

$$\frac{d\mu}{dt} + \frac{dx'}{dt} = \mu + x',$$

If we average (bar) in the ensemble-sense, we get

$$\overline{\frac{d\mu}{dt}} + \overline{\frac{dx'}{dt}} = \overline{\mu} + \overline{x'}$$

and since $\overline{x'} = d\overline{x'}/dt = 0$, this reduces to

$$\frac{d\mu}{dt} = \mu\,.$$

To form the second-moment equation, we multiply the dynamical law by x' and average. The result is

$$\frac{\overline{d(x')^2/2}}{dt} = \overline{(x')^2}$$

or

$$\frac{d\sigma}{dt} = 2\sigma.$$

Thus, we have

$$\mu(t) = \mu(0)e^t$$
$$\sigma(t) = \sigma(0)\sigma^{2t}.$$

Since the initial probability distribution is normal (no moments higher than 2), the third and higher moments will never appear (consequence of the linear dynamics). At later times, the probability distribution is given by

$$\frac{1}{\sqrt{2\pi}e^{2t}} \exp\left\{-\frac{(x-\mu e^t)^2}{2e^{4t}}\right\}.$$

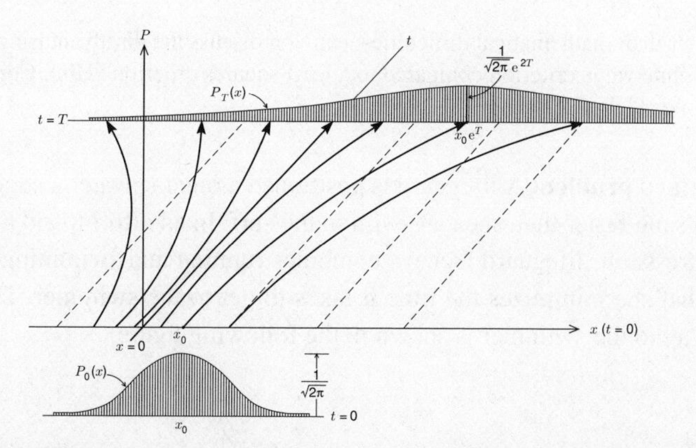

Fig. 2.9.1 Probability density functions (pdf) of the stochastic/dynamic solution to $\mathrm{d}x/\mathrm{d}t = x$. $P_0(x)$ and $P_T(x)$ are the pdf's at $t = 0$ and $t = T$, respectively.

We have graphically displayed the model output from a probabilistic viewpoint in Figure 2.9.1.

From the figure it becomes clear that the probability density function spreads outward along x, of course remaining greatest at the position of the mean value. Neighboring states ("analogues") at the initial time will diverge where the separation distance is proportional to e^t. The probability of a state far from the mean becomes significant in this case.

If the forecast from this dynamical system were to be combined with observations to improve the estimate of the system state, it is prudent to have knowledge of the initial uncertainty and its spread with time. In meteorology, this growth of the uncertainty is labelled the background error variance (or covariance as is typical for the many-variable system).

Exercises

2.1 **David Blackwell**, noted combinatorialist and U.C.-Berkeley professor, generally introduced undergraduates to the least squares principle by posing the following problem (see Blackwell 1969):

> Randomly choose a word from a phrase (such as) GO ON A HIKE. Predict the number of letters in the word you will choose. Your penalty will be the squared difference between your prediction and the number of letters in the word chosen. What is your best prediction? Set up the function J that measures the penalty. Hint: Assume that each word has equal probability of being chosen.
>
> By completing the square for this expression J, determine the best prediction by inspection.
>
> If, instead of a least squares penalty, the penalty is given by the absolute value of the difference between your prediction and the number of letters in the word chosen, what is your best estimate?

Other than mathematical difficulties, can you discuss the disadvantage of an absolute value criterion compared to a least squares criterion? Hint: Consider the phrase: IT IS GARGANTUAN.

2.2 **Lifeguard problem** A lifeguard is positioned along the water's edge. She can run in sand faster than she can swim in the surf. In an effort to aid a swimmer in distress, our lifeguard Geneva combines running and swimming in such a way that she minimizes the time it takes to get to the swimmer. The typical pathway to the swimmer is shown in the following figure.

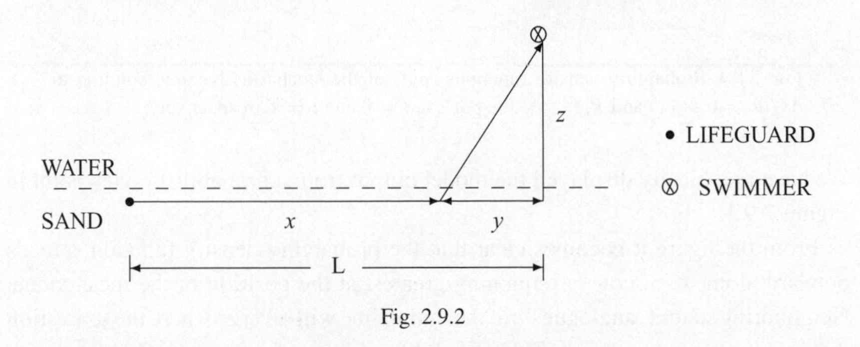

Fig. 2.9.2

Assume speed of swimming is "s" and speed of running is "r".

(a) Write the expression for the time T it takes the lifeguard to get to the swimmer. Write this as a function of x, s, r, L, and z. These elements constitute the control vector (5 elements).

(b) To minimize T, we require the derivative (with respect to x) to vanish. Since s and r are assumed constant, minimization of sT or rT will give the same answer as minimization of T. Using this fact, the control vector can be reduced by 1 where instead of using the elements s and r separately, we can use the one element s/r or vice versa, r/s. Take the required derivative with respect to the unknown x to find the optimal x and then find the expression for optimal time.

(c) If $L = 10$ units, $z = 2$ units, and $r/s = 3$, find the optimal x.

(d) For each execution of a distress event, observations of x, z, L, r and T are made. The estimated error variances in the observations are: $(1m)^2$, $(3m)^2$, $(1m)^2$, $(2ms^{-1})^2$ and $(0sec)^2$, respectively. Set up a functional to be minimized under the constraint of minimal T, and find the governing equations that must be satisfied to determine "s".

2.3 Consider an estimate at $t = 0$ given by a weighted combination of the forecast from $t = -T$ and the analysis at $t = 0$. That is,

$$\widehat{X} = W_0 X_0 + (1 - W_0)\, X^{\mathrm{f}}_{-1}.$$

Further assume that the dynamical model is

$$X_{k+1} = aX_k, \qquad a = \text{constant}.$$

Assume the errors of analysis at $t = -T$ and $t = 0$ are uncorrelated. If X_T is the true state at $t = 0$, find the value of W_0 that minimizes

$$J = \overline{(\widehat{X} - X_T)^2}$$

If $a = 2$, discuss the weighting and find the error variance of the estimate. Do the same for $a = 1/2$.

2.4 Consider analysis at three times given by z_0, z_1, and z_2. Assume the dynamical law that connects the state of the system at these times is

$$X_1 = aX_0$$
$$X_2 = aX_1.$$

Find $\widehat{X}_0$, an estimate at the initial time, that minimizes

$$J = (\widehat{X}_0 - z_0)^2 + (\widehat{X}_1 - z_1)^2 + (\widehat{X}_2 - z_2)^2$$

where $\widehat{X}_1$ and $\widehat{X}_2$ are forecasts from the optimal initial condition $\widehat{X}_0$. Assume the errors of analysis at each time are uncorrelated but equal.

Find the optimal $\widehat{X}_0$.

Find the error variances at $t = 0$, 1, and 2. Calculate these error variances for the cases when $a = 1/2$, 1, and 2. Discuss results.

2.5 An object is tracked on a radar screen as discussed in Section 2.5. The first observations of the object are registered as follows:

t	x	y
0	0.0	0.0
1	0.5	1.0

where x, y are the rectangular coordinates. Dead reckoning is used to exrapolate into the future. Assume the observation of the subsequent positions exhibit an error variance $\sigma_z^2 = 0.1$ (in both x and y). Further, the error variance associated with the dead reckoning is $\sigma_f^2 = 0.4$ (again, in both x and y). Extrapolate to time increment 5 in steps of 1 time unit by using the stochastic/dynamic approach discussed in Section 2.5, when the observations are:

t	x	y
2	1.8	2.4
3	4.7	2.6
4	8.1	3.7
5	12.0	5.2

Calculate the error variance of the estimate. Prove that it is always less than the smallest of the observational and forecast error variances.

Notes and references

Section 2.2 An excellent discussion of the use of characteristics in solution of differential equations is found in Carrier and Pearson (1976).
Section 2.3 For further explorations refer to Miyakoda and Talagrand (1971).
Section 2.4 Platzman (1964) contains a thorough introduction to Burgers equation.
Section 2.5 See Part IV and References for details.
Section 2.6 Refer to Thompson (1969) for more details.
Section 2.7 Refer to Sanders and Burpee (1968). The historical review of hurricane track forecasting by Mark Demaria is an informative and stimulating account of forecasting practice throughout the twentieth century. Demaria (1996). See Lewis et al. (2001) for details on sensitivity displayed in Figure 2.7.4.
Section 2.9 Saaty's book (Saaty (1967)), especially Chapter 8, is a solid introduction to the ideas associated with stochastic dynamic prediction written at a level that is accessible to students who have had a course in ordinary differential equations. And the above mentioned paper by Kikuro Miyakoda and Olivier Talagrand stimulates our thought about melding predictability with data assimilation. The short paper by du Plessis (1967) on Kalman filters gives the reader a set of interesting examples that are conceptually easy to follow.

3

Applications

In this chapter we introduce a variety of models, and in some cases associated data. Some of these models/data will be used at various junctures in the book. We further invite the teachers and students to generate their own problems based on these examples. Alongside the description of the models, we identify the data assimilation problems that can be explored with the models. Some of the examples are pedagogical in nature, such as the straight line problem, but others are identified with specific branches of science.

3.1 Straight line problem

The phrase "fitting model to data" generally conjures up the idea of fitting a straight line to a set of observations. Indeed, its widespread use in virtually every quantitative discipline makes it the most common example of data assimilation under constraint. It has intuitive appeal since the goodness of fit is easily ascertained by visual inspection of the plotted line in relation to the observations. We investigate this data assimilation problem in several guises.

3.1.1 Slope–Intercept form: static/deterministic, off-line problem

In this classic approach, the slope and intercept of the assumed line are the two fixed but unknown constants. We represent these unknowns as

$$\mathbf{x} = \begin{bmatrix} \beta \\ \alpha \end{bmatrix}, \quad \mathbf{x} \in \mathbb{R}^2 .$$

In the data assimilation literature this unknown vector is called the control vector. These unknowns are related to the observation z as follows:

$$z = \beta + \alpha t + v$$

where t is the independent variable (time, e.g.) and v is the noise. In discrete form,

$$z_k = \beta + \alpha t_k + v_k$$

where v_k is the noise at time t_k. Given m such observations, we can succinctly represent this relation in a matrix form

$$\mathbf{z} = \mathbf{H}\mathbf{x} + \mathbf{v} \tag{3.1.1}$$

where $\mathbf{z}, \mathbf{v} \in \mathbb{R}^m$, and $\mathbf{H} \in \mathbb{R}^{m\times 2}$, and $\mathbf{H}$ has the form

$$\mathbf{H} = \begin{bmatrix} 1 & t_1 \\ 1 & t_2 \\ \vdots & \vdots \\ 1 & t_m \end{bmatrix}. \tag{3.1.2}$$

It is assumed that the noise vector $\mathbf{v}$ is such that $E(\mathbf{v}) = 0$ and $\mathrm{Cov}(\mathbf{v}) = E(\mathbf{v}\,\mathbf{v}^{\mathrm{T}}) = \mathbf{R}$, a real symmetric and a positive definite matrix. The criterion used to find the elements of $\mathbf{x}$ takes the form:

$$J(\mathbf{x}) = \frac{1}{2}\|\,\mathbf{z} - \mathbf{H}\,\mathbf{x}\,\|_2^2 = \frac{1}{2}(\mathbf{z} - \mathbf{H}\,\mathbf{x})^{\mathrm{T}}(\mathbf{z} - \mathbf{H}\,\mathbf{x}), \tag{3.1.3}$$

which is viewed as a deterministic criterion, i.e., without account for the characteristics of the noise. If the noise is accounted, we have the modified criterion as follows:

$$J(\mathbf{x}) = \frac{1}{2}\|\,\mathbf{z} - \mathbf{H}\,\mathbf{x}\,\|_{\mathbf{R}^{-1}}^2 = \frac{1}{2}(\mathbf{z} - \mathbf{H}\,\mathbf{x})^{\mathrm{T}}\mathbf{R}^{-1}(\mathbf{z} - \mathbf{H}\,\mathbf{x}). \tag{3.1.4}$$

In either case, the static, deterministic, off-line estimation problem is stated as follows: given $\mathbf{z}$, $\mathbf{H}$ and $\mathbf{R}$, find the $\mathbf{x} \in \mathbb{R}^2$ that minimizes $J(\mathbf{x})$ in (3.1.4) (Exercise 3.1). If $\mathbf{R}$ is diagonal with equal value of these diagonal elements, the modified version reduces to the deterministic version of the criterion. If $\mathbf{R}$ is diagonal, but the elements are unequal, then the latter version of the criterion reduces to a minimization where each squared departure term in the functional J is weighted differently – the weight is the inverse of the error variance. Thus, an observation with relatively large error receives less weight than an observation with smaller error variance. Generally, a diagonal form of $\mathbf{R}$ implies that the observational errors are uncorrelated. Correlation between observations implies the existence of off-diagonal elements, yet $\mathbf{R}$ is symmetric. These problems are solved in Part II.

3.1.2 Initial value problem: dynamic/deterministic, off-line problem

The straight line problem can be viewed as a dynamic problem governed by the following differential equation:

$$\frac{\mathrm{d}x}{\mathrm{d}t} = \hat{\alpha}, \quad \hat{\alpha} \text{ a constant}.$$

Discretizing the above equation using the forward Euler scheme, we obtain the discrete time counterpart given by

$$x_k = x_{k-1} + \alpha = x_0 + k\alpha \tag{3.1.5}$$

where $\alpha = \hat{\alpha}(\Delta t)$. Let us assume there are m observations of the state at a subset of the discrete times given by

$$z_i = x_i + v_i, \qquad i = 1 \text{ to } m \tag{3.1.6}$$

where v_i's are the random observation errors. We wish to determine the initial condition x_0 and the parameter (the slope) α such that the functional J is minimized,

$$J(x_0, \alpha) = \frac{1}{2}\sum_{i=1}^{m}(x_i - z_i)^2. \tag{3.1.7}$$

This class of problems is solved using the variational method in Part VI. (Exercise 3.2)

3.1.3 Boundary value problem: static/deterministic off-line problem

Here we assume that the straight line is given by

$$\frac{\mathrm{d}^2 x}{\mathrm{d}t^2} = 0\,. \tag{3.1.8}$$

Using the standard central difference approximation for the second derivative, we obtain the following discrete form:

$$x_{k-1} - 2x_k + x_{k+1} = 0\,, \tag{3.1.9}$$

for $k = 1, 2, \ldots, N-1$. In this case, the two boundary values x_0 and x_N are the unknowns. Rewriting (3.1.9) in matrix form, we obtain

$$\mathbf{Ax} = \mathbf{b} \tag{3.1.10}$$

where $\mathbf{x} = (x_1, x_2, \ldots, x_{N-1})^{\mathrm{T}}$, $\mathbf{b} = (-x_0, 0, 0, \ldots, 0, x_N)^{\mathrm{T}}$ and

$$\mathbf{A} = \begin{bmatrix} -2 & 1 & 0 & \cdots & 0 & 0 & 0 & 0 \\ 1 & -2 & 1 & \cdots & 0 & 0 & 0 & 0 \\ \vdots & \vdots & \vdots & \cdots & \vdots & \vdots & \vdots & \vdots \\ 0 & 0 & 0 & \cdots & 0 & 1 & -2 & 1 \\ 0 & 0 & 0 & \cdots & 0 & 0 & 1 & -2 \end{bmatrix} \tag{3.1.11}$$

is a symmetric, tridiagonal matrix of size $(N-1)$. It can be verified that $\mathbf{A}$ is **non-singular**.

Let $\mathbf{z} \in \mathbb{R}^m$ be the given set of observations where

$$\mathbf{z} = \mathbf{H}\mathbf{x} + \mathbf{v} \tag{3.1.12}$$

Following (3.1.4) we state the problem as follows: given **z**, **H**, **R**, and the matrix **A**, find the **b** that minimizes

$$J(\mathbf{b}) = \frac{1}{2}(\mathbf{z} - \overline{\mathbf{H}}\mathbf{b})^{\mathrm{T}}\mathbf{R}^{-1}(\mathbf{z} - \overline{\mathbf{H}}\mathbf{b}) \tag{3.1.13}$$

where $\overline{\mathbf{H}} = \mathbf{H}\mathbf{A}^{-1}$. This class of problems is solved in Part V. (Exercise 3.3)

3.1.4 Initial value problem: stochastic/dynamic, online problem

Following (3.1.5), consider the stochastic version

$$x_{k+1} = M(x_k) + w_{k+1} \tag{3.1.14}$$

where $M(x_k) = x_k + \alpha$ and $\{w_k\}$ is the model error with

$$E[w_k] = 0 \quad \text{and} \quad E[w_k^2] = Q_k \in \mathbb{R}$$

and $\{w_k\}$ is serially uncorrelated. Let

$$z_k = x_k + v_k \tag{3.1.15}$$

denote the sequence of observations. The sequential or the online problem calls for generating a sequence $\widehat{\mathbf{x}}_k$ of minimum variance estimate along with its variance $\widehat{\mathbf{P}}_k$. This problem is solved using the Kalman filtering algorithms in Part VII.

3.2 Celestial dynamics

A problem that has stimulated great interest through the past two centuries is the special three-body problem – determination of the orbit of an infinitesimally small body in the presence of two bodies of finite mass. The American astronomer and mathematician George William Hill posed the problem in mid-nineteenth century in his efforts to understand the motion of the moon in the presence of the Sun and Earth. Henri Poincaré explored it more completely. Poincaré demonstrated that the general three-body problem could not be solved analytically, but this specialized problem is solvable and exploration of its solution under various initial conditions has proved fruitful.

Let us state the problem. Three heavenly bodies move under the action of gravitational attraction. One of the bodies is of infinitesimally small mass and exerts no appreciable force on the objects of finite mass. The mass of the objects are μ and $1 - \mu$ where $\mu \leq 1/2$. These larger bodies move in a circular path about their center of mass, where the separation of these objects is unity (nondimensional). It is further assumed that the three bodies move in a plane, and the location of the

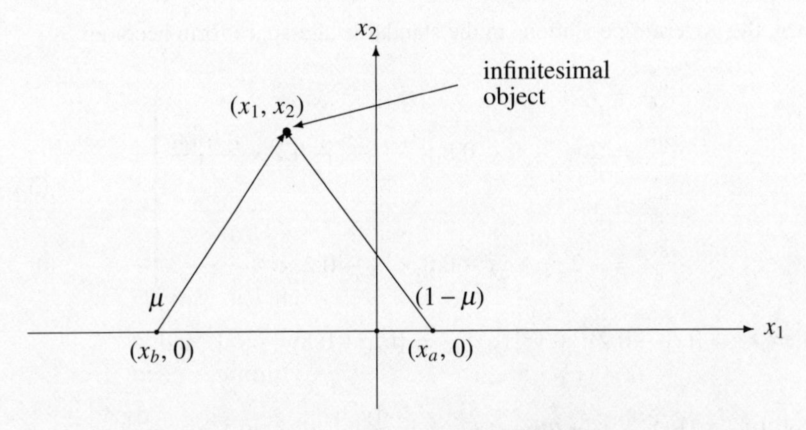

Fig. 3.2.1 Three-body problem: a special case.

infinitesimal object is found relative to the rotating coordinate system defined by the rotation of the finite objects.

The finite objects are positioned along the x_1-axis as shown in Figure 3.2.1. The $x_1 x_2$ plane rotates, the finite masses remain fixed on the x_1 axis, and the position of the infinitesimal object is (x_1, x_2). The larger mass is four times greater than the smaller mass in the case shown. That is, (center of mass at origin)

$$\mu = 0.2, \qquad 1 - \mu = 0.8$$
$$x_a = 0.2, \qquad x_b = -0.8.$$

The governing equations for the infinitesimal object are:

$$\left.\begin{aligned} \frac{d^2x_1}{dt^2} - 2\frac{dx_2}{dt} &= x_1 - 0.8 \times \frac{(x_1-0.2)}{r_1^3} - 0.2 \times \frac{(x_1+0.8)}{r_2^3} \\ \frac{d^2x_2}{dt^2} + 2\frac{dx_1}{dt} &= x_2 - 0.8 \times \frac{x_2}{r_1^3} - 0.2 \times \frac{x_2}{r_2^3} \end{aligned}\right\} \tag{3.2.1}$$

where we choose initial conditions as follows:

$$x_1(0) = -1/2, \qquad x_2(0) = 1/2, \qquad \left.\frac{dx_1}{dt}\right|_{t=0} = -0.1, \qquad \left.\frac{dx_2}{dt}\right|_{t=0} = 0$$
$$r_1 = [(x_1 - 0.2)^2 + x_2^2]^{\frac{1}{2}}, \qquad r_2 = [(x_1 + 0.8)^2 + x_2^2]^{\frac{1}{2}} \tag{3.2.2}$$

where r_1 and r_2 are the distances from the infinitesimal object to the larger and smaller finite objects, respectively.

It is often convenient to reduce the set of second-order differential equations to a set of first-order equations. This is accomplished by letting

$$y_1 = x_1, \quad y_2 = \frac{dx_1}{dt}, \quad y_3 = x_2, \quad y_4 = \frac{dx_2}{dt}. \tag{3.2.3}$$

Then, the governing equations in the standard state-space form become:

$$\left.\begin{aligned} \frac{dy_1}{dt} &= y_2 \\ \frac{dy_2}{dt} &= 2\,y_4 + y_1 - 0.8 \times \frac{(y_1-0.2)}{r_1^3} - 0.2 \times \frac{(y_1+0.8)}{r_2^3} \\ \frac{dy_3}{dt} &= y_4 \\ \frac{dy_4}{dt} &= -2\,y_2 + y_3 - 0.8 \times \frac{y_3}{r_1^3} - 0.2 \times \frac{y_3}{r_2^3} \end{aligned}\right\} \tag{3.2.4}$$

where $r_1 = [(y_1 - 0.2)^2 + y_3^2]^{\frac{1}{2}}$, $\quad r_2 = [(y_1 + 0.8)^2 + y_3^2]^{\frac{1}{2}}$ and

$$y_1(0) = -1/2, \qquad y_3(0) = 1/2, \qquad \left.\frac{dy_1}{dt}\right|_{t=0} = -0.1, \qquad \left.\frac{dy_3}{dt}\right|_{t=0} = 0.$$

From a data assimilation viewpoint, several interesting problems can be posed in the context of this special three-body problem. The first issue of interest is the predictability of the object (the infinitesimal object). That is, if there is slight error in the initial condition, is this inaccuracy "forgiven" – i.e., does the uncertainty in the objects future position increase, decrease or remain the same. What Poincaré found is that the predictability is "flow dependent" which means that the evolution of the uncertainty critically depends on the intrinsic stability properties of the underlying dynamics and the initial conditions (where the object is initially located, its initial velocity, and the uncertainty in those initial conditions).(Exercise 3.4 and 3.5)

Thus, the growth of error in the prediction system must be determined. For example, how long does it take the initial position error to double for a given set of initial conditions. Then, if observations of the object are given at various epochs (points in time), how can these observations (with known error) be combined with the forecast to yield an optimal state under the least squares criteria?

3.3 Fluid dynamics

J. M. Burgers, a quantum physicist turned fluid dynamist, was known for his ability to simplify the complex problems of fluid dynamics including turbulence. In his reminiscences (Burgers, 1975), he said:

> Scientific problems may come forward from things heard, or read in books and papers, and sometimes they seem to arise from nowhere, but the background of the entire society is always there and effective. An important influence is the often felt need to reduce a scientific problem to its most essential and simple points, in order to make clear to others what can be done and what would be beyond reach, or to defend one's manner of thinking and one's way of approach.

His most famous equations take the following two forms:

$$\frac{\partial u}{\partial t} + u\frac{\partial u}{\partial x} = 0 \tag{3.3.1}$$

and, appending the diffusion term,

$$\frac{\partial u}{\partial t} + u\frac{\partial u}{\partial x} = \frac{1}{\mathrm{Re}}\frac{\partial^2 u}{\partial x^2} \tag{3.3.2}$$

where Re is the Reynolds number (nondimensional),

$$\mathrm{Re} = \frac{uL}{\nu},$$

the ratio of inertia to viscous force where u is velocity scale, L is length scale, and ν is viscosity. The first of these two equations has been used extensively by meteorologists because of its similarity to the nonlinear barotropic constraint, the governing equation for the large-scale transient waves of mid-latitudes (see Section 2.6). The second equation pits nonlinear advective processes against turbulent dissipation.

Platzman's (1964) study of the spectral solution to (3.3.1) has been one of the most fruitful testbeds for problems in prediction and data assimilation. We state the problem as follows (following Section 2.2):

Assume $u(x, t)$ is periodic in the domain $[0, 2\pi]$ and that the initial condition is given by the sine wave that spans this domain, wavenumber 1,

$$u(x, 0) = \sin x, \qquad 0 \leq x \leq 2\pi. \tag{3.3.3}$$

The analytic solution, found by the method of characteristics is

$$u(x, t) = \sin(x - ut). \tag{3.3.4}$$

Following Platzman, a solution of the form

$$u = \sum_{n-1}^{\infty} u_n(t)\sin(nx) \tag{3.3.5}$$

is sought, an odd-function Fourier expansion. This leads to a system of ordinary differential equations given by

$$\begin{aligned}
\frac{2}{1}\frac{\mathrm{d}u_1}{\mathrm{d}t} &= (u_1u_2 + u_2u_3 + u_3u_4 + \cdots) \\
\frac{2}{2}\frac{\mathrm{d}u_2}{\mathrm{d}t} &= -\frac{1}{2}u_1^2 + (u_1u_3 + u_2u_4 + \cdots) \\
\frac{2}{3}\frac{\mathrm{d}u_3}{\mathrm{d}t} &= -u_1u_2 + (u_1u_4 + u_2u_5 + \cdots) \\
\frac{2}{4}\frac{\mathrm{d}u_4}{\mathrm{d}t} &= -u_1u_3 - \frac{1}{2}u_2^2 + (u_1u_5 + u_2u_6 + \cdots).
\end{aligned}$$

In general,

$$\frac{2}{n}\frac{\mathrm{d}u_n}{\mathrm{d}t} = -\frac{1}{2}\sum_{k=1}^{n-1} u_k u_{n-k} + \sum_{k=1}^{\infty} u_k u_{n+k}. \tag{3.3.6}$$

See Platzman (1964) for interesting details, including the comparison of the analytic solution to truncated forms of the dynamics.

If the advective term is absent from (3.3.2), we obtain the diffusion equation:

$$\frac{\partial u}{\partial t} = \frac{1}{\mathrm{Re}}\frac{\partial^2 u}{\partial x^2} \tag{3.3.7}$$

where initial and boundary condition must be specified. If we take the same initial condition and periodic boundaries mentioned above, we can find a solution by specification of the initial condition only. If the Reynolds number is constant, then the analytic solution is

$$u(x,t) = \mathrm{e}^{-\frac{t}{\mathrm{Re}}} \sin x\,.$$

In finite difference form, the equation can be expressed as

$$\frac{u_j^{k+1} - u_j^k}{\Delta t} = \frac{1}{\mathrm{Re}}\frac{u_{j+1}^k - 2u_j^k + u_{j-1}^k}{(\Delta x)^2}$$

where $u(x,t) = u(j\Delta x, k\Delta t) \equiv u_j^k$.

If $\Delta x = 2\pi/8$ and $\sigma \equiv \frac{\Delta t}{\mathrm{Re}(\Delta x)^2} = 1/4$, we get

$$\begin{aligned} u_j^{k+1} &= u_j^k + \frac{1}{4}(u_{j+1}^k - 2u_j^k + u_{j-1}^k) \\ &= \frac{1}{2}u_j^k + \frac{1}{4}(u_{j+1}^k + u_{j-1}^k). \end{aligned}$$

Another form of the advection/diffusion equation of Burgers can be derived by a parameterization of the diffusion term. Using the Guldberg-Mohn hypothesis, named after two late-nineteenth-century meteorologists, we have

$$\frac{\partial^2 u}{\partial x^2} = -\kappa u, \qquad \kappa > 0$$

which is seen to be a form of spectral representation, i.e., u expressed as sines and cosines. Of course, the κ is generally empirical and dependent on spatial scale. In this case, we get

$$\frac{\partial u}{\partial t} + u\frac{\partial u}{\partial x} = -\sigma^2 u$$

where $\sigma^2 = \kappa/\mathrm{Re}$.

The analytic solution to Burger's equation (3.3.1) is

$$u(x,t) = f(x - tu(x,t))$$

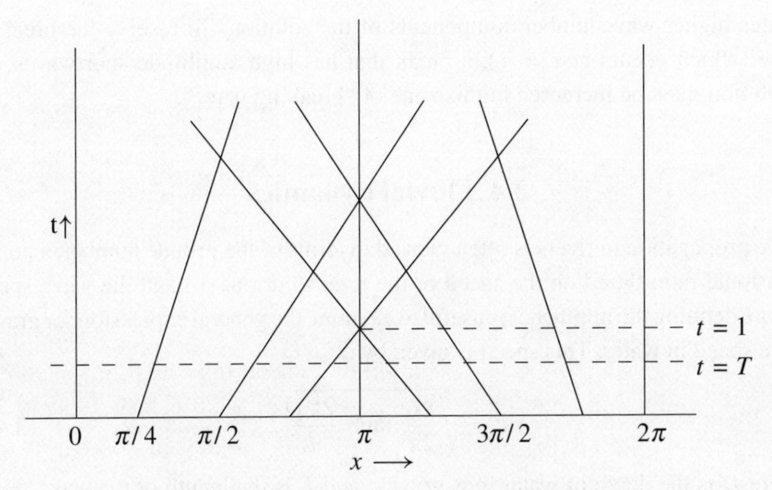

Fig. 3.3.1 Characteristics for the Burgers equation (3.3.1).

where $u(x, 0) = f(x)$ is the initial condition. This solution is generally found by the method of characteristics. If the function f is such that this equation can be solved for u, then we have the analytic solution for $u(x, t)$. This can only be accomplished in special circumstances (e.g., see Platzman (1964)).

It is still instructive to examine the nature of the solution in terms of the characteristics (discussed earlier in Section 2.2). The characteristics are the straight lines in the $x - t$ plane given by

$$x - ut = \text{constant}.$$

The slopes of this family of lines is $\mathrm{d}t/\mathrm{d}x = 1/u(x, 0)$, i.e., the inverse of the value of u on the initial line determines the slope of the characteristic line. The associated constant is x_0, the value of x where the characteristic passes through the line of initial values of u, i.e., the value of x on the $t = 0$ line. It follows that $u = u(x, 0)$ at all points along the characteristic. For Platzman's case of $u = \sin x$, $0 \leq x \leq 2\pi$, at $t = 0$, with periodicity, the characteristics converge toward $x = \pi$ as t increases. And, indeed, there comes a point where the characteristics cross and the solution is multivalued (beyond $t = 1$). The schematic of the characteristics in this case is shown in Figure 3.3.1.

From a data assimilation viewpoint this problem is most interesting. In the classic case of using observations spread over time and space to determine the optimal initial condition, it is clear that there must be at least one observation on each characteristic line. Furthermore, if observations are only available at times close to $t = T < 1$ (as seen in figure 3.3.1), a uniform distribution of observations is not the most desirable. It will be more advantageous to have observations bunched together near $x = \pi$, with coarser resolution near the boundaries. In effect, the evolution takes the initial wave (wavenumber 1), and through nonlinear interaction, creates higher wavenumber components of the solution. To resolve the breaking wave (which occurs at $x = \pi$), a break that has high amplitude shortwaves, the resolution must be increased in this zone of "breaking wave".

3.4 Fluvial dynamics

Wave propagation in rivers is often viewed in terms of the Froude number, a nondimensional ratio based on the speed of the river's current (v) and the wave speed. Before defining the number, it is useful to examine the general expression for gravity wave speed in water. This speed is given by

$$c = \sqrt{\frac{gL}{2\pi} \tanh \frac{2\pi D}{L}}, \tag{3.4.1}$$

where D is the depth of water, g is gravity, and L is the length of the wave. Since $\tanh x \approx x$ for small and positive x and $\tanh x \approx 1$ for large and positive x, it follows that for sufficiently long waves ($L/D \gg 1$), the speed reduces to $\sqrt{gD}$, and for sufficiently short waves ($L/D \ll 1$), the speed reduces to $\sqrt{\frac{gL}{2\pi}}$. Now, the Froude number, F, is defined as

$$F \equiv \frac{v^2}{gD}, \tag{3.4.2}$$

where gD is the square of the speed of "shallow water" ($L \gg D$) waves. It is well known that gravity waves cannot move upstream if the flow is supercritical, i.e., ($F > 1$). For subcritical flow ($F < 1$), there is a particular wavelength for which the wave is stationary. The governing law for this stationary wavelength can be expressed as

$$\tanh(\sigma) = \sigma F \tag{3.4.3}$$

where $\sigma = 2\pi D/L$. From a series of measurements of the current v and knowledge of the depth D of the river, the wavelength of this stationary wave can be determined using the static/deterministic formulation.

3.5 Oceanography

The most ubiquitous force in the geosphere is gravity. For a mass distribution in stable equilibrium, disturbances are resisted by gravitational restoring forces and result in oscillations which may take the form of standing waves or propagating waves. Examples in the hydrosphere – ocean, lakes, and rivers – are: long waves (shallow water waves), such as seiche and tsunami, and short waves such as ship-induced waves and wind-generated waves. In this section, we will discuss the dynamics of the shallow water waves.

Table 3.5.1 *Observations on Tsunami*

Station	Distance from epicenter (stat.mi.)	Mean observed velocity (stat.mi./hr)	Observed travel time (hr min)
Honolulu	2141	490	4 34
San Francisco	2197	398	5 31
La Jolla	2646	428	6 11

Table is extracted from the information in C. K. Green (1946).

Another force that impacts the hydrosphere on the larger scales of motion – scales influenced by the earth's rotation – is the inertia force. The classic demonstration of this force is found by viewing the Foucault pendulum, oscillations of a bob on the end of a wire ideally suspended from the top of a high-ceilinged building. Depending on the latitude, this pendulum precesses and executes a rotation about the zenith in $24/\sin\phi$ hours (ϕ the latitude). The dynamics of this motion stems from the Coriolis force, and we will examine motion in the ocean under the action of this inertia force.

3.5.1 Shallow water equations

As mentioned in Section 3.4, the speed of propagation of gravity waves in shallow water, i.e., depth of water less, considerably less, than the length of the generated wave, is $\sqrt{gD}$. The most notable shallow water waves are those generated by under-ocean disturbances that stem from earthquakes. These waves, called Tsunami, are generally undetected by the naked eye in the open sea, but when these waves approach land, and especially in the presence of a dramatic rise in height of the ocean floor (and associated decreasing depth of the water), these waves attain great height and can spread inland.

Entries in Table 3.5.1 and Figure 3.5.1 roughly indicate the travel time of the Tsunami generated near the Aleutian Chain on April 1, 1946. The travel times shown in this table are along the great circle routes from the epicenter of the quake to the various coastal stations.

The momentum and mass conservation equations that govern these gravity waves in an (x, t) plane are:

$$\left.\begin{aligned} \frac{\partial u}{\partial t} + u\frac{\partial u}{\partial x} &= -g\frac{\partial h}{\partial x} \\ \frac{\partial h}{\partial t} + \frac{\partial}{\partial x}(Dh) &= 0 \end{aligned}\right\} \tag{3.5.1}$$

where u is the speed of the water particles and h is the height of the water above the equilibrium height D. If we neglect the nonlinear advection and assume a constant

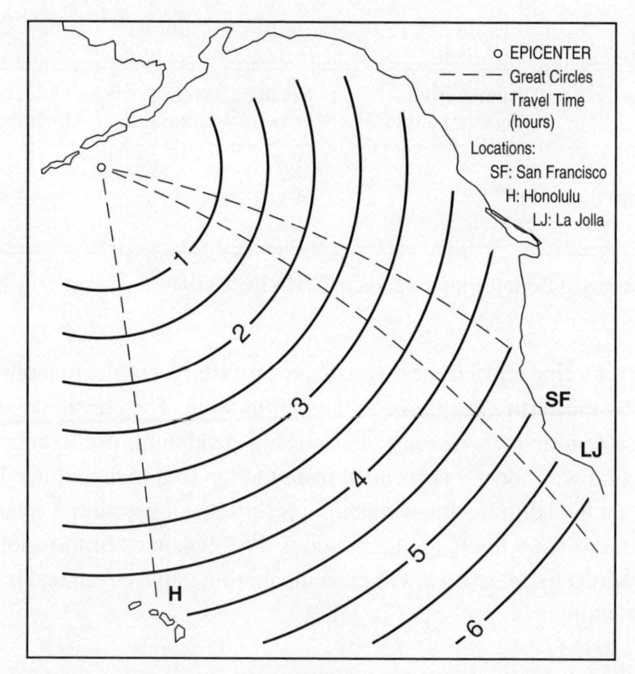

Fig. 3.5.1 The progression of the tsunami generated near the Aleutian Chain in April, 1946. The isolines show the position of the tsunami from 1 to 6 hours after its generation.

depth D, then the equations assume the simplified form:

$$\left.\begin{aligned}\frac{\partial u}{\partial t} &= -g\frac{\partial h}{\partial x}\\ \frac{\partial h}{\partial t} &= -D\frac{\partial u}{\partial x}\end{aligned}\right\} \tag{3.5.2}$$

In this form it is apparent that the phase speed is $\sqrt{gD}$. The wave equation is formed (for either variable) by differentiation (with respect to t followed by substitution), i.e.,

$$\frac{\partial^2 h}{\partial t^2} = gD\frac{\partial^2 h}{\partial x^2} \tag{3.5.3}$$

(similarly for u). Assumption of a propagating wave solution, e.g., $\sin k(x - ct)$, where k is wavenumber and c propagation speed, leads to $c = \sqrt{gD}$. If an underwater disturbance raises the water surface as shown in Figure 3.5.2, the hydrostatic pressure gradient force, $-g\,\partial h/\partial x$, will create accelerated particle motion aside the center line of raised water (shown by arrows). This divergence of water along this centerline, $-D\,\partial u/\partial x$, will lead to a fall of water level along this centerline and an associated rise on both sides of it. In effect, this action describes the propagation of a water wave in the positive and negative directions at speed $\sqrt{gD}$.

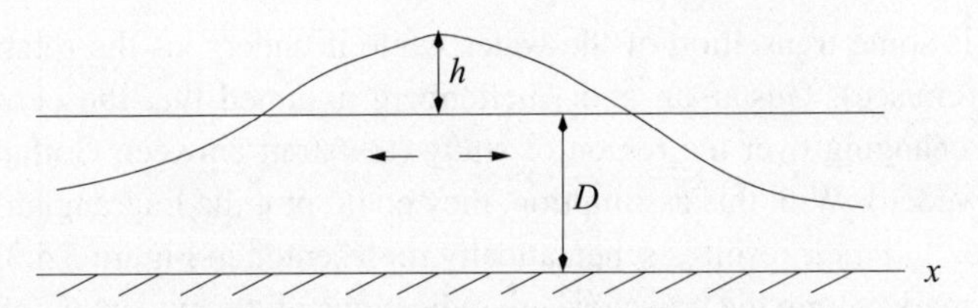

Fig. 3.5.2 Propagation of water waves.

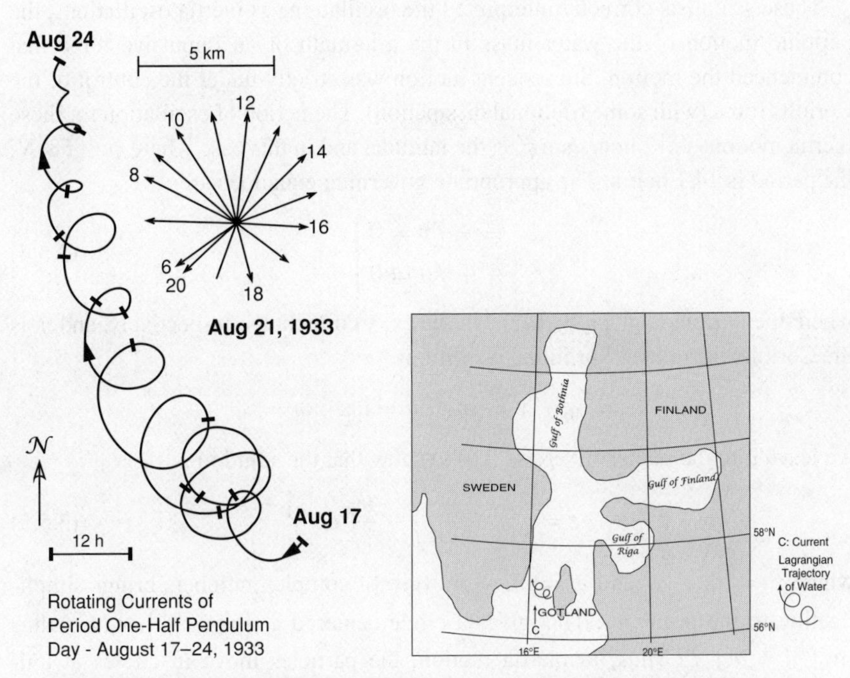

Fig. 3.5.3 The spoke diagram shows the direction and magnitude of the surface current on August 21, 1933, starting at 0600 (6) local time and extending to 2000 (20). These currents over the period August 17–24 were used to trace the water movement as shown by the trochoid where the short bars are placed at 12 hour intervals, and the space scale atop the diagram applies to this trajectory. Location of the current meter is shown on the inset.

3.5.2 Inertia Oscillations

In the late 1930s, an interesting response of the ocean to an impulsive force was noticed by Gustafson and Kuellenberg in the body of water between the coast of Sweden and the island of Gotland. A schematic roughly indicating the current in the upper layer of ocean appeared as Figure 3.5.3.

The current is shown at hourly intervals during the day, starting with hour 6. The length of the vector is the magnitude of the current and the direction is given by the arrow. As seen, the current makes a complete 360° rotation in slightly more than 14 hours. The magnitude of the current, given by the length of the vectors, indicates that there is some translation of the water while it undergoes the rotation (order of several cm/sec). Gustafson and Kuellenberg assumed that the currents were uniformly changing over the region of study (the strait between Gotland and the coast of Sweden). With this assumption, they could plot the Lagrangian trajectory of the water and their result is schematically represented in Figure 3.5.3.

The tic marks along the trajectory are indications of the movement of the water over 12 hour intervals (6 days shown).

These scientists correctly interpreted the oscillations as inertia oscillations, the periodic motion of the water mass in the aftermath of an impulsive force that commenced the motion. Subsequent motion was strictly under the control of the Coriolis force (with some frictional dissipation). The period of oscillation for these inertia motions is 12 hours/sin ϕ, ϕ the latitude, and in this case where $\phi \sim 58^\circ N$, the period is 14.1 hours. The appropriate governing equations are

$$\left.\begin{aligned} \frac{\partial u}{\partial t} - fv = 0 \\ \frac{\partial v}{\partial t} + fu = 0 \end{aligned}\right\} \tag{3.5.4}$$

where the current is given by (u, v) in the (x, y) directions, respectively, and f is the Coriolis parameter. For initial conditions

$$x = x_0, \quad y = y_0, \quad u = u_0, \quad v = v_0\,,$$

we leave it to the reader (Exercise 3.6) to show that the solution is

$$z = \left(z_0 + \frac{w_0}{\mathrm{i}f}\right) - \frac{w_0}{\mathrm{i}f}\,\mathrm{e}^{-\mathrm{i}ft} \tag{3.5.5}$$

where $z_0 = x_0 + \mathrm{i}y_0$ and $w_0 = u_0 + \mathrm{i}v_0$ (use of complex numbers brings simplification to the problem). This gives a circle centered at $\left(z_0 + \frac{w_0}{\mathrm{i}f}\right)$, with radius $|w_0|/f = |w|/f$. Thus, in inertia motion, the particles move in circles at uniform speed, the radius of the circle being = speed$/f$. The inertia period is $2\pi/f = 2\pi/2\Omega \sin\phi = 12/\sin\phi$ hours.

If we assume a uniform current in the x-direction, U, this leads to the solution:

$$\left.\begin{aligned} x &= x_0 + Ut + \tfrac{u_0}{f}\sin(ft) \\ y &= y_0 - \tfrac{u_0}{f}(1 - \cos(ft)) \end{aligned}\right\} \quad \text{assuming } v_0 = 0 \tag{3.5.6}$$

This trajectory is a trochoid. If $U < u_0$, the trochoid is prolate; if $U > u_0$, it is curate. Refer to Figure 3.5.4.

With the Guldberg–Mohn linear friction (frictional force opposite to direction of motion and proportional to the components, i.e., $-ku$ (x-direction) and $-kv$ (y-direction)), the solution is

$$\left.\begin{aligned} x &= \tfrac{u_0}{(k^2+f^2)}\mathrm{e}^{-kt}[k(1 - \cos(ft)) + f\sin(ft)] \\ y &= -\tfrac{u_0}{(k^2+f^2)}\mathrm{e}^{-kt}[f(1 - \cos(ft)) - k\sin(ft)] \end{aligned}\right\} \tag{3.5.7}$$

when $x_0 = y_0 = 0$ and $v_0 = 0$.

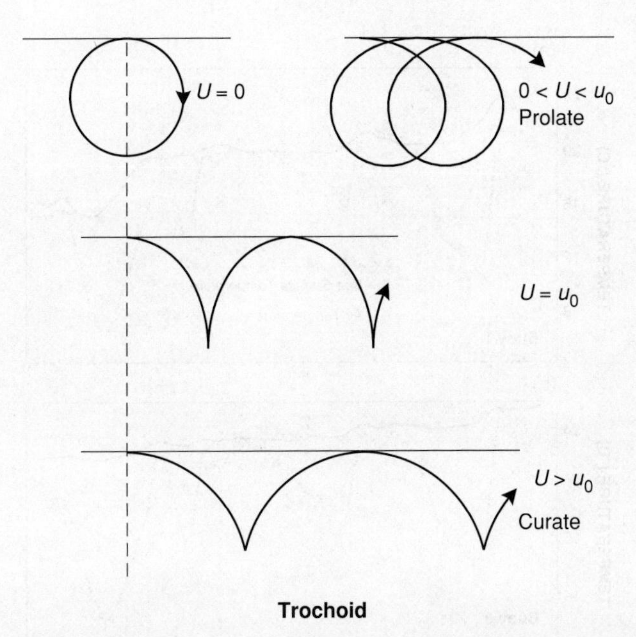

Fig. 3.5.4 Trochoids governed by the relative size of the perturbation (u_0) and base-state current (U).

3.5.3 Cooling of shelf water

During the "cool season" when weather disturbances move out over the Gulf of Mexico (GofM), typically November through April, the shelf water cools in steps associated with each cold air outbreak. This can be seen by examination of the data collected on moored buoys residing over the continental shelf. In Figure 3.5.5 we have plotted the time series of air temperature (T_a), sea-surface temperature (T_s), and wind speed ($|V|$), from two buoys off the GofM coast (see Figure 3.5.6). Buoy#3 is atop water that is 50 m deep while buoy#1 is in water less than 20 m deep. Four cold fronts passed over these buoys during the month of November 1992. The sudden drop in T_u occurs at the following times: 3 Nov, 12 Nov, 21 Nov, and 25 Nov. The air temperature exhibits an oscillatory trace where each cold outbreak event is typified by a sudden drop in T_a that is followed by a recovery. This oscillatory trace is associated with wind reversals as shown in Figure 3.5.7. Northerly wind comes with the outbreak of cold air, but the wind typically turns clockwise with time and a modified (warmed and moistened) air mass returns several days later as shown in the lower panel of Figure 3.5.7. As the result of these cold air passages, the sea surface temperature drops from 24°C to 16°C at buoy#1, and from 24°C to 19°C at buoy#3 during November. The smaller volume or mass of water in the shallower regions permit a more-pronounced cooling as the result of heat transfer from the relatively warm ocean to the adjoining air.

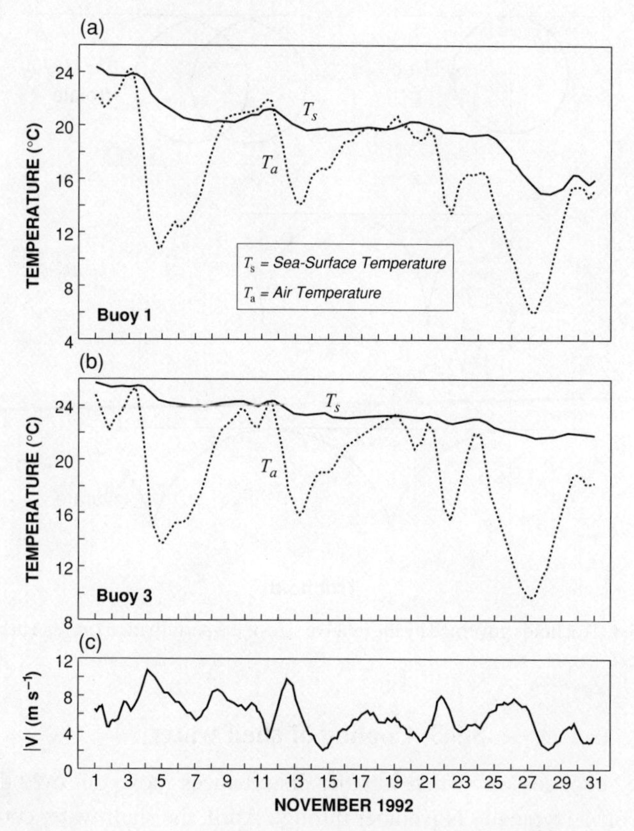

Fig. 3.5.5 Traces of air temperature (T_a) and sea surface temperature (T_s) during November 1992 as measured by instruments aboard two buoys moored in shelf water over the Gulf of Mexico. The wind speed trace applies to buoy 1.

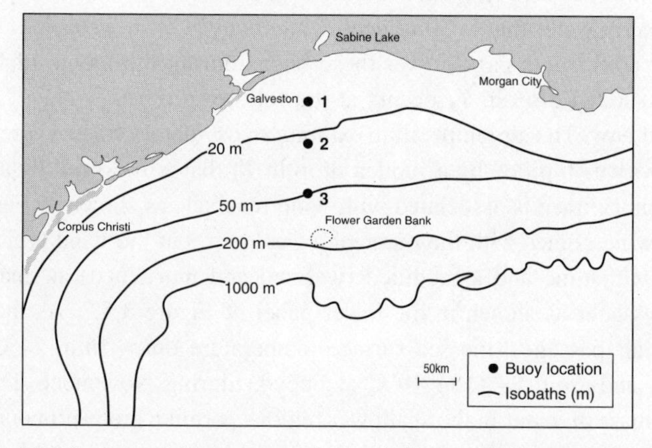

Fig. 3.5.6 Locations of buoys moored over shelf water off the coast of Texas.

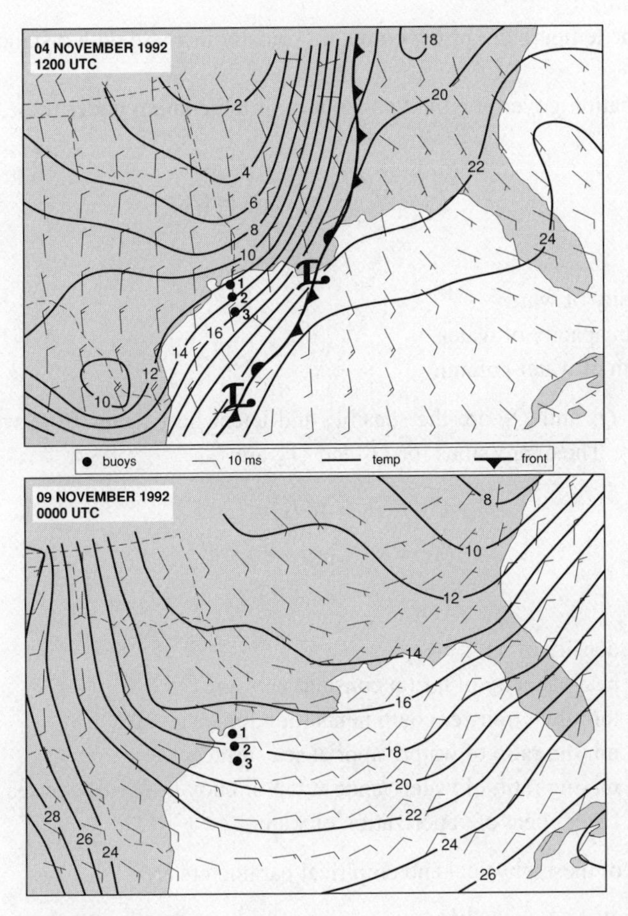

Fig. 3.5.7 Surface winds and weather front in relation to the three moored buoys off the coast of Texas. The top map shows the surface flow shortly after the frontal passage and the bottom map shows the flow five days later. A full barb represents 10 knots and the direction of flow is from tail "feathers" to head of wind barb.

The dynamics of air/sea interaction that leads to this drop in sea temperature can be modeled by assuming that the loss of heat from the ocean column is due to the turbulent transfer of sensible heat and latent heat (the evaporation of water). Under the conditions of cold air outbreak (where differences in air and sea surface temperature are the order of 5–10°C), the other energy sources/sinks (solar radiation input and longwave radiation output and transport within the ocean, e.g.) are an order of magnitude less than the turbulent transfers at the air/sea interface. We further assume that the temperature of the sea water is constant throughout the individual columns. Over the shelf, where the water depths are about 50 m or less, the strong winds associated with the cold outflows mix the water to the bottom. The temperature of the columns typically increase with distance from the coastline.

The equation governing the temperature of the column under these conditions is:

$$\rho_{\mathrm{w}} c_{\mathrm{w}} H \frac{\partial T_{\mathrm{s}}}{\partial t} = -Q_{\mathrm{s}} - Q_{\mathrm{E}}$$

where

ρ_{w} : density of water
c_{w} : heat capacity of water
H : depth of water column

and where Q_{s} and Q_{E} are the sensible and latent heat fluxes from water to air, respectively. The expressions for Q_s and Q_E are:

$$Q_s = \rho c_p C_{\mathrm{H}}(T_{\mathrm{s}} - T_{\mathrm{a}})|\vec{V}|$$

$$Q_E = L\rho C_{\mathrm{E}}(q_{\mathrm{s}} - q_{\mathrm{a}})|\vec{V}|$$

where

ρ : density of air
c_p : heat capacity of air (at constant pressure)
$C_{\mathrm{H}}, C_{\mathrm{E}}$: turbulent transfer coefficients (nondimensional)
q_{s} : mixing ratio of water vapor at sea surface
q_{a} : mixing ratio of water vapor at top of buoy (~3 m above sea surface)
L : Latent heat of vaporization of water

The values of these physical and empirical parameters are:

c_p : 1004 Joule $\mathrm{kg}^{-1}\,{}^{\circ}\mathrm{C}^{-1}$
ρ : 1.28 kg m^{-3}
$T_s - T_a$: (°C)
$q_s - q_a$: (g/kg) (nondimensional $\sim 10^{-3}$)
$|\vec{V}|$: m s^{-1}
L : $2.5 \cdot 10^6$ Joules kg^{-1}
ρ_w : $10^3\ \mathrm{kg} \cdot \mathrm{m}^{-3}$
c_w : 4183 Joule $\mathrm{kg}^{-1}\,{}^{\circ}\mathrm{C}^{-1}$
H : (m)
C_H, C_E : $\sim 1.5 \cdot 10^{-3}$ (nondimensional)

The calculation of the water vapor mixing ratio is found empirically from the following formulae (refer to Table 3.5.2):

If $T_{\mathrm{s}} = 20$°C, $p = 1020$ mb, $e_{\mathrm{s}} = 23.4$ mb, then $q_{\mathrm{s}} = 0.014$ or 14 g kg^{-1}.

Now,

$$\frac{\partial T_{\mathrm{s}}}{\partial t} = -\frac{1}{\rho_{\mathrm{w}} c_{\mathrm{w}} H} \cdot (Q_{\mathrm{s}} + Q_{\mathrm{E}})$$

Table 3.5.2 *Mixing ratios*

Unsaturated air (use T_d: dew point temp)	Saturated air (at sea surface) (use T_s: sea temp)
Vapor pressure, $e(mb)$, where $e = 6.11 \times 10^{\alpha}$	Saturated vapor pressure, $e_s(mb)$ $e_s = 6.11 \times 10^{\hat{\alpha}}$
where $\alpha = \frac{7.5 \times T_d(°C)}{237.3 + T_d(°C)}$	where $\hat{\alpha} = \frac{7.5 \times T_s(°C)}{237.3 + T_s(°C)}$
Air pressure, p(mb)	Air pressure, p(mb)
mixing ratio, q $q = 0.662 \frac{e}{p}$	saturated mixing ratio, q_s $q_s = 0.662 \frac{e_s}{p}$

Table 3.5.3

hour	wind speed (m/s)	pressure (mb)	T_s (°C)	T_a (°C)	T_d (°C)
0	8	1027	18.5	11.7	0.7
1	9	1028	18.5	12.6	−0.6
2	10	1029	18.4	12.1	−1.6
3	12	1030	18.2	11.7	−1.9
4	12	1030	18.2	11.2	−3.1
5	12	1031	18.1	10.5	−2.1
6	10	1031	18.1	10.4	−2.3
7	11	1031	18.0	10.5	−1.2
8	11	1031	17.9	10.6	−0.9
9	11	1030	17.9	10.6	−1.0
10	11	1030	17.8	10.7	−0.5
11	11	1030	17.7	10.7	0.3
12	11	1030	17.7	10.3	0.0
13	11	1030	MSG	10.2	0.5
14	11	1031	17.6	9.8	0.3
15	10	1031	17.6	10.0	0.0
16	9	1032	17.6	10.0	−1.2
17	8	1032	17.6	10.2	−2.5
18	8	1031	17.6	11.3	−2.0
19	6	1030	17.5	11.7	−1.6
20	6	1030	17.5	12.7	−1.0
21	7	1029	17.5	13.7	−1.2
22	8	1028	17.5	14.5	−1.0
23	9	1028	17.4	14.5	−0.8
24	9	1029	17.4	14.0	−0.6

For these outflows, $Q_s \sim 150$ watts m^{-2} and $Q_E \sim 400$ watts m^{-2}, or the turbulent transfer total is ~ 550 W m^{-2}. For a water column that is 13 m deep, the rate of decrease of sea column temperature is ~ -0.04°C/hr. Thus, in 48h, T_s would drop ~ 1.9°C.

The data from the buoy off Biloxi, MS, given in Table 3.5.3, exhibits the effect of water column cooling (13 m deep water).

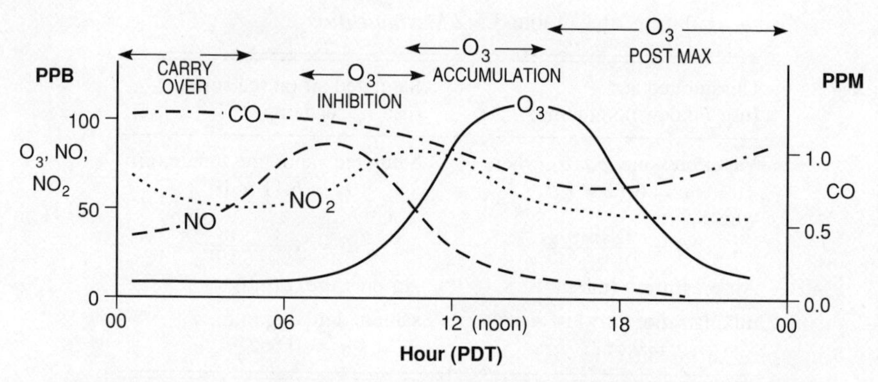

Fig. 3.6.1 Diurnal cycle of ozone (O_3) concentration in the southern California basin along with other reactants (CO: Carbon monoxide, and NO, NO_2: oxides of nitrogen). PPB and PPM represent parts per billion and million, respectively.

3.6 Atmospheric chemistry

In summer, the production of ozone (O_3) in large metropolitan areas such as the Southern California Basin (SOCAB), has become a paramount concern[‡]. The photochemistry, i.e., photolysis of chemical species by the action of sunlight (photon absorption), is a maximum during summer, and when combined with the sources of nitrogen compounds, nitric and nitrous oxide (NO and N_2O, respectively), the daily evolution of ozone follows a time series similar to that shown in Figure 3.6.1.

3.6.1 Nonlinear stable chain reaction

The chemical reactions that give rise to this daily distribution of O_3 are classified as stable chain reactions, specifically, the oxidation of carbon monoxide (CO). The governing equations follow where the quantum mechanical symbol ($h\nu$) represents the photolysis process.

$$\left.\begin{aligned} O_3\,[(+h\nu)+H_2O] &\longrightarrow 2HO+O_2 \\ HO+CO(+O_2) &\longrightarrow HO_2+CO_2 \\ HO_2+NO &\longrightarrow HO+NO_2 \\ HO_2+HO_2 &\longrightarrow H_2O_2+O_2 \end{aligned}\right\} \tag{3.6.1}$$

where HO is the hydroxyl radical, O_2 is oxygen, HO_2 is the hydroperoxy radical, and H_2O_2 is hydrogen peroxide. We symbolically represent this stable chain reaction

[‡] We are grateful to Professor William Stockwell, Howard University, for his help in writing this section.

for both the oxidation of CO (3.6.1) [and oxidation of sulfur dioxide discussed below] as

$$\left.\begin{array}{lll} \mathrm{A} \longrightarrow \mathrm{R1} & & K_1 \\ \mathrm{R1} + \mathrm{S1} \longrightarrow \mathrm{R2} + \mathrm{P1} & & \mathrm{K}_2 \\ \mathrm{R2} + \mathrm{S2} \longrightarrow \mathrm{R1} + \mathrm{P2} & & \mathrm{K}_3 \\ \mathrm{R1} + \mathrm{R1} \longrightarrow \mathrm{P3} & & \mathrm{K}_4 \end{array}\right\} \tag{3.6.2}$$

where A is the initial reactant, S is the stable reactant, R is the fast reacting intermediate, P is the product and where the K_i are the reaction rates. For the oxidation of sulfur dioxide (SO_2) and production of sulfuric acid (H_2SO_4) – i.e., the "acid rain" process – we only need replace the second reaction in (3.6.1) with

$$\mathrm{HO} + \mathrm{SO_2}(+\mathrm{O_2}, +\mathrm{H_2O}) \longrightarrow \mathrm{H_2SO_4} + \mathrm{HO_2}\,.$$

Otherwise, the reactions are the same. The differential equations governing these processes (oxidation of CO or oxidation of SO_2) can be written as:

$$\left.\begin{array}{l} \frac{\mathrm{d}}{\mathrm{d}t}[\mathrm{A}] = -K_1[\mathrm{A}] \\ \frac{\mathrm{d}}{\mathrm{d}t}[\mathrm{R1}] = \begin{cases} K_1[\mathrm{A}] + K_3[\mathrm{R2}]\cdot[\mathrm{S2}] \\ -K_2[\mathrm{R1}]\cdot[\mathrm{S1}] - 2K_4[\mathrm{R1}]^2 \end{cases} \\ \frac{\mathrm{d}}{\mathrm{d}t}[\mathrm{R2}] = K_2[\mathrm{R1}]\cdot[\mathrm{S1}] - K_3[\mathrm{R2}]\cdot[\mathrm{S2}] \\ \frac{\mathrm{d}}{\mathrm{d}t}[\mathrm{S1}] = -K_2[\mathrm{R1}]\cdot[\mathrm{S1}] \\ \frac{\mathrm{d}}{\mathrm{d}t}[\mathrm{P1}] = K_2[\mathrm{R1}]\cdot[\mathrm{S1}] \\ \frac{\mathrm{d}}{\mathrm{d}t}[\mathrm{S2}] = -K_3[\mathrm{R2}]\cdot[\mathrm{S2}] \\ \frac{\mathrm{d}}{\mathrm{d}t}[\mathrm{P2}] = K_3[\mathrm{R2}]\cdot[\mathrm{S2}] \\ \frac{\mathrm{d}}{\mathrm{d}t}[\mathrm{P3}] = K_4[\mathrm{R1}]^2 \end{array}\right\} \tag{3.6.3}$$

Meaningful dimensionless values of the various species and the reaction rates are given below:

K_1	K_2	K_3	K_4	$[A]_0$	$[R1]_0$	$[R2]_0$	$[S1]_0$	$[S2]_0$	$[P1]_0$	$[P2]_0$
0.1	1.0	1.0	0.01	1	0	0	10	10	0	0

(3.6.4)

where $[\ ]_0$ indicates the initial value ($t = 0$).

In both types of oxidation, the photolysis of ozone (O_3) leads to the production of hydroxyl radicals (HO)[§] which are considered to be "detergents" (cleansers) in the atmosphere. The first reaction in the oxidation process is known as "initiation". Then the hydroxyl radicals react with carbon monoxide (CO) [or sulfur dioxide (SO_2)] to produce products (CO_2 and H_2SO_4) and hydroperoxy radicals (HO_2).

§ These radicals are **fast** reacting products in such a stable chain reaction.

${}^{238}_{92}\text{U} \longrightarrow {}^{234}_{90}\text{Th} + \alpha$	${}^{226}_{88}\text{Ra} \longrightarrow {}^{222}_{86}\text{Rn} + \alpha$	${}^{214}_{84}\text{Po} \longrightarrow {}^{210}_{82}\text{Pb} + \alpha$
${}^{234}_{90}\text{Th} \longrightarrow {}^{234}_{91}\text{Pa} + \beta$	${}^{222}_{86}\text{Rn} \longrightarrow {}^{218}_{84}\text{Po} + \alpha$	${}^{210}_{82}\text{Pb} \longrightarrow {}^{210}_{83}\text{Bi} + \beta$
${}^{234}_{91}\text{Pa} \longrightarrow {}^{234}_{92}\text{U} + \beta$	${}^{218}_{84}\text{Po} \longrightarrow {}^{214}_{82}\text{Pb} + \alpha$	${}^{210}_{83}\text{Bi} \longrightarrow {}^{210}_{84}\text{Po} + \beta$
${}^{234}_{92}\text{U} \longrightarrow {}^{230}_{90}\text{Th} + \alpha$	${}^{214}_{82}\text{Pb} \longrightarrow {}^{214}_{83}\text{Bi} + \beta$	${}^{210}_{84}\text{Po} \longrightarrow {}^{206}_{82}\text{Pb} + \alpha$
${}^{230}_{90}\text{Th} \longrightarrow {}^{226}_{88}\text{Ra} + \alpha$	${}^{214}_{83}\text{Bi} \longrightarrow {}^{214}_{84}\text{Po} + \beta$	

Fig. 3.6.2 Transmutation of Uranium-238.

The hydroperoxy radicals react with nitric oxide (NO) to reproduce hydroxyl radicals (HO). These reactions are called "chain propagating" reactions that reproduce the oxidants HO and HO_2 and could continue indefinitely except for the existence of "termination reactions". The final reaction involving two hydroperoxy radicals produces hydrogen peroxide (H_2O_2) and is known as a "radical terminating" reaction. This reaction prevents the "runaway" production of radicals in the atmosphere.

3.6.2 Simple linear decay

Nuclei of atoms are composed of protons and neutrons, and nuclei that change their structure spontaneously and emit radiation are termed radioactive. Many such unstable nuclei occur naturally and among them is Uranium-238, denoted by ${}^{238}_{92}\text{U}$ where 238 represents the mass (the mass number given by the number of protons and neutrons), and 92 represents the atomic number (number of protons).

The sequential transformation from Uranium-238 to Lead-206 (${}^{206}_{82}Pb$) is shown in Figure 3.6.2, where α, β represents the ejection of alpha and beta particles from the various nuclei. This sequence indicates that nuclear reactions result in the formation of an element not initially present – called nuclear transmutation.

The decay of one element that leads to the production of others follows a process we label as simple linear decay. Symbolically,

$$\begin{array}{ll} \text{A} \longrightarrow \text{R1} & K_1 \\ \text{R1} \longrightarrow \text{P1} & K_2 \end{array}$$

with associated differential equations:

$$\left.\begin{aligned} \tfrac{\mathrm{d}}{\mathrm{d}t}[\text{A}] &= -K_1[\text{A}] \\ \tfrac{\mathrm{d}}{\mathrm{d}t}[\text{R1}] &= K_1[\text{A}] - K_2[\text{R1}] \\ \tfrac{\mathrm{d}}{\mathrm{d}t}[\text{P1}] &= K_2[\text{R1}] \end{aligned}\right\} \qquad (3.6.5)$$

where K_1 and K_2 are the reaction rates and nondimensional values of rate parameters and initial concentrations are given as follows:

$$\begin{array}{ccccc} K_1 & K_2 & [A]_0 & [R1]_0 & [P1]_0 \\ 0.01 & 0.1 & 10 & 0 & 0 \end{array} \qquad (3.6.6)$$

One of the environmental concerns related to the transmutation of Uranium-235 is the production of radium, ${}^{236}_{88}$ Ra, which in turn transmutes to radon, ${}^{222}_{86}$ Rn. Radon is an inert radioactive gas that is emitted from soils that contain uranium. It can accumulate in homes and long term exposure has been linked to cancer.

3.7 Meteorology

The weather affecting us is generally associated with processes in the lowest 10 km of the atmosphere, the troposphere. Viewed from afar, this layer of gases constitutes a small fraction of the global dimension – the solid earth, the hydrosphere, and adjoining gaseous envelope. This atmospheric "shell" has vertical dimension $< 1/100$ in ratio to the Earth's radius. A visual analogy is the thickness of an orange peel in relation to the radius of the sphere that approximates the orange. It should then not be surprising that the wind, the horizontal motion of the air relative to the rotating earth, has become central to our exploration of the atmosphere. That is not to say that the vertical motion of the air is secondary. On the contrary, this upward and downward motion of air, although difficult to measure, is supremely important in understanding atmospheric phenomena, not the least of which is the energy exchange and energy conservation principles that span the spectrum from convective storms to the global circulation.

Some of the most important relationships and equations in meteorology are the static laws that relate wind to pressure gradient. These are approximations of the wind, and they are faithful representations to the wind under special circumstances governed by assumptions that stem from dominance of certain terms in the horizontal momentum equations. We discuss these wind laws and submit them as excellent examples of static constraints in geophysical data assimilation.

Advection is a phenomenon and a constraint that is pervasive in meteorology. It was conceptually discussed in Chapter 2 and its nonlinear form, idealized in Burgers' equation (Section 3.3), has provided an excellent testbed for studying advective processes.

3.7.1 Wind in the presence of friction

Friction can be parameterized as a force opposite to the wind direction and proportional to its speed (addressed earlier). A balance between the pressure gradient

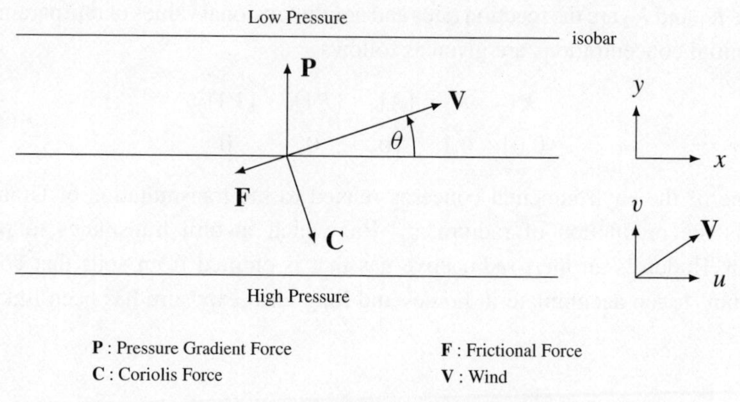

Fig. 3.7.1 **An illustration of the balance.**

force (**P**), the Coriolis force (**C**), and this frictional force (**F**) can only be achieved when the wind cuts across the isobars in the direction from high to low pressure as shown in Figure 3.7.1 (northern hemisphere).

The component equations can be written

$$\left.\begin{aligned} -fv &= -fv_{\mathrm{g}} - \kappa u \\ fu &= fu_{\mathrm{g}} - \kappa v \end{aligned}\right\} \tag{3.7.1}$$

where $\kappa(>0)$ is the empirical coefficient of friction and $\mathbf{V}_{\mathrm{g}} = (u_{\mathrm{g}}, v_{\mathrm{g}})^{\mathrm{T}} = \frac{1}{f}(-P_y, P_x)^{\mathrm{T}}$ where $\mathbf{P} = (P_x, P_y)^{\mathrm{T}}$, and f is the Coriolis parameter. $\mathbf{V}_{\mathrm{g}}$ is the geostrophic or "earth turning" wind (directed 90° *cum sole* from the pressure gradient force). The magnitude of this balanced wind $\mathbf{V} = (u, v)^{\mathrm{T}}$ can be expressed in terms of the geostrophic wind as follows:

$$|\mathbf{V}| = \frac{|\mathbf{V}_{\mathrm{g}}|}{\sqrt{1 + (\kappa/f)^2}} < |\mathbf{V}_{\mathrm{g}}| \tag{3.7.2}$$

The angle θ is given by $\tan^{-1}(\kappa/f)$.

The view of surface winds over the Pacific Ocean is shown in Figure 3.7.2. It is clear that the winds generally exhibit a component or flow from high pressure to low pressure where the angle θ is generally in the range of 20–40°.

3.7.2 Lorenz's "minimum equations"

In much the same way that Burgers' equation has been used by meteorologists to study nonlinear advection, Lorenz's "maximum simplification" or "minimum" equations have been used to test forecasting and data assimilation (Lorenz (1960); for tests using these equations see Epstein (1969b) and Lakshmivarahan et al. (2003)). We will use these equations at several junctures in the text.

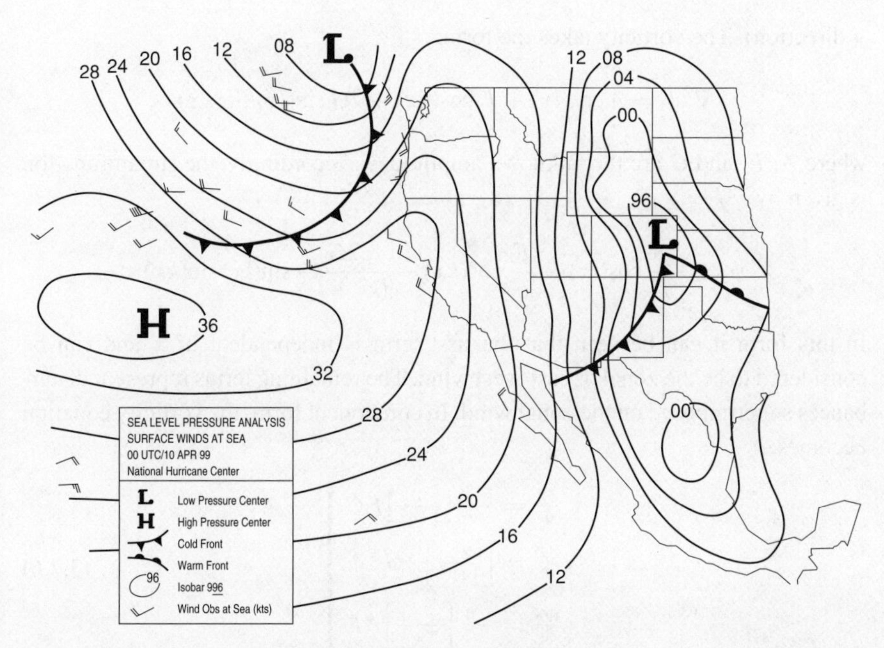

Fig. 3.7.2 Surface winds over Pacific Ocean.

Lorenz's idea stemmed from his desire to examine the equations for large-scale weather prediction under the simplest possible circumstances. The dynamical law is the conservation of vorticity at the nondivergent level of the troposphere, typically near 500 mb (see Section 2.6). The streamfunction ψ can thus be used to represent the flow, where

$$\frac{\partial \psi}{\partial x} = v, \qquad -\frac{\partial \psi}{\partial y} = u \tag{3.7.3}$$

where (u, v) are the velocity components in the (positive east, positive north) directions respectively and where (x, y) are coordinates in the (positive east, positive north) direction. Thus, the vorticity ζ is given by

$$\zeta = \frac{\partial v}{\partial x} - \frac{\partial u}{\partial y} = \nabla^2 \psi. \tag{3.7.4}$$

The governing dynamical law is expressed as

$$\frac{\partial}{\partial t}\nabla^2\psi - \frac{\partial \psi}{\partial y}\frac{\partial}{\partial x}\nabla^2\psi + \frac{\partial \psi}{\partial x}\frac{\partial}{\partial y}\nabla^2\psi = 0 \tag{3.7.5}$$

and indeed this has similarity to Burgers' equation – a nonlinear advection of vorticity in this case.

Lorenz found a spectral solution to this equation by truncating the Fourier series form of solution and using a variety of symmetry properties. The only waves admitted have wavenumber k (in the x-direction) and l (in the

y-direction). The vorticity takes the form:

$$\nabla^2\psi = A\cos(ly) + F\cos(kx) + 2G\sin(ly)\sin(kx)$$

where A, F, and G are the unknown amplitudes. Accordingly, the streamfunction is given by

$$\psi = -\frac{A}{l^2}\cos(ly) - \frac{F}{k^2}\cos(kx) - \frac{2G}{(k^2+l^2)}\sin(ly)\sin(kx)$$

In this form it can be seen that the first term is independent of x and can be considered to be the zonal or east/west wind. The remaining terms represent disturbances superimposed on the zonal wind. In component form, the vorticity equation becomes:

$$\left.\begin{aligned}\frac{dA}{dt} &= -\frac{1}{\alpha}\left[\frac{1}{1+\alpha^2}\right]FG\\ \frac{dF}{dt} &= \left[\frac{\alpha^3}{1+\alpha^2}\right]AG\\ \frac{dG}{dt} &= -\frac{1}{2}\left[\frac{\alpha^2-1}{\alpha}\right]AF\end{aligned}\right\} \tag{3.7.6}$$

where $\alpha = k/l$.

These equations can be expressed in nondimensional form by choosing a time scale T. In Lorenz's case $T = 3$ hours (dictated by his desire to scale such that T^{-1} is the order of the atmosphere's large-scale vorticity in mid-latitudes). We use this scaling to define the following nondimensional variables:

$$x_1 = TA, \quad x_2 = TF, \quad x_3 = TG$$

and

$$\hat{t} = t/T \qquad \text{(nondimensional time } \hat{t}\text{)},$$

then,

$$\left.\begin{aligned}\frac{dx_1}{d\hat{t}} &= -\frac{1}{\alpha}\left[\frac{1}{1+\alpha^2}\right]x_2x_3\\ \frac{dx_2}{d\hat{t}} &= \left[\frac{\alpha^3}{1+\alpha^2}\right]x_1x_3\\ \frac{dx_3}{d\hat{t}} &= -\frac{1}{2\alpha}\left[\alpha^2-1\right]x_1x_2\end{aligned}\right\} \tag{3.7.7}$$

For a 24-hour forecast, we integrate from $\hat{t} = 0$ to $\hat{t} = 8$. Lorenz examined solutions for $\alpha = 0.95$ and 1.05, respectively labeled "unstable" and "stable". In the unstable case, the disturbances (amplitudes x_2 and x_3) grow at the expense of the zonal flow (amplitude x_1). In the stable case, no such energy transfer occurs.

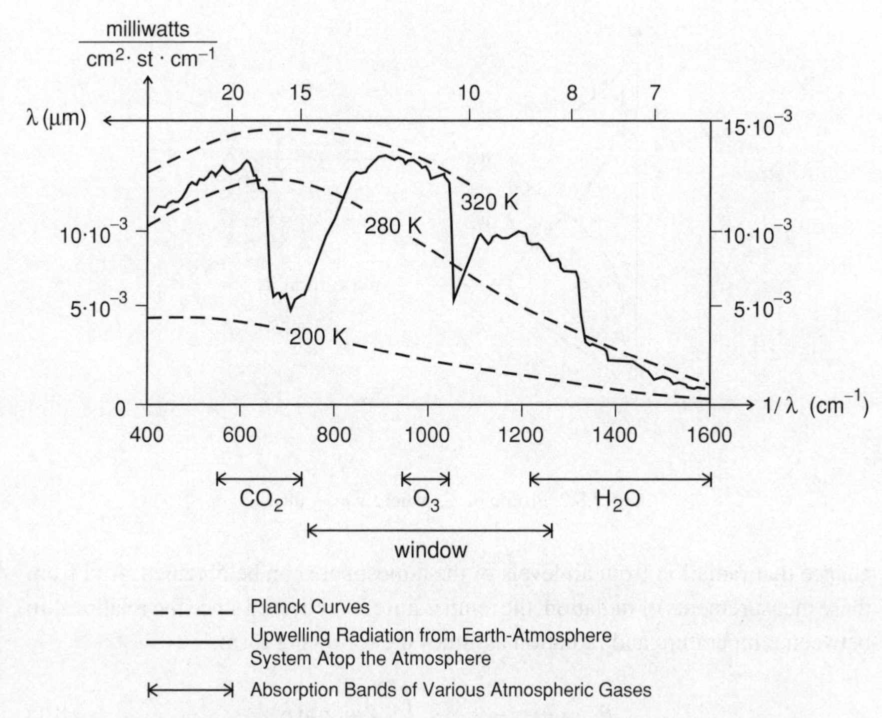

Fig. 3.8.1 The spectrum of radiation emanating from the Earth's surface and atmosphere as viewed from a location high in the atmosphere.

3.8 Atmospheric physics (an inverse problem)

In the late 1950s, in the aftermath of Sputnik's launch, military planners and scientists gave serious thought to surveillance and measurement from artificial satellites that would circle the earth. In this milieu, Lewis Kaplan (1959) proposed measuring the radiance from instruments aboard these satellites in an effort to reconstruct the temperature structure of the atmosphere. The idea rested on the fact that the polyatomic gases in the atmosphere – H_2O (water vapor), CO_2 (carbon dioxide), O_3 (ozone), among others – absorb and emit infrared radiation that comes from the earth's surface as well as from the layers of air above the surface.

Kaplan proposed that the radiation from CO_2 be used for this reconstruction. Whereas water vapor exhibits great variations in the atmosphere, CO_2 has a nearly constant mixing ratio, i.e., the number of grams of CO_2 per grams of other gases is nearly unvarying throughout the atmosphere and so the absorber mass for this constituent is treated as constant. This simplifies the problem significantly. Kaplan suggested that the $15\mu m$ band be used. In Figure 3.8.1 note the rich structure of the radiant energy near the top of the atmosphere in the $15\mu m$ band. There is strong absorption near the center of the band but relatively little at the edges. If radiation is measured from the various lines in this spectrum, there is a good

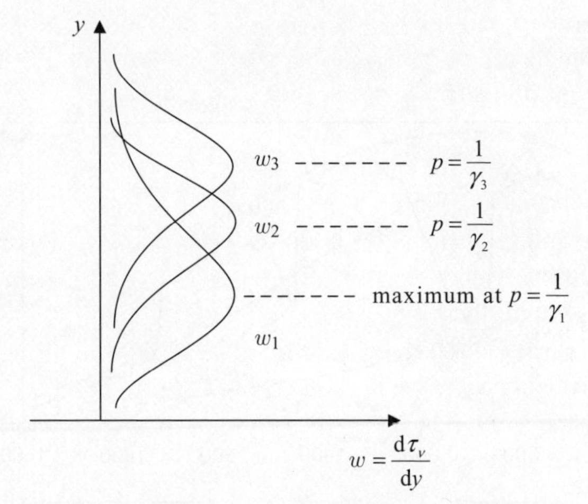

Fig. 3.8.2 Profile of $\frac{d\tau_\nu}{dy}$ where $y = -\ln(p)$.

chance that radiation from all levels of the atmosphere can be obtained. And from these measurements of radiation, the temperature is recovered since the relationship between temperature and radiation assumes the following form:

$$R_\nu = \exp(-\gamma_\nu) + \int_0^1 T w(p) dp \tag{3.8.1}$$

where we have taken the luxury of simplifying the general radiative transfer equation by assuming the Rayleigh–Jeans form of Planck's law (most appropriate for the microwave region of the spectrum) and further expressed the variables as nondimensional numbers, all the order of 1. Now, the variables in the expression are:

p : pressure
ν : wavenumber (inverse wavelength)
$w(p)$: weight function $= p\,\gamma_\nu \exp(-p\,\gamma_\nu)$
γ_ν : transmissivity parameter (highly dependent on wavenumber)

The point to be made here is that the weight functions generally overlap as shown in Figure 3.8.2

If these weight functions had structures more reminiscent of Dirac's delta functions, i.e., spiked at a given value of p and nearly zero elsewhere, then recovery of $T(p)$ from R_ν would be much more straightforward. However, since this is not the case, the redundancy of information from the various radiation measurements presents challenging problems in data assimilation. For example, if we stratified the atmosphere into three layers: $0 \le p \le 0.2$, $0.2 \le p \le 0.5$, and $0.5 \le p \le 1.0$ where the mean temperatures in the layers were represented by T_1, T_2, and T_3, respectively, we should in principle be able to find these temperatures from measurements of radiation at three spectral wavenumbers. However, since the radiation measurements are subject to error and the weighting possess the strong overlap as shown in Figure 3.8.2, the three governing equations do not exhibit a "strong" independence. To overcome these difficulties, we generally use more than the minimum requisite set of observations (more than three in this case). To further help overcome the difficulties, a "guess" profile of temperature from standard upper-air observation is included as "prior" information. One of the most difficult aspects of this recovery problem is that thick cloud effectively makes it impossible to recover temperatures below the cloud top – in effect, the cloud top is a "surrogate" ground surface and it behaves as a black body and exhibits the full spectrum of infrared radiation in accord with Planck's law.

In the formulation of this recovery of temperature problem, the determination of T from measurements of R_ν is called the "inverse" problem while the calculation of R_ν from temperature is referred to as the "forward" problem. In analogy with computation of area under a curve in calculus, we are usually given the integrand and compute the area. The "inverse" is to find the integrand given the area (generally not unique). The forward calculations are typically used to find the model counterpart to the observation, i.e., radiation is typically not one of the model variables but temperature is one of the variables.

Exercises

3.1 Compute the gradient and the Hessian of the functional $J(\mathbf{x})$ in (3.1.4).

3.2 Substituting $x_i = x_0 + i\alpha$ in (3.1.7), compute an explicit expression for $J(x_0, \alpha)$. Then compute $\partial J/\partial x_0$ and $\partial J/\partial \alpha$ and compute the minimizer for $J(x_0, \alpha)$.

3.3 Compute the gradient and the Hessian of $J(\mathbf{b})$ in (3.1.13).

3.4 Using the 4th-order Runge–Kutta method solve the system (3.2.4) and plot $y_1(t)$, $y_2(t)$, $y_3(t)$, $y_4(t)$ as a function of t for various initial conditions.

3.5 Draw the plot that depicts the evolution of the position $(y_1(t), y_3(t))^{\mathrm{T}}$ of the infinitesimal object in the $y_1 - y_3$ plane as t evolves from 0 to 10. Repeat this for various initial conditions.

3.6 Verify that (3.5.5) is the solution of (3.5.4).

Notes and references

The following books and articles elaborate on the applications found in this chapter:

Section 3.2 The dynamics of the special three-body problem is found in Moulton (1902) while Lorenz (1993) and Wolfram (2002) give interesting views on particular solutions.

Section 3.3 Refer to Platzman (1964), Benton and Platzman (1972) and Burgers (1975). Burgers' reminiscences (Burgers 1975) are especially interesting and informative.
Section 3.4 Volume 6 of the Course in Theoretical Physics (Fluid Mechanics) by Landau and Lifshitz (1959) is an excellent, yet concise, review of fluid dynamics including the various forms of gravity waves.
Section 3.5 Proudman's *Dynamical Oceanography*, Gill's *Atmosphere Ocean Dynamics*, and Pedlosky's *Geophysical Fluid Dynamics*, present the governing equations for a variety of problems in oceanography. They also incorporate historical developments that add flavor to the study. The investigation of the Tsunami of April 1, 1946 is found in Green (1946). Discussion of inertia oscillations is found in Sverdrup et al. (1942). The historical review of Coriolis' work by Anders Persson is most informative (Persson (1998)).
Section 3.6 Chapters 11–13 in Jacobson (2005) are an excellent source of background information for this section.
Section 3.7 Lorenz (1960) develops the maximum simplification equations. Epstein (1969b) and Lakshmivarahan et al. (2003) use them to test stochastic/dynamic prediction and data assimilation, respectively.
Section 3.8 Jacobson (2005) and Houghton (2002) introduce the reader to the radiative transfer equation.

4
Brief history of data assimilation

In this chapter we provide an overview of the historical developments that led to the vast and rich discipline called dynamic data assimilation.

4.1 Where do we begin the history?

In Chapters 1 and 2 we have established our philosophy of dynamic data assimilation. Central to this philosophy is the existence of data and governing equations or model dynamics. Thus, the Herculean efforts by scientists like Galileo and Kepler and Newton, efforts that made use of observations to formulate theory, fall outside our scope. Their monumental contributions established some of the governing equations (also known as constraints) upon which dynamic data assimilation depends.

The mathematicians and astronomers of the seventeenth and eighteenth centuries who made use of the Newtonian laws to calculate the orbits of comets were the first data assimilators in our sense of the definition. Newton was among them and he discussed the problem in *Principia* (Book III, Prop. XLI). Regarding the problem of determining the orbit of comets, he said: "This being a problem of very great difficulty, I tried many methods of resolving it." Among the early investigators of this problem were Leonard Euler, Louis Lagrange, Pierre-Simon Laplace, and lesser known amateur astronomers like Heinrich Olbers. The task of finding the path of a comet, of course, relied on the coupled set of nonlinear differential equations that described its path under the assumption of two-body celestial mechanics – the motion controlled by the gravitational attraction of the comet to the Sun.

What is the requisite number of observations to determine the comet's path? Further, what observations are made? As we know from experience, viewing a celestial object in the heavens gives us no information on its distance from us. In short, we express the position by two angular measurements – azimuth and elevation. Since we are ignorant of its distance from us, we are unable to estimate its velocity from successive observations. And from our experience in solving problems in classical physics where the equation(s) of motion are generally second-order ordinary differential equations, the initial conditions are ideally given in the form of position and velocity at the initial time.

It becomes immediately clear that the observations of celestial bodies that are available to us, no matter how many, are not easily translated into the "standard" initial conditions, velocity and position. At each epoch we can obtain two observations (the angles). Thus, in principle, the six constants that arise from the integration of the governing differential equations should be determined from three complete observations (i.e., two angular measurements at each of the three instants of time). If the object is known to be a comet in an assumed parabolic orbit, then the number of requisite observations can be reduced to five since the eccentricity is then known.

As you can imagine, expressing the observed angles as functions of the elements of the orbit (eccentricity, orientation of orbital plane with respect to the ecliptic, period of the motion [if elliptical path]), and time of observations, leads to transcendental functions of some complexity. There is no direct/analytical solution to these equations; yet some of the great minds of the eighteenth century, most notably Lagrange and Laplace, were able to simplify the problem through a number of ingenious transformations that are presented in Forrest Moulton's pedagogical and thorough text on celestial mechanics (Moulton 1902).

4.2 Laplace's strategy for orbital determination

It is not unusual to have more than the minimum requisite set of observations. As shown by Moulton (1902), this allows better approximations in the power series expansions of the basic variables. Yet, Laplace realized that this advantage still fell short of maximum utilization of the observations. In the late 1700s, he established a data assimilation strategy that rested upon the use of residuals, the measure of the dissatisfaction of the governing equations when observations and by-products of observations are substituted into the equations. The constraints he placed on the problem were:

(a) the algebraic sum of the residuals should vanish, and
(b) the sum of the absolute values of the residuals should be a minimum.

The first constraint is a statement of the assumption that the positive and the negative residuals should balance out. This constraint reduces the number of control variables (the elements of the orbit) by one. The second constraint, an optimality condition, established conditions that the control variables must satisfy. Namely, the derivatives of the sum of the absolute values with respect to each control variable should vanish. These minimization principles had been established through the earlier work of Euler and Lagrange in the development of the calculus of variations. The intuitive difficulty with Laplace's methodology is that the extreme values of residuals are not sufficiently penalized. Furthermore, the mathematical operations, such as taking derivatives of the absolute values of complicated expressions, transcendental expressions for example, is complicated, laborious, and certainly nontrivial.

4.3 The search for Ceres

Near the end of the eighteenth century, the German astronomer Johann Bode predicted the existence of a planet between Mars and Jupiter (at 2.8 A.U. from the sun†). The prediction was based on an empirical law of planetary distances – "... a regularity first noted by J. D. Titius in 1766, but discussed more fully and applied with great daring by the director of the Berlin Observatory, J. E. Bode (1747–1826) ..." (Phelps and Stein (1962)). Astronomers began to search for the "missing planet". Guiseppe (Joseph) Piazzi (1746–1826) is credited with the first sighting of this "planet" on January 1, 1801, at $\sim$ 2.8 A.U. from the Sun. The sighting occurred in Palermo, Italy. Piazzi had devoted his career to cataloging stars from the favorable observing station at Palermo on the north coast of the Mediterranean island of Sicily. On January 1, 1801, he was trying to confirm the location of a faint star (7th magnitude) in the constellation *Taurus* (the Bull) when he noticed another celestial object that was unknown to him. He initially believed it to be a star but further observation indicated that it was moving, albeit extremely slowly, among the fixed stars. Its initial motion (until January 10–11) was retrograde (westward), but it eventually began to move eastward and over a 42-day period of observation, the object traversed a small celestial arc of only 3°. In the twilight of February 11, the object disappeared in the strong solar rays (conjunction with the sun). Among others, Piazzi communicated his finding to Bode at the Berlin Observatory, and in this correspondence it is clear that Piazzi was uncertain whether the object was a star, planet, or comet. This fact is reflected in the title of his booklet *Risultate delle Osservazioni di Palermo* (Result of the observations at Palermo).

Piazzi's observations were made available to astronomers and mathematicians in Europe, at first incompletely in May, 1801, and then in detail by September. The race was on to predict the place and time of reappearance of this celestial body. At some point after the publication of the booklet, it was determined that the object was a planet and Piazzi named it *Ceres Ferdinadea* in honor of the patron goddess of Sicily (*Ceres*–goddess of agriculture in Roman mythology) and Ferdinand IV (1751–1825) (King of Naples and Sicily). During the summer 1801, Carl Gauss, at that time a 24 year old mathematician who had just received his doctoral degree after completing studies at the Universities of Göttingen and Helmstedt, decided to enter the "race".

By November of 1801, Gauss had calculated the future path of *Ceres* and published the results a month later in the booklet titled: *Fortgesetzte Nachrichten über*

† The minimum distance between Sun and Earth is designated as 1 astronomical unit (A.U.)

den längst vermutheten neuen Haupt-Planeten unseres Sonnen-Systems (Continuing news on the long-time suspected new major planet in our solar system). Gauss's predicted position of reappearance and path of *Ceres* was confirmed simultaneously on January 1, 1802, by Franz Zach at the Gotha Observatory in central Germany and by Heinrich Olbers in Bremen (northern Germany). It was truly one of the most spectacular exhibitions of mathematical prowess in the history of astronomy. Through his tracking of *Ceres*, Olbers detected another "planet" on 2 March 1802 at approximately 2.8 A.U. He called it *Pallas*. It was then realized that these smaller celestial objects moving in elliptical paths around the sun at $\sim$ 2.8 A.U. were much smaller than the other planets and they have since become known as planetoids (asteroids). Since that time, many thousands of planetoids have been discovered in this belt.

4.4 Gauss's method: least squares

In Gauss's book, *Theoria Modus Corporum Coelestium* (Theory of the Motion of Heavenly Bodies), written in 1809 (8 years after his calculations that led to the location of Ceres after its conjunction with the sun), we have the record of his views and development of least squares[‡]. It is not an easy book to read, but we should have expected nothing else in view of Gauss's approach to discussion of his contributions. Eric Temple Bell, Caltech mathematician and historian of mathematics described Gauss's writing as follows:

> A cathedral is not a cathedral, he [Gauss] said, until the last scaffolding is down and out of sight. Working with this ideal before him, Gauss preferred to polish one masterpiece several times rather than to publish the broad outlines of many as he might easily have done.
>
> (Bell 1937)

Although the mathematics is difficult to follow in Gauss's treatise, the verbal statements are often quite clear. Regarding the optimal use of all observations in the calculation of orbits, he says:

> If the astronomical observations and other quantities on which the computation of orbits is based were absolutely correct, the elements also, whether deduced from three or four observations, would be strictly accurate (so far indeed as the motion is supposed to take place exactly according to the laws of Kepler) and, therefore, if other observations were used, they might be confirmed but not corrected. But since all our observations and measurements are nothing more than approximations to the truth, the same must be true of all calculations resting on them, and the highest aim of all computations made concerning concrete phenomena must be to approximate, as nearly as practicable, to the truth. But this can be accomplished in no other way than by suitable combination of

[‡] Gauss's book was reprinted by Dover in 1963

more observations than the number absolutely requisite for the determination of the unknown quantities. This problem can only be properly understood when an approximate knowledge of the orbit has been already attained, which is afterwards to be corrected so as to satisfy all the observations in the most accurate manner possible.

(Gauss 1963)

In this statement, Gauss lays the groundwork for the method of least squares – "... to satisfy all the observations in the most accurate manner." When the optimal condition involves the squared departure between estimate and observation, the associated mathematical operations are ever so much more straightforward than when dealing with absolute values. Further, the penalty for extreme departures is intuitively satisfying. Gauss also implies that linearization about the current operating estimate is necessary – an iterative process that makes use of the latest estimate to linearize the dynamical constraints.

4.5 Gauss's problem: a simplified version

In the preface of *Theoria Motus*, Gauss reconstructs the events that led to the discovery of Uranus (in 1781, four years after Gauss's birth). It indeed is a dramatic story that we mentioned earlier (Hoyle 1962). Gauss's discussion focuses on the special circumstances that surrounded the discovery. Among these circumstances were: assumption of a circular orbit ("By a happy accident the orbit of this planet has a small eccentricity") (Gauss, 1963, xiii), the slow motion of the planet, the very small inclination of the orbit to the plane of the ecliptic, and its brilliant light. Gauss goes on to say:

To determine the orbit of a heavenly body without any hypothetical assumption, from observations not embracing a great period of time, and not allowing a selection with a view to the application of special methods, was almost wholly neglected up to the beginning of the present century [the nineteenth century].

(Gauss 1963, xiv)

With this background, he sets the stage for his work on the method of least squares applied to the planetoid Ceres. He ends his discussion with the statement.

Could I ever have found a more seasonable opportunity to test the practical value of my conceptions, than now in employing them for the determination of the orbit of the planet Ceres, which during these forty-one days had described a geocentric arc of only 3 degrees, and after the lapse of a year must be looked for in the region of the heavens very remote from that in which it was last seen? This first application of the method was made in the month of October, 1801, and the first clear night, when the planet was sought for as deduced by the numbers deduced from it, restored the fugitive to observation.

(Gauss 1963, xv)

In this chapter, we present issues faced by Gauss in the determination of the orbit of a heavenly body rotating about the sun. We take the liberty of introducing several

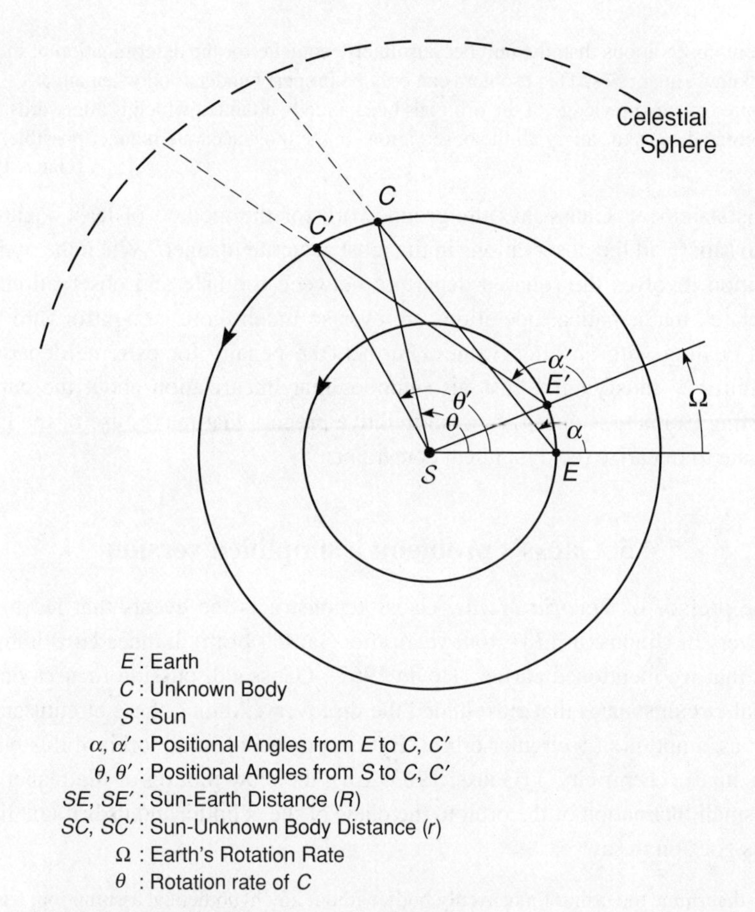

Fig. 4.5.1 Measurement symbols associated with tracking of a planetoid (C) from earth (E).

assumptions that allow us to solve the problem meaningfully yet more simply than the general problem discussed by Gauss in *Theoria*. We first solve our hypothetical problem with the minimum set of observations and afterwards outline the method of solution in the presence of more than the minimum requisite set.

We assume that the orbit of the planetoid (denoted by C) and that of the earth (denoted by E) are circular about the sun (denoted by $\mathcal{S}$). The conceptual configuration is shown in Figure 4.5.1. We will further assume that the orbits of the earth and planetoid are co-planar, i.e., the objects $\mathcal{S}$, E and C lie in the same plane.

Angular observations of C are made from E. These angles are denoted by α and α' in Figure 4.5.1. The rotation rate of E about $\mathcal{S}$ is known ($\Omega = 2\pi$ radians per year). We assume that at $t = 0$, we sight C from E at angle $\alpha(0)$, i.e., $t = 0$. C can be anywhere on this line of sight. At a later time, E will have moved around the circle of radius R and the angular movement will be $\frac{2\pi}{1y} \cdot \Delta t$, where Δt is the later time expressed in fraction of a year. Of course, C will have moved in its circle an angular distance $\dot{\theta}\,\Delta t$ where $\dot{\theta}$ is the uniform angular rate of rotation of C about $\mathcal{S}$. Sighting C from E at Δt will give us a new $\alpha(\Delta t)$ $(= \alpha')$ and the time interval between is known.

Assuming the measurements are exact and that we believe Kepler's law to be perfect, two observations will be sufficient to find the orbital elements of C. Intuitively, we can see that this is the case if we invoke Kepler's 3rd law that says

$$\frac{r^3}{T^2} = \text{constant}$$

for each body in rotation around the sun (the circle being a special case of the more general elliptical orbits) where r is the radius and T is the period of rotation. We know the value of the constant since $T = 1$ year (y) and $r = 1\ A.U.$ for the Earth, i.e.,

$$1 = \text{constant} = \frac{(1\ A.U.)^3}{(1\ y)^2}.$$

Assume we site C at the initial time ($t = 0$) along the line extending from earth (E) to the point on the celestial sphere marked "observation 0" (Figure 4.5.2). We will make another observation one month later. In anticipation of this observation, let us find the displacement of objects moving about the sun that lie on this initial line. The displacements over the one month period are shown for objects at radii = 1.5, 2.0, 3.0, and 3.5 A.U. (arrows show relative displacement). One month after the initial observation, earth is at E'. The object is cited along the line extending from E' to the point marked "observation 1" on the celestial sphere. By inspection of the schematic diagram, it is seen that the only displacement that satisfies Kepler's law is the one at 2 A.U.

For a more rigorous solution, we derive formulas for finding r and $\dot{\theta}$ (radius and rotation rate of C, respectively) from the measurements of α, where Figure 4.5.1 defines notation. In this approach we will assume a measurement of α and the rate of change of α near the time of initial measurement (i.e., $\dot{\alpha}|_{t=0}$).

It can be shown that

$$\alpha(t) = \tan^{-1}\frac{r\sin(\theta + \dot{\theta}\,t) - R\sin\Omega t}{r\cos(\theta + \dot{\theta}\,t) - R\cos\Omega t}.$$

Let

$$x(t) = -R\cos\Omega t + r\cos(\theta + \dot{\theta}\,t)$$

and

$$y(t) = -R\sin\Omega t + r\sin(\theta + \dot{\theta}\,t).$$

Then

$$\tan\alpha = \frac{y}{x}.$$

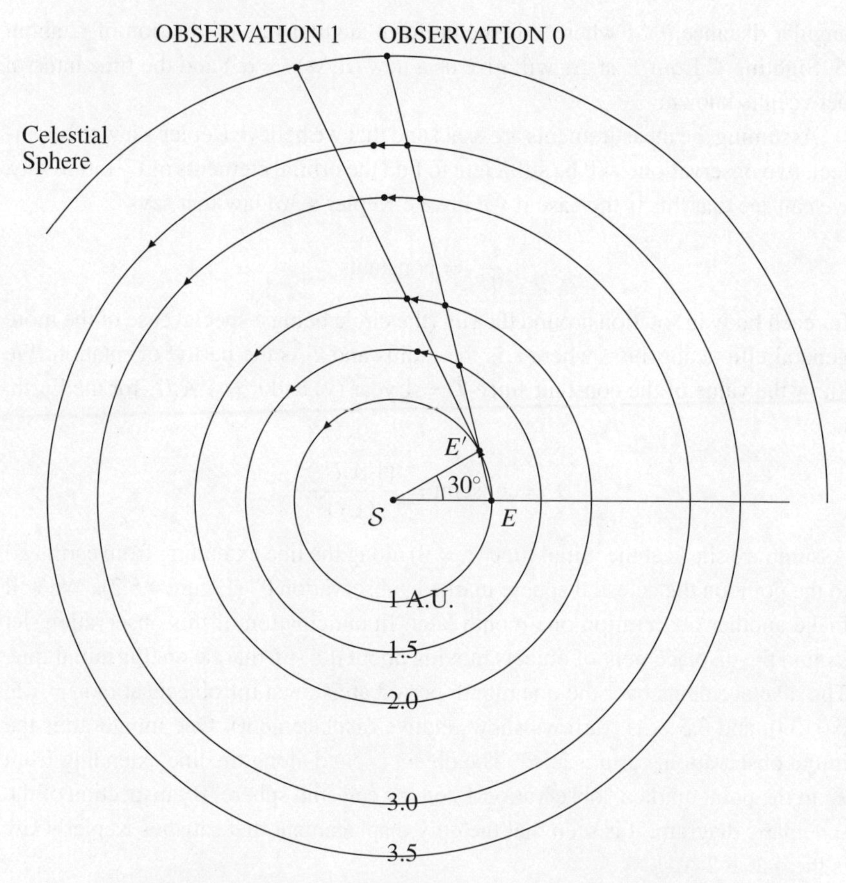

Fig. 4.5.2 Rotation rates determined from Kepler's 3rd Law. An object at unknown distance from earth is observed at two times – "0" and "1" (earth at E and E', respectively). Use of Kepler's law yields its distance from the Sun (2.0 A.U.).

Taking the derivative of α with respect to time, we get

$$\dot{\alpha} = \frac{\dot{y} - \dot{x}\ \tan\alpha}{x\ \sec^2\alpha}.$$

Now, at $t = 0$ (where θ is the unknown initial angle), we have

$$\tan\alpha = \frac{r\ \sin\theta}{r\ \cos\theta - R}.$$

Also, if we evaluate $\dot{\alpha}$ at $t = 0$, we get

$$\dot{\alpha}|_{t=0} = \frac{\dot{y} - \dot{x}\ \tan\alpha}{x\ \sec^2\alpha}|_{t=0}.$$

From Kepler's 3rd Law, we also know

$$\dot{\theta}^2 = \frac{4\pi^2}{r^3}.$$

Thus, if we examine these last three equations in terms of r, $\dot{\theta}$, θ, and the known quantities, we have

$$\tan\alpha = \frac{r\,\sin\theta}{r\,\cos\theta - R} = \frac{\sin\theta}{\cos\theta - [R/r]} \tag{4.5.1}$$

$$\dot{\alpha}|_{t=0} = \left[\frac{\dot{y}(0) - \dot{x}(0)\tan\alpha}{x(0)\sec^2\alpha}\right]_{t=0} \tag{4.5.2}$$

$$\dot{\theta}^2 = \frac{4\pi^2}{r^3} \tag{4.5.3}$$

where

$$x(0) = -R + r\,\cos\theta, \qquad y(0) = r\,\sin\theta$$

$$\dot{x}(0) = -r\dot{\theta}\sin\theta$$

$$\dot{y}(0) = r\dot{\theta}\cos\theta - R\Omega\,.$$

Remembering that α, $\dot{\alpha}|_{t=0}$, R and Ω are known, equations (4.5.1)–(4.5.3) constitute three equations in the three unknowns r, $\dot{\theta}$, and θ. They are not easily solved, and an iterative method is dictated.

Since we know the planetoid is positioned near 3 A.U. (Bode's Law), we begin our iteration by "guessing" at r, say $\hat{r} = 3$. We have measured α, so (4.5.1) will give us a guess at θ, say $\hat{\theta}$. If we substitute from (4.5.3) into (4.5.2), eliminating $\dot{\theta}$, then (4.5.2) contains θ and r. We have measured $\dot{\alpha}|_{t=0}$, and we can solve for an improved value of r by assuming $r = \hat{r} + p$, linearizing (4.5.2) as a function only of p, $\hat{\theta}$, and then solve for p.

We then return to (4.5.1) to get a new estimate of θ using $(\hat{r} + p)$ in place of r. Continue to iterate until p becomes vanishing small. This is Newton's original idea of solving these equations, later augmented by Joseph Raphson and subsequently called the Newton-Raphson method.

For a more meaningful problem formulation, assume observations of angle $\alpha(t)$ are subject to error. How would we accommodate more than two observations? For example, assume we have three measurements, now denoted by $\tilde{\alpha}_0$, $\tilde{\alpha}_1$, and $\tilde{\alpha}_2$. We also assume we know the time that each observation is made, say $t_0 = 0$, t_1, and t_2. As an extension of the method just discussed, we could take the observations in three sets of two each. These sets would give different solutions and we could average them; yet we intuitively know that this approach does not account for all three observations in a unified manner. What Gauss did was to construct a measure of the fit of the model (Kepler's law) to the observations by using the least squares principle. Let us call this expression, J, and define it as follows:

$$J = (\alpha_0 - \tilde{\alpha}_0)^2 + (\alpha_1 - \tilde{\alpha}_1)^2 + (\alpha_2 - \tilde{\alpha}_2)^2\,.$$

The α_0, α_1, and α_2 are the angular measurements we desire to find, and we wish to find them in a way that will minimize J under the constraint that the α's must satisfy Kepler's law. In essence, we find the set $(r, \dot{\theta}, \theta)$ that gives us the α's that minimize J.

We can think of the triad r, $\dot{\theta}$, and θ as elements of the (control) vector $\mathbf{C} = (r, \dot{\theta}, \theta)$. This vector controls the evolution of the path of the planetoid. J is determined by this vector. In the space of these elements, J has a distinct distribution. The job is to find the point in this space that will yield the smallest value of J, i.e., the least-squares fit between the model-derived state (α_0, α_1 and α_2) and the observations ($\tilde{\alpha}_0$, $\tilde{\alpha}_1$, and $\tilde{\alpha}_2$, respectively).

We can start as we did in the earlier example, namely, we make a guess at the elements. Let us assume that our initial guesses are:

$$\alpha_0^{(1)} = \tilde{\alpha}_0$$
$$r^{(1)} = 3$$

and $\theta_0^{(1)}$ the solution to equation (1) where α and r are given by the guesses. Now we can make predictions of α_1 (call it $\alpha_1^{(1)}$) and α_2 (call it $\alpha_2^{(1)}$) by using

$$\dot{\alpha} = \frac{\dot{y} - \dot{x}\ \tan\alpha}{x\ \sec^2\alpha}$$

evaluated at $t = 0$ as shown above. The position coordinate x and y are known as functions of r, $\dot{\theta}$, and θ and known quantities. Thus, we have a forecast equation to get values of the α's at the various times downstream. To accomplish this, we would generally use some finite-difference approximation to the forecast equation.

Is our initial guess the minimum? We would be most fortunate if this was the case and we would never expect it. Generally, we are tasked with the problem of finding an improvement to our guess. We know the value of J associated with this guess, and we can find the derivative of J with respect to each element of the control vector. This is not trivial, but there are a variety of ways to find these derivatives and this is an important component of our methodologies of data assimilation. If we could find the second derivatives of J with respect to these elements, we would be well-posed to determine a better estimate. In some cases, this is possible but generally we make improvements knowing only the first derivatives. Let us proceed along these lines.

If we know the first derivatives at our operating point (the initial guess), we can hope to find an improved estimate by moving along the direction of the negative gradient of J, i.e., along the direction of the vector – $-(\partial J/\partial r, \partial J/\partial\dot{\theta}, \partial J/\partial\theta)$. Just how far we move in that direction is still undetermined; yet there are strategies to find these "step lengths". Assuming we find the gradient and an appropriate step length, we get a new estimate of the minimum of J. The iterative process is continued until we satisfy some empirical criterion related to the size of the gradient (near the minimum the gradient approaches zero) or differences in successive values of J are infinitesimal (the successive iterates are smaller than some predetermined value).

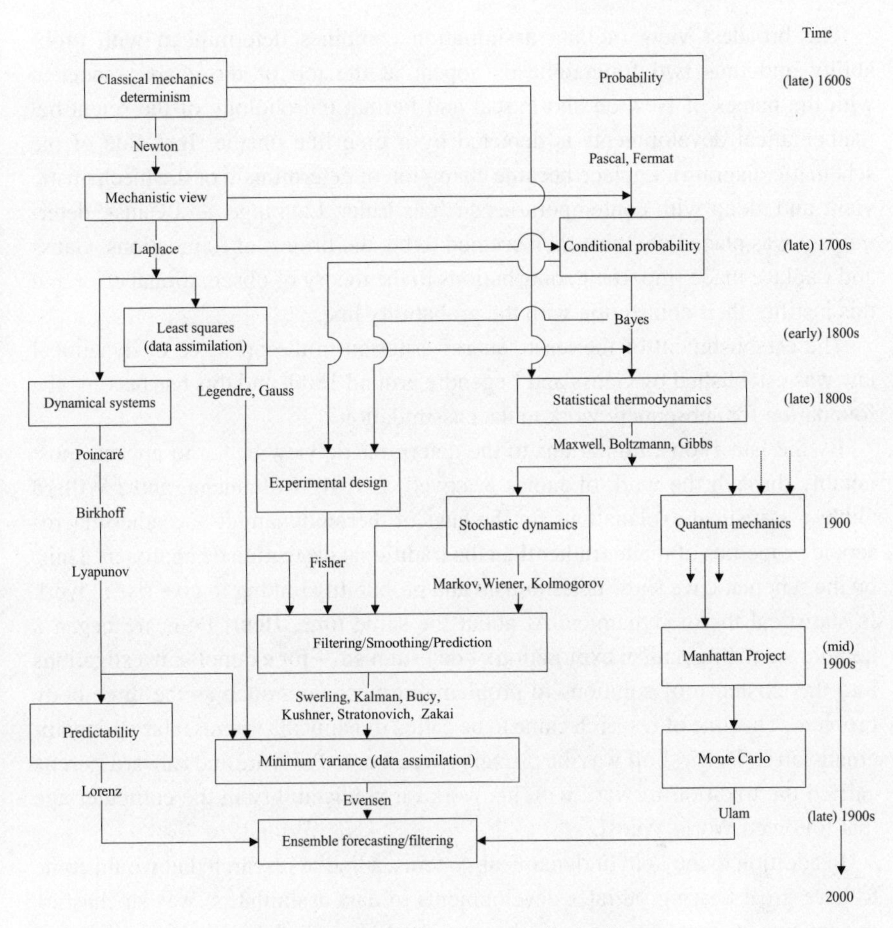

Fig. 4.6.1 Schematic for data assimilation history.

In essence, Gauss used methodology similar to the one we have outlined. The structure of J in the space of the control vector is fundamental to dynamic data assimilation and its geometric complexity governs the ease or difficulty we experience when searching for the minimum. Observation density and character of the dynamics, especially the degree of nonlinearity, are the primary factors that control this complexity.

4.6 Probability enters data assimilation

The development of dynamic data assimilation that couples both dynamical law and probability can be traced to several fundamental lines of research in the history of science. The schematic diagram in Figure 4.6.1 depicts these lines.

The broadest view of data assimilation combines determinism with probability and thus two fountainheads appear at the top of the chart associated with the names of Newton and Pascal and Fermat (chronology of the scientific/ mathematical developments is depicted by a time line on the right side of the schematic diagram). Laplace became champion of determinism or the mechanistic view, and along with contemporaries such as Euler, Lagrange, and Gauss, determinism was placed on, what then seemed to be, the firmest of foundations. Gauss and Laplace made important contributions to the theory of observational error and this justifies their connection with the probability line.

The establishment of the least squares criterion in the presence of dynamical law was established by Gauss and Legendre around 1800, and this has become the foundation for subsequent work in data assimilation.

By the late 1800s, limitations to the deterministic view began to appear, most notably through the work of James Maxwell, Ludwig Boltzmann, and J.Willard Gibbs – statistical explanations for the laws of thermodynamics and other macroscopic properties of matter rather than the traditional deterministic approach. Thus, on the schematic we show determinism and probability linking to give rise to work in statistical thermodynamics. At about the same time, Henri Poincaré began a heightened mathematical exploration of determinism – for example, investigations into the existence of solutions to problems in mechanics such as the three-body problem. This line of research came to be called dynamical systems. Harvard mathematician G. D. Birkhoff was the primary successor to Poincaré and Edward Lorenz carried the tradition forward with his work on predictability in the computer age that followed World War II.

In addition to the field of dynamical systems, a line of research that would come to have great bearing on later developments in data assimilation was stochastic–dynamic prediction, an approach that stemmed from mathematical issues related to Brownian motion and statistical thermodynamics, among other processes that exhibited randomness. The foundations of this field of study are associated with the name of Alexei Markov, Norbert Wiener, and Andrei Kolmogorov.

Aside the work in dynamical systems and stochastic–dynamic processes, we have a statistical vein that stems from the Rev. Thomas Bayes' and Ronald Fisher's work in experimental design and hypothesis testing. By mid-nineteenth century, this line of attack and the work in stochastic–dynamic processes laid the groundwork for the theory of filtering, smoothing, and prediction – analysis of random processes.

With a base in deterministic least squares, dynamical systems, stochastic–dynamic processes, and random processes, the stage was set for the sequential or online approach to the minimum variance approach to data assimilation. This work is primarily associated with the names – Swerling, Kalman, Bucy, Stratanovich, Kushner, and Zakai.

Finally, we view the current practice of ensemble forecasting as a culmination of the work in predictability, data analysis via minimum variance, and Monte Carlo, the

Fig. 4.6.2

probabilistic approach that found wide application in the investigation of branching processes associated with nuclear bombardment. It, of course, is traced back to the Manhattan Project and ultimately to quantum mechanics, the fundamental break with Newtonian mechanics. Stanislaw Ulam, mathematician and colleague of von Neumann and Fermi, was the originator of this probabilistic approach.

Several of the pioneers of data assimilation are pictured in Figure 4.6.2 (Early Period) and Figure 4.6.3 (Later Period) [Portraits by the author, J. L.]. In the early

Fig. 4.6.3

period, we have (clockwise from upper-left): Isaac Newton, Carl Gauss, Heinrich Olbers, and Pierre-Simon Laplace. In the later period, we have: Henri Poincaré, Andrei Kolmogorov, Edward Lorenz, and Norbert Wiener.

Exercises

4.1 Referring to Figure 4.5.1, show that

$$\tan\alpha = \frac{r\sin(\theta + \dot{\theta}t) - R\sin(\Omega t)}{r\cos(\theta + \dot{\theta}t) - R\cos(\Omega t)}.$$

Hint: Use complex number notation and associated rules.

4.2 A planetoid is sited from Earth. Following the notation and graphics in Figure 4.5.1, the observations are as follows:

$$\alpha = 1.82 \text{ radians} \qquad \text{at } t = 0$$
$$\dot{\alpha} = 1.10 \text{ (radians/year)} \quad \text{at } t = 0$$

Find the distance of the planetoid from the Sun by following the iterative procedure outlined in the text. As a first guess, assume $r = 3.5\, A.U.$

4.3 A planet is observed from Earth. Following the notation and graphics in Figure 4.5.1, the three observations are:

$$\widetilde{\alpha}_{-1} = 1.03\,\pi, \quad \widetilde{\alpha}_0 = \pi\,, \quad \widetilde{\alpha}_1 = 0.97\,\pi$$

These observations are taken at $t_{-1} = -2 \times 10^{-2}$, $t_0 = 0$, and $t_1 = 2 \times 10^{-2}$ (in fraction of 1 year) [roughly 1 week apart]. As a first guess for r (distance of planet from Sun), use the geometry of the observation angles to express r in A.U. Use this guess for r to estimate the rotation rate of the planet (Kepler's law). We will assume that θ is known exactly, $\theta = \alpha_0 = \pi$ radians. Thus, the control vector is $(r, \dot{\theta})$. Given the control vector, the planet's position can be found. And from the planet's position at a given time, the angles α can be calculated – and the value of the functional determined. We write this functional as:

$$J = (\alpha_{-1} - \widetilde{\alpha}_{-1})^2 + (\alpha_0 - \widetilde{\alpha}_0)^2 + (\alpha_1 - \widetilde{\alpha}_1)^2\,.$$

Using a range of values around the estimated r and $\dot{\theta}$, calculate the corresponding values of J. From this plot, determine the value of r and $\dot{\theta}$ that minimize J.

Notes and references

Section 4.2–4.3 In addition to Gauss (1963), the following books discuss the discovery of Ceres: Hall (1970), Dunnington (1955) and Reich (1985). An interesting discussion of Bode's work is found in Phelps and Stein (1962).
Section 4.4 Eric Temple Bell's book on the history of math makes for enjoyable reading – very opinionated!
Section 4.5 Discussion of Kepler's laws that set the stage for the problem in this section is found in Cohen (1960).
Section 4.6 Harold Sorenson's article on least squares and Kalman filter in light of Gauss' work is a splendid historical view of data assimilation Sorenson (1970). The article by Lewis (2005) links dynamics and probability with the current practice of ensemble forecasting in meteorology.

PART II

Data assimilation: deterministic/static models

5

Linear least squares estimation: method of normal equations

In this chapter our goal is to describe the classical method of linear least squares estimation as a deterministic process wherein the estimation problem is recast as an optimization (minimization) problem. This approach is quite fundamental to data assimilation and was originally developed by Gauss in the nineteenth century (refer to Part I). The primary advantage of this approach is that it requires no knowledge of the properties of the observational errors which is an integral part of any measurement system. A statistical approach to the estimation, on the other hand, relies on a probabilistic model for the observational errors. One of the important facets of the statistical approach is that under appropriate choice of the probabilistic model for the observational errors, we can indeed reproduce the classical deterministic least squares solution described in this chapter. Statistical methods for estimation are reviewed in Part IV.

The opening Section 5.1 introduces the basic "trails of thought" leading to the first formulation of the linear least squares estimation using a very simple problem called the straight line problem (see Chapter 3 for details). This problem involves estimation of two parameters – the *intercept* and the *slope* of the straight line that is being "fitted" to a swarm of m points (that align themselves very nearly along a line) in a two-dimensional plane. An extension to the general case of linear models – m points in n dimensions ($n \geqq 2$) is pursued in Section 5.2. Thanks to the beauty and the power of the vector-matrix notation, the derivation of this extension is no more complex than the simple two-dimensional example discussed in Section 5.1.

For concreteness, in Sections 5.1 and 5.2, it is assumed that the number m of observations is greater than n, the number of unknowns – the *inconsistent* or the *over-determined* problem. The dual case of $m < n$, known as the *under-determined* problem is developed in Section 5.3.

A unified treatment of both the over- and under-determined cases using Tikhonov regularization is described in Section 5.4. The last Section 5.5 contains several concluding observations and provides links to the vast literature on this topic.

5.1 The straight line problem

Consider an object travelling in a straight line at a constant velocity. We can observe the position z_i of this object at time t_i, for $i = 1, 2, \ldots, m$, where $t_1 < t_2 < \cdots < t_m$. Given the pairs $\{(z_i, t_i) | i = 1, 2, \ldots, m\}$, the problem of interest is to *estimate* the *unknown velocity* ν and the *initial position* z_0. From the first principles, we readily see that

$$
\begin{aligned}
z_1 &= z_0 + \nu t_1 \\
z_2 &= z_0 + \nu t_2 \\
&\vdots \\
z_m &= z_0 + \nu t_m .
\end{aligned}
\tag{5.1.1}
$$

In expressing (5.1.1) using matrix/vector notation (see Appendices A and B for details), we introduce the following. Let $\mathbf{z} \in \mathbb{R}^m$ with $\mathbf{z} = (z_1, z_2, \ldots, z_m)^{\mathrm{T}}$, $\mathbf{x} \in \mathbb{R}^2$ with $\mathbf{x} = (z_0, \nu)^{\mathrm{T}}$, and $\mathbf{H} \in \mathbb{R}^{m \times 2}$ with

$$
\mathbf{H} = \begin{bmatrix} 1 & t_1 \\ 1 & t_2 \\ \vdots & \vdots \\ 1 & t_m \end{bmatrix}
\tag{5.1.2}
$$

where T denotes the transpose operation. Then (5.1.1) can be written as

$$
\begin{bmatrix} z_1 \\ z_2 \\ \vdots \\ z_m \end{bmatrix} = \begin{bmatrix} 1 & t_1 \\ 1 & t_2 \\ \vdots & \vdots \\ 1 & t_m \end{bmatrix} \begin{bmatrix} z_0 \\ \nu \end{bmatrix}
\tag{5.1.3}
$$

or more succinctly as

$$
\mathbf{z} = \mathbf{H}\mathbf{x}. \tag{5.1.4}
$$

Remark 5.1.1 In the parlance of data assimilation, the matrix $\mathbf{H}$ represents the measurement system that relates the *unknown state vector* $\mathbf{x}$ to the *observation vector*, $\mathbf{z}$. In this example, the observation is *linearly* related to the unknown state, and hence the title linear least squares estimation. In general, the observation may be a *non-linear* function of the elements of the state vector. For example, satellites measure the energy radiated which is a non-linear function of the temperature, and radars measure the radiance which is non-linearly related to the diameter of the rain droplets. In this opening section, since our aim is to get a traction on the basic principles and techniques of least squares, we confine our attention to the linear case. Non-linear least squares estimation is described in Chapter 8.

The equation (5.1.4) denotes a system of $m(\geq 2)$ linear equations in two unknowns. Let $\mathbf{h}_{*1}$ and $\mathbf{h}_{*2}$ be the first and the second columns of the matrix $\mathbf{H}$. Then, range space of $\mathbf{H}$, defined by

$$\text{Range}(\mathbf{H}) = \{\mathbf{y} | \mathbf{y} = a\mathbf{h}_{*1} + b\mathbf{h}_{*2}, \text{ where } a \text{ and } b \text{ are real numbers}\}$$

denotes the two-dimensional subspace of $\mathbb{R}^m$ defined by the hyperplane that contains $\mathbf{h}_{*1}$ and $\mathbf{h}_{*2}$ and passes through the origin. In other words, Range($\mathbf{H}$) denotes the set of all linear combinations of the columns of $\mathbf{H}$. If the observation vector $\mathbf{z} \in$ Range($\mathbf{H}$), then (5.1.4) is called a *consistent* system, otherwise it is called an *inconsistent* system. Thus, unless $\mathbf{z}$ lies in this range space of $\mathbf{H}$, we *cannot* find an $\mathbf{x} \in \mathbb{R}^2$ that solves (5.1.4) in the usual sense of the solution, namely, $\mathbf{z} = \mathbf{Hx}$. In the following, it is assumed that $\mathbf{z}$ does not belong to the range of $\mathbf{H}$, and (5.1.4) is called the *over-determined* and *inconsistent* system of equations. In this case, there is a need to redefine the solution of (5.1.4).

To this end, the notion of *residual* vector

$$\mathbf{r} = \mathbf{r}(\mathbf{x}) = \mathbf{z} - \mathbf{Hx} \tag{5.1.5}$$

is introduced, where $\mathbf{r} = (r_1, r_2, \ldots, r_m)^{\mathrm{T}}$ and $r_i = z_i - (z_0 + vt_i)$ for $i = 1, 2, \ldots, m$. The *length* of this residual vector, denoted by $\|\mathbf{r}\|$, is often taken as a measure of the "goodness" of the solution. Since the ideal solution of $\mathbf{r}(\mathbf{x}) = \mathbf{0}$, the null vector, is impossible to achieve in an inconsistent system, a useful characterization of the solution is to seek that vector $\mathbf{x} \in \mathbb{R}^2$ for which the length $\|\mathbf{r}(\mathbf{x})\|$ attains a minimum value. This discussion now leads to the following.

(A) Statement of the linear least squares problem: Given a measurement system denoted by $\mathbf{H} \in \mathbb{R}^{m\times 2}$ and the observation vector $\mathbf{z} \in \mathbb{R}^m$, find an $\mathbf{x} \in \mathbb{R}^2$, such that

$$f(x) = \|\mathbf{r}(\mathbf{x})\|^2 = \|\mathbf{z} - \mathbf{Hx}\|^2 \tag{5.1.6}$$

attains the minimum value.

Notice that the unknown vector $\mathbf{x}$ to be estimated becomes the independent variable over which this minimization problem is defined. In this case since $\mathbf{x}$ is allowed to take any value in $\mathbb{R}^2$ without any restriction, this minimization problem is often known as the *unconstrained minimization* problem.

Clearly, $\|\mathbf{r}\|^2 = f : \mathbb{R}^2 \longrightarrow \mathbb{R}$ is a scalar valued function of the vector $\mathbf{x}$ and is known as a *functional*. By invoking the first principles of multivariate optimization (Appendix D), it follows that minimizing $\mathbf{x}$ is obtained as a solution to the following sufficient conditions:

$$\left.\begin{array}{l}\text{first-order condition: } \nabla f(\mathbf{x}) = 0 \\ \text{second-order condition: } \nabla^2 f(\mathbf{x}) \text{ is positive definite}\end{array}\right\} \tag{5.1.7}$$

where

$$\nabla f(\mathbf{x}) = \left(\frac{\partial f}{\partial z_0}, \frac{\partial f}{\partial v}\right)^{\mathrm{T}} \tag{5.1.8}$$

is the *gradient* vector of $f(\mathbf{x})$, and

$$\nabla^2 f(\mathbf{x}) = \begin{bmatrix} \frac{\partial^2 f}{\partial {z_0}^2} & \frac{\partial^2 f}{\partial z_0 \partial v} \\ \frac{\partial^2 f}{\partial v \partial z_0} & \frac{\partial^2 f}{\partial v^2} \end{bmatrix} \tag{5.1.9}$$

is the Hessian which is a *symmetric* matrix of second partial derivatives of $f(\mathbf{x})$ (Appendix C).

While the above framework provides a broad brush approach for solving the original estimation problem, still there are a few critical choices to be made before arriving at an algorithm for solving it.

The first choice to make is the measure for the length of the residual vector. Referring to Appendix A, there are at least three very useful ways to define this measure:

$$\left.\begin{aligned} &\text{Euclidean/2-norm}: \|\mathbf{r}\|_2 = (r_1^2 + r_2^2 + \cdots + r_m^2)^{1/2} \\ &\text{Manhattan/1-norm}: \|\mathbf{r}\|_1 = |r_1| + |r_2| + \cdots + |r_m| \\ &\text{Chebychev}/\infty\text{-norm}: \|\mathbf{r}\|_\infty = \max\{|r_1| + |r_2| + \cdots + |r_m|\} \end{aligned}\right\} \tag{5.1.10}$$

Using these three norms, we now describe three useful versions of our minimization problem.

(1) Least sum of squares of the errors By choosing the Euclidean norm, (5.1.6) becomes

$$\begin{aligned} f(\mathbf{x}) &= \|\mathbf{r}(\mathbf{x})\|_2^2 = \sum_{i=1}^{m} r_i^2(\mathbf{x}) \\ &= \sum_{i=1}^{m} [z_i - (z_0 + v t_i)]^2 = (\mathbf{z} - \mathbf{H}\mathbf{x})^{\mathrm{T}}(\mathbf{z} - \mathbf{H}\mathbf{x}). \end{aligned} \tag{5.1.11}$$

This is the popular least squares error (LSE) criterion.

(2) Least sum of the absolute errors By choosing the Manhattan/1-norm, (5.1.6) becomes

$$\begin{aligned} f(\mathbf{x}) &= \|\mathbf{r}(\mathbf{x})\|_1 = |r_1| + |r_2| + \cdots + |r_m| \\ &= \sum_{i=1}^{m} |[z_i - (z_0 + v t_i)]|. \end{aligned} \tag{5.1.12}$$

This is known as the least absolute error criterion.

(3) Least maximum of the absolute errors By choosing the Chebychev/ ∞-norm, (5.1.6) becomes:

$$f(\mathbf{x}) = \|\mathbf{r}(\mathbf{x})\|_\infty = \max_{1 \le i \le m} [|r_i|] = \max_{1 \le i \le m} \{|z_i - (z_0 + v t_i)|\}. \tag{5.1.13}$$

This is often called the *min-max* criterion.

These norms are *equivalent* (in the sense that if a vector has finite length in any one norm, then it has finite length in all the other norms) and the choice of a particular norm is often controlled by convenience and ease of analysis. For our analysis, conditions (5.1.7) essentially dictate our choice of norms. While all the three norms are continuous functions, only the Euclidean norm has continuous derivatives at the origin (see Exercise 5.1) and hence is the choice for our analysis. In other words, the method of least squares is defined as the minimization of the square of the Euclidean length of the residual vector which is the sum of the squares of its components.

(B) The least squares method We now describe a method to minimize $f(\mathbf{x})$ in (5.1.11). On rewriting, we obtain

$$\begin{aligned} f(\mathbf{x}) &= (\mathbf{z} - \mathbf{Hx})^{\mathrm{T}}(\mathbf{z} - \mathbf{Hx}) = (\mathbf{z}^{\mathrm{T}} - (\mathbf{Hx})^{\mathrm{T}})(\mathbf{z} - \mathbf{Hx}) \\ &= (\mathbf{z}^{\mathrm{T}} - \mathbf{x}^{\mathrm{T}}\mathbf{H}^{\mathrm{T}})(\mathbf{z} - \mathbf{Hx}) \\ &= \mathbf{z}^{\mathrm{T}}\mathbf{z} - \mathbf{z}^{\mathrm{T}}\mathbf{Hx} - \mathbf{x}^{\mathrm{T}}\mathbf{H}^{\mathrm{T}}\mathbf{z} + \mathbf{x}^{\mathrm{T}}\mathbf{H}^{\mathrm{T}}\mathbf{Hx} \\ &= \mathbf{z}^{\mathrm{T}}\mathbf{z} - 2\mathbf{z}^{\mathrm{T}}\mathbf{Hx} + \mathbf{x}^{\mathrm{T}}(\mathbf{H}^{\mathrm{T}}\mathbf{H})\mathbf{x}. \end{aligned} \tag{5.1.14}$$

Note that in obtaining the last expression in (5.1.14), we have used the fact that the transpose of a scalar is the scalar itself, and that $\mathbf{z}^{\mathrm{T}}\mathbf{Hx}$ is a scalar.

The gradient $\nabla f(\mathbf{x})$ and the Hessian $\nabla^2 f(\mathbf{x})$ are given by (Appendix C)

$$\nabla f(\mathbf{x}) = -2\mathbf{H}^{\mathrm{T}}\mathbf{z} + 2(\mathbf{H}^{\mathrm{T}}\mathbf{H})\mathbf{x} \tag{5.1.15}$$

and

$$\nabla^2 f(\mathbf{x}) = 2(\mathbf{H}^{\mathrm{T}}\mathbf{H}). \tag{5.1.16}$$

The first-order condition in (5.1.7) when applied to (5.1.15) defines the minimizing $\mathbf{x}$ as the solution of

$$(\mathbf{H}^{\mathrm{T}}\mathbf{H})\mathbf{x} = \mathbf{H}^{\mathrm{T}}\mathbf{z} \tag{5.1.17}$$

which is a linear system of simultaneous equations, which often goes by the name *normal* equations. This approach is now classical and has come to be known as the normal equation method. Assuming that the 2×2 matrix $\mathbf{H}^{\mathrm{T}}\mathbf{H}$ is *non-singular*, we get the solution

$$\mathbf{x}^* = (\mathbf{H}^{\mathrm{T}}\mathbf{H})^{-1}\mathbf{H}^{\mathrm{T}}\mathbf{z}. \tag{5.1.18}$$

Again, referring to the second-order condition in (5.1.7), it follows that the solution of (5.1.18) is not a minimizer of $f(\mathbf{x})$ in (5.1.14) unless the matrix $\mathbf{H}^{\mathrm{T}}\mathbf{H}$ is *positive definite*.

Recall that the matrix $\mathbf{H}$ represents the measurement system and two of the properties of $\mathbf{H}^{\mathrm{T}}\mathbf{H}$ – non-singularity and positive definiteness – are key to defining the minimizing solution for the least squares estimation problem in question. In other words, analysis of the properties of $\mathbf{H}^T\mathbf{H}$ is vital to our overall mission and in the following we take up this analysis.

(C) Properties of $\mathbf{H}^T\mathbf{H}$ Referring to (5.1.2), we readily see that

$$\mathbf{H}^T\mathbf{H} = \begin{bmatrix} 1 & 1 & \cdots & 1 \\ t_1 & t_2 & \cdots & t_m \end{bmatrix} \begin{bmatrix} 1 & t_1 \\ 1 & t_2 \\ \vdots & \vdots \\ 1 & t_m \end{bmatrix} = \begin{bmatrix} m & \sum_{i=1}^m t_i \\ \sum_{i=1}^m t_i & \sum_{i=1}^m t_i^2 \end{bmatrix}. \tag{5.1.19}$$

(a) $\mathbf{H}^T\mathbf{H}$ is called a *Grammian* matrix and is always a symmetric matrix, since $(\mathbf{H}^T\mathbf{H})^T = \mathbf{H}^T\mathbf{H}$.

(b) $\mathbf{H}^+ = (\mathbf{H}^T\mathbf{H})^{-1}\mathbf{H}^T$ is called the *generalized inverse* of $\mathbf{H}$. The reader can readily verify the following properties of the generalized inverse $\mathbf{H}^+$ (Exercise 5.6).
   (a) $\mathbf{H}\mathbf{H}^+\mathbf{H} = \mathbf{H}$
   (b) $\mathbf{H}^+\mathbf{H}\mathbf{H}^+ = \mathbf{H}^+$
   (c) $(\mathbf{H}\mathbf{H}^+)^T = \mathbf{H}\mathbf{H}^+$ i.e. $\mathbf{H}\mathbf{H}^+$ is symmetric.
   (d) $(\mathbf{H}^+\mathbf{H})^T = \mathbf{H}^+\mathbf{H}$, i.e. $\mathbf{H}^+\mathbf{H}$ is symmetric.

(c) $\mathbf{H}^T\mathbf{H}$ is positive definite if for any $\mathbf{H} \in \mathbb{R}^2$ and $\mathbf{y} \neq \mathbf{0}$,

$$\mathbf{0} < \mathbf{y}^T(\mathbf{H}^T\mathbf{H})\mathbf{y} = (\mathbf{y}^T\mathbf{H}^T)(\mathbf{H}\mathbf{y}) = (\mathbf{H}\mathbf{y})^T(\mathbf{H}\mathbf{y}) = \|\mathbf{H}\mathbf{y}\|_2^2. \tag{5.1.20}$$

Since the norm of a non-null vector is always positive, (5.1.20) will hold only when $\mathbf{H}\mathbf{y} \neq \mathbf{0}$, for $\mathbf{y} \neq \mathbf{0}$. In other words, (5.1.20) requires that $\mathbf{H}$ maps non-null vectors into non-null vectors. To further examine this condition, denote $\mathbf{H}$ as

$$\mathbf{H} = [\mathbf{h}_{*1}, \mathbf{h}_{*2}]$$

where $\mathbf{h}_{*j}$ is the jth column of the matrix $\mathbf{H}$. If $\mathbf{y} = (y_1, y_2)^T$, then

$$\mathbf{H}\mathbf{y} = [\mathbf{h}_{*1}, \mathbf{h}_{*2}] \begin{bmatrix} y_1 \\ y_2 \end{bmatrix} = y_1\mathbf{h}_{*1} + y_2\mathbf{h}_{*2}$$

denotes the linear combination of the columns of $\mathbf{H}$, with the elements of $\mathbf{y}$ as the coefficients. From the definition of linear independence (Appendix A), we can guarantee that $\mathbf{H}\mathbf{y} \neq \mathbf{0}$ when $\mathbf{y} \neq \mathbf{0}$ exactly when the *columns of* $\mathbf{H}$ *are linearly independent.*

The above discussion leads to the following *requirement on* $\mathbf{H}$. In order for the linear least squares problem to be well-defined, the measurement system represented by the matrix $\mathbf{H}$ must be carefully designed to render the columns of $\mathbf{H}$ to be linearly independent. Recall that the Rank($\mathbf{H}$) $= \min(m, 2) \leq 2$, and linear independence of the columns of $\mathbf{H}$ would imply that Rank($\mathbf{H}$) $= 2$, that is, $\mathbf{H}$ is of maximal rank.

(d) The question now is what factors determine the rank of $\mathbf{H}$. To examine this, recall that $\mathbf{H} \in \mathbb{R}^{m\times 2}$ and Rank($\mathbf{H}$) ≤ 2. Consider the case when all the $t_i = t$, that is, all the m measurements are taken at only one time epoch. Then

$$\mathbf{H} = \begin{bmatrix} 1 & t \\ 1 & t \\ \vdots & \vdots \\ 1 & t \end{bmatrix}$$

and for $\mathbf{y} = (-t, 1)^{\mathrm{T}} \neq \mathbf{0}$, we immediately have $\mathbf{Hy} = \mathbf{0}$, that is, columns of $\mathbf{H}$ are linearly dependent and Rank($\mathbf{H}$) $= 1$. In this case, ($\mathbf{H}^{\mathrm{T}}\mathbf{H}$) is *not* positive definite and the second-order condition does not hold. Also, in this case,

$$\mathbf{H}^{\mathrm{T}}\mathbf{H} = \begin{bmatrix} m & mt \\ mt & mt^2 \end{bmatrix}$$

and $\mathbf{H}^{\mathrm{T}}\mathbf{H}$ is a *singular* matrix, since det($\mathbf{H}^{\mathrm{T}}\mathbf{H}$) $= 0$. That is, there is no minimizing solution to (5.1.17), which represents the first-order condition for a minimum.

We now examine the case when all the measurements are made at *two distinct* time epochs. Consider an extreme case where the first measurement is made at time t_1, and the rest of the $m-1$ measurements at time $t_2 > t_1$. Then (see Exercise 5.2)

$$\mathbf{H} = \begin{bmatrix} 1 & t_1 \\ 1 & t_2 \\ 1 & t_2 \\ \vdots & \vdots \\ 1 & t_2 \end{bmatrix}.$$

and

$$\mathbf{H}^{\mathrm{T}}\mathbf{H} = \begin{bmatrix} m & t_1 + (m-1)t_2 \\ t_1 + (m-1)t_2 & t_1^2 + (m-1)t_2^2 \end{bmatrix}.$$

It can be verified that the det($\mathbf{H}^{\mathrm{T}}\mathbf{H}$) $\neq 0$, Rank($\mathbf{H}$) $= 2$, and the minimizing solution exists and is unique.

A fundamental and an inescapable conclusion is that unless we have measurements of positions of the moving object at least at two different instances in time, the stated minimization and hence the LSE problem is *not* well defined, and *cannot* be solved.

This conclusion is also intuitively appealing since we have two unknown components of $\mathbf{x} = (\mathbf{z}_0, \nu)^{\mathrm{T}}$ to be estimated and we need at least two observations at distinct epochs. A larger import of the analysis of this simple problem is that great care must be exercised in the planning and the design of observational systems to render the underlying estimation problem solvable.

(D) Explicit solution Having isolated the conditions under which the solution to (5.1.17) is defined, we now provide expressions for the explicit solution. To simplify the notation, we introduce the following.

$$
\begin{aligned}
\bar{t} &= \tfrac{1}{m}\textstyle\sum_{i=1}^{m} t_i \\
\bar{t^2} &= \tfrac{1}{m}\textstyle\sum_{i=1}^{m} t_i^2 \\
\bar{z} &= \tfrac{1}{m}\textstyle\sum_{i=1}^{m} z_i \\
\bar{tz} &= \tfrac{1}{m}\textstyle\sum_{i=1}^{m} t_i z_i .
\end{aligned}
\tag{5.1.21}
$$

Using (5.1.19) and (5.1.21), we can write (5.1.17) as

$$
\begin{bmatrix} 1 & \bar{t} \\ \bar{t} & \bar{t^2} \end{bmatrix}
\begin{bmatrix} z_0 \\ v \end{bmatrix}
=
\begin{bmatrix} \bar{z} \\ \bar{tz} \end{bmatrix}
\tag{5.1.22}
$$

from which we immediately obtain (see Exercise 5.3)

$$
v^* = \frac{\bar{tz} - \bar{t}\bar{z}}{\bar{t^2} - (\bar{t})^2} \tag{5.1.23}
$$

$$
= \frac{\sum_{i=1}^{m}(z_i - \bar{z})(t_i - \bar{t})}{\sum_{i=1}^{m}(t_i - \bar{t})^2} \tag{5.1.24}
$$

and

$$
z_0^* = \bar{z} - \bar{t}v^* . \tag{5.1.25}
$$

Statistical interpretation of the quantities in (5.1.21) and (5.1.24) are given in Part III. Hence, the linear model that predicts the position z_t of the moving object at time t is given by

$$
z_t = z_o^* + v^* t. \tag{5.1.26}
$$

Using $\mathbf{x}^*$ in (5.1.18), we can readily compute the minimum value of the sum of the squared residual or error (SSE) which denotes the error between the estimated linear model and the observations as follows:

$$
\begin{aligned}
\mathbf{r}(\mathbf{x}^*) = \mathbf{z} - \mathbf{H}\mathbf{x}^* &= \mathbf{z}^{\mathrm{T}} - \mathbf{H}(\mathbf{H}^{\mathrm{T}}\mathbf{H})^{-1}\mathbf{H}^{\mathrm{T}}\mathbf{z} \\
&= [\mathbf{I} - \mathbf{H}(\mathbf{H}^{\mathrm{T}}\mathbf{H})^{-1}\mathbf{H}^{\mathrm{T}}]\mathbf{z}
\end{aligned}
\tag{5.1.27}
$$

where $\mathbf{I} \in \mathbb{R}^{m\times m}$ is an identity matrix, and (see Exercise 5.4)

$$
\begin{aligned}
SSE &= \left\|\mathbf{r}(\mathbf{x}^*)\right\|^2 \\
&= \mathbf{z}^{\mathrm{T}}[\mathbf{I} - \mathbf{H}(\mathbf{H}^{\mathrm{T}}\mathbf{H})^{-1}\mathbf{H}^{\mathrm{T}}]^{\mathrm{T}}[\mathbf{I} - \mathbf{H}(\mathbf{H}^{\mathrm{T}}\mathbf{H})^{-1}\mathbf{H}^{\mathrm{T}}]\mathbf{z} \\
&= \mathbf{z}^{\mathrm{T}}[\mathbf{I} - \mathbf{H}(\mathbf{H}^{\mathrm{T}}\mathbf{H})^{-1}\mathbf{H}^{\mathrm{T}}]\mathbf{z}.
\end{aligned}
\tag{5.1.28}
$$

(Note $\mathbf{H}^{\mathrm{T}}\mathbf{H}$ is non-singular and $\mathbf{A}\mathbf{A}^{-1} = \mathbf{A}^{-1}\mathbf{A} = \mathbf{I}$.)

In component form, SSE can also be expressed as

$$\mathrm{SSE} = \sum_{i=1}^{m} [(z_o^* + \nu^* t_i) - z_i]^2 \tag{5.1.29}$$

where $\mathbf{x}^* = (z_0^*, \nu^*)$ is given in (5.1.24) and (5.1.25). The square root of the average value of SSE, called the root mean square error (RMSE) given by

$$\mathrm{RMSE} = \left(\frac{\mathrm{SSE}}{m}\right)^{1/2}$$

is often used as a measure of the fit.

(E) Examples We now present several illustrative examples of this methodology.

Example 5.1.1 Let $m = 4$, and the (t_i, z_i) pairs are given in the following table.

	$i = 1$	$i = 2$	$i = 3$	$i = 4$
t_i	0.0	1.0	2.0	3.0
z_i	1.0	3.0	2.0	3.0

Then, $\bar{t} = 1.5$, $\overline{t^2} = 3.5$, $\bar{z} = 2.25$, and $\overline{tz} = 4$. Hence (5.1.22) becomes

$$\begin{bmatrix} 1 & 1.5 \\ 1.5 & 3.5 \end{bmatrix} \begin{bmatrix} z_0 \\ \nu \end{bmatrix} = \begin{bmatrix} 2.25 \\ 4 \end{bmatrix}$$

and $\nu^* = 0.5$ and $z_0^* = 1.5$, and the model equation becomes

$$z_t = 1.5 + 0.5t.$$

Using this, we obtain $SSE = 1.5$ and $RMSE = 0.6124$.

Example 5.1.2 Suppose you want to estimate your own weight, say, w. Since the measured weight may vary depending on the type of clothes you wear, the food you ate, the scale you use, etc., a good strategy would be to measure your weight under various conditions. Given this strategy, if $z_1, z_2, \ldots, z_m$ are the m such measurements, what is the best estimate of z? Let $\mathbf{z} = (z_1, z_2, \ldots, z_m)^{\mathrm{T}}$, and let $\mathbf{H} = [1, 1, \ldots, 1]^{\mathrm{T}} \in \mathbb{R}^{m\times 1}$, be a column vector of all 1's of size m. The problem is to find w such that

$$f(\mathbf{w}) = \|\mathbf{r}(\mathbf{w})\|^2 = \|\mathbf{z} - \mathbf{H}\mathbf{w}\|^2$$

is a minimum. From (5.1.17), we immediately obtain

$$(\mathbf{H}^{\mathrm{T}}\mathbf{H})\mathbf{w} = \mathbf{H}^{\mathrm{T}}\mathbf{z}$$

which reduces to

$$w = \frac{1}{m}\sum_{i=1}^{m} z_i.$$

That is, the numerical average of the m measured weights is the best (in the sense of least squares) estimate of your weight.

Example 5.1.3 In this example we examine the effect of *multiple observations* on the process of estimation. Let $k \geq 2$ and let

$$t_1 < t_2 < \cdots < t_k$$

be a set of k increasing time instances at which observations of positions are made. Let $m_1, m_2, \ldots, m_k$ be the number of observations of the position of the object at these k times, where $m_1 + m_2 + \cdots + m_k = m$. For definiteness, we assume the following notation for these m observations.

Time	Measurements of Position
t_1	$z_{11}, z_{12}, \ldots, z_{1m_1}$
t_2	$z_{21}, z_{22}, \ldots, z_{2m_2}$
$\vdots$	$\vdots$
t_i	$z_{i1}, z_{i2}, \ldots, z_{im_i}$
$\vdots$	$\vdots$
t_k	$z_{k1}, z_{k2}, \ldots, z_{km_k}$

The $\mathbf{H}$ matrix and $\mathbf{z}$ vector take a block-partitioned structure given by

$$\mathbf{H}^{\mathrm{T}} = \begin{bmatrix} 1 & 1 & \cdots & 1 & \vdots & 1 & 1 & \cdots & 1 & \vdots & \cdots & \vdots & 1 & 1 & \cdots & 1 \\ t_1 & t_1 & \cdots & t_1 & \vdots & t_2 & t_2 & \cdots & t_2 & \vdots & \cdots & \vdots & t_k & t_k & \cdots & t_k \end{bmatrix}$$

$$\mathbf{z}^{\mathrm{T}} = \begin{bmatrix} z_{11} & \cdots & z_{1m_1} & \vdots & z_{21} & \cdots & z_{2m_2} & \vdots & \cdots & \vdots & z_{k1} & \cdots & z_{km_k} \end{bmatrix}$$

It can be verified that

$$\mathbf{H}^{\mathrm{T}}\mathbf{H} = \begin{bmatrix} m & \sum_{i=1}^{k} m_i t_i \\ \sum_{i=1}^{k} m_i t_i & \sum_{i=1}^{k} m_i t_i^2 \end{bmatrix}$$

and

$$\mathbf{H}^{\mathrm{T}}\mathbf{z} = \begin{bmatrix} \sum_{i=1}^{k} \sum_{j=1}^{m_i} z_{ij} \\ \sum_{i=1}^{k} t_i \sum_{j=1}^{m_i} z_{ij} \end{bmatrix}$$

and the equation (5.1.17) becomes

$$\begin{bmatrix} m & \sum_{i=1}^{k} m_i t_i \\ \sum_{i=1}^{k} m_i t_i & \sum_{i=1}^{k} m_i t_i^2 \end{bmatrix} \begin{bmatrix} z_0 \\ v \end{bmatrix} = \begin{bmatrix} \sum_{i=1}^{k} \sum_{j=1}^{m_i} z_{ij} \\ \sum_{i=1}^{k} t_i \sum_{j=1}^{m_i} z_{ij} \end{bmatrix}. \tag{5.1.30}$$

Dividing both sides by m, and defining

$$\alpha_i = \frac{m_i}{m}, \quad \sum_{i=1}^{k} \alpha_i = 1,$$

$$\bar{z}_i = \frac{1}{m_i} \sum_{j=1}^{m_i} z_{ij} = \text{Average of the measurements at time } t_i,$$

we can rewrite (5.1.30) as

$$\begin{bmatrix} 1 & \sum_{i=1}^{k} \alpha_i t_i \\ \sum_{i=1}^{k} \alpha_i t_i & \sum_{i=1}^{k} \alpha_i t_i^2 \end{bmatrix} \begin{bmatrix} z_0 \\ v \end{bmatrix} = \begin{bmatrix} \sum_{i=1}^{k} \alpha_i \bar{z}_i \\ \sum_{i=1}^{k} t_i \alpha_i \bar{z}_i \end{bmatrix}. \tag{5.1.31}$$

We now introduce the following notation:

$$\bar{\mathbf{H}}^{\mathrm{T}} = \begin{bmatrix} 1 & 1 & \cdots & 1 \\ t_1 & t_2 & \cdots & t_k \end{bmatrix}, \quad \bar{\mathbf{z}}^{\mathrm{T}} = \begin{bmatrix} \bar{z}_1 & \bar{z}_2 & \cdots & \bar{z}_k \end{bmatrix}$$

and a *diagonal weight matrix*

$$\mathbf{W} = \begin{bmatrix} \alpha_1 & 0 & \cdots & 0 \\ 0 & \alpha_2 & \cdots & 0 \\ \vdots & \vdots & \vdots & \vdots \\ 0 & 0 & \cdots & \alpha_k \end{bmatrix}.$$

It can be verified that (5.1.31) can be represented as

$$(\bar{\mathbf{H}}^{\mathrm{T}} \mathbf{W} \bar{\mathbf{H}}) \mathbf{x} = (\bar{\mathbf{H}}^{\mathrm{T}} \mathbf{W}) \bar{\mathbf{z}}$$

$$\mathbf{x} = (\bar{\mathbf{H}}^{\mathrm{T}} \mathbf{W} \bar{\mathbf{H}})^{-1} (\bar{\mathbf{H}}^{\mathrm{T}} \mathbf{W}) \bar{\mathbf{z}}. \tag{5.1.32}$$

That is, if there are multiple observations, then the fraction of the observation α_i taken at time t_i plays the role of a weight factor that determines the contribution of the observations at time t_i. Larger (smaller) the value of α_i, the larger (smaller) is the importance of the data at time t_i in computing the overall estimate.

As a special case, if there are the same number of observations at each time, that is, $m_i = m/k$ and $\alpha_i = 1/k$ for $i = 1, 2, \ldots, k$, then (5.1.31) becomes

$$\begin{bmatrix} 1 & \frac{1}{k} \sum_{i=1}^{k} t_i \\ \frac{1}{k} \sum_{i=1}^{k} t_i & \frac{1}{k} \sum_{i=1}^{k} t_i^2 \end{bmatrix} \begin{bmatrix} z_0 \\ v \end{bmatrix} = \begin{bmatrix} \frac{1}{k} \sum_{i=1}^{k} \bar{z}_i \\ \frac{1}{k} \sum_{i=1}^{k} t_i \bar{z}_i \end{bmatrix} \tag{5.1.33}$$

which is the standard least squares formulation with observation at time t_i replaced by the average of the observations at that time.

From (5.1.31) and (5.1.32), it can be verified that

$$
\begin{aligned}
\det(\bar{\mathbf{H}}^{\mathrm{T}}\mathbf{W}\bar{\mathbf{H}}) &= \left[\sum_{i=1}^{k} \alpha_i t_i^2\right] - \left[\sum_{i=1}^{k} \alpha_i t_i\right]^2 \\
&= \sum_{i=1}^{k} t_i^2 \alpha_i (1-\alpha_i) \\
&= -2 \sum_{1 \le i < j \le k} (\alpha_i t_i)(\alpha_j t_j).
\end{aligned}
$$

As another special case, let $k = 2$, and

$$\det[\bar{\mathbf{H}}^{\mathrm{T}}\mathbf{W}\bar{\mathbf{H}}] = -2\alpha_1\alpha_2 t_1 t_2.$$

Since $\alpha_1 + \alpha_2 = 1$, it can be seen that the matrix $\bar{\mathbf{H}}^{\mathrm{T}}\mathbf{W}\bar{\mathbf{H}}$ becomes singular as $\alpha_1 \longrightarrow 0$ or as $\alpha_2 \longrightarrow 0$. That is, if we take multiple observations, unless we distribute them wisely, either it may lead to singularity of the matrices (when $k = 2$) or have an effect of not treating all the observations alike by inducing an implicit weight that determines their relative importance which may have undesirable consequences.

5.2 Generalized least squares

In this section we generalize the normal equation method for the basic linear least squares problem discussion in Section 5.1 in two directions. First, it is assumed that each observation z_i depends on n (state) variables (instead of two) which are the components of the vector $\mathbf{x} = (x_1, x_2, \ldots, x_n)^{\mathrm{T}} \in \mathbb{R}^n$. Thus, the observation z_i depends linearly on the n variables as

$$z_i = h_{i1}x_1 + h_{i2}x_2 + \cdots + h_{in}x_n \tag{5.2.1}$$

for $i = 1, 2, \ldots, m$, where h_{ij} denotes the characteristics of the measurement system. If $\mathbf{z} = (z_1, z_2, \ldots, z_m)^{\mathrm{T}} \in \mathbb{R}^m$ and $\mathbf{H} = [h_{ij}] \in \mathbb{R}^{m \times n}$, then (5.2.1) can be written succinctly as

$$\mathbf{z} = \mathbf{H}\mathbf{x}. \tag{5.2.2}$$

Second, we would like to introduce an *explicit weighting scheme* in defining the residuals which is conceptually different from the *implicit weights* induced by the fraction of the multiple observations as in Example 5.1.3. This is done using an extension of the Euclidean norm called the *energy norm*, which is a quadratic form of the residual vector $\mathbf{r}(\mathbf{x})$ and defined as follows. Let $\mathbf{W} \in \mathbb{R}^{m \times m}$ be a *symmetric* and *positive definite* matrix, and recall that the residual $\mathbf{r}(\mathbf{x}) = (\mathbf{z} - \mathbf{H}\mathbf{x}) \in \mathbb{R}^m$. Define

$$
\begin{aligned}
f(\mathbf{x}) = \|\mathbf{r}(\mathbf{x})\|_{\mathrm{W}}^2 &= \mathbf{r}^{\mathrm{T}}(\mathbf{x})\mathbf{W}\mathbf{r}(\mathbf{x}) \\
&= ((\mathbf{z} - \mathbf{H}\mathbf{x}))^{\mathrm{T}}\mathbf{W}((\mathbf{z} - \mathbf{H}\mathbf{x})) \\
&= \mathbf{z}^{\mathrm{T}}\mathbf{W}\mathbf{z} - 2\mathbf{z}^{\mathrm{T}}\mathbf{W}\mathbf{H}\mathbf{x} + \mathbf{x}^{\mathrm{T}}(\mathbf{H}^{\mathrm{T}}\mathbf{W}\mathbf{H})\mathbf{x}.
\end{aligned} \tag{5.2.3}
$$

A number of observations are in order. First, this expression representing the sum of the weighted squares of the residuals is a generalization of that in (5.1.14) in that we obtain the latter from (5.2.3) when $\mathbf{W} = \mathbf{I}$, the identity matrix. Second, the difference between this weighting scheme and the one in Example 5.1.3 is that while the matrix $\mathbf{W}$ in (5.1.30) is necessarily a *diagonal matrix* (with non-negative entries along the diagonal that add up to unity, and hence is a symmetric and positive definite matrix), the matrix $\mathbf{W}$ in (5.2.3) is not required to be diagonal and allows a wider choice. Third, thanks to the beauty of matrix-vector notation, there is virtually no difference in the algebra as we go from (5.1.14) where $\mathbf{x} \in \mathbb{R}^2$ to (5.2.3) where $\mathbf{x} \in \mathbb{R}^n$.

In minimizing $f(\mathbf{x})$ in (5.2.3), the gradient and Hessian of $f(\mathbf{x})$ in (5.2.3) are given by

$$\nabla f(\mathbf{x}) = -2\mathbf{H}^{\mathrm{T}}\mathbf{W}\mathbf{z} + 2(\mathbf{H}^{\mathrm{T}}\mathbf{W}\mathbf{H})\mathbf{x} \tag{5.2.4}$$

and

$$\nabla^2 f(\mathbf{x}) = 2(\mathbf{H}^{\mathrm{T}}\mathbf{W}\mathbf{H}). \tag{5.2.5}$$

The first-order condition when applied to (5.2.4) gives rise to the *normal equations*

$$(\mathbf{H}^{\mathrm{T}}\mathbf{W}\mathbf{H})\mathbf{x} = \mathbf{H}^{\mathrm{T}}\mathbf{W}\mathbf{z}. \tag{5.2.6}$$

The minimizing solution is given by

$$\left.\begin{aligned} \mathbf{x}^* &= (\mathbf{H}^{\mathrm{T}}\mathbf{W}\mathbf{H})^{-1}\mathbf{H}^{\mathrm{T}}\mathbf{W}\mathbf{z} \\ \text{and} \qquad & \\ \mathbf{z}^* &= \mathbf{H}\mathbf{x}^* = \mathbf{H}(\mathbf{H}^{\mathrm{T}}\mathbf{W}\mathbf{H})^{-1}\mathbf{H}^{\mathrm{T}}\mathbf{W}\mathbf{z} \end{aligned}\right\} \tag{5.2.7}$$

Again, referring to the second order condition, it is required that $(\mathbf{H}^{\mathrm{T}}\mathbf{W}\mathbf{H})$ be positive definite. That is for any $\mathbf{y} \in \mathbb{R}^m$, and $\mathbf{y} \neq \mathbf{0}$,

$$\begin{aligned} \mathbf{0} < \mathbf{y}^{\mathrm{T}}(\mathbf{H}^{\mathrm{T}}\mathbf{W}\mathbf{H})\mathbf{y} &= (\mathbf{y}^{\mathrm{T}}\mathbf{H}^{\mathrm{T}})\mathbf{W}(\mathbf{H}\mathbf{y}) \\ &= (\mathbf{H}\mathbf{y})^{\mathrm{T}}\mathbf{W}(\mathbf{H}\mathbf{y}) \\ &= \|\mathbf{H}\mathbf{y}\|_{\mathbf{W}}^2 . \end{aligned} \tag{5.2.8}$$

This inequality will hold only if $\mathbf{H}\mathbf{y} \neq \mathbf{0}$ when $\mathbf{y} \neq \mathbf{0}$ which happens precisely when the columns of $\mathbf{H}$ are linearly independent – once again reaffirming a fundamental requirement in the theory of linear least squares estimation.

Remark 5.2.1 The key question that still remains is what is the basic guideline for the choice of the weight matrix in the generalized least squares theory. In this chapter and indeed in this Part II, we have chosen to take a *deterministic* view of the world. However, in practice it is very difficult, if not impossible, to make precise measurements and actual measurements always have a *random* (additive) error component embedded in them. Depending on the instruments and the physical quantities being measured, these random errors may exhibit a spatial and/or temporal correlation. In such cases, the observation vector $\mathbf{z}$ is decomposed into

$$\mathbf{z} = \mathbf{H}\mathbf{x} + \mathbf{v} \tag{5.2.9}$$

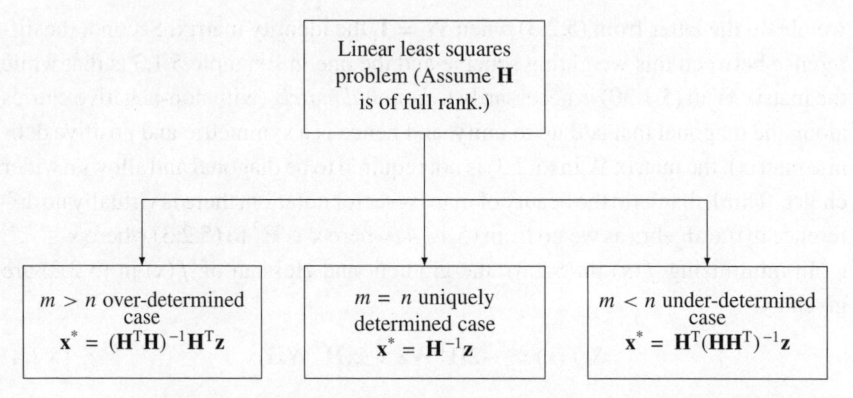

Fig. 5.3.1 A classification of linear least squares problems.

where $\mathbf{Hx}$ is the model that relates the model variables $\mathbf{x}$ to the observations $\mathbf{z}$ and $\mathbf{v} \in \mathbb{R}^m$ is a non-observable random error vector. A standard model for $\mathbf{v}$ is that it is a multivariate Gaussian noise with mean zero and covariance matrix $\mathbf{R}$, that is, $\mathbf{v} \sim N(0, \mathbf{R})$. When $\mathbf{v}$ is modelled in this fashion, an appropriate choice for the weight matrix is $\mathbf{W} = \mathbf{R}^{-1}$. This choice of the weight has an effect of normalizing the noise variance. For more details refer to Part III on Statistical Estimation.

5.3 Dual problem: $m < n$

Analysis of the linear least squares problem with respect to m, *the number of observations* and n, *the number of unknowns* has three different versions – *over-determined* ($m > n$), *uniquely determined* ($m = n$), and *under-determined* ($m < n$) cases as shown in Figure 5.3.1. Having covered the over-determined case in Sections 5.1 and 5.2, in this section, we take up the analysis of the dual case of the under-determined problem when $m < n$. For completeness, we first consider the rather simple case when $m = n$.

(A) Uniquely determined problem When $m = n$ and if $\mathbf{H}$ is of full rank, that is Rank($\mathbf{H}$) $= m = n$, then the linearly independent columns of $\mathbf{H}$ span space $\mathbb{R}^n$. Hence, any vector $\mathbf{z} \in \mathbb{R}^n$ must be uniquely expressible as a linear combination of the columns of $\mathbf{H}$. In other words, there exists a unique $\mathbf{x}^* \in \mathbb{R}^n$ such that

$$\mathbf{Hx}^* = \mathbf{z} \quad \text{or} \quad \mathbf{x}^* = \mathbf{H}^{-1}\mathbf{z}. \tag{5.3.1}$$

It can be verified that in this case the residual $\mathbf{r} = (\mathbf{z} - \mathbf{Hx}) = \mathbf{z} - \mathbf{HH}^{-1}\mathbf{z} = \mathbf{0}$.

(B) Dual Problem $m < n$ In this case Rank($\mathbf{H}$) $= \min(m, n) = m$. In other words, only m of the n columns of $\mathbf{H}$ are linearly independent. For definiteness, assume without loss of generality that the first m columns of $\mathbf{H}$ are linearly independent. (Otherwise, we can permute the columns which is essentially a relabelling of the components of the unknown vector $\mathbf{x}$ so that the first m columns of $\mathbf{H}$ are linearly independent.) We can then partition $\mathbf{H}$ as

$$\mathbf{H} = [\mathbf{H}_1 \vdots \mathbf{H}_2] \tag{5.3.2}$$

where $\mathbf{H}_1 \in \mathbb{R}^{m \times m}$ consisting of the first m columns and $\mathbf{H}_2 \in \mathbb{R}^{m \times (n-m)}$ has the rest of the $(n-m)$ columns of $\mathbf{H}$. By assumption, $\mathbf{H}_1$ is non-singular. Similarly, induce a compatible partition of $\mathbf{x}$ as

$$\mathbf{x} = \begin{bmatrix} \mathbf{x}_1 \\ \cdots \\ \mathbf{x}_2 \end{bmatrix} \tag{5.3.3}$$

where $\mathbf{x}_1 \in \mathbb{R}^m$ has the first m components and $\mathbf{x}_2 \in \mathbb{R}^{n-m}$ has the rest of the $(n-m)$ components of $\mathbf{x}$.

From $\mathbf{z} = \mathbf{H}\mathbf{x}$, we get

$$\mathbf{z} = [\mathbf{H}_1 \vdots \mathbf{H}_2] \begin{bmatrix} \mathbf{x}_1 \\ \cdots \\ \mathbf{x}_2 \end{bmatrix} = \mathbf{H}_1\mathbf{x}_1 + \mathbf{H}_2\mathbf{x}_2 \tag{5.3.4}$$

which, on rewriting, becomes

$$\mathbf{x}_1 = \mathbf{H}_1^{-1}[\mathbf{z} - \mathbf{H}_2\mathbf{x}_2]. \tag{5.3.5}$$

The $(n-m)$ components of $\mathbf{x}_2$ are *free* variables and hence there are infinitely many $\mathbf{x}_1$ from (5.3.5), one for each $\mathbf{x}_2$. Using (5.3.5), we now verify that the residual

$$\begin{aligned}
\mathbf{r} = \mathbf{r}(\mathbf{x}) = (\mathbf{z} - \mathbf{H}\mathbf{x}) &= \mathbf{z} - [\mathbf{H}_1 \vdots \mathbf{H}_2] \begin{bmatrix} \mathbf{x}_1 \\ \cdots \\ \mathbf{x}_2 \end{bmatrix} \\
&= \mathbf{z} - [\mathbf{H}_1 \vdots \mathbf{H}_2] \begin{bmatrix} \mathbf{H}_1^{-1}[\mathbf{z} - \mathbf{H}_2\mathbf{x}_2] \\ \cdots \\ \mathbf{x}_2 \end{bmatrix} \\
&= \mathbf{z} - [(\mathbf{z} - \mathbf{H}_2\mathbf{x}_2) + \mathbf{H}_2\mathbf{x}_2] = 0
\end{aligned} \tag{5.3.6}$$

that is, each of the infinitely many $\mathbf{x} = \begin{bmatrix} \mathbf{x}_1 \\ \cdots \\ \mathbf{x}_2 \end{bmatrix}$ with $\mathbf{x}_1$ defined by (5.3.5) and $\mathbf{x}_2$ as free variable is a "solution" to the original problem. Thus, the earlier formulation as the minimization of the square of the Euclidean norm of the residual is *not* an option in this case. An overwhelming question now is: how to pick the "right" or "meaningful" estimate $\mathbf{x}$ from these infinitely many consistent choices? An interesting way is to pick that $\mathbf{x} \in \mathbb{R}^n$ such that (a) $\mathbf{x}$ is a solution in the sense of (5.3.6) and (b) has the least length. Mathematically, this can be formulated as a constrained minimization problem as follows (Appendix D).

A constrained minimization problem:

$$\begin{gathered}\text{Among all the values of } \mathbf{x} \in \mathbb{R}^n \text{ that satisfy } \mathbf{z} = \mathbf{H}\mathbf{x} \\ \textit{via } (5.3.5) \text{ find that } \mathbf{x} \text{ which minimizes } f(\mathbf{x}) = \|\mathbf{x}\|_2^2 .\end{gathered}$$

Remark 5.3.1 The set of all $\mathbf{x} \in \mathbb{R}^n$ that satisfy the linear relation $\mathbf{z} = \mathbf{H}\mathbf{x}$ for the given $\mathbf{z}$ and $\mathbf{H}$ is clearly a subset S of $\mathbb{R}^n$ and is denoted by $S \subseteq \mathbb{R}^n$. This subset is called the *feasible* set for the minimization problem. The linear relation $\mathbf{z} = \mathbf{H}\mathbf{x}$ that defines S is called the *linear constraint*.

One of the basic ideas in constrained minimization is to reformulate it as an equivalent unconstrained minimization problem using the method of *Lagrangian multipliers* as demonstrated below.

Let $\lambda = (\lambda_1, \lambda_2, \ldots, \lambda_m)^{\mathrm{T}} \in \mathbb{R}^m$ be a vector of m unknowns called the Lagrangian multipliers. Using this λ, define a new function $L(\lambda, \mathbf{x})$, called the Lagrangian as

$$L(\lambda, \mathbf{x}) = \|\mathbf{x}\|_2^2 + \lambda^{\mathrm{T}}(\mathbf{z} - \mathbf{H}\mathbf{x}) = \mathbf{x}^{\mathrm{T}}\mathbf{x} + \lambda^{\mathrm{T}}(\mathbf{z} - \mathbf{H}\mathbf{x}) \tag{5.3.7}$$

which is a function of $(m + n)$ variables. A fundamental fact in the theory of constrained minimization is that the $\mathbf{x}^*$ that minimizes (5.3.7) in $\mathbb{R}^{m+n}$ also minimizes the constrained problem (5.3.7) in $\mathbb{R}^n$. So, at the expense of increasing the dimensionality of the search space, we have converted a constrained problem to an equivalent unconstrained minimization problem. In view of this basic fact, in the following we concentrate on solving the following problem.

$$\begin{gathered}\text{Minimize } L(\lambda, \mathbf{x}) \text{ defined in (5.3.7)} \\ \text{over } \lambda \in \mathbb{R}^m \text{ and } \mathbf{x} \in \mathbb{R}^n\end{gathered} \tag{5.3.8}$$

The first-order condition for the minimum of (5.3.8) is given by

$$\left.\begin{aligned}\nabla_{\mathbf{x}} L(\lambda, \mathbf{x}) &= 2\mathbf{x} - \mathbf{H}^{\mathrm{T}}\lambda = \mathbf{0} \\ \nabla_{\lambda} L(\lambda, \mathbf{x}) &= \mathbf{z} - \mathbf{H}\mathbf{x} = \mathbf{0}\end{aligned}\right\} \tag{5.3.9}$$

where $\nabla_{\mathbf{x}}$ and ∇_{λ} denote the *gradient* operators with respect to the vector variables $\mathbf{x}$ and λ respectively. Solving (5.3.9), we get $\mathbf{x} = \frac{1}{2}\mathbf{H}^{\mathrm{T}}\lambda$, and

$\mathbf{z} = \frac{1}{2}(\mathbf{H}\mathbf{H}^{\mathrm{T}})\lambda$. Since $(\mathbf{H}\mathbf{H}^{\mathrm{T}}) \in \mathbb{R}^{m\times m}$ is non-singular (recall that $\mathbf{H}$ is of full rank and Rank($\mathbf{H}$) $= m$), it follows that $\lambda = 2(\mathbf{H}\mathbf{H}^{\mathrm{T}})^{-1}\mathbf{z}$. Combining this with $\mathbf{x} = \frac{1}{2}\mathbf{H}^{\mathrm{T}}\lambda$, we immediately get

$$\mathbf{x}^* = \mathbf{H}^{\mathrm{T}}(\mathbf{H}\mathbf{H}^{\mathrm{T}})^{-1}\mathbf{z} \tag{5.3.10}$$

as the solution to the original constrained problem. The length of this minimum length solution $\mathbf{x}^*$ in (5.3.10) is given by

$$\begin{aligned} \|\mathbf{x}^*\|_2^2 &= \|\mathbf{H}^{\mathrm{T}}(\mathbf{H}\mathbf{H}^{\mathrm{T}})^{-1}\mathbf{z}\|_2^2 \\ &= \mathbf{z}^{\mathrm{T}}(\mathbf{H}\mathbf{H}^{\mathrm{T}})^{-1}\mathbf{H}\mathbf{H}^{\mathrm{T}}((\mathbf{H}\mathbf{H}^{\mathrm{T}})^{-1})\mathbf{z} \\ &= \mathbf{z}^{\mathrm{T}}(\mathbf{H}\mathbf{H}^{\mathrm{T}})^{-1}\mathbf{z}. \end{aligned} \tag{5.3.11}$$

A generalized (weighted) version of this under-determined version of the least squares problem is pursued in Exercise 5.5. In this case $\mathbf{H}^+ = \mathbf{H}^{\mathrm{T}}(\mathbf{H}\mathbf{H}^{\mathrm{T}})^{-1}$ is called the generalized inverse of $\mathbf{H}$ (see Exercise 5.6).

Remark 5.3.2 This dual case where m, the number of observations, is less than n, the number of unknowns, often occurs in geophysical data assimilation problems (see Part V on 3DVAR problems). Within the context of meteorological data assimilation, the uniqueness is achieved *not* by seeking the minimum norm solution but often by clever probabilistic reasoning using the Bayesian framework. More specifically, prior distribution of the unknowns (which is derived from the previous forecast or climatology, etc.) is assumed. This, when combined with the new information contained in the observation $\mathbf{z}$ using the Bayes' rule, provides a framework for obtaining the unique solution.

5.4 A unified approach: Tikhonov regularization

Having solved the over-determined and under-determined problems separately, the question: "is there a **unified** formulation where both of these cases can be rolled into one?" becomes interesting. The answer lies in using a variation of the formulation in Section 5.3 as follows. Define a new objective function†

$$f(\mathbf{x}) = \frac{\alpha}{2}\mathbf{x}^{\mathrm{T}}\mathbf{x} + \frac{1}{2}\,\|(\mathbf{z} - \mathbf{H}\mathbf{x})\|^2 \tag{5.4.1}$$

to be minimized for some real constant $\alpha > 0$. The basic idea is, instead of enforcing the **strong** requirement of equality $\mathbf{z} = \mathbf{H}\mathbf{x}$ or the vanishing of the residual as in (5.3.7), in here we settle for a **weaker** requirement of the reduction of the norm

† Recall that multiplying a function $f(\mathbf{x})$ by a constant does not alter the location of the minimum or maximum of $f(\mathbf{x})$.

of residual, $(\mathbf{z} - \mathbf{Hx})$. Stated in other words, (5.4.1) seeks a compromise between obtaining the **minimum norm** solution (as in Section 5.3) and the **minimum residual** solution (as in Section 5.1). The nature and degree of this compromise or trade-off is decided by the value of the arbitrarily chosen constant α. Clearly, the gradient $\nabla f(\mathbf{x})$ of $f(\mathbf{x})$ in (5.4.1) given by

$$\nabla f(\mathbf{x}) = (\mathbf{H}^{\mathrm{T}}\mathbf{H} + \alpha\mathbf{I})\mathbf{x} - \mathbf{H}^{\mathrm{T}}\mathbf{z}$$

vanishes when

$$\mathbf{x} = (\mathbf{H}^{\mathrm{T}}\mathbf{H} + \alpha\mathbf{I})^{-1}\mathbf{H}^{\mathrm{T}}\mathbf{z} \tag{5.4.2}$$

which reduces to the minimum residual solution in (5.1.19) when $\alpha = 0$ as it should. To see its relation to the minimum norm solution in Section 5.3, we first invoke the following matrix identity (refer to Appendix B)

$$[\mathbf{A}^{\mathrm{T}}\mathbf{B}^{-1}\mathbf{A} + \mathbf{D}^{-1}]\mathbf{A}^{\mathrm{T}}\mathbf{B}^{-1} = \mathbf{D}\mathbf{A}^{\mathrm{T}}[\mathbf{B} + \mathbf{A}\mathbf{D}\mathbf{A}^{\mathrm{T}}]^{-1} \tag{5.4.3}$$

Setting $\mathbf{A} = \mathbf{H}$, $\mathbf{B} = \mathbf{I}$, and $\mathbf{D}^{-1} = \alpha\mathbf{I}$, the above identity becomes

$$\begin{aligned}
[\mathbf{H}^{\mathrm{T}}\mathbf{H} + \alpha\mathbf{I}]^{-1}\mathbf{H}^{\mathrm{T}} &= \alpha^{-1}\mathbf{I}\mathbf{H}^{\mathrm{T}}[\mathbf{I} + \alpha^{-1}\mathbf{H}\mathbf{H}^{\mathrm{T}}]^{-1} \\
&= \alpha^{-1}\mathbf{H}^{\mathrm{T}}[\alpha^{-1}(\alpha\mathbf{I} + \mathbf{H}\mathbf{H}^{\mathrm{T}})]^{-1} \\
&= \mathbf{H}^{\mathrm{T}}[\alpha\mathbf{I} + \mathbf{H}\mathbf{H}^{\mathrm{T}}]^{-1}.
\end{aligned} \tag{5.4.4}$$

Substituting (5.4.4) into the r.h.s. of (5.4.2), the latter becomes

$$\mathbf{x} = \mathbf{H}^{\mathrm{T}}[\alpha\mathbf{I} + \mathbf{H}\mathbf{H}^{\mathrm{T}}]^{-1}\mathbf{z} \tag{5.4.5}$$

which clearly includes the minimum norm solution in (5.3.10) as a special case when $\alpha = 0$.

Several observations are in order.

(a) This unified framework was first introduced by Tikhonov and in his honor it has come to be known as **Tikhonov regularization** (Tikhonov and Arsenin (1977)). The addition of $\frac{\alpha}{2}\mathbf{x}^{\mathrm{T}}\mathbf{x}$ in (5.4.1) has a smoothing or dampening effect and the solutions (5.4.2) and (5.4.5) are called **damped solutions.**

(b) **Conversion of ill-posed to well-posed problems** Given that $\mathbf{H} \in \mathbb{R}^{m\times n}$, then the Rank($\mathbf{H}$) $= \min(m, n)$ if $\mathbf{H}$ is of **full rank**. In this case of the two Grammian matrices $\mathbf{H}^{\mathrm{T}}\mathbf{H}$ or $\mathbf{H}\mathbf{H}^{\mathrm{T}}$, only one is of full rank, and, hence, nonsingular. When $\mathbf{H}$ is **rank deficient**, then Rank($\mathbf{H}$) $= k < \min(m, n)$, and in this case both the Grammian matrices $\mathbf{H}^{\mathrm{T}}\mathbf{H}$ and $\mathbf{H}\mathbf{H}^{\mathrm{T}}$ are rank deficient and hence singular. However, the addition of the dampening term with suitable α, renders both the matrices $(\mathbf{H}^{\mathrm{T}}\mathbf{H} + \alpha\mathbf{I})$ and $(\mathbf{H}\mathbf{H}^{\mathrm{T}} + \alpha\mathbf{I})$ non-singular. In other words, the least squares problem in this unified framework for a suitable $\alpha > 0$ is always well-posed.

(c) The Hessian of $f(\mathbf{x})$ in (5.4.1) is given by

$$\nabla^2 f(\mathbf{x}) = (\mathbf{H}^{\mathrm{T}}\mathbf{H} + \alpha\mathbf{I}).$$

For any $\mathbf{y} \in \mathbb{R}^n$,

$$\mathbf{y}^{\mathrm{T}}(\mathbf{H}^{\mathrm{T}}\mathbf{H} + \alpha\mathbf{I})\mathbf{y} = (\mathbf{H}\mathbf{y})^{\mathrm{T}}(\mathbf{H}\mathbf{y}) + \alpha\mathbf{y}^{\mathrm{T}}\mathbf{y}. \tag{5.4.6}$$

Since $\mathbf{y}^{\mathrm{T}}\mathbf{y} > 0$ for all non-null vectors $\mathbf{y}$, the r.h.s. of (5.4.6) is positive for all $\mathbf{y} \neq 0$ whenever $\alpha > 0$. Hence the Hessian for this unified framework is always positive definite if $\alpha > 0$, and the solution of (5.4.2) or (5.4.5) is the minimizer of $f(\mathbf{x})$ in (5.4.1).

(d) On the down side, this framework does not provide any guidelines for the choice of α except that it be positive. Since α decides the degree of trade-off between the minimum norm and minimum residual solutions, one may have to solve the problem for different sets of values of α and evaluate the goodness of the associated solution.

Exercises

5.1 Let $\mathbf{x} = (x_1, x_2)^{\mathrm{T}}$ and define $g(\mathbf{x})$ as follows:

(a) $g(\mathbf{x}) = x_1^2 + x_2^2$

(b) $g(\mathbf{x}) = |x_1| + |x_2|$

(c) $g(\mathbf{x}) = \max\{|x_1|, |x_2|\}$

Compute $\partial g/\partial x_1$ and $\partial g/\partial x_2$ in each case, and analyze their continuity at $\mathbf{x} = (0, 0)^{\mathrm{T}}$. Plot $g(\mathbf{x})$ in each case.

5.2 Investigate the linear independence of the columns of the following $\mathbf{H}$ matrices and compute their rank.

$$\begin{bmatrix} 1 & t \\ 1 & t \\ 1 & t \\ 1 & t \end{bmatrix}, \begin{bmatrix} 1 & t_1 \\ 1 & t_2 \\ 1 & t_2 \\ 1 & t_2 \end{bmatrix}, \begin{bmatrix} 1 & t_1 \\ 1 & t_1 \\ 1 & t_2 \\ 1 & t_2 \end{bmatrix}, \begin{bmatrix} 1 & t_1 \\ 1 & t_1 \\ 1 & t_1 \\ 1 & t_2 \end{bmatrix}.$$

In each case compute $\mathbf{H}^{\mathrm{T}}\mathbf{H}$, and $\det(\mathbf{H}^{\mathrm{T}}\mathbf{H})$.

5.3 (a) Verify that equation (5.1.17) reduces to equation (5.1.22) using (5.1.21).

(b) Verify that (5.1.23) and (5.1.25) solve the system (5.1.22).

(c) Verify that (5.1.23) can be rewritten as (5.1.24).

5.4 Referring to (5.1.27), define $\mathbf{P_H} = \mathbf{H}(\mathbf{H}^{\mathrm{T}}\mathbf{H})^{-1}\mathbf{H}^{\mathrm{T}}$ and $\mathbf{P}_{\mathbf{H}}^{\perp} = \mathbf{I} - \mathbf{P_H}$

(a) Prove that $\mathbf{P}_{\mathbf{H}}^{\perp}$ and $\mathbf{P_H}$ are symmetric matrices.

(b) Prove that $\mathbf{P}_{\mathbf{H}}^2 = \mathbf{P_H}$ and $(\mathbf{I} - \mathbf{P_H})^2 = \mathbf{I} - \mathbf{P_H}$, that is, both $\mathbf{P_H}$ and $\mathbf{I} - \mathbf{P_H}$ are idempotent matrices.

5.5 Consider the following formulation of the weighted version of the underdetermined case: Minimize $f(\mathbf{x}) = \|\mathbf{x}\|_{\mathbf{W}^{-1}} = \mathbf{x}^{\mathrm{T}}\mathbf{W}^{-1}\mathbf{x}$ where $\mathbf{x}$ is required to satisfy $\mathbf{z} = \mathbf{H}\mathbf{x}$, where $\mathbf{W}$ is a symmetric, positive definite matrix. (Recall that the inverse of a positive definite matrix is also positive definite.) By forming the Lagrangian function $L(\lambda, \mathbf{x})$ similar to (5.1.14), show that the optimal $\mathbf{x}^* = \mathbf{W}\mathbf{H}^{\mathrm{T}}(\mathbf{H}\mathbf{W}\mathbf{H}^{\mathrm{T}})^{-1}\mathbf{z}$.

5.6 Let $\mathbf{H} \in \mathbb{R}^{m\times n}$ and $\mathbf{H}^+ \in \mathbb{R}^{n\times m}$. If $\mathbf{H}$ and $\mathbf{H}^+$ satisfy the following four properties, then $\mathbf{H}^+$ is called the Moore–Penrose (generalized) inverse of $\mathbf{H}$.
(a) $\mathbf{H}\mathbf{H}^+\mathbf{H} = \mathbf{H}$
(b) $\mathbf{H}^+\mathbf{H}\mathbf{H}^+ = \mathbf{H}^+$
(c) $(\mathbf{H}\mathbf{H}^+)^{\mathrm{T}} = \mathbf{H}\mathbf{H}^+$
(d) $(\mathbf{H}^+\mathbf{H})^{\mathrm{T}} = \mathbf{H}^+\mathbf{H}$
In Section 5.1 (for the case when $m > n$), $\mathbf{H}^+ = (\mathbf{H}^{\mathrm{T}}\mathbf{H})^{-1}\mathbf{H}^{\mathrm{T}}$ has been called the generalized inverse of $\mathbf{H}$ and in Section 5.3 (for the case $m < n$), $\mathbf{H}^+ = \mathbf{H}^{\mathrm{T}}(\mathbf{H}\mathbf{H}^{\mathrm{T}})^{-1}$ has been called the generalized inverse of $\mathbf{H}$. Verify that $\mathbf{H}^+$ defined in each of these cases satisfies the four conditions given above.

5.7 Following the development of temperature determination from radiance measurements (Section 3.8), we divide the atmosphere into 3 layers as shown below:

		Pressure	Layer
	————	0 mb	
	T_3		Layer 3
	————	200 mb	
	T_2		Layer 2
	————	500 mb	
	T_1		Layer 1
T_0	————	1000 mb	

where T_0 is the temperature of the earth's surface, and T_1, T_2, and T_3 are the mean temperatures of the layers. The layers are bounded by $\hat{p} = 1, 0.5, 0.2, 0$. Assume the true state of the atmosphere is given by $T_1 = 0.9$, $T_2 = 0.85$, and $T_3 = 0.875$. The radiances are given by

$$R_\nu = \exp(-\gamma_\nu) + \int_0^1 T\, w(p, \gamma_\nu)\, \mathrm{d}p$$

where all variables are nondimensional and

$$w(p, \gamma_\nu) = p\, \gamma_\nu \exp[-\gamma_\nu\, p].$$

The problem is to find T by measurements of R_ν, assuming γ_ν is known. We choose the following wavenumbers (nondimensional) as follows:

i	ν_i	γ_ν	$w(p, \gamma_\nu)$	$\exp[-\gamma_\nu]$
1	0.9	(1/0.9)	$(1/0.9)p \exp(-p/0.9)$	0.329
2	1.0	(1/0.7)	$(1/0.7)p \exp(-p/0.7)$	0.240
3	1.1	(1/0.5)	$(1/0.5)p \exp(-p/0.5)$	0.135
4	1.2	(1/0.3)	$(1/0.3)p \exp(-p/0.3)$	0.036
5	1.3	(1/0.2)	$(1/0.2)p \exp(-p/0.2)$	0.007

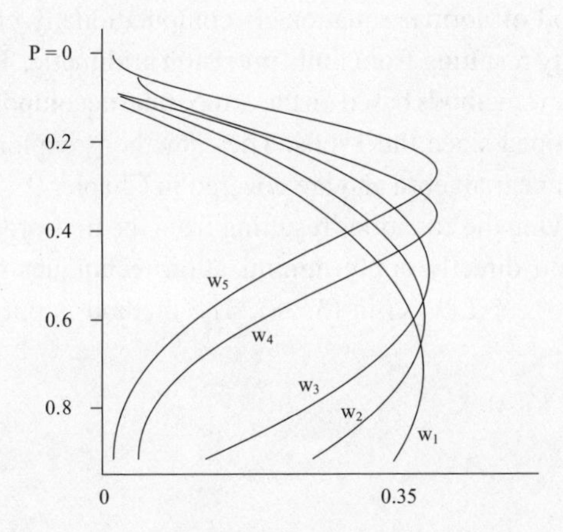

Calculate true values of R_ν through "forward" calculation. Contaminate R_ν with random noise (call these $\widetilde{R}_\nu$). Using only $\widetilde{R}_1$, $\widetilde{R}_2$, and $\widetilde{R}_3$, recover T_1, T_2, and T_3. Using $\widetilde{R}_1$, $\widetilde{R}_2$, $\widetilde{R}_3$, $\widetilde{R}_4$, and $\widetilde{R}_5$, recover T_1, T_2, and T_3.

Notes and references

In this chapter, we have described the classical method for least squares estimation using the method of normal equations.

The method described in this chapter is quite standard and goes by several names: method of linear regression, curve fitting, etc., Draper and Smith (1966). The basic ideas behind this method goes back to Gauss. This deterministic approach to least squares estimation is covered in detail in several classics, including Lawson and Hanson (1995), Golub and van Loan (1989). For a comprehensive coverage of the theory and applications of optimization refer to Luenberger (1973), Dennis and Schnabel (1996), and Nash and Sofer (1996). Refer to Appendix A for a discussion of various vector norms and their properties.

The notion of weighted least squares is central to data assimilation and is used extensively in Part IV on 3DVAR. A good discussion of deterministic weighted least squares is contained in Golub and van Loan (1989). Both ordinary and weighted least squares are extensively used in econometric and finance literature, for details refer to Johnston and DiNardo (1997), Pindyck and Rubinfeld (1998), Greene (2000), and Hamilton (1994).

The concept of generalized inverses of matrices arise naturally in the least squares context. For a comprehensive review of the properties of generalized inverses and methods for computing them refer to Albert (1972), Basilevshy (1983), and Rao and Mitra (1971).

Numerical methods for solving the normal equations are described in Part III. While the method of normal equations is computationally efficient, it may show signs of instability resulting from finite precision arithmetic. To alleviate this instability problem, new methods based on the orthogonal decomposition of the **H** matrix have been developed since the 1960s. These methods cxploit various techniques from numerical linear algebra and are covered in Chapter 9.

Instead of solving the equations resulting from the first-order/second-order conditions, one could directly apply minimization techniques to minimize $f(\mathbf{x})$ in (5.1.14) or (5.2.3), or $L(\lambda, \mathbf{x})$ in (5.3.8). The method is pursued in Chapters 10 through 12.

6

A geometric view: projection and invariance

In this chapter we revisit the linear least squares estimation problem and solve it using the method of *orthogonal projection*. This geometric view is quite fundamental and has guided the development and extension of least squares solutions in several directions. In Section 6.1 we describe the basic principles of orthogonal projections, namely, projecting a vector $\mathbf{z}$ onto a single vector $\mathbf{h}$. In Section 6.2, we discuss the extension of this idea of projecting a given vector $\mathbf{z}$ onto the subspace spanned by the columns of the measurement matrix $\mathbf{H} \in \mathbb{R}^{m \times n}$. An interesting outcome of this exercise is that the set of linear equations defining the optimal estimate by this geometric method are identical to those derived from the method of normal equations. This invariance of the least squares solution with respect to the methods underscores the importance of this class of solutions. Section 6.3 develops the geometric equivalent of the weighted or generalized linear least squares problem. It is shown that the optimal solution is given by an *oblique* projection as opposed to an orthogonal projection. In Section 6.4 we derive conditions for the invariance of least squares solutions under linear transformations of both the **model space** $\mathbb{R}^n$ and the **observation space** $\mathbb{R}^m$. It turns out invariance is achievable within the framework of generalized or weighted least squares formulation.

6.1 Orthogonal projection: basic idea

Let $\mathbf{h} = (h_1, h_2, \ldots, h_m)^{\mathrm{T}} \in \mathbb{R}^m$ be the given vector representing the measurement system, and let $\mathbf{z} = (z_1, z_2, \ldots, z_m)^{\mathrm{T}} \in \mathbb{R}^m$ be a set of m observations, where it is assumed that $\mathbf{z}$ is *not* a multiple of $\mathbf{h}$. Refer to Figure 6.1.1.

Let $x \in \mathbb{R}^m$, that is x be a real scalar which is not known, and let $x\mathbf{h}$ denote the unknown multiple of $\mathbf{h}$. The question is what is the "best" representation of $\mathbf{z}$ along $\mathbf{h}$. In answering the question, first recall the following basic result from the first course in Euclidean geometry: "the shortest distance between a line (say h) and a point (say z) not on the line is the length of the perpendicular line segment from the point (z) to the line (h)". This fact holds the key to

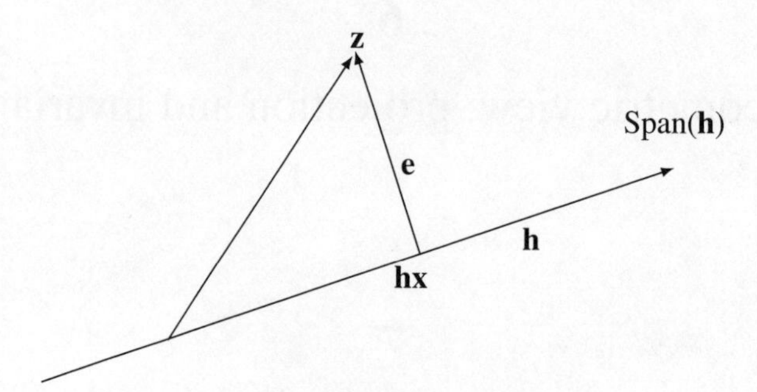

Fig. 6.1.1 An illustration of orthogonal projection.

finding the best representation we are seeking. That is, given $\mathbf{h}$ and $\mathbf{z}$ find the scalar x such that the magnitude of the error $\mathbf{e} = (\mathbf{z} - \mathbf{h}x)$ in this representation is such that $\|\mathbf{e}\|$ is as small as possible. (In this case and in almost all the developments in this chapter, unless specified otherwise, $\|\mathbf{e}\|$ denotes the Euclidean norm.) Accordingly, $\|\mathbf{e}\|$ is minimum when the error vector $\mathbf{e}$ is orthogonal to $\mathbf{h}$. That is,

$$0 = \mathbf{h}^{\mathrm{T}}\mathbf{e} = \mathbf{h}^{\mathrm{T}}(\mathbf{z} - \mathbf{h}x) = \mathbf{h}^{\mathrm{T}}\mathbf{z} - (\mathbf{h}^{\mathrm{T}}\mathbf{h})x \tag{6.1.1}$$

from which we obtain the optimal value for x to be

$$x^* = \frac{\mathbf{h}^{\mathrm{T}}\mathbf{z}}{\mathbf{h}^{\mathrm{T}}\mathbf{h}} = (\mathbf{h}^{\mathrm{T}}\mathbf{h})^{-1}\mathbf{h}^{\mathrm{T}}\mathbf{z} \tag{6.1.2}$$

and the optimal representation $\mathbf{z}^*$ is given by

$$\mathbf{z}^* = \mathbf{h}x^* = \frac{\mathbf{h}(\mathbf{h}^{\mathrm{T}}\mathbf{z})}{\mathbf{h}^{\mathrm{T}}\mathbf{h}} = \mathbf{h}(\mathbf{h}^{\mathrm{T}}\mathbf{h})^{-1}\mathbf{h}^{\mathrm{T}}\mathbf{z} = \mathbf{h}\,\mathbf{h}^{+}\mathbf{z} \tag{6.1.3}$$

where

$$\mathbf{h}^{+} = (\mathbf{h}^{\mathrm{T}}\mathbf{h})^{-1}\mathbf{h}^{\mathrm{T}} \tag{6.1.4}$$

is called the generalized inverse of $\mathbf{h}$. The vector $\mathbf{z}^*$ given by (6.1.3) is the *orthogonal projection of* $\mathbf{z}$ *onto* $\mathbf{h}$.

If $x \neq x^*$, then $\mathbf{h}x$ still represents a projection, but is called an *oblique* projection. In the same way orthogonal projections and ordinary least squares are intimately related, oblique projections and weighted or generalized least squares are close to each other. See Section 6.3 for details. Several comments and observations are in order.

Remark 6.1.1 Given a vector $\mathbf{h}$, Span($\mathbf{h}$) = $\{\mathbf{y} \mid \mathbf{y} = \alpha\mathbf{h}$ for any $\alpha \in \mathbb{R}\}$, that is, span of $\mathbf{h}$ denotes the set of all scalar multiples of $\mathbf{h}$. Span($\mathbf{h}$) is also called the subspace generated by $\mathbf{h}$ and geometrically it denotes the line that coincides with $\mathbf{h}$ that extends from $-\infty$ to $+\infty$.

Remark 6.1.2 The formula (6.1.2) is structurally very similar to that obtained in (5.1.18). This similarity should not be surprising, since the error vector $\mathbf{e} = (\mathbf{z} - \mathbf{h}x)$ is, in fact, the residual $\mathbf{r}(\mathbf{x}) = (\mathbf{z} - \mathbf{h}x)$. It is an easy exercise to verify that $\mathbf{h}^+$ in (6.1.4) is indeed the generalized inverse of $\mathbf{h}$ (see Exercise 6.1).

Remark 6.1.3 A justification for calling $\mathbf{z}^*$ an orthogonal projection can also be seen from another basic fact of analytical geometry. Let

$$\hat{\mathbf{h}} = \frac{\mathbf{h}}{\|\mathbf{h}\|} \tag{6.1.5}$$

be the unit vector in the direction of $\mathbf{h}$. It is well known that the inner product of $\mathbf{z}$ with $\hat{\mathbf{h}}$, namely, $\mathbf{z}^{\mathrm{T}}\hat{\mathbf{h}}$ denotes the *magnitude* of the (orthogonal) projection of $\mathbf{z}$ onto $\mathbf{h}$. This magnitude times the unit vector $\hat{\mathbf{h}}$ then denotes the projection (vector) of $\mathbf{z}$ onto $\mathbf{h}$. Thus,

$$\begin{aligned}
\mathbf{z}^* &= (\mathbf{z}^{\mathrm{T}}\hat{\mathbf{h}})\hat{\mathbf{h}} = \hat{\mathbf{h}}(\mathbf{z}^{\mathrm{T}}\hat{\mathbf{h}}) = \hat{\mathbf{h}}(\hat{\mathbf{h}}^{\mathrm{T}}\mathbf{z}) \\
&= (\hat{\mathbf{h}}\hat{\mathbf{h}}^{\mathrm{T}})\mathbf{z} = \left(\frac{\mathbf{h}}{\|\mathbf{h}\|}\frac{\mathbf{h}^{\mathrm{T}}}{\|\mathbf{h}\|}\right)\mathbf{z} \\
&= \frac{1}{\|\mathbf{h}\|^2}\mathbf{h}\mathbf{h}^{\mathrm{T}}\mathbf{z} = \frac{1}{\mathbf{h}^{\mathrm{T}}\mathbf{h}}\mathbf{h}\mathbf{h}^{\mathrm{T}}\mathbf{z} \\
&= \mathbf{h}(\mathbf{h}^{\mathrm{T}}\mathbf{h})^{-1}\mathbf{h}^{\mathrm{T}}\mathbf{z}
\end{aligned} \tag{6.1.6}$$

which is identical to (6.1.3).

Now define

$$\mathbf{P_h} = \hat{\mathbf{h}}\hat{\mathbf{h}}^{\mathrm{T}} = \frac{1}{\mathbf{h}^{\mathrm{T}}\mathbf{h}}\mathbf{h}\mathbf{h}^{\mathrm{T}} \tag{6.1.7}$$

which is the *outer-product* matrix of the unit vector $\hat{\mathbf{h}}$ with itself. The matrix $\mathbf{P_h}$ is called the *orthogonal projection matrix (or operator)* onto the subspace generated by $\mathbf{h}$. The orthogonal projection of $\mathbf{z}$ onto Span$\{\mathbf{h}\}$ is obtained simply by multiplying $\mathbf{z}$ by $\mathbf{P_h}$ on the left, i.e. $\mathbf{z}^* = \mathbf{P_h}\mathbf{z}$.

Properties of the orthogonal projection matrix $\mathbf{P_h}$ Let $\mathbf{h} = (h_1, h_2, \ldots, h_m)^{\mathrm{T}}$ and $\hat{\mathbf{h}} = (\hat{h}_1, \hat{h}_2, \ldots, \hat{h}_m)^{\mathrm{T}}$. Then

$$\begin{aligned}
\mathbf{P_h} = \hat{\mathbf{h}}\hat{\mathbf{h}}^{\mathrm{T}} &= \begin{bmatrix} \hat{h}_1 \\ \hat{h}_2 \\ \vdots \\ \hat{h}_m \end{bmatrix} [\hat{h}_1, \hat{h}_2, \ldots, \hat{h}_m] \\
&= \begin{bmatrix} \hat{h}_1^2 & \hat{h}_1\hat{h}_2 & \hat{h}_1\hat{h}_3 & \cdots & \hat{h}_1\hat{h}_m \\ \hat{h}_2\hat{h}_1 & \hat{h}_2^2 & \hat{h}_2\hat{h}_3 & \cdots & \hat{h}_2\hat{h}_m \\ \vdots & \vdots & \vdots & \vdots & \vdots \\ \hat{h}_m\hat{h}_1 & \hat{h}_m\hat{h}_2 & \hat{h}_m\hat{h}_3 & \cdots & \hat{h}_m^2 \end{bmatrix}.
\end{aligned} \tag{6.1.8}$$

Verification of the following properties is left as an exercise. (See Exercise 6.2.)

(a) $\mathbf{P}_{\mathbf{h}}^{\mathrm{T}} = \mathbf{P}_{\mathbf{h}}$, that is, $\mathbf{P}_{\mathbf{h}}$ is a *symmetric* matrix.
(b) $\mathbf{P}_{\mathbf{h}}^{2} = \mathbf{P}_{\mathbf{h}}$, that is, $\mathbf{P}_{\mathbf{h}}$ is *idempotent*.
(c) Since each column of $\mathbf{P}_{\mathbf{h}}$ is a multiple of $\hat{\mathbf{h}}$, $\mathrm{Rank}(\mathbf{P}_{\mathbf{h}}) = 1$.
(d) $\det(\mathbf{P}_{\mathbf{h}}) = 0$, that is, $\mathbf{P}_{\mathbf{h}}$ is *singular*.
(e) 1 is the only non-zero *eigenvalue* of $\mathbf{P}_{\mathbf{h}}$.
(f) $\mathbf{P}_{\mathbf{h}}^{\mathrm{T}} \neq \mathbf{P}_{\mathbf{h}}^{-1}$, that is, $\mathbf{P}_{\mathbf{h}}$ is *not an orthogonal matrix* even though it produces an orthogonal projection.

6.2 Ordinary least squares estimation: orthogonal projection

In generalizing the above development, consider now a measurement matrix $\mathbf{H} \in \mathbb{R}^{m\times n}$ $(m > n)$ with

$$\mathbf{H} = [\mathbf{h}_{*1}, \mathbf{h}_{*2}, \ldots, \mathbf{h}_{*n}] \tag{6.2.1}$$

where $\mathbf{h}_{*j}$ denotes the jth column of $\mathbf{H}$, $j = 1, \ldots, n$. Clearly,

$$\begin{aligned}\mathrm{Span}(\mathbf{H}) &= \left\{\mathbf{y}|\mathbf{y} = \alpha_1\mathbf{h}_{*1} + \alpha_2\mathbf{h}_{*2} + \cdots + \alpha_n\mathbf{h}_{*n},\ \text{with } \alpha_i' s \text{ scalars}\right\}\\ &= \left\{\mathbf{y}|\mathbf{y} = \mathbf{H}\alpha, \alpha = (\alpha_1, \alpha_2, \ldots, \alpha_n)^{\mathrm{T}} \in \mathbb{R}^n\right\}\end{aligned} \tag{6.2.2}$$

denotes the subspace of $\mathbb{R}^n$ generated by the n columns of $\mathbf{H}$. Let $\mathbf{z} \in \mathbb{R}^m$ where it is assumed that $\mathbf{z} \notin \mathrm{Span}(\mathbf{H})$. The question is: what is the best representation of $\mathbf{z}$ in $\mathrm{Span}(\mathbf{H})$?

Let $\mathbf{x} = (x_1, x_2, \ldots, x_n)^{\mathrm{T}} \in \mathbb{R}^n$ and let $\mathbf{Hx}$ denote a representation of $\mathbf{z}$ in $\mathrm{Span}(\mathbf{H})$, where recall

$$\mathbf{Hx} = x_1\mathbf{h}_{*1} + x_2\mathbf{h}_{*2} + \cdots + x_n\mathbf{h}_{*n}. \tag{6.2.3}$$

Then, the error $\mathbf{e}$ in this representation is given by

$$\mathbf{e} = (\mathbf{z} - \mathbf{Hx}). \tag{6.2.4}$$

From Section 6.1 it follows that $\mathbf{Hx}$ will be the best representation for $\mathbf{z}$ in $\mathrm{Span}(\mathbf{H})$ in the sense of $\|\mathbf{e}\|$ is a minimum exactly when $\mathbf{e}$ is orthogonal to the $\mathrm{Span}(\mathbf{H})$. That is,

$$\begin{aligned}\mathbf{0} &= (\mathbf{Hx})^{\mathrm{T}}\mathbf{e} = (\mathbf{Hx})^{\mathrm{T}}(\mathbf{z} - \mathbf{Hx})\\ &= \left(\textstyle\sum_{j=1}^{m} x_j\mathbf{h}_{*j}\right)^{\mathrm{T}}(\mathbf{z} - \mathbf{Hx}) = \left(\textstyle\sum_{j=1}^{m} x_j\mathbf{H}_{*j}^{\mathrm{T}}\right)(\mathbf{z} - \mathbf{Hx})\\ &= \textstyle\sum_{j=1}^{m} x_j\left[\mathbf{h}_{*j}^{\mathrm{T}}(\mathbf{z} - \mathbf{Hx})\right].\end{aligned} \tag{6.2.5}$$

Since x_j's are *not* known, (6.2.5) can be true only when

$$\mathbf{0} = \mathbf{h}_{*j}^{\mathrm{T}}[(\mathbf{z} - \mathbf{Hx})] \text{ for } j = 1, 2, \ldots, n. \tag{6.2.6}$$

That is, when the error $\mathbf{e} = [(\mathbf{z} - \mathbf{H}\mathbf{x})]$ is orthogonal to each column of $\mathbf{H}$. By stacking up all these n conditions, we get

$$\begin{bmatrix} \mathbf{h}_{*1}^{\mathrm{T}} \\ \mathbf{h}_{*2}^{\mathrm{T}} \\ \vdots \\ \mathbf{h}_{*n}^{\mathrm{T}} \end{bmatrix} [(\mathbf{z} - \mathbf{H}\mathbf{x})] = \begin{bmatrix} 0 \\ 0 \\ \vdots \\ 0 \end{bmatrix} \tag{6.2.7}$$

or

$$\mathbf{H}^{\mathrm{T}}[(\mathbf{z} - \mathbf{H}\mathbf{x})] = \mathbf{0}$$

which leads to

$$(\mathbf{H}^{\mathrm{T}}\mathbf{H})\mathbf{x} = \mathbf{H}^{\mathrm{T}}\mathbf{z}$$

or

$$\mathbf{x}^* = (\mathbf{H}^{\mathrm{T}}\mathbf{H})^{-1}\mathbf{H}^{\mathrm{T}}\mathbf{z} \tag{6.2.8}$$

which is the same as (5.1.18). Again, this similarity is to be expected since the error vector $\mathbf{e}$ is indeed the residual vector $\mathbf{r}(\mathbf{x})$ considered in Chapter 2. The optimal representation $\mathbf{z}^*$ is given by

$$\mathbf{z}^* = \mathbf{H}\mathbf{z}^* = \mathbf{H}(\mathbf{H}^{\mathrm{T}}\mathbf{H})^{-1}\mathbf{H}^{\mathrm{T}}\mathbf{z} = \mathbf{P}_{\mathbf{H}}\mathbf{z} \tag{6.2.9}$$

where

$$\mathbf{P}_{\mathbf{H}} = \mathbf{H}(\mathbf{H}^{\mathrm{T}}\mathbf{H})^{-1}\mathbf{H} \tag{6.2.10}$$

is an $m \times m$ matrix called the *(orthogonal) projection matrix* (operator) onto the Span($\mathbf{H}$).

Properties of $\mathbf{P}_{\mathbf{H}}$

(a) $\mathbf{P}_{\mathbf{H}}^{\mathrm{T}} = \mathbf{P}_{\mathbf{H}}$, that is, $\mathbf{P}_{\mathbf{H}}$ is an $m \times m$ *symmetric* matrix.
(b) $\mathbf{P}_{\mathbf{H}}^2 = \mathbf{P}_{\mathbf{H}}$, that is, $\mathbf{P}_{\mathbf{H}}$ is an *idempotent* matrix.
(c) Assuming Rank($\mathbf{H}$) $= n$, Rank($\mathbf{P}_{\mathbf{H}}$) $= n$
(d) det($\mathbf{P}_{\mathbf{H}}$) $= 0$, that is, $\mathbf{P}_{\mathbf{H}}$ is singular.
(e) There are exactly n non-zero eigenvalues of $\mathbf{P}_{\mathbf{H}}$.
(f) $\mathbf{P}_{\mathbf{H}}$ is *not* an orthogonal matrix.

Given $\mathbf{P}_{\mathbf{H}}$ defined by (6.2.10), we can define a new $m \times m$ matrix (operator)

$$\mathbf{P}_{\mathbf{H}}^{\perp} = \mathbf{I} - \mathbf{P}_{\mathbf{H}}. \tag{6.2.11}$$

The following properties of $\mathbf{P}_{\mathbf{H}}^{\perp}$ can be easily verified.

(a) $\mathbf{P}_{\mathbf{H}}^{\perp}$ is an $m \times m$ symmetric matrix, that is $(\mathbf{P}_{\mathbf{H}}^{\perp})^{\mathrm{T}} = \mathbf{P}_{\mathbf{H}}^{\perp}$.
(b) $\mathbf{P}_{\mathbf{H}}^{\perp}$ is idempotent, that is, $(\mathbf{P}_{\mathbf{H}}^{\perp})^2 = \mathbf{P}_{\mathbf{H}}^{\perp}$.

(c) $\mathbf{P}_{\mathbf{H}}^{\perp}\mathbf{z} = (\mathbf{I} - \mathbf{P}_{\mathbf{H}})\mathbf{z} = \mathbf{z} - \mathbf{P}_{\mathbf{H}}\mathbf{z} = \mathbf{z} - \mathbf{z}^* = \mathbf{e}^*$, that is, $\mathbf{P}_{\mathbf{H}}^{\perp}$ when applied to $\mathbf{z}$ gives the optimal error $\mathbf{e}^*$ in representing $\mathbf{z}$ by $\mathbf{z}^*$. Hence,

$$\mathbf{z} = \mathbf{P}_{\mathbf{H}}\mathbf{z} + (\mathbf{I} - \mathbf{P}_{\mathbf{H}})\mathbf{z} = \mathbf{z}^* + \mathbf{e}^* \tag{6.2.12}$$

which represents an orthogonal decomposition of $\mathbf{z}$ induced by the projection operator $\mathbf{P}_{\mathbf{H}}$, where $\mathbf{z}^* \in \text{Span}(\mathbf{H})$ and $\mathbf{e}^*$ is orthogonal to the Span($\mathbf{H}$).

(d) Rank($\mathbf{P}_{\mathbf{H}}^{\perp}$) $= n - m$ and det($\mathbf{P}_{\mathbf{H}}^{\perp}$) $= 0$; hence, $\mathbf{P}_{\mathbf{H}}^{\perp}$ is also singular.

(e) $\mathbf{P}_{\mathbf{H}} + \mathbf{P}_{\mathbf{H}}^{\perp} = \mathbf{I}$.

6.3 Generalized least squares estimation: oblique projection

In this section we examine the geometric interpretation of the weighted least squares solution derived in Section 5.2.

Let $\mathbf{H} \in \mathbb{R}^{m\times n}$ denote the given measurement system and $\mathbf{z} \in \mathbb{R}^m$ be the given observation where the matrix $\mathbf{H}$ and the vector $\mathbf{z}$ are obtained with respect to a given coordinate system, for definiteness, called the system A. Let $\mathbf{W} \in \mathbb{R}^{m\times m}$ be the given symmetric, positive definite matrix to be used as the weight matrix. The idea here is to transform the given coordinate system A to a new coordinate system B so that the given weighted least squares problem in the coordinate system A becomes the ordinary least squares problem in the new coordinate system B. This transformation from the system A to B is to be accomplished using the given weight matrix. Once the ordinary least squares solution in system B is obtained using the method of orthogonal projection described in Section 6.2, then the required solution to the weighted least squares problem is obtained by inverse transformation of the solution from system B to system A. In the following we provide an implementation of this strategy.

To this end, first recall (Appendix B) that any symmetric positive definite matrix $\mathbf{W}$ can be factored as

$$\mathbf{W} = \mathbf{C}^{\mathrm{T}}\mathbf{C} \tag{6.3.1}$$

where $\mathbf{C} \in \mathbb{R}^{m\times m}$ is a non-singular matrix. We now use $\mathbf{C}$ as the matrix that transforms the original coordinate system A into the new coordinate system B. Define

$$\bar{\mathbf{H}} = \mathbf{C}\mathbf{H} \tag{6.3.2}$$

where

$$\bar{\mathbf{H}} = [\bar{\mathbf{h}}_{*1}, \bar{\mathbf{h}}_{*2}, \ldots, \bar{\mathbf{h}}_{*n}].$$

That is, the jth column of $\bar{\mathbf{H}}$ in the coordinate system B is related to the jth column of $\mathbf{H}$ in the coordinate system A as in

$$\bar{\mathbf{h}}_{*j} = \mathbf{C}\mathbf{h}_{*j}, \ \text{ for } j = 1, 2, \ldots, n \tag{6.3.3}$$

Similarly, let

$$\bar{\mathbf{z}} = \mathbf{C}\mathbf{z} \tag{6.3.4}$$

denote the observations in the coordinate system B. That is, we now have the observation vector $\bar{\mathbf{z}}$ based on the measurement system denoted by $\bar{\mathbf{H}}$ and the problem is to find the optimal $\bar{\mathbf{x}}^*$ using the method of orthogonal projection in Section 6.2. First compute the orthogonal projection matrix in the coordinate system B as:

$$\mathbf{P}_{\bar{\mathbf{H}}} = \bar{\mathbf{H}}[\bar{\mathbf{H}}^{\mathrm{T}}\bar{\mathbf{H}}]^{-1}\bar{\mathbf{H}}^{\mathrm{T}} \tag{6.3.5}$$

and

$$\bar{\mathbf{z}}^* = \mathbf{P}_{\bar{\mathbf{H}}}\bar{\mathbf{z}}. \tag{6.3.6}$$

Now the projection $\mathbf{z}^*$ in the original coordinate system A is obtained by using (6.3.4):

$$\mathbf{z}^* = \mathbf{C}^{-1}\bar{\mathbf{z}}^* = \mathbf{C}^{-1}\mathbf{P}_{\bar{\mathbf{H}}}\bar{\mathbf{z}} = \mathbf{C}^{-1}\bar{\mathbf{H}}[\bar{\mathbf{H}}^{\mathrm{T}}\bar{\mathbf{H}}]^{-1}\bar{\mathbf{H}}^{\mathrm{T}}\bar{\mathbf{z}} \tag{6.3.7}$$

Now using (6.3.2) and (6.3.4), we get a representation for $\mathbf{z}^*$ in the original coordinate system A as

$$\mathbf{z}^* = \mathbf{C}^{-1}(\mathbf{CH})[\mathbf{H}^{\mathrm{T}}\mathbf{C}^{\mathrm{T}}\mathbf{CH}](\mathbf{H}^{\mathrm{T}}\mathbf{C}^{\mathrm{T}})(\mathbf{Cz}) = \mathbf{H}[\mathbf{H}^{\mathrm{T}}\mathbf{WH}]\mathbf{H}^{\mathrm{T}}\mathbf{Wz} \tag{6.3.8}$$

which is the same as (5.2.7). In other words, our original plan to convert the given weighted least squares in coordinate system A into an ordinary least squares in coordinate system B has indeed worked well.

In analogy with (6.2.9), express (6.3.8) as $\mathbf{z}^* = \mathbf{P}_{\mathbf{H}}\mathbf{z}$ where

$$\mathbf{P}_{\mathbf{H}} = \mathbf{H}[\mathbf{H}^{\mathrm{T}}\mathbf{WH}]^{-1}\mathbf{H}^{\mathrm{T}}\mathbf{Wz} \tag{6.3.9}$$

plays the role of the *projection* matrix in coordinate system A.

Interestingly enough, $\mathbf{P}_{\mathbf{H}}$ in (6.3.9) does *not* share the properties of $\mathbf{P}_{\mathbf{H}}$ listed in Section 6.2. In particular, it can be verified (Exercise 6.5) that $\mathbf{P}_{\mathbf{H}}$ in (6.3.9) is idempotent($\mathbf{P}_{\mathbf{H}}^2 = \mathbf{P}_{\mathbf{H}}$), but it is *not* symmetric ($\mathbf{P}_{\mathbf{H}}^{\mathrm{T}} \neq \mathbf{P}_{\mathbf{H}}$). Since a matrix is an orthogonal projection matrix only when it is *idempotent* and symmetric, the projection matrix $\mathbf{P}_{\mathbf{H}}$ in (6.3.9) corresponding to the weighted least squares problem denotes an *oblique* projection and *not* an orthogonal projection matrix.

6.4 Invariance under linear transformation

Once a least squares problem is formulated before attempting to solve it numerically, there may often arise a need to change the **scales** of the variables so as to make the components of the observation vector **z** and/or the state vector **x** to be of similar order of magnitude. This is usually achieved by multiplying selected components of $\mathbf{z}$ and/or $\mathbf{x}$ by suitable scaling factors. As an example, consider the case when $m = 3$ and $\mathbf{z} = (z_1, z_2, z_3)^T$. Let $\mathbf{B}$ be a 3×3 diagonal matrix given by $\mathbf{B} = \text{Diag}(b_1, b_2, b_3)$. Then $\bar{\mathbf{z}} = \mathbf{Bz} = (b_1 z_1, b_2 z_2, b_3 z_3)^T$ is the new set of scaled observations. Thus, scaling is obtained by multiplying the vector to be scaled on the left by a suitable matrix. This process of transforming the vector $\mathbf{z}$ to $\bar{\mathbf{z}} = \mathbf{Bx}$ by multiplying $\mathbf{z}$ by a matrix is a **linear transformation** (Appendix B). A basic requirement is that any linear transformation used in scaling must be **invertible**.

In this section we examine the conditions under which the solution of the least squares problem is invariant under linear transformation of both the **model space** $\mathbb{R}^n$ where $\mathbf{x}$ resides and the **observation space** $\mathbb{R}^m$ that contains $\mathbf{z}$.

Let $\mathbf{A} \in \mathbb{R}^{n \times n}$ and $\mathbf{B} \in \mathbb{R}^{m \times m}$ be two non-singular matrices representing the linear transformations in the model space $\mathbb{R}^n$ and the observation space $\mathbb{R}^m$. Let $\bar{\mathbf{x}}$ and $\bar{\mathbf{z}}$ be the new (scaled) state vector and the observation vector, where

$$\mathbf{x} = \mathbf{A}\bar{\mathbf{x}} \quad \text{and} \quad \mathbf{z} = \mathbf{B}\bar{\mathbf{z}}. \tag{6.4.1}$$

Then from the old residual $r(\mathbf{x}) = (\mathbf{z} - \mathbf{Hx})$, we obtain the new transformed residual $r(\bar{\mathbf{x}})$ as (since the residual lies in the observation space)

$$\begin{aligned} r(\bar{\mathbf{x}}) &= \mathbf{B}^{-1} r(\mathbf{x}) = \mathbf{B}^{-1}((\mathbf{z} - \mathbf{Hx})) \\ &= \mathbf{B}^{-1}\mathbf{z} - (\mathbf{B}^{-1}\mathbf{HA})\mathbf{A}^{-1}\mathbf{x} = \bar{\mathbf{z}} - \bar{\mathbf{H}}\bar{\mathbf{x}} \end{aligned} \tag{6.4.2}$$

where

$$\bar{\mathbf{H}} = \mathbf{B}^{-1}\mathbf{HA} \tag{6.4.3}$$

denotes the new measurement system.

Case A: $m > n$ Invoking the results in Section 5.1, the transformed least squares solution minimizing $\|r(\bar{\mathbf{x}})\|$ in (6.4.2) is then given by

$$\bar{\mathbf{x}}_{\text{LS}} = (\bar{\mathbf{H}}^T\bar{\mathbf{H}})^{-1}\bar{\mathbf{H}}^T\bar{\mathbf{z}}. \tag{6.4.4}$$

Substituting for $\bar{\mathbf{H}}$ and $\bar{\mathbf{z}}$, we get

$$\begin{aligned} \bar{\mathbf{x}}_{\text{LS}} &= (\mathbf{A}^T\mathbf{H}^T\mathbf{B}^{-T}\mathbf{B}^{-1}\mathbf{HA})^{-1}\mathbf{A}^T\mathbf{H}^T\mathbf{B}^{-T}(\mathbf{B}^{-1}\mathbf{z}) \\ &= \mathbf{A}^{-1}[\mathbf{H}^T(\mathbf{BB}^T)^{-1}\mathbf{H}]^{-1}\mathbf{H}^T(\mathbf{BB}^T)^{-1}\mathbf{z}. \end{aligned}$$

Converting back to the original variables, we get

$$\bar{\mathbf{x}}_{\text{LS}} = \mathbf{A}\bar{\mathbf{x}}_{\text{LS}} = [\mathbf{H}^T(\mathbf{BB}^T)^{-1}\mathbf{H}]^{-1}\mathbf{H}^T(\mathbf{BB}^T)^{-1}\mathbf{z}. \tag{6.4.5}$$

Notice first that the right-hand side of (6.4.5) is independent of $\mathbf{A}$, which in turn implies that the classical least squares solution in (5.1.18) is **invariant** under (non-singular) linear transformation of the model space. The story is quite different, however, with respect to the linear transformation of the observation space. From (6.4.5) it follows that the classical solution is invariant only when $\mathbf{BB}^T = \mathbf{I}$ which can happen exactly when $\mathbf{B}$ is an **orthogonal** transformation

(Appendix B). This is a rather natural requirement since $r(\mathbf{x})$ resides in $\mathbb{R}^m$ and the length of vectors in $\mathbb{R}^m$ remains invariant under orthogonal transformation of $\mathbb{R}^m$.

Case B: $m < n$ In this case by minimizing

$$f(\bar{\mathbf{x}}) = \|\bar{\mathbf{x}}\|^2 + \lambda(\bar{\mathbf{z}} - \bar{\mathbf{H}}\bar{\mathbf{x}}) \tag{6.4.6}$$

which is the analog of (5.3.8), we obtain from (5.3.11) the new least squares solution

$$\bar{\mathbf{x}}_{\mathrm{LS}} = \bar{\mathbf{H}}^{\mathrm{T}}(\bar{\mathbf{H}}\bar{\mathbf{H}}^{\mathrm{T}})^{-1}\bar{\mathbf{z}}. \tag{6.4.7}$$

Converting back to the original variables, we obtain after simplifying

$$\mathbf{x}_{\mathrm{LS}} = \mathbf{A}\bar{\mathbf{x}}_{\mathrm{LS}} = (\mathbf{A}\mathbf{A}^{\mathrm{T}})\mathbf{H}^{\mathrm{T}}[\mathbf{H}(\mathbf{A}\mathbf{A}^{\mathrm{T}})\mathbf{H}^{\mathrm{T}}]^{-1}\mathbf{z}. \tag{6.4.8}$$

Clearly, this is independent of $\mathbf{B}$ and hence invariant under the transformation of the observation space. However, this solution (6.4.8) is invariant only when $\mathbf{A}\mathbf{A}^{\mathrm{T}} = \mathbf{I}$ or $\mathbf{A}$ is an orthogonal matrix.

Notice the duality between these two cases – in Case A, minimization is performed in the observation space $\mathbb{R}^m$ while in Case B, minimization is performed in the model space $\mathbb{R}^n$. This fact is reflected by the reversal of the conditions on the matrices $\mathbf{A}$ and $\mathbf{B}$ to obtain overall invariance of solutions.

Case C Consider the minimization of the combined objective function resulting from Tikhonov regularization (Section 5.4)

$$f(\bar{\mathbf{x}}) = \frac{\alpha}{2}\bar{\mathbf{x}}^{\mathrm{T}}\bar{\mathbf{x}} + \frac{1}{2}(\bar{\mathbf{z}} - \bar{\mathbf{H}}\mathbf{x})^{\mathrm{T}}(\bar{\mathbf{z}} - \bar{\mathbf{H}}\mathbf{x}) \tag{6.4.9}$$

which is the analog of the function in (5.4.1). It can be verified (Exercise 6.6) that the equation defining the least squares solution of (6.4.9) given by

$$(\bar{\mathbf{H}}^{\mathrm{T}}\bar{\mathbf{H}} + \alpha\mathbf{I})\bar{\mathbf{x}} = \bar{\mathbf{H}}^{\mathrm{T}}\bar{\mathbf{z}} \tag{6.4.10}$$

which when converted back into the original variables becomes

$$[\mathbf{H}^{\mathrm{T}}(\mathbf{B}\mathbf{B}^{\mathrm{T}})\mathbf{H} + \alpha(\mathbf{A}\mathbf{A}^{\mathrm{T}})^{-1}]\mathbf{x} = \mathbf{H}^{\mathrm{T}}(\mathbf{B}\mathbf{B}^{\mathrm{T}})^{-1}\mathbf{z}. \tag{6.4.11}$$

That is, while the combined formulation using the Tikhonov regularization unifies the dual formulations into one, it does **not** lead to invariance of the solution unless both the matrices $\mathbf{A}$ and $\mathbf{B}$ are orthogonal, which is too restrictive.

The above development leads to an inescapable conclusion that the classical ordinary least squares formulation of Section 5.1 does **not** have desirable invariance property. In search for an invariance under general (non-singular) linear transformations of both the model and the observation spaces, we now turn to the generalized (weighted) least squares formulation of Section 5.2.

Case D Generalized (weighted) least squares. Let $\bar{\mathbf{W}}_{\mathbf{x}} \in \mathbb{R}^{n\times n}$ and $\bar{\mathbf{W}}_{\mathbf{d}} \in \mathbb{R}^{m\times m}$ be two real, symmetric and positive definite matrices. Consider the combined weighted least squares criterion using the new (scaled) variables given by

$$f(\bar{\mathbf{x}}) = \frac{1}{2}\bar{\mathbf{x}}^{\mathrm{T}}\bar{\mathbf{W}}_{\mathbf{x}}\bar{\mathbf{x}} + \frac{1}{2}(\bar{\mathbf{z}} - \bar{\mathbf{H}}\bar{\mathbf{x}})^{\mathrm{T}}\bar{\mathbf{W}}_{\mathbf{d}}(\mathbf{z} - \bar{\mathbf{H}}\bar{\mathbf{x}}). \tag{6.4.12}$$

Comparing this with (5.2.3) and (5.4.1), it follows that this function is a result of combining the **Tikhonov regularization** idea (Section 5.4) and the idea of using the **energy** or the **weighted norm**.

The gradient and the Hessian of (6.4.12) are given by

$$\nabla f(\bar{\mathbf{x}}) = [\bar{\mathbf{H}}^{\mathrm{T}}\bar{\mathbf{W}}_{\mathbf{d}}\bar{\mathbf{H}} + \bar{\mathbf{W}}_{\mathbf{x}}]\mathbf{x} - \bar{\mathbf{H}}^{\mathrm{T}}\bar{\mathbf{W}}_{\mathbf{d}}\bar{\mathbf{z}}, \tag{6.4.13}$$

and

$$\nabla^2 f(\bar{\mathbf{x}}) = [\bar{\mathbf{H}}^{\mathrm{T}}\bar{\mathbf{W}}_{\mathbf{d}}\bar{\mathbf{H}} + \bar{\mathbf{W}}_{\mathbf{x}}]. \tag{6.4.14}$$

Since $\bar{\mathbf{W}}_{\mathbf{d}}$ and $\bar{\mathbf{W}}_{\mathbf{x}}$ are positive definite, and $\bar{\mathbf{H}}$ is of full rank if $\mathbf{H}$ is (why?), it follows that this Hessian is positive definite. Hence the minimizer is obtained by setting the gradient to zero leading to the solution of the linear system

$$[\bar{\mathbf{H}}^{\mathrm{T}}\bar{\mathbf{W}}_{\mathbf{d}}\bar{\mathbf{H}} + \bar{\mathbf{W}}_{\mathbf{x}}]\bar{\mathbf{x}} = \bar{\mathbf{H}}^{\mathrm{T}}\bar{\mathbf{W}}_{\mathbf{d}}\bar{\mathbf{z}}. \tag{6.4.15}$$

Substituting for $\bar{\mathbf{H}}$, $\bar{\mathbf{x}}$ and $\bar{\mathbf{z}}$, and simplifying, we obtain (Exercise 6.7)

$$[\mathbf{H}^{\mathrm{T}}(\mathbf{B}^{-\mathrm{T}}\bar{\mathbf{W}}_{\mathbf{d}}\mathbf{B}^{-1})\mathbf{H} + \mathbf{A}^{-\mathrm{T}}\bar{\mathbf{W}}_{\mathbf{x}}\mathbf{A}^{-1}]\mathbf{x} = \mathbf{H}^{\mathrm{T}}(\mathbf{B}^{-\mathrm{T}}\bar{\mathbf{W}}_{\mathbf{d}}\mathbf{B}^{-1})\mathbf{z}. \tag{6.4.16}$$

Setting

$$\mathbf{W}_{\mathbf{d}} = (\mathbf{B}^{-\mathrm{T}}\bar{\mathbf{W}}_{\mathbf{d}}\mathbf{B}^{-1}) \quad \text{or} \quad \bar{\mathbf{W}}_{\mathbf{d}} = \mathbf{B}^{\mathrm{T}}\mathbf{W}_{\mathbf{d}}\mathbf{B}$$

and

$$\mathbf{W}_{\mathbf{x}} = \mathbf{A}^{-\mathrm{T}}\bar{\mathbf{W}}_{\mathbf{x}}\mathbf{A}^{-1} \quad \text{or} \quad \bar{\mathbf{W}}_{\mathbf{x}} = \mathbf{A}^{\mathrm{T}}\mathbf{W}_{\mathbf{x}}\mathbf{A}, \tag{6.4.17}$$

the above equation becomes

$$[\mathbf{H}^{\mathrm{T}}\mathbf{W}_{\mathbf{d}}\mathbf{H} + \mathbf{W}_{\mathbf{x}}]\mathbf{x} = \mathbf{H}^{\mathrm{T}}\mathbf{W}_{\mathbf{d}}\mathbf{z}. \tag{6.4.18}$$

In other words, the combined weighted least squares solution is invariant under the linear transformation of both the model space and the observation space exactly when the old weight matrices $\mathbf{W}_{\mathbf{x}}$ and $\mathbf{W}_{\mathbf{d}}$ are related to the new weight matrices $\bar{\mathbf{W}}_{\mathbf{x}}$ and $\bar{\mathbf{W}}_{\mathbf{d}}$ via the **congruence transformation** (Appendix B) in (6.4.17).

Thus, the search for the invariance of the least squares solution reduces to finding a class of weight matrices that transform according to the rule in (6.4.17). Indeed, as shown in Appendix F, the answer lies in picking the original weight matrix to be the inverse of an appropriate covariance matrix. For completeness in the following, we provide a quick verification of this claim.

Let $\eta \in \mathbb{R}^n$ be a random vector with Σ_η as its covariance matrix. That is,

$$\Sigma_\eta = E[(\eta - E(\eta))(\eta - E(\eta))^{\mathrm{T}}]. \tag{6.4.19}$$

Let $\mathbf{D} \in \mathbb{R}^{n\times n}$ be a non-singular matrix and define a new random vector ξ using the linear transformation

$$\eta = \mathbf{D}\xi. \tag{6.4.20}$$

Then, if Σ_ξ is the covariance matrix of ξ, then

$$E(\xi) = E[\mathbf{D}^{-1}\eta] = \mathbf{D}^{-1}E(\eta)$$

and

$$\begin{aligned}\Sigma_\xi &= E[(\xi - E(\xi))(\xi - E(\xi))^{\mathrm{T}}]\\ &= \mathbf{D}^{-1}E[(\eta - E(\eta))(\eta - E(\eta))^{\mathrm{T}}]\mathbf{D}^{-\mathrm{T}}\\ &= \mathbf{D}^{-1}\Sigma_\eta\mathbf{D}^{-\mathrm{T}}\end{aligned}$$

or

$$\Sigma_\xi^{-1} = \mathbf{D}^{\mathrm{T}}\Sigma_\eta^{-1}\mathbf{D}, \tag{6.4.21}$$

which is exactly of the same form as required in (6.4.17).

We now state the conditions for the invariance of the least squares solution.

(a) Choose $\mathbf{W_d}$ to be the inverse of the **observational error covariance matrix**.
(b) Choose $\mathbf{W_x}$ to be the inverse of the **background error covariance matrix**.

This is the main reason for the widespread use of the inverse of the covariance matrix as the weight matrices in the formulation of the data assimilation problems of interest in Parts V and VI.

Exercises

6.1 Verify that $\mathbf{h}^+ = (\mathbf{h}^{\mathrm{T}}\mathbf{h})^{-1}\mathbf{h}^{\mathrm{T}}$ defined in (6.1.4) is indeed the Moore–Penrose generalized inverse of $\mathbf{h}$. That is, verify the following:
   (a) $\mathbf{hh^+h} = \mathbf{h}$
   (b) $\mathbf{h^+hh^+} = \mathbf{h^+}$
   (c) $(\mathbf{h^+h})^{\mathrm{T}} = \mathbf{h^+h}$
   (d) $(\mathbf{hh^+})^{\mathrm{T}} = \mathbf{h}\,\mathbf{h^+}$

6.2 Verify the following properties of $\mathbf{P_h}$.
   (a) $\mathbf{P_h^{\mathrm{T}}} = \mathbf{P_h}$, that is, $\mathbf{P_h}$ is a *symmetric* matrix.
   (b) $\mathbf{P}^2 = \mathbf{P_h}$, that is, $\mathbf{P_h}$ is *idempotent*.
   (c) Since each column of $\mathbf{P_h}$ is a multiple of $\hat{\mathbf{h}}$, Rank$(\mathbf{P_h}) = 1$.
   (d) $\det(\mathbf{P_h}) = 0$, that is, $\mathbf{P_h}$ is *singular*.
   (e) 1 is the only non-zero *eigenvalue* of $\mathbf{P_h}$.
   (f) $\mathbf{P_h^{\mathrm{T}}} \neq \mathbf{P_h^{-1}}$, that is, $\mathbf{P_h}$ is *not an orthogonal matrix* even though it produces an orthogonal projection.

6.3 Verify the following properties of $\mathbf{P_H}$.
   (a) $\mathbf{P_H^{\mathrm{T}}} = \mathbf{P_H}$, that is $\mathbf{P_H}$ is an $m \times m$ *symmetric* matrix.
   (b) $\mathbf{P_H^2} = \mathbf{P_H}$, that is $\mathbf{P_H}$ is an *idempotent* matrix.
   (c) Assuming Rank$(\mathbf{H}) = n$, Rank$(\mathbf{P_H}) = n$.
   (d) $\det(\mathbf{P_H}) = 0$, that is, $\mathbf{P_H}$ is singular.
   (e) There are exactly n non-zero eigenvalues of $\mathbf{P_H}$.
   (f) $\mathbf{P_H}$ is *not* an orthogonal matrix.

6.4 Define $\mathbf{P}_{\mathbf{H}}^{\perp} = \mathbf{I} - \mathbf{P}_{\mathbf{H}}$ where $\mathbf{P}_{\mathbf{H}} = \mathbf{H}(\mathbf{H}^{\mathrm{T}}\mathbf{H})^{\mathrm{T}}\mathbf{H}^{\mathrm{T}}$. Verify that (a) $\mathbf{P}_{\mathbf{H}}^{\perp}$ is symmetric, (b) $\mathbf{P}_{\mathbf{H}}^{\perp}$ is idempotent (c) $\mathrm{Rank}(\mathbf{P}_{\mathbf{H}}^{\perp}) = n - m$, and (d) $\det(\mathbf{P}_{\mathbf{H}}^{\perp}) = 0$.

6.5 If $\mathbf{P}_{\mathbf{H}} = \mathbf{H}[\mathbf{H}^{\mathrm{T}}\mathbf{W}\mathbf{H}]^{-1}\mathbf{H}^{\mathrm{T}}\mathbf{W}$ where $\mathbf{W}$ is a symmetric positive definite matrix, verify that $\mathbf{P}_{\mathbf{H}}^{2} = \mathbf{P}_{\mathbf{H}}$ and that $\mathbf{P}_{\mathbf{H}}^{\mathrm{T}} \neq \mathbf{P}_{\mathbf{H}}$, that is, $\mathbf{P}_{\mathbf{H}}$ is *idempotent* but *not* symmetric, and hence is not an orthogonal projection matrix.

6.6 Verify the correctness of (6.4.11).

6.7 Derive (6.4.16) from (6.4.15).

Notes and references

Orthogonal projection theorem is the basis of all the orthogonal projection methods described within the framework of abstract Hilbert space (Friedman 1956). Projection methods have played a central role in the development of many branches of applied mathematics – Galerkin and Petrov–Galerkin methods in *finite element methods* (Zienkiewicz (2000), Reddy and Gartling (2001)), Krylov subspace methods for *solving linear systems* (Greenbaum (1997)), and *least squares estimation theory* (Catlin (1989), Kailath (1974)), to mention a few.

Basilevsky (1983) and Meyer (2000) contain a very readable and a thorough exposition of the properties of projection matrices relevant to our development in this chapter.

For another readable account of invariance results covered in Section 6.4, refer to Chapter 2 on **Data Analysis Methods in Geodesy** by Dermanis and Rummel, and Chapter 3 on **Linear and Nonlinear Inverse Problems** by R. Snieder and J. Trampert in Dermanis *et al.* (2000).

7

Nonlinear least squares estimation

In this chapter our aim is to provide an introduction to the nonlinear least squares problem. In practice many of the problems of interest are nonlinear in nature. These include several problems of interest in radar and satellite meteorology, exploration problems in geology, and tomography, to mention a few. In Section 7.1, we describe the first-order method which in many ways is a direct extension of the ideas developed in Chapters 5 and 6. This method is based on a classical idea from numerical analysis – replacing $h(x)$ by its *linear* approximation at a given operating point x_c, and solving a linear problem to obtain a new operating point, x_{new}, which is *closer* to the target state, x^*, than the original starting point. By repeatedly applying this idea, we can get as close to the target state as needed. The second-order counterpart of this idea is to replace $h(x)$ by its *quadratic* approximation and, except for a few algebraic details, this method essentially follows the above iterative paradigm. This second-order method is described in Section 7.2.

7.1 A first-order method

Let $\mathbf{x} \in \mathbb{R}^n$ denote the *state* of a system under observation. Let $\mathbf{z} \in \mathbb{R}^m$ denote a set of *observables* which depend on the state $\mathbf{x}$, and let $\mathbf{z} = \mathbf{h}(\mathbf{x})$ be a representation of the physical laws that relate the underlying state to the observables – *temperature* to the *energy* radiated measured by a satellite or *rain* to the *reflectivity* measured by a doppler radar, where $\mathbf{h}(\mathbf{x}) = (h_1(\mathbf{x}), h_2(\mathbf{x}), \ldots, h_m(\mathbf{x}))^{\mathrm{T}}$ is a m-vector-valued function of the vector $\mathbf{x}$, and $h_i(\mathbf{x}) : \mathbb{R}^n \longrightarrow \mathbb{R}$ is the scalar valued function which is the ith component of $\mathbf{h}(\mathbf{x})$ for $i = 1, 2, \ldots, m$.

Given a set $\mathbf{z}$ of observations and knowing the functional form of $\mathbf{h}$, the problem is to find $\mathbf{x} \in \mathbb{R}^n$ that may be responsible for the observation. Recognizing that the system

$$\mathbf{z} = \mathbf{h}(\mathbf{x}) \tag{7.1.1}$$

may not be consistent (in the sense that there may not be an $\mathbf{x}$ for the given $\mathbf{z}$ satisfying (7.1.1)), our aim is to look for an $\mathbf{x}$ that will minimize the square of the norm of the residual vector

$$\mathbf{r}(\mathbf{x}) = \mathbf{z} - \mathbf{h}(\mathbf{x}) \tag{7.1.2}$$

For purposes of later reference, in this chapter, we consider the energy norm (Appendix A). Let $\mathbf{W} \in \mathbb{R}^{m \times m}$ be a symmetric positive definite matrix. Then

$$f(\mathbf{x}) = \frac{1}{2} \|\mathbf{r}(\mathbf{x})\|^2_{\mathbf{W}} = \frac{1}{2}(\mathbf{z} - \mathbf{h}(\mathbf{x}))^{\mathrm{T}}\mathbf{W}(\mathbf{z} - \mathbf{h}(\mathbf{x})) \tag{7.1.3}$$

where the factor $1/2$ is introduced† to cancel out the factor 2 that would otherwise arise in differentiation of quadratic forms.

The first-order method described in this section begins by approximating the nonlinear function $\mathbf{h}(\mathbf{x})$ locally by its linear counterpart obtained by using the first-order Taylor expansion of $\mathbf{h}(\mathbf{x})$ around an operating point, say, $\mathbf{x}_c$ (Appendix C). Accordingly, at any point $\mathbf{x}$ in a small neighborhood of $\mathbf{x}_c$, $\mathbf{h}(\mathbf{x})$ can be represented as

$$\mathbf{h}(\mathbf{x}) = \mathbf{h}(\mathbf{x}_c) + \mathcal{D}_{\mathbf{h}}(\mathbf{x}_c)(\mathbf{x} - \mathbf{x}_c) \tag{7.1.4}$$

where $\mathcal{D}_{\mathbf{h}}(\mathbf{x})$ denotes the *Jacobian* matrix of $\mathbf{h}$ which is an $m \times n$ matrix given by

$$\mathcal{D}_{\mathbf{h}}(\mathbf{x}) = \left[\frac{\partial h_i}{\partial x_j}\right], 1 \le i \le m; 1 \le j \le n. \tag{7.1.5}$$

It can be verified that the ith row of $\mathcal{D}_{\mathbf{h}}(\mathbf{x})$ is the *gradient* vector of $h_i(\mathbf{x})$ for $i = 1, 2, \ldots, m$. Substituting (7.1.4) into (7.1.3), and simplifying the notation by defining

$$\mathbf{g}(\mathbf{x}) = (\mathbf{z} - \mathbf{h}(\mathbf{x})) \tag{7.1.6}$$

we obtain

$$\mathbf{Q}_1(\mathbf{x}) = \frac{1}{2}[\mathbf{g}(\mathbf{x}_c) - \mathcal{D}_{\mathbf{h}}(\mathbf{x}_c)(\mathbf{x} - \mathbf{x}_c)]^{\mathrm{T}}\mathbf{W}[\mathbf{g}(\mathbf{x}_c) - \mathcal{D}_{\mathbf{h}}(\mathbf{x}_c)(\mathbf{x} - \mathbf{x}_c)] \tag{7.1.7}$$

where $\mathbf{g}(\mathbf{x}_c)$ is known and is independent of $\mathbf{x}$.

The idea is $\mathbf{Q}_1(\mathbf{x})$ has a much simple quadratic structure and is a respectable approximation to $f(\mathbf{x})$ in a small neighborhood around $\mathbf{x}_c$. Thus, we minimize $\mathbf{Q}_1(\mathbf{x})$ instead of $f(\mathbf{x})$.

Expanding the r.h.s. of (7.1.7), we readily see that

$$\begin{aligned}\mathbf{Q}_1(\mathbf{x}) = \frac{1}{2}\{&\mathbf{g}^{\mathrm{T}}(\mathbf{x}_c)\mathbf{W}\mathbf{g}(\mathbf{x}_c) - 2\mathbf{g}^{\mathrm{T}}(\mathbf{x}_c)\mathbf{W}\mathcal{D}_{\mathbf{h}}(\mathbf{x}_c)(\mathbf{x} - \mathbf{x}_c)\\ &+(\mathbf{x} - \mathbf{x}_c)^{\mathrm{T}}[\mathcal{D}^{\mathrm{T}}_{\mathbf{h}}(\mathbf{x}_c)\mathbf{W}\mathcal{D}_{\mathbf{h}}(\mathbf{x}_c)](\mathbf{x} - \mathbf{x}_c)\}.\end{aligned} \tag{7.1.8}$$

† Recall that multiplying a function $f(x)$ by a constant $a > 0$ represents a uniform magnification and it does *not* alter the location of the *critical points* such as the maximum, minimum, etc.

Given $\mathbf{z}$, $\mathbf{h}(\mathbf{x})$, find $\mathbf{x}^*$ iteratively that minimizes $f(\mathbf{x})$ in (7.1.3).

Step 1 Pick an initial operating point $\mathbf{x}_c$.

Step 2 Evaluate the vectors $\mathbf{h}(\mathbf{x}_c)$ and $\mathbf{g}(\mathbf{x}_c) = [\mathbf{z} - \mathbf{h}(\mathbf{x}_c)]$ and the matrix $\mathcal{D}_{\mathbf{h}}(\mathbf{x}_c)$.

Step 3 Compute the matrix $[\mathcal{D}_{\mathbf{h}}^{\mathrm{T}}(\mathbf{x}_c)\mathbf{W}\mathcal{D}_{\mathbf{h}}(\mathbf{x}_c)]$ and the vector $\mathcal{D}_{\mathbf{h}}^{\mathrm{T}}(\mathbf{x}_c)\mathbf{W}\mathbf{g}(\mathbf{x}_c)$.

Step 4 Solve the linear system (7.1.11) for the increment $(\mathbf{x} - \mathbf{x}_c)$.

Step 5 If $\|\mathbf{x} - \mathbf{x}_c\| < \epsilon$, a pre-specified tolerance limit, then $\mathbf{x}^* = \mathbf{x}$. Else, redefine $\mathbf{x}_c \longleftarrow \mathbf{x}$, and go to Step 2.

Fig. 7.1.1 First-order algorithm: nonlinear least squares.

Hence, the gradient of $\mathbf{Q}_1(\mathbf{x})$ is

$$\nabla\mathbf{Q}_1(\mathbf{x}) = -\mathcal{D}_{\mathbf{h}}^{\mathrm{T}}(\mathbf{x}_c)\mathbf{W}\mathbf{g}(\mathbf{x}_c) + [\mathcal{D}_{\mathbf{h}}^{\mathrm{T}}(\mathbf{x}_c)\mathbf{W}\mathcal{D}_{\mathbf{h}}(\mathbf{x}_c)](\mathbf{x} - \mathbf{x}_c) \tag{7.1.9}$$

and its Hessian is

$$\nabla^2\mathbf{Q}_1(\mathbf{x}) = \mathcal{D}_{\mathbf{h}}^{\mathrm{T}}(\mathbf{x}_c)\mathbf{W}\mathcal{D}_{\mathbf{h}}(\mathbf{x}_c). \tag{7.1.10}$$

By setting the gradient to zero, we immediately obtain

$$[\mathcal{D}_{\mathbf{h}}^{\mathrm{T}}(\mathbf{x}_c)\mathbf{W}\mathcal{D}_{\mathbf{h}}(\mathbf{x}_c)](\mathbf{x} - \mathbf{x}_c) = \mathcal{D}_{\mathbf{h}}^{\mathrm{T}}(\mathbf{x}_c)\mathbf{W}\mathbf{g}(\mathbf{x}_c) \tag{7.1.11}$$

or

$$(\mathbf{x} - \mathbf{x}_c) = [\mathcal{D}_{\mathbf{h}}^{\mathrm{T}}(\mathbf{x}_c)\mathbf{W}\mathcal{D}_{\mathbf{h}}(\mathbf{x}_c)]^{-1}\mathcal{D}_{\mathbf{h}}^{\mathrm{T}}(\mathbf{x}_c)\mathbf{W}[\mathbf{z} - \mathbf{h}(\mathbf{x}_c)]. \tag{7.1.12}$$

This process is now repeated by redefining $\mathbf{x}$ from (7.1.12) to be the new operating point until such time when $\|\mathbf{x} - \mathbf{x}_c\|$ is below a prescribed threshold.

In other words, this iterative approach minimizes a sequence of quadratic approximations $\mathbf{Q}_1(\mathbf{x})$ to $f(\mathbf{x})$. An iterative framework for implementing the first-order algorithm is given in Figure 7.1.1. It can be verified that $[\mathcal{D}_{\mathbf{h}}^{\mathrm{T}}(\mathbf{x}_c)\mathbf{W}\mathcal{D}_{\mathbf{h}}(\mathbf{x}_c)]$ is a symmetric matrix and we could, in principle, use Cholesky decomposition (Chapter 9) to solve (7.1.12).

Remark 7.1.1 When $\mathbf{h}(\mathbf{x})$ is a linear function, then there exists a matrix $\mathbf{H} \in \mathbb{R}^{m\times n}$, such that $\mathbf{h}(\mathbf{x}) = \mathbf{H}\mathbf{x}$. In this case, it can be verified that $\mathcal{D}_{\mathbf{h}}(\mathbf{x}) = \mathbf{H}$. In this case, we can choose the initial operating point $\mathbf{x}_c = \mathbf{0}$, and (7.1.12) becomes

$$\mathbf{x}^* = (\mathbf{H}^{\mathrm{T}}\mathbf{W}\mathbf{H})^{-1}\mathbf{H}^{\mathrm{T}}\mathbf{W}\mathbf{z} \tag{7.1.13}$$

which is the same as the one derived in Chapter 5. Thus, when $\mathbf{h}(\mathbf{x})$ is linear, there is no need to iterate and the optimal $\mathbf{x}^*$ is found in one step by solving (7.1.13).

Remark 7.1.2 For any iterative scheme to be useful, it must have desirable convergence properties. A typical convergence result may be stated as follows. Let $S \subseteq \mathbb{R}^n$ be a closed subset that contains the minimum of $f(\mathbf{x})$ that is sought. Then under mild conditions on $\mathbf{h}(\mathbf{x})$, the sequence of iterates defined by the first-order algorithm given in Figure 7.1.1 converges to the minimum $\mathbf{x}^*$. Once convergence is guaranteed, then the interest shifts to improving the rate of convergence. See Chapter 10 for details.

Example 7.1.1 Let $m = n = 2$, $\mathbf{x} = (x_1, x_2)^T$, $\mathbf{h}(\mathbf{x}) = (h_1(\mathbf{x}), h_2(\mathbf{x}))^T$, with $h_1(\mathbf{x}) = ax_1x_2$, and $h_2(\mathbf{x}) = bx_1^2$. Let $\mathbf{z} = (z_1, z_2)^T$ and $\mathbf{W} = \mathbf{I}$. Then,

$$\mathcal{D}_\mathbf{h}(\mathbf{x}) = \begin{bmatrix} ax_2 & ax_1 \\ 2bx_1 & 0 \end{bmatrix}$$

$$\mathcal{D}_\mathbf{h}^T(\mathbf{x})\mathcal{D}_\mathbf{h}(\mathbf{x}) = \begin{bmatrix} 4b^2x_1^2 + a^2x_2^2 & a^2x_1x_2 \\ a^2x_1x_2 & a^2x_1^2 \end{bmatrix}$$

$$\mathcal{D}_\mathbf{h}^T(\mathbf{x})\mathbf{g}(\mathbf{x}_c) = \begin{bmatrix} ax_2g_1(\mathbf{x}_c) + 2bx_1g_2(\mathbf{x}_c) \\ ax_1g_1(\mathbf{x}_c) \end{bmatrix}$$

where $\mathbf{g}(\mathbf{x}_c) = (g_1(\mathbf{x}), g_2(\mathbf{x}_c))^T$ and $g_i(\mathbf{x}) = (z_i - h_i(\mathbf{x}_c))$ for $i = 1, 2$. Then, (7.1.11) takes the form

$$[\mathcal{D}_\mathbf{h}^T(\mathbf{x}_c)\mathcal{D}_\mathbf{h}(\mathbf{x}_c)](\mathbf{x} - \mathbf{x}_c) = \mathcal{D}_\mathbf{h}^T(\mathbf{x}_c)\mathbf{g}(\mathbf{x}_c)$$

which can be solved by using methods in Chapter 9.

7.2 A second-order method

The framework for this method is exactly the same as that for the first-order method described in Section 7.1. However, the main difference between the first-order and the second-order methods lies in the details of the approximation of $\mathbf{h}(\mathbf{x})$ around the operating point $\mathbf{x}_c$. In the second-order method, we use the second-order Taylor expansion of $\mathbf{h}(\mathbf{x})$ around $\mathbf{x}_c$ (Appendix C) as described below. Thus,

$$\mathbf{h}(\mathbf{x}) \simeq \mathbf{h}(\mathbf{x}_c) + \mathcal{D}_\mathbf{h}(\mathbf{x}_c)(\mathbf{x} - \mathbf{x}_c) + \psi(\mathbf{x} - \mathbf{x}_c) \tag{7.2.1}$$

where as before $\mathcal{D}_\mathbf{h}(\mathbf{x}_c)$ is the *Jacobian* of $\mathbf{h}(\mathbf{x})$ and $\psi(\mathbf{y})$ is a vector representing the contributions from the second-order terms. It can be verified (Appendix C) that

$$\psi(\mathbf{y}) = (\psi_1(\mathbf{y}), \psi_2(\mathbf{y}), \ldots, \psi_m(\mathbf{y}))^T$$

$$\psi_k(\mathbf{y}) = \tfrac{1}{2}\mathbf{y}^T[\nabla^2\mathbf{h}_k(\mathbf{x}_c)]\mathbf{y},\ 1 \le k \le m \tag{7.2.2}$$

and

$$\nabla^2\mathbf{h}_k(\mathbf{x}) = \left[\frac{\partial^2\mathbf{h}_k(\mathbf{x})}{\partial x_i \partial x_j}\right],\ 1 \le i, j \le n$$

is the Hessian of $\mathbf{h}_k(\mathbf{x})$. Thus, $\psi(\mathbf{y})$ is a vector each of whose components is a quadratic form in $\mathbf{y} = (\mathbf{x} - \mathbf{x}_c)$ where the matrices of the quadratic form are the Hessian of the components of $\mathbf{h}(\mathbf{x})$.

Now substitute (7.2.1) into (7.1.3) to get

$$f(\mathbf{x}) \approx \frac{1}{2}[\mathbf{g}(\mathbf{x}_c) - \mathcal{D}_\mathbf{h}(\mathbf{x}_c)\mathbf{y} - \psi(\mathbf{y})]^T\mathbf{W}[\mathbf{g}(\mathbf{x}_c) - \mathcal{D}_\mathbf{h}(\mathbf{x}_c)\mathbf{y} - \psi(\mathbf{y})] \tag{7.2.3}$$

which is another approximation to $f(\mathbf{x})$ around $\mathbf{x} = \mathbf{x}_c$. Since the components of $\psi(\mathbf{y})$ are quadratic in $\mathbf{y}$, the r.h.s. of (7.2.3) when multiplied represents a fourth degree polynomial in $\mathbf{y} = (\mathbf{x} - \mathbf{x}_c)$. Expanding the r.h.s. of (7.2.3) and keeping only the terms up to the second degree in $\mathbf{y}$, we get

$$\begin{aligned}\mathbf{Q}_2(\mathbf{x}) = &\tfrac{1}{2}\mathbf{g}^{\mathrm{T}}(\mathbf{x}_c)\mathbf{W}\mathbf{g}(\mathbf{x}_c) - \mathbf{g}^{\mathrm{T}}(\mathbf{x}_c)\mathbf{W}\mathcal{D}_{\mathbf{h}}(\mathbf{x}_c)(\mathbf{x} - \mathbf{x}_c)\\ &+ \tfrac{1}{2}(\mathbf{x} - \mathbf{x}_c)^{\mathrm{T}}[\mathcal{D}_{\mathbf{h}}^{\mathrm{T}}(\mathbf{x}_c)\mathbf{W}\mathcal{D}_{\mathbf{h}}(\mathbf{x}_c)](\mathbf{x} - \mathbf{x}_c) - \mathbf{g}^{\mathrm{T}}(\mathbf{x}_c)\mathbf{W}\psi(\mathbf{x} - \mathbf{x}_c)\end{aligned} \tag{7.2.4}$$

which is a new quadratic approximation to $f(\mathbf{x})$ around $\mathbf{x}_c$. Again, the idea is to minimize $\mathbf{Q}_2(\mathbf{x})$ instead of $f(\mathbf{x})$.

Remark 7.2.1 A comparison of $\mathbf{Q}_2(\mathbf{x})$ in (7.2.4) with the $\mathbf{Q}_1(\mathbf{x})$ in (7.1.8) immediately reveals that $\mathbf{Q}_2(\mathbf{x})$ has the extra quadratic term $\mathbf{g}^{\mathrm{T}}(\mathbf{x}_c)\mathbf{W}\psi(\mathbf{x} - \mathbf{x}_c)$. Thus, $\mathbf{Q}_2(\mathbf{x})$ is a *full* quadratic approximation to $f(\mathbf{x})$ while $\mathbf{Q}_1(\mathbf{x})$ is only a *partial* quadratic approximation. It is this term involving $\psi(\mathbf{y})$ that underscores the difference between the first- and the second-order methods.

As a preparation for computing the gradient of $\mathbf{Q}_2(\mathbf{x})$, let us first compute that of the last term in (7.2.4). To further simplify notation, define

$$\mathbf{b}(\mathbf{x}) = \mathbf{W}\mathbf{g}(\mathbf{x}) = (b_1(\mathbf{x}), b_2(\mathbf{x}), \ldots, b_m(\mathbf{x}))^{\mathrm{T}}. \tag{7.2.5}$$

Combining this with (7.2.3), we have

$$\begin{aligned}\nabla[\mathbf{g}^{\mathrm{T}}(\mathbf{x}_c)\mathbf{W}\psi(\mathbf{x} - \mathbf{x}_c)] &= \nabla[\mathbf{b}^{\mathrm{T}}(\mathbf{x}_c)\psi(\mathbf{x} - \mathbf{x}_c)]\\ &= \nabla[\textstyle\sum_{k=1}^m b_k(\mathbf{x})\psi(\mathbf{x} - \mathbf{x}_c)]\\ &= \nabla[\tfrac{1}{2}\textstyle\sum_{k=1}^m b_k(\mathbf{x}_c)[(\mathbf{x} - \mathbf{x}_c)^{\mathrm{T}}\nabla^2 h_k(\mathbf{x}_c)(\mathbf{x} - \mathbf{x}_c)]\\ &= \tfrac{1}{2}\textstyle\sum_{k=1}^m b_k(\mathbf{x}_c)\nabla[(\mathbf{x} - \mathbf{x}_c)^{\mathrm{T}}\nabla^2 h_k(\mathbf{x}_c)(\mathbf{x} - \mathbf{x}_c)]\\ &= \textstyle\sum_{k=1}^m b_k(\mathbf{x}_c)\nabla^2 h_k(\mathbf{x}_c)(\mathbf{x} - \mathbf{x}_c)\end{aligned} \tag{7.2.6}$$

Using (7.2.6), we now compute the *gradient* of $\mathbf{Q}_2(\mathbf{x})$ as

$$\begin{aligned}\nabla\mathbf{Q}_2(\mathbf{x}) = &-\mathcal{D}_{\mathbf{h}}^{\mathrm{T}}(\mathbf{x}_c)\mathbf{W}(\mathbf{g}(\mathbf{x}_c)) + [\mathcal{D}_{\mathbf{h}}^{\mathrm{T}}(\mathbf{x}_c)\mathbf{W}\mathcal{D}_{\mathbf{h}}(\mathbf{x}_c)](\mathbf{x} - \mathbf{x}_c)\\ &+ \sum_{k=1}^m b_k(\mathbf{x}_c)\nabla^2 h_k(\mathbf{x}_c)(\mathbf{x} - \mathbf{x}_c)\end{aligned} \tag{7.2.7}$$

and the *Hessian* of $\mathbf{Q}_2(\mathbf{x})$ is

$$\nabla^2\mathbf{Q}_2(\mathbf{x}) = [\mathcal{D}_{\mathbf{h}}^{\mathrm{T}}(\mathbf{x}_c)\mathbf{W}\mathcal{D}_{\mathbf{h}}(\mathbf{x}_c) + \sum_{k=1}^m b_k(\mathbf{x}_c)\nabla^2 h_k(\mathbf{x}_c)] \tag{7.2.8}$$

Setting the gradient to zero, we obtain

$$\begin{aligned}&[\mathcal{D}_{\mathbf{h}}^{\mathrm{T}}(\mathbf{x}_c)\mathbf{W}\mathcal{D}_{\mathbf{h}}(\mathbf{x}_c) + \textstyle\sum_{k=1}^m b_k(\mathbf{x}_c)\nabla^2 h_k(\mathbf{x}_c)](\mathbf{x} - \mathbf{x}_c)\\ &\qquad = \mathcal{D}_{\mathbf{h}}^{\mathrm{T}}(\mathbf{x}_c)\mathbf{W}[\mathbf{z} - \mathbf{h}(\mathbf{x}_c)]\end{aligned} \tag{7.2.9}$$

where, recall that, $\mathbf{g}(\mathbf{x}_c) = \mathbf{z} - \mathbf{h}(\mathbf{x}_c)$.

The second-order algorithm may be described as in Figure 7.2.1.

Given $\mathbf{z}$, $\mathbf{h}(\mathbf{x})$, find $\mathbf{x}^*$ iteratively that minimizes $f(\mathbf{x})$ in (7.1.3).

Step 1 Pick an initial operating point $\mathbf{x}_c$.

Step 2 Evaluate the vectors $\mathbf{h}(\mathbf{x}_c)$ and $\mathbf{g}(\mathbf{x}_c) = [\mathbf{z} - \mathbf{h}(\mathbf{x}_c)]$ and $\mathbf{b}(\mathbf{x}_c) = \mathbf{W}(\mathbf{z} - \mathbf{h}(\mathbf{x}_c))$ and the matrices $\mathcal{D}_{\mathbf{h}}(\mathbf{x}_c)$ and $\nabla^2 h_k(\mathbf{x}_c)$ for $1 \le k \le m$.

Step 3 Assemble the matrix on the l.h.s. of (7.2.9).

Step 4 Compute $[\mathcal{D}_{\mathbf{h}}^{\mathrm{T}}(\mathbf{x}_c)\mathbf{W}\mathbf{g}(\mathbf{x}_c)]$, the r.h.s. of (7.2.9).

Step 5 Solve (7.2.9) for the increment $(\mathbf{x} - \mathbf{x}_c)$.

Step 6 If $\|\mathbf{x} - \mathbf{x}_c\| < \epsilon$, a pre-specified threshold, then $\mathbf{x}^* = \mathbf{x}$. Else, redefine $\mathbf{x}_c \longleftarrow \mathbf{x}$, and go to Step 2.

Fig. 7.2.1 Second-order algorithm: nonlinear least squares.

Remark 7.2.2 Equation (7.2.9) is known as the Newton's equation and its solution $\mathbf{x} - \mathbf{x}_c$ is called the Newton direction (Chapter 12). Equation (7.1.11) arising from the partial quadratic approximation is called the Gauss–Newton equation.

Remark 7.2.3 If $\mathbf{h}(\mathbf{x})$ is linear, that is $\mathbf{h}(\mathbf{x}) = \mathbf{H}\mathbf{x}$, where $\mathbf{H} \in \mathbb{R}^{m\times n}$, then $\mathcal{D}_{\mathbf{h}}(\mathbf{x}) = \mathbf{H}$, and $\nabla^2 \mathbf{h}_k(\mathbf{x}) = 0$ for all $1 \le k \le m$. In this case (7.2.9) reduces to the well-known formula for the linear least squares treated in Chapter 2.

Example 7.2.1 Continuing the Example 7.1.1., we have $h_1(\mathbf{x}) = ax_1x_2$ and $h_2(\mathbf{x}) = bx_1^2$, and

$$\nabla h_1(\mathbf{x}) = \begin{pmatrix} \frac{\partial h_1}{\partial x_1} \\ \frac{\partial h_1}{\partial x_2} \end{pmatrix} = \begin{pmatrix} ax_2 \\ ax_1 \end{pmatrix}$$

with

$$\nabla h_2(\mathbf{x}) = \begin{pmatrix} \frac{\partial h_2}{\partial x_1} \\ \frac{\partial h_2}{\partial x_2} \end{pmatrix} = \begin{pmatrix} 2bx_1 \\ 0 \end{pmatrix}$$

Hence,

$$\nabla^2 h_1(\mathbf{x}) = \begin{pmatrix} \frac{\partial^2 h_1}{\partial x_1^2} & \frac{\partial^2 h_1}{\partial x_1 \partial x_2} \\ \frac{\partial h_1}{\partial x_1 \partial x_2} & \frac{\partial^2 h_1}{\partial x_2^2} \end{pmatrix} = \begin{pmatrix} 0 & a \\ a & 0 \end{pmatrix}$$

and

$$\nabla^2 h_2(\mathbf{x}) = \begin{pmatrix} \frac{\partial^2 h_2}{\partial x_1^2} & \frac{\partial^2 h_2}{\partial x_1 \partial x_2} \\ \frac{\partial h_2}{\partial x_1 \partial x_2} & \frac{\partial^2 h_2}{\partial x_2^2} \end{pmatrix} = \begin{pmatrix} 2b & 0 \\ 0 & 0 \end{pmatrix}$$

Hence, the matrix on the l.h.s. of (7.2.9) can be readily obtained using these Hessians of the components of $\mathbf{h}(\mathbf{x})$.

Exercises

7.1 Let $h(t, \mathbf{x}) = e^{tx_1} + e^{tx_2}$ be the sum of two exponential functions, where t is a real parameter and $\mathbf{x} = (x_1, x_2)^T$. Let z_i denote the observation of $h(t_i, \mathbf{x})$ for $i = 1, 2, \ldots, m$. Define the residual

$$r_i(\mathbf{x}) = [z_i - h(t_i, \mathbf{x})], \text{ for } i = 1, 2, \ldots, m$$

Let $\mathbf{r}(\mathbf{x}) = (r_1(\mathbf{x}), r_2(\mathbf{x}), \ldots, r_m(\mathbf{x}))^T$, $\mathbf{z} = (z_1, z_2, \ldots, z_m)^T$, and $h(\mathbf{x}) = (h(t_1, \mathbf{x}), h(t_2, \mathbf{x}), \ldots, h(t_m, \mathbf{x}))^T$. Then, consider

$$f(\mathbf{x}) = \frac{1}{2}\|\mathbf{r}(\mathbf{x})\|^2 = \frac{1}{2}(\mathbf{z} - h(\mathbf{x}))^T(\mathbf{z} - h(\mathbf{x})).$$

Derive explicit expressions for the first-order and the second-order approximations for $f(\mathbf{x})$ around the operating point $\mathbf{x} = \mathbf{x}_c$.

7.2 Repeat the above exercise for the case when

$$h(t, \mathbf{x}) = x_1 + x_2 e^{-(t+x_3)^2/x_4}$$

where $\mathbf{x} = (x_1, x_2, x_3, x_4)^T$ and t is a real parameter.

7.3 Following the development in Section 3.8, assume we know the functional form of the temperature curve between $p = 0, 1$. Assume it to be parabolic:

$$T(p) = x_1(p - x_2)^2 + x_3, \quad 0 \leqq p \leqq 1.$$

Our observations ("radiation") are measures of overlapping fractions of the area under the curve. Generally,

$$Z_{ij} = \int_{p_i}^{p_j} T(p)\mathrm{d}p.$$

The observations follow:

p_i	p_j	Z_{ij}
0.00	0.25	0.21
0.20	0.50	0.15
0.30	0.70	0.51
0.60	0.80	0.11

(a) Define $\vec{h}$ vector.
(b) Derive elements of Jacobian matrix.
(c) With $\vec{X} = (x_1, x_2, x_3)^T$ and $\vec{X}_c = (0.5, 1.0, 0.4)^T$ iterate to find the optimal $\vec{X}$ in the solution of (7.1.3) when the weight matrix is the identity matrix.

Notes and references

Full quadratic approximation of non-linear functions based on the second-order method is the basis for the modern theory of non-linear optimization (Nash and Sofer (1996), Dennis and Schanabel (1996)). Second-order method is the basis for Newton's algorithm for finding zeros of non-linear functions (Ortega and Rheinboldt (1970)) and the first-order methods give rise to the so called secant methods for finding the zeros of non-linear functions.

Developments in this chapter follow Lakshmivarahan, Honda and Lewis (2003). As explained in this paper, the prudent strategy is a hybrid approach that uses the first-order method in the early stages and then switches over to the second-order method. The first-order method is somctimcs superior to the second-order method when the operating point is far from the minimum; here the word "far" is used to indicate that the full quadratic approximation afforded by the second-order method is not representative of the curvature near the minimum in these cases.

8

Recursive least squares estimation

So far in Chapters 5 through 7, it was assumed that the number m of observations is fixed and is known in advance. This treatment has come to be known as the *fixed sample* or *off-line* version of the least squares problem. In this chapter, we introduce the rudiments of the dual problem wherein the data or the observations are *not* known in advance and arrive *sequentially* in time. The challenge is to keep updating the optimal estimates as the new observations arrive on the scene. A naive way would be to repeatedly solve a sequence of least squares problems after the arrival of every new observation using the methods described in Chapters 5 through 7. A little reflection will, however, reveal that this is inefficient and computationally very expensive. The real question is: knowing the optimal estimate $\mathbf{x}^*(m)$ based on the m samples, can we compute $\mathbf{x}^*(m+1)$, the optimal estimate for $(m+1)$ samples, recursively by computing an increment or a correction to $\mathbf{x}^*(m)$ that reflects the new information contained in the new $(m+1)$th observation? The answer is indeed "yes", and leads to the *sequential* or *recursive method* for least squares estimation which is the subject of this chapter.

Section 8.1 provides an introduction to the deterministic recursive linear least squares estimation.

8.1 A recursive framework

Let $\mathbf{x} \in \mathbb{R}^n$ denote the state of the system under observation where n is fixed. Let $\mathbf{z} \in \mathbb{R}^m$ denote a set of m observations where it is assumed that $\mathbf{x}$ and $\mathbf{z}$ are related linearly as

$$\mathbf{z} = \mathbf{H}\mathbf{x} \tag{8.1.1}$$

and $\mathbf{H} \in \mathbb{R}^{m\times n}$ denotes the measurement matrix. Let $\mathbf{x}^*(m)$ denote the *optimal linear least squares estimate* (refer to Chapter 5)

$$\mathbf{x}^*(m) = (\mathbf{H}^{\mathrm{T}}\mathbf{H})^{-1}\mathbf{H}^{\mathrm{T}}\mathbf{z} \tag{8.1.2}$$

where we have introduced the parameter m in $\mathbf{x}^*(m)$ to denote its dependence on the number of observations used in arriving at this estimate. Let $z_{m+1} \in \mathbb{R}$ be the new observation. Then (8.1.1) can be expanded in the form of a *partitioned* matrix-vector relation as

$$\begin{bmatrix} \mathbf{z} \\ \\ z_{m+1} \end{bmatrix} = \begin{bmatrix} \mathbf{H} \\ \\ \mathbf{h}_{m+1}^{\mathrm{T}} \end{bmatrix} \mathbf{x} \tag{8.1.3}$$

where z_{m+1} denotes the $(m+1)$th element of the new or expanded observation vector and $\mathbf{h}_{m+1} \in \mathbb{R}^n$, that is, $\mathbf{h}_{m+1}^{\mathrm{T}}$ denotes the $(m+1)$th row of the new or expanded $(m+1) \times n$ measurement matrix. Then,

$$\begin{aligned} r_{m+1}(\mathbf{x}) &= \begin{bmatrix} \mathbf{z} \\ \\ z_{m+1} \end{bmatrix} - \begin{bmatrix} \mathbf{H} \\ \\ \mathbf{h}_{m+1}^{\mathrm{T}} \end{bmatrix} \mathbf{x} = \begin{bmatrix} (\mathbf{z} - \mathbf{H}\mathbf{x}) \\ \\ z_{m+1} - \mathbf{h}_{m+1}^{\mathrm{T}}\mathbf{x} \end{bmatrix} \\ &= \begin{bmatrix} r_m(\mathbf{x}) \\ \\ z_{m+1} - \mathbf{h}_{m+1}^{\mathrm{T}}\mathbf{x} \end{bmatrix} \end{aligned} \tag{8.1.4}$$

denotes the *new* residual vector as a function of the *old* residual vector $r_m(\mathbf{x}) = (\mathbf{z} - \mathbf{H}\mathbf{x})$. Recall that $\mathbf{x}^*(m)$ in (8.1.2) minimizes $\|r_m(\mathbf{x})\|^2$, and our aim is to find $\mathbf{x}^*(m+1)$ that minimizes $\|r_{m+1}(\mathbf{x})\|^2$. To this end, define

$$\begin{aligned} f_{m+1}(\mathbf{x}) &= \|r_{m+1}(\mathbf{x})\|^2 \\ &= f_m(\mathbf{x}) + \left(z_{m+1} - \mathbf{h}_{m+1}^{\mathrm{T}}\mathbf{x}\right)^{\mathrm{T}} \left(z_{m+1} - \mathbf{h}_{m+1}^{\mathrm{T}}\mathbf{x}\right) \end{aligned} \tag{8.1.5}$$

where, again, by definition $f_m(\mathbf{x}) = \|r_m(\mathbf{x})\|^2$. This additive recursive relation is quite basic, and it relates to the evolution of the square of the norm of the residual as a function of observations.

Computing the gradient of $f_{m+1}(\mathbf{x})$, we obtain

$$\nabla f_{m+1}(\mathbf{x}) = \nabla f_m(\mathbf{x}) + 2(\mathbf{h}_{m+1}\mathbf{h}_{m+1}^{\mathrm{T}})\mathbf{x} - 2\mathbf{h}_{m+1}z_{m+1} \tag{8.1.6}$$

where, recall that (Chapter 5)

$$\nabla f_m(\mathbf{x}) = 2(\mathbf{H}^{\mathrm{T}}\mathbf{H})\mathbf{x} - 2\mathbf{H}^{\mathrm{T}}\mathbf{z}.$$

Combining these and setting $\nabla f_{m+1}(\mathbf{x})$ to zero, we obtain

$$(\mathbf{H}^{\mathrm{T}}\mathbf{H} + \mathbf{h}_{m+1}\mathbf{h}_{m+1}^{\mathrm{T}})\mathbf{x} = (\mathbf{H}^{\mathrm{T}}\mathbf{z} + \mathbf{h}_{m+1}z_{m+1}). \tag{8.1.7}$$

That is,

$$\mathbf{x}^*(m+1) = (\mathbf{H}^{\mathrm{T}}\mathbf{H} + \mathbf{h}_{m+1}\mathbf{h}_{m+1}^{\mathrm{T}})^{-1}(\mathbf{H}^{\mathrm{T}}\mathbf{z} + \mathbf{h}_{m+1}z_{m+1}). \tag{8.1.8}$$

A little reflection would reveal that (8.1.8) is related to (8.1.3) in exactly the same way as (8.1.2) is related to (8.1.1). So, what is the net gain? Nothing! In fact, we could have easily obtained (8.1.8) by substituting

$$\begin{pmatrix} \mathbf{H} \\ \mathbf{h}_{m+1}^{\mathrm{T}} \end{pmatrix} \text{ for } \mathbf{H} \text{ and } \begin{pmatrix} \mathbf{z} \\ z_{m+1} \end{pmatrix} \text{ for } \mathbf{z}$$

in (8.1.2). The reason for us taking a little circuitous path is to emphasize the *recursive* relations (8.1.5) and (8.1.6).

Our stated goal is to be able to compute $\mathbf{x}^*(m+1)$ from $\mathbf{x}^*(m)$ without having to invert the matrix $(\mathbf{H}^{\mathrm{T}}\mathbf{H} + \mathbf{h}_{m+1}\mathbf{h}_{m+1}^{\mathrm{T}})$ all over again, since $\mathbf{x}^*(m)$ involves the inverse of $(\mathbf{H}^{\mathrm{T}}\mathbf{H})$. This is accomplished by invoking a very useful result from matrix theory called the *Sherman–Morrison* formula (Appendix B)

$$(\mathbf{P} + \mathbf{h}\mathbf{h}^{\mathrm{T}})^{-1} = \mathbf{P}^{-1} - \frac{\mathbf{P}^{-1}\mathbf{h}\mathbf{h}^{\mathrm{T}}\mathbf{P}^{-1}}{1 + \mathbf{h}^{\mathrm{T}}\mathbf{P}^{-1}\mathbf{h}} \tag{8.1.9}$$

which relates the inverse of $(\mathbf{P} + \mathbf{h}\mathbf{h}^{\mathrm{T}})$ to that of $\mathbf{P}$, where $\mathbf{P} \in \mathbb{R}^{n\times n}$ is non-singular and $\mathbf{h} \in \mathbb{R}^n$. The matrix $\mathbf{h}\mathbf{h}^{\mathrm{T}}$ is called the *outer product* matrix and is of rank one and $(\mathbf{P} + \mathbf{h}\mathbf{h}^{\mathrm{T}})$ is called the *rank-one update* of the matrix $\mathbf{P}$. By identifying $\mathbf{P}$ with $\mathbf{H}^{\mathrm{T}}\mathbf{H}$ and $\mathbf{h}$ with $\mathbf{h}_{m+1}$, we readily see the use of (8.1.9) in our goal to obtain a recursive relation for $\mathbf{x}^*(m+1)$. Substituting (8.1.9) in (8.1.8), we get (where $\mathbf{P} = \mathbf{H}^{\mathrm{T}}\mathbf{H}$ and $\mathbf{h} = \mathbf{h}_{m+1}$ for simplicity in notation)

$$\begin{aligned} \mathbf{x}_{m+1}^* &= (\mathbf{P} + \mathbf{h}\mathbf{h}^{\mathrm{T}})^{-1}(\mathbf{H}^{\mathrm{T}}\mathbf{z} + \mathbf{h}z_{m+1}) \\ &= [\mathbf{P}^{-1} - \mathbf{P}^{-1}\mathbf{h}\alpha^{-1}\mathbf{h}^{\mathrm{T}}\mathbf{P}^{-1}](\mathbf{H}^{\mathrm{T}}\mathbf{z} + \mathbf{h}z_{m+1}) \end{aligned} \tag{8.1.10}$$

where, again, to simplify the notation, we set the scalar

$$\alpha = (1 + \mathbf{h}^{\mathrm{T}}\mathbf{P}^{-1}\mathbf{h}). \tag{8.1.11}$$

Multiplying the terms on the right-hand side of (8.1.10) and recognizing that $\mathbf{P}^{-1}\mathbf{H}^{\mathrm{T}}\mathbf{z} = (\mathbf{H}^{\mathrm{T}}\mathbf{H})^{-1}\mathbf{H}^{\mathrm{T}}\mathbf{z} = \mathbf{x}^*(m)$, we get

$$\begin{aligned} \mathbf{x}_{m+1}^* = \mathbf{x}^*(m) &+ \mathbf{P}^{-1}\mathbf{h}z_{m+1} - \mathbf{P}^{-1}\mathbf{h}\alpha^{-1}\mathbf{h}^{\mathrm{T}}\mathbf{x}^*(m) \\ &- \mathbf{P}^{-1}\mathbf{h}\alpha^{-1}\mathbf{h}^{\mathrm{T}}\mathbf{P}^{-1}\mathbf{h}z_{m+1}. \end{aligned} \tag{8.1.12}$$

But from (8.1.11)

$$\begin{aligned} \mathbf{P}^{-1}\mathbf{h}\alpha^{-1}(\mathbf{h}^{\mathrm{T}}\mathbf{P}^{-1}\mathbf{h})z_{m+1} &= \mathbf{P}^{-1}\mathbf{h}\alpha^{-1}(\alpha - 1)z_{m+1} \\ &= \mathbf{P}^{-1}\mathbf{h}z_{m+1} - \mathbf{P}^{-1}\mathbf{h}\alpha^{-1}z_{m+1}. \end{aligned} \tag{8.1.13}$$

Substituting (8.1.13) into (8.1.12) and simplifying, we get

$$\mathbf{x}^*(m+1) = \mathbf{x}^*(m) + \mathbf{P}^{-1}\mathbf{h}\alpha^{-1}[z_{m+1} - \mathbf{h}^{\mathrm{T}}\mathbf{x}^*(m)].$$

Substituting again for $\mathbf{P} = \mathbf{H}^{\mathrm{T}}\mathbf{H}$ and $\mathbf{h} = \mathbf{h}_{m+1}$, we obtain the final recursive formula as

$$\mathbf{x}^*(m+1) = \mathbf{x}^*(m) + \frac{(\mathbf{H}^{\mathrm{T}}\mathbf{H})^{-1}\mathbf{h}_{m+1}}{1 + \mathbf{h}_{m+1}^{\mathrm{T}}(\mathbf{H}^{\mathrm{T}}\mathbf{H})^{-1}\mathbf{h}_{m+1}}[z_{m+1} - \mathbf{h}_{m+1}^{\mathrm{T}}\mathbf{x}^*(m)]. \tag{8.1.14}$$

The scalar $\mathbf{h}_{m+1}^{\mathrm{T}}\mathbf{x}^*(m)$ may be thought of as the *prediction* of the $(m+1)$th observation based on the current optimal estimate $\mathbf{x}^*(m)$ and the difference $(z_{m+1} - \mathbf{h}_{m+1}^{\mathrm{T}}\mathbf{x}^*(m))$ is called *innovation* which is the new information contained in the $(m+1)$th observation beyond what was predicted. The vector

$$\mathbf{g} = \frac{(\mathbf{H}^{\mathrm{T}}\mathbf{H})^{-1}\mathbf{h}_{m+1}}{1 + \mathbf{h}_{m+1}^{\mathrm{T}}(\mathbf{II}^{\mathrm{T}}\mathbf{H})^{-1}\mathbf{h}_{m+1}} \in \mathbb{R}^n \tag{8.1.15}$$

that multiplies the innovation term, is often called the *gain*. Notice that this gain does not depend on the observation and is purely a function of the characteristics of the measurement system. It is instructive to rewrite this expression for the gain in a recursive form. To this end, let

$$\begin{aligned} K_m^{-1} &= \mathbf{H}^{\mathrm{T}}\mathbf{H} \\ \textit{and} \\ K_{m+1}^{-1} &= K_m^{-1} + \mathbf{h}_{m+1}\mathbf{h}_{m+1}^{\mathrm{T}}. \end{aligned} \tag{8.1.16}$$

Then, using the Sherman–Morrison formula (8.1.9), it follows that

$$\begin{aligned} K_{m+1}\mathbf{h}_{m+1} &= (\mathbf{H}^{\mathrm{T}}\mathbf{H} + \mathbf{h}_{m+1}\mathbf{h}_{m+1}^{\mathrm{T}})^{-1}\mathbf{h}_{m+1} \\ &= (\mathbf{H}^{\mathrm{T}}\mathbf{H})^{-1}\mathbf{h}_{m+1} - \frac{(\mathbf{H}^{\mathrm{T}}\mathbf{H})^{-1}\mathbf{h}_{m+1}\mathbf{h}_{m+1}^{\mathrm{T}}(\mathbf{H}^{\mathrm{T}}\mathbf{H})^{-1}\mathbf{h}_{m+1}}{1 + \mathbf{h}_{m+1}^{\mathrm{T}}(\mathbf{H}^{\mathrm{T}}\mathbf{H})^{-1}\mathbf{h}_{m+1}} \\ &= (\mathbf{H}^{\mathrm{T}}\mathbf{H})^{-1}\mathbf{h}_{m+1}[1 - \frac{\mathbf{h}_{m+1}^{\mathrm{T}}(\mathbf{H}^{\mathrm{T}}\mathbf{H})^{-1}\mathbf{h}_{m+1}}{\mathbf{I} + \mathbf{h}_{m+1}^{\mathrm{T}}(\mathbf{H}^{\mathrm{T}}\mathbf{H})^{-1}\mathbf{h}_{m+1}}] \\ &= \frac{(\mathbf{H}^{\mathrm{T}}\mathbf{H})^{-1}\mathbf{h}_{m+1}}{1 + \mathbf{h}_{m+1}^{\mathrm{T}}(\mathbf{H}^{\mathrm{T}}\mathbf{H})^{-1}\mathbf{h}_{m+1}} = \mathbf{g}. \end{aligned} \tag{8.1.17}$$

Thus, we can recast (8.1.14) as

$$\begin{aligned} \mathbf{x}^*(m+1) &= \mathbf{x}^*(m) + K_{m+1}\mathbf{h}_{m+1}[z_{m+1} - \mathbf{h}_{m+1}^{\mathrm{T}}\mathbf{x}^*(m)] \\ K_{m+1}^{-1} &= K_m^{-1} + \mathbf{h}_{m+1}\mathbf{h}_{m+1}^{\mathrm{T}} \\ \textit{where } K_m^{-1} &= \mathbf{H}^{\mathrm{T}}\mathbf{H}. \end{aligned} \tag{8.1.18}$$

We now illustrate this recursive formulation using the simple example of estimating one's own weight considered in Example 5.1.2.

Example 8.1.1 Consider the case when $n = 1$, $\mathbf{H} = [1, 1, \ldots, 1]^{\mathrm{T}} \in \mathbb{R}^m$ and $\mathbf{h}_{m+1} = 1$. If $\mathbf{z} = (z_1, z_2, \ldots, z_m)^{\mathrm{T}} \in \mathbb{R}^m$ is the set of m observations, and z_{m+1}

is the new observation, then we have

$$\begin{pmatrix} \mathbf{z} \\ z_{m+1} \end{pmatrix} = \begin{pmatrix} \mathbf{H} \\ 1 \end{pmatrix} \mathbf{x}.$$

Thus,

$$K_m^{-1} = \mathbf{H}^{\mathrm{T}}\mathbf{H} = m \quad \text{and} \quad K_m^{-1} = (m+1).$$

Substituting into (8.1.18), we get

$$\mathbf{x}^*(m+1) = \mathbf{x}^*(m) + \tfrac{1}{m+1}[z_{m+1} - \mathbf{x}^*(m)]$$

$$= \tfrac{m}{m+1}\mathbf{x}^*(m) + \tfrac{1}{m+1} z_{m+1}.$$

As $m \longrightarrow \infty$, then $K_m \longrightarrow 0$ and the contributions from the innovation term becomes increasingly smaller. This indicates stability and convergence of the sequence of recursive estimates as m increases.

Exercises

8.1 Let $x_1, x_2, \ldots, x_m$ be a set of observations of an unknown scalar x. The *sample mean* and the *sample variance* are given by

$$\bar{x}_m = \frac{1}{m}\sum_{i=1}^{m} x_i \quad \text{and} \quad \bar{\sigma}_m^2 = \frac{1}{m-1}\sum_{i=1}^{m}(x_i - \bar{x}_m)^2,$$

respectively. Recast these expressions in the recursive form.

8.2 In this exercise we provide a recursive formulation of the weighted linear least squares problem treated in Section 5.2. From (5.2.7), we have

$$\mathbf{x}^*(m) = (\mathbf{H}_m^{\mathrm{T}}\mathbf{W}_m\mathbf{H}_m)^{-1}\mathbf{H}_m^{\mathrm{T}}\mathbf{W}_m\mathbf{z}_m$$

where we have added the index m to emphasize the fact that this expression is based on m observations. Recall $\mathbf{x}^*(m) \in \mathbb{R}^n$, $\mathbf{H}_m \in \mathbb{R}^{m\times n}$, $\mathbf{W}_m \in \mathbb{R}^{m\times m}$ is symmetric positive definite, and $\mathbf{z}_m \in \mathbb{R}^m$.

Let

$$\mathbf{h}_{m+1} = \begin{bmatrix} \mathbf{H}_m \\ \cdots \\ \mathbf{h}_{m+1}^{\mathrm{T}} \end{bmatrix}, \mathbf{z}_{m+1} = \begin{bmatrix} \mathbf{z}_m \\ \cdots \\ z_{m+1} \end{bmatrix}, \mathbf{h}_{m+1} \in \mathbb{R}^n, \mathbf{z}_{m+1} \in \mathbb{R}$$

$$\mathbf{W}_{m+1} = \begin{bmatrix} \mathbf{W}_m & \vdots & \mathbf{0} \\ \ldots & \vdots & \ldots \\ \mathbf{0} & \vdots & w_{m+1} \end{bmatrix}, w_{m+1} \in \mathbb{R} \text{ and } w_{m+1} > 0.$$

Then, clearly,

$$\mathbf{x}^*(m+1) = (\mathbf{H}_{m+1}^{\mathrm{T}}\mathbf{W}_{m+1}\mathbf{h}_{m+1})^{-1}\mathbf{h}_{m+1}^{\mathrm{T}}\mathbf{W}_{m+1}\mathbf{z}_{m+1}$$

denotes the optimal estimate using the $(m+1)$ observations. By following the developments in Section 5.2 and Section 8.1, verify that $\mathbf{x}^*(m+1)$ can be computed recursively as

$$\mathbf{x}(m+1)^* = \mathbf{x}^*(m) + \mathbf{K}_{m+1}h_{m+1}w_{m+1}[z_{m+1} - \mathbf{h}_{m+1}^{\mathrm{T}}\mathbf{x}^*(m)]$$

where

$$\mathbf{K}_{m+1}^{-1} = \mathbf{K}_m^{-1} + h_{m+1}w_{m+1}\mathbf{h}_{m+1}^{\mathrm{T}}$$

and

$$\mathbf{K}_m^{-1} = \mathbf{H}_m^{\mathrm{T}}\mathbf{W}_m\mathbf{H}_m.$$

8.3 Let $\mathbf{x_b} \in \mathbb{R}^n$, and $\mathbf{B}^{-1} \in \mathbb{R}^{n\times n}$ be a symmetric and positive definite matrix. Consider a functional $J_{\mathbf{b}} : \mathbb{R}^n \longrightarrow \mathbb{R}$,

$$J_{\mathbf{b}}(\mathbf{x}) = \frac{1}{2}(\mathbf{x} - \mathbf{x_b})^{\mathrm{T}}\mathbf{B}^{-1}(\mathbf{x} - \mathbf{x_b}).$$

Then, clearly $\mathbf{x} = \mathbf{x_b}$ minimizes $J_{\mathbf{b}}(\mathbf{x})$. Now, suppose we have an observation $\mathbf{z} \in \mathbb{R}^m$ and $\mathbf{H} \in \mathbb{R}^{m\times n}$ denotes the measurement system that relates $\mathbf{z}$ to $\mathbf{x}$ *via* $\mathbf{z} = \mathbf{H}\mathbf{x}$. Let

$$J_o(\mathbf{x}) = \frac{1}{2}((\mathbf{z} - \mathbf{H}\mathbf{x}))^{\mathrm{T}}\mathbf{R}^{-1}((\mathbf{z} - \mathbf{H}\mathbf{x}))$$

where $\mathbf{R}^{-1} \in \mathbb{R}^{m\times m}$ is a symmetric positive definite matrix. Let $J(\mathbf{x})$ be a new combined functional where

$$J(\mathbf{x}) = J_{\mathbf{b}}(\mathbf{x}) + J_o(\mathbf{x}).$$

If $\mathbf{x}^*$ is the minimizer of $J(\mathbf{x})$, our interest is in recursively computing $\mathbf{x}^*$ from $\mathbf{x_b}$, the minimizer of $J_{\mathbf{b}}(\mathbf{b})$. The term $\mathbf{x_b}$ represents the a priori optimal estimate and $\mathbf{x}^*$ is the a posteriori optimal estimate. Verify that $\mathbf{x}^*$ can be recursively computed as

$$\mathbf{x}^* = \mathbf{x_b} + \mathbf{K}\mathbf{H}^{\mathrm{T}}\mathbf{R}^{-1}[\mathbf{z} - \mathbf{H}\mathbf{x_b}]$$

where

$$\mathbf{K}^{-1} = (\mathbf{B}^{-1} + \mathbf{H}^{\mathrm{T}}\mathbf{R}^{-1}\mathbf{H}).$$

Notes and references

Abraham Wald (1947) pioneered the introduction of sequential or recursive techniques in statistical estimation and decision making. The developments in this chapter are quite elementary and serve as a precursor to the derivation of the Kalman filters in Part V of this book. Refer to Gelb (1974), Sorenson (1970), Schweppe (1973), and Sage and Melsa (1971) for further details.

PART III

Computational techniques

9

Matrix methods

Recall from Chapters 5 and 6 that the optimal linear estimate $\mathbf{x}^*$ is given by the solution of the normal equation

$$(\mathbf{H}^T\mathbf{H})\mathbf{x}^* = \mathbf{H}^T\mathbf{z} \quad \text{when } m > n$$

and

$$(\mathbf{H}^T\mathbf{H})\mathbf{y} = \mathbf{z} \quad \text{and} \quad \mathbf{x}^* = \mathbf{H}\mathbf{y} \quad \text{when } m < n$$

where $\mathbf{H} \in \mathbb{R}^{m\times n}$ and is of full rank. In either case $\mathbf{H}^T\mathbf{H} \in \mathbb{R}^{n\times n}$ and $\mathbf{H}\mathbf{H}^T \in \mathbb{R}^{m\times m}$, called the Grammian, is a symmetric and positive definite matrix. In the opening Section 9.1, we describe the classical *Cholesky decomposition* algorithm for solving linear systems with symmetric and positive definite matrices. This algorithm is essentially an adaptation of the method of *LU* decomposition for general matrices. This method of solving the normal equations using the Cholesky decomposition is computationally very efficient, but it may exhibit instability resulting from finite precision arithmetic. To alleviate this problem, during the 1960s a new class of methods based directly on the orthogonal decomposition of the (rectangular) measurement matrix $\mathbf{H}$ have been developed. In this chapter we describe two such methods. The first of these is based on the *QR-decomposition* in Section 9.2 and the second, called the *singular value decomposition*(SVD) is given in Section 9.3. Section 9.4 provides a comparison of the amount of *work* measured in terms of the number of floating point operations (FLOPs) to solve the linear least squares problem by these methods.

9.1 Cholesky decomposition

We begin by describing the classical LU-decomposition. Consider the generic linear system

$$\mathbf{A}\mathbf{x} = \mathbf{b} \tag{9.1.1}$$

Given $\mathbf{A} \in \mathbb{R}^{n\times n}$ and $\mathbf{b} \in \mathbb{R}^n$. Solve $\mathbf{Ax} = \mathbf{b}$.
Step 1 Decompose $\mathbf{A}$ as $\mathbf{A} = \mathbf{LU}$ with $\mathbf{L}$ lower triangular matrix with unity along the principal diagonal and $\mathbf{U}$ an upper triangular matrix (Exercise 9.2). Then $\mathbf{Ax} = (\mathbf{LU})\mathbf{x} = \mathbf{b}$.
Step 2 Solve the lower triangular system $\mathbf{Lg} = \mathbf{b}$ (Exercise 9.2).
Step 3 Solve the upper triangular system $\mathbf{Ux} = \mathbf{g}$ (Exercise 9.4).

Fig. 9.1.1 LU-decomposition algorithm.

to be solved where $\mathbf{A} \in \mathbb{R}^{n\times n}$, a non-singular matrix and $\mathbf{b} \in \mathbb{R}^n$ are given. Perhaps the most fundamental idea in all of numerical linear algebra is the concept that relates to the multiplicative factorization/decomposition of the matrix $\mathbf{A}$ as

$$\mathbf{A} = \mathbf{LU} \tag{9.1.2}$$

where $\mathbf{L}$ and $\mathbf{U}$ are both $n \times n$ matrices, with $\mathbf{L}$ a lower triangular with unit element along the principal diagonal and $\mathbf{U}$ an upper triangular matrix. It is instructive to rewrite (9.1.2) in component form as a matrix identity as follows:

$$\begin{bmatrix} a_{11} & a_{12} & \cdots & a_{1n} \\ a_{21} & a_{22} & \cdots & a_{2n} \\ \vdots & \vdots & \vdots & \vdots \\ a_{n1} & a_{n2} & \cdots & a_{nn} \end{bmatrix} = \begin{bmatrix} 1 & 0 & \cdots & 0 \\ l_{21} & 1 & \cdots & 0 \\ \vdots & \vdots & \vdots & \vdots \\ l_{n1} & l_{n2} & \cdots & 1 \end{bmatrix} \times \begin{bmatrix} u_{11} & u_{12} & \cdots & u_{1n} \\ 0 & u_{22} & \cdots & u_{2n} \\ \vdots & \vdots & \vdots & \vdots \\ 0 & 0 & \cdots & u_{nn} \end{bmatrix}. \tag{9.1.3}$$

Notice that the $\mathbf{L}$ matrix has $n(n-1)/2$ unknown elements and $\mathbf{U}$ has $n(n+1)/2$ unknown elements, which together add up to a total of n^2 unknowns. By multiplying the right hand side and equating the elements of this product matrix element by element with the left hand side matrix, we get a system of n^2 (non-linear) equations in n^2 unknowns. By exploring the inherent structure (Exercise 9.1), we can explicitly solve for these n^2 unknowns which leads to the factors $\mathbf{L}$ and $\mathbf{U}$. An algorithm is given in Exercise 9.2.

In the light of this decomposition, a general framework for solving (9.1.1) can be stated as in Figure 9.1.1, where the complete details of the algorithm for each step are pursued in Exercises 9.2 through 9.4.

Example 9.1.1 Consider the case when

$$\mathbf{A} = \begin{bmatrix} 1 & 3/2 \\ 3/2 & 7/2 \end{bmatrix}.$$

Using (9.1.2), we have the following:

$$\begin{bmatrix} 1 & 3/2 \\ 3/2 & 7/2 \end{bmatrix} = \begin{bmatrix} 1 & 0 \\ l_{21} & 1 \end{bmatrix} \begin{bmatrix} u_{11} & u_{12} \\ 0 & u_{22} \end{bmatrix} = \begin{bmatrix} u_{11} & u_{12} \\ l_{21}u_{11} & l_{21}u_{12} + u_{22} \end{bmatrix}.$$

From the definition of the equality of matrices, we immediately get $u_{11} = 1$, $u_{12} = 3/2$, $l_{21} = 3/2$ and $u_{22} = 5/4$. Thus, we have

$$\mathbf{L} = \begin{bmatrix} 1 & 0 \\ 3/2 & 1 \end{bmatrix} \quad \text{and} \quad \mathbf{U} = \begin{bmatrix} 1 & 3/2 \\ 0 & 5/4 \end{bmatrix}.$$

In going from **LU** to Cholesky decomposition, we further decompose **U** as

$$\mathbf{U} = \mathbf{DM} \tag{9.1.4}$$

where

$$\mathbf{D} = \text{Diag}(u_{11}, u_{22}, \ldots, u_{nn})$$

is a diagonal matrix formed by the elements along the principal diagonal of **U**. Clearly the ith row of the upper triangular matrix **M** is obtained by dividing the ith row of **U** by u_{ii} for $i = 1, 2, \ldots, n$. Combining (9.1.4) with (9.1.2) we get the following decomposition:

$$\mathbf{A} = \mathbf{LDM} \tag{9.1.5}$$

Now, if **A** is symmetric, then it can be verified that $\mathbf{M} = \mathbf{L}^{\mathrm{T}}$. In addition, if we further require **A** to be positive definite, then it follows that the diagonal elements of **D** are all positive. Thus, when **A** is symmetric and positive definite, we have

$$\mathbf{A} = \mathbf{LDL}^{\mathrm{T}} = \mathbf{L}(\mathbf{D}^{1/2}\mathbf{D}^{1/2})\mathbf{L}^{\mathrm{T}} = (\mathbf{LD}^{1/2})(\mathbf{D}^{1/2}\mathbf{L}^{\mathrm{T}}) = \mathbf{GG}^{\mathrm{T}} \tag{9.1.6}$$

where $\mathbf{G} = \mathbf{LD}^{1/2}$ and $\mathbf{D}^{1/2} = \text{Diag}(u_{11}^{1/2}, u_{22}^{1/2}, \ldots, u_{nn}^{1/2})$ is the diagonal matrix whose diagonal elements are the square roots of the corresponding elements of **D**. The lower triangular matrix **G** is called the *Cholesky factor* of **A**. It is instructive to rewrite (9.1.6) in explicit component form as

$$\begin{bmatrix} a_{11} & a_{12} & \cdots & a_{1n} \\ a_{21} & a_{22} & \cdots & a_{2n} \\ \vdots & \vdots & \vdots & \vdots \\ a_{n1} & a_{n}2 & \cdots & a_{nn} \end{bmatrix} = \begin{bmatrix} g_{11} & 0 & \cdots & 0 \\ g_{21} & g_{22} & \cdots & 0 \\ \vdots & \vdots & \vdots & \vdots \\ g_{n1} & g_{n2} & \cdots & g_{nn} \end{bmatrix} \times \begin{bmatrix} g_{11} & g_{21} & \cdots & g_{n1} \\ 0 & g_{22} & \cdots & g_{n2} \\ \vdots & \vdots & \vdots & \vdots \\ 0 & 0 & \cdots & g_{nn} \end{bmatrix}. \tag{9.1.7}$$

Notice that **G** has $n(n+1)/2$ unknown elements. Since **A** is symmetric, there are $n(n+1)/2$ distinct elements in **A** as well. By equating these elements we get a

Given $\mathbf{H} \in \mathbb{R}^{m \times n}$ and $\mathbf{z} \in \mathbb{R}^m$, with $m > n$. Solve $(\mathbf{H}^T\mathbf{H})\mathbf{x} = \mathbf{H}^T\mathbf{z}$.

Step 1 Compute the $n \times n$ $\mathbf{H}^T\mathbf{H}$ symmetric matrix – (matrix-matrix multiplication).
Step 2 Compute the $n \times 1$ vector $\mathbf{H}^T\mathbf{z}$ – (matrix-vector multiplication).
Step 3 Compute the Cholesky factor $\mathbf{G}$ such that $(\mathbf{H}^T\mathbf{H}) = \mathbf{G}\mathbf{G}^T$(Exercise 9.6).
Step 4 Solve the lower triangular system $\mathbf{G}\mathbf{g} = (\mathbf{H}^T\mathbf{z})$ (Exercise 9.3).
Step 5 Solve the upper triangular system $\mathbf{G}^T\mathbf{x}^* = \mathbf{g}$ (Exercise 9.4).

Fig. 9.1.2 Cholesky method for normal equations: over-determined system.

system of $n(n+1)/2$ (nonlinear) equations in as many unknowns. By exploiting the inherent structure (Exercise 9.5), we can explicitly solve for these unknowns. An algorithm for this Cholesky decomposition is given in Exercise 9.6.

Example 9.1.2 For the matrix $\mathbf{A}$ in Example 9.1.1, the $\mathbf{U}$ factor can be written as

$$\mathbf{U} = \begin{bmatrix} 1 & 3/2 \\ 0 & 5/4 \end{bmatrix} = \mathbf{DM}$$

where

$$\mathbf{D} = \begin{bmatrix} 1 & 0 \\ 0 & 5/4 \end{bmatrix}, \mathbf{M} = \begin{bmatrix} 1 & 3/2 \\ 0 & 1 \end{bmatrix}.$$

Then $\mathbf{D}^{1/2} = \begin{bmatrix} 1 & 0 \\ 0 & \sqrt{5}/2 \end{bmatrix}$ and

$$\begin{aligned} \mathbf{A} &= \begin{bmatrix} 1 & 0 \\ 3/2 & 1 \end{bmatrix} \begin{bmatrix} 1 & 0 \\ 0 & \sqrt{5}/2 \end{bmatrix} \begin{bmatrix} 1 & 0 \\ 0 & \sqrt{5}/2 \end{bmatrix} \begin{bmatrix} 1 & 3/2 \\ 0 & 1 \end{bmatrix} \\ &= \begin{bmatrix} 1 & 0 \\ 3/2 & \sqrt{5}/2 \end{bmatrix} \begin{bmatrix} 1 & 3/2 \\ 0 & \sqrt{5}/2 \end{bmatrix} = \mathbf{G}\mathbf{G}^T. \end{aligned}$$

Against this backdrop, we now describe a framework for the Cholesky decomposition based algorithm to solve the normal equation. Recall that $\mathbf{H}^T\mathbf{H}$ is symmetric and positive definite when $\mathbf{H}$ is of full rank and likewise for $\mathbf{H}\mathbf{H}^T$. The two algorithms are given in Figure 9.1.2 and Figure 9.1.3.

Remark 9.1.1 Symmetric square root of a matrix Let $\mathbf{A} \in \mathbb{R}^{n \times n}$ be a symmetric, positive definite matrix. Then, there exists a symmetric, positive definite matrix $\mathbf{S}$ such that $\mathbf{A} = \mathbf{S}^2$. This matrix $\mathbf{S}$ is called the **square root** of $\mathbf{A}$.

One method of computing $\mathbf{S}$ is to use the matrix analog of the standard Newton's iterative method for finding the square root of a positive real number. Let $\mathbf{x}_0 = \mathbf{I}$ and define

$$\mathbf{x}_{k+1} = \frac{1}{2}(\mathbf{x}_k + \mathbf{A}\,\mathbf{x}_k^{-1}).$$

Given $\mathbf{H} \in \mathbb{R}^{m \times n}$ and $\mathbf{z} \in \mathbb{R}^m$, with $m < n$. Solve $\mathbf{x}^* = \mathbf{H}^{\mathrm{T}}(\mathbf{H}\mathbf{H}^{\mathrm{T}})^{-1}\mathbf{z}$.

Step 1 Compute the $m \times m$ symmetric matrix $\mathbf{H}\mathbf{H}^{\mathrm{T}}$ – (matrix-matrix multiplication).

Step 2 Compute the Cholesky factor $\mathbf{G}$ such that $(\mathbf{H}\mathbf{H}^{\mathrm{T}}) = \mathbf{G}\mathbf{G}^{\mathrm{T}}$(Exercise 9.6).

Step 3 Solve the lower triangular system $\mathbf{G}\mathbf{g} = \mathbf{z}$ (Exercise 9.3).

Step 4 Solve the upper triangular system $\mathbf{G}^{\mathrm{T}}\mathbf{y} = \mathbf{g}$ (Exercise 9.4).

Step 5 Compute $\mathbf{x}^* = \mathbf{H}^{\mathrm{T}}y$ – (Matrix-vector multiplication.)

Fig. 9.1.3 Cholesky method for normal equations: under-determined system.

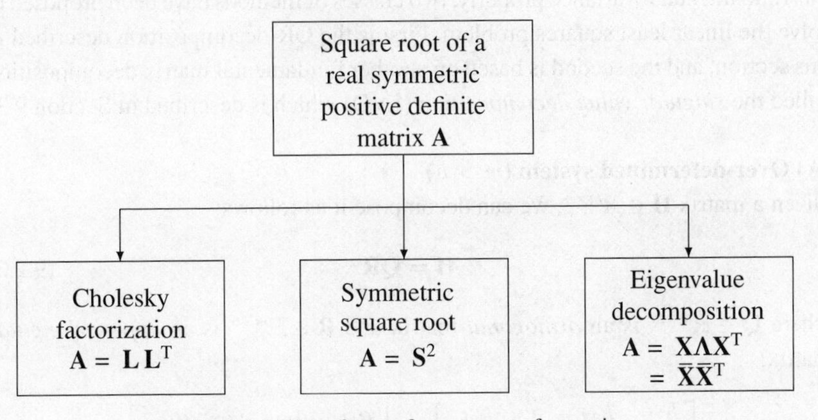

Fig. 9.1.4 Three forms of square root of a matrix.

Then, $\mathbf{x}_k$ converges to $\mathbf{S}$ quadratically (Chapter 12).

As an example, if

$$\mathbf{A} = \begin{bmatrix} 1 & 3/2 \\ 3/2 & 7/2 \end{bmatrix}, \quad \text{then} \quad \mathbf{S} = \begin{bmatrix} 0.8161 & 0.5779 \\ 0.5779 & 1.7793 \end{bmatrix}.$$

Remark 9.1.2 Three forms of square root of a matrix In analogy with the definition of the square root of a positive real number, while $\mathbf{A} = \mathbf{S}^2$ resembles the conventional definition of the square root of a matrix, in the literature on Kalman filtering (Chapters 28 and 30) the concept of a square root of a real symmetric and positive definite matrix is used in an extended sense to include the following three factorizations: (*a*) Cholesky factorization $\mathbf{A} = \mathbf{L}\mathbf{L}^{\mathrm{T}}$ where $\mathbf{L}$ is a lower triangular matrix, (*b*) Symmetric square root factorization $\mathbf{A} = \mathbf{S}^2$ where $\mathbf{S}$ is symmetric positive definite matrix and (*c*) eigen decomposition $\mathbf{A} = \mathbf{X}\mathbf{\Lambda}\mathbf{X}^{\mathrm{T}} = \bar{\mathbf{X}}\bar{\mathbf{X}}^{\mathrm{T}}$ where $\mathbf{X}$ is the matrix of eigenvectors and $\mathbf{\Lambda}$ is the diagonal matrix of eigenvalues of $\mathbf{A}$ (Chapter 28). Refer to Figure 9.1.4.

9.2 QR-decomposition

A matrix $\mathbf{Q} \in \mathbb{R}^{m \times m}$ is said to be *orthogonal* if its transpose is its inverse, that is $\mathbf{Q}\mathbf{Q}^{\mathrm{T}} = \mathbf{Q}^{\mathrm{T}}\mathbf{Q} = \mathbf{I}$. Thus, the columns and the rows constitute a complete orthonormal basis for $\mathbb{R}^m$. One of the basic properties of the orthogonal matrices is that as a linear transformation of vectors in $\mathbb{R}^m$ to $\mathbb{R}^m$, that is $\mathbf{Q} : \mathbb{R}^m \longrightarrow \mathbb{R}^m$, it preserves the Euclidean norm. Let $\mathbf{y} \in \mathbb{R}^m$. Then

$$\|\mathbf{Q}\mathbf{y}\|^2 = (\mathbf{Q}\mathbf{y})^{\mathrm{T}}(\mathbf{Q}\mathbf{y}) = \mathbf{y}^{\mathrm{T}}(\mathbf{Q}^{\mathrm{T}}\mathbf{Q})\mathbf{y} = \mathbf{y}^{\mathrm{T}}\mathbf{y} = \|\mathbf{y}\|^2 \tag{9.2.1}$$

where $\mathbf{Q}$ is orthogonal. Similarly, it can be verified that $\left\|\mathbf{Q}^{\mathrm{T}}\mathbf{y}\right\|^2 = \|\mathbf{y}\|^2$. Exploiting this fundamental invariance property, two classes of methods have been proposed to solve the linear least squares problem. First is the QR-decomposition described in this section, and the second is based on another fundamental matrix decomposition called the *singular value decomposition* (SVD) which is described in Section 9.3.

(A) Over-determined system $(m > n)$

Given a matrix $\mathbf{H} \in \mathbb{R}^{m \times n}$, we can decompose it as follows:

$$\mathbf{H} = \mathbf{Q}\mathbf{R} \tag{9.2.2}$$

where $\mathbf{Q} \in \mathbb{R}^{m \times m}$ is an *orthogonal* matrix and $\mathbf{R} \in \mathbb{R}^{m \times n}$ is an *upper triangular* matrix:

$$\begin{bmatrix} h_{11} & h_{12} & \cdots & h_{1n} \\ h_{21} & h_{22} & \cdots & h_{2n} \\ \vdots & \vdots & \vdots & \vdots \\ h_{m1} & h_{m2} & \cdots & h_{mn} \end{bmatrix} = \begin{bmatrix} q_{11} & q_{12} & \cdots & q_{1m} \\ q_{21} & q_{22} & \cdots & q_{2m} \\ \vdots & \vdots & \vdots & \vdots \\ q_{m1} & q_{m2} & \cdots & q_{mm} \end{bmatrix} \times \begin{bmatrix} r_{11} & r_{12} & \cdots & r_{1n} \\ 0 & r_{22} & \cdots & r_{2n} \\ \vdots & \vdots & \vdots & \vdots \\ 0 & 0 & \cdots & r_{nn} \\ 0 & 0 & \cdots & 0 \\ \vdots & \vdots & \vdots & \vdots \\ 0 & 0 & \cdots & 0 \end{bmatrix} \tag{9.2.3}$$

called the full QR-decomposition.

We begin by partitioning the matrices on the right-hand side of (9.2.3) as follows. Let

$$\mathbf{Q} = \left[\mathbf{Q}_1 \vdots \mathbf{Q}_2\right] \quad \text{and} \quad \mathbf{R} = \begin{bmatrix} \mathbf{R}_1 \\ \cdots \\ \mathbf{R}_2 \end{bmatrix} = \begin{bmatrix} \mathbf{R}_1 \\ \cdots \\ \mathbf{0} \end{bmatrix} \tag{9.2.4}$$

where $\mathbf{Q}_1 \in \mathbb{R}^{m\times n}$ contains the leftmost n columns with $\mathbf{Q}_2 \in \mathbb{R}^{m\times(m-n)}$ has the rest of the $m-n$ columns of $\mathbf{Q}$; and $\mathbf{R}_1 \in \mathbb{R}^{n\times n}$ contains the topmost n rows of $\mathbf{R}$ with $\mathbf{R}_2 \in \mathbb{R}^{(m-n)\times n}$ contains the rest of the $m-n$ rows of all zeros, and hence, is a *null* or *zero* matrix. Since $\mathbf{Q}$ is orthogonal, it follows that

$$\mathbf{Q}_1^{\mathrm{T}}\mathbf{Q}_1 = \mathbf{I}_n$$

$$\text{and} \quad \mathbf{P}_1 \in \mathbb{R}^{m\times m} \quad \text{given by}$$

$$\mathbf{P}_1 = \mathbf{Q}_1(\mathbf{Q}_1^{\mathrm{T}}\mathbf{Q}_1)^{-1}\mathbf{Q}_1^{\mathrm{T}} = \mathbf{Q}_1\mathbf{Q}_1^{\mathrm{T}} \tag{9.2.5}$$

is an *orthogonal projection* (Chapter 6) on to the subspace spanned by the columns of $\mathbf{Q}_1$.

Combining (9.2.2) with (9.2.4), we obtain another related decomposition:

$$\mathbf{H} = [\mathbf{Q}_1\ \mathbf{Q}_2]\begin{bmatrix}\mathbf{R}_1\\ \mathbf{0}\end{bmatrix} = \mathbf{Q}_1\mathbf{R}_1 \tag{9.2.6}$$

called the *reduced QR-decomposition*.

We now return to our linear least squares problem, where $\mathbf{r}(\mathbf{x}) = (\mathbf{z} - \mathbf{H}\mathbf{x})$ is the residual vector. In the light of (9.2.1), we have

$$\begin{aligned} f(\mathbf{x}) &= \|\mathbf{r}(\mathbf{x})\|^2 = \left\|\mathbf{Q}^{\mathrm{T}}\mathbf{r}(\mathbf{x})\right\|^2 \\ &= \left\|\mathbf{Q}^{\mathrm{T}}(\mathbf{z} - \mathbf{H}\mathbf{x})\right\|^2 = \left\|\mathbf{Q}^{\mathrm{T}}\mathbf{z} - \mathbf{Q}^{\mathrm{T}}\mathbf{H}\mathbf{x}\right\|^2. \end{aligned} \tag{9.2.7}$$

But

$$\mathbf{Q}^{\mathrm{T}}\mathbf{z} = \begin{bmatrix}\mathbf{Q}_1^{\mathrm{T}}\\ \mathbf{Q}_2^{\mathrm{T}}\end{bmatrix}\mathbf{z} = \begin{bmatrix}\mathbf{Q}_1^{\mathrm{T}}\mathbf{z}\\ \mathbf{Q}_2^{\mathrm{T}}\mathbf{z}\end{bmatrix}$$

and from (9.2.2)

$$\mathbf{Q}^{\mathrm{T}}\mathbf{H}\mathbf{x} = \mathbf{Q}^{\mathrm{T}}\mathbf{Q}\mathbf{R}\mathbf{x} = \mathbf{R}\mathbf{x} = \begin{bmatrix}\mathbf{R}_1\\ \mathbf{0}\end{bmatrix}\mathbf{x} = \begin{bmatrix}\mathbf{R}_1\mathbf{x}\\ \mathbf{0}\end{bmatrix}.$$

Substituting these into (9.2.7),

$$\begin{aligned} f(\mathbf{x}) &= \left\|\begin{bmatrix}\mathbf{Q}_1^{\mathrm{T}}\mathbf{z}\\ \mathbf{Q}_2^{\mathrm{T}}\mathbf{z}\end{bmatrix} - \begin{bmatrix}\mathbf{R}_1\mathbf{x}\\ \mathbf{0}\end{bmatrix}\right\|^2 \\ &= \left\|\mathbf{Q}_1^{\mathrm{T}}\mathbf{z} - \mathbf{R}_1\mathbf{x}\right\|^2 + \left\|\mathbf{Q}_2^{\mathrm{T}}\mathbf{z}\right\|^2. \end{aligned} \tag{9.2.8}$$

Thus, $f(\mathbf{x})$ is minimum when

$$\mathbf{R}_1\mathbf{x} = \mathbf{Q}_1^{\mathrm{T}}\mathbf{z} \quad \text{or} \quad \mathbf{x}^* = \mathbf{R}_1^{-1}\mathbf{Q}_1^{\mathrm{T}}\mathbf{z} \tag{9.2.9}$$

and the minimum value of the least squares error is given by

$$\left\|\mathbf{r}(\mathbf{x}^*)\right\|^2 = \left\|\mathbf{Q}_2^{\mathrm{T}}\mathbf{z}\right\|^2. \tag{9.2.10}$$

Refer to Figure 9.2.1 for a description of this approach.

Given $\mathbf{H} \in \mathbb{R}^{m\times n}$ and $\mathbf{z} \in \mathbb{R}^m$, $m > n$, solve $\mathbf{R}_1\mathbf{x} = \mathbf{Q}_1^{\mathrm{T}}\mathbf{z}$.

Step 1 Compute the factors $\mathbf{Q}_1$ and $\mathbf{R}_1$ such that $\mathbf{H} = \mathbf{Q}_1\mathbf{R}_1$, where $\mathbf{Q}_1 \in \mathbb{R}^{m\times n}$ has orthonormal columns and $\mathbf{R}_1 \in \mathbb{R}^{n\times n}$, an upper triangular matrix – (use modified Gram–Schmidt algorithm given below).

Step 2 Compute $\mathbf{Q}_1^{\mathrm{T}}\mathbf{z}$ – matrix-vector product.

Step 3 Solve the upper triangular system $\mathbf{R}_1\mathbf{x} = \mathbf{Q}_1^{\mathrm{T}}\mathbf{z}$ (Exercise 9.4).

Fig. 9.2.1 QR-decomposition: over-determined system.

Remark 9.2.1 One could derive (9.2.8) alternatively by using the reduced QR-decomposition as follows:

$$\begin{aligned} f(\mathbf{x}) = \|\mathbf{r}(\mathbf{x})\|^2 &= \|(\mathbf{z} - \mathbf{Hx})\|^2 = \|\mathbf{z} - \mathbf{Q}_1\mathbf{R}_1\mathbf{x}\|^2 \\ &= \mathbf{z}^{\mathrm{T}}\mathbf{z} - 2\mathbf{z}^{\mathrm{T}}\mathbf{Q}_1\mathbf{R}_1\mathbf{x} + \mathbf{x}^{\mathrm{T}}\mathbf{R}_1^{\mathrm{T}}\mathbf{R}_1\mathbf{x} \end{aligned} \tag{9.2.11}$$

from which, we have

$$\nabla f(\mathbf{x}) = -2\mathbf{R}_1^{\mathrm{T}}\mathbf{Q}_1^{\mathrm{T}}\mathbf{z} + 2\mathbf{R}_1^{\mathrm{T}}\mathbf{R}_1\mathbf{x} \tag{9.2.12}$$

and $\nabla^2 f(\mathbf{x}) = 2\mathbf{R}_1^{\mathrm{T}}\mathbf{R}_1$. Setting (9.2.12) to zero, we have

$$\mathbf{R}_1^{\mathrm{T}}\mathbf{R}_1\mathbf{x} = \mathbf{R}_1^{\mathrm{T}}\mathbf{Q}_1^{\mathrm{T}}\mathbf{z}.$$

Since $\mathbf{R}_1$ is non-singular when $\mathbf{H}$ is of full rank, multiplying both sides by $\mathbf{R}_1^{-\mathrm{T}} = (R^{\mathrm{T}})^{-1}$, we immediately get (9.2.9).

(B) Under-determined system: $m < n$

The above development can be readily adapted to the under-determined case when $m < n$. Since $\mathbf{H}^{\mathrm{T}} \in \mathbb{R}^{n\times m}$, with $n > m$, we can obtain the full QR-decomposition using (9.2.2) with n and m interchanged. Thus, we immediately get

$$\mathbf{H}^{\mathrm{T}} = \mathbf{QR} \tag{9.2.13}$$

where $\mathbf{Q} \in \mathbb{R}^{n\times n}$ and $\mathbf{R} \in \mathbb{R}^{n\times m}$ is an upper triangular matrix. Partitioning $\mathbf{Q}$ and $\mathbf{R}$, we get

$$\mathbf{Q} = \begin{bmatrix} \mathbf{Q}_1 \vdots \mathbf{Q}_2 \end{bmatrix} \quad \text{and} \quad \mathbf{R} = \begin{bmatrix} \mathbf{R}_1 \\ \cdots \\ 0 \end{bmatrix} \tag{9.2.14}$$

where $\mathbf{Q}_1 \in \mathbb{R}^{n\times m}$ has the first m columns with $\mathbf{Q}_2 \in \mathbb{R}^{n\times(n-m)}$ has the rest of the columns of $\mathbf{Q}$ and $\mathbf{R}_1 \in \mathbb{R}^{m\times m}$ is the upper triangular matrix. Again, $\mathbf{Q}_1^{\mathrm{T}}\mathbf{Q}_1 = \mathbf{I}_m$ and $\mathbf{Q}_1\mathbf{Q}_1^{\mathrm{T}}$ is the orthogonal projection matrix on the sub-space spanned by the first m columns of $\mathbf{Q}$.

Now using (9.2.13), we immediately get

$$\begin{aligned} f(\mathbf{x}) &= \|\mathbf{r}(\mathbf{x})\|^2 = \|(\mathbf{z}-\mathbf{H}\mathbf{x})\|^2 = \left\|\mathbf{z}-\mathbf{R}^{\mathrm{T}}\mathbf{Q}^{\mathrm{T}}\mathbf{x}\right\|^2 \\ &= \mathbf{z}^{\mathrm{T}}\mathbf{z} - 2\mathbf{z}^{\mathrm{T}}\mathbf{R}^{\mathrm{T}}\mathbf{Q}^{\mathrm{T}}\mathbf{x} + \mathbf{x}^{\mathrm{T}}(\mathbf{Q}\mathbf{R}\mathbf{R}^{\mathrm{T}}\mathbf{Q}^{\mathrm{T}})\mathbf{x} \end{aligned} \tag{9.2.15}$$

whose gradient and Hessian are given by

$$\begin{aligned} \nabla f(\mathbf{x}) &= -2\mathbf{Q}\mathbf{R}\mathbf{z} + 2(\mathbf{Q}\mathbf{R}\mathbf{R}^{\mathrm{T}}\mathbf{Q}^{\mathrm{T}})\mathbf{x} \\ \text{and} \quad & \\ \nabla^2 f(\mathbf{x}) &= 2(\mathbf{Q}\mathbf{R}\mathbf{R}^{\mathrm{T}}\mathbf{Q}^{\mathrm{T}}). \end{aligned} \tag{9.2.16}$$

Setting the gradient to zero, and since $\mathbf{Q}$ is an orthogonal matrix, the minimizing $\mathbf{x}$ is the solution of

$$\mathbf{R}\mathbf{R}^{\mathrm{T}}(\mathbf{Q}^{\mathrm{T}}\mathbf{x}) = \mathbf{R}\mathbf{z}. \tag{9.2.17}$$

Let

$$\mathbf{Q}^{\mathrm{T}}\mathbf{x} = \begin{bmatrix} \mathbf{Q}_1^{\mathrm{T}} \\ \mathbf{Q}_2^{\mathrm{T}} \end{bmatrix} \mathbf{x} = \begin{bmatrix} \mathbf{Q}_1^{\mathrm{T}}\mathbf{x} \\ \mathbf{Q}_2^{\mathrm{T}}\mathbf{x} \end{bmatrix} = \begin{bmatrix} \mathbf{y}_1 \\ \mathbf{y}_2 \end{bmatrix} \tag{9.2.18}$$

where $\mathbf{y}_1 \in \mathbb{R}^m$ and $\mathbf{y}_2 \in \mathbb{R}^{n-m}$. Now combining (9.2.14), (9.2.17), and (9.2.18), we get

$$\begin{bmatrix} \mathbf{R}_1 \\ \cdots \\ \mathbf{0} \end{bmatrix} \begin{bmatrix} \mathbf{R}_1^{\mathrm{T}} \vdots \mathbf{0} \end{bmatrix} \begin{bmatrix} \mathbf{y}_1 \\ \mathbf{y}_2 \end{bmatrix} = \begin{bmatrix} \mathbf{R}_1 \\ \cdots \\ \mathbf{0} \end{bmatrix} \mathbf{z} \tag{9.2.19}$$

or

$$\begin{bmatrix} \mathbf{R}_1\mathbf{R}_1^{\mathrm{T}} & \mathbf{0} \\ \mathbf{0} & \mathbf{0} \end{bmatrix} \begin{bmatrix} \mathbf{y}_1 \\ \mathbf{y}_2 \end{bmatrix} = \begin{bmatrix} \mathbf{R}_1\mathbf{z} \\ \mathbf{0} \end{bmatrix}$$

from which since $\mathbf{R}_1$ is non-singular, we get $\mathbf{y}_1$ as the solution of

$$\mathbf{R}_1\mathbf{R}_1^{\mathrm{T}}\mathbf{y}_1 = \mathbf{R}_1\mathbf{z}$$

or

$$\mathbf{R}_1^{\mathrm{T}}\mathbf{y}_1 = \mathbf{z} \tag{9.2.20}$$

and $\mathbf{y}_2$ is arbitrary.

Solving the lower triangular system (9.2.20) for $\mathbf{y}_1$, we can build the required solution of $\mathbf{x}$ using (9.2.18) as

$$\mathbf{x} = \mathbf{Q}\begin{bmatrix} \mathbf{y}_1 \\ \mathbf{y}_2 \end{bmatrix} = [\mathbf{Q}_1\ \mathbf{Q}_2]\begin{bmatrix} \mathbf{y}_1 \\ \mathbf{y}_2 \end{bmatrix} = \mathbf{Q}_1\mathbf{y}_1 + \mathbf{Q}_2\mathbf{y}_2. \tag{9.2.21}$$

Several observations are in order: (1) Since $\mathbf{y}_2$ is arbitrary, there are clearly many solutions which is to be expected as we are dealing with an under-determined

Given $\mathbf{H} \in \mathbb{R}^{m\times n}$ and $\mathbf{z} \in \mathbb{R}^m$, $m < n$, solve $\mathbf{R}_1^{\mathrm{T}}\mathbf{Q}_1^{\mathrm{T}}\mathbf{x} = \mathbf{z}$.

Step 1 Compute the factors $\mathbf{H}^{\mathrm{T}} = \mathbf{Q}_1\mathbf{R}_1$ as in (9.2.14), where $\mathbf{Q} \in \mathbb{R}^{n\times m}$ has orthonormal columns and $\mathbf{R}_1 \in \mathbb{R}^{m\times m}$ is an upper triangular matrix, using the modified Gram–Schmidt algorithm.

Note: Build $\mathbf{Q} = \begin{bmatrix} \mathbf{Q}_1 \vdots \mathbf{Q}_2 \end{bmatrix} \in \mathbb{R}^{n\times n}$, where $\mathbf{Q}_2 \in \mathbb{R}^{n\times(n-m)}$ by adding $(n-m)$ new orthonormal vectors so that $\mathbf{Q}$ is orthogonal.

Step 2 Solve the lower triangular system $\mathbf{R}_1^{\mathrm{T}}\mathbf{y}_1 = \mathbf{z}$ for $\mathbf{y}_1$.

Step 3 Compute $\mathbf{x}^* = \mathbf{Q}_1\mathbf{y}_1$ as the minimum norm solution.

Note: Any arbitrary solution $\mathbf{x} = \mathbf{x}^* + \mathbf{Q}_2\mathbf{y}_2$, where $\mathbf{y}_2$ is arbitrary.

Fig. 9.2.2 QR-decomposition: under-determined system.

system. (2) The solution $\mathbf{x}$ of minimum norm is given by

$$\mathbf{x}^* = \mathbf{Q}_1\mathbf{y}_1 \tag{9.2.22}$$

since

$$\|\mathbf{x}\|^2 = \|\mathbf{Q}_1\mathbf{y}_1\|^2 + \|\mathbf{Q}_2\mathbf{y}_2\|^2 = \|\mathbf{y}_1\|^2 + \|\mathbf{y}_2\|^2 \geq \|\mathbf{y}_1\|^2 = \left\|\mathbf{x}^*\right\|^2.$$

Figure 9.2.2 contains a version of this algorithm.

The above development is predicated on the existence of the QR-decomposition of a matrix. For completeness, we now turn to providing a very simple and elegant algorithm based on the classical *Gram–Schmidt* orthogonalization method for computing this decomposition.

(C) Gram–Schmidt algorithm: basic idea

Let $S_1 = \{\mathbf{h}_1, \mathbf{h}_2, \ldots, \mathbf{h}_n\}$ be a set of n linearly independent vectors in $\mathbb{R}^m$, where $m > n$. The problem is to generate $S_2 = \{\mathbf{q}_1, \mathbf{q}_2, \ldots, \mathbf{q}_n\}$, the set of n orthonormal vectors in $\mathbb{R}^m$ from S_1.

The idea behind this algorithm may be described as follows. First choose $\mathbf{q}_1$ such that

$$\mathbf{q}_1 = \frac{\mathbf{h}_1}{r_{11}} \quad \text{and} \quad r_{11} = \|\mathbf{h}_1\|. \tag{9.2.23}$$

Then, let

$$\mathbf{q}_2 = \frac{\mathbf{h}_2 - r_{12}\mathbf{q}_1}{r_{22}}. \tag{9.2.24}$$

Taking inner product of both sides with $\mathbf{q}_1$ and requiring that $\mathbf{q}_2$ is orthogonal to $\mathbf{q}_1$, since $\|\mathbf{q}_1\| = 1$, we obtain

$$\mathbf{0} = \mathbf{q}_1^{\mathrm{T}}\mathbf{q}_2 = \frac{1}{r_{22}}[\mathbf{q}_1^{\mathrm{T}}\mathbf{h}_2 - r_{12}]$$

or

$$r_{12} = \mathbf{q}_1^{\mathrm{T}} \mathbf{h}_2. \tag{9.2.25}$$

Now, normalizing $\mathbf{q}_2$, we get

$$r_{22} = \|\mathbf{h}_2 - r_{12}\mathbf{q}_1\|. \tag{9.2.26}$$

Generalizing this, we obtain for $1 \le j \le n$,

$$\mathbf{q}_j = \frac{\mathbf{h}_j - \sum_{i=1}^{j-1} r_{ij}\mathbf{q}_i}{r_{jj}} \tag{9.2.27}$$

where

$$\begin{aligned} r_{ij} &= \mathbf{q}_i^{\mathrm{T}} \mathbf{h}_j, \quad 1 \le i \le j-1. \\ r_{jj} &= \left\| \mathbf{h}_j - \sum_{j-1}^{i=1} r_{ij}\mathbf{q}_i \right\|. \end{aligned} \tag{9.2.28}$$

Now, we can rewrite (9.2.23), (9.2.24) and (9.2.27) succinctly in matrix notation as

$$[\mathbf{h}_1, \mathbf{h}_2, \ldots, \mathbf{h}_n] = [\mathbf{q}_1, \mathbf{q}_2, \ldots, \mathbf{q}_n] \begin{bmatrix} r_{11} & r_{12} & \cdots & r_{1j} & \cdots & r_{1n} \\ 0 & r_{22} & \cdots & r_{2j} & \cdots & r_{2n} \\ 0 & 0 & \vdots & \vdots & \vdots & \vdots \\ \cdots & \cdots & \vdots & r_{jj} & \cdots & r_{jn} \\ \vdots & \vdots & \vdots & \vdots & \vdots & \vdots \\ 0 & 0 & \cdots & 0 & \cdots & r_{nn} \end{bmatrix}$$

or

$$\mathbf{H} = \mathbf{Q}\mathbf{R} \tag{9.2.29}$$

where

$$\mathbf{H} = [\mathbf{h}_1, \mathbf{h}_2, \ldots, \mathbf{h}_n] \in \mathbb{R}^{m \times n}$$

$$\mathbf{Q} = [\mathbf{q}_1, \mathbf{q}_2, \ldots, \mathbf{q}_n] \in \mathbb{R}^{m \times n}$$

has n orthonormal columns and

$$\mathbf{R} = \left[r_{ij}\right] \in \mathbb{R}^{m \times n}$$

is the upper triangular matrix, which gives the required reduced QR-decomposition.

The algorithm is summarized in Figure 9.2.3.

While this classical algorithm is very simple and elegant, this is not known to be numerically stable. A stable version of this algorithm based on the principles of orthogonal projection (Chapter 6), called the modified Gram–Schmidt algorithm, is developed in Exercises 9.7 and 9.8. It is an interesting computational exercise to implement this classical and the modified versions of this algorithm.

Given $S_1 = \{\mathbf{h}_1, \mathbf{h}_2, \ldots, \mathbf{h}_n\}$, where $\mathbf{h}_i \in \mathbb{R}^m$ are linearly independent, find $S_2 = \{\mathbf{q}_1, \mathbf{q}_2, \ldots, \mathbf{q}_n\}$, where $\mathbf{q}_i \in \mathbb{R}^m$ are orthonormal.

Step 1 Repeat steps 2 through 5 for $j = 1, \ldots, n$.

Step 2 Set $v_j = \mathbf{h}_j$.

Step 3 For $i = 1, \ldots, j-1$

Compute the inner product $r_{ij} = \mathbf{q}_i^{\mathrm{T}} \mathbf{h}_j$

Update $v_j = v_j - r_{ij}\mathbf{q}_i$.

Step 4 Compute the norm of $v_j : r_{jj} = \|v_j\| = (v_j^{\mathrm{T}} v_j)^{1/2}$

Step 5 Compute $\mathbf{q}_j = \frac{v_j}{r_{jj}}$.

Fig. 9.2.3 Classical Gram–Schmidt algorithm.

Remark 9.2.2 QR-decomposition is one of the most basic tools in numerical linear algebra. In addition to linear least squares problems, this decomposition is widely used in the computation of eigenvalues of symmetric matrices. Besides the Gram-Schmidt algorithm, another competing method for obtaining this decomposition is based on Householder's transformation which geometrically is a reflection. It is beyond our scope to take up the description of this important and useful idea. We refer the reader to excellent text books for details.

9.3 Singular value decomposition

This method is based on the eigen decomposition of the Grammian matrices $\mathbf{H}^{\mathrm{T}}\mathbf{H} \in \mathbb{R}^{n \times n}$ and $\mathbf{H}\mathbf{H}^{\mathrm{T}} \in \mathbb{R}^{m \times m}$. Recall that a Grammian, by definition, is a symmetric and positive definite matrix (assuming that $\mathbf{H}$ is of full rank) and, hence, its eigenvalues are real and positive. Let (λ_i, v_i) for $i = 1, 2, \ldots, n$ be the eigenvalue/vector pair for $\mathbf{H}^{\mathrm{T}}\mathbf{H}$. Then,

$$(\mathbf{H}^{\mathrm{T}}\mathbf{H})v_i = \lambda_i v_i, \quad v_i \in \mathbb{R}^n \tag{9.3.1}$$

for $i = 1, 2, \ldots, n$. By collecting all these n relations, we get

$$\mathbf{H}^{\mathrm{T}}\mathbf{H}[v_1, v_2, \ldots, v_n] = [v_1, v_2, \ldots, v_n] \begin{bmatrix} \lambda_1 & 0 & \cdots & 0 \\ 0 & \lambda_2 & \cdots & 0 \\ \vdots & \vdots & & \vdots \\ 0 & 0 & \cdots & \lambda_n \end{bmatrix} \tag{9.3.2}$$

Denoting

$$\mathbf{V} = [v_1, v_2, \ldots, v_n] \in \mathbb{R}^{m \times n}$$

and

$$\mathbf{\Lambda} = \mathrm{Diag}[\lambda_1, \lambda_2, \ldots, \lambda_n] \in \mathbb{R}^{n \times n}$$

we can rewrite (9.3.2) succinctly as

$$\mathbf{H}^{\mathrm{T}}\mathbf{H}\mathbf{V} = \mathbf{V}\boldsymbol{\Lambda}. \tag{9.3.3}$$

Since the eigenvectors of a real symmetric matrix are orthogonal, it follows that the columns of **V** are orthogonal. Without loss of generality, we can assume that the columns of **V** are also normalized. Hence, in the following **V** is taken to be an orthogonal matrix, that is,

$$\mathbf{V}^{\mathrm{T}}\mathbf{V} = \mathbf{V}\mathbf{V}^{\mathrm{T}} = \mathbf{I} \in \mathbb{R}^{n\times n}. \tag{9.3.4}$$

Now, define a new system of vectors,

$$\mathbf{u}_i = \frac{1}{\sqrt{\lambda_i}}\mathbf{H}\upsilon_i \tag{9.3.5}$$

where $\mathbf{u}_i \in \mathbb{R}^m$ for $i = 1, 2, \ldots, n$. Then,

$$\begin{aligned}(\mathbf{H}\mathbf{H}^{\mathrm{T}})\mathbf{u}_i &= \tfrac{1}{\sqrt{\lambda_i}}(\mathbf{H}\mathbf{H}^{\mathrm{T}})\mathbf{H}\upsilon_i = \tfrac{1}{\sqrt{\lambda_i}}\mathbf{H}(\mathbf{H}^{\mathrm{T}}\mathbf{H})\upsilon_i \\ &= \tfrac{1}{\sqrt{\lambda_i}}\mathbf{H}(\lambda_i\upsilon_i) \text{ from (9.3.1)} \\ &= \sqrt{\lambda_i}\mathbf{H}\upsilon_i = \lambda_i\mathbf{u}_i \text{ from (9.3.5)}\end{aligned} \tag{9.3.6}$$

that is, $(\lambda_i, \mathbf{u}_i)$ is an eigenvalue/vector pair for $\mathbf{H}\mathbf{H}^{\mathrm{T}}$, for $i = 1, 2, \ldots, n$. Define

$$\mathbf{U} = [\mathbf{u}_1, \mathbf{u}_2, \ldots, \mathbf{u}_n] \in \mathbb{R}^{m\times n}. \tag{9.3.7}$$

We first verify that the columns of **U** are orthonormal, that is, $\mathbf{U}^{\mathrm{T}}\mathbf{U} = \mathbf{I} \in \mathbb{R}^{n\times n}$. To this end, rewrite (9.3.5) as

$$\mathbf{u}_i\sqrt{\lambda_i} = \mathbf{H}\upsilon_i \quad \text{for} \quad i = 1, 2, \ldots, n$$

which, when rewritten in matrix form, becomes

$$\mathbf{U}\sqrt{\boldsymbol{\Lambda}} = \mathbf{H}\mathbf{V} \tag{9.3.8}$$

where $\sqrt{\boldsymbol{\Lambda}} = \mathrm{Diag}[\sqrt{\lambda_1}, \sqrt{\lambda_2}, \ldots, \sqrt{\lambda_n}]$. Hence,

$$\begin{aligned}\mathbf{U}^{\mathrm{T}}\mathbf{U} &= (\mathbf{H}\mathbf{V}\boldsymbol{\Lambda}^{-1/2})^{\mathrm{T}}(\mathbf{H}\mathbf{V}\boldsymbol{\Lambda}^{-1/2}) = \boldsymbol{\Lambda}^{-1/2}\mathbf{V}^{\mathrm{T}}(\mathbf{H}^{\mathrm{T}}\mathbf{H}\mathbf{V})\boldsymbol{\Lambda}^{-1/2} \\ &= \boldsymbol{\Lambda}^{-1/2}\mathbf{V}^{\mathrm{T}}\mathbf{V}\boldsymbol{\Lambda}^{1/2} \text{ (from (9.3.3))} \\ &= \mathbf{I} \quad \text{(from (9.3.4))}.\end{aligned} \tag{9.3.9}$$

An important observation is that both $\mathbf{H}^{\mathrm{T}}\mathbf{H}$ and $\mathbf{H}\mathbf{H}^{\mathrm{T}}$ have the same set of non-zero eigenvalues. Now, rewrite (9.3.5) as

$$\mathbf{H}\mathbf{u}_i = \sqrt{\lambda_i}\mathbf{u}_i$$

or

$$\mathbf{H}[v_1, v_2, \ldots, v_n] = [\mathbf{u}_1, \mathbf{u}_2, \ldots, \mathbf{u}_n]\begin{bmatrix} \sqrt{\lambda_1} & 0 & \cdots & 0 \\ 0 & \sqrt{\lambda_2} & \cdots & 0 \\ \vdots & \vdots & & \vdots \\ 0 & 0 & 0 & \sqrt{\lambda_n} \end{bmatrix}$$

which becomes

$$\mathbf{HV} = \mathbf{U}\sqrt{\mathbf{\Lambda}}$$

or

$$\mathbf{H} = \mathbf{U}\sqrt{\mathbf{\Lambda}}\mathbf{V}^{\mathrm{T}}. \tag{9.3.10}$$

This is called the *reduced singular value decomposition* of **H**. Expanding the r.h.s. of (9.3.10), we obtain

$$\begin{aligned} \mathbf{H} &= [\mathbf{u}_1, \mathbf{u}_2, \ldots, \mathbf{u}_n]\begin{bmatrix} \sqrt{\lambda_1} & 0 & \cdots & 0 \\ 0 & \sqrt{\lambda_2} & \cdots & 0 \\ \vdots & \vdots & & \vdots \\ 0 & 0 & 0 & \sqrt{\lambda_n} \end{bmatrix}\begin{bmatrix} v_1^{\mathrm{T}} \\ v_2^{\mathrm{T}} \\ \vdots \\ v_n^{\mathrm{T}} \end{bmatrix} \\ &= \textstyle\sum_{i=1}^{n} \sqrt{\lambda_i}\mathbf{u}_i v_i^{\mathrm{T}} \end{aligned} \tag{9.3.11}$$

where the outer product $\mathbf{u}_i v_i^{\mathrm{T}}$ is a *rank-one* matrix. That is, the measurement matrix **H** can be thought of as the sum of n rank-one matrices as in (9.3.11).

It is customary to call the columns of **U** the left and those of **V** the right *singular vectors* of **H** and $\sqrt{\lambda_i}$ are called the *singular values*.

Remark 9.3.1 When $\mathbf{H} \in \mathbb{R}^{n\times n}$ is symmetric, then $\mathbf{H} = \mathbf{H}^{\mathrm{T}}$ and $\mathbf{H}^{\mathrm{T}}\mathbf{H} = \mathbf{H}\mathbf{H}^{\mathrm{T}} = \mathbf{H}^2$. In this case $\mathbf{u}_i = v_i$ and λ_i is the eigenvalue of $\mathbf{H}^2$, and $\sqrt{\lambda_i}$ that of **H**. Hence

$$\mathbf{H} = \sum_{i=1}^{n} \sqrt{\lambda_i}\mathbf{u}_i\mathbf{u}_i^{\mathrm{T}}$$

is the well-known spectral decomposition of the symmetric matrix **H**. In other words, SVD is an extension of this idea of spectral expansion for rectangular matrices.

Now returning to the least squares problem on hand, we get

$$\begin{aligned} f(\mathbf{x}) &= \|\mathbf{r}(\mathbf{x})\|^2 = \|(\mathbf{z} - \mathbf{Hx})\|^2 \\ &= \left\|(\mathbf{z} - \mathbf{U}\sqrt{\mathbf{\Lambda}}\mathbf{V}^{\mathrm{T}}\mathbf{x})\right\|^2 \quad \text{(from (9.3.10))} \\ &= (\mathbf{z} - \mathbf{U}\sqrt{\mathbf{\Lambda}}\mathbf{V}^{\mathrm{T}}\mathbf{x})^{\mathrm{T}}(\mathbf{z} - \mathbf{U}\sqrt{\mathbf{\Lambda}}\mathbf{V}^{\mathrm{T}}\mathbf{x}) \\ &= \mathbf{z}^{\mathrm{T}}\mathbf{z} - 2\mathbf{z}^{\mathrm{T}}\mathbf{U}\sqrt{\mathbf{\Lambda}}\mathbf{V}^{\mathrm{T}}\mathbf{x} + \mathbf{x}^{\mathrm{T}}\mathbf{V}\mathbf{\Lambda}\mathbf{V}^{\mathrm{T}}\mathbf{x} \end{aligned} \tag{9.3.12}$$

and

$$0 = \nabla f(\mathbf{x}) = -2\mathbf{V}\sqrt{\mathbf{\Lambda}}\mathbf{U}^{\mathrm{T}}\mathbf{z} + 2\mathbf{V}\mathbf{\Lambda}\mathbf{V}^{\mathrm{T}}\mathbf{x}$$

leading to the minimizer as the solution of

$$(\mathbf{V}\mathbf{\Lambda}\mathbf{V}^{\mathrm{T}})\mathbf{x} = \mathbf{V}\sqrt{\mathbf{\Lambda}}\mathbf{U}^{\mathrm{T}}\mathbf{z}. \tag{9.3.13}$$

Multiplying both sides of (9.3.13) on the left successively by $\mathbf{V}^{\mathrm{T}}$, $\mathbf{\Lambda}^{-1}$, and $\mathbf{V}$, we get

$$\mathbf{x}^* = \mathbf{V}\mathbf{\Lambda}^{-1/2}\mathbf{U}^{\mathrm{T}}\mathbf{z}. \tag{9.3.14}$$

This lends itself to a natural geometric interpretation: $\mathbf{U}^{\mathrm{T}}$ as a linear transformation transforms $\mathbf{z} \in \mathbb{R}^m$ to, say, $\mathbf{y} = \mathbf{U}^{\mathrm{T}}\mathbf{z} \in \mathbb{R}^m$. The matrix $\mathbf{\Lambda}^{-1/2}$ being a diagonal matrix *stretches* the components of $\mathbf{y}$, that is, $\mathbf{w} = \mathbf{\Lambda}^{-1/2}\mathbf{y} \in \mathbb{R}^n$, where $w_i = \lambda_i^{-1/2} y_i$, $i = 1, 2, \ldots, n$. Then, $\mathbf{V}$ being an orthogonal transformation in $\mathbb{R}^n$, *rotates* $\mathbf{w}$ to obtain $\mathbf{x}^* = \mathbf{V}\mathbf{w}$.

Remark 9.3.2 The above development is based on the reduced SVD. Technically, one could also use the so called full SVD and in the following, we indicate the major steps. The full SVD for $\mathbf{H}$ is given by

$$\mathbf{H} = \bar{\mathbf{U}}\Sigma\bar{\mathbf{V}} \tag{9.3.15}$$

where

$$\mathbf{U} = [\mathbf{U}_1, \mathbf{U}_2] \in \mathbb{R}^{m\times m}$$

is an *orthogonal* matrix with $\mathbf{U}_1$ containing the first n columns and $\mathbf{U}_2$ with $(m - n)$ columns of $\bar{\mathbf{U}}$,

$$\Sigma = \begin{bmatrix} \Sigma_1 \\ \cdots \\ \Sigma_2 \end{bmatrix} \in \mathbb{R}^{m\times n} \tag{9.3.16}$$

where $\Sigma_1 \in \mathbb{R}^{n\times n}$ is the *diagonal matrix* of singular values of $\mathbf{H}$ and $\Sigma_2 \in \mathbb{R}^{(m-n)\times n}$ matrix of *all zeros* and $\bar{\mathbf{V}} \in \mathbb{R}^{n\times n}$ is an orthogonal matrix. Substituting these into (9.3.15), we get

$$\mathbf{H} = [\mathbf{U}_1, \mathbf{U}_2] \begin{bmatrix} \Sigma_1 \\ \cdots \\ \Sigma_2 \end{bmatrix} \mathbf{V}^{\mathrm{T}} = \mathbf{U}_1\Sigma_1\bar{\mathbf{V}}^{\mathrm{T}}$$

which is the reduced SVD given in (9.3.10), where $\mathbf{U}_1 = \mathbf{V}$, $\Sigma_1 = \sqrt{\mathbf{\Lambda}}$ and $\bar{\mathbf{V}} = \mathbf{V}$.

The key steps of the algorithm based on SVD are given in Figure 9.3.1.

Remark 9.3.3 This method based on SVD is very general in the sense that it is applicable even in cases when $\mathbf{H}$ is *not* of full rank. Let Rank$(\mathbf{H}) = r$, where $r < \min\{m, n\}$. Then, the full SVD in (9.3.15) takes the following form:

$$\mathbf{H} = \bar{\mathbf{U}}\Sigma\bar{\mathbf{V}}^{\mathrm{T}} \tag{9.3.17}$$

where $\bar{\mathbf{U}} \in \mathbb{R}^{m\times m}$ and $\bar{\mathbf{V}} \in \mathbb{R}^{n\times n}$ are both orthogonal matrices, and $\Sigma \in \mathbb{R}^{m\times n}$ is a matrix with only the first r non-zero singular values across its main diagonal. We

Given $\mathbf{H} \in \mathbb{R}^{m\times n}$ and $\mathbf{z} \in \mathbb{R}^m$, compute $\mathbf{x}^* = \mathbf{V}\mathbf{\Lambda}^{-1/2}\mathbf{U}^{\mathrm{T}}\mathbf{z}$.

Step 1 Compute the reduced SVD $\mathbf{H} = \mathbf{U}\sqrt{\mathbf{\Lambda}}\mathbf{V}^{\mathrm{T}}$ as in (9.3.10), where $\mathbf{V} \in \mathbb{R}^{m\times n}$ has orthonormal columns, $\mathbf{\Lambda}^{-1/2}$ is the diagonal matrix of singular values and $\mathbf{V} \in \mathbb{R}^{n\times m}$ is an orthogonal matrix.

Step 2 Compute $\mathbf{y}_1 = \mathbf{U}^{\mathrm{T}}\mathbf{z}$ – matrix-vector product.

Step 3 Compute $\mathbf{y}_2 = \mathbf{\Lambda}^{-1/2}\mathbf{y}_1$. – this is a simple scaling.

Step 4 Compute $\mathbf{x}^* = \mathbf{V}\mathbf{y}_2$ – matrix-vector product.

Fig. 9.3.1 SVD algorithm.

now partition $\mathbf{U} = [\mathbf{U}_1\ \mathbf{U}_2]$, where $\mathbf{U}_1$ has the first r columns of $\mathbf{U}$, and $\mathbf{U}_2$ has the rest of the $(m-r)$ columns. Similarly, let

$$\Sigma = \begin{bmatrix} \Sigma_{11} & \Sigma_{12} \\ \Sigma_{21} & \Sigma_{22} \end{bmatrix} \tag{9.3.18}$$

where $\Sigma_{11} \in \mathbb{R}^{r\times r}$ is the uppermost principal submatrix which is a diagonal matrix with the r singular values across its main diagonal. It can be verified that $\Sigma_{12} \in \mathbb{R}^{r\times(n-r)}$, $\Sigma_{21} \in \mathbb{R}^{(m-r)\times r}$, and $\Sigma_{22} \in \mathbb{R}^{(m-r)\times(n-r)}$ are all null matrices. Finally, partitioning $\mathbf{V}$ as

$$\mathbf{V}^{\mathrm{T}} = \begin{bmatrix} \mathbf{V}_1^{\mathrm{T}} \\ \mathbf{V}_2^{\mathrm{T}} \end{bmatrix}$$

where $\mathbf{V}_1^{\mathrm{T}} \in \mathbb{R}^{r\times n}$ has the first r columns and $\mathbf{V}_2^{\mathrm{T}} \in \mathbb{R}^{(n-r)\times n}$ has the rest of the $(n-r)$ columns of $\mathbf{V}^{\mathrm{T}}$. Hence,

$$\mathbf{H} = [\mathbf{U}_1\,\mathbf{U}_2]\begin{bmatrix} \sqrt{\Sigma_{11}} & \Sigma_{12} \\ \Sigma_{21} & \Sigma_{22} \end{bmatrix}\begin{bmatrix} \mathbf{V}_1^{\mathrm{T}} \\ \mathbf{V}_2^{\mathrm{T}} \end{bmatrix} = \mathbf{U}_1\sqrt{\Sigma_{11}}\mathbf{V}_1^{\mathrm{T}} \tag{9.3.19}$$

which is the reduced SVD of the rank deficient $\mathbf{H}$. Using this, we can now apply the SVD algorithm.

Exercises

9.1 Consider the following 3×3 matrix identity:

$$\begin{bmatrix} a_{11} & a_{12} & a_{13} \\ a_{21} & a_{22} & a_{23} \\ a_{31} & a_{32} & a_{33} \end{bmatrix} = \begin{bmatrix} 1 & 0 & 0 \\ l_{21} & 1 & 0 \\ l_{31} & l_{32} & 1 \end{bmatrix}\begin{bmatrix} u_{11} & u_{12} & u_{13} \\ 0 & u_{22} & u_{23} \\ 0 & 0 & u_{33} \end{bmatrix}.$$

By multiplying the matrices on the right-hand side and equating the produce matrix element-wise with the one on the left-hand side, write out the set of (non-linear) equations in l_{ij} and u_{ij}. Verify that you can compute the elements of the first row of $\mathbf{U}$ and $\mathbf{L}$ alternately, first computing the elements of the first row of $\mathbf{U}$, then the elements of the first column of $\mathbf{L}$, followed by the second row of **U**, and then the second column of **L**, and so on. Translate the pattern into an algorithm, and verify its correctness.

9.2 The following is an explicit algorithm for computing the factor matrices **L** and **U** from the given matrix **A**. Verify its correctness. (Refer (9.1.3) for the notation.)

for $r = 1$ to n
  for $i = r$ to n
    $u_{ri} = a_{ri} - \sum_{j=1}^{r-1} l_{rj} u_{ji}$ /* Recover rows of **U** */
  end {for}
  for $i = r + 1$ to n
    $l_{ir} = \frac{a_{ir} - \sum_{j=1}^{r-1} l_{ij} u_{jr}}{u_{rr}}$ /* Recover columns of **L** */
  end {for}
end {for}

Note: This version of the **LU**-decomposition is also known as the *Doolittle* reduction.

9.3 The following is an algorithm for solving the *lower triangular* system:

$$\begin{bmatrix} l_{11} & 0 & 0 & \cdots & 0 \\ l_{21} & l_{22} & 0 & \cdots & 0 \\ l_{31} & l_{32} & l_{31} & \cdots & 0 \\ \vdots & \vdots & \vdots & \vdots & \vdots \\ l_{n1} & l_{n2} & l_{n3} & \cdots & l_{nn} \end{bmatrix} \begin{bmatrix} g_1 \\ g_2 \\ g_3 \\ \vdots \\ g_n \end{bmatrix} = \begin{bmatrix} b_1 \\ b_2 \\ b_3 \\ \vdots \\ b_n \end{bmatrix}.$$

$g_1 = b_1 / l_{11}$
for $i = 2$ to n
  $g_i = \frac{b_i - \sum_{j=1}^{i-1} l_{ij} g_j}{l_{ii}}$
end {for}

Verify the correctness of this algorithm.

9.4 The following is an algorithm for solving the *upper triangular* system:

$$\begin{bmatrix} u_{11} & u_{12} & u_{13} & \cdots & u_{1n} \\ 0 & u_{22} & u_{23} & \cdots & u_{2n} \\ 0 & 0 & u_{33} & \cdots & u_{3n} \\ \vdots & \vdots & \vdots & \vdots & \vdots \\ 0 & 0 & 0 & \cdots & u_{nn} \end{bmatrix} \begin{bmatrix} x_1 \\ x_2 \\ x_3 \\ \vdots \\ x_n \end{bmatrix} = \begin{bmatrix} g_1 \\ g_2 \\ g_3 \\ \vdots \\ g_n \end{bmatrix}.$$

$x_n = g_n / u_{nn}$

for $i = n - 1$ to 1
  $x_i = \frac{g_i - \sum_{j=i+1}^{n} u_{ij} x_j}{u_{ii}}$
end

Verify the correctness of this algorithm.

9.5 Consider the following 3×3 matrix identity, where $\mathbf{A}$ is symmetric:

$$\begin{bmatrix} a_{11} & a_{12} & a_{13} \\ a_{12} & a_{22} & a_{23} \\ a_{13} & a_{23} & a_{33} \end{bmatrix} = \begin{bmatrix} g_{11} & 0 & 0 \\ g_{21} & g_{22} & 0 \\ g_{31} & g_{32} & g_{33} \end{bmatrix} \begin{bmatrix} g_{11} & g_{12} & g_{13} \\ 0 & g_{22} & g_{23} \\ 0 & 0 & g_{33} \end{bmatrix}.$$

By multiplying the matrices on the right-hand side and equating the product matrix element-wise with the symmetric matrix on the left-hand side, write out the set of equations in six unknowns g_{ij}'s. Verify that you can compute the g_{ij}'s in a systematic manner and identify the pattern for the recovery of g_{ij}'s. Translate the pattern into an algorithm and verify your result.

9.6 The following is an algorithm for computing the *Cholesky factor* for a given symmetric and positive definite matrix $\mathbf{A}$.

for $j = 1$ to n

$g_{jj} = [a_{jj} - \sum_{k=1}^{j-1} g_{jk}^2]^{1/2}$ /* compute diagonal elements */

for $i = j + 1$ to n

$g_{ij} = \frac{a_{ij} - \sum_{k=1}^{j-1} g_{ij} g_{jk}}{g_{jj}}$ /*Recover jth columns of $\mathbf{g}$ */

end

Verify the correctness of the algorithm.

9.7 In this exercise, we recast the computation of the Gram–Schmidt algorithm using the idea of *orthogonal projections* described in Chapter 6. Let $\mathbf{q}_1$ be the unit vector defined in (9.2.23). We can then rewrite (9.2.24) as

$$\begin{aligned} r_{22}\mathbf{q}_2 &= \mathbf{h}_2 - r_{12}\mathbf{q}_1 = \mathbf{h}_2 - (\mathbf{q}_1^T\mathbf{h}_2)\mathbf{q}_1 \\ &= \mathbf{h}_2 - (\mathbf{q}_1\mathbf{q}_1^T)\mathbf{h}_2 = (\mathbf{I} - \mathbf{P}_1)\mathbf{h}_2 \end{aligned}$$

where $\mathbf{P}_1 = \mathbf{q}_1\mathbf{q}_1^T$ is a rank-one orthogonal projection matrix onto the subspace Span($\mathbf{q}_1$). Thus, $(\mathbf{I} - \mathbf{P}_1)\mathbf{h}_2$ is the projection of $\mathbf{h}_2$ onto the subspace that is orthogonal to Span($\mathbf{q}_1$). Hence, the unit vector $\mathbf{q}_2$ is obtained by first computing the projection $(\mathbf{I} - \mathbf{P}_1)\mathbf{h}_2$, and then normalizing it using (9.2.27). Consider now

$$\begin{aligned} r_{33}\mathbf{q}_3 &= \mathbf{h}_3 - r_{13}\mathbf{q}_1 - r_{23}\mathbf{q}_2 = \mathbf{h}_3 - (\mathbf{q}_1^T\mathbf{h}_3)\mathbf{q}_1 - (\mathbf{q}_2^T\mathbf{h}_3)\mathbf{q}_2 \\ &= \mathbf{h}_3 - (\mathbf{q}_1\mathbf{q}_1^T)\mathbf{h}_3 - (\mathbf{q}_2\mathbf{q}_2^T)\mathbf{h}_3 = (\mathbf{I} - \mathbf{P}_1 - \mathbf{P}_2)\mathbf{h}_3 \end{aligned}$$

where $\mathbf{P}_i = \mathbf{q}_i\mathbf{q}_i^T$ is the rank-one orthogonal projection matrix onto Span($\mathbf{q}_i$).

(a) Since $\{\mathbf{q}_1, \mathbf{q}_2\}$ are orthonormal, verify

$$(\mathbf{I} - \mathbf{P}_1 - \mathbf{P}_2) = (\mathbf{I} - \mathbf{P}_1)(\mathbf{I} - \mathbf{P}_2).$$

Hint: $\mathbf{P}_1\mathbf{P}_2 = (\mathbf{q}_1\mathbf{q}_1^T)(\mathbf{q}_2\mathbf{q}_2^T) = \mathbf{q}_1(\mathbf{q}_1^T\mathbf{q}_2)\mathbf{q}_2^T = 0.$

(b) Also verify that

$$(\mathbf{I} - \mathbf{P}_1 - \mathbf{P}_2) = (\mathbf{I} - \mathbf{P}_2 - \mathbf{P}_1) = (\mathbf{I} - \mathbf{P}_2)(\mathbf{I} - \mathbf{P}_1)$$

and that

$$r_{33}\mathbf{q}_3 = (\mathbf{I} - \mathbf{P}_2)(\mathbf{I} - \mathbf{P}_1)\mathbf{h}_3$$

that is, $\mathbf{q}_3$ is obtained by nested projections of $\mathbf{h}_3$ and then normalizing the result.

(c) Verify that $\mathbf{q}_j$ in (9.2.27) can be calculated as

$$(r_{jj}\mathbf{q}_j) = (\mathbf{I} - \mathbf{P}_{j-1})(\mathbf{I} - \mathbf{P}_{j-2}) \cdots (\mathbf{I} - \mathbf{P}_2)(\mathbf{I} - \mathbf{P}_1)\mathbf{h}_j$$

where $\mathbf{P}_i = \mathbf{q}_i\mathbf{q}_i^{\mathrm{T}}$ for $i = 1, 2, \ldots, j-1$, is a rank-one orthogonal projection onto Span($\mathbf{q}_i$).

9.8 Exercise 9.7 is the basis for the *modified Gram–Schmidt* algorithm which is described in here. We first rewrite the result in the above exercise as follows.

$$\mathbf{q}_1 = \frac{\mathbf{h}_1}{r_{11}},\; r_{11} = \|\mathbf{h}_1\|$$
$$(r_{22}\mathbf{q}_2) = (\mathbf{I} - \mathbf{P}_1)\mathbf{h}_2$$
$$(r_{33}\mathbf{q}_3) = (\mathbf{I} - \mathbf{P}_2)(\mathbf{I} - \mathbf{P}_1)\mathbf{h}_3$$
$$(r_{44}\mathbf{q}_4) = (\mathbf{I} - \mathbf{P}_3)(\mathbf{I} - \mathbf{P}_2)(\mathbf{I} - \mathbf{P}_1)\mathbf{h}_3$$
$$\vdots$$
$$(r_{nn}\mathbf{q}_n) = (\mathbf{I} - \mathbf{P}_{n-1})(\mathbf{I} - \mathbf{P}_{n-2}) \cdots (\mathbf{I} - \mathbf{P}_2)(\mathbf{I} - \mathbf{P}_1)\mathbf{h}_n$$

Modified Gram–Schmidt algorithm:

for $i = 1$ to n
  $v_i = \mathbf{h}_i$
end
for $j = 1$ to n
  $r_{jj} = \|v_j\|$
  $\mathbf{q}_j = v_j / r_{jj}$
  for $i = j + 1$ to n
    $r_{ji} = \mathbf{q}_j^{\mathrm{T}} v_i$
    $v_i = v_i - r_{ji}\mathbf{q}_j$
  end
end

Verify that this algorithm is correct.

Notes and references

The topics discussed in this chapter lie at the heart of numerical linear algebra. There are several excellent text books on these topics – Trefethen and Bau (1997), Golub and van Loan (1989), and Higham (1996).

There are basically two classes of methods for solving linear systems – *direct* and *iterative* methods. Direct methods are based on the multiplicative decomposition and the **LU**, Cholesky, **QR**, and singular value decomposition belong to this category. Ortega (1988), and Golub and Ortega (1993) provide a comprehensive coverage of both serial and parallel versions of direct methods.

The classical iterative techniques are based on the *additive decomposition* and the Jacobi, Gauss–Siedel, successive over-relaxation (SOR), symmetric successive over-relaxation (SSOR) belong to this category. Varga (2000), Young (1971), Hageman and Young (1981) provide a thorough and a comprehensive treatment of iterative methods.

Exploiting the intrinsic relation between the quadratic minimization problem and the solution of symmetric positive definite linear systems, Hestenes and Stiefel (1952) developed the conjugate gradient method (Chapter 12), which was originally considered as a direct method. The revival of the classical conjugate gradient method as an iterative technique has spurred enormous interest in generalizing this class of methods. This effort has led to the modern theory of iterative methods based on Krylov subspace techniques. Refer to Greenbaum (1997) and Brauset (1995).

10

Optimization: steepest descent method

In Chapters 5 and 7 the least squares problem – minimization of the residual norm, $f(\mathbf{x}) = \|\mathbf{r}(\mathbf{x})\|$ where $\mathbf{r}(\mathbf{x}) = (\mathbf{z} - \mathbf{H}\mathbf{x})$ in (5.1.11) and (7.1.3) with respect to the state variable $\mathbf{x}$ is formulated. There are essentially two mathematically equivalent approaches to this minimization. In the first, compute the gradient $\nabla f(\mathbf{x})$ and obtain (the minimizer) $\mathbf{x}^*$ by solving $\nabla f(\mathbf{x}) = \mathbf{0}$. We then check if the Hessian $\nabla^2 f(\mathbf{x}^*)$ is positive definite to guarantee that $\mathbf{x}^*$ is indeed a local minimum. In the linear least squares problem in Chapter 5, $f(\mathbf{x})$ is a quadratic function $\mathbf{x}$ and hence $\nabla f(\mathbf{x}) = \mathbf{0}$ leads to the solution of a linear system of the type $\mathbf{A}\mathbf{x} = \mathbf{b}$ with $\mathbf{A}$ a symmetric and positive definite matrix (refer to (5.1.17)) which can be solved by the methods described in Chapter 9. In the nonlinear least squares problem, $f(\mathbf{x})$ may be highly nonlinear (far beyond the quadratic nonlinearity). In this case, we can compute $\mathbf{x}^*$ by solving a nonlinear algebraic system given by $\nabla f(\mathbf{x}) = \mathbf{0}$, and then checking for the positive definiteness of the Hessian $\nabla^2 f(\mathbf{x}^*)$. Alternatively, we can approximate $f(\mathbf{x})$ locally around a current operating point, say, $\mathbf{x}_c$ by a quadratic form $\mathbf{Q}(y)$ (using either the first-order or the second-order method described in Chapter 7) where $\mathbf{y} = (\mathbf{x} - \mathbf{x}_c)$. We then find the minimizer $\mathbf{y}^*$ for this approximating quadratic form much like the linear least squares problem. This process is repeated by resetting $\mathbf{x}_c \longleftarrow \mathbf{x}_c + \mathbf{y}^*$ until convergence to $\mathbf{x}^*$, the minimum for the original $f(\mathbf{x})$.

The second approach which is the topic of this chapter is a direct iterative approach to minimizing $f(\mathbf{x})$ in which we generate a sequence $\mathbf{x}_0, \mathbf{x}_1, \mathbf{x}_2, \ldots$ converging in the limit to $\mathbf{x}^*$, the minimizer of $f(\mathbf{x})$ with the property that for $k = 0, 1, 2, \ldots, (\mathbf{x}_{k+1} - \mathbf{x}_k)$ is a **descent** direction that is, $f(\mathbf{x}_{k+1}) < f(\mathbf{x}_k)$ for all $k \geq 0$. A little reflection would reveal that the descent direction must have a non-zero projection (see Appendices A and B for concepts related to projection) with the negative gradient. There is a vast body of literature dealing with this class of algorithms for minimization. Our aim in this chapter is to provide an introduction to the basic ideas leading to the design of these algorithms. Conditions characterizing the minima are developed in Appendix C.

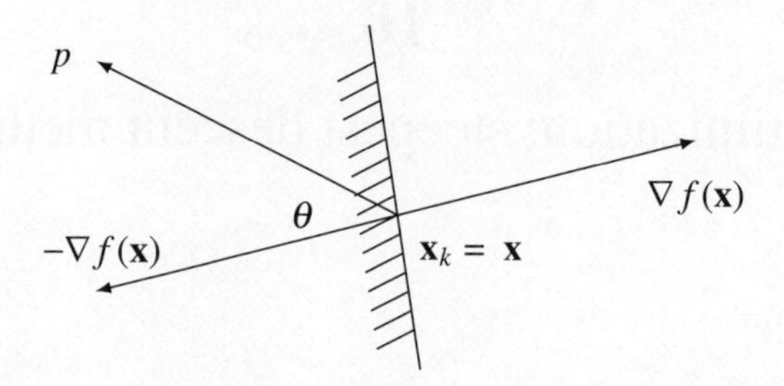

Fig. 10.1.1 A descent direction.

10.1 An iterative framework for minimization

Let $f : \mathbb{R}^n \longrightarrow \mathbb{R}$ be the scalar-valued function of a vector (also called a **functional**) to be minimized, and let $\mathbf{x}^*$ be a (local) minimizer of $f(\mathbf{x})$. The iterative framework for minimization seeks to generate a sequence $\mathbf{x}_0, \mathbf{x}_1, \mathbf{x}_2, \ldots, \mathbf{x}_k, \ldots$ in $\mathbb{R}^n$ satisfying the following two conditions

$$f(\mathbf{x}_{k+1}) < f(\mathbf{x}_k) \tag{10.1.1}$$

and

$$\lim_{k\to\infty} \mathbf{x}_k = \mathbf{x}^*. \tag{10.1.2}$$

That is, the value of the function $f(\mathbf{x})$ **monotonically decreases** along the sequence which in the limit **converges** to the minimizer. Any mechanism for generating such a sequence is said to be based on a **greedy strategy**.

Let $\mathbf{x}_k = \mathbf{x}$ be such that $\nabla f(\mathbf{x}) \neq \mathbf{0}$. That is, $\mathbf{x}_k$ is **not** a minimizer of $f(\mathbf{x})$. Let $\mathbf{p} \in \mathbb{R}^n$ be a direction such that

$$\langle \mathbf{p}, \nabla f(\mathbf{x})\rangle = \mathbf{p}^{\mathrm{T}}\nabla f(\mathbf{x}) < 0. \tag{10.1.3}$$

That is, the direction $\mathbf{p}$ has a non-zero **projection** (Appendix A) on the negative of the gradient of $f(\mathbf{x})$. See Figure 10.1.1 for an illustration. Since $\mathbf{p}^{\mathrm{T}}\nabla f(\mathbf{x})$ is proportional to the **directional derivative** of $f(\mathbf{x})$ (Appendix C) in the direction $\mathbf{p}$, (10.1.3) implies that we can **reduce** the value of $f(\mathbf{x})$ by moving along this direction $\mathbf{p}$. Hence, such a direction $\mathbf{p}$ has come to be known as the **descent** direction. Given a descent direction $\mathbf{p}$ at $\mathbf{x}_k = \mathbf{x}$, we can now find a (sufficiently small) positive constant α, called the **step length** parameter such that

$$\mathbf{x}_{k+1} = \mathbf{x}_k + \alpha\mathbf{p} \tag{10.1.4}$$

satisfies the inequality (10.1.1). Bravo! we now have a framework for iterative minimization in place which is described in Figure 10.1.2.

We now elaborate on the components that make up this iterative framework.

(a) Specification and properties of function to be minimized The function $f(\mathbf{x})$ to be minimized can be specified in two distinct ways. In the first, $f(\mathbf{x})$ is

Given $f(\mathbf{x})$ and an initial point $\mathbf{x}_0$, such that $\nabla f(\mathbf{x}_0) \neq \mathbf{0}$.
For $k = 0, 1, 2, 3, \ldots$
Step 1 Find a descent direction $\mathbf{p}$ at $\mathbf{x}_k$, that is,
$\mathbf{p}$ satisfying (10.1.3) at $\mathbf{x}_k = \mathbf{x}$.
Step 2 Find a suitable value of the parameter α such that
$\mathbf{x}_{k+1} = \mathbf{x}_k + \alpha\mathbf{p}$ satisfies (10.1.1)
Step 3 Test for convergence. If satisfied, set $\mathbf{x}^* = \mathbf{x}_{k+1}$
and Exit, else go to Step 1.

Fig. 10.1.2 A general iterative framework.

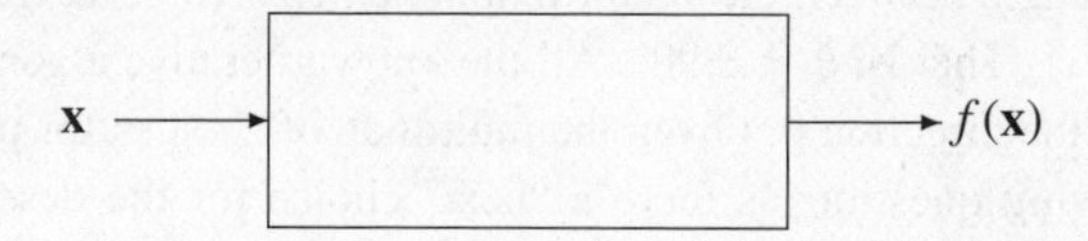

Fig. 10.1.3 A black box representing $f(\mathbf{x})$.

given explicitly, say, for example, as a quadratic form

$$f(\mathbf{x}) = \frac{1}{2}\mathbf{x}^{\mathrm{T}}\mathbf{A}\mathbf{x} - \mathbf{b}^{\mathrm{T}}\mathbf{x} \tag{10.1.5}$$

where $\mathbf{A} \in \mathbb{R}^{n\times n}$ is a symmetric and positive definite matrix and $\mathbf{b} \in \mathbb{R}^n$. In this case, we can pre-compute the quantities of interest, namely the gradient, $\nabla f(\mathbf{x})$ and the Hessian $\nabla^2 f(\mathbf{x})$, etc. This greatly facilitates the numerical evaluation of these quantities during the course of the algorithm. In the second method, $f(\mathbf{x})$ may be given only as a black box, as in Figure 10.1.3.

That is, we can only get the value of $f(\mathbf{x})$ as the output for a given input $\mathbf{x}$ and there is no recourse to obtaining the functional form of $f(\mathbf{x})$. In this case, the quantities of interest such as gradient and Hessian are calculated numerically by invoking one of the many **finite difference** approximations. For example,

$$\frac{\partial f}{\partial x_i} = \frac{f(x_i + h) - f(x_i)}{h}$$

and

$$\frac{\partial^2 f}{\partial x_i^2} = \frac{f(x_i + h) - 2f(x_i) + f(x_i - h)}{h^2}. \tag{10.1.6}$$

In addition, the function $f(x)$, in either specification may possess **multiple minima**. The iterative framework described above is only suitable for finding a **local minimum**. There are several promising frameworks for finding a global minimum in a multi-minima case. Discussion of these techniques is beyond our scope and we refer the reader to the literature for details.

(b) Choice of initial conditions The general rule is that the iterates generated by the general framework given above converge to a minimum that is **closest** to the initial starting point. Hence, the choice of the initial condition $\mathbf{x}_0$ is very critical in determining the limit to which the iterates converge as well as the number of iterations needed for such a convergence. A clever choice of $\mathbf{x}_0$ must be based on all the a priori information we have about $f(\mathbf{x})$.

(c) Choice of descent direction Looking at the picture in Figure 10.1.1, we can infer that any vector $\mathbf{p}$ that lies in the left half of the perpendicular to $\nabla f(\mathbf{x})$ at $\mathbf{x}_k = \mathbf{x}$ (shown as the hatched region) is a candidate for the descent direction. Clearly, a sufficient condition for a descent direction is that

$$-\frac{\mathbf{p}^{\mathrm{T}}\nabla f(\mathbf{x})}{\|\mathbf{p}\|\,\|\nabla f(\mathbf{x})\|} = \cos\theta \geq \delta > 0 \tag{10.1.7}$$

where θ is the angle between the vector $\mathbf{p}$ and the negative of the gradient $\nabla f(\mathbf{x})$, see Figure 10.1.1. That is, $\theta < \pm 90°$. All the known iterative algorithms differ in their choice of the direction $\mathbf{p}$. Given the multitude of choices for $\mathbf{p}$, we are faced with the following question: Is there a "best" choice for the descent direction? where best is to be interpreted in the sense of enabling a maximum reduction in the value of $f(\mathbf{x})$. The answer is "yes" and to this end recall that the gradient $\nabla f(\mathbf{x})$ represents the direction of **maximum rate of change** in $f(\mathbf{x})$. Thus, $\mathbf{p} = -\nabla f(\mathbf{x})$ would guarantee a maximum reduction locally. Accordingly,

$$\mathbf{x}_{k+1} = \mathbf{x}_k - \alpha\nabla f(\mathbf{x}) \tag{10.1.8}$$

has come to be known as the **steepest descent** or the **gradient** algorithm. The steepest descent algorithm was developed by Cauchy in 1847.

There are numerous other choices for $\mathbf{p}$ – the **conjugate gradient algorithm**, **Newton's algorithm**, a whole host of **quasi-Newton algorithms, trust region methods**, to mention a few, all characterized by the special choice of this direction $\mathbf{p}$. Some of these will be described in Chapters 11 and 12.

(d) Line search and step length Given a descent direction $\mathbf{p}$, the emphasis then shifts to finding the **best** value (in the sense of bringing the **maximum reduction** in the value of the function $f(\mathbf{x})$) of the step length parameter α at the current operating point $\mathbf{x}_k$ and along the chosen direction $\mathbf{p}$. Given $\mathbf{x}_k = \mathbf{x}$, and $\mathbf{p}$ define $g : \mathbb{R} \longrightarrow \mathbb{R}$ where

$$g(\alpha) = f(\mathbf{x} + \alpha\mathbf{p}). \tag{10.1.9}$$

Thus, finding the best value of α reduces to solving a one-dimensional minimization problem – minimization of $g(\alpha)$ in (10.1.9). Refer to Figure 10.1.4. Clearly, the minimizing α_k is obtained as the solution of

$$\frac{\mathrm{d}g}{\mathrm{d}\alpha} = [\nabla f(\mathbf{x} + \alpha\mathbf{p})]^{\mathrm{T}}\mathbf{p} = 0. \tag{10.1.10}$$

That is, the best value of α is one that renders the current descent direction $\mathbf{p}$ orthogonal to $\nabla f(\mathbf{x}_{k+1})$ where $\mathbf{x}_{k+1} = \mathbf{x}_k + \alpha\mathbf{p}$. Herein lies the fundamental principle of

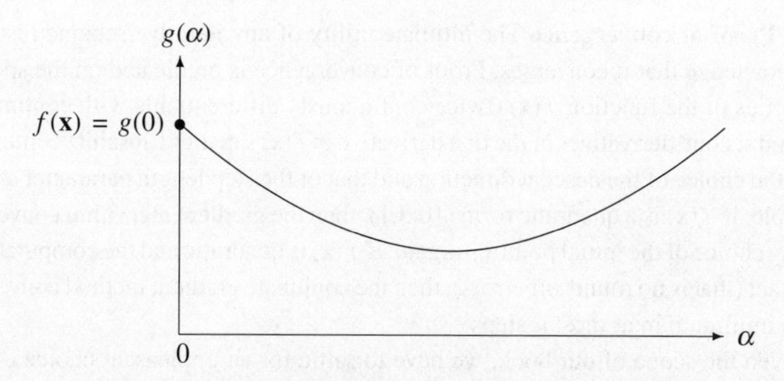

Fig. 10.1.4 Linear search at $\mathbf{x}_k = \mathbf{x}$ along the descent direction $\mathbf{p}$.

the design of minimization algorithms (based on the **divide and conquer strategy**) in which a minimization in the multidimensional space $\mathbb{R}^n$ is reduced to a sequence of one-dimensional minimization problems which are far easier to tackle than the original problem.

If α is a **constant** across the iterations, then it is called a **stationary** iteration. In a **non-stationary** algorithm, α changes with k. In the gradient algorithm (10.1.8), the value of α that minimizes $g(\alpha)$ in (10.1.9) depends on $\mathbf{x}_k = \mathbf{x}$, $\mathbf{p}$ and the properties of $f(\mathbf{x})$ along $\mathbf{p}$ and hence is a non-stationary algorithm. In fact, most of the best-known minimization algorithms are non-stationary. Specific details of minimization of $g(\alpha)$ are discussed in Section 10.4.

(e) Test for convergence and scaling Almost all the tests for convergence are essentially derived from the necessary and sufficient conditions for the minimum, namely $\nabla f(\mathbf{x}) = 0$, and $\nabla^2 f(\mathbf{x})$ is positive definite (Appendix C). We enlist several choices for the convergence test. Let $\epsilon > 0$ be a pre-specified tolerance – a good choice is the machine precision: $\epsilon = 10^{-7}$ for the single and $\epsilon = 10^{-15}$ for the double precision arithmetic.

(i) $\|\nabla f(\mathbf{x})\| \leq \epsilon$
(ii) $\|\mathbf{x}_{k+1} - \mathbf{x}_k\| \leq \epsilon$
(iii) $\|f(\mathbf{x}_{k+1}) - f(\mathbf{x}_k)\| \leq \epsilon$
(iv) $\nabla^2 f(\mathbf{x}^*)$ is positive definite, where $\mathbf{x}^*$ is an estimate of the minimum

Since calculation of the norm involves square root operation, it is often convenient to use the square of the norm instead of the norm itself. In the design of reliable software one may want to combine several of these conditions,

$$(v)\ \|\nabla f(\mathbf{x})\| \leq [1 + |f(\mathbf{x})|]\epsilon.$$

Recall that the norm of the gradient is sensitive to the scaling of both the independent variable $\mathbf{x}$ and the dependent quantity $f(\mathbf{x})$. So, great care must be exercised in checking for the effect of the scaling on the convergence tests.

(f) Proof of convergence The ultimate utility of any iterative scheme rests on the knowledge that it converges. Proof of convergence is predicated on the special properties of the function $f(\mathbf{x})$ (twice continuously differentiable with continuous first and second derivatives or the first derivative of $f(\mathbf{x})$ satisfies Lipschitz condition, etc.), the choice of the descent direction and that of the step length parameter α. For example, if $f(\mathbf{x})$ is a quadratic form (10.3.1), then the gradient algorithm converges for any choice of the initial point $\mathbf{x}_0$. Again, if $f(\mathbf{x})$ is quadratic and the computations are exact (that is no round-off errors), then the conjugate gradient method converges to the minimum in at most n steps.

Given the scope of our book, we have to settle for an unpleasant choice of not indulging in the details of the proof of convergence. We refer the reader to many readable texts given at the end of this chapter for details.

(g) Rate of convergence Once the convergence is guaranteed, our curiosity shifts to quantifying the **rate of convergence**. The usefulness of this rate information is that we can pre-compute the number of iterations, n^*, needed to achieve a pre-specified tolerance used in the stopping condition. Detailed discussion of the rate of convergence is given in Section 10.2.

(h) Time–space requirements The amount of **time** required per iteration is often characterized by the amount of **work** (measured in terms of the **number of basic operations** – add/subtract, multiply/divide) to be done in each iteration. This when multiplied by n^*, the number of iterations needed to achieve a pre-specified tolerance obtained using the rate information, provides an estimate of the cost (often measured in megaflops) of minimization. Also a good estimate of the memory space requirement (measured in megabytes) during each iteration would help the choice of computer configuration needed in the successful implementation of these algorithms.

(i) Serial vs. parallel computers With ever-increasing speed of the underlying hardware and available RAM (random access memory) and decreasing cost of the computer hardware, problems that dictated the use of supercomputers (costing multi-millions of dollars) can now be solved on a desk/laptop computer. This unprecedented growth in the computer technology has pushed the envelope so much that the class of problems requiring truly large and expensive machines is continuously changing. The data assimilation problems of interest in this book are some of the few examples that require truly large machines. Parallel computing is the only answer to numerically solving large-scale problems of interest in weather forecasting. In fact, most of the national centers across the globe have already switched over to using large distributed memory architectures to produce their daily forecasts.

Implementing large-scale minimization problems of interest in data assimilation in a distributed memory environment with a view to speed up the overall computation provides interesting problems of its own. Again, given our scope and limitation of space, we shall not indulge in this direction.

10.2 Rate of convergence

Let $\mathbf{x}_0, \mathbf{x}_1, \mathbf{x}_2, \ldots$ be a sequence of iterates in $\mathbb{R}^n$ converging to $\mathbf{x}^* \in \mathbb{R}^n$. That is $\mathbf{x}^*$ is known as the **limit** of this sequence denoted by

$$\lim_{k\to\infty} \mathbf{x}_k = \mathbf{x}^*. \tag{10.2.1}$$

Let $\mathbf{e}_k = \mathbf{x}_k - \mathbf{x}^*$ denote the **error** in the kth iterate $\mathbf{x}_k$. If the sequence $\mathbf{e}_k$ is such that for some real constants $p > 0$ and $0 \le q < \infty$,

$$\lim_{k\to\infty} \frac{\|\mathbf{e}_{k+1}\|}{\|\mathbf{e}_k\|^p} = q, \tag{10.2.2}$$

then $\mathbf{x}_k$ is said to converge to $\mathbf{x}^*$ at a **rate** (or **order**) p with a **rate constant** q. This requirement implies that there exists a constant k^* such that for all $k > k^*$ (that is, the tail of the sequence)

$$\|\mathbf{e}_{k+1}\| \approx q\, \|\mathbf{e}_k\|^p\,. \tag{10.2.3}$$

In other words, while definition (10.2.2) does **not** restrict the **initial** (**transient**) behavior of $\mathbf{x}_k$, it prescribes a monotonic behavior for the **tail** of the sequence $\mathbf{x}_k$.

In the following, we isolate and describe three important convergence classes of interest in practical minimization.

(A) Linear convergence Any sequence $\mathbf{x}_k$ satisfying (10.2.2) with $p = 1$ and $q < 1$, for $k > k^*$, that is,

$$\|\mathbf{e}_{k+1}\| = q\, \|\mathbf{e}_k\| \text{ for all } k > k^* \tag{10.2.4}$$

is said to exhibit **linear rate** of convergence. To fix the ideas, we illustrate using two examples.

Example 10.2.1 Let $x_k = a^k$ for some $0 < a < 1$. Then, $x^* = 0$ and $k^* = 0$, with $e_k = x_k$. Thus,

$$x_{k+1} = a^{k+1} = a.a^k = ax_k \text{ for all } k > 0$$

from which we obtain $q = a < 1$, and $p = 1$. Hence, this sequence converges to zero at a **linear rate** or exhibits **order 1** convergence. Since this is also a **geometric sequence**, it is often conceptually beneficial to view the **linear convergence** as the proverbial **geometric convergence**.

Example 10.2.2 Let $x_k = \frac{1}{k}$. Here again, $x^* = 0$ and $x_k = e_k$. From

$$\frac{x_{k+1}}{x_k} = \frac{k}{k+1} \longrightarrow 1 \text{ as } k \longrightarrow \infty,$$

it follows that $p = 1$ and $q = 1$. While this **harmonic** sequence converges to zero, its convergence rate is **not** linear.

Table 10.2.1 *Convergence analysis*

q	n^*
0.1	7
0.2	11
0.4	18
0.6	32
0.8	73
0.9	153
0.99	1,604
0.999	16,111

One can use (10.2.4) to pre-compute the number of iterations needed to achieve a desirable tolerance limit. Let $\epsilon = 10^{-d}$ for some integer $d > 1$ ($d = 7$ for single precision and $d = 15$ for double precision arithmetic). By iterating (10.2.4), we obtain

$$\frac{\|e_k\|}{\|e_0\|} = q^k.$$

By requiring this ratio to be less than or equal to $\epsilon = 10^{-d}$, we obtain

$$\frac{\|e_k\|}{\|e_0\|} = q^k \leq 10^{-d} = \epsilon. \tag{10.2.5}$$

By taking the logarithm and remembering $q < 1$, we obtain†

$$k \geq \lceil \frac{d}{\log_{10}(q^{-1})} \rceil = n^*. \tag{10.2.6}$$

Clearly, the right-hand side of (10.2.6) gives the **minimum** number, n^*, of iterations needed to achieve the required tolerance limit. Table 6.2.1 gives typical values of q for $d = 7$.

Thus, when $q = 0.99$, we would require over 1600 iterations to obtain single precision accuracy.

The steepest descent method for minimization is known to converge at a linear rate as will be shown in Section 10.3.

(B) Quadratic convergence Any sequence satisfying (10.2.2) with $p = 2$ for any $k > k^*$ is said to exhibit **quadratic convergence** or convergence of **order 2**.

Example 10.2.3 Let $x_k = a^{2^k}$ where $0 < a < 1$. Then, x^* and $k^* = 0$ with $x_k = e_k$. Hence

$$x_{k+1} = a^{2^{k+1}} = a^{2^k + 2^k} = a^{2^k}.a^{2^k} = (x_k)^2 \text{ for all } k > 0,$$

† $\lceil x \rceil$ is called the ceiling of x which is the smallest integer larger than or equal to x. Thus, $\lceil 3.7 \rceil = 4$ and $\lceil -3.7 \rceil = -3$.

from which we get $q = 1$ and $p = 2$. Hence, this sequence exhibits quadratic convergence or order 2 convergence.

To fix the ideas, let $a = 10^{-1}$, that is $x_0 = a = 10^{-1}$. Then, the sequence generated becomes

$$10^{-1},\ 10^{-2},\ 10^{-4},\ 10^{-8},\ 10^{-16},\ldots$$

That is, we should be able to surpass the tolerance of $\epsilon = 10^{-6}$ in no more than 3 iterations. Another illustrative explanation of this sequence is that the number of correct digits doubles after each iteration.

It is well known that the classical Newton's method for finding square root of a real number $a > 0$ converges at a quadratic rate.

(C) Superlinear convergence Any sequence that satisfies (10.2.2) with $p = 1$ and $q = 0$ is said to attain the superlinear convergence rate.

Example 10.2.4 Let $x = (1/k)^k$. Again, in this case, we have $x^* = 0$. From

$$\frac{x_{k+1}}{x_k} = \frac{k^k}{(k+1)^{k+1}} = \frac{1}{k+1}\left(\frac{k}{k+1}\right)^k \longrightarrow 0 \text{ as } k \to \infty,$$

we get $q = 0$ and $p = 1$. Hence, this sequence exhibits superlinear convergence.

Remark 10.2.1 There is a built-in ambiguity in this definition of superlinear convergence. For example, consider the sequence in Example 10.2.3. From

$$x_{k+1} = a^{2^{k+1}} = a^{2^k}.a^{2^k} = a^{2^k} x_k$$

or

$$\frac{x_{k+1}}{x_k} = a^{2^k} \longrightarrow 0 \text{ as } k \longrightarrow \infty.$$

Thus, we get $p = 1$ and $q = 0$ and hence this also exhibits superlinear convergence. In other words, sequences exhibiting higher order convergence rate ($p > 1$) can be shown to possess superlinear convergence as well.

10.3 Steepest descent algorithm

In this section we provide an analysis of the steepest descent algorithm using a model problem of quadratic minimization.

A model problem Let $\mathbf{A} \in \mathbb{R}^{n\times n}$ be a symmetric and positive definite matrix and $\mathbf{b} \in \mathbb{R}^n$. Define $f : \mathbb{R}^n \longrightarrow \mathbb{R}$ as

$$f(\mathbf{x}) = \frac{1}{2}\mathbf{x}^{\mathrm{T}}\mathbf{A}\mathbf{x} - \mathbf{b}^{\mathrm{T}}\mathbf{x}. \tag{10.3.1}$$

This $f(\mathbf{x})$ is twice continuously differentiable and convex (Appendix C) and hence it has a unique global minimum at say, $\mathbf{x}^*$. This minimizer $\mathbf{x}^*$ is obtained as the

Given $\mathbf{x}_0$ and $r_0 = \mathbf{b} - \mathbf{A}\mathbf{x}_0$
For $k = 0, 1, 2, 3, \ldots$
Step 1 Compute $\alpha_k = \frac{r_k^T r_k}{r_k^T \mathbf{A} r_k}$
Step 2 Compute $\mathbf{x}_{k+1} = \mathbf{x}_k + \alpha_k r_k$
Step 3 Test for convergence. If yes, **EXIT**
Step 4 Compute $r_{k+1} = r_k - \alpha_k \mathbf{A} r_k$.

Fig. 10.3.1 Steepest descent algorithm.

solution of

$$\nabla f(\mathbf{x}) = \mathbf{A}\mathbf{x} - \mathbf{b} = 0 \tag{10.3.2}$$

that is, $\mathbf{x}^* = \mathbf{A}^{-1}\mathbf{b}$.

Let $\mathbf{x}_k$ be the current operating point. Then the **residual** vector (Chapter 5)

$$r_k = r(\mathbf{x}_k) = \mathbf{b} - \mathbf{A}\mathbf{x}_k = -\nabla f(\mathbf{x}_k) \tag{10.3.3}$$

represents the steepest descent direction for $f(\mathbf{x})$ at $\mathbf{x}_k$.

Line search and optimal step length Let the next iterate be given by

$$\mathbf{x}_{k+1} = \mathbf{x}_k + \alpha r_k \tag{10.3.4}$$

where α is chosen to minimize

$$\begin{aligned} \mathbf{g}(\alpha) = f(\mathbf{x}_{k+1}) &= \tfrac{1}{2}[\mathbf{x}_k + \alpha r_k]^{\mathrm{T}} \mathbf{A} [\mathbf{x}_k + \alpha r_k] - \mathbf{b}^{\mathrm{T}}[\mathbf{x}_k + \alpha r_k] \\ &= \tfrac{1}{2}[r_k^{\mathrm{T}} \mathbf{A} r_k]\alpha^2 + [(r_k^{\mathrm{T}} \mathbf{A}\mathbf{x}_k) - r_k^{\mathrm{T}}\mathbf{b}]\alpha \\ &\quad + \tfrac{1}{2}\mathbf{x}_k^{\mathrm{T}} \mathbf{A}\mathbf{x}_k - \mathbf{b}^{\mathrm{T}}\mathbf{x}_k. \end{aligned} \tag{10.3.5}$$

From

$$\frac{\mathrm{d}\mathbf{g}}{\mathrm{d}\alpha} = [r_k^{\mathrm{T}} \mathbf{A} r_k]\alpha + r_k^{\mathrm{T}}[\mathbf{A}\mathbf{x}_k - \mathbf{b}] = 0 \tag{10.3.6}$$

we obtain the minimizer α_k of $\mathbf{g}(\alpha)$ as

$$\alpha_k = \frac{r_k^{\mathrm{T}}[\mathbf{b} - \mathbf{A}\mathbf{x}_k]}{r_k^{\mathrm{T}} \mathbf{A} r_k} = \frac{r_k^{\mathrm{T}} r_k}{r_k^{\mathrm{T}} \mathbf{A} r_k}. \tag{10.3.7}$$

That is, the subproblem of the one-dimensional minimization is solved exactly in this case. Clearly, (10.3.4) and (10.3.7) together define the steepest descent algorithm stated in Figure 10.3.1.

Space/time requirements We first quantify the space/time requirements of this algorithm. In the initialization step, the **matrix-vector product**, $\mathbf{A}\mathbf{x}_0$ requires $2n^2$ operations (Appendix B) and **vector add/subtract** $\mathbf{b} - \mathbf{A}\mathbf{x}_0$ requires n operations. Then, in each iteration, we need to compute the **matrix-vector product** $\mathbf{A}r_k$ and store it since it is needed twice – once in Step 1 and again in Step 4. This would need $2n^2$ operations. In Step 1, computation of α_k requires two **inner products** – $r_k^{\mathrm{T}} r_k$

and $r_k^{\mathrm{T}}(\mathbf{A}r_k)$ and a division amounting to a total of $4n+1$ operations. Step 2 then performs a **scalar times a vector plus a vector** operation needing 2n operations. Testing in Step 3 would generally require **computation of the norm** which requires n operations. Finally, Step 4, like Step 2 performs a **scalar times a vector plus a vector** operation requiring $2n$ operations. Thus, except for the matrix-vector product, the rest of the computations require a linear $O(n)$ number of operations.

In each step we need to store three vectors, r_k, $\mathbf{A}r_k$, and $\mathbf{x}_k$. The storage and the time required to compute $\mathbf{A}r_k$ depends critically on the structure and sparsity of $\mathbf{A}$.

Orthogonality of residuals The new residual at $\mathbf{x}_{k+1}$ is given by

$$r_{k+1} = \mathbf{b} - \mathbf{A}\mathbf{x}_{k+1} = \mathbf{b} - \mathbf{A}[\mathbf{x}_k + \alpha_k r_k] = r_k - \alpha_k \mathbf{A} r_k \tag{10.3.8}$$

Now taking the inner product of both sides with r_k and using (10.3.7), we obtain

$$r_k^{\mathrm{T}} r_{k+1} = r_k^{\mathrm{T}} r_k - \alpha_k r_k^{\mathrm{T}} \mathbf{A} r_k = 0 \tag{10.3.9}$$

that is, successive residuals are orthogonal. Since r_k is also the descent direction at Step k, we can also interpret (10.3.9) as follows: the new search direction r_{k+1} is orthogonal to the earlier residual r_k.

Convergence To understand the convergence of the steepest descent algorithm, recall first that the global minimizer of $f(\mathbf{x})$ is

$$\mathbf{x}^* = \mathbf{A}^{-1}\mathbf{b}. \tag{10.3.10}$$

Let

$$\mathbf{e}_k = \mathbf{x}_k - \mathbf{x}^* \tag{10.3.11}$$

denote the **error** in the kth iterate $\mathbf{x}_k$. Since

$$\lim_{k\to\infty} \mathbf{x}_k = \mathbf{x}_0, \text{ exactly when } \lim_{k\to\infty} \mathbf{e}_k = 0$$

in the following we base the convergence analysis on $\mathbf{e}_k$. To this end, define (using $\mathbf{b} = \mathbf{A}\mathbf{x}^*$)

$$\begin{aligned} E(\mathbf{x}_k) &= f(\mathbf{x}_k) - f(\mathbf{x}^*) \\ &= [\tfrac{1}{2}\mathbf{x}_k^{\mathrm{T}}\mathbf{A}\mathbf{x}_k - \mathbf{x}^{*\mathrm{T}}\mathbf{A}\mathbf{x}_k] - [\tfrac{1}{2}\mathbf{x}^{*\mathrm{T}}\mathbf{A}\mathbf{x}^* - \mathbf{x}^{*\mathrm{T}}\mathbf{A}\mathbf{x}^*] \\ &= \tfrac{1}{2}\mathbf{x}_k^{\mathrm{T}}\mathbf{A}\mathbf{x}_k - \mathbf{x}^{*\mathrm{T}}\mathbf{A}\mathbf{x}_k + \tfrac{1}{2}\mathbf{x}^{*\mathrm{T}}\mathbf{A}\mathbf{x}^* \\ &= \tfrac{1}{2}(\mathbf{x}_k - \mathbf{x}^*)^{\mathrm{T}}\mathbf{A}(\mathbf{x}_k - \mathbf{x}^*) \\ &= \tfrac{1}{2}\mathbf{e}_k^{\mathrm{T}}\mathbf{A}\mathbf{e}_k. \end{aligned} \tag{10.3.12}$$

It can be verified that $E(\mathbf{x}_k)$ is zero exactly when $\mathbf{e}_k = 0$ (why?). Hence, we can analyze the convergence of steepest descent algorithm by analyzing the convergence of $E(\mathbf{x}_k)$. The following recursive relation for $E(\mathbf{x}_k)$ is easy to verify (Exercise 10.2):

$$E(\mathbf{x}_{k+1}) = \left[1 - \frac{[r_k^{\mathrm{T}} r_k]^2}{[r_k^{\mathrm{T}}\mathbf{A} r_k][r_k^{\mathrm{T}}\mathbf{A}^{-1} r_k]}\right] E(\mathbf{x}_k). \tag{10.3.13}$$

Now, since **A** is symmetric and positive definite, the n eigenvalues $\lambda_i, i = 1$ to n of **A** satisfy the following relation (Appendix B):

$$\lambda_1 \geq \lambda_2 \geq \lambda_3 \geq \cdots \geq \lambda_n > 0. \tag{10.3.14}$$

Against this background, we now state without proof a very basic result that is key to the convergence proof called the **Kantrovich inequality**. If **A** is a symmetric and positive definite matrix, then for any $0 \neq \mathbf{y} \in \mathbb{R}^n$, it can be shown that

$$\frac{[\mathbf{y}^T\mathbf{y}]^2}{[\mathbf{y}^T\mathbf{A}\mathbf{y}][\mathbf{y}^T\mathbf{A}^{-1}\mathbf{y}]} \geq 1 - \left[\frac{\lambda_1 - \lambda_n}{\lambda_1 + \lambda_n}\right]^2 \tag{10.3.15}$$

where λ_1 and λ_n are the maximum and the minimum eigenvalues of the matrix **A**.

Now combining (10.3.15) and (10.3.13), we get

$$E(\mathbf{x}_{k+1}) \leq \beta E(\mathbf{x}_k) \tag{10.3.16}$$

where the **rate constant**

$$\beta = \left[\frac{\lambda_1 - \lambda_n}{\lambda_1 + \lambda_n}\right]^2 = \left[\frac{\kappa_2(\mathbf{A}) - 1}{\kappa_2(\mathbf{A}) + 1}\right]^2 < 1, \tag{10.3.17}$$

where

$$\kappa_2(\mathbf{A}) = \frac{\lambda_1}{\lambda_n} \tag{10.3.18}$$

is called the **spectral condition number** of the matrix **A**. Comparing (10.3.16) with (10.2.4), it immediately follows that $E(\mathbf{x}_k)$ **converges linearly** with β as its rate constant. Iterating (10.3.16), we obtain

$$E(\mathbf{x}_k) \leq \beta^k E(\mathbf{x}_0). \tag{10.3.19}$$

Hence

$$\frac{E(\mathbf{x}_k)}{E(\mathbf{x}_0)} \leq \beta^k \leq \epsilon = 10^{-d}. \tag{10.3.20}$$

We get

$$k^* = \left\lceil \frac{d}{\log_{10} \beta^{-1}} \right\rceil \tag{10.3.21}$$

to be the minimum number of iterations needed to reduce $E(\mathbf{x}_0)$ to a desired threshold. Table 10.3.1 provides sample values for k^* for various $\kappa_2(\mathbf{A})$ for $d = 7$. Thus, for example, when $\kappa_2(\mathbf{A}) = 1000$, it would require more than 4000 iterations to converge.

The level surfaces of $f(\mathbf{x})$ are hyper-ellipsoids in $\mathbb{R}^n$. When $n = 2$, these become ellipses. In this case, the length of the semi-major and semi-minor axes are proportional to $1/\sqrt{\lambda_1}$ and $1/\sqrt{\lambda_n}$. Thus, when $\lambda_1 >> \lambda_n$, these ellipses are elongated in one direction. These ellipses become near circular when $\lambda_1 \approx \lambda_n$. This is the reason

Table 10.3.1 *Condition number, for $d = 7$*

$\kappa_2(\mathbf{A})$	β	κ^*
1	0	—
10	0.669421	40
10^2	0.960788	403
10^3	0.996008	4030
10^4	0.999600	40,288

that it takes a large number of iterations to converge when $\kappa_1(\mathbf{A})$ is quite large as illustrated in the following example.

Example 10.3.1 Let $\lambda \geq 1$ and let

$$\mathbf{A} = \begin{bmatrix} 1 & 0 \\ 0 & \lambda \end{bmatrix}.$$

Consider the minimization of

$$f(\mathbf{x}) = \frac{1}{2}\mathbf{x}^{\mathrm{T}}\mathbf{A}\mathbf{x} = \frac{1}{2}(x_1^2 + \lambda x_2^2).$$

It can be verified that

$$\nabla f(\mathbf{x}) = (x_1, \lambda x_2)^{\mathrm{T}} = -\mathbf{r}(x)$$

and that the minimizer $\mathbf{x}^* = (0, 0)^{\mathrm{T}}$ with $f(\mathbf{x}^*) = 0$. Starting from $\mathbf{x}_0 = (\lambda, 1)^{\mathrm{T}}$, apply the steepest descent algorithm. It can be verified that

$$\alpha_0 = \frac{\mathbf{r}_0^{\mathrm{T}}\mathbf{r}_0}{\mathbf{r}_0^{\mathrm{T}}\mathbf{A}\mathbf{r}_0} = \frac{2}{1+\lambda},$$

and that

$$\mathbf{x}_1 = \mathbf{x}_0 - \alpha_0 \nabla f(\mathbf{x}_0) = \frac{(\lambda - 1)}{(\lambda + 1)}(\lambda, -1)^{\mathrm{T}}.$$

Continuing these calculations, it can be verified that

$$\mathbf{x}_k = \left(\frac{\lambda - 1}{\lambda + 1}\right)^k (\lambda, (-1)^k)^{\mathrm{T}}.$$

Thus, when $\lambda = 4$, $\mathbf{x}_k = (0.6)^k(4, (-1)^k)^{\mathrm{T}}$. The actual trajectory is shown in Figure10.3.2 from which the zig-zag behavior of this algorithm is rather obvious.

Remark 10.3.1 The reason for slow convergence To understand the reason for the slow convergence rate of the steepest descent algorithm, recall from (10.3.9) that the successive search directions (which are the residuals) are orthogonal to each

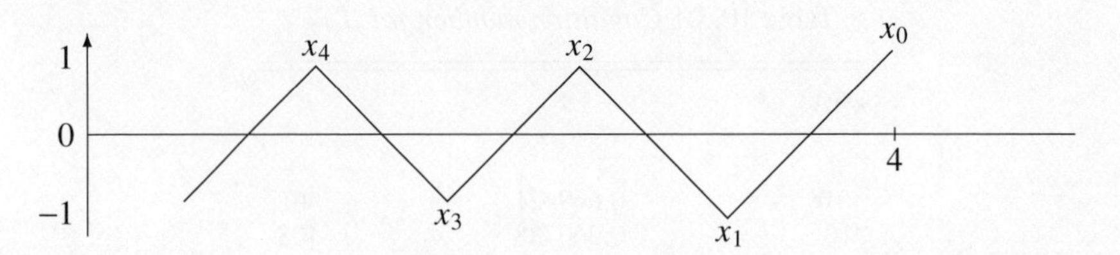

Fig. 10.3.2 Zig-zag behavior of the steepest descent algorithm.

other. But, in the above example, something more is happening. It can be verified

$$\mathbf{r}_k = \nabla f(\mathbf{x}_k) = \left(\frac{\lambda - 1}{\lambda + 1}\right)^k \lambda(1, (-1)^k)^{\mathrm{T}},$$

from which it follows that while $\mathbf{r}_k \perp \mathbf{r}_{k+1}$ for each k, it also happens that for k even the vectors $\mathbf{r}_0, \mathbf{r}_2, \mathbf{r}_4, \ldots$ all represent the same direction, and likewise for k odd, the vectors $\mathbf{r}_1, \mathbf{r}_3, \mathbf{r}_5, \ldots$ represent the same direction. Consequently, the odd iterates $\mathbf{x}_1, \mathbf{x}_3, \mathbf{x}_5, \ldots$ lie along the direction $\mathbf{r}_1$ and the even iterates $\mathbf{x}_0, \mathbf{x}_2, \mathbf{x}_4, \ldots$ lie along the direction $\mathbf{r}_0$ (Exercises 10.5 and 10.6). Formally stated, when the condition number is large, the sequence generated by the steepest descent algorithm lie in the affine subspace

$$\mathbf{x}_0 + \mathrm{Span}\{\mathbf{r}_0, \mathbf{r}_1\} \tag{10.3.22}$$

which is essentially determined by $\mathbf{x}_0$. Thus, while this algorithm nicely **divides** the n-dimensional problems into a sequence of one-dimensional problems, because the iterates are "caged" in a two-dimensional subspace, it does not exploit the full n-degrees of freedom inherent to $\mathbb{R}^n$ and hence does not conquer this space easily. The conjugate gradient algorithm described in Chapter 12 overcomes this difficulty by requiring (a) the successive search directions are gradient related and (b) they are also linearly independent.

10.4 One-dimensional search

In this section we describe a broad set of guidelines for computing the step length parameter α. Referring to (10.1.9), we see that α minimizes

$$\mathbf{g}(\alpha) = f(\mathbf{x} + \alpha\mathbf{p}) \tag{10.4.1}$$

where $\mathbf{x}_k = \mathbf{x}$ is the current operating point and $p_k = p$ is the current descent direction at $\mathbf{x}$. In principle, the best value of α is the one that minimizes $\mathbf{g}(\alpha)$ and is obtained by solving

$$\frac{\mathrm{d}g}{\mathrm{d}\alpha} = [\nabla f(\mathbf{x} + \alpha\mathbf{p})]^{\mathrm{T}}\mathbf{p} = 0. \tag{10.4.2}$$

Table 10.4.1

k	0	1	2	3	4	5
x_k	2	$-\frac{3}{2}$	$\frac{5}{4}$	$-\frac{9}{8}$	$\frac{17}{16}$	$-\frac{33}{32}$
$f(x_k)$	4	2.25	1.5625	1.2656	1.1289	1.0635

Except in special cases, where we may have complete prior knowledge of the properties of $f(\mathbf{x})$ (witness, $f(\mathbf{x})$ is a quadratic form), (10.4.2) is often solved numerically and could take considerable effort. Since this one-dimensional minimization problem is to be repeated in every step until convergence, the cost of this step is a major component in deciding the overall cost of minimization. Our overall aim is to obtain a provably convergent iterative method for minimization whose total cost is not prohibitively large. This desire to obtain guaranteed convergence at a lower cost promotes considerations for **trade-off** between **speed** and **accuracy**. That is, can we afford to settle for an easily computable but a near optimum value of α instead of the true minimum α which could take considerable effort? It turns out that a good near-optimal value of α is sufficient to guarantee overall convergence of many minimization algorithms. While our scope prevents us from indulging in the proof of these claims, in the following we summarize the algorithmic aspects of this theory.

Given $\mathbf{x}_k = \mathbf{x}$ and $\mathbf{p}_k = \mathbf{p}$, an obvious necessary condition on α is that

$$f(\mathbf{x} + \alpha \mathbf{p}) < f(\mathbf{x}). \tag{10.4.3}$$

However, this is **not** sufficient for convergence as the following two examples illustrate.

Example 10.4.1 Let $f(\mathbf{x}) = x^2$, $x_0 = 2$, $\alpha_k = 2 + 3 * 2^{-(k+1)}$, and $p_k = (-1)^{k+1}$. The first few iterates of the algorithm

$$x_{k+1} = x_k + \alpha_k p_k$$

are given in Table 10.4.1.

Several observations are in order. Notice that 0 is the minimum of $f(\mathbf{x})$. From the fact that p_k is -1 when x_k is positive and $p_k = +1$ when x_k is negative, it follows that p_k is a descent direction. It can be verified that

$$x_k = (-1)^k \left[1 + \frac{1}{2^k}\right] \quad \text{and} \quad f(x_k) = \left(1 + \frac{1}{2^k}\right)^2.$$

Hence

$$\lim_{k\to\infty} |x_k| = |x^*| = 1 \quad \text{and} \quad \lim_{k\to\infty} f(x_k) = 1,$$

which implies that the algorithm does **not** converge.

Table 10.4.2

k	0	1	2	3	4	5
x_k	2	$\frac{3}{2}$	$\frac{5}{4}$	$\frac{9}{8}$	$\frac{17}{16}$	$\frac{33}{32}$
$f(x_k)$	4	2.25	1.5625	1.2656	1.1289	1.0635

The reason is that the reduction $|f(x_{k+1}) - f(x_k)|$ is much smaller than the step length $|x_{k+1} - x_k|$ (Exercise 10.7). That is, the decrease in $f(x)$ is not commensurate with the step length.

Example 10.4.2 Let $f(x) = x^2$, $x_0 = 2$, $p_k = -1$, and $\alpha_k = 2^{-(k+1)}$. Table 10.4.2 provides values of the first few iterates of the algorithm

$$x_{k+1} = x_k + \alpha_k p_k$$

Clearly, $\lim_{k\to\infty} x_k = 1$, and $\lim_{k\to\infty} f(x_k) = 1$, while the true minimum is at 0 – no convergence! Here the issue is the step lengths are too small compared to the initial rate of decrease.

The import of these two counter examples is that the step length cannot be too large nor too small, and, in some sense, must be related to the initial rate of decrease of $f(\mathbf{x})$ in the direction $\mathbf{p}$. We now state two conditions that guarantee suitable upper and lower bounds on the step length parameter.

- Condition A. For some $a \in (0, 1)$, pick an $\alpha > 0$ that satisfies

$$f(x_{k+1}) \le f(x_k) + a[\nabla f(\mathbf{x}_k)]^{\mathrm{T}}(\alpha\mathbf{p}) \tag{10.4.4}$$

  which can be rewritten as

$$\frac{|f(\mathbf{x}_{k+1}) - f(\mathbf{x}_k)|}{\mathbf{x}_{k+1} - \mathbf{x}_k} < a\,\|\nabla f(\mathbf{x}_k)\|$$

  which implies that the average rate of decrease of $f(\mathbf{x})$ is a prescribed fraction of the initial rate of decrease. The inequality (10.4.4) is graphically represented in Figure 10.4.1. This condition defines an upper bound on α.
- Condition G. For some $b \in (a, 1)$ (where a is defined in condition A), pick an $\alpha > 0$ such that

$$f(\mathbf{x}_{k+1}) \ge f(\mathbf{x}_k) + b[\nabla f(\mathbf{x}_k)]^{\mathrm{T}}(\alpha\mathbf{p}). \tag{10.4.5}$$

  This condition is graphically illustrated in Figure 10.4.1, and it defines a lower bound on α.

Since $b > a$, these two conditions can coexist, and together define a suitable range for α. The key conclusion is that if α is chosen to lie in the region shown in Figure 10.4.1, then it is sufficient to guarantee convergence of the overall algorithm.

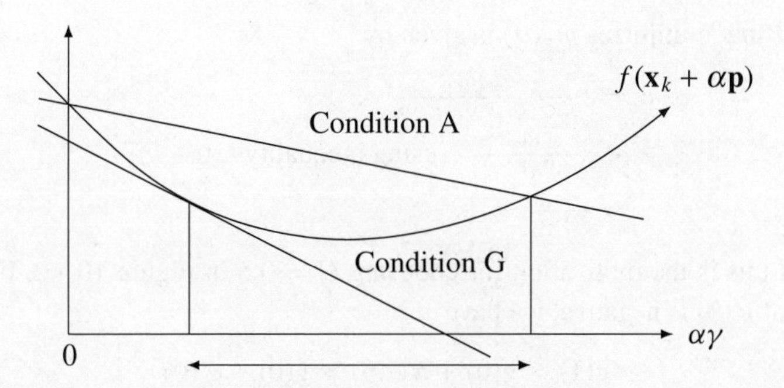

Fig. 10.4.1 Combined conditions A and G.

Choose $a \in (0, 0.5)$. Pick constants L and U, such that $0 < L < U < 1$.
Let $\mathbf{p}_k$ be the descent direction
/* suggested values are: $a = 10^{-4}$, $L = 0.1$, and $U = 0.5$. */

$\alpha_k = 1.0$
while $\{f(\mathbf{x}_{k+1}) > f(\mathbf{x}_k) + a[\nabla f(\mathbf{x}_k)]^T(\alpha_k \mathbf{p}_k)\}$
/* Condition A is violated */
$\alpha_k \leftarrow \rho\alpha_k$ for some $\rho \in [L, U]$
/* reduce step length */
end while
$\mathbf{x}_{k+1} = \mathbf{x}_k + \alpha_k \mathbf{p}_k$.

Fig. 10.4.2 Backtracking algorithm.

There are countless ways to implement these ideas and we conclude this section with a discussion of a simple and yet elegant **backtracking** algorithm. The basic idea is to first choose $\alpha_k = 1$ and compute $\mathbf{x}_{k+1} = \mathbf{x}_k + \alpha_k \mathbf{p}_k$. If this $\mathbf{x}_{k+1}$ is not acceptable in the sense of Condition A, then backtrack by reducing α_k. Since α_k is reduced from a larger value, the problem of too small a step will not arise and so the Condition G is not explicitly used in this approach. A conceptual version of the backtracking algorithm is given in Figure 10.4.2.

We now describe a practical version of this idea.

- Step 1 Let $\alpha = 1.0$. Given $\mathbf{x}_k$ and $\mathbf{p}_k$, compute $g(1) = f(\mathbf{x}_k + \mathbf{p}_k)$. If $f(\mathbf{x}_k + \mathbf{p}_k)$ satisfies Condition A, we are done. Otherwise, we have

$$g(1) = f(\mathbf{x}_k + \mathbf{p}_k) > f(\mathbf{x}_k) + a[\nabla f(\mathbf{x}_k)]^{\mathrm{T}}\mathbf{p} = g(0) + ag'(0) \qquad (10.4.6)$$

where $g'(\alpha)$ denotes the derivative of $g(\alpha)$. In this latter case, using the three pieces of information – $g(0)$, $g(1)$, and $g'(0)$ – model $g(\alpha)$ using a quadratic function of α as follows.

$$m_g(\alpha) = (g(1) - g(0) - g'(0))\alpha^2 + (g'(0))\alpha + g(0). \qquad (10.4.7)$$

Then $\hat{\alpha}$ that minimizes $m_g(\alpha)$ is given by

$$\begin{aligned}\hat{\alpha} &= -\frac{g'(0)}{2(g(1)-g(0)-g'(0))}\\ &< \frac{g'(0)}{2g'(0)[1-\alpha]} \quad \text{(using inequality (10.4.6))}\\ &= \frac{1}{2(1-\alpha)} < \frac{1}{2}. \end{aligned} \tag{10.4.8}$$

Indeed this is the motivation for choosing $U = 0.5$ in Figure 10.4.2. From the fact that $g'(0)$ is negative, we have

$$g(1) > g(0) + \alpha g'(0) > g(0) + g'(0). \tag{10.4.9}$$

Thus,

$$m_g''(\alpha) = (g(1) - g(0) - g'(0)) > 0$$

which, in turn, implies that $\hat{\alpha}$ in (10.4.8) is the minimum of $m_g(\alpha)$. In this case, choose $\alpha_k = \hat{\alpha}$ and go to step 2.

Remark 10.4.1 There is a need for one more test. If $g(1) > g(0)$, then $\hat{\alpha}$ determined in (10.4.8) may be very small which implies that the quadratic function is **not** a good model for $g(\alpha)$ in this region. To avoid too small values for $\hat{\alpha}$, we require that $\alpha > L = 0.1$. Thus, if in the first backtrack step $\hat{\alpha} < 0.1$, we then set $\hat{\alpha} = 0.1$.

- Step 2 Using α_k obtained in Step 1, test Condition A. If true, we are done, otherwise we need to backtrack again. In this latter case, either we can repeat the quadratic analysis of Step 1, or else we can fit a cubic polynomial using four pieces of information – $g(0)$, $g'(0)$, $g(1)$, and $g(\hat{\alpha})$. (See Exercise 10.8.)

Exercises

10.1 Compute the order of convergence and rate constant for the following sequences:
(a) $\mathbf{x}_k = \frac{1}{k^2}$
(b) $\mathbf{x}_k = \frac{1}{2^k}$
(c) $\mathbf{x}_k = \frac{1}{\log_e^k}$
(d) $\mathbf{x}_k = \frac{1}{k!}$
(e) $\mathbf{x}_k = \frac{1}{k \log k}$

10.2 Verify (10.3.13).

10.3 If $F(\mathbf{x}) = a_1 f(\mathbf{x})$, where a_1 is a positive real constant, verify that $F(\mathbf{x})$ and $f(\mathbf{x})$ have the same set of minimizers.

10.4 Let

$$\mathbf{A} = \begin{bmatrix} \lambda & a \\ a & 1 \end{bmatrix} \text{ and } \mathbf{b} = \begin{pmatrix} 1 \\ 1 \end{pmatrix}$$

and let $f(\mathbf{x}) = \frac{1}{2}\mathbf{x}^{\mathrm{T}}\mathbf{A}\mathbf{x} - \mathbf{b}^{\mathrm{T}}\mathbf{x}$ for $\mathbf{x} \in \mathbb{R}^2$.

(a) Draw the contours of $f(\mathbf{x})$ for $\lambda = 1, 10, 50, 100, 500$ for $a = 0.0, \pm 0.25, \pm 0.5$ and ± 0.75.

(b) Discuss the effect of increasing/decreasing values of a on the orientation of the ellipses.

(c) For each combination of λ and a given above, compute the spectral condition number of $\mathbf{A}$.

(d) Run the steepest descent algorithm and experimentally determine the number of iterations needed to obtain single precision accuracy.

10.5 Let $f(\mathbf{x}) = \frac{1}{2}\mathbf{x}^{\mathrm{T}}\mathbf{A}\mathbf{x} - \mathbf{b}^{\mathrm{T}}\mathbf{x}$, and $\nabla f(\mathbf{x}) = \mathbf{A}\mathbf{x} - \mathbf{b}$, and the residual $\mathbf{r}_k = \mathbf{b} - \mathbf{A}\mathbf{x}_k$. If $\mathbf{x}_k$'s are defined using the steepest descent algorithm, verify that $\mathbf{r}_{k+1} \perp \mathbf{r}_k$, and $\mathbf{r}_k \perp \mathbf{r}_{k-1}$ does **not** imply $\mathbf{r}_{k+1} \perp \mathbf{r}_{k-1}$, that is, orthogonality of the residuals is **not transitive**.

10.6 **Conjugate direction** Let $f(\mathbf{x}) = \frac{1}{2}\mathbf{x}^{\mathrm{T}}\mathbf{A}\mathbf{x} - \mathbf{b}^{\mathrm{T}}\mathbf{x}$. We say that a point $\mathbf{y} \in \mathbb{R}^n$ is **optimal for $f(\mathbf{x})$ with respect to a (non-null) direction $\mathbf{p} \in \mathbb{R}^n$**, if

$$f(\mathbf{y}) \leq f(\mathbf{y} + \alpha\mathbf{p}) \text{ for any } \alpha \in \mathbb{R}.$$

In Appendix D, it is proved that this condition is equivalent to requiring $\mathbf{p} \perp \mathbf{r}_{\mathbf{y}} = \mathbf{b} - \mathbf{A}\mathbf{y} = -\nabla f(\mathbf{y})$. Let $\mathbf{z} = \mathbf{y} + \mathbf{q}$, where $\mathbf{y}$ is optimal for $f(\mathbf{x})$ with respect to the direction $\mathbf{p}$, that is $\mathbf{p} \perp \mathbf{r}_{\mathbf{y}}$, and $\mathbf{q} \neq \mathbf{p}$. Define $\mathbf{r}_{\mathbf{z}} = \mathbf{b} - \mathbf{A}\mathbf{z} = \mathbf{b} - \mathbf{A}\mathbf{y} - \mathbf{A}\mathbf{q} = \mathbf{r}_{\mathbf{y}} - \mathbf{A}\mathbf{q}$. Verify that $\mathbf{z}$ is optimal for $f(\mathbf{x})$ with respect to the direction $\mathbf{p}$ if $\mathbf{p} \perp \mathbf{r}_{\mathbf{z}} = \mathbf{r}_{\mathbf{y}} - \mathbf{A}\mathbf{q}$. Since, $\mathbf{p} \perp \mathbf{r}_{\mathbf{y}}$, verify that this can happen if $\mathbf{p} \perp \mathbf{A}\mathbf{q}$, that is $\mathbf{p}^{\mathrm{T}}\mathbf{A}\mathbf{q} = 0$.

Note: Given a matrix $\mathbf{A}$, two directions $\mathbf{p}$ and $\mathbf{q}$, are said to be $\mathbf{A}$-conjugate if $\mathbf{p}^{\mathrm{T}}\mathbf{A}\mathbf{q} = 0$. Thus, $\mathbf{A}$-conjugacy is an extension of the notion of orthogonality. Thus, conjugacy implies transitivity. From Remark 10.3.1, it follows that the gradient method does **not** possess this transitivity property.

10.7 Compute $|f(\mathbf{x}_{k+1} - \mathbf{x}_k)|$ and $|\mathbf{x}_{k+1} - \mathbf{x}_k|$ for the Example 10.4.2 and plot their ratio as a function of k.

10.8 Define a cubic polynomial

$$g(\alpha) = a_3\alpha^3 + a_2\alpha^2 + g'(0)\alpha + g(0).$$

(a) Verify that $(a_3, a_2)^{\mathrm{T}}$ is obtained as (where $\alpha_1 = \hat{\alpha}$ and $\alpha_2 = 1$)

$$\begin{pmatrix} a_3 \\ a_2 \end{pmatrix} = \frac{1}{\alpha_1 - \alpha_2}\begin{bmatrix} \frac{1}{\alpha_1^2} & -\frac{1}{\alpha_2^2} \\ -\frac{\alpha_2}{\alpha_1^2} & +\frac{\alpha_1}{\alpha_2^2} \end{bmatrix}\begin{bmatrix} g(\alpha_1) - g(0) - \alpha_1 g'(0) \\ g(\alpha_2) - g(0) - \alpha_2 g'(0) \end{bmatrix}.$$

(b) Verify that the minimizer of $g(\alpha)$ is given by

$$\hat{\alpha} = \frac{-a_2 + \sqrt{a_2^2 - 3a_3 g'(0)}}{3a_3}.$$

10.9 Let $\mathbf{x} = (x_1, x_2)^{\mathrm{T}}$ and $\mathbf{A} = \begin{bmatrix} 1 & 0 \\ 0 & 2 \end{bmatrix}$. Consider

$$f(\mathbf{x}) = \frac{1}{2}\mathbf{x}^{\mathrm{T}}\mathbf{A}\mathbf{x}.$$

(a) Apply the steepest descent algorithm in Figure 10.3.1, and show that

$$\mathbf{x}_k = \left(\frac{1}{3}\right)^k \begin{bmatrix} 2 \\ (-1)^k \end{bmatrix}$$

where $\mathbf{x}_0 = (2, 1)^{\mathrm{T}}$.

(b) Show that $f(\mathbf{x}_{k+1}) = f(\mathbf{x}_k)/9$.

(c) Compare the rate of convergence given in (10.3.16) with the actual convergence obtained.

10.10 Consider the minimization of $f(\mathbf{x}) = \frac{1}{2}(x_1^2 + \lambda x_2^2)$, the same function considered in Example 10.3.1.

(a) Plot the contours of $f(\mathbf{x})$, where $\lambda = 4$, 9, and 50. Apply the steepest descent algorithm, and plot the trajectories for the following cases.

(b) $\mathbf{x}_0 = (\lambda, 1)^{\mathrm{T}}$, and $\lambda = 9$, and $\lambda = 50$.

(c) $\mathbf{x}_0 = (1, 1)^{\mathrm{T}}$, and $\lambda = 4, 9$ and 50.

10.11 **Generation of descent direction for a given gradient vector** Let $\nabla f(\mathbf{x})$ be the gradient of a function $f : \mathbb{R}^n \longrightarrow \mathbb{R}$. Let $\mathbf{B} \in \mathbb{R}^{n\times n}$ be a **symmetric and positive definite** matrix and define a vector $\mathbf{p}$ as the solution of $\mathbf{B}\mathbf{p} = -\nabla f(\mathbf{x})$.

(a) Verify that $\mathbf{p}$ is a descent direction for $f(\mathbf{x})$.

Hint: The inverse of a symmetric and positive definite matrix is also symmetric and positive definite.

Let $\nabla f(\mathbf{x}) = (1, 1)^{\mathrm{T}}$, and $\mathbf{B} = \begin{bmatrix} a & b \\ b & c \end{bmatrix}$, such that $b < \sqrt{ac}$. Compute the solution $\mathbf{p}$ of the linear system $\mathbf{B}\mathbf{p} = -\nabla f(\mathbf{x})$ and verify that indeed $\mathbf{p}$ is a descent direction for $f(\mathbf{x})$.

10.12 **Curvature of a function** Let $f : \mathbb{R}^n \longrightarrow \mathbb{R}$. The direction $\mathbf{p} \in \mathbb{R}$ is called the direction of **positive** or **negative curvature** if $\mathbf{p}^{\mathrm{T}}\nabla^2 f(\mathbf{x})\mathbf{p} > 0$ or < 0, respectively, where $\nabla^2 f(\mathbf{x})$ is the Hessian of $f(\mathbf{x})$. Verify that a direction of negative curvature exists exactly when at least one of the eigenvalues of $\nabla^2 f(\mathbf{x})$ is negative.

10.13 **Saddle point of a function** A point that is simultaneously the maximum and the minimum of a function in two different directions is called a saddle point of the function, Let $f(\mathbf{x}) = \frac{1}{2}(x_1^2 - x_2^2)$.

(a) Verify that the origin is a saddle point for $f(\mathbf{x})$ since it is a minimum along the x_1-axis and a maximum along the x_2-axis.

(b) Verify that any vector $\mathbf{p} = (a, 0)^T$ is a direction of positive curvature of $f(\mathbf{x})$ for any real number a. Similarly, $\mathbf{p} = (0, b)^T$ is a direction of negative curvature of $f(\mathbf{x})$ for any real number b.

(c) Let $\mathbf{p} = (1, 1+\epsilon)^T$. For what values of ϵ can we have $\mathbf{p}^T \nabla^2 f(\mathbf{x}) \mathbf{p} < 0$ and $\mathbf{p}^T \nabla f(\mathbf{x}) < 0$, that is, $\mathbf{p}$ is simultaneously a direction of negative curvature and descent direction, when $\mathbf{x} = (1, 1)^T$?

(d) Pick other values for $\mathbf{x}$ and repeat the computations in part (c). Summarize your findings.

10.14 Let $f(\mathbf{x}) = \frac{1}{3}x_1^3 + \frac{1}{2}x_1^2 + 2x_1x_2 + \frac{1}{2}x_2^2 - x_2 + 9$.

(a) Verify that $\nabla f(\mathbf{x})$ vanishes for two distinct values of $\mathbf{x}$, say $\mathbf{x}_a$ and $\mathbf{x}_b$.

(b) Evaluate the eigenvalues of the Hessian at $\mathbf{x}_a$ and $\mathbf{x}_b$.

(c) Identify which one of $\mathbf{x}_a$ and $\mathbf{x}_b$ is a maximum, minimum, or the saddle point.

(d) Plot the contours of $f(\mathbf{x})$.

10.15 Let $f(\mathbf{x}) = ax_1^2 + x_2^2 - 2x_1x_2 - 2x_2$. Compute the **stationary points**, that is, maximum, minimum and the saddle points of $f(\mathbf{x})$ for various values of a.

Notes and references

The material covered in this chapter is a part of the classic folklore in optimization. While the gradient-based algorithms are seldom used in practice, the ideas are very intuitive and help to motivate the reason for more sophisticated ideas and algorithms covered in Chapters 11 and 12.

Our coverage is an adaptation from Nash and Sofer (1996) and Dennis and Schnabel (1996). Ortega and Rhienboldt (1970) and the above two books contain an extensive coverage of one-dimensional optimization. Many of the original ideas relating to the one-dimensional search are due to Armijo (1966) and Goldstein (1967).

11

Conjugate direction/gradient methods

The major impetus for the development of conjugate direction/gradient methods stems from the weakness of the steepest descent method (Chapter 10). Recall that while the search directions which are the negative of the gradient of the function being minimized can be computed rather easily, the convergence of the steepest descent method can be annoyingly slow. This is often exhibited by the zig-zag or oscillatory behavior of the iterates. To use an analogy, there is lot of talk with very little substance. The net force that drives the iterates towards the minimum becomes very weak as the problem becomes progressively ill-conditioned (see Remark 10.3.1). The reason for this undesirable behavior is largely a result of the absence of transitivity of the orthogonality of the successive search directions (Exercise 10.5). Consequently the iterates are caged up in a smaller (two) dimensional subspace and the method is unable to exploit the full n degrees of freedom that are available at our disposal. Conjugate direction method was designed to remedy this situation by requiring that the successive search directions are mutually **A**-Conjugate (Exercise 10.6). **A**-Conjugacy is a natural extension of the classical orthogonality. It can be shown that if a set of vectors are **A**-Conjugate, then they are also linearly independent. Thus, as the iteration proceeds conjugate direction/gradient method guarantees that the iterates minimize the given function in subspaces of increasing dimension. It is this **expanding subspace** property which is a hallmark of this method that guarantees convergence in almost n steps provided that the arithmetic is exact. Conjugate gradient (CG) method is a special class of conjugate direction (CD) method where the mutually **A**-Conjugate directions are recursively derived using the gradient of the function being minimized.

As a technique for minimization CG lies somewhere in between the steepest descent methods (Chapter 10) and the Newton's family of methods (Chapter 12), in that they are faster than the steepest descent methods and do not require as much computational effort to generate the next search direction as the Newton's method. The CD/CG method was developed by Hestenes and Stiefel in 1952 and has become one of the major methods for optimization especially for quadratic problems.

In section 11.1, we describe the conjugate direction method and derive several of its properties. The classical conjugate gradient method is developed in Section 11.2. Extension of this classical algorithm for the minimization of nonlinear functions is covered in Section 11.3. Section 11.4 provides an introduction to the concept of preconditioning in the context of the classical CG method.

11.1 Conjugate direction method

Let $\mathbf{A} \in \mathbb{R}^{n\times n}$ be a symmetric, positive definite matrix and $\mathbf{b} \in \mathbb{R}^n$. Let $f : \mathbb{R}^n \to \mathbb{R}$ be a quadratic form given by

$$f(x) = \frac{1}{2}\mathbf{x}^{\mathrm{T}}\mathbf{A}\mathbf{x} - \mathbf{b}^{\mathrm{T}}\mathbf{x}. \tag{11.1.1}$$

Conjugate direction method provides an elegant framework for the minimization of the quadratic from $f(x)$.

Let $\mathcal{S} = \{\mathbf{p}_0, \mathbf{p}_1, \mathbf{p}_2, \ldots, \mathbf{p}_{n-1}\}$ be a set of n non-null vectors in $\mathbb{R}^n$. This set is said to be (mutually) **A-Conjugate** if

$$\mathbf{p}_k^{\mathrm{T}}\mathbf{A}\mathbf{p}_j = 0 \quad \text{for } k \neq j. \tag{11.1.2}$$

When $\mathbf{A} = \mathbf{I}$, the identity matrix, **A**-Conjugacy reduces to the well-known **orthogonality** of vectors. It can be verified that if a set of vectors is **A**-Conjugate, then they are also independent (Exercise 11.1), and it constitutes a basis for $\mathbb{R}^n$.

We begin by establishing the power and the import of the notion of **A**-Conjugacy. Let $\mathbf{x}_0 \in \mathbb{R}^n$ be given. Then, for any $\mathbf{x} \in \mathbb{R}^n$, we can express $(\mathbf{x} - \mathbf{x}_0)$ uniquely as a linear combination of the elements of $\mathcal{S}$ as follows.

$$\mathbf{x} - \mathbf{x}_0 = \alpha_0\mathbf{p}_0 + \alpha_1\mathbf{p}_1 + \alpha_2\mathbf{p}_2 + \cdots + \alpha_{n-1}\mathbf{p}_{n-1}. \tag{11.1.3}$$

To determine the coefficients α_i in (11.1.3), we multiply both sides of (11.1.3) on the left first by **A** and then $\mathbf{p}_i^{\mathrm{T}}$. By **A**-Conjugacy, we obtain

$$\mathbf{p}_k^{\mathrm{T}}\mathbf{A}(\mathbf{x} - \mathbf{x}_0) = \sum_{j=0}^{n-1}\alpha_j\mathbf{p}_k^{\mathrm{T}}\mathbf{A}\mathbf{p}_j = \alpha_k\mathbf{p}_k^{\mathrm{T}}\mathbf{A}\mathbf{p}_k$$

from which it follows that, for $k = 0, 1, \ldots, n-1$,

$$\alpha_k = \frac{\mathbf{p}_k^{\mathrm{T}}\mathbf{A}(\mathbf{x} - \mathbf{x}_0)}{\mathbf{p}_k^{\mathrm{T}}\mathbf{A}\mathbf{p}_k}. \tag{11.1.4}$$

If $\mathbf{x} = \mathbf{x}_*$ is the solution of $\mathbf{A}\mathbf{x} = \mathbf{b}$, then

$$\alpha_k = \frac{\mathbf{p}_k^{\mathrm{T}}\mathbf{A}(\mathbf{x} - \mathbf{x}_0)}{\mathbf{p}_k^{\mathrm{T}}\mathbf{A}\mathbf{p}_k} = \frac{\mathbf{p}_k^{\mathrm{T}}\mathbf{r}_0}{\mathbf{p}_k^{\mathrm{T}}\mathbf{A}\mathbf{p}_k} \tag{11.1.5}$$

where $\mathbf{r}_0 = \mathbf{b} - \mathbf{A}\mathbf{x}_0$, the residual at $\mathbf{x}_0$. Herein lies the power of **A**-Conjugacy – the solution of the linear system $\mathbf{A}\mathbf{x} = \mathbf{b}$ can be expressed as a linear combination of the **A**-Conjugate vectors and the coefficients of this linear combination can be calculated based on **A**, **b**, $\mathbf{x}_0$, and the **A**-Conjugate vectors.

Let us take one more look at the impact of **A**-Conjugacy on the minimization problem of interest to us. To this end, define

$$\mathbf{P} = \left[\mathbf{p}_0\ \mathbf{p}_1\ \cdots\ \mathbf{p}_{n-1}\right] \in \mathbb{R}^{n\times n} \tag{11.1.6}$$

the matrix whose columns are the **A**-Conjugate vectors in $\mathcal{S}$. Then, in view of (11.1.2), we readily see that

$$\begin{aligned}
\mathbf{P}^{\mathrm{T}}\mathbf{A}\mathbf{P} &= \begin{bmatrix} \mathbf{p}_0^{\mathrm{T}} \\ \mathbf{p}_1^{\mathrm{T}} \\ \vdots \\ \mathbf{p}_{n-1}^{\mathrm{T}} \end{bmatrix} \mathbf{A}\left[p_0\ p_1\ \cdots\ p_{n-1}\right] \\
&= \begin{bmatrix} d_0 & 0 & 0 & \cdots & 0 \\ 0 & d_1 & 0 & \cdots & 0 \\ \vdots & \vdots & \vdots & & \vdots \\ 0 & 0 & 0 & \cdots & d_{n-1} \end{bmatrix} \\
&= \mathrm{Diag}(d_0, d_1, d_2, \cdots, d_{n-1}).
\end{aligned} \tag{11.1.7}$$

Let $\alpha = (\alpha_0, \alpha_1, \ldots, \alpha_{n-1})^{\mathrm{T}} \in \mathbb{R}^n$. Then, (11.1.3) can be succinctly written as

$$\mathbf{x} = \mathbf{x}_0 + \mathbf{P}\alpha. \tag{11.1.8}$$

Now, define $G : \mathbb{R}^n \to \mathbb{R}$ as

$$\begin{aligned}
G(\alpha) &= f(\mathbf{x}_0 + \mathbf{P}\alpha) \\
&= \tfrac{1}{2}(\mathbf{x}_0 + \mathbf{P}\alpha)^{\mathrm{T}}\mathbf{A}(\mathbf{x}_0 + \mathbf{P}\alpha) - \mathbf{b}^{\mathrm{T}}(\mathbf{x}_0 + \mathbf{P}\alpha) \\
&= \left(\tfrac{1}{2}\mathbf{x}_0^{\mathrm{T}}\mathbf{A}\mathbf{x}_0 - \mathbf{b}^{\mathrm{T}}\mathbf{x}_0\right) + \tfrac{1}{2}\alpha^{\mathrm{T}}(\mathbf{P}^{\mathrm{T}}\mathbf{A}\mathbf{P})\alpha - (\mathbf{b} - \mathbf{A}\mathbf{x}_0)^{\mathrm{T}}\mathbf{P}\alpha \\
&= f(\mathbf{x}_0) + \tfrac{1}{2}\alpha^{\mathrm{T}}\mathbf{D}\alpha - \mathbf{r}_0^{\mathrm{T}}\mathbf{P}\alpha \\
&= f(\mathbf{x}_0) + \tfrac{1}{2}\textstyle\sum_{k=0}^{n-1}\alpha_k^2 d_k - \sum_{k=0}^{n-1}\mathbf{r}_0^{\mathrm{T}}\mathbf{p}_k\alpha_k \\
&= f(\mathbf{x}_0) + \textstyle\sum_{k=0}^{n-1}\left[\tfrac{1}{2}\alpha_k^2 d_k - \mathbf{r}_0^{\mathrm{T}}\mathbf{p}_k\alpha_k\right] \\
&= f(\mathbf{x}_0) + \textstyle\sum_{k=0}^{n-1} g_k(\alpha_k)
\end{aligned} \tag{11.1.9}$$

where $\mathbf{D} = \mathbf{P}^{\mathrm{T}}\mathbf{A}\mathbf{P}$ and $\mathbf{r}_0 = \mathbf{b} - \mathbf{A}\mathbf{x}_0$ and

$$g_k(\alpha) = \frac{1}{2}d_k\alpha^2 - \left(\mathbf{r}_0^{\mathrm{T}}\mathbf{p}_k\right)\alpha. \tag{11.1.10}$$

Thus, the **linear transformation** $\mathbf{x} - \mathbf{x}_0 = \mathbf{P}\alpha$ in (11.1.8), in view of the **A**-Conjugacy property of the columns of **P**, reduces the general positive definite quadratic form to its **canonical** form where the matrix **A** is reduced to a diagonal form **D**. The import of transformation is that it reduces the n-dimensional minimization of $f(\mathbf{x})$ in (11.1.1) to a collection of n (decoupled) one-dimensional

Given $f(\mathbf{x})$ in (11.1.1) and a set of **A**-Conjugate vectors $\mathcal{S}$ satisfying (11.1.2). Choose $\mathbf{x}_0 \in \mathbb{R}^n$, and compute $\mathbf{r}_0 = \mathbf{b} - \mathbf{A}\mathbf{x}_0$.

For $k = 0$ to $n - 1$

Step 1 $\alpha_k = \frac{\mathbf{p}_k^T \mathbf{r}_k}{\mathbf{p}_k^T \mathbf{A}\mathbf{p}_k}$.

Step 2 $\mathbf{x}_{k+1} = \mathbf{x}_k + \alpha_k \mathbf{p}_k$.

Step 3 $\mathbf{r}_{k+1} = \mathbf{r}_k - \alpha_k \mathbf{A}\mathbf{p}_k$.

Step 4 If $\mathbf{r}_{k+1} = 0$ then $\mathbf{x}_* = \mathbf{x}_{k+1}$.

Fig. 11.1.1 Conjugate direction method.

minimization problems as shown below:

$$\begin{aligned}
\min_{\mathbf{x} \in \mathbb{R}^n} f(\mathbf{x}) &= \min_{\alpha \in \mathbb{R}^n} f(\mathbf{x}_0 + \mathbf{P}\alpha) \\
&= \min_{\alpha \in \mathbb{R}^n} G(\alpha) \\
&= \min_{\alpha \in \mathbb{R}^n} \sum_{k=0}^{n-1} g_k(\alpha_k) \\
&= \sum_{k=0}^{n-1} \min_{\alpha_k \in \mathbb{R}} g_k(\alpha_k)
\end{aligned} \tag{11.1.11}$$

since each $g_k(\alpha_k)$ depends only on α_k and not on any other component of α. From (8.1.10), the minimizer of $g_k(\alpha_k)$ is given by the solution of

$$\frac{\mathrm{d}g_k(\alpha_k)}{\mathrm{d}\alpha_k} = d_k\alpha_k - \mathbf{p}_k^T \mathbf{r}_0 = 0$$

or

$$\alpha_k = \frac{\mathbf{p}_k^T \mathbf{r}_0}{d_k} \tag{11.1.12}$$

which not surprisingly, is the same as (11.1.5), since at the minimum $\nabla f(\mathbf{x}) = \mathbf{A}\mathbf{x} - \mathbf{b} = 0$.

Against this backdrop, we now describe a framework for the conjugate direction method for the minimization of $f(\mathbf{x})$ in (11.1.1) in Figure 11.1.1.

We establish a series of results relating to the behavior of this algorithm.

(a) First, we prove that the choice of α_k in step 1 minimizes $f(\mathbf{x})$ at the point $\mathbf{x}_k$ in the direction $\mathbf{p}_k$.

From

$$\begin{aligned}
g(\alpha) &= f(\mathbf{x}_k + \alpha\mathbf{p}_k) \\
&= \tfrac{1}{2}(\mathbf{x}_k + \alpha\mathbf{p}_k)^T \mathbf{A}(\mathbf{x}_k + \alpha\mathbf{p}_k) - \mathbf{b}^T(\mathbf{x}_k + \alpha\mathbf{p}_k) \\
&= \left(\tfrac{1}{2}\mathbf{x}_k^T\mathbf{A}\mathbf{x}_k - \mathbf{b}^T\mathbf{x}_k\right) + \tfrac{1}{2}\left(\mathbf{p}_k^T\mathbf{A}\mathbf{p}_k\right)\alpha^2 - (\mathbf{b} - \mathbf{A}\mathbf{x}_k)^T\mathbf{p}_k\alpha \\
&= f(\mathbf{x}_k) + \tfrac{1}{2}\left(\mathbf{p}_k^T\mathbf{A}\mathbf{p}_k\right)\alpha^2 - \left(\mathbf{r}_k^T\mathbf{p}_k\right)\alpha
\end{aligned} \tag{11.1.13}$$

we get the minimizing α as the solution of

$$\frac{dg(\alpha)}{d\alpha} = (\mathbf{p}_k^T \mathbf{A} \mathbf{p}_k)\alpha - (\mathbf{p}_k^T \mathbf{r}_k) = 0$$

which gives the value of α_k in step 1.

(b) The value of the function $f(\mathbf{x})$ decreases monotonically along the trajectory generated by this algorithm. From (11.1.13), by substituting for α_k from step 1, we get

$$\begin{aligned} f(\mathbf{x}_{k+1}) - f(\mathbf{x}_k) &= \tfrac{1}{2}(\mathbf{p}_k^T \mathbf{A} \mathbf{p}_k)\alpha^2 - (\mathbf{r}_k^T \mathbf{p}_k)\alpha_k \\ &= -\frac{1}{2}\frac{\left(\mathbf{r}_k^T \mathbf{p}_k\right)^2}{\mathbf{p}_k^T \mathbf{A} \mathbf{p}_k} \\ &< 0. \end{aligned} \tag{11.1.14}$$

(c) Consider the expression for $\mathbf{r}_{k+1}$ in step 3. Taking inner product of both sides with $\mathbf{p}_k$, in light of **A**-Conjugacy of $\mathbf{p}_i$'s and the expression of α_k in step 1, we obtain

$$\mathbf{p}_k^T \mathbf{r}_{k+1} = \mathbf{p}_k^T \mathbf{r}_k - \alpha_k \mathbf{p}_k^T \mathbf{A} \mathbf{p}_k = 0. \tag{11.1.15}$$

That is, $\mathbf{r}_{k+1}$ is orthogonal to $\mathbf{p}_k$. Since $\mathbf{r}_{k+1} = \mathbf{b} - \mathbf{A}\mathbf{x}_{k+1} = -\nabla f(\mathbf{x}_{k+1})$, (11.1.15) implies that the negative gradient of $f(\mathbf{x})$ at $\mathbf{x}_{k+1}$ is orthogonal to $\mathbf{p}_k$. Hence, by invoking the result in Appendix D, it follows that $\mathbf{x}_{k+1}$ minimizes $f(\mathbf{x})$ along the line $\mathbf{x}_k + \alpha \mathbf{p}_k$.

(d) From step 3, in view of **A**-Conjugacy we get

$$\begin{aligned} \mathbf{p}_k^T \mathbf{r}_k &= \mathbf{p}_k^T \mathbf{r}_{k-1} - \alpha_{k-1} \mathbf{p}_k^T \mathbf{A} \mathbf{p}_{k-1} \\ &= \mathbf{p}_k^T \mathbf{r}_{k-1}. \end{aligned}$$

Applying step 3 repeatedly to the r.h.s., we obtain

$$\mathbf{p}_k^T \mathbf{r}_k = \mathbf{p}_k^T \mathbf{r}_{k-1} = \cdots = \mathbf{p}_k^T \mathbf{r}_1 = \mathbf{p}_k^T \mathbf{r}_0. \tag{11.1.16}$$

This in turn implies that the value of α_k in step 1 and that given in (11.1.12) are indeed the same.

(e) Using step 3, **A**-Conjugacy and (11.1.15) it follows that

$$\begin{aligned} \mathbf{p}_k^T \mathbf{r}_{k+2} &= \mathbf{p}_k^T (\mathbf{r}_{k+1} - \alpha_{k+1} \mathbf{A} \mathbf{p}_{k+1}) \\ &= \mathbf{p}_k^T \mathbf{r}_{k+1} - \alpha_{k+1} \mathbf{p}_k^T \mathbf{A} \mathbf{p}_{k+1} \\ &= 0. \end{aligned}$$

By repeating this argument, we get

$$0 = \mathbf{p}_k^T \mathbf{r}_{k+1} = \mathbf{p}_k^T \mathbf{r}_{k+2} = \cdots = \mathbf{p}_k^T \mathbf{r}_{n-1} = \mathbf{p}_k^T \mathbf{r}_n. \tag{11.1.17}$$

(f) **Expanding subspace property** By iterating step 2, we get

$$\mathbf{x}_{k+1} = \mathbf{x}_0 + \alpha_0 \mathbf{p}_0 + \alpha_1 \mathbf{p}_1 + \cdots + \alpha_k \mathbf{p}_k. \tag{11.1.18}$$

Then,

$$\begin{aligned}\mathbf{r}_{k+1} &= \mathbf{b} - \mathbf{A}\mathbf{x}_{k+1} \\ &= \mathbf{r}_0 - \alpha_0\mathbf{A}\mathbf{p}_0 - \alpha_1\mathbf{A}\mathbf{p}_1 - \cdots - \alpha_k\mathbf{A}\mathbf{p}_k.\end{aligned}$$

Taking inner product of both sides with $\mathbf{p}_j$ for $0 \le j \le k-1$ and using the value of α_j in step 1, and (11.1.16), we get

$$\mathbf{p}_j^{\mathrm{T}}\mathbf{r}_{k+1} = \mathbf{p}_j^{\mathrm{T}}\mathbf{r}_0 - \alpha_j\mathbf{p}_j^{\mathrm{T}}\mathbf{A}\mathbf{p}_j = 0.$$

That is, $\mathbf{r}_{k+1} = -\nabla f(\mathbf{x}_{k+1})$ is orthogonal to $\mathbf{p}_j$ for all $j = 0, 1, 2, \ldots, k-1$. Hence by invoking the standard result in constrained minimization given in Appendix D, we conclude that $\mathbf{x}_{k+1}$ minimizes $f(\mathbf{x})$ over all

$$\mathbf{x} \in \mathbf{x}_0 + \mathrm{Span}\{\mathbf{p}_0, \mathbf{p}_1, \ldots, \mathbf{p}_k\}. \tag{11.1.19}$$

Thus, $\mathbf{x}_{k+1}$ in addition to minimizing $f(\mathbf{x})$ along the line $\mathbf{x}_k + \alpha\mathbf{p}_k$ also minimizes $f(\mathbf{x})$ in the **expanding subspace** given by (11.1.19). This is the special feature that distinguishes CD method from other methods, especially, the steepest descent method of Chapter 10. (Remark 10.3.1)

(g) An immediate consequence of this minimization over expanding subspaces is that it guarantees convergence in no more than n-step assuming that the arithmetic is exact.

11.2 Conjugate gradient method

The conjugate gradient method is a conjugate direction method wherein the search directions are not given a priori but are generated iteratively based on the gradient of the objective function at the current operating point. The classical CG method is stated in Figure 11.2.1

Comparing this with the CD method in Section 11.1, it follows that the first three steps – steps 1, 2, and 3 of both the methods are the same. The steps 5 and 6 of the CG method together define the process of generating the new search direction $\mathbf{p}_{k+1}$ as a linear combination of the new residual $\mathbf{r}_{k+1}$ (which is in fact the negative of the gradient of $f(\mathbf{x})$ at $\mathbf{x}_{k+1}$) and the previous search direction $\mathbf{p}_k$. Initially, $\mathbf{p}_0 = \mathbf{r}_0 = -\nabla f(\mathbf{x}_0)$. Thus, the CG method starts like the steepest descent method in Chapter 6. We now state several properties of this algorithm.

(1) **Choice of** α_k It is required that the new iterate $\mathbf{x}_{k+1}$ is the minimizer of $f(\mathbf{x})$ in the direction $\mathbf{x}_k + \alpha\mathbf{p}_k$. This in turn requires (Appendix D) that the negative of the gradient of $f(\mathbf{x})$ at $\mathbf{x}_{k+1}$, namely $-\nabla f(\mathbf{x}_{k+1}) = \mathbf{b} - \mathbf{A}\mathbf{x}_{k+1} = \mathbf{r}_{k+1}$ is orthogonal to $\mathbf{p}_k$. Thus

$$0 = \mathbf{p}_k^{\mathrm{T}}\mathbf{r}_{k+1} = \mathbf{p}_k^{\mathrm{T}}(\mathbf{r}_k - \alpha\mathbf{A}\mathbf{p}_k) \tag{11.2.1}$$

Given $f(\mathbf{x}) = \frac{1}{2}\mathbf{x}^{\mathrm{T}}\mathbf{A}\mathbf{x} - \mathbf{b}^{\mathrm{T}}\mathbf{x}$ and the initial choice $\mathbf{x}_0 \in \mathbb{R}^n$. Compute $\mathbf{r}_0 = \mathbf{b} - \mathbf{A}\mathbf{x}_0$ and let $\mathbf{p}_0 = \mathbf{r}_0$

For $k = 0$ to $n-1$ do the following

Step 1 $\alpha_k = \frac{\mathbf{p}_k^{\mathrm{T}}\mathbf{r}_k}{\mathbf{p}_k^{\mathrm{T}}\mathbf{A}\mathbf{p}_k} = \frac{\mathbf{r}_k^{\mathrm{T}}\mathbf{r}_k}{\mathbf{p}_k^{\mathrm{T}}\mathbf{A}\mathbf{p}_k}$

Step 2 $\mathbf{x}_{k+1} = \mathbf{x}_k + \alpha_k\mathbf{p}_k$

Step 3 $\mathbf{r}_{k+1} = \mathbf{r}_k - \alpha_k\mathbf{A}\mathbf{p}_k$

Step 4 Test for convergence: If $\mathbf{r}_{k+1}^{\mathrm{T}}\mathbf{r}_{k+1} < \epsilon$, then exit

Step 5 $\beta_k = -\frac{\mathbf{r}_{k+1}^{\mathrm{T}}\mathbf{A}\mathbf{p}_k}{\mathbf{p}_k^{\mathrm{T}}\mathbf{A}\mathbf{p}_k} = \frac{\mathbf{r}_{k+1}^{\mathrm{T}}\mathbf{r}_{k+1}}{\mathbf{r}_k^{\mathrm{T}}\mathbf{r}_k}$

Step 6 **New search direction:** $\mathbf{p}_{k+1} = \mathbf{r}_{k+1} + \beta_k\mathbf{p}_k$

Fig. 11.2.1 Conjugate gradient method.

from which we obtain the minimizing α as

$$\alpha_k = \frac{\mathbf{p}_k^{\mathrm{T}}\mathbf{r}_k}{\mathbf{p}_k^{\mathrm{T}}\mathbf{A}\mathbf{p}_k} \tag{11.2.2}$$

which agrees with step 1.

(2) **An alternate choice of α_k** We now derive an alternate formula for α_k. At $k = 0$, since $\mathbf{r}_0 = \mathbf{p}_0$, we readily see that

$$\alpha_o = \frac{\mathbf{p}_0^{\mathrm{T}}\mathbf{r}_0}{\mathbf{p}_0^{\mathrm{T}}\mathbf{A}\mathbf{p}_0} = \frac{\mathbf{r}_0^{\mathrm{T}}\mathbf{r}_0}{\mathbf{p}_0^{\mathrm{T}}\mathbf{A}\mathbf{p}_0}.$$

Now from

$$\begin{aligned}\mathbf{p}_k^{\mathrm{T}}\mathbf{r}_k &= (\mathbf{r}_k + \beta_k\mathbf{p}_{k-1})^{\mathrm{T}}\mathbf{r}_k \\ &= \mathbf{r}_k^{\mathrm{T}}\mathbf{r}_k + \beta_k\mathbf{p}_{k-1}^{\mathrm{T}}\mathbf{r}_k \\ &= \mathbf{r}_k^{\mathrm{T}}\mathbf{r}_k,\end{aligned}$$

since $\mathbf{p}_{k-1}^{\mathrm{T}}\mathbf{r}_k = 0$ by the same argument that leads to (11.2.1). Combining these, we get

$$\alpha_k = \frac{\mathbf{r}_k^{\mathrm{T}}\mathbf{r}_k}{\mathbf{p}_k^{\mathrm{T}}\mathbf{A}\mathbf{p}_k}, \tag{11.2.3}$$

which is the second formula for α_k in step 1.

(3) **Choice of β_k** The constant β_k is chosen to enforce the condition that $\mathbf{p}_{k+1}$ is $\mathbf{A}$-Conjugate to $\mathbf{p}_k$. Thus

$$0 = \mathbf{p}_k^{\mathrm{T}}\mathbf{A}\mathbf{p}_{k+1} = \mathbf{p}_k^{\mathrm{T}}\mathbf{A}(\mathbf{r}_{k+1} + \beta_k\mathbf{p}_k) \tag{11.2.4}$$

from which it follows that

$$\beta_k = -\frac{\mathbf{p}_k^{\mathrm{T}}\mathbf{A}\mathbf{r}_{k+1}}{\mathbf{p}_k^{\mathrm{T}}\mathbf{A}\mathbf{p}_k} \tag{11.2.5}$$

which is the first formula for it in step 5.

(4) **A-Conjugacy** Consider

$$\begin{aligned}\mathbf{p}_{k+1}^{\mathrm{T}}\mathbf{A}\mathbf{p}_{k+1} &= \mathbf{p}_{k+1}^{\mathrm{T}}\mathbf{A}(\mathbf{r}_{k+1} + \beta_k\mathbf{p}_k)\\ &= \mathbf{p}_{k+1}^{\mathrm{T}}\mathbf{A}\mathbf{r}_{k+1} + \beta_k\mathbf{p}_{k+1}^{\mathrm{T}}\mathbf{A}\mathbf{p}_k\\ &= \mathbf{p}_{k+1}^{\mathrm{T}}\mathbf{A}\mathbf{r}_{k+1},\end{aligned} \tag{11.2.6}$$

since $\mathbf{p}_{k+1}$ is **A**-Conjugate to $\mathbf{p}_k$ by (11.2.4). This reformulation is useful in deriving several more properties.

(5) $\mathbf{r}_{k+1}$ **is orthogonal to** $\mathbf{r}_k$ Much like the steepest descent algorithm, the gradient of $f(\mathbf{x})$ at $\mathbf{x}_{k+1}$ is orthogonal to that at $\mathbf{x}_k$. For

$$\begin{aligned}\mathbf{r}_k^{\mathrm{T}}\mathbf{r}_{k+1} &= \mathbf{r}_k^{\mathrm{T}}(\mathbf{r}_k - \alpha_k\mathbf{A}\mathbf{p}_k)\\ &= \mathbf{r}_k^{\mathrm{T}}\mathbf{r}_k - \alpha_k\,\mathbf{r}_k^{\mathrm{T}}\mathbf{A}\mathbf{p}_k\\ &= \mathbf{r}_k^{\mathrm{T}}\mathbf{r}_k - \frac{(\mathbf{r}_k^{\mathrm{T}}\mathbf{r}_k)}{\mathbf{p}_k^{\mathrm{T}}\mathbf{A}\mathbf{p}_k}(\mathbf{r}_k^{\mathrm{T}}\mathbf{A}\mathbf{p}_k)\\ &= 0\end{aligned} \tag{11.2.7}$$

in view of (11.2.3).

(6) **A new formula for** β_k Using (11.2.3), (11.2.5), and (11.2.7), we obtain

$$\begin{aligned}\mathbf{r}_{k+1}^{\mathrm{T}}\mathbf{r}_{k+1} &= (\mathbf{r}_k - \alpha_k\mathbf{A}\mathbf{p}_k)^{\mathrm{T}}\mathbf{r}_{k+1}\\ &= \mathbf{r}_k^{\mathrm{T}}\mathbf{r}_{k+1} - \alpha_k\big(\mathbf{p}_k^{\mathrm{T}}\mathbf{A}\mathbf{r}_{k+1}\big)\\ &= \alpha_k\beta_k\big(\mathbf{p}_k^{\mathrm{T}}\mathbf{A}\mathbf{p}_k\big)\\ &= \beta_k\big(\mathbf{r}_k^{\mathrm{T}}\mathbf{r}_k\big)\end{aligned}$$

and

$$\beta_k = \frac{\mathbf{r}_{k+1}^{\mathrm{T}}\mathbf{r}_{k+1}}{\mathbf{r}_k^{\mathrm{T}}\mathbf{r}_k}. \tag{11.2.8}$$

(7) $\mathbf{r}_k$ **is orthogonal to all** $\mathbf{p}_j$ **for** $0 \le j \le k$ By iterating step 2, we obtain

$$\mathbf{x}_k = \mathbf{x}_0 + \alpha_0\mathbf{p}_0 + \alpha_1\mathbf{p}_1 + \cdots + \alpha_{k-1}\mathbf{p}_{k-1}.$$

Thus

$$\mathbf{x}_k \in \mathbf{x}_0 + \mathrm{Span}\{\mathbf{p}_0, \mathbf{p}_1, \ldots .\mathbf{p}_{k-1}\}. \tag{11.2.9}$$

Let

$$\mathbf{P} = \big[\,\mathbf{p}_0\ \mathbf{p}_1\ \cdots\ \mathbf{p}_{k-1}\,\big] \in \mathbb{R}^{n\times k}.$$

Thus, for $\alpha = (\alpha_0, \alpha_1, \alpha_2, \ldots, \alpha_{k-1})^{\mathrm{T}} \in \mathbb{R}^k$, we get

$$\mathbf{x}_k = \mathbf{x}_0 + \mathbf{P}\alpha \tag{11.2.10}$$

and (using 11.1.14)

$$f(\mathbf{x}_k) = f(\mathbf{x}_0) + \frac{1}{2}\alpha^{\mathrm{T}}(\mathbf{P}^{\mathrm{T}}\mathbf{A}\mathbf{P})\alpha - \mathbf{r}_0^{\mathrm{T}}\mathbf{P}\alpha. \tag{11.2.11}$$

Hence the α that minimizes $f(\mathbf{x}_k)$ is given by

$$0 = -\nabla f(\mathbf{x}_k) = (\mathbf{P}^{\mathrm{T}}\mathbf{A}\mathbf{P})\alpha - \mathbf{P}^{\mathrm{T}}\mathbf{r}_0$$

or

$$\alpha = (\mathbf{P}^{\mathrm{T}}\mathbf{A}\mathbf{P})^{-1}\mathbf{P}^{\mathrm{T}}\mathbf{r}_0. \tag{11.2.12}$$

Thus

$$\mathbf{x}_k = \mathbf{x}_0 + \mathbf{P}(\mathbf{P}^{\mathrm{T}}\mathbf{A}\mathbf{P})^{-1}\mathbf{P}^{\mathrm{T}}\mathbf{r}_0$$

and

$$\begin{aligned} \mathbf{r}_k &= \mathbf{b} - \mathbf{A}\mathbf{x}_k \\ &= \mathbf{r}_0 - \mathbf{A}\mathbf{P}(\mathbf{P}^{\mathrm{T}}\mathbf{A}\mathbf{P})^{-1}\mathbf{P}^{\mathrm{T}}\mathbf{r}_0. \end{aligned} \tag{11.2.13}$$

Multiplying both sides by $\mathbf{P}^{\mathrm{T}}$, it follows that

$$\mathbf{P}^{\mathrm{T}}\mathbf{r}_k = \mathbf{P}^{\mathrm{T}}\mathbf{r}_0 - (\mathbf{P}^{\mathrm{T}}\mathbf{A}\mathbf{P})(\mathbf{P}^{\mathrm{T}}\mathbf{A}\mathbf{P})^{-1}\mathbf{P}^{-1}\mathbf{r}_0 = 0 \tag{11.2.14}$$

Stated in other words, (11.2.14) implies that $\mathbf{r}_k$ which is the negative of the gradient of $f(\mathbf{x})$ at $\mathbf{x}_k$ is orthogonal to each of the previous search directions $\mathbf{p}_0, \mathbf{p}_1, \ldots, \mathbf{p}_{k-1}$. That is,

$$\mathbf{p}_j^{\mathrm{T}}\mathbf{r}_k = 0 \ \text{ for all } 0 \leq j \leq k. \tag{11.2.15}$$

(8) **The residuals are mutually orthogonal and the search directions are A-Conjugate** In (11.2.7) we prove that $\mathbf{r}_{k+1}$ is orthogonal to $\mathbf{r}_k$, and in (11.2.4) we proved that $\mathbf{p}_{k+1}$ is **A**-Conjugate to $\mathbf{p}_k$. We now simultaneously prove by induction that, indeed, for all $j < k$

$$\mathbf{r}_k^{\mathrm{T}}\mathbf{r}_j = 0 \ \text{ and } \ \mathbf{p}_k^{\mathrm{T}}\mathbf{A}\mathbf{p}_j = 0. \tag{11.2.16}$$

When $k = 0$, since $\mathbf{r}_j = 0$ and $\mathbf{p}_j = 0$ for $j < k$, we verify (11.2.16) which is the basis for the inductive argument. We now hypothesize that (11.2.16) is true for k, and we wish to extend it for $k + 1$.

(a) Since $j = k$ is already covered in (11.2.7) we only need to consider the case $j < k$. Thus

$$\begin{aligned}
\mathbf{r}_j^{\mathrm{T}}\mathbf{r}_{k+1} &= \mathbf{r}_j^{\mathrm{T}}(\mathbf{r}_k - \alpha_k \mathbf{A}\mathbf{p}_k) \\
&= \mathbf{r}_j^{\mathrm{T}}\mathbf{r}_k - \alpha_k \mathbf{r}_j^{\mathrm{T}}\mathbf{A}\mathbf{p}_k \\
&= \mathbf{r}_j^{\mathrm{T}}\mathbf{r}_k - \alpha_k (\mathbf{p}_j - \beta_{j-1}\mathbf{p}_{j-1})^{\mathrm{T}}\mathbf{A}\mathbf{p}_k \\
&= \mathbf{r}_j^{\mathrm{T}}\mathbf{r}_k - \alpha_k \mathbf{p}_j^{\mathrm{T}}\mathbf{A}\mathbf{p}_k + \alpha_k\beta_{j-1}\mathbf{p}_{j-1}^{\mathrm{T}}\mathbf{A}\mathbf{p}_k \\
&= 0.
\end{aligned}$$

(b) Again, since $j = k$ is covered in (11.2.4), we only consider $j < k$. Accordingly,

$$\begin{aligned}
\mathbf{p}_{k+1}^{\mathrm{T}}\mathbf{A}\mathbf{p}_j &= (\mathbf{r}_{k+1} + \beta_k \mathbf{p}_k)^{\mathrm{T}}\mathbf{A}\mathbf{p}_j \\
&= \mathbf{r}_{k+1}^{\mathrm{T}}\mathbf{A}\mathbf{p}_j + \beta_k \mathbf{p}_k^{\mathrm{T}}\mathbf{A}\mathbf{p}_j \\
&= \mathbf{r}_{k+1}^{\mathrm{T}}\tfrac{(\mathbf{r}_j - \mathbf{r}_{j+1})}{\alpha_j} + \beta_k \mathbf{p}_k^{\mathrm{T}}\mathbf{A}\mathbf{p}_j \\
&= \frac{1}{\alpha_j}[\mathbf{r}_{k+1}^{\mathrm{T}}\mathbf{r}_j - \mathbf{r}_{k+1}^{\mathrm{T}}\mathbf{r}_{j+1}] + \beta_k \mathbf{p}_k^{\mathrm{T}}\mathbf{A}\mathbf{p}_j \\
&= 0.
\end{aligned}$$

Since each of these vanish by inductive hypothesis.

Clearly, this property (11.2.16) distinguishes CG from the steepest descent algorithm. Again, since all the $\mathbf{p}_j$'s are **A**-Conjugate by (11.2.16), it immediately follows that CG is indeed a conjugate direction method and hence must converge in no more than n steps when the arithmetic is exact.

(9) **Relation between various subspaces** We state yet another important property whose verification is left as Exercise 11.4.

$$\begin{aligned}
\mathrm{Span}\{\mathbf{p}_0, \mathbf{p}_1, \ldots, \mathbf{p}_{k-1}\} &= \mathrm{Span}\{\mathbf{r}_0, \mathbf{r}_1, \ldots, \mathbf{r}_{k-1}\} \\
&= \mathrm{Span}\{\mathbf{r}_0, \mathbf{A}\mathbf{r}_0, \mathbf{A}^2\mathbf{r}_0, \ldots, \mathbf{A}^{k-1}\mathbf{r}_0\}. \qquad (11.2.17)
\end{aligned}$$

(10) **Krylov framework** Given a nonsingular matrix $\mathbf{A} \in \mathbb{R}^{n\times n}$ and a vector $\mathbf{y} \in \mathbb{R}^n$, we can generate a sequence of vectors $\mathbf{y}, \mathbf{A}\mathbf{y}, \mathbf{A}^2\mathbf{y}, \ldots$. Since $\mathbb{R}^n$ is of dimension n, there exists an integer $m \le n$ such that $\mathbf{A}^m\mathbf{y}$ is a **linear combination** of $\{\mathbf{y}, \mathbf{A}\mathbf{y}, \mathbf{A}^2\mathbf{y}, \ldots, \mathbf{A}^{m-1}\mathbf{y}\}$. That is, $\mathbf{A}^m\mathbf{y}$ belongs to the Span$\{\mathbf{y}, \mathbf{A}\mathbf{y}, \mathbf{A}^2\mathbf{y}, \ldots, \mathbf{A}^{m-1}\mathbf{y}\}$. The vector space

$$\mathcal{K}_m(\mathbf{y}, \mathbf{A}) = \mathrm{Span}\{\mathbf{y}, \mathbf{A}\mathbf{y}, \mathbf{A}^2\mathbf{y}, \ldots, \mathbf{A}^{m-1}\mathbf{y}\} \qquad (11.2.18)$$

defined by **A** and **y** is of dimension m is called the **Krylov Subspace**. Now, combining this with (11.2.9) and (11.2.17), we readily see that the iterates of the CG algorithm are such that

$$\mathbf{x}_k \in \mathbf{x}_0 + \mathcal{K}_k(\mathbf{r}_0, \mathbf{A}). \tag{11.2.19}$$

That is, there exist constants $a_0, a_1, a_2, \ldots, a_{k-1}$ such that

$$\begin{aligned} \mathbf{x}_k &= \mathbf{x}_0 + (a_0\mathbf{I} + a_1\mathbf{A} + a_2\mathbf{A}^2 + \cdots + a_{k-1}\mathbf{A}^{k-1})\mathbf{r}_0 \\ &= \mathbf{x}_0 + q_{k-1}(\mathbf{A})\mathbf{r}_0 \end{aligned} \tag{11.2.20}$$

where

$$q_{k-1}(x) = a_0 + a_1 x + a_2 x^2 + \cdots + a_{k-1}x^{k-1} \tag{11.2.21}$$

is a $(k-1)$ degree polynomial. Accordingly, $q_{k-1}(\mathbf{A})$ is called a **matrix-polynomial**. Indeed, there exists an intimate relation between the properties of matrix-polynomials and Krylov Subspace.

Since $\mathbf{r}_0 = \mathbf{b} - \mathbf{A}\mathbf{x}_0 = \mathbf{A}(\mathbf{x}_* - \mathbf{x}_0)$, we can rewrite (11.2.21) as

$$\begin{aligned} \mathbf{x}_* - \mathbf{x}_k &= (\mathbf{x}_* - \mathbf{x}_0) - q_{k-1}(\mathbf{A})\mathbf{A}(\mathbf{x}_* - \mathbf{x}_0) \\ &= [\mathbf{I} - q_{k-1}(\mathbf{A})\mathbf{A}](\mathbf{x}_* - \mathbf{x}_0) \end{aligned} \tag{11.2.22}$$

where $(\mathbf{x}_* - \mathbf{x}_k)$ denotes the error in the kth iterate $\mathbf{x}_k$. This is a very basic relation from which we can quantify the convergence of the CG algorithm as shown below.

(11) **CG as an optimal process** Define a quadratic function with the A-norm of the error $\mathbf{e}_k = (\mathbf{x}_* - \mathbf{x}_0)$ in the kth iterate as follows:

$$\begin{aligned} E(\mathbf{x}_k) &= \frac{1}{2}(\mathbf{x}_* - \mathbf{x}_k)^{\mathrm{T}}\mathbf{A}(\mathbf{x}_* - \mathbf{x}_k) \\ &= \frac{1}{2}\|\mathbf{x}_* - \mathbf{x}_k\|^2\mathbf{A} \end{aligned} \tag{11.2.23}$$

where recall that $\|\mathbf{x}\|_{\mathbf{A}}$ is the A-norm of $\mathbf{x}$ (Appendix A). Substituting (11.2.22), we get after simplifying (Exercise 11.5)

$$E(\mathbf{x}_k) = \frac{1}{2}(\mathbf{x}_* - \mathbf{x}_0)^{\mathrm{T}}\mathbf{A}[\mathbf{I} - \mathbf{A}q_{k-1}(\mathbf{A})]^2(\mathbf{x}_* - \mathbf{x}_0). \tag{11.2.24}$$

One of the important consequences of the expanding subspace property in Section 11.1 is that the polynomials that define the iterates generated by the CG method (11.2.20) provide the solution to the following minimization problem:

$$E(\mathbf{x}_k) = \min_{g_{k-1}} \frac{1}{2}(\mathbf{x}_* - \mathbf{x}_0)^{\mathrm{T}}\mathbf{A}[\mathbf{I} - \mathbf{A}g_{k-1}(\mathbf{A})]^2(\mathbf{x}_* - \mathbf{x}_0) \tag{11.2.25}$$

where the minimum is taken over all polynomials of degree $k-1$. In this sense, the CG algorithm has a natural optimal property associated with it.

(12) **Rate of convergence of CG algorithm** It might sound odd at first sight to discuss the rate of convergence of the CG method since it is known to converge in at most n steps. This claim is, however, true only if the arithmetic is exact.

Because of round-off errors resulting from finite precision arithmetic, the conjugacy of search directions may be lost. In view of this, in practice we may not get convergence in n steps, and may need to iterate longer. It is often convenient to perform the convergence analysis by transforming the standard coordinate space into one that is defined by the eigenvectors of the matrix $\mathbf{A}$, called the **eigenspace** of the matrix $\mathbf{A}$ and representing the iterates in this new space. To this end, we begin by introducing some useful notation.

For $i = 1, 2, \ldots, n$, let (λ_i, η_i) denote the eigenvalue-vector pair of $\mathbf{A}$, that is, $\mathbf{A}\eta_i = \lambda_i \eta_i$, for $1 \leq i \leq n$. Since $\mathbf{A}$ is symmetric, it is well known that (Appendix B) λ_i's are real and η_i's are mutually orthogonal, that is, $\eta_i \perp \eta_j$ for $i \neq j$. Without loss of generality, it is assumed that

$$\lambda_1 \geq \lambda_2 \geq \lambda_3 \geq \cdots \geq \lambda_n > 0 \tag{11.2.26}$$

and that η_i's are normalized, that is, $\|\eta_i\| = 1$.

From this it can be verified that

$$\mathbb{R}^n = \text{Span}\{\eta_1, \eta_2, \ldots, \eta_n\}. \tag{11.2.27}$$

Let the linear combination

$$(\mathbf{x}_* - \mathbf{x}_0) = c_1\eta_1 + c_2\eta_2 + \cdots + c_n\eta_n \tag{11.2.28}$$

denote the new representation of the initial error $\mathbf{e}_0 = (\mathbf{x}_* - \mathbf{x}_0)$ in the eigenspace of $\mathbf{A}$.

Then

$$\begin{aligned}
E(\mathbf{x}_0) &= \tfrac{1}{2}(\mathbf{x}_* - \mathbf{x}_0)^{\mathrm{T}}\mathbf{A}(\mathbf{x}_* - \mathbf{x}_0) \\
&= \tfrac{1}{2}\left(\textstyle\sum_{i=1}^n c_i\eta_i\right)^{\mathrm{T}}\mathbf{A}\left(\textstyle\sum_{i=1}^n c_i\eta_i\right) \\
&= \tfrac{1}{2}\textstyle\sum_{i=1}^n c_i^2\left(\eta_i^{\mathrm{T}}\mathbf{A}\eta_i\right) && [\text{since } \eta_i \perp \eta_j] \\
&= \tfrac{1}{2}\textstyle\sum_{i=1}^n c_i^2\left(\lambda_i\eta_i^{\mathrm{T}}\eta_i\right) && [\text{since } \mathbf{A}\eta_i = \lambda_i\eta_i] \\
&= \tfrac{1}{2}\textstyle\sum_{i=1}^n c_i^2\lambda_i && [\text{since } \|\eta_i\| = 1].
\end{aligned} \tag{11.2.29}$$

Similarly, from (11.2.25) it can be verified (Exercise 11.6) that for any polynomial $g_{k-1}(x)$ of degree $(k-1)$, it follows that

$$\begin{aligned}
E(\mathbf{x}_k) &\leq \tfrac{1}{2}\textstyle\sum_{i=1}^n [1 - \lambda_i g_{k-1}(\lambda_i)]^2 \lambda_i c_i^2 \\
&\leq \max_{\lambda_i}(1 - \lambda_i g_{k-1}(\lambda_i))^2\left(\tfrac{1}{2}\textstyle\sum_{i=1}^n \lambda_i c_i^2\right) \\
&= \max_{\lambda_i}(1 - \lambda_i g_{k-1}(\lambda_i))^2 E(\mathbf{x}_0)
\end{aligned} \tag{11.2.30}$$

where the maximum is taken over the n eigenvalues of $\mathbf{A}$. By invoking the standard results relating to the properties of the class of orthogonal polynomials, called the **Chebyshev polynomials** of the first kind [Hageman and Young (1981)], it can be shown that the **relative error**

$$\frac{E(\mathbf{x}_k)}{E(\mathbf{x}_0)} = \frac{\|\mathbf{x}_* - \mathbf{x}_k\|_{\mathbf{A}}}{\|\mathbf{x}_* - \mathbf{x}_0\|_{\mathbf{A}}} \leq 2\left[\frac{\sqrt{\kappa_2(\mathbf{A})} - 1}{\sqrt{\kappa_2(\mathbf{A})} + 1}\right]^k \tag{11.2.31}$$

Table 11.2.1 $\epsilon = 10^{-7}$

$\kappa_2(\mathbf{A})$	k^*
1	4
10	24
10^2	74
10^3	231
10^4	730

where

$$\kappa_2(\mathbf{A}) = \frac{\lambda_1}{\lambda_n}$$

is called the **spectral condition number** of the matrix **A**.

Given a prespecified tolerance $\epsilon > 0$, let k^* denote the number of iterations needed for the relative error to be less than or equal to ϵ. Then

$$2\left[\frac{\sqrt{\kappa_2(\mathbf{A})} - 1}{\sqrt{\kappa_2(\mathbf{A})} + 1}\right]^{k^*} < \epsilon \tag{11.2.32}$$

from which we get an estimate of k^* as

$$k^* \approx \frac{\sqrt{\kappa_2(\mathbf{A})}}{2}\left|log_e \frac{\epsilon}{2}\right| \tag{11.2.33}$$

which is proportional to $\sqrt{\kappa_2(\mathbf{A})}$. Table 11.2.1 contains the values of k^* for typical values of the condition number when $\epsilon = 10^{-7}$. Comparing this with Table 10.2.1, it follows that CG is a lot faster than the steepest descent method.

11.3 Nonlinear conjugate gradient method

In developing the basic conjugate gradient method, it was assumed that the function $f(\mathbf{x})$ to be minimized is a quadratic form. However, the basic principles are indeed applicable when $f(\mathbf{x})$ is any nonlinear function. In this section, we present such an extension. A version of this nonlinear CG algorithm is given in Figure 11.3.1

The algorithm in this form was introduced by Fletcher and Reeves in 1964, where it was shown to converge under suitable conditions on $f(\mathbf{x})$. We now list several of the key features of this algorithm.

(a) This does not require any second derivative information.
(b) Compared to the quasi-Newton methods described in Chapter 12, computation of the search directions are simple and straightforward.
(c) The storage requirements are only linear in n since it requires only a few vectors to be stored. Hence this is highly suitable for large-scale problems.

Given $f : \mathbb{R}^n \to \mathbb{R}$ and $\nabla f(\mathbf{x})$. Let $\mathbf{x}_0$ be an initial starting position, and let $\mathbf{p}_0 = -\nabla f(\mathbf{x}_0)$.

For $i = 0, 1, 2, \ldots$.

Step 1 Perform the one-dimensional minimization of $g(\alpha) = f(\mathbf{x}_i + \alpha \mathbf{p}_i)$ and let α_i be the minimizer of $g(\alpha)$. [see Section 10.4.]

Step 2 Compute $\mathbf{x}_{i+1} = \mathbf{x}_i + \alpha_i \mathbf{p}_i$.

Step 3 Test for convergence. If YES, STOP.

Step 4 Define $\beta_i = \frac{[\nabla f(\mathbf{x}_i)]^{\mathrm{T}}[\nabla f(\mathbf{x}_i)]}{[\nabla f(\mathbf{x}_{i-1})]^{\mathrm{T}}[\nabla f(\mathbf{x}_{i-1})]}$.

Step 5 Compute $\mathbf{p}_{i+1} = -\nabla f(\mathbf{x}_{i+1}) + \beta_i \mathbf{p}_i$ and go to step 1.

Fig. 11.3.1 Nonlinear conjugate gradient method.

(d) There are a couple of other ways for computing the step-length parameter β_i in step 4. To this end, define $\mathbf{y}_{i-1} = \nabla f(\mathbf{x}_i) - \nabla f(\mathbf{x}_{i-1})$. Then

$$\beta_i = \frac{\mathbf{y}_{i-1}^{\mathrm{T}} \nabla f(\mathbf{x}_i)}{[\nabla f(\mathbf{x}_{i-1})]^{\mathrm{T}}[\nabla f(\mathbf{x}_{i-1})]} \quad \text{– **Polak–Ribiere** formula}$$

and

$$\beta_i = \frac{\mathbf{y}_{i-1}^{\mathrm{T}} \nabla f(\mathbf{x}_i)}{\mathbf{y}_{i-1}^{\mathrm{T}} \mathbf{p}_{i-1}} \quad \text{– **Hestenes and Stiefel** formula}$$

could be used in place of step 4 in Figure 11.3.1 where the **Fletcher–Reeves** formula is used. It turns out, when $f(\mathbf{x})$ is a quadratic form all these three formulae become identical and coincide with that given in step 5 in Figure 11.2.1.

11.4 Preconditioning

In Section 11.2, it was shown that the rate of convergence of the CG method for solving the linear system

$$\mathbf{A}\mathbf{x} = \mathbf{b} \tag{11.4.1}$$

depends critically on $\kappa_2(\mathbf{A})$, the **condition number** of the system matrix $\mathbf{A}$ – **the larger the condition number, the slower is the rate of convergence**. This fact calls for examining ways to accelerate the convergence by reducing the condition number. The standard approach to accomplish this goal is to transform the given set of equations to a new system

$$\hat{\mathbf{A}}\hat{\mathbf{x}} = \hat{\mathbf{b}} \tag{11.4.2}$$

such that

$$\kappa_2(\hat{\mathbf{A}}) < \kappa_2(\mathbf{A}). \tag{11.4.3}$$

The transformation often used is called the **congruence transformation** (Appendix B) which we now describe. Let $\mathbf{S}$ be a **non singular** matrix. Then the process of transforming

$$\mathbf{A} \rightarrow \hat{\mathbf{A}} = \mathbf{S}\mathbf{A}\mathbf{S}^{\mathrm{T}} \tag{11.4.4}$$

is called the congruence transformation. To see how this transformation is used, multiply both sides of (11.4.1) by $\mathbf{S}$ leading to

$$\mathbf{S}\mathbf{A}\mathbf{x} = \mathbf{S}\mathbf{b}. \tag{11.4.5}$$

Now define a new variable $\hat{\mathbf{x}}$ as a solution of

$$\mathbf{S}^{\mathrm{T}}\hat{\mathbf{x}} = \mathbf{x}. \tag{11.4.6}$$

Setting $\hat{\mathbf{b}} = \mathbf{S}\mathbf{b}$ and (11.4.4) can be rewritten using the new variable $\hat{\mathbf{x}}$ as

$$(\mathbf{S}\mathbf{A}\mathbf{S}^{\mathrm{T}})(\mathbf{S}^{-\mathrm{T}}\mathbf{x}) = \mathbf{S}\mathbf{b}$$

or

$$\hat{\mathbf{A}}\hat{\mathbf{x}} = \hat{\mathbf{b}}. \tag{11.4.7}$$

This process of converting (11.4.1) into (11.4.7) has come to be known as the **preconditioning** using congruence transformation and the matrix $\mathbf{S}$ is called the **preconditioner**.

Several observations are in order:

(1) The best choice for the preconditioner is $\mathbf{S} = \mathbf{A}^{-\frac{1}{2}}$ in which case $\hat{\mathbf{A}} = \mathbf{I}$. But this would beg the question since if we know $\mathbf{A}^{-1}$, there is no need for the preconditioner.
(2) In many cases, the use of $\mathbf{S} = \mathbf{D}$, a diagonal matrix can help to achieve (11.4.3).
(3) Actual conversion of (11.4.1) to (11.4.7) involves heavy computational burden of matrix-matrix multiplications requiring $O(n^3)$ operations. So, for this framework to have practical significance, we need to avoid computing these quantities explicitly. Herein lies the challenge in the design of practically useful preconditioning processes.
(4) In developing such practical schemes, define $\mathbf{M} = (\mathbf{S}^{\mathrm{T}}\mathbf{S})^{-1}$. Now using the **similarity transformation** (Appendix B) of $\hat{\mathbf{A}}$ using $\mathbf{S}^{\mathrm{T}}$, we get

$$\mathbf{S}^{\mathrm{T}}\hat{\mathbf{A}}\mathbf{S}^{-\mathrm{T}} = \mathbf{S}^{\mathrm{T}}(\mathbf{S}\mathbf{A}\mathbf{S}^{\mathrm{T}})\mathbf{S}^{-\mathrm{T}} = (\mathbf{S}^{\mathrm{T}}\mathbf{S})\mathbf{A} = \mathbf{M}^{-1}\mathbf{A}. \tag{11.4.8}$$

That is, $\hat{\mathbf{A}}$ is similar to $\mathbf{M}^{-1}\mathbf{A}$ and hence they have the same set of eigenvalues and the same condition number. This implies that we could work with $\mathbf{M}^{-1}\mathbf{A}$ in place of $\hat{\mathbf{A}}$. Since $\mathbf{S}$ is non singular, the Grammian matrices $\mathbf{S}^{\mathrm{T}}\mathbf{S}$ and its inverse $\mathbf{M} = (\mathbf{S}^{\mathrm{T}}\mathbf{S})^{-1}$ are both symmetric and positive definite matrices. The primary advantage of using $\mathbf{M}^{-1}\mathbf{A}$ is that if $\mathbf{M} = \mathbf{A}$, then $\mathbf{M}^{-1}\mathbf{A} = \mathbf{I}$. Thus, if $\mathbf{M}$ is an approximation to $\mathbf{A}$, so will $\mathbf{M}^{-1}$ to $\mathbf{A}^{-1}$, and $\mathbf{M}^{-1}\mathbf{A}$ will be closer to the identity matrix which is what is desired.

Given $\hat{\mathbf{A}}\hat{\mathbf{x}} = \mathbf{b}$. Choose $\hat{\mathbf{x}}_0$ and let $\hat{\mathbf{r}}_0 = \hat{\mathbf{b}} - \hat{\mathbf{A}}\hat{\mathbf{x}}_0$ and $\hat{\mathbf{p}}_0 = \hat{\mathbf{r}}_0$.

For $k = 0, 1, 2, \ldots$

Step 1 $\hat{\alpha}_k = \frac{(\hat{\mathbf{r}}_k)^{\mathrm{T}}\hat{\mathbf{r}}_k}{(\hat{\mathbf{p}}_k)^{\mathrm{T}}\hat{\mathbf{A}}\hat{\mathbf{p}}_k}$.

Step 2 $\hat{\mathbf{x}}_{k+1} = \hat{\mathbf{x}}_k + \hat{\alpha}_k\hat{\mathbf{p}}_k$.

Step 3 $\hat{\mathbf{x}}_{k+1} = \hat{\mathbf{r}}_k - \hat{\alpha}_k\hat{\mathbf{A}}\hat{\mathbf{p}}_k$.

Step 4 Test for convergence using $\|\hat{\mathbf{r}}_{k+1}\|$.

Step 5 $\hat{\beta}_k = \frac{(\hat{\mathbf{r}}_{k+1})^{\mathrm{T}}\hat{\mathbf{r}}_{k+1}}{(\hat{\mathbf{r}}_k)^{\mathrm{T}}\hat{\mathbf{r}}_k}$.

Step 6 $\hat{\mathbf{p}}_{k+1} = \hat{\mathbf{r}}_{k+1} + \hat{\beta}_k\hat{\mathbf{p}}_k$.

Fig. 11.4.1 Classical CG for $\hat{\mathbf{A}}\hat{\mathbf{x}} = \hat{\mathbf{b}}$.

In the light of this interpretation, we will use $\mathbf{M}$ instead of $\mathbf{S}$ as the preconditioner.

In the following, we illustrate the use of preconditioning in the conjugate gradient method. Instead of simply quoting the final version of this algorithm using the matrix $\mathbf{M}$, we first start with the classical CG algorithm for the system (11.4.2) and then illustrate the process that renders the conversion to the final form using the matrix $\mathbf{M}$ as the preconditioner.

The classical CG algorithm for solving (11.4.2) is given in Figure 11.4.1.

As observed earlier this version carries a heavy computational cost and is not usable as such. We now turn to transforming this algorithm into a form where as much of the computations can be performed with the original matrix $\mathbf{A}$ and the vector $\mathbf{b}$. To this end, we develop a series of relations.

(1) If $\hat{\mathbf{p}}_j$'s are $\hat{\mathbf{A}}$-Conjugate, then $\mathbf{p}_j$'s defined by

$$\mathbf{p}_j = \mathbf{S}^{\mathrm{T}}\hat{\mathbf{p}}_j \tag{11.4.9}$$

are $\mathbf{A}$-Conjugate. For,

$$(\hat{\mathbf{p}}_j)^{\mathrm{T}}\hat{\mathbf{A}}\hat{\mathbf{p}}_i = (\hat{\mathbf{p}}_j)^{\mathrm{T}}\mathbf{SAS}^{\mathrm{T}}\hat{\mathbf{p}}_i = (\mathbf{S}^{\mathrm{T}}\hat{\mathbf{p}}_j)^{\mathrm{T}}\mathbf{A}(\mathbf{S}^{\mathrm{T}}\hat{\mathbf{p}}_i) = \mathbf{p}_j^{\mathrm{T}}\mathbf{A}\mathbf{p}_i$$

from which the claim follows.

(2) Consider the computations at step 0. Using (11.4.4) and (11.4.6), we have

$$\begin{aligned}\hat{\mathbf{r}}_0 &= \hat{\mathbf{b}} - \hat{\mathbf{A}}\hat{\mathbf{x}}_0 \\ &= \mathbf{Sb} - (\mathbf{SAS}^{\mathrm{T}})(\mathbf{S}^{-\mathrm{T}}\mathbf{x}_0) \\ &= \mathbf{S}(\mathbf{b} - \mathbf{A}\mathbf{x}_0) = \mathbf{S}\mathbf{r}_0.\end{aligned} \tag{11.4.10a}$$

Hence

$$(\hat{\mathbf{r}}_0)^{\mathrm{T}}\hat{\mathbf{r}}_0 = (\mathbf{S}\mathbf{r}_0)^{\mathrm{T}}(\mathbf{S}\mathbf{r}_0) = \mathbf{r}_0^{\mathrm{T}}(\mathbf{S}^{\mathrm{T}}\mathbf{S}\mathbf{r}_0) = \mathbf{r}_0^{\mathrm{T}}\tilde{\mathbf{r}}_0 \tag{11.4.10b}$$

where

$$\tilde{\mathbf{r}}_0 = (\mathbf{S}^{\mathrm{T}}\mathbf{S})\mathbf{r}_0 = \mathbf{M}^{-1}\mathbf{r}_0. \tag{11.4.11}$$

(3) Also, from (11.4.9), we get

$$(\hat{\mathbf{p}}_0)^{\mathrm{T}}\hat{\mathbf{A}}\hat{\mathbf{p}}_0 = (\mathbf{S}^{-\mathrm{T}}\mathbf{p}_0)^{\mathrm{T}}(\mathbf{SAS}^{\mathrm{T}})(\mathbf{S}^{-\mathrm{T}}\mathbf{p}_0) = \mathbf{p}_0^{\mathrm{T}}\mathbf{A}\mathbf{p}_0. \tag{11.4.12}$$

(4) Combining (11.4.10) and (11.4.12), we get

$$\hat{\alpha}_0 = \frac{(\hat{\mathbf{r}}_0)^{\mathrm{T}}\hat{\mathbf{r}}_0}{(\hat{\mathbf{p}}_0)^{\mathrm{T}}\hat{\mathbf{A}}\hat{\mathbf{p}}_0} = \frac{\mathbf{r}_0^{\mathrm{T}}\tilde{\mathbf{r}}_0}{\mathbf{p}_0^{\mathrm{T}}\mathbf{A}\mathbf{p}_0}. \tag{11.4.13}$$

That is, $\hat{\alpha}_0$ can be computed using the old variables $\mathbf{r}_0$, $\mathbf{A}$, suitably chosen $\mathbf{p}_0$, and the new variable $\tilde{\mathbf{r}}_0$ defined in (11.4.11) using $\mathbf{M}$.

(5) Multiplying both sides of step 2 in Figure 11.4.1 on the left by $\mathbf{S}^{\mathrm{T}}$ to get

$$\mathbf{S}^{\mathrm{T}}\hat{\mathbf{x}}_1 = \mathbf{S}^{\mathrm{T}}\hat{\mathbf{x}}_0 + \hat{\alpha}_0\mathbf{S}^{\mathrm{T}}\hat{\mathbf{p}}_0.$$

Using (11.4.9) and (11.4.6) this becomes

$$\mathbf{x}_1 = \mathbf{x}_0 + \hat{\alpha}_0\mathbf{p}_0. \tag{11.4.14}$$

(6) Likewise multiply both sides of step 3 in Figure 11.4.1 on the left by $\mathbf{S}^{-1}$ to get

$$\mathbf{S}^{-1}\hat{\mathbf{r}}_1 = \mathbf{S}^{-1}\hat{\mathbf{r}}_0 - \hat{\alpha}_0\mathbf{S}^{-1}\hat{\mathbf{A}}\hat{\mathbf{p}}_0.$$

From (11.4.10) and (11.4.9) this reduces to

$$\mathbf{r}_1 = \mathbf{r}_0 - \hat{\alpha}_0\mathbf{A}\mathbf{p}_0 \tag{11.4.15}$$

which again involves the original variables.

(7) Consider step 5, using (11.4.10) and (11.4.11)

$$\begin{aligned}\hat{\beta}_0 &= \frac{(\hat{\mathbf{r}}_1)^{\mathrm{T}}\hat{\mathbf{r}}_1}{(\hat{\mathbf{r}}_0)^{\mathrm{T}}\hat{\mathbf{r}}_0} \\ &= \frac{(\mathbf{S}\mathbf{r}_1)^{\mathrm{T}}(\mathbf{S}\mathbf{r}_1)}{(\mathbf{S}\mathbf{r}_0)^{\mathrm{T}}(\mathbf{S}\mathbf{r}_0)} \\ &= \frac{\mathbf{r}_1^{\mathrm{T}}(\mathbf{S}^{\mathrm{T}}\mathbf{S}\mathbf{r}_1)}{\mathbf{r}_0^{\mathrm{T}}(\mathbf{S}^{\mathrm{T}}\mathbf{S}\mathbf{r}_0} \\ &= \frac{\mathbf{r}_1^{\mathrm{T}}\tilde{\mathbf{r}}_1}{\mathbf{r}_0^{\mathrm{T}}\tilde{\mathbf{r}}_0}.\end{aligned} \tag{11.4.16}$$

(8) Consider step 6. Multiply both sides on the left by $\mathbf{S}^{\mathrm{T}}$, we obtain

$$\mathbf{S}^{\mathrm{T}}\hat{\mathbf{p}}_1 = \mathbf{S}^{\mathrm{T}}\hat{\mathbf{r}}_1 + \hat{\beta}_0\mathbf{S}^{\mathrm{T}}\hat{\mathbf{p}}_0.$$

Using (11.4.9) and (11.4.10) this reduces to

$$\begin{aligned}\mathbf{p}_1 &= (\mathbf{S}^{\mathrm{T}}\mathbf{S})\mathbf{r}_1 + \hat{\beta}_0\mathbf{p}_0 \\ &= \mathbf{M}^{-1}\mathbf{r}_1 + \hat{\beta}_0\mathbf{p}_0 = \tilde{\mathbf{r}}_0 + \tilde{\beta}_0\mathbf{p}_0.\end{aligned} \tag{11.4.17}$$

This completes the transformation of the computations in step 0. Since these are to be repeated for each k, similar relations hold at every step. This leads to a practical version of the **preconditioned conjugate gradient**(PCG) algorithm given in Figure 11.4.2.

Given $\mathbf{Ax} = \mathbf{b}$. Pick a preconditioner $\mathbf{M}$ which is a symmetric and positive definite matrix. Choose $\mathbf{x}_0$ and let $\mathbf{r}_0 = \mathbf{b} - \mathbf{A}\mathbf{x}_0$. Solve $\mathbf{M}\tilde{\mathbf{r}}_0 = \mathbf{r}_0$ and set $\mathbf{p}_0 = \tilde{\mathbf{r}}_0$, the initial search direction.

For $k = 0, 1, 2, \ldots$

Step 1 $\alpha_k = \frac{\mathbf{r}_k^{\mathrm{T}}\tilde{\mathbf{r}}_k}{\mathbf{p}_k^{\mathrm{T}}\mathbf{A}\mathbf{p}_k}$.

Step 2 $\mathbf{x}_{k+1} = \mathbf{x}_k + \alpha_k \mathbf{p}_k$.

Step 3 $\mathbf{r}_{k+1} = \mathbf{r}_k - \alpha_k \mathbf{A}\mathbf{p}_k$.

Step 4 Test for convergence.

Step 5 Solve $\mathbf{M}\tilde{\mathbf{r}}_{k+1} = \mathbf{r}_{k+1}$.

Step 6 $\beta_k = \frac{\mathbf{r}_{k+1}^{\mathrm{T}}\tilde{\mathbf{r}}_{k+1}}{\mathbf{r}_k^{\mathrm{T}}\tilde{\mathbf{r}}_k}$.

Step 7 $\mathbf{p}_{k+1} = \tilde{\mathbf{r}}_{k+1} + \beta_k \mathbf{p}_k$.

Fig. 11.4.2 Preconditioned conjugate gradient method.

By way of further simplifying the notation, we have dropped the super-scripts in α and β. If $\mathbf{M} = \mathbf{I}$, this algorithm reduces to the classical CG method. The primary difference between the CG and PCG is that the latter incurs additional computational cost in solving $\mathbf{M}\tilde{\mathbf{r}}_k = \mathbf{r}_k$ for each k. This in turn requires that $\mathbf{M}$ must have a simple structure such as diagonal matrix. But for $\mathbf{M}$ to be effective as a preconditioner, it must be a close approximation to $\mathbf{A}$. Herein lies the conflict between the cost vs. the effectiveness of the preconditioner, which often leads to seeking good trade-off between these opposing goals. While there are several general guidelines for the design of preconditioners, the choice of $\mathbf{M}$ ultimately has to be dependent on the structure and properties of the given matrix $\mathbf{A}$.

Exercises

11.1 Verify the claim that if vectors of a set $\mathcal{S} = \{\mathbf{p}_0, \mathbf{p}_1, \ldots, \mathbf{p}_{k-1}\}$ are (mutually) $\mathbf{A}$-Conjugate, then they are also linearly independent.

11.2 Let $\mathbf{p}_1, \mathbf{p}_2, \ldots, \mathbf{p}_n$ be the set of orthogonal eigenvectors of a real symmetric and positive definite matrix $\mathbf{A}$ corresponding the n eigenvalues $\lambda_1 \geq \lambda_2 \geq \lambda_3 \geq \cdots \geq \lambda_n > 0$. That is, $\mathbf{A}\mathbf{p}_i = \lambda_i \mathbf{p}_i$ for $i = 1$ to n.

(a) Verify that $\{\mathbf{p}_1, \mathbf{p}_2, \ldots, \mathbf{p}_n\}$ are $\mathbf{A}$-Conjugate, that is, $\mathbf{p}_i^{\mathrm{T}}\mathbf{A}\mathbf{p}_j = 0$ and $\mathbf{p}_i^{\mathrm{T}}\mathbf{A}\mathbf{p}_i = \lambda_i > 0$ for $i = 1$ to n

(b) If $\mathbf{P} = [\mathbf{p}_1\, \mathbf{p}_2 \cdots \mathbf{p}_n] \in \mathbb{R}^{n\times n}$ and $\Lambda = \mathrm{Diag}[\lambda_1, \lambda_2, \ldots, \lambda_n]$ from $\mathbf{AP} = \mathbf{P}\Lambda$ verify that

$$\mathbf{A} = \mathbf{P}\Lambda\mathbf{P}^{\mathrm{T}} = \sum_{i=1}^{n} \lambda_i \mathbf{p}_i \mathbf{p}_i^{\mathrm{T}}$$

and

$$\mathbf{A}^{-1} = \mathbf{P}\Lambda^{-1}\mathbf{P}^{\mathrm{T}} = \sum_{i=1}^{n} \frac{\mathbf{p}_i \mathbf{p}_i^{\mathrm{T}}}{\lambda_i}.$$

11.3 From (11.1.9), consider the minimization of $G(\alpha) = \frac{1}{2}\alpha^{\mathrm{T}}\mathbf{D}\alpha - (\mathbf{b}^{\mathrm{T}}\mathbf{P})\alpha$. Clearly, the minimizer $\alpha = \mathbf{D}^{-1}\mathbf{P}^{\mathrm{T}}\mathbf{b}$. From this we obtain $\mathbf{x}_* = \mathbf{P}\mathbf{D}^{-1}\mathbf{P}^{-1}\mathbf{b}$ as the minimizer for $f(\mathbf{x})$
(a) Comparing this expression with $\mathbf{x}_* = \mathbf{A}^{-1}\mathbf{b}$ verify that

$$\mathbf{A}^{-1} = \mathbf{P}\mathbf{D}^{-1}\mathbf{P}^{\mathrm{T}} = \sum_{i=1}^{n} \frac{\mathbf{p}_i\mathbf{p}_i^{\mathrm{T}}}{\mathbf{d}_i} = \sum_{i=1}^{n} \frac{\mathbf{p}_i\mathbf{p}_i^{\mathrm{T}}}{\mathbf{p}_i^{\mathrm{T}}\mathbf{A}\mathbf{p}_i}$$

(b) Compare and comment on the expression for $\mathbf{A}^{-1}$ obtained in Exercise (11.2 (b)).

11.4 Verify the relation in (11.2.17).

11.5 Verify the correctness of (11.2.24).

11.6 (a) Using $\mathbf{A}\eta = \lambda\eta$, first verify that $\mathbf{A}^2\eta = \lambda^2\eta$ and $\mathbf{A}^k\eta = \lambda^k\eta$ for any integer $k \geq 1$.
(b) Let $g(\mathbf{x}) = a_0 + a_1\mathbf{x} + a_2\mathbf{x}^2$. Then verify that $g(\mathbf{A})\eta = g(\lambda)\eta$ where $g(\mathbf{A}) = a_0\mathbf{I} + a_1\mathbf{A} + a_2\mathbf{A}^2$ is called the matrix polynomial of $\mathbf{A}$ of degree 2.

Notes and references

Conjugate gradient method was originally developed by Hestenes and Stiefel in their seminal paper in (1952) . The book by Hestenes (1980) provides an authoritative account of this class of methods. Books by Luenberger (1973) and Nash and Sofer (1996) provide a very readable account of the theory of this algorithm.

Lanczos around the same time in the early 1950s developed independently a class of methods that bears his name. His approach was based on the intrinsic properties of orthogonal polynomials and he developed a class of algorithm for solving linear systems based on the three-term recurrence for a class of orthogonal polynomials. On further analysis, it turned out that there is a one to one correspondence between the method of Lanczos and those of Hestenes and Stiefel. For a readable account of Lanczos method and its relation to CG method refer to monographs by Greenbaum (1997), Golub and van Loan (1989), Hanke (1995), Brauset (1995), Hageman and Young (1981), and Trefethen and Bau (1997).

Barrett et al. (1994) contains an extremely useful collection of pseudocodes for many algorithms including the CG method.

CG method is but one member of an increasing family of methods called the Krylov subspace methods. Refer to Greenbaum (1997) and Brauset (1995).

Nonlinear conjugate gradient methods are discussed in Luenberger (1973) and Nash and Sofer (1996).

Ortega (1988) contains a comprehensive discussion of various methods for designing preconditioners. Also refer to Brauset (1995) for details.

12

Newton and quasi-Newton methods

It was around 1660 Newton discovered the method for solving nonlinear equations that bears his name. Shortly thereafter – around 1665 – he also developed the **secant** method for solving nonlinear equations. Since then these methods have become a part of the folklore in numerical analysis (see Exercises 12.1 and 12.2). In addition to solving nonlinear equations, these methods can also be applied to the problem of minimizing a nonlinear function. In this chapter we provide an overview of the classical Newton's method and many of its modern relatives called quasi-Newton methods for unconstrained minimization. The major advantage of the Newton's method is its quadratic convergence (Exercise 12.3) but in finding the next descent direction it requires solution of a linear system which is often a bottleneck. Quasi-Newton methods are designed to preserve the good convergence properties of the Newton's method while they provide considerable relief from this computational bottleneck. Quasi-Newton methods are extensions of the **secant** method. Davidon was the first to revive the modern interest in quasi-Newton methods in 1959 but his work remained unpublished till 1991. However, Fletcher and Powell in 1963 published Davidon's ideas and helped to revive this line of approach to designing efficient minimization algorithms.

The philosophy and practice that underlie the design of quasi-Newton methods underscore the importance of the trade - off between rate of convergence and computational cost and storage. The classical Newton's method has quadratic convergence but requires $O(n^2)$ storage and $O(n^3)$ times. quasi-Newton methods while settling for super linear convergence have reduced the space requirements to $O(n)$ and time to $O(n^2)$ and/or $O(n)$. Because of these attractive features, they are often the method of choice for large-scale problems of interest in the geophysical domain. Ready to use programs based on quasi-Newton methods are available in several software libraries including IMSL, MATLAB and NETLIB.

In Section 12.1 we describe the classical Newton's method and many of its properties. Quasi-Newton's methods are described in Sections 12.2. Strategies for reducing space requirements in quasi-Newton methods are described in Section 12.3.

12.1 Newton's method

let $f : \mathbb{R}^n \rightarrow \mathbb{R}$ be the function to be minimized and let $\mathbf{x}^*$ be a local minimum for $f(\mathbf{x})$. The basic idea behind the Newton's method may be stated as follows: let $\mathbf{x}_k$ be a current operating point. First approximate $f(\mathbf{x})$ around $\mathbf{x}_k$ by a **quadratic** function using the second-order Taylor expansion (Appendix C) and minimize this quadratic function. This minimizer then defines the new operating point and the cycle is repeated until convergence is obtained.

More formally, let $\mathbf{p} \in \mathbb{R}^n$ and define

$$\begin{aligned} m(\mathbf{p}) &= f(\mathbf{x}_k + \mathbf{p}) \\ &= f(\mathbf{x}_k) + [\nabla f(\mathbf{x}_k)]^{\mathrm{T}}\mathbf{p} + \tfrac{1}{2}\mathbf{p}^{\mathrm{T}}[\nabla^2 f(\mathbf{x}_k)]\mathbf{p} \end{aligned} \tag{12.1.1}$$

the second-order quadratic approximation for $f(\mathbf{x})$ in a small enough neighborhood around $\mathbf{x}_k$, where $\nabla f(\mathbf{x}_k)$ is the gradient and $\nabla^2 f(\mathbf{x}_k)$ is the Hessian of $f(\mathbf{x})$ at $\mathbf{x}_k$ (Appendix C). Then

$$\nabla m(\mathbf{p}) = \nabla f(\mathbf{x}_k) + \nabla^2 f(\mathbf{x}_k)\mathbf{p} \tag{12.1.2}$$

and

$$\nabla^2 m(\mathbf{p}) = \nabla^2 f(\mathbf{x}_k). \tag{12.1.3}$$

Setting the gradient of $m(\mathbf{p})$ to zero, we get a system of linear equations

$$\nabla^2 f(\mathbf{x}_k)\mathbf{p} = -\nabla f(\mathbf{x}_k), \tag{12.1.4}$$

whose solution $\mathbf{p}^*$ is the minimizer of $m(\mathbf{p})$ provided that the Hessian $\nabla^2 m(\mathbf{p}^*)$ is positive definite (Exercise 12.7). Condition (12.1.4) is often known as the **Newton's equation** and its solution $\mathbf{p}^*$ is called the **Newton direction**. Once $\mathbf{p}^*$ is known, define a new operating point

$$\mathbf{x}_{k+1} = \mathbf{x}_k + \mathbf{p}^*$$

and the process is repeated until $\mathbf{x}_{k+1}$ is sufficiently close to $\mathbf{x}^*$, the true local minimum of $f(\mathbf{x})$.

To understand the full power of the Newton method, consider the case where $f(\mathbf{x})$ is a quadratic form. That is,

$$f(\mathbf{x}) = \frac{1}{2}\mathbf{x}^{\mathrm{T}}\mathbf{A}\mathbf{x} - \mathbf{b}^{\mathrm{T}}\mathbf{x} \tag{12.1.5}$$

where $\mathbf{A} \in \mathbb{R}^{n\times n}$ is symmetric and positive definite.

Then, it can be verified that

$$\begin{aligned} m(\mathbf{p}) &= f(\mathbf{x}_k + \mathbf{p}) \\ &= \tfrac{1}{2}(\mathbf{x}_k + \mathbf{p})^{\mathrm{T}}\mathbf{A}(\mathbf{x}_k + \mathbf{p}) - \mathbf{b}^{\mathrm{T}}(\mathbf{x}_k + \mathbf{p}) \\ &= \tfrac{1}{2}\mathbf{p}^{\mathrm{T}}\mathbf{A}\mathbf{p} + (\mathbf{A}\mathbf{x}_k - \mathbf{b})^{\mathrm{T}}\mathbf{p} + f(\mathbf{x}_k). \end{aligned}$$

Given $f : \mathbb{R}^n \to \mathbb{R}$ and $\mathbf{x}_0$. Let $\nabla f(\mathbf{x})$ and $\nabla^2 f(\mathbf{x})$ be the gradient and Hessian of $f(\mathbf{x})$ respectively.
For $k = 0, 1, 2, \ldots$ until convergence do:

Step 1 Compute $\nabla f(\mathbf{x}_k)$ and $\nabla^2 f(\mathbf{x}_k)$.

Step 2 Solve $\nabla^2 f(\mathbf{x}_k)\mathbf{p} = -\nabla f(\mathbf{x}_k)$ and let $\mathbf{p}_k$ be the solution.

Step 3 Compute $\mathbf{x}_{k+1} = \mathbf{x}_k + \mathbf{p}_k$.

Step 4 Check for convergence. If YES stop. Else go to step1.

Fig. 12.1.1 Classical full Newton's method.

Now $\mathbf{p}^*$ is obtained by solving

$$\nabla m(\mathbf{p}) = \mathbf{A}\mathbf{p} + (\mathbf{A}\mathbf{x}_k - \mathbf{b}) = 0$$

where $\mathbf{p}^* = -\mathbf{A}^{-1}(\mathbf{A}\mathbf{x}_k - \mathbf{b}) = -\mathbf{x}_k + \mathbf{A}^{-1}\mathbf{b}$. Then $\mathbf{x}_n = \mathbf{x}_k + \mathbf{p}^* = \mathbf{A}^{-1}\mathbf{b} = \mathbf{x}^*$, the true minimum.

That is, we obtain convergence in one step irrespective of where the current operating point $\mathbf{x}_k$ is. The classical Newton's algorithm is given in Figure 12.1.1.

Several observations are in order.

(1) Under ideal conditions Newton's method converges quadratically. However, Newton's method is seldom used in this form in practice for, it may not converge and even if it did converge, it may not converge to the minimizer. To make it more robust several modifications are needed and we describe two of the very useful ones below.

(a) The first modification calls for changing step 3 in Figure 12.1.1 by introducing the step length parameter. More formally, once the Newton direction is available in Step 2, then the one dimensional minimization of $f(\mathbf{x})$ along the chosen direction $\mathbf{p}_k$ is performed. That is if

$$g(\alpha) = f(\mathbf{x}_k + \alpha\mathbf{p}_k), \tag{12.1.6}$$

then α_k that minimizes $g(\alpha)$ is obtained using the method described in Section 10.4. We then replace Step 3 with the following:

Step 3′ Compute $\mathbf{x}_{k+1} = \mathbf{x}_k + \alpha_k\mathbf{p}_k$ where α_k minimizes $g(\alpha)$ in equation (12.1.6).

(b) The second modification deals with the Newton's equation in Step 2. Assume that $\nabla^2 f(\mathbf{x}_k)$ is indefinite (that is, some of the eigenvalues are negative and others are positive), then while

$$\nabla^2 f(\mathbf{x}_k)\mathbf{p} = -\nabla f(\mathbf{x}_k) \tag{12.1.7}$$

can be solved, the solution $\mathbf{p}$ may **not** be a descent direction. In such cases, to improve the robustness of this algorithm, at the very least, we need to guarantee that the solution $\mathbf{p}$ is a descent direction. This is often done by modifying the Hessian as follows. Let $\mathbf{E}_k = \mu_k \mathbf{I}$ be a diagonal matrix where all the diagonal entries are μ_k. The goal is to choose the scalar μ_k to be positive and sufficiently large such that the modified Hessian

$$(\nabla^2 f(\mathbf{x}_k) + \mu_k \mathbf{I})$$

is positive definite. Then, it can be verified the solution $\mathbf{p}$ of

$$(\nabla^2 f(\mathbf{x}_k) + \mu_k \mathbf{I})\mathbf{p} = -\nabla f(\mathbf{x}_k) \tag{12.1.8}$$

is indeed a descent direction for $f(\mathbf{x})$ at $\mathbf{x}_k$ (Exercise 10.11). Thus, we modify Step 2 as follows:
Step 2′**:** Solve $(\nabla^2 f(\mathbf{x}_k) + \mu_k \mathbf{I})\mathbf{p} = -\nabla f(\mathbf{x}_k)$ and let $\mathbf{p}_k$ be the solution where μ_k is chosen to force the modified Hessian $(\nabla^2 f(\mathbf{x}_k) + \mu_k \mathbf{I})$ to be positive definite.

Remark 12.1.1 The constant $\mu_k \geq 0$ that is needed in equation 12.1.8 can be easily found in the course of solving for $\mathbf{p}$ using the Cholesky decomposition algorithm (see Chapter 9). It is well known that the symmetric matrix $\nabla^2 f(\mathbf{x}_k)$ can be factored as

$$\nabla^2 f(\mathbf{x}_k) = \mathbf{LDL}^{\mathrm{T}} \tag{12.1.9}$$

where $\mathbf{L}$ is a lower triangular matrix with unit diagonal entry and $\mathbf{D}$ is a diagonal matrix. The diagonal elements of $\mathbf{D}$ are positive exactly when $\nabla^2 f(\mathbf{x}_k)$ is positive definite.

Thus, in the course of the Cholesky decomposition, if it turns out that any element $d_{ii} \leq 0$, then it signals that $\nabla^2 f(\mathbf{x}_k)$ is not positive definite. We can then decide on the value of μ_k on the fly to force positive definiteness of $(\nabla^2 f(\mathbf{x}_k) + \mu_k \mathbf{I})$. Refer to Dennis and Schnabel (1996) for details of the mechanics of this implementation.

(2) **Truncated Newton's method** While the above modifications induce a much needed robustness to the Newton's method, Step 2 or its modification, Step 2′ calls for solving a large symmetric and positive definite linear system and constitutes a major computational bottleneck.

While there are numerous algorithms for solving such special class of linear systems – Cholesky decomposition (Chapter 9), conjugate gradient method (Chapter 11), to mention a few – in the worst case when $\mathbf{A}$ is dense, this could take $O(n^3)$ operations. To ease the burden of this computational bottleneck, one often invokes the trade-off between speed of convergence and accurate determination of the Newton direction. Recall that minimization algorithms are pretty robust with respect to the search of the descent direction $\mathbf{p}$ and the choice of the step length parameter α. Thus, instead of solving the Newton's equation exactly, we may only seek an acceptable approximate solution. This can be done by using a whole host of ideas and tools from the Krylov subspace

Given $f : \mathbb{R}^n \to \mathbb{R}$ and $\mathbf{x}_0$. Let $\nabla f(\mathbf{x})$ and $\nabla^2 f(\mathbf{x})$ denote the gradient and Hessian of $f(\mathbf{x})$.

Outer iteration For $k = 0, 1, 2, \ldots$

Step 1 Compute $\nabla f(\mathbf{x})$ and $\nabla^2 f(\mathbf{x})$.

Inner iteration:

Step 2 Approximately solve $\nabla^2 f(\mathbf{x}_k)\mathbf{p} = -\nabla f(\mathbf{x}_k)$ using iterative methods such as, say, the conjugate gradient method of Chapter 11 and let $\mathbf{p}_k$ denote such an approximate solution.

Step 3 Perform a one-dimensional minimization of $g(\alpha) = f(\mathbf{x}_k + \alpha \mathbf{p}_k)$ and let α_k be the step length described in Chapter 10.

End Inner iteration

Step 4 Define $\mathbf{x}_{k+1} = \mathbf{x}_k + \alpha_k \mathbf{p}_k$.

Step 5 Test for Convergence. If YES stop. Else go to step 1.

End Outer iteration

Fig. 12.1.2 Truncated Newton's method.

projection methods where an iterative method (such as the conjugate gradient method), used in solving the Newton's equation, is truncated once a "good" approximate solution is obtained. Variation of the Newton's algorithm where the search direction is obtained by only approximately solving the Newton's equation has come to be known as **truncated Newton's method**.

For later reference, we now describe a framework for the truncated Newton's method in Figure 12.1.2.

(3) **Prior Knowledge of Gradient and Hessian** The classical Newton's algorithm and its variations described above tacitly assume that the functional forms of gradient and the Hessian of $f(\mathbf{x})$ are known a priori. This is possible only in rare cases where $f(\mathbf{x})$ is known explicitly in advance. In many practical problems $f(\mathbf{x})$ may **not** be known in advance but often specified only as a black box where we can obtain the value of $f(\mathbf{x})$ for any input $\mathbf{x}$ (see Figure 10.1.3). In such cases, both the gradient vector $\nabla f(\mathbf{x})$ and the Hessian matrix $\nabla^2 f(\mathbf{x})$ have to be estimated on the fly by clever numerical approximations. Even granting that such approximations are feasible, it might happen that the Hessian $\nabla^2 f(\mathbf{x})$ at the current operating point may not be positive definite. This loss of positive definiteness would often cause difficulty in solving the Newton's equation. The whole family of quasi-Newton algorithms are meant to address and avoid this loss of positive definiteness of the Hessian, as well as reducing the cost of computing Newton direction by settling for a good approximation to it.

12.2 Quasi-Newton methods

Let $f : \mathbb{R}^n \to \mathbb{R}$ be the function to be minimized. It is assumed that $f(\mathbf{x})$ is not known explicitly but its value can be obtained for any point $\mathbf{x}$. Recall that the centerpiece of the Newton's method is the quadratic model for $f(\mathbf{x})$ at the current operating point $\mathbf{x}_k$. This requires the knowledge of the gradient $\nabla f(\mathbf{x})$. It is assumed that the ith element, $[\nabla f(\mathbf{x})]_i$ of the gradient is computed using the well known central difference approximation:

$$[\nabla f(\mathbf{x}_k)]_i \approx \frac{f(\mathbf{x}_k + ae_i) - f(\mathbf{x}_k - ae_i)}{2a} \tag{12.2.1}$$

for some small real constant $a > 0$ and e_i is the ith unit vector. Without getting into the details (of how to) let us assume for now that an approximation $\mathbf{B}_k$ to the Hessian of $f(\mathbf{x}_k)$ is known. Then, let

$$m(\mathbf{p}) = f(\mathbf{x}_k) + \mathbf{p}^{\mathrm{T}}\nabla f(\mathbf{x}_k) + \frac{1}{2}\mathbf{p}^{\mathrm{T}}\mathbf{B}_k\mathbf{p} \tag{12.2.2}$$

be the resulting quadratic model for $f(\mathbf{x}_k)$ at $\mathbf{x}_k$. If the gradient of $f(\mathbf{x})$ is really available then we could use the actual gradient in (12.2.2) instead of the approximation obtained using (12.2.1). The next search direction $\mathbf{p}_k$ is obtained as the solution of the following analog of the Newton's equation

$$\nabla m(\mathbf{p}) = \mathbf{B}_k\mathbf{p} + \nabla f(\mathbf{x}_k) = 0$$

or

$$\mathbf{p}_k = -\mathbf{B}_k^{-1}\nabla f(\mathbf{x}_k). \tag{12.2.3}$$

Clearly, $\mathbf{p}_k$ is an approximation to the actual Newton direction and the goodness of this approximation is directly related to that of $\mathbf{B}_k$ to the actual Hessian.

Before stating the specific proposals for the choice of $\mathbf{B}_k$, we list many of the conditions and properties required of $\mathbf{B}_k$.

(1) **Recursive update** Since $\mathbf{B}_k$ is to be computed in every iterative step, it is desirable to define $\mathbf{B}_k$ recursively as

$$\mathbf{B}_{k+1} = \mathbf{B}_k + \mathbf{E}_k \tag{12.2.4}$$

where $\mathbf{E}_k$ is called the matrix update. All the known quasi-Newton methods differ essentially in the way the matrix $\mathbf{E}_k$ is specified.

(2) **Symmetry** Since the actual Hessian is always symmetric, it is required that $\mathbf{B}_k$ is symmetric for all k. This can be guaranteed by specifying $\mathbf{B}_0$, the initial approximation to the Hessian a symmetric matrix and by requiring that $\mathbf{E}_k$ is symmetric for all k.

(3) **Positive definiteness** The Hessian of $f(\mathbf{x})$ near the minimum is positive definite. Hence it is required that $\mathbf{B}_k$ is positive definite. This guarantees that $\mathbf{p}_k$ defined in (12.2.3) is unique.

(4) **Secant condition** The question now is where to begin for computing the second derivatives based on the assumption that the first derivatives are available. The answer lies in an age old tradition in numerical analysis of using the so

called **secant formula** which in the univariate case may be stated as follows: If $g : \mathbb{R} \to \mathbb{R}$, and $g'(\mathbf{x})$ denotes the first derivative of $g(\mathbf{x})$, then

$$g''(\mathbf{x}_n) \approx \frac{g'(\mathbf{x}_n) - g'(\mathbf{x}_c)}{\mathbf{x}_n - \mathbf{x}_c} \tag{12.2.5}$$

where $|\mathbf{x}_n - \mathbf{x}_c|$ is small. We directly invoke the multidimensional generalization of this device which can be stated as follows:

$$\nabla^2 f(\mathbf{x}_{k+1})[\mathbf{x}_{k+1} - \mathbf{x}_k] \approx \nabla f(\mathbf{x}_{k+1}) - \nabla f(\mathbf{x}_k). \tag{12.2.6}$$

It is natural to require that any approximation $\mathbf{B}_k$ to the Hessian $\nabla^2 f(\mathbf{x}_{k+1})$ also satisfy this basic relation. In imposing this requirement, on $\mathbf{B}_k$, let us simplify the notation by defining

$$\mathbf{s}_k = \mathbf{x}_{k+1} - \mathbf{x}_k = \alpha \mathbf{p}, \text{ the search direction,}$$

and

$$\mathbf{y}_k = \nabla f(\mathbf{x}_{k+1}) - \nabla f(\mathbf{x}_k). \tag{12.2.7}$$

It is now required that any $\mathbf{B}_k$ must satisfy the following **secant condition**

$$\mathbf{B}_k \mathbf{s}_k = \mathbf{y}_k. \tag{12.2.8}$$

The import of this condition is that the approximation $\mathbf{B}_k$ behaves like the Hessian with respect to the current search direction $\mathbf{p}$. (Exercise 12.6)

(5) **Ease of computation** The ultimate test of this approximation lies in the computational efforts needed to solve (12.2.3). Notice that our interest in $\mathbf{B}_k$ is only through its inverse to obtain $\mathbf{p}_k = -\mathbf{B}_k^{-1} \nabla f(\mathbf{x}_k)$ which in turn implies whatever approximation we choose for $\mathbf{B}_k$, it must be easily invertible. By invoking the **Sherman–Morrison–Woodbury** formula (Appendix B), it is immediate that we can recursively compute $\mathbf{B}_{k+1}^{-1}$ from $\mathbf{B}_k^{-1}$ using only $O(n^2)$ computations provided that the rank of $\mathbf{E}_k$ is very small (say one or two) and the initial choice of $\mathbf{B}_0$ is readily invertible.

Having laid the ground rules for the design of the Hessian approximation, we now describe the first proposal for $\mathbf{B}_k$.

(1) **Broyden's Formula** Let $\mathbf{B}_0 = \mathbf{I}$, the identity matrix and

$$\mathbf{B}_{k+1} = \mathbf{B}_k + \frac{(\mathbf{y}_k - \mathbf{B}_k \mathbf{s}_k)(\mathbf{y}_k - \mathbf{B}_k \mathbf{s}_k)^{\mathrm{T}}}{(\mathbf{y}_k - \mathbf{B}_k \mathbf{s}_k)^{\mathrm{T}} \mathbf{s}_k}. \tag{12.2.9}$$

Here the matrix update $\mathbf{E}_k$ is given by the second term on the r.h.s. of (12.2.9) which is a constant multiple of the outer product of $(\mathbf{y}_k - \mathbf{B}_k \mathbf{s}_k)$ with itself and hence it is symmetric, and is of rank one. It is easy to verify that $\mathbf{B}_{k+1}$

also satisfies the secant condition (Exercise 12.8) and that $\mathbf{B}_{k+1}^{-1}$ can be computed recursively from $\mathbf{B}_k^{-1}$ using Sherman–Morrison–Woodbury's formula (Exercise 12.9).

Broyden (1965) was the first to propose update formulae of the above type in the context of solving nonlinear equations using quasi-Newton methods and proved their intrinsic properties. Since then the update formula of the type (12.2.9) has come to be known as the Broyden class of formulae. Notwithstanding its use in the solution of nonlinear equations, since our goal is the minimization of $f(\mathbf{x})$, we in addition require that $\mathbf{B}_{k+1}$ in (12.2.9) is also positive definite. It turns out that rank one updates do not guarantee positive definiteness of $\mathbf{B}_{k+1}$ even if $\mathbf{B}_k$ is. To remedy this problem, several modifications of Broyden's formula were proposed. In the following, we mention only the two important ones.

(2) **Broyden–Fletcher–Goldfarb–Shanno (BFGS) formula** In this approach $\mathbf{B}_{k+1}$ is obtained from $\mathbf{B}_k$ using a **rank-two** update as follows:

$$\mathbf{B}_{k+1} = \mathbf{B}_k - \frac{(\mathbf{B}_k\mathbf{s}_k)^{\mathrm{T}}(\mathbf{B}_k\mathbf{s}_k)}{(\mathbf{s}_k^{\mathrm{T}}\mathbf{B}_k\mathbf{s}_k)} + \frac{\mathbf{y}_k\mathbf{y}_k^{\mathrm{T}}}{\mathbf{y}_k^{\mathrm{T}}\mathbf{s}_k}. \tag{12.2.10}$$

Clearly, the update matrix $\mathbf{E}_k$ is the sum of two symmetric, rank-one, outer product matrices.

Hence $\mathbf{E}_k$ is symmetric and is of rank two. It can be shown that if $\mathbf{B}_k$ is positive definite then so is $\mathbf{B}_{k+1}$ defined by (12.2.10) if and only if $\mathbf{y}_k^{\mathrm{T}}\mathbf{s}_k > 0$, which when expanded becomes

$$\mathbf{p}^{\mathrm{T}}\nabla f(\mathbf{x}_k) < \mathbf{p}^{\mathrm{T}}\nabla f(\mathbf{x}_{k+1}). \tag{12.2.11}$$

That is, the directional derivative of $f(\mathbf{x})$ in the direction $\mathbf{p}$ evaluated at $\mathbf{x}_{k+1}$ is larger than its value at $\mathbf{x}_k$. Recall that for $\mathbf{p}$ to be a descent direction at $\mathbf{x}_k$ it is required that $\mathbf{p}^{\mathrm{T}}\nabla f(\mathbf{x}_k) < 0$. Given $\mathbf{x}_k$ and the search direction $\mathbf{p}$, the next iterate $\mathbf{x}_{k+1}$ is obtained by the dimensional minimization (Section 10.4). Recall that $\mathbf{x}_{k+1}$ is optimal for $\mathbf{x}_k$ in the direction $\mathbf{p}$ only if $\mathbf{p}^{\mathrm{T}}\nabla f(\mathbf{x}_{k+1}) = 0$ (see Exercise 10.6 and Appendix D). Hence we can readily ensure the condition (12.2.11) for the positive definiteness of $\mathbf{B}_{k+1}$ by suitable one-dimensional minimization to compute $\mathbf{x}_{k+1}$. Further, $\mathbf{B}_{k+1}^{-1}$ can be computed readily using the Sherman-Morrison-Woodbury formula.

The quasi-Newton method using $\mathbf{B}_{k+1}$ given by the BFGS formula (12.2.10) has become a standard method for multidimensional minimization.

(3) **Generalized Davidon–Fletcher–Powell (DFP) Method** In this $\mathbf{B}_k$ recurrence is given by

$$\mathbf{B}_{k+1} = \mathbf{B}_k - \frac{(\mathbf{B}_k\mathbf{s}_k)(\mathbf{B}_k\mathbf{s}_k)^{\mathrm{T}}}{\mathbf{s}_k^{\mathrm{T}}\mathbf{B}_k\mathbf{s}_k} + \frac{\mathbf{y}_k\mathbf{y}_k^{\mathrm{T}}}{\mathbf{y}_k^{\mathrm{T}}\mathbf{s}_k} + (\mathbf{s}_k^{\mathrm{T}}\mathbf{B}_k\mathbf{s}_k)\delta\boldsymbol{\eta}_k\boldsymbol{\eta}_k^{\mathrm{T}}, \tag{12.2.12}$$

where δ is a real constant and

$$\eta_k = \frac{\mathbf{y}_k}{\mathbf{y}_k^{\mathrm{T}}\mathbf{s}_k} - \frac{\mathbf{B}_k\mathbf{s}_k}{\mathbf{s}_k^{\mathrm{T}}\mathbf{B}_k\mathbf{s}_k}. \tag{12.2.13}$$

When $\delta = 0$, we get the BFGS scheme and when $\delta = 1$, it is called the Davidon–Fletcher–Powell scheme. It can be verified that this family of update schemes preserve the positive definiteness of $\mathbf{B}_k$.

Remark 12.2.1 Given this level of approximations, one might wonder whether quasi-Newton algorithms converge at all. Indeed, it can be shown algorithm based on the BFGS scheme converges at a super linear rate. For details, refer to Dennis and Schnabel (1996).

12.3 Limiting space requirement in quasi-Newton method

All the improvements made thus far – introduction of the step length parameter, recursive estimation of the Hessian using BFGS method or other equivalent scheme, the notion of truncated Newton method – all have contributed to the overall robustness and reduction in computation time. But it still requires $O(n^2)$ words of memory for storing the Hessian approximation. For large scale problems of interest in geophysical sciences, the value of n could easily be in the range 10^6–10^8. Such large problems arise in the context of dynamic data assimilation using the 4DVAR method described in Part IV. To accommodate problems of this size, we need to look for ways to reduce the space requirement. Any such attempt may have an undesirable effect on the rate of convergence, but this is a small price to pay to make such large-scale problems feasible in today's technology. It is worth remembering that the nonlinear conjugate gradient method (Section 11.3) does not require any matrix storage and is a viable algorithm for large scale problems. In the following, we describe two such modifications.

(1) **Approximation Hessian-Vector Product**
In this approach, the special nature of the Newton's equation

$$\nabla^2 f(\mathbf{x}_k)\mathbf{p} = -\nabla f(\mathbf{x}_k) \tag{12.3.1}$$

is exploited. Since it is assumed that $f(\mathbf{x})$ is twice continuously differentiable, we can use the first-order Taylor series expansion for the gradient to express $\nabla f(\mathbf{x}_k + \alpha\mathbf{p})$ as

$$\nabla f(\mathbf{x}_k + a\mathbf{p}) \approx \nabla f(\mathbf{x}_k) + a\nabla^2 f(\mathbf{x}_k)\mathbf{p} \tag{12.3.2}$$

for any real constant $a > 0$. On rearranging, we get an approximate expression for the Hessian-vector product

$$\nabla^2 f(\mathbf{x}_k)\mathbf{p} \approx \frac{1}{a}[\nabla f(\mathbf{x}_k + a\mathbf{p}) - \nabla f(\mathbf{x}_k)]. \tag{12.3.3}$$

Since $\nabla f(\mathbf{x}_k)$ is already known, we can compute the r.h.s. of this expression with one more evaluation of the gradient at $(\mathbf{x}_k + a\mathbf{p})$.

The basic idea is to integrate this computation of the Hessian-vector product with the truncated Newton's method described in Figure 12.1.2 where the inner iteration solves the Newton's equation using the conjugate gradient method. A quick review of the conjugate gradient method (Figure 11.2.1) for solving $\mathbf{Ax} = \mathbf{b}$ reveals that this algorithm uses the matrix A only through the matrix-vector product $(\mathbf{Ax}_0)$ and $(\mathbf{Ap}_k)$ for each $k = 0, 1, 2, \ldots, n-1$ and does **not** need the matrix $\mathbf{A}$ explicitly.[†]

To see the final connection; recall that the kth outer iteration of the truncated Newton's method in Figure 12.1.2, the inner iteration is called to solve the Newton's equation (12.3.1) using the conjugate gradient method using the following association:

$$\left.\begin{aligned} \nabla^2 f(\mathbf{x}_k) &\leftrightarrow \mathbf{A} \\ -\nabla f(\mathbf{x}_k) &\leftrightarrow \mathbf{b} \\ \mathbf{p} &\leftrightarrow \mathbf{x} \end{aligned}\right\} \tag{12.3.4}$$

Indeed, whenever the matrix-vector product involving $\mathbf{A}$ occurs, it is to be replaced by the Hessian–vector product (12.3.3).

A complete integrated view of the resulting minimization algorithm is given in Figure 12.3.1. To avoid confusion, we have made appropriate changes to the notation in describing the CG method and this modified version is given in Figure 12.3.2. A program for computing the Hessian-vector product is given in Figure 12.3.3. It can be readily verified that this algorithm requires only $O(n)$ storage and $O(n)$ time per iteration of the inner loops. It is for this reason this is one of the recommended methods for large scale problems.

Remark 12.3.1 Within the context of 4DVAR method for dynamic data assimilation, we can obtain the gradient and the Hessian-vector product using the first-order and the second-order adjoint methods respectively. Given this information using the tools of this chapter, we can design a wide variety of algorithms for large-scale data assimilation problems.

(2) **Limiting space using a restart strategy**

In looking for another strategy to reduce space requirements, recall from Section 12.2 that the BFGS formula for the Hessian is given by

$$\mathbf{B}_k = \mathbf{B}_{k-1} - \frac{(\mathbf{B}_{k-1}\mathbf{s}_{k-1})(\mathbf{B}_{k-1}\mathbf{s}_{k-1})^{\mathrm{T}}}{(\mathbf{s}_{k-1}^{\mathrm{T}}\mathbf{B}_{k-1}\mathbf{s}_{k-1})} + \frac{\mathbf{y}_{k-1}\mathbf{y}_{k-1}^{\mathrm{T}}}{\mathbf{y}_{k-1}^{\mathrm{T}}\mathbf{y}_{k-1}} \tag{12.3.5}$$

[†] It is important to realize that the $\mathbf{p}_k$ used in the CG method in Figure (11.2.1) and $\mathbf{p}$'s used in the Newton's method are different. To avoid any unintended confusion when rewriting the CG method in Figure 12.3.2, we use a different set of notation.

Given $f : \mathbb{R}^n \to \mathbb{R}$ and $\nabla f(\mathbf{x})$. $\mathbf{x}_0$ is the starting vector.

Iteration: $k = 0, 1, 2, \ldots$

Step 1 Compute $\nabla f(\mathbf{x}_k)$.

Step 2 **Call the Conjugate Gradient routine** in Figure 12.3.2 with $\nabla f(\mathbf{x}_k)$ as the input. This routine will deliver a descent direction $\mathbf{p}_k$ which is an approximate solution to the Newton's equation (12.3.1).

Step 3 Given $\mathbf{x}_k$ and $\mathbf{p}_k$, compute α_k, an approximate minimizer for $g(\alpha) = f(\mathbf{x}_k + \alpha\mathbf{p})$ using the one-dimensional minimization method in Section 10.4.

Step 4 Compute $\mathbf{x}_{k+1} = \mathbf{x}_k + \alpha_k\mathbf{p}_k$.

Step 5 Test for convergence. If YES, exit. Else go to step 1.

Fig. 12.3.1 Truncated Newton – MAIN PROGRAM.

- This Conjugate gradient routine computes an approximate solution to the Newton's equation using the Hessian-vector product approximation in (12.3.3).
- Let $\mathbf{b} = -\nabla f(\mathbf{x}_k)$ where $\nabla f(\mathbf{x}_k)$ is received as an input from the Main Program in Figure 12.3.1.
- This program solves $\mathbf{By} = \mathbf{b}$ where $\mathbf{B}$ is the Hessian $\nabla^2 f(\mathbf{x}_k)$ and $\mathbf{y}$ is the Newton direction. Since this routine does **not** have access to $\mathbf{B}$, whenever the the Hessian-vector product of the form ($\mathbf{Bz}$) is needed for any vector $\mathbf{z}$, it calls another routine in Figure 12.3.3, that computes an approximation to this Hessian-vector product using (12.3.3).
- In solving $\mathbf{By} = \mathbf{b}$, let $\mathbf{y}_0$ be the initial approximation to $\mathbf{y}$. That is, pick $\mathbf{y}_0$.
- Compute $\mathbf{r}_0 = \mathbf{b} - (\mathbf{By}_0)$ and let $\mathbf{q}_0 = \mathbf{r}_0$. Notice that the product ($\mathbf{By}_0$) is obtained by a call to the program in Figure 12.3.3, with $\mathbf{y}_0$ as the input.

For $k = 0, 1, 2, \ldots$

Step 1 **Compute step length**

$$\delta_k = \frac{\mathbf{r}_k^{\mathrm{T}}\mathbf{r}_k}{\mathbf{q}_k^{\mathrm{T}}(\mathbf{Bq}_k)}$$

/*This step needs the Hessian-vector product $\mathbf{Bq}_k$.*/

Step 2 **Update the iterate**

$$\mathbf{y}_{k+1} = \mathbf{y}_k + \delta_k\mathbf{q}_k.$$

Step 3 **Update the residual**

$$\mathbf{r}_{k+1} = \mathbf{r}_k - \delta_k(\mathbf{Bq}_k).$$

/*In this step, we can reuse the ($\mathbf{Bq}_k$) from Step 1.*/

Step 4 Test for convergence.

Step 5 Compute step length

$$\eta_k = \frac{\mathbf{r}_{k+1}^{\mathrm{T}}\mathbf{r}_{k+1}}{\mathbf{r}_k^{\mathrm{T}}\mathbf{r}_k}$$

Step 6 Update the search direction

$$\mathbf{q}_{k+1} = \mathbf{r}_{k+1} + \eta_k\mathbf{q}_k$$

Fig. 12.3.2 Conjugate Gradient method using the Hessian-vector Product.

- This routine computes an approximation to the Hessian-vector product.
- Needs access to the formula for $\nabla f(\mathbf{x})$. If this is not available, need access to $f(\mathbf{x})$ where $\nabla f(\mathbf{x})$ is computed using finite-difference approximation.
- Computes an approximation to $\mathbf{Bz}$

$$\mathbf{Bz} \approx \frac{1}{a}[\nabla f(\mathbf{x}_k + a\mathbf{z}) - \nabla f(\mathbf{x}_k)]$$

for some real constant $a > 0$ where $\mathbf{z}$ is an input.

Fig. 12.3.3 The Hessian-vector product.

where $\mathbf{B}_0 = \mathbf{I}$, the identity matrix, $\mathbf{s}_k = (\mathbf{x}_{k-1} - \mathbf{x}_k)$ and $\mathbf{y}_k = [\nabla f(\mathbf{x}_{k+1}) - \nabla f(\mathbf{x}_k))]$. At each step, the aim is to solve the approximate Newton's equation

$$\mathbf{B}_k\mathbf{p} = -\nabla f(\mathbf{x}_k) \tag{12.3.6}$$

and use the solution $\mathbf{p}_k$ as the next descent direction. It turns out that we could indeed derive an update formula for $\mathbf{H}_k = \mathbf{B}_k^{-1}$ using the Sherman–Morrison–Woodbury formula. It can be shown that

$$\mathbf{H}_k = \mathbf{H}_{k-1} + \frac{(\mathbf{y}_{k-1} - \mathbf{H}_{k-1}\mathbf{s}_{k-1})\mathbf{y}_{k-1}^{\mathrm{T}}}{\mathbf{y}_{k-1}^{\mathrm{T}}\mathbf{s}_{k-1}} - \left[\frac{(\mathbf{y}_{k-1} - \mathbf{H}_{k-1}\mathbf{s}_{k-1})^{\mathrm{T}}\mathbf{s}_{k-1}}{(\mathbf{y}_{k-1}^{\mathrm{T}}s_{k-1})^2}\right]\mathbf{y}_{k-1}\mathbf{y}_{k-1}^{\mathrm{T}}. \tag{12.3.7}$$

Given this, the required search direction $\mathbf{p}_k$ can be computed using

$$\mathbf{p}_k = -\mathbf{H}_k\nabla f(\mathbf{x}_k). \tag{12.3.8}$$

This is the starting point for introducing various approximations to save space.

First observe that (12.3.7) is a first-order matrix recurrence relation. Consequently, $\mathbf{H}_k$ depends on all the past values of $\mathbf{H}_j$, $j = 0, 1, 2, \ldots, k-1$. Since $\mathbf{H}_0 = \mathbf{I}$, we can express $\mathbf{H}_1$ as (using $k = 1$ in (12.3.7))

$$\mathbf{H}_1 = \mathbf{I} + \frac{(\mathbf{y}_0 - \mathbf{s}_0)\mathbf{y}_0^{\mathrm{T}}}{\mathbf{y}_0^{\mathrm{T}}\mathbf{s}_0} - \left[\frac{(\mathbf{y}_0 - \mathbf{s}_0)^{\mathrm{T}}\mathbf{s}_0}{(\mathbf{y}_0^{\mathrm{T}}\mathbf{s}_0)^2}\right]\mathbf{y}_0\mathbf{y}_0^{\mathrm{T}} \tag{12.3.9}$$

where the r.h.s. of (12.3.8) depends only on $(\mathbf{s}_0, \mathbf{y}_0)$. We can now substitute this formula for $\mathbf{H}_1$ into $\mathbf{H}_2$ and express $\mathbf{H}_2$ as a function of the pairs $(\mathbf{s}_1, \mathbf{y}_1)$ and $(\mathbf{s}_0, \mathbf{y}_0)$ (Exercise 12.10). Continuing this, we can readily see that $\mathbf{H}_k$ is a function of all the pairs $\{(\mathbf{s}_j, \mathbf{y}_j) | j = 0, 1, 2, \ldots, k-1\}$.

One way to reduce the space requirement is to artificially reduce this dependence of $\mathbf{H}_k$ on the entire past by limiting its dependence to only m pairs $\{(\mathbf{s}_j, \mathbf{y}_j) | j = k-1, k-2, \ldots, k-m\}$ for some prespecified small value of m. That is, we limit the extent of this dependence to a **moving window** of size m and hence the name "restart strategy." The following is an illustration.

Restart at every step, m = 1 In this case, it is assumed that $\mathbf{H}_{k-1} = \mathbf{I}$ and then (12.3.7) becomes

$$\mathbf{H}_k = \mathbf{I} + \frac{(\mathbf{y}_{k-1} - \mathbf{s}_{k-1})\mathbf{y}_{k-1}^{\mathrm{T}}}{\mathbf{y}_{k-1}^{\mathrm{T}}\mathbf{s}_{k-1}} - \left[\frac{(\mathbf{y}_{k-1} - \mathbf{s}_{k-1})^{\mathrm{T}}\mathbf{s}_{k-1}}{(\mathbf{y}_{k-1}^{\mathrm{T}} s_{k-1})^2}\right]\mathbf{y}_{k-1}\mathbf{y}_{k-1}^{\mathrm{T}}. \qquad (12.3.10)$$

Combining this with (12.3.8), we get, after simplication,

$$\mathbf{p}_k = -\nabla f(\mathbf{x}_k) - \frac{(\mathbf{y}_{k-1}^{\mathrm{T}}\nabla f(\mathbf{x}_k))}{\mathbf{y}_{k-1}^{\mathrm{T}}\mathbf{s}_{k-1}}(\mathbf{y}_{k-1} - \mathbf{s}_{k-1}) + \frac{(\mathbf{y}_{k-1}\mathbf{s}_{k-1})^{\mathrm{T}}\mathbf{s}_{k-1}}{(\mathbf{y}_{k-1}^{\mathrm{T}}\mathbf{s}_{k-1})^2}[\mathbf{y}_{k-1}^{\mathrm{T}}\nabla f(\mathbf{x})]\mathbf{y}_{k-1}. \qquad (12.3.11)$$

Notice that the r.h.s. of (12.3.11) uses only three vectors $\mathbf{s}_{k-1}$, $\mathbf{y}_{k-1}$, and $\nabla f(\mathbf{x}_k)$ and no explicit matrix storage or matrix-vector product is needed. Bravo!

Clearly, while this idea has drastically reduced the storage, for sure we have also compromised on the quality of the resulting descent direction. It is obvious that the quality of this approximation improves with m.

Restart with m = 2 Assuming $\mathbf{H}_{k-2} = \mathbf{I}$, we can obtain an update formula for $\mathbf{H}_k$ as a function of $(\mathbf{s}_{k-1}, \mathbf{y}_{k-1})$ and $(\mathbf{s}_{k-2}, \mathbf{y}_{k-2})$ (see Exercise 12.10). In this case, the formula $\mathbf{H}_k$ and hence $\mathbf{p}_k$ will be much more complex than the ones in (12.3.10) and (12.3.11) respectively. Herein lies the saving – instead of storing the matrix $\mathbf{H}_k$, we need only store five vectors $(\mathbf{s}_{k-1}, \mathbf{y}_{k-1})$, $(\mathbf{s}_{k-2}, \mathbf{y}_{k-2})$, and $\nabla f(\mathbf{x}_k)$ with an increased cost of computing $\mathbf{p}_k$ which uses more information compared to when $m = 1$ and hopefully of a better quality.

Notice that the window size m effectively controls the amount of information used in obtaining $\mathbf{p}_k$. The ultimate effectiveness of this strategy may have to be settled by careful experimental study. Past experience has shown that a window size of three to five has been found adequate in many problems.

Exercises

12.1 **Newton's Method Solution of Equation Scalar Case** Let $f : \mathbb{R} \to \mathbb{R}$. Let $\mathbf{x}_c$ be the current operating point. The next approximation is obtained by using the first-order Taylor expansion:

$$f(\mathbf{x}_c + \mathbf{p}) \approx f(\mathbf{x}_c) + \mathbf{p}f'(\mathbf{x}_c) = 0$$

that is

$$\mathbf{p} = -\frac{f(\mathbf{x}_c)}{f'(\mathbf{x}_c)}$$

provided $f'(\mathbf{x}_c) \neq 0$. Then the Newton's method is defined by

$$\mathbf{x}_{k+1} = \mathbf{x}_k - \frac{f(\mathbf{x}_k)}{f'(\mathbf{x}_k)}.$$

(a) Apply this algorithm to compute the solution of $\mathbf{x}^2 = a$, that is, the square root of a.

12.2 **Newton's Method Solution of Equation Vector Case** Let $f(\mathbf{x}) = (f_1(\mathbf{x}), f_2(\mathbf{x}), \ldots, f_n(\mathbf{x}))^{\mathrm{T}}$ where $f_i : \mathbb{R}^n \to \mathbb{R}$ with $\mathbf{x} = (\mathbf{x}_1, \mathbf{x}_2, \ldots, \mathbf{x}_n)^{\mathrm{T}}$. The first-order Taylor expansion for $f(\mathbf{x})$ is given by

$$f(\mathbf{x} + \mathbf{p}) \approx f(\mathbf{x}) + \mathbf{D}_f(\mathbf{x})\mathbf{p} = 0$$

from which we obtain the following algorithm

$$\mathbf{x}_{k+1} = \mathbf{x}_k - \mathbf{D}_f^{-1}(\mathbf{x}) f(\mathbf{x})$$

where $\mathbf{D}_f(\mathbf{x})$ is the Jacobian of $f(\mathbf{x})$

(a) Apply this algorithm to solve for the zeros of $f(\mathbf{x}) = (f_1(\mathbf{x}), f_2(\mathbf{x}))^{\mathrm{T}}$ where $f_1(\mathbf{x}) = \mathbf{x}_1^2 - 17$ and $f_2(\mathbf{x}) = \mathbf{x}_2^2 - 11$.

12.3 **Quadratic Convergence of Newton's Algorithm** Consider the scalar case in Exercise(12.1). Let $\mathbf{e}_k = \mathbf{x}_k - \mathbf{x}^*$, the error in the kth iterate $\mathbf{x}_k$.

Then, using the second-order Taylor series, we get

$$\begin{aligned} 0 = f(\mathbf{x}^*) &= f(\mathbf{x}_k - (\mathbf{x}_k - \mathbf{x}^*)) \\ &= f(\mathbf{x}_k) - \mathbf{e}_k f'(\mathbf{x}_k) + \tfrac{1}{2}\mathbf{e}_k^2 f''(\xi) \end{aligned}$$

for some $\mathbf{x}_k - \mathbf{x}^* < \xi < \mathbf{x}^*$.

(a) Rearrange the above equation to obtain

$$\mathbf{x}_{k+1} - \mathbf{x}^* = \frac{1}{2}(\mathbf{x}_k - \mathbf{x}^*)^2 \frac{f''(\xi)}{f'(\mathbf{x}_k)}$$

from which conclude that

$$|\mathbf{x}_{k+1} - \mathbf{x}^*| \leq c_k |\mathbf{x}_k - \mathbf{x}^*|^2$$

for some $c_k = |\frac{f''(\xi)}{2f'(\mathbf{x}_k)}|$.

That is, the Newton's algorithm converges quadratically.

12.4 Consider the minimization of $f(\mathbf{x}) = 2\mathbf{x}_1^2 + \mathbf{x}_2^2 - 2\mathbf{x}_1\mathbf{x}_2 + 2\mathbf{x}_1^3 + \mathbf{x}_1^4$. Derive the Newton's equation and solve for the Newton direction.

12.5 Show that the Newton's method is invariant under the transformation $\mathbf{y} = \mathbf{B}\mathbf{x} + \mathbf{b}$ where $\mathbf{B}$ is a non-singular matrix and $\mathbf{b}$ is a vector.

12.6 Let $f(\mathbf{x}) = \frac{1}{2}\mathbf{x}^{\mathrm{T}}\mathbf{A}\mathbf{x} - \mathbf{b}^{\mathrm{T}}\mathbf{x}$. Then verify that the Hessian of the quadratic form satisfies the secant condition (12.2.8) exactly.

12.7 Under what condition, the Newton direction $\mathbf{p}^* = -[\nabla^2 f(\mathbf{x}_c)]\nabla f(\mathbf{x}_c)$ defined by (12.1.4) is a descent direction.

Hint: Recall that for $\mathbf{p}$ to be a descent direction $\mathbf{p}^{\mathrm{T}}[\nabla f(\mathbf{x})] < 0$.

12.8 Verify that $\mathbf{B}_{k+1}$ defined in (12.2.9) satisfies the secant condition (12.2.8).

12.9 Compute the recurrence formula for the inverse of $\mathbf{B}_{k+1}$ in (12.2.9) using the Sherman–Morrison formula.

12.10 Using the expression for $\mathbf{H}_1$ in (12.3.9). Compute an explicit for $\mathbf{H}_2$ using $(\mathbf{s}_0, \mathbf{y}_0)$ and $(\mathbf{s}_1, \mathbf{y}_1)$. Repeat it for $\mathbf{H}_3$. Do you see any pattern in these formulae? If so can you generalize to $\mathbf{H}_m$?

Notes and references

The book by Dennis and Schnabel (1996) provide a thorough analysis of Newton and quasi-Newton methods (Chapter 12) for both the solution of nonlinear equations and minimization. Also refer to Ortega and Rhienboldt (1970). Nash and Sofer (1996) provide a very readable and thorough overview of these methods.

PART IV

Statistical estimation

13

Principles of statistical estimation

This opening chapter of Part IV provides an introduction to basic concepts and definitions that are germane to the statistical theory of estimation. Section 13.1 provides an overview of the underpinnings of the various formulations of the estimation problem namely the **deterministic** vs. **Fisher's** vs. **Bayesian** framework. Many of desirable attributes of a "good" estimate are characterized in Section 13.2.

13.1 Statement and formulation of the estimation problem

There is a **variable** or **parameter** $\mathbf{x} \in \mathbb{R}^n (n \geq 1)$ representing an **unknown** quantity, often called the **true state** of the underlying system, to be estimated. This unknown $\mathbf{x}$ is **not** directly observable but we can measure a quantity $\mathbf{z} \in \mathbb{R}^m (m \geq 1)$ called the **observation**, that depends on this unknown $\mathbf{x}$. Stated in simple terms, the estimation problem of interest to us is: given $\mathbf{z}$, obtain the "best" estimate $\hat{\mathbf{x}}$ of $\mathbf{x}$. To build the bridge between the observation $\mathbf{z}$ and the estimate $\hat{\mathbf{x}}$ of $\mathbf{x}$, we need to develop several building blocks to which we now turn.

(a) **Measurement system** First and foremost is a mathematical model that relates the observation $\mathbf{z}$ to the state $\mathbf{x}$. Let $h : \mathbb{R}^n \to \mathbb{R}^m$ where $\mathbf{z} = \mathbf{h}(\mathbf{x}) = (h_1(\mathbf{x}), h_2(\mathbf{x}), \ldots, h_m(\mathbf{x}))^{\mathrm{T}}$. This function $\mathbf{h}(\cdot)$ represents the physical relation between $\mathbf{x}$ and $\mathbf{z}$. As observed in Chapter 1, this $\mathbf{h}(\cdot)$ could be based on **Faraday's law** that relates ($\mathbf{x} =$) the speed of a car and ($\mathbf{z} =$) the voltage generated. Given the fixed properties of the electrical generator, voltage is directly proportional to the speed. As a second example, $\mathbf{z}$ could denote the reflectivity as observed by the radar and $\mathbf{x}$ could represent the rate of rain. This function $\mathbf{h}$ is based on the physical or empirical laws that are used by the transducers that make up the measurement system. In general $\mathbf{h}(\cdot)$ could be a nonlinear function of $\mathbf{x}$. In the special case when it is linear, we use the notation $\mathbf{z} = \mathbf{H}\mathbf{x}$ where $\mathbf{H} \in \mathbb{R}^{m \times n}$ is a real $m \times n$ matrix.

(b) **Model for the observation** Many a time, observations are corrupted or contaminated by noise. Let $\mathbf{v} \in \mathbb{R}^m$ denote a vector that represents the actual noise corrupting the observation. It is assumed that this noise is **additive** in nature in that $\mathbf{z}$ is given by

$$\mathbf{z} = \mathbf{h}(\mathbf{x}) + \mathbf{v}. \tag{13.1.1}$$

We would like to emphasize that in this relation $\mathbf{z}$ and $\mathbf{h}(\cdot)$ are known but $\mathbf{x}$ and $\mathbf{v}$ are **not**.

To meaningfully formulate the estimation problem, we need further assumptions about $\mathbf{x}$ and $\mathbf{v}$.

(c) **Model for x** There are essentially two schools of thought relating to the handling of the unknown $\mathbf{x}$. First is the one introduced by **Sir Ronald Fisher** in the early 1920s that championed the idea of treating the **unknown x as an unknown constant** μ. Based on this idea, he developed a fundamental method called the **maximum likelihood technique** thereby erecting the first cornerstone for the modern statistical theory of **point estimation**. The second is the **Bayesian approach** wherein the unknown $\mathbf{x}$ itself is treated as a random variable/vector whose distribution $p(\mathbf{x})$ is known a priori. This latter distribution models the uncertainty in $\mathbf{x}$ by summarizing all the available (subjective) information about it well before the arrival of any observation. In this Bayesian approach, it is usually assumed that the a priori distribution $p(\mathbf{x})$ is centered at μ, that is $E(\mathbf{x}) = \mu$ where the expectation (Appendix F) is taken with respect to $p(\mathbf{x})$.

(d) **Model for noise** While the actual additive noise $\mathbf{v}$ is **not** directly observable, for tractability, we still need a good mathematical model for it. The standard assumptions are:
   (a) $\mathbf{v}$ has **mean zero**, that is, $E(\mathbf{v}) = 0$
   (b) The **covariance matrix** R of $\mathbf{v}$ is known and is positive definite, that is, $E(\mathbf{v}\mathbf{v}^{\mathrm{T}}) = R$, a symmetric and positive definite matrix, and
   (c) The observation noise $\mathbf{v}$ and the unknown state $\mathbf{x}$ are uncorrelated. In Fisher's framework since $\mathbf{x} = \mu$, this condition reduces to $E[\mathbf{v}\mu^{\mathrm{T}}] = E[\mathbf{v}]\mu^{\mathrm{T}} = 0$. However, in the Bayesian framework, this condition implies $E[\mathbf{v}(\mathbf{x} - \mu)^{\mathrm{T}}] = 0$. Notice that these standard assumptions relate only to the first two moments of $\mathbf{v}$ and do not require the knowledge of the distribution of $\mathbf{v}$. In special cases we will assume that $\mathbf{v}$ has multivariate normal distribution, that is, $\mathbf{v} \sim N(0, R)$.

Given these four components, let $\phi : \mathbb{R}^m \rightarrow \mathbb{R}^n$ where the **estimate $\hat{\mathbf{x}}$ of x** is given by

$$\hat{\mathbf{x}} = \phi(\mathbf{z}) \tag{13.1.2}$$

and the function ϕ depends only on the measurement system $\mathbf{h}(\cdot)$, and the models for both $\mathbf{x}$ and $\mathbf{v}$. This function ϕ is called the **estimator** and its value evaluated at $\mathbf{z}$ gives an estimate $\hat{\mathbf{x}}$ of $\mathbf{x}$. Since $\mathbf{z}$ is random, the estimate $\hat{\mathbf{x}}$ is also a random vector. The goal of the estimation theory is to quantify the properties of the probability distribution of

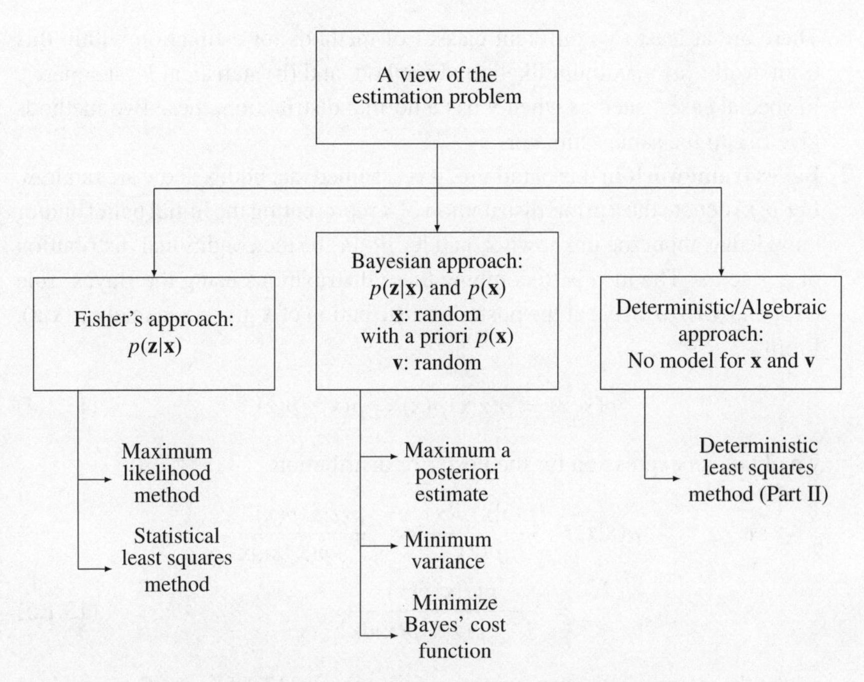

Fig. 13.1.1 A global view of the statistical estimation problem.

$\hat{\mathbf{x}}$ which depends on the structure of the estimator $\phi(\cdot)$, the measurement system $\mathbf{h}(\cdot)$ and on the statistical properties of $\mathbf{v}$ and $\mathbf{x}$. If ϕ is such that $\hat{\mathbf{x}}$ is a linear function of $\mathbf{z}$, then $\hat{\mathbf{x}}$ is called a **linear estimate** and $\phi(\cdot)$ is called a linear estimator. Otherwise, $\hat{\mathbf{x}}$ is called a **nonlinear estimate**, and ϕ is a **nonlinear estimator**.

The best estimate is one that is defined in terms of minimizing a scalar quantity based on the **error in the estimate** defined by

$$\tilde{\mathbf{x}} = \hat{\mathbf{x}} - \mathbf{x}. \tag{13.1.3}$$

Since $\mathbf{x}$ is **not** known, we must arrange matters in such a way that the statistical properties of $\tilde{\mathbf{x}}$ depend only on $\mathbf{h}(\,)$, the models for both $\mathbf{x}$ and $\mathbf{v}$ and the chosen estimator. Further it is desirable to ensure that the statistical properties of the error $\tilde{\mathbf{x}}$ do **not** depend on the knowledge of the values of the observation $\mathbf{z}$. In this case we can enjoy the luxury of performing the error analysis even before the arrival of the first observation. Indeed, the well-known Kalman filtering algorithm permits such an error analysis.

Against this background, we now describe three useful frameworks for statistical estimation. Refer to Figure 13.1.1.

(1) **Fisher's framework** It is assumed that $\mathbf{x}$ is an **unknown constant** μ and that $\mathbf{v}$ is random. Hence $\mathbf{z} = \mathbf{h}(\mathbf{x}) + \mathbf{v}$ is random. This approach exploits the properties of the multivariate probability distribution of $\mathbf{z}$ conditioned on $\mathbf{x}$, namely $p(\mathbf{z}|\mathbf{x})$.

There are at least two different classes of methods for estimation within this framework: (a) maximum likelihood method, and (b) statistical least squares. In special cases, such as when $\mathbf{v}$ has a normal distribution, these two methods give rise to the same estimator.

(2) **Bayes framework** In this paradigm, it is assumed that both $\mathbf{x}$ and $\mathbf{v}$ are random. Let $p(\mathbf{x})$ denote the a priori distribution of $\mathbf{x}$ representing the initial belief and/or knowledge about the unknown $\mathbf{x}$, and let $p(\mathbf{z}|\mathbf{x})$ be the conditional distribution of $\mathbf{z}$ given $\mathbf{x}$. The idea is to combine these distributions using the Bayes' rule (Appendix F) to arrive at the posterior distribution of $\mathbf{x}$ given $\mathbf{z}$, namely $p(\mathbf{x}|\mathbf{z})$. From

$$p(\mathbf{x}, \mathbf{z}) = p(\mathbf{z}|\mathbf{x})p(\mathbf{x}) = p(\mathbf{x}|\mathbf{z})p(\mathbf{z}) \tag{13.1.4}$$

we obtain an expression for the posterior distribution

$$\begin{aligned} p(\mathbf{x}|\mathbf{z}) &= \frac{p(\mathbf{z}|\mathbf{x})p(\mathbf{x})}{p(\mathbf{z})} = \frac{p(\mathbf{z}|\mathbf{x})p(\mathbf{x})}{\int_{-\infty}^{\infty} p(\mathbf{x}, \mathbf{z})d\mathbf{x}} \\ &= \frac{p(\mathbf{z}|\mathbf{x})p(\mathbf{x})}{\int_{-\infty}^{\infty} p(\mathbf{z}|\mathbf{x})p(\mathbf{x})d\mathbf{x}} \end{aligned} \tag{13.1.5}$$

where the integral in the denominator of the r.h.s. in (13.1.5) gives the **marginal distribution** $p(\mathbf{z})$ of $\mathbf{z}$.

Once $p(\mathbf{x}|\mathbf{z})$ is computed, we could use this information in a variety of ways to select an estimator. This is often done by defining a Bayes cost function and selecting an estimator that minimizes this cost function. The well-known **maximum a posteriori** (MAP) **estimate** and the **minimum variance** (MV) estimates are some of the examples resulting from this framework.

(3) **Deterministic least squares framework** In cases when no acceptable models for $\mathbf{x}$ and $\mathbf{v}$ exist, the idea is to resort to a pure **algebraic approach** leading to the **deterministic least squares** method described in Part II.

13.2 Properties of estimates

Before moving on to defining the notion of and methods for optimal estimation, we need to first guarantee that the estimate $\hat{\mathbf{x}}$ as a random variable satisfies some of the quite basic and natural requirements.

(A) **Unbiasedness** The first requirement relates to the relative location of the mean of the conditional distribution $p(\hat{\mathbf{x}}|\mathbf{x})$ with respect to the true value of $\mathbf{x}$. It stands to reason to expect that the mean of $p(\hat{\mathbf{x}}|\mathbf{x})$ must be the same as $\mathbf{x}$. That is,

$$E(\hat{\mathbf{x}}|\mathbf{x}) = \mathbf{x} \tag{13.2.1}$$

if $\mathbf{x}$ is a constant. In the case when $\mathbf{x}$ has a priori distribution $p(\mathbf{x})$, it is natural to expect that

$$E_{\mathbf{x}}\{E(\hat{\mathbf{x}}|\mathbf{x})\} = E(\hat{\mathbf{x}}) = E(\mathbf{x}) \tag{13.2.2}$$

where the first expectation operator $E_{\mathbf{x}}$ on the l.h.s. of (13.2.2) is w.r. to the prior distribution of $\mathbf{x}$. Any estimate $\hat{\mathbf{x}}$ that satisfies (13.2.1) or (13.2.2) is called an **unbiased estimate**. If $\hat{\mathbf{x}}$ is not unbiased, then the difference $(E(\hat{\mathbf{x}}) - \mathbf{x})$ or $(E(\hat{\mathbf{x}}) - E(\mathbf{x}))$ is called the **bias**.

We now illustrate this attribute using two standard examples.

Example 13.2.1 Consider a coin that falls head with probability p and tail with probability $q = 1 - p$. It is assumed that p is not known but is a fixed constant. Our aim is to estimate p.

Let us first translate this problem into our notation. Define, for $i = 1, 2, \ldots, m$, observations z_i given by

$$z_i = p + v_i \tag{13.2.3}$$

where the random variables v_i are **independent and identically distributed** as follows:

$$v_i = \begin{cases} (1-p) & \text{with probability } p \\ -p & \text{with probability } q = 1-p \end{cases} \tag{13.2.4a}$$

Hence

$$E(v_i) = 0 \text{ and } \mathrm{Var}(v_i) = pq. \tag{13.2.4b}$$

Combining this with (13.2.2), it can be verified that z_i is a Bernoulli random variable:

$$z_i = \begin{cases} 1 & \text{with probability } p \text{ – coin falls head} \\ 0 & \text{with probability } q \text{ – coin falls tail} \end{cases}$$

and that

$$E(z_i) = p \quad \text{and} \quad \mathrm{Var}(z_i) = pq. \tag{13.2.5}$$

Recall that the sample mean is a good estimator of p. Thus, define

$$\hat{p} = \frac{1}{m}\sum_{i=1}^{m} z_i.$$

If it can be verified that (since z_i's are independent)

$$E(\hat{p}) = \frac{1}{m}\sum_{i=1}^{m} E(z_i) = p \tag{13.2.6a}$$

and

$$\begin{aligned}\operatorname{Var}(\hat{p}) &= E\left[\frac{1}{m}\sum_{i=1}^{m} z_i - p\right]^2 \\ &= \frac{1}{m^2}\sum_{i=1}^{m} E(z_i - p)^2 \\ &= \frac{pq}{m}.\end{aligned} \tag{13.2.6b}$$

Indeed, $\hat{p}$ is an **unbiased** estimate for p. This is **not** the only one, however. It follows from (13.2.5) indeed each z_i is also an **unbiased estimate for** p.

To understand the import of unbiasedness, consider the **mean squared error** in the estimate $\hat{\mathbf{x}}$ when the unknown $\mathbf{x}$ is assumed to be a constant. Then

$$\begin{aligned}E(\hat{\mathbf{x}} - \mathbf{x})^2 &= E[\hat{\mathbf{x}} - E(\hat{\mathbf{x}}) + E(\hat{\mathbf{x}}) - \mathbf{x}]^2 \\ &= E(\hat{\mathbf{x}} - E(\hat{\mathbf{x}}))^2 + E(E(\hat{\mathbf{x}}) - \mathbf{x})^2 \\ &\quad + 2E[(\hat{\mathbf{x}} - E(\hat{\mathbf{x}}))(E(\hat{\mathbf{x}}) - \mathbf{x})].\end{aligned} \tag{13.2.7}$$

Since $(E(\hat{\mathbf{x}}) - \mathbf{x})$ is a constant, the last term on the r.h.s. becomes

$$2[E(\hat{\mathbf{x}}) - \mathbf{x}][E(\hat{\mathbf{x}}) - E(\hat{\mathbf{x}})] = 0.$$

Combining these, the new expression for the mean squared error becomes

$$MSE(\hat{\mathbf{x}}) = E(\hat{\mathbf{x}} - \mathbf{x})^2 = \operatorname{Var}(\hat{\mathbf{x}}) + [\operatorname{Bias}(\hat{\mathbf{x}})]^2. \tag{13.2.8}$$

Thus, when $\hat{\mathbf{x}}$ is unbiased, the mean squared error in $\hat{\mathbf{x}}$ reduces to its Variance. Let $\mathcal{U}$ denote the set of all unbiased estimates for $\hat{\mathbf{x}}$. For estimators in this class, the problem of minimizing the mean squared error and that of minimizing the variance are one and the same.

It is, however, possible to obtain even lower mean squared error if we are willing to allow a small bias as illustrated in the following.

Example 13.2.2 Let $z_i = \mu + v_i$ where μ is a constant and v_i are independent and identically distributed (iid) normal random variables with mean zero and variance σ^2. Hence z_i are also iid random variables with mean μ and variance σ^2.

A well-known estimator for μ is the sample mean $\bar{z} = \frac{1}{m}\sum_{i=1}^{m} z_i$. It can be verified that this estimate is unbiased, that is $E(\bar{z}) = \mu$ and

$$\operatorname{Var}(\bar{z}) = \operatorname{Var}\left(\frac{1}{m}\sum_{i=1}^{m} z_i\right) = E\left[\frac{1}{m}\sum_{i=1}^{m}(z_i - \mu)\right]^2 = \frac{\sigma^2}{m}.$$

Consider now the problem of estimating σ^2. There are two cases to consider – when the mean μ is known and μ is **not** known. In the first case, since μ is known, an obvious estimator for σ^2 is

$$\hat{\sigma}^2 = \frac{1}{m}\sum_{i=1}^{m}(z_i - \mu)^2 \tag{13.2.9}$$

which is unbiased since

$$E(\hat{\sigma}^2) = \frac{1}{m}\sum_{i=1}^{m} E(z_i - \mu)^2 = \sigma^2.$$

It can be shown that the variance of $\hat{\sigma}^2$ (Exercise 13.1) is

$$\mathrm{Var}(\hat{\sigma}^2) = \frac{2\sigma^4}{m}. \tag{13.2.10}$$

In the second case when μ is not known we are forced to use its estimate in estimating σ^2 using a variation of (13.2.9):

$$s^2 = \frac{1}{m}\sum_{i=1}^{m}(z_i - \bar{z})^2.$$

Then

$$\begin{aligned} E\left(\sum\nolimits_{i=1}^{m}(z_i - \bar{z})^2\right) &= E\left[\sum\nolimits_{i=1}^{m}\left(z_i^2 - 2z_i\bar{z} + \bar{z}^2\right)\right] \\ &= E\left[\sum\nolimits_{i=1}^{m}\left(z_i^2\right) - m\bar{z}^2\right] \\ &= \sum\nolimits_{i=1}^{m} E\left(z_i^2\right) - mE\left(\bar{z}^2\right). \end{aligned}$$

But

$$\begin{aligned} E\left(z_i^2\right) &= \sigma^2 + \mu^2 \quad \text{and} \\ E(\bar{z}^2) &= \mathrm{Var}(\bar{z}) + [E(\bar{z})]^2 = \frac{\sigma^2}{m} + \mu^2. \end{aligned}$$

Combining these, we obtain

$$E(s^2) = \frac{1}{m}[m\sigma^2 + m\mu^2 - \sigma^2 - m\mu^2] = \left(\frac{m-1}{m}\right)\sigma^2 < \sigma^2.$$

Hence s^2 is slightly biased where the bias is $E(s^2) - \sigma^2 = -\sigma^2/m$.

It can be shown that the variance of s^2 is given by (Exercise 13.2)

$$\mathrm{Var}(s^2) = \frac{2(m-1)\sigma^4}{m^2}. \tag{13.2.11}$$

Comparing (13.2.10) and (13.2.11), we see that

$$\mathrm{Var}(\sigma^2) = \frac{2\sigma^4}{m} > \frac{2\sigma^4}{m-1}\left(\frac{m-1}{m}\right)^2 = \mathrm{Var}(s^2).$$

But from (13.2.8), it follows that the mean squared error (MSE) in $\hat{\sigma}^2$ is less than the MSE in s^2 (Exercise 13.3):

$$\mathrm{MSE}(\hat{\sigma}^2) = \frac{2\sigma^4}{m} > \frac{2(m-1)}{m^2}\sigma^4 + \frac{\sigma^4}{m^2} = \mathrm{MSE}(s^2). \tag{13.2.12}$$

That is, a slightly biased estimate is more precise.

(B) **Relative efficiency** Let $\hat{\mathbf{x}}_a$ and $\hat{\mathbf{x}}_b$ be two estimates of an unknown parameter $\mathbf{x}$. We say that the estimate $\hat{\mathbf{x}}_a$ is **more efficient** relative to $\hat{\mathbf{x}}_b$ if

$$\mathrm{Var}(\hat{\mathbf{x}}_a) \leq \mathrm{Var}(\hat{\mathbf{x}}_b).$$

The ratio $\mathrm{Var}(\hat{\mathbf{x}}_b)/\mathrm{Var}(\hat{\mathbf{x}}_a)$ is a measure of the relative efficiency of these estimates.

In the Example 13.2.1 , while both z_i and $\hat{p}$ are estimates, since

$$\mathrm{Var}(\hat{p}) = \frac{pq}{m} < pq = \mathrm{Var}(z_i)$$

it follows that the sample mean is **more efficient** compared to z_i. This definition naturally leads us to seeking estimators with least variance. Since unbiasedness is a very basic requirement, indeed we are seeking for unbiased estimates with least variance. This search is guided by one of the most fundamental results in the theory of point estimation called the **Cramer–Rao lower bound** which we now quote without proof. First we introduce some relevant concepts and notations.

Recall that $p(\mathbf{z}|\mathbf{x})$ is the conditional distribution of $\mathbf{z}$ given $\mathbf{x}$. Since $\mathbf{x}$ occurs as a parameter in $p(\mathbf{z}|\mathbf{x})$, it is useful to consider this as a function of $\mathbf{x}$. Then

$$L(\mathbf{x}|\mathbf{z}) = p(\mathbf{z}|\mathbf{x}) \tag{13.2.13}$$

as a function of $\mathbf{x}$ defines the **likelihood function**. As an example, let

$$p(\mathbf{z}|\mathbf{x}) = \frac{1}{\sqrt{2\pi}\sigma} \exp\left[-\frac{(\mathbf{z}-\mu)^2}{2\sigma^2}\right]$$

where $\mathbf{x} = (\mu, \sigma)^{\mathrm{T}}$ is the parameter. When considered as a function of $\mathbf{z}$ for a given $\mathbf{x}$, this is called the normal density and when considered as function of $\mathbf{x}$ for a given $\mathbf{z}$, it is called the likelihood function for $\mathbf{z}$. The natural logarithm of the likelihood function $\ln L(\mathbf{x}|\mathbf{z})$ plays a very basic role in the definition of the Cramer–Rao lower bound.

Cramer–Rao bound (scalar case). Let x be a scalar and $\hat{x}$ be an unbiased estimate for x. Then the conditional variance of $\hat{x}$ is bounded below and is given by

$$\mathrm{Var}(\hat{x}|x) \geq \left\{E\left[\frac{\partial}{\partial x}\ln L(x|z)\right]^2\right\}^{-1}$$
$$= -\left\{E\left[\frac{\partial^2}{\partial x^2}\ln L(x|z)\right]\right\}^{-1} \tag{13.2.14}$$

where $\ln L(x|z)$ is the **log likelihood function**. Notice that there are two equivalent expressions for this lower bound – one involving only the **first derivative** and the other involving the **second derivative** of the likelihood function.

In extending this lower bound to the case when $\mathbf{x}$ is a vector, we first introduce a relation among symmetric positive definite matrices. Let $\mathbf{A}$ and $\mathbf{B}$ be symmetric positive definite matrices. Then, we say

$$\mathbf{A} \geq \mathbf{B} \tag{13.2.15}$$

exactly, when $\mathbf{A} - \mathbf{B}$ is a symmetric and positive semidefinite matrix.

Let $\nabla_{\mathbf{x}} \ln L(\mathbf{x}|\mathbf{z})$ be the gradient and $\nabla^2_{\mathbf{x}} \ln L(\mathbf{x}|\mathbf{z})$ be the Hessian of the likelihood function with respect to $\mathbf{x}$. Then the Cramer–Rao bound can be stated as follows:

$$\mathrm{Cov}(\hat{\mathbf{x}}|\mathbf{x}) \geq (E(([\nabla_{\mathbf{x}} \ln L(\mathbf{x}|\mathbf{z})][\nabla_{\mathbf{x}} \ln L(\mathbf{x}|\mathbf{z})]^{\mathrm{T}})))^{-1}$$
$$= -(E[\nabla^2_{\mathbf{x}} \ln L(\mathbf{x}|\mathbf{z})])^{-1} \tag{13.2.16}$$

where the first expression on the r.h.s. is the inverse of the expected value of the square of the outer product matrix of the gradient of log likelihood function with itself and the second expression is the negative of the inverse of the expected value of the Hessian of the likelihood function.

At this juncture, it is useful to introduce the notion of an **information** matrix $\mathbf{I}(\mathbf{x})$ for the sample defined by

$$\mathbf{I}(\mathbf{x}) = -E[\nabla^2_{\mathbf{x}} \ln L(\mathbf{x}|\mathbf{z})]$$
$$= E([\nabla_{\mathbf{x}} \ln L(\mathbf{x}|\mathbf{z})][\nabla_{\mathbf{x}} \ln L(\mathbf{x}|\mathbf{z})]^{\mathrm{T}}) \tag{13.2.17}$$

which summarizes the amount of information in the observation. Using this, we could restate the Cramer–Rao inequality as

$$\mathrm{Cov}(\hat{\mathbf{x}}|\mathbf{x}) \geq \mathbf{I}^{-1}(\mathbf{x}). \tag{13.2.18}$$

It is instructive to compute an example of the information matrix.

Example 13.2.3 Let $z_i = \mu + v_i$ where μ is an unknown constant and v_i are iid random variables from a common normal distribution with zero mean and variance σ^2, that is, $v_i \sim$ iid $N(0, \sigma^2)$. Then, $z_i \sim$ iid $N(\mu, \sigma^2)$. Let $\mathbf{x} = (\mu, \sigma^2)^{\mathrm{T}}$ be the vector of unknown parameters. Since $\mathbf{z} = (z_1\ z_2\ \cdots\ z_m)^{\mathrm{T}}$ is jointly normal, its likelihood function is given by

$$L(\mathbf{x}|\mathbf{z}) = p(\mathbf{z}|\mathbf{x}) = \Pi_{i=1}^{m} f(z_i|\mathbf{x})$$

$$= (2\pi\sigma^2)^{-\frac{m}{2}} \exp[-\tfrac{1}{2\sigma^2} \textstyle\sum_{i=1}^{m}(z_i - \mu)^2].$$

The log likelihood function is given by

$$\ln L(\mathbf{x}|\mathbf{z}) = -\frac{m}{2} \ln 2\pi - \frac{m}{2} \ln \sigma^2 - \frac{1}{2\sigma^2} \sum_{i=1}^{m} (z_i - \mu)^2.$$

Hence

$$\frac{\partial \ln L(\mathbf{x}|\mathbf{z})}{\partial \mu} = \frac{1}{\sigma^2} \sum_{i=1}^{m} (z_i - \mu), \qquad \frac{\partial^2 \ln L(\mathbf{x}|\mathbf{z})}{\partial \mu^2} = -\frac{m}{\sigma^2}$$

$$\frac{\partial \ln L(\mathbf{x}|\mathbf{z})}{\partial (\sigma^2)} = -\frac{m}{2\sigma^2} + \frac{1}{2\sigma^4} \sum_{i=1}^{m} (z_i - \mu)^2$$

$$\frac{\partial^2 \ln L(\mathbf{x}|\mathbf{z})}{\partial (\sigma^2)^2} = \frac{m}{2\sigma^4} - \frac{1}{\sigma^6} \sum_{i=1}^{m} (z_i - \mu)^2$$

$$\frac{\partial^2 \ln L(\mathbf{x}|\mathbf{z})}{\partial \mu \partial (\sigma^2)} = -\frac{1}{\sigma^4} \sum_{i=1}^{m} (z_i - \mu).$$

Using the facts

$$E\left[\sum_{i=1}^{m} (z_i - \mu)^2\right] = m\sigma^2 \quad \text{and} \quad E\left[\sum_{i=1}^{m} (z_i - \mu)\right] = 0$$

we get

$$E[-\nabla^2 \ln L(\mathbf{x}|\mathbf{z})] = \begin{bmatrix} \frac{m}{\sigma^2} & 0 \\ 0 & \frac{m}{2\sigma^4} \end{bmatrix}$$

and hence

$$\mathbf{I}^{-1}(\mathbf{x}) = \begin{bmatrix} \frac{\sigma^2}{m} & 0 \\ 0 & \frac{2\sigma^4}{m} \end{bmatrix}.$$

Combining this with the example (13.2.2), we can now verify the Cramer–Rao inequality as follows:

Let $\hat{\mathbf{x}} = (\bar{\mathbf{z}}, \hat{\mathbf{s}}^2)^{\mathrm{T}}$ where

$$\hat{s}^2 = \left(\frac{m}{m-1}\right) \mathbf{s}^2 = \frac{1}{(m-1)} \sum_{i=1}^{m} (z_i - \bar{z})^2. \tag{13.2.19}$$

It can be verified (Exercise 13.4) that $\hat{\mathbf{x}}$ is an unbiased estimate for $\hat{\mathbf{x}} = (\mu, \sigma^2)^{\mathrm{T}}$. Further, it can be shown (Exercise 13.5) that $\bar{z}$ and $\hat{s}^2$ are independent and that

$$\mathrm{Var}(\bar{z}) = \frac{\sigma^2}{m} \quad \text{and} \quad \mathrm{Var}(\hat{s}^2) = \frac{2\sigma^4}{m-1}.$$

Hence

$$\mathrm{Cov}(\hat{x}) = \begin{bmatrix} \frac{\sigma^2}{m} & 0 \\ 0 & \frac{2\sigma^4}{m-1} \end{bmatrix}.$$

Now

$$\mathrm{Cov}(\hat{x}) - \mathbf{I}^{-1}(x) = \begin{bmatrix} 0 & 0 \\ 0 & \frac{2\sigma^4}{m(m-1)} \end{bmatrix}$$

which is clearly symmetric and non-negative definite.

An immediate consequence of this Cramer–Rao inequality is that it naturally leads to the definition of an efficient estimate.

(C) **Efficient estimate** An efficient estimate is an unbiased estimate whose conditional variance is equal to the lower bound dictated by the Cramer–Rao bound.

In general, there is no guarantee that an efficient estimate exists for a given estimation problem. However, when it does, it turns out that the maximum likelihood estimate is an efficient estimate. Herein lies the importance of the maximum likelihood estimate introduced by Fisher.

Example 13.2.4 Let $z_i = p + v_i$ be as defined in Example 13.2.1 with $x = p$, where recall this z_i is Bernoulli random variable whose distribution is given by

$$f(z_i) = p(z_i|x) = p^{z_i}(1-p)^{1-z_i}.$$

Since z_i are independent, letting $\mathbf{z} = (z_1, z_2, \ldots, z_m)^{\mathrm{T}}$, it follows that

$$L(x|\mathbf{z}) = p(\mathbf{z}|x) = p(z_1|x)\, p(z_2|x) \cdots p(z_m|x)$$

$$= p^{Y_m}(1-p)^{m-Y_m}$$

where

$$Y_m = \sum_{i=1}^{m} z_i \quad \text{and} \quad E[Y_m] - mp.$$

Hence

$$\ln L(\mathbf{z}|x) = Y_m \log p + (m - Y_m)\log(1-p).$$

Differentiating w.r. to p, we get (with $x = p$)

$$\frac{\partial}{\partial x} \ln L(\mathbf{z}|x) = \frac{Y_m}{p} - \frac{(m - Y_m)}{(1-p)}$$

and

$$\frac{\partial^2}{\partial x^2} \ln L(\mathbf{z}|x) = -\frac{Y_m}{p^2} - \frac{(m - Y_m)}{(1-p)^2}$$

from which the Cramer–Rao lower bound on the variance of Y_m is pq/m. Comparing this with the variance of the sample mean, in (13.2.6a) and (13.2.6b) it follows that the sample mean is indeed an efficient estimate for p.

There are two more desirable attributes for the estimates called **consistency** and **sufficiency**.

Consistency of an estimate relates to the behavior of the distribution of the estimate $\hat{x}$ of x as the number m of observations grows without bound. In such a case, it is natural to expect this distribution of $\hat{\mathbf{x}}$ to be increasingly clustered around the true value $\mathbf{x}$ as m increases.

(D) **Consistency** An estimate $\hat{\mathbf{x}}$ of $\mathbf{x}$ is said to be consistent if for any $\epsilon > 0$

$$\text{Prob}[|\hat{\mathbf{x}} - \mathbf{x}| > \epsilon] \rightarrow 0 \text{ as } m \rightarrow \infty. \tag{13.2.20}$$

That is, $\hat{x}$ **converges in probability** to x as $m \rightarrow \infty$. The interest in consistency essentially stems from two of the important consequences of it, namely if $\hat{x}$ is consistent then $\hat{x}$ is **asymptotically unbiased** and $\text{Var}(\hat{x})$ asymptotically converges to zero.

(E) **Sufficiency** Relates to guaranteeing conditions under which a random sample of observations will have enough or sufficient information to obtain an estimate for x. A formal condition on the conditional density $p(\mathbf{z}|x)$ for sufficiency was developed by Fisher and Neyman. This condition leads to a natural factorization of this conditional density function. Using this result, we can guarantee the existence of unbiased estimates for x.

Exercises

13.1 Verify the correctness of the expression for $\text{Var}(\hat{\sigma}^2)$ in (13.2.10).

Hint: Since $z_i \sim \text{iid } N(\mu, \sigma^2)$, it follows (Appendix F) that $(m/\sigma^2)\,\hat{\sigma}^2 = \sum_{i=1}^{m}(z_i - \mu)^2/\sigma^2 \sim \chi^2(m)$. The mean and variance of $\chi^2(m)$ are m and $2m$, respectively. Also $\text{Var}(a\mathbf{x}) = a^2\,\text{Var}(\mathbf{x})$.

13.2 Verify the correctness of the expression for $\text{Var}(s^2)$ in (13.2.11).

Hint: $(ms^2/\sigma^2) = \sum_{i=1}^{m}(z_i - \bar{z})^2 \sim \chi^2(m-1)$.

13.3 Verify the inequality in (13.2.12).

13.4 Show that $\hat{s}^2$ defined in (13.2.19) is an unbiased estimate for σ^2 and that its variance is $(\frac{2\sigma^4}{m-1})$.

Hint: $((m-1)s^2/\sigma^2) = \sum_{i=1}^{m}(z_i - \bar{z})^2 \sim \chi^2(m-1)$.

13.5 Prove that the estimates $\bar{z} = \frac{1}{m}\sum_{i=1}^{m} z_i$ and $\hat{s}^2 = \frac{1}{m-1}\sum_{i=1}(z_i - \bar{z})^2$ are independent.

Hint: Refer to Appendix F, and independence of quadratic forms of multivariate normal random vectors.

Notes and references

The material covered in this chapter is quite basic and can be found in just about any book dealing with statistical estimation, such as Deutsch (1965), Melsa and Cohn (1978), and Sorenson (1980). Jazwinski(1970) and Schweppe (1973) provide an extensive coverage of estimation within the context of Stochastic Dynamics. Rao (1945) is one of the early papers on efficiency of estimation wherein a derivation of the lower bound is presented. For an expanded view of the contents of this part covering Chapters 13–17 refer to the classic book by Rao (1973). The survey papers by Kailath (1974) and Cohn (1997) provide an illuminating discussion and a rather comprehensive coverage of the literature.

14

Statistical least squares estimation

This chapter provides an introduction to the principles and techniques of statistical least squares estimation of an unknown vector $\mathbf{x} \in \mathbb{R}^n$ when the observations are corrupted by additive random noise. While the techniques and developments in this chapter parallel those of Chapter 5, the key assumption relative to the random nature of the observation sets this chapter apart. An immediate consequence is that the estimates are random variables and we now need to contend with the additional challenge of quantifying its mean, variance and many of the other desirable attributes such as unbiasedness, efficiency, consistency, to mention a few.

Section 14.1 contains the derivation of the statistical least squares estimate. An analysis of the quality of the fit between the linear model and the data is presented in Section 14.2. The **Gauss–Markov theorem** and its implications of optimality of the linear least squares estimates are covered in Section 14.3. A discussion of the model error and its impact on the quality of the least squares estimate is presented in Section 14.4.

14.1 Statistical least squares estimate

Consider the **linear estimation** problem where the **unknown** $\mathbf{x} \in \mathbb{R}^n$ and the **known** observation $\mathbf{z} \in \mathbb{R}^m$ are related as

$$\mathbf{z} = \mathbf{H}\mathbf{x} + \mathbf{v} \tag{14.1.1}$$

where $\mathbf{H} \in \mathbb{R}^{m \times n}$ is a **known** matrix and $\mathbf{v}$ is the **additive random noise** corrupting the observations. For definiteness, it is assumed that $m > n$. This noise vector $\mathbf{v}$ is **not** observable and to render the problem tractable, the following assumptions are made.

(1) $E(\mathbf{v}) = 0$,
(2) $E(\mathbf{v}\mathbf{v}^{\mathrm{T}}) = \mathbf{R}$, symmetric and positive definite,
(3) $\mathbf{v}$ and $\mathbf{x}$ are uncorrelated.

Define the residual $\mathbf{r}(\mathbf{x}) = \mathbf{z} - \mathbf{H}\mathbf{x}$. Since the covariance matrix $\mathbf{R}$ is symmetric and positive definite, so is $\mathbf{R}^{-1}$. Recall from Chapter 13 (Remark 13.2.1) that this inverse $\mathbf{R}^{-1}$ is also known as the **information matrix** or **precision matrix** and is often used as a weight matrix in formulating the least squares problem. Then

$$\begin{aligned} f(\mathbf{x}) &= \frac{1}{2}\mathbf{r}^{\mathrm{T}}(\mathbf{x})\mathbf{R}^{-1}\mathbf{r}(\mathbf{x}) = \frac{1}{2}\|\mathbf{r}(\mathbf{x})\|^2_{\mathbf{R}^{-1}} \\ &= \frac{1}{2}(\mathbf{z} - \mathbf{H}\mathbf{x})^{\mathrm{T}}\mathbf{R}^{-1}(\mathbf{z} - \mathbf{H}\mathbf{x}) \end{aligned} \tag{14.1.2}$$

denotes the weighted sum of the squares of the residuals or the **energy norm** of the residual vector $\mathbf{r}(\mathbf{x})$.

Remark 14.1.1 To get a feel for the effect of using $\mathbf{R}^{-1}$ as the weight matrix, consider the special case when $\mathbf{R}$ is a diagonal matrix which happens when the elements of $\mathbf{v}$ are uncorrelated. Let $\mathbf{R} = \mathrm{Diag}(\sigma_1^2, \sigma_2^2, \ldots, \sigma_n^2)$. Then $\mathbf{R}^{-1} = \mathrm{Diag}(\sigma_1^{-2}, \sigma_2^{-2}, \ldots, \sigma_n^{-2})$ and

$$f(\mathbf{x}) = \frac{1}{2}\sum_{i=1}^{m}\frac{(z_i - \mathbf{H}_{i*}\mathbf{x})^2}{\sigma_i^2} = \frac{1}{2}\sum_{i=1}^{m}\frac{v_i^2}{\sigma_i^2} \tag{14.1.3}$$

which is the sum of the squares of the normalized random variables (v_i/σ_i) with mean zero and unit variance where $\mathbf{H}_{i*}$ denotes the ith row of $\mathbf{H}$. Since the variance is sensitive to scaling, this normalization eliminates the impact of scaling on the analysis and conclusions. (Refer to Chapter 6.)

Our aim is to minimize $f(\mathbf{x})$ w.r. to $\mathbf{x}$. To this end compute the gradient and Hessian of $f(\mathbf{x})$ in (14.1.2):

$$\nabla f(\mathbf{x}) = (\mathbf{H}^{\mathrm{T}}\mathbf{R}^{-1}\mathbf{H})\mathbf{x} - (\mathbf{H}^{\mathrm{T}}\mathbf{R}^{-1})\mathbf{z} \tag{14.1.4}$$

and

$$\nabla^2 f(\mathbf{x}) = \mathbf{H}^{\mathrm{T}}\mathbf{R}^{-1}\mathbf{H}. \tag{14.1.5}$$

Setting (14.1.4) to zero, we obtain the least squares estimate

$$\hat{\mathbf{x}}_{\mathrm{LS}} = (\mathbf{H}^{\mathrm{T}}\mathbf{R}^{-1}\mathbf{H})^{-1}\mathbf{H}^{\mathrm{T}}\mathbf{R}^{-1}\mathbf{z}. \tag{14.1.6}$$

Notice that this formula is identical to the weighted least squares discussed in Chapter 5.

If $\mathbf{H}$ is of full rank $(\mathrm{Rank}(\mathbf{H}) = n)$, then $\mathbf{H}\mathbf{y} \neq 0$ for any $\mathbf{y} \neq 0$. This when combined with the positive definiteness of $\mathbf{R}^{-1}$ gives

$$\mathbf{y}^{\mathrm{T}}(\mathbf{H}^{\mathrm{T}}\mathbf{R}^{-1}\mathbf{H})\mathbf{y} = (\mathbf{H}\mathbf{y})^{\mathrm{T}}\mathbf{R}^{-1}(\mathbf{H}\mathbf{y}) > 0 \quad \text{for all} \quad \mathbf{y} \neq 0.$$

Thus, the Hessian of $f(\mathbf{x})$ at $\mathbf{x} = \hat{\mathbf{x}}_{\mathrm{LS}}$ is positive definite and hence $\hat{\mathbf{x}}_{\mathrm{LS}}$ is indeed the minimizer of $f(\mathbf{x})$.

Several observations are in order.

(1) **Unbiasedness** Notice that the least squares estimate $\hat{\mathbf{x}}_{\text{LS}}$ in (14.1.6) is a linear function of the observation. Since $\mathbf{z}$ is random, so is $\hat{\mathbf{x}}_{\text{LS}}$. Then, using (14.1.1) it follows that

$$\begin{aligned}\hat{\mathbf{x}}_{\text{LS}} &= (\mathbf{H}^{\text{T}}\mathbf{R}^{-1}\mathbf{H})^{-1}\mathbf{H}^{\text{T}}\mathbf{R}^{-1}\mathbf{z}\\ &= \mathbf{x} + (\mathbf{H}^{\text{T}}\mathbf{R}^{-1}\mathbf{H})^{-1}\mathbf{H}^{\text{T}}\mathbf{R}^{-1}\mathbf{v}\end{aligned} \tag{14.1.7}$$

from which we obtain

$$\begin{aligned}E(\hat{\mathbf{x}}_{\text{LS}}) &= \mathbf{x} + (\mathbf{H}^{\text{T}}\mathbf{R}^{-1}\mathbf{H})^{-1}\mathbf{H}^{\text{T}}\mathbf{R}^{-1}E(\mathbf{v})\\ &= \mathbf{x}.\end{aligned} \tag{14.1.8}$$

That is, $\hat{\mathbf{x}}_{\text{LS}}$ is an **unbiased** estimate of $\mathbf{x}$.

(2) **Covariance of the estimate** The covariance of $\hat{\mathbf{x}}_{\text{LS}}$ is given by

$$\begin{aligned}\text{Cov}(\hat{\mathbf{x}}_{\text{LS}}) &= E[(\hat{\mathbf{x}}_{\text{LS}} - \mathbf{x})(\hat{\mathbf{x}}_{\text{LS}} - \mathbf{x})^{\text{T}}]\\ &= (\mathbf{H}^{\text{T}}\mathbf{R}^{-1}\mathbf{H})^{-1}\mathbf{H}^{\text{T}}\mathbf{R}^{-1}E(\mathbf{v}\mathbf{v}^{\text{T}})\mathbf{R}^{-1}\mathbf{H}(\mathbf{H}^{\text{T}}\mathbf{R}^{-1}\mathbf{H})^{-1}\\ &= (\mathbf{H}^{\text{T}}\mathbf{R}^{-1}\mathbf{H})^{-1}\\ &= [\nabla^2 f(\mathbf{x})]^{-1}.\end{aligned} \tag{14.1.9}$$

(3) **Relation to projection** Define

$$\begin{aligned}\hat{\mathbf{z}} &= \mathbf{H}\hat{\mathbf{x}}_{\text{LS}} \qquad (14.1.10)\\ &= \mathbf{H}(\mathbf{H}^{\text{T}}\mathbf{R}^{-1}\mathbf{H})^{-1}\mathbf{H}^{\text{T}}\mathbf{R}^{-1}\mathbf{z}\\ &= \mathbf{P}\mathbf{z}\end{aligned}$$

where

$$\mathbf{P} = \mathbf{H}(\mathbf{H}^{\text{T}}\mathbf{R}^{-1}\mathbf{H})^{-1}\mathbf{H}^{\text{T}}\mathbf{R}^{-1}. \tag{14.1.11}$$

It can be verified that $\mathbf{P}^2 = \mathbf{P}$, that is $\mathbf{P}$ is **idempotent** and that $\mathbf{P}$ is **not** symmetric. Hence $\mathbf{P}$ represents an **oblique** projection operator projecting $\mathbf{z}$ obliquely on to the space spanned by the columns of $\mathbf{H}$ (refer to Chapter 6). This projection, $\hat{\mathbf{z}}$, is often called the model counterpart of the observation.

(4) **A special case** Consider the case in which the components of the noise vector $\mathbf{v}$ are **uncorrelated** and share a common variance, σ^2. In this case $\mathbf{R} = \sigma^2\mathbf{I}$, a diagonal matrix with the constant value of σ^2 along the diagonal. From (14.1.6), it follows that

$$\hat{\mathbf{x}}_{\text{LS}} = (\mathbf{H}^{\text{T}}\mathbf{H})^{-1}\mathbf{H}^{\text{T}}\mathbf{z}, \text{ and } \text{Cov}(\hat{\mathbf{x}}_{\text{LS}}) = \sigma^2(\mathbf{H}^{\text{T}}\mathbf{H})^{-1}. \tag{14.1.12}$$

Since $(\mathbf{H}^{\text{T}}\mathbf{H})$ is symmetric, there exists an orthogonal matrix $\mathbf{Q}$ of eigenvectors of $(\mathbf{H}^{\text{T}}\mathbf{H})$ such that (Appendix B)

$$(\mathbf{H}^{\text{T}}\mathbf{H})\mathbf{Q} = \mathbf{Q}\Lambda$$

where Λ is the diagonal matrix of eigenvalues of $(\mathbf{H}^{\mathrm{T}}\mathbf{H})$. Then

$$(\mathbf{H}^{\mathrm{T}}\mathbf{H}) = \mathbf{Q}\Lambda\mathbf{Q}^{\mathrm{T}} \quad \text{or} \quad (\mathbf{H}^{\mathrm{T}}\mathbf{H})^{-1} = \mathbf{Q}\,\Lambda^{-1}\mathbf{Q}^{\mathrm{T}}. \tag{14.1.13}$$

Recall that the sum of the variances of the components of $\hat{\mathbf{x}}_{\mathrm{LS}}$ is given by

$$\begin{aligned}
\mathrm{tr}[\mathrm{Cov}(\hat{\mathbf{x}}_{\mathrm{LS}})] &= \mathrm{tr}[\sigma^2(\mathbf{H}^{\mathrm{T}}\mathbf{H})^{-1}] \\
&= \sigma^2\mathrm{tr}[(\mathbf{H}^{\mathrm{T}}\mathbf{H})^{-1}] \\
&= \sigma^2\mathrm{tr}[\mathbf{Q}\Lambda^{-1}\mathbf{Q}^{\mathrm{T}}] \text{ (using 14.1.13)} \\
&= \sigma^2\mathrm{tr}[\mathbf{Q}^{\mathrm{T}}\mathbf{Q}\,\Lambda^{-1}]\text{(Exercise 14.1)} \\
&= \sigma^2\mathrm{tr}[\Lambda^{-1}]\ (\mathbf{Q}^{\mathrm{T}}\mathbf{Q} = \mathbf{Q}\mathbf{Q}^{\mathrm{T}} = \mathbf{I}) \\
&= \sigma^2 \sum_{i=1}^{n} \frac{1}{\lambda_i}.
\end{aligned} \tag{14.1.14}$$

In other words, the total variance of the estimate is proportional to the sum of the reciprocals of the eigenvalues of $(\mathbf{H}^{\mathrm{T}}\mathbf{H})^{-1}$. Thus, if $\mathbf{H}^{\mathrm{T}}\mathbf{H}$ is nearly singular, then at least one of the λ_i is close to zero and the variance in $\hat{\mathbf{x}}_{\mathrm{LS}}$ would be excessively large. In this case, the projection matrix $\mathbf{P}$ in (14.1.11) becomes

$$\mathbf{P} = \mathbf{H}(\mathbf{H}^{\mathrm{T}}\mathbf{H})^{-1}\mathbf{H}^{\mathrm{T}} \tag{14.1.15}$$

which is **idempotent** ($\mathbf{P}^2 = \mathbf{P}$) and **symmetric** ($\mathbf{P}^{\mathrm{T}} = \mathbf{P}$), that is, $\mathbf{P}$ is an **orthogonal projection** matrix and $\hat{\mathbf{z}} = \mathbf{P}\mathbf{z}$ is the orthogonal projection of $\mathbf{z}$ on to the space spanned by the columns of $\mathbf{H}$. This special case is quite similar to the standard least squares described in Chapters 5 and 6.

(5) **Estimation of σ^2** In the above analysis, it was tacitly assumed that the noise covariance matrix $\mathbf{R}$ is known. We now address the special case $\mathbf{R} = \sigma^2\mathbf{I}$ when σ^2 is **not** known. To estimate σ^2, first define the **residual e**, using (14.1.12) and (14.1.13), as

$$\begin{aligned}
\mathbf{e} = \mathbf{z} - \hat{\mathbf{z}} &= \mathbf{z} - \mathbf{H}\hat{\mathbf{x}}_{\mathrm{LS}} \\
&= (\mathbf{I} - \mathbf{P})\mathbf{z} \\
&= (\mathbf{I} - \mathbf{P})(\mathbf{H}\mathbf{x} + \mathbf{v}) \\
&= (\mathbf{I} - \mathbf{P})\mathbf{v},
\end{aligned} \tag{14.1.16}$$

since it can be verified that $(\mathbf{I} - \mathbf{P})\mathbf{H} = 0$. Hence the mean of $\mathbf{e}$ is given by

$$E(\mathbf{e}) = E[(\mathbf{I} - \mathbf{P})\mathbf{v}] = (\mathbf{I} - \mathbf{P})E(\mathbf{v}) = 0. \tag{14.1.17}$$

From

$$
\begin{aligned}
E(\mathbf{e}^{\mathrm{T}}\mathbf{e}) &= E[\mathbf{v}^{\mathrm{T}}(\mathbf{I}-\mathbf{P})(\mathbf{I}-\mathbf{P})\mathbf{v}] \\
&= E[\mathbf{v}^{\mathrm{T}}(\mathbf{I}-\mathbf{P})\mathbf{v}] && ((\mathbf{I}-\mathbf{P}) \text{ is idempotent}) \\
&= E[\mathrm{tr}(\mathbf{v}^{\mathrm{T}}(\mathbf{I}-\mathbf{P})\mathbf{v})] && (\mathrm{tr}(a) = a \text{ for scalar } a) \\
&= E[\mathrm{tr}(\mathbf{v}\mathbf{v}^{\mathrm{T}}(\mathbf{I}-\mathbf{P}))] && (\mathrm{tr}(\mathbf{ABC}) = \mathrm{tr}(\mathbf{CBA})) \\
&= \sigma^2 \mathrm{tr}(\mathbf{I}-\mathbf{P}) && (\text{Exercises 14.1–14.3}) \\
&= \sigma^2[\mathrm{tr}(\mathbf{I}) - \mathrm{tr}(\mathbf{P})] && (\text{Exercise 14.4}) \\
&= \sigma^2(m-n)
\end{aligned}
\tag{14.1.18}
$$

it follows that

$$
\hat{\sigma}^2 = \frac{\mathbf{e}^{\mathrm{T}}\mathbf{e}}{(m-n)} \tag{14.1.19}
$$

is an **unbiased** estimate for σ^2.

14.2 Analysis of the quality of the fit

In this section, we provide some further insight into the linear least squares estimation problem by analyzing the quality of the fit between the linear mathematical model and the available observation.

We begin by rewriting (14.1.1) as

$$
\mathbf{z} = \sum_{j=1}^{n} \mathbf{H}_{*j}\mathbf{x}_j + \mathbf{v} \tag{14.2.1}
$$

where $\mathbf{H}_{*j}$ denotes the jth column of $\mathbf{H}$. The physical variables representing the columns of $\mathbf{H}$ are often known as the **independent** variables or **regressors** and the model (14.2.1) is known as the **multivariate** ($n \geq 2$) linear regression model. This linear model in general can allow for a $\mathbf{z}$-intercept term which corresponds to choosing the first column $\mathbf{H}_{*1}$ of $\mathbf{H}$ to be all 1's. For example, when $n = 2$, this model takes the familiar from

$$
\begin{bmatrix} z_1 \\ z_2 \\ \cdot \\ \cdot \\ z_m \end{bmatrix} = \begin{bmatrix} 1 & h_{12} \\ 1 & h_{22} \\ \cdot & \cdot \\ \cdot & \cdot \\ 1 & h_{m2} \end{bmatrix} \begin{bmatrix} x_1 \\ x_2 \end{bmatrix} + \begin{bmatrix} v_1 \\ v_2 \\ \cdot \\ \cdot \\ v_m \end{bmatrix} \tag{14.2.2}
$$

where x_1 denotes the $\mathbf{z}$-intercept and x_2 the slope of the straight line being fitted.

We now define the model counterpart of the observation

$$
\hat{\mathbf{z}} = \mathbf{H}\hat{\mathbf{x}}_{\mathrm{LS}} = \mathbf{H}(\mathbf{H}^{\mathrm{T}}\mathbf{R}^{-1}\mathbf{H})^{-1}\mathbf{H}^{\mathrm{T}}\mathbf{R}^{-1}\mathbf{z}. \tag{14.2.3}
$$

The model **residual e** is then given by

$$\mathbf{e} = \mathbf{z} - \hat{\mathbf{z}} = (\mathbf{I} - \mathbf{P})\mathbf{z} \tag{14.2.4}$$

where **P** is the **oblique** projection matrix given in (14.1.11), from which we obtain

$$\mathbf{H}^{\mathrm{T}}\mathbf{e} = (\mathbf{H}^{\mathrm{T}} - \mathbf{H}^{\mathrm{T}}\mathbf{P})\mathbf{z}.$$

In the special cases, when $\mathbf{R} = \sigma^2\mathbf{I}$, it can be verified using (14.1.15) that $\mathbf{H}^{\mathrm{T}} = \mathbf{H}^{\mathrm{T}}\mathbf{P}$ and hence

$$\mathbf{H}^{\mathrm{T}}\mathbf{e} = 0. \tag{14.2.5}$$

That is, when $\mathbf{R} = \sigma^2\mathbf{I}$, the model residual vector is orthogonal to the columns of **H**. This should not come as any surprise since **P** is an orthogonal projection matrix on to the columns of **H** when $\mathbf{R}^2 = \sigma^2\mathbf{I}$. Combining (14.2.5) with the special case when $\mathbf{H}_{*1}$ is all 1's, it follows that

$$\sum_{i=1}^{m} e_i = 0 \tag{14.2.6}$$

that is, the components of the model residual add up to zero. Since

$$\mathbf{z} = \hat{\mathbf{z}} + \mathbf{e}$$

using (14.2.6), we immediately obtain

$$\frac{1}{m}\sum_{i=1}^{m} z_i = \frac{1}{m}\sum_{i=1}^{m} \hat{z}_i \tag{14.2.7}$$

that is, the mean of the actual observations and that of the model counterpart or the fitted value are the same when the model allows for an intercept term. Again, rewriting

$$\mathbf{z} = \mathbf{H}\hat{\mathbf{x}}_{\mathrm{LS}} + \mathbf{e}$$

in the component form, we get

$$z_i = \sum_{j=1}^{n} h_{ij}(\hat{x}_{\mathrm{LS}})_j + e_i.$$

Hence, using (14.2.6)

$$\begin{aligned}\bar{\mathbf{z}} = \frac{1}{m}\sum\nolimits_{i=1}^{m} z_i &= \frac{1}{m}\sum\nolimits_{i=1}^{m}\sum\nolimits_{j=1}^{n} h_{ij}(\hat{x}_{\mathrm{LS}})_j \\ &= \sum\nolimits_{j=1}^{n}\Big(\frac{1}{m}\sum\nolimits_{i=1}^{m} h_{ij}\Big)(\hat{x}_{\mathrm{LS}})_j\end{aligned}$$

that is,

$$\bar{\mathbf{z}} = \sum_{j=1}^{n} \bar{\mathbf{h}}_{*j}(\hat{x}_{\mathrm{LS}})_j \tag{14.2.8}$$

where $\bar{\mathbf{z}}$ is the average of $\mathbf{z}$ and $\bar{\mathbf{h}}_{*j}$ is the average of the jth column of $\mathbf{H}$. This immediately implies that the regression line passes through the mean of both the dependent variable $\mathbf{z}$ and the independent variables denoted by the columns of $\mathbf{H}$. We hasten to add that the derivation of the properties (14.2.6) through (14.2.8) is conditioned on the linear model having an intercept term. If such an intercept term is not present in the model one or more of these properties may not hold.

We now examine another consequence of the orthogonality property (14.2.5). Consider

$$\begin{aligned}\mathbf{z}^{\mathrm{T}}\mathbf{z} &= (\hat{\mathbf{z}} + \mathbf{e})^{\mathrm{T}}(\hat{\mathbf{z}} + \mathbf{e}) \\ &= (\mathbf{H}\hat{\mathbf{x}}_{\mathrm{LS}} + \mathbf{e})^{\mathrm{T}}(\mathbf{H}\hat{\mathbf{x}}_{\mathrm{LS}} + \mathbf{e}) \\ &= (\hat{\mathbf{x}}_{\mathrm{LS}})^{\mathrm{T}}\mathbf{H}^{\mathrm{T}}\mathbf{H}\hat{\mathbf{x}}_{\mathrm{LS}} + \mathbf{e}^{\mathrm{T}}\mathbf{e} + 2(\hat{\mathbf{x}}_{\mathrm{LS}})^{\mathrm{T}}\mathbf{H}^{\mathrm{T}}\mathbf{e} \\ &= (\hat{\mathbf{x}}_{\mathrm{LS}})^{\mathrm{T}}\mathbf{H}^{\mathrm{T}}\mathbf{H}\hat{\mathbf{x}}_{\mathrm{LS}} + \mathbf{e}^{\mathrm{T}}\mathbf{e} \quad \text{(from 14.2.5)}.\end{aligned}$$

The ratio

$$\frac{(\hat{\mathbf{x}}_{\mathrm{LS}})^{\mathrm{T}}\mathbf{H}^{\mathrm{T}}\mathbf{H}\hat{\mathbf{x}}_{\mathrm{LS}}}{\mathbf{z}^{\mathrm{T}}\mathbf{z}} = 1 - \frac{\mathbf{e}^{\mathrm{T}}\mathbf{e}}{\mathbf{z}^{\mathrm{T}}\mathbf{z}} \tag{14.2.9}$$

is known as the **uncentered $\mathbf{R}^2$** that indicates the **goodness** of the model fit. Clearly, the smaller the value of $\mathbf{e}^{\mathrm{T}}\mathbf{e}$, the better is the fit between the model and the data. Sometimes it is convenient to work with variance instead of the raw second moments. If $\mathbf{i} = (1, 1, \ldots, 1)^{\mathrm{T}} \in \mathbb{R}^m$ and $\bar{\mathbf{z}}$ denotes the mean of $\mathbf{z}$ then a measure of the variance of $\mathbf{z}$ is given by

$$\begin{aligned}\textstyle\sum_{i=1}^{m}(z_i - \bar{z})^2 &= (\mathbf{z} - \mathbf{i}\bar{z})^{\mathrm{T}}(\mathbf{z} - \mathbf{i}\bar{z}) \\ &= \mathbf{z}^{\mathrm{T}}\mathbf{z} - m(\bar{z})^2.\end{aligned}$$

Then, from

$$\mathbf{z}^{\mathrm{T}}\mathbf{z} - m(\bar{z})^2 = (\hat{\mathbf{x}}_{\mathrm{LS}})^{\mathrm{T}}\mathbf{H}^{\mathrm{T}}\mathbf{H}\hat{\mathbf{x}}_{\mathrm{LS}} - m(\bar{z})^2 + \mathbf{e}^{\mathrm{T}}\mathbf{e}$$

we get the centered $\mathbf{R}^2$ as

$$\frac{(\hat{\mathbf{x}}_{\mathrm{LS}})^{\mathrm{T}}\mathbf{H}^{\mathrm{T}}\mathbf{H}\hat{\mathbf{x}}_{\mathrm{LS}} - m(\bar{z})^2}{[\mathbf{z}^{\mathrm{T}}\mathbf{z} - m(\bar{z})^2]} = 1 - \frac{\mathbf{e}^{\mathrm{T}}\mathbf{e}}{[\mathbf{z}^{\mathrm{T}}\mathbf{z} - m(\bar{z})^2]} \tag{14.2.10}$$

which is another indicator of the quality of the fit.

14.3 Optimality of least squares estimates

In this section, we derive a natural and one of the fundamental optimality properties of the linear least squares estimates. Let $\mathcal{U}$ denote the class of all **unbiased** estimates and $\mathcal{L}$ denote the class of all **linear** estimates of the state variable $\mathbf{x}$ in the linear model 14.1.1. Refer to Figure 14.3.1. The intersection of these two classes contains the family of linear, unbiased estimates. For example, recall that the least squares estimate $\hat{\mathbf{x}}_{\mathrm{LS}}$ in (14.1.6) is both linear and unbiased. The **best linear unbiased**

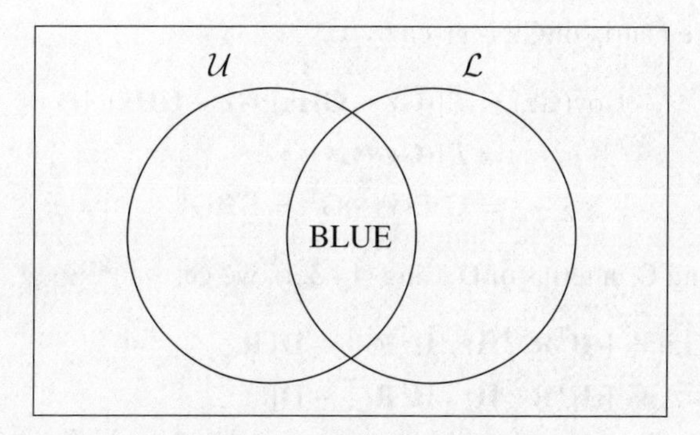

Fig. 14.3.1 A view of the set of all estimators.

estimate (BLUE) is defined as an estimate whose sample variance is the minimum among all the linear unbiased estimates. It turns out that $\hat{\mathbf{x}}_{\text{LS}}$ also enjoys the property that it is a BLUE and this result has come to be known as the **Gauss–Markov Theory.** As a prelude to establishing this basic result, we introduce a useful concept relating matrices. If **A** and **B** are two real symmetric and positive definite matrices, then we say that $\mathbf{A} \geq \mathbf{B}$ if there exists a symmetric and positive semi-definite matrix **C** such that

$$\mathbf{A} = \mathbf{B} + \mathbf{C}. \tag{14.3.1}$$

In the following, we apply this concept to covariance matrices of linear estimates.

Gauss–Markov theorem: version I Let $\hat{\mathbf{x}}$ be any linear unbiased estimator of **x** in (14.1.1) and $\hat{\mathbf{x}}_{\text{LS}}$ in (14.1.6) be the least squares estimate of the same **x**. Then

$$\text{Cov}(\hat{\mathbf{x}}) \geq \text{Cov}(\hat{\mathbf{x}}_{\text{LS}}) \tag{14.3.2}$$

Let $\mathbf{G} \in \mathbb{R}^{n\times m}$ and $\hat{\mathbf{x}} = \mathbf{Gz}$ denote an arbitrary linear estimate of **x**. Then, from $\mathbf{Gz} = \mathbf{GHx} + \mathbf{Gv}$ and

$$\begin{aligned} E(\mathbf{Gz}) &= \mathbf{GH}E(\mathbf{x}) + \mathbf{G}E(\mathbf{v}) \\ &= \mathbf{GHx} \end{aligned}$$

it follows that $\hat{\mathbf{x}}$ is unbiased exactly when $\mathbf{GH} = \mathbf{I}_n$, the identity matrix of order n. Consider now the difference

$$\begin{aligned} \hat{\mathbf{x}}_{\text{LS}} - \mathbf{Gz} &= [(\mathbf{H}^{\text{T}}\mathbf{R}^{-1}\mathbf{H})^{-1}\mathbf{H}^{\text{T}}\mathbf{R}^{-1} - \mathbf{G}]\mathbf{z} \\ &= \mathbf{Dz} \end{aligned}$$

where $\mathbf{D} \in \mathbb{R}^{n\times m}$ and is given by

$$\mathbf{D} = [(\mathbf{H}^{\text{T}}\mathbf{R}^{-1}\mathbf{H})^{-1}\mathbf{H}^{\text{T}}\mathbf{R}^{-1} - \mathbf{G}]. \tag{14.3.3}$$

The covariance matrix of $\mathbf{Gz}$ is given by

$$\begin{aligned}\text{Cov}(\mathbf{Gz}) &= E[(\mathbf{Gz} - \mathbf{GHx})(\mathbf{Gz} - \mathbf{GHx})^{\mathrm{T}}] \\ &= E[(\mathbf{Gv})(\mathbf{Gv})^{\mathrm{T}}] \\ &= \mathbf{G}\, E(\mathbf{vv}^{\mathrm{T}})\mathbf{G}^{\mathrm{T}} = \mathbf{GRG}^{\mathrm{T}}. \end{aligned} \tag{14.3.4}$$

Now expressing $\mathbf{G}$ in terms of $\mathbf{D}$ using (14.3.3), we get

$$\begin{aligned}\text{Cov}(\mathbf{Gz}) &= [(\mathbf{H}^{\mathrm{T}}\mathbf{R}^{-1}\mathbf{H})^{-1}\mathbf{H}^{\mathrm{T}}\mathbf{R}^{-1} - \mathbf{D}]\,\mathbf{R} \\ &\quad [(\mathbf{H}^{\mathrm{T}}\mathbf{R}^{-1}\mathbf{H})^{-1}\mathbf{H}^{\mathrm{T}}\mathbf{R}^{-1} - \mathbf{D}]^{\mathrm{T}} \\ &= (\mathbf{H}^{\mathrm{T}}\mathbf{R}^{-1}\mathbf{H})^{-1} + \mathbf{DRD}^{\mathrm{T}} - [(\mathbf{H}^{\mathrm{T}}\mathbf{R}^{-1}\mathbf{H})^{-1}\mathbf{H}^{\mathrm{T}}\mathbf{R}^{-1}]\,\mathbf{RD}^{\mathrm{T}} \\ &\quad - \mathbf{DR}\,[(\mathbf{H}^{\mathrm{T}}\mathbf{R}^{-1}\mathbf{H})^{-1}\mathbf{H}^{\mathrm{T}}\mathbf{R}^{-1}]^{\mathrm{T}}. \end{aligned} \tag{14.3.5}$$

But using the condition for unbiasedness, we get

$$\begin{aligned}\mathbf{I} = \mathbf{GH} &= [(\mathbf{H}^{\mathrm{T}}\mathbf{R}^{-1}\mathbf{H})^{-1}\mathbf{H}^{\mathrm{T}}\mathbf{R}^{-1} - \mathbf{D}]\,\mathbf{H} \\ &= \mathbf{I} - \mathbf{DH}\end{aligned}$$

from which it follows that $\mathbf{DH} = 0$. Combining this with (14.3.5) and simplifying we immediately obtain

$$\begin{aligned}\text{Cov}(\mathbf{Gz}) &= (\mathbf{H}^{\mathrm{T}}\mathbf{R}^{-1}\mathbf{H})^{-1} + \mathbf{DRD}^{\mathrm{T}} \\ &= \text{Cov}(\hat{\mathbf{x}}_{\text{LS}}) + \mathbf{DRD}^{\mathrm{T}}. \end{aligned} \tag{14.3.6}$$

Recall that $\mathbf{R}$ is symmetric and positive definite and $\mathbf{DRD}^{\mathrm{T}}$ is symmetric. For any $\mathbf{y} \in \mathbb{R}^n$

$$\mathbf{y}^{\mathrm{T}}\mathbf{DRD}^{\mathrm{T}}\mathbf{y} = (\mathbf{D}^{\mathrm{T}}\mathbf{y})^{\mathrm{T}}\,\mathbf{R}\,(\mathbf{D}^{\mathrm{T}}\mathbf{y})$$

where $\mathbf{D}^{\mathrm{T}}\mathbf{y} \in \mathbb{R}^m$. Depending on the rank of $\mathbf{D}$, since it is possible for $\mathbf{D}^{\mathrm{T}}\mathbf{y}$ to be zero for a non-zero vector $\mathbf{y}$, it follows that $\mathbf{DRD}^{\mathrm{T}}$ is symmetric and positive semi-definite. This when combined with (14.3.1), the theorem follows.

For completeness, we now establish a second and a slightly stronger version of the optimality of least squares estimates.

Gauss–Markov theorem: version II Pick a vector $\mu \in \mathbb{R}^n$ and fix it. Define a **linear functional** of x, namely

$$\phi(\mathbf{x}) = \mu^{\mathrm{T}}\mathbf{x}. \tag{14.3.7}$$

Now consider the problem of estimating $\phi(\mathbf{x})$. We are seeking a linear and unbiased estimate for $\phi(\mathbf{x})$. To this end, let $\mathbf{a} \in \mathbb{R}^m$ and let $\mathbf{a}^{\mathrm{T}}\mathbf{z}$ be an estimator for $\phi(\mathbf{x})$, which is clearly a linear function of $\mathbf{z}$. From

$$\begin{aligned}E(\mathbf{a}^{\mathrm{T}}\mathbf{z}) &= E[\mathbf{a}^{\mathrm{T}}(\mathbf{Hx} + \mathbf{v})] = \mathbf{a}^{\mathrm{T}}\mathbf{H}\,E(\mathbf{x}) + \mathbf{a}^{\mathrm{T}}E(\mathbf{v}) \\ &= \mathbf{a}^{\mathrm{T}}\mathbf{Hx}\end{aligned}$$

it follows that this linear estimate is **unbiased** if

$$\mu^{\mathrm{T}}\mathbf{x} = \mathbf{a}^{\mathrm{T}}\mathbf{H}\mathbf{x} \quad \text{or} \quad \mu = \mathbf{H}^{\mathrm{T}}\mathbf{a}. \tag{14.3.8}$$

The variance of this linear unbiased estimate is given by

$$\begin{aligned} \mathrm{Var}(\mathbf{a}^{\mathrm{T}}\mathbf{z}) &= E[\mathbf{a}^{\mathrm{T}}\mathbf{z} - E(\mathbf{a}^{\mathrm{T}}\mathbf{z})]^2 = E[(\mathbf{a}^{\mathrm{T}}\mathbf{v})^2] \\ &= E[\mathbf{a}^{\mathrm{T}}\mathbf{v}\mathbf{v}^{\mathrm{T}}\mathbf{a}] = \mathbf{a}^{\mathrm{T}}\mathbf{R}\mathbf{a}. \end{aligned} \tag{14.3.9}$$

We now examine the problem of minimizing the variance of $\mathbf{a}^{\mathrm{T}}\mathbf{z}$ when $\mathbf{a}$ is subjected to the constraint (14.3.8). This is solved by minimizing the Lagrangian

$$L(\mathbf{a}, \lambda) = \mathbf{a}^{\mathrm{T}}\mathbf{R}\mathbf{a} - \lambda^{\mathrm{T}}(\mathbf{H}^{\mathrm{T}}\mathbf{a} - \mu) \tag{14.3.10}$$

where $\lambda \in \mathbb{R}^n$ and $\mathbf{a} \in \mathbb{R}^m$. Hence

$$\left.\begin{aligned} \nabla_a L(\mathbf{a}, \lambda) &= 2\mathbf{R}\mathbf{a} - \mathbf{H}\lambda \\ \nabla_\lambda L(\mathbf{a}, \lambda) &= \mathbf{H}^{\mathrm{T}}\mathbf{a} - \mu \end{aligned}\right\} \tag{14.3.11}$$

Equating these gradients to zero and solving, we get

$$\mathbf{a} = \frac{1}{2}\mathbf{R}^{-1}\mathbf{H}\lambda \quad \text{and} \quad \frac{1}{2}(\mathbf{H}^{\mathrm{T}}\mathbf{R}^{-1}\mathbf{H})\lambda = \mu$$

which when combined gives

$$\mathbf{a} = \mathbf{R}^{-1}\mathbf{H}(\mathbf{H}^{\mathrm{T}}\mathbf{R}^{-1}\mathbf{H})^{-1}\mu.$$

Hence, the **linear, unbiased, minimum variance** estimates $\mathbf{a}^{\mathrm{T}}\mathbf{z}$ of $\phi(\mathbf{x}) = \mu^{\mathrm{T}}\mathbf{x}$ is given by

$$\begin{aligned} \mathbf{a}^{\mathrm{T}}\mathbf{z} &= \mu^{\mathrm{T}}(\mathbf{H}^{\mathrm{T}}\mathbf{R}^{-1}\mathbf{H})^{-1}\mathbf{H}^{\mathrm{T}}\mathbf{R}^{-1}\mathbf{z} \\ &= \mu^{\mathrm{T}}\hat{\mathbf{x}}_{\mathrm{LS}} \ [\text{using}(14.1.6)]. \end{aligned} \tag{14.3.12}$$

In other words, the best linear unbiased estimate of a linear combination of $\mathbf{x}$ is the same linear combination of $\hat{\mathbf{x}}_{\mathrm{LS}}$.

Now by picking $\mu = (1, 0, \ldots, 0)^{\mathrm{T}}$, it follows that the first component of $\hat{\mathbf{x}}_{\mathrm{LS}}$ is BLUE for the first component of $\mathbf{x}$. Likewise, by picking μ to be any other standard unit vector, we see that each component of $\hat{\mathbf{x}}_{\mathrm{LS}}$ is BLUE for the corresponding component of $\mathbf{x}$. Hence, $\hat{\mathbf{x}}_{\mathrm{LS}}$ is a BLUE for $\mathbf{x}$.

Remark 14.3.1 It is possible to construct nonlinear estimates whose variance is smaller than the linear estimates. However, if we restrict the noise vector $\mathbf{v}$ in (14.1.1) to have a multivariate normal distribution then it can be shown that the linear least squares estimate $\hat{\mathbf{x}}_{\mathrm{LS}}$ in (14.1.6) has minimum variance among all the linear and nonlinear unbiased estimates for $\mathbf{x}$. This latter result is known as the **Rao-Blackwell Theorem**. For details refer to Rao (1973).

14.4 Model error and sensitivity

In this section, we briefly discuss the effect of model error on the quality of the linear estimate of the state $\mathbf{x}$ in the linear model (14.1.1). To make the basic ideas transparent, first consider the unperturbed scalar case (when $n = 1$ and $m = 1$), that is,

$$z = hx + v \tag{14.4.1}$$

where the scalar noise v is such that $E(v) = 0$ and $E(v^2) = \sigma^2$. From (14.1.7), the linear, unbiased, least squares estimate for x is given by

$$\hat{\mathbf{x}}_{\text{LS}} = \frac{z}{h} \tag{14.4.2}$$

where

$$E(\hat{\mathbf{x}}_{\text{LS}}) = \frac{1}{h}E(z) = x \tag{14.4.3}$$

and

$$\begin{aligned} \text{Var}(\hat{\mathbf{x}}_{\text{LS}}) = E(\hat{\mathbf{x}}_{\text{LS}} - x)^2 &= E(\tfrac{z}{h} - x)^2 \\ &= \tfrac{1}{h^2}E(z - hx)^2 = \tfrac{1}{h^2}E(v^2) \\ &= \tfrac{\sigma^2}{h^2}. \end{aligned} \tag{14.4.4}$$

Now consider the perturbed system

$$z = \bar{h}x + v \tag{14.4.5}$$

where $\bar{h} = h + \epsilon$, for some $\epsilon \neq 0$. The aim of the following analysis is to discover the difference between the reality and the assumption. While in reality the system parameter h may have changed to $\bar{h}$, unaware of this change we might still continue to assume that the system parameter has the value h.

Thus, the actual value of the least squares estimate $\hat{x}$ is given by

$$\hat{x} = \frac{z}{\bar{h}} = \frac{z}{(h + \epsilon)}.$$

From

$$\begin{aligned} (h + \epsilon)^{-1} = [h(1 + h^{-1}\epsilon)]^{-1} &= (1 + h^{-1}\epsilon)^{-1}h^{-1} \\ &\approx (1 - h^{-1}\epsilon)h^{-1} \\ &= h^{-1} - h^{-1}\epsilon h^{-1} \end{aligned}$$

we get

$$\begin{aligned} \hat{x} &= (1 - h^{-1}\epsilon)h^{-1}z \\ &= (1 - h^{-1}\epsilon)h^{-1}[hx + v] \ (\because \text{we are unaware of the change}) \\ &= (1 - h^{-1}\epsilon)x + (1 - h^{-1}\epsilon)h^{-1}v \end{aligned}$$

from which it follows that

$$E(\hat{x}) = x - h^{-1}\epsilon x. \tag{14.4.6}$$

Hence, $\hat{x}$ is a biased estimate with the bias given by

$$|x - E(\hat{x})| = h^{-1}\epsilon x.$$

Thus, the model errors show up as the bias in the least squares estimates. Similarly,

$$\mathrm{Var}(\hat{x}) = \frac{\sigma^2}{\bar{h}^2} = \frac{\sigma^2}{(h+\epsilon)^2} = \frac{\sigma^2}{h^2}[1 - 2h^{-1}\epsilon] \tag{14.4.7}$$

which could be smaller or larger than the variance in (14.4.4).

We now present an extension of this result to the general linear case. To this end, let

$$\mathbf{z} = \bar{\mathbf{H}}\mathbf{x} + \mathbf{v} \tag{14.4.8}$$

where $\bar{\mathbf{H}} = \mathbf{H} + \epsilon\mathbf{E}$ with $\mathbf{E} \in \mathbb{R}^{m \times n}$ and $\epsilon > 0$ denote the perturbed system. It is assumed that the noise vector continues to satisfy the standard assumptions in Section 14.1. The least squares estimate for $\mathbf{x}$ using this model (14.4.8) is given by

$$\hat{\mathbf{x}} = (\bar{\mathbf{H}}^{\mathrm{T}}\mathbf{R}^{-1}\bar{\mathbf{H}})^{-1}\bar{\mathbf{H}}^{\mathrm{T}}\mathbf{R}^{-1}\mathbf{z}. \tag{14.4.9}$$

Let us now compute the difference between $(\bar{\mathbf{H}}^{\mathrm{T}}\mathbf{R}^{-1}\bar{\mathbf{H}})$ and $(\mathbf{H}^{\mathrm{T}}\mathbf{R}^{-1}\mathbf{H})$ by using only the first degree terms in ϵ which is justified since ϵ is assumed to be small.

$$\begin{aligned}
\bar{\mathbf{H}}^{\mathrm{T}}\mathbf{R}^{-1}\bar{\mathbf{H}} &= (\mathbf{H} + \epsilon\mathbf{E})^{\mathrm{T}}\mathbf{R}^{-1}(\mathbf{H} + \epsilon\mathbf{E}) \\
&= \mathbf{H}^{\mathrm{T}}\mathbf{R}^{-1}\mathbf{H} + \epsilon\mathbf{H}^{\mathrm{T}}\mathbf{R}^{-1}\mathbf{E} + \epsilon\mathbf{E}^{\mathrm{T}}\mathbf{R}^{-1}\mathbf{H} + \epsilon^2\mathbf{E}^{\mathrm{T}}\mathbf{R}^{-1}\mathbf{E} \\
&\approx \mathbf{H}^{\mathrm{T}}\mathbf{R}^{-1}\mathbf{H} + \epsilon[\mathbf{H}^{\mathrm{T}}\mathbf{R}^{-1}\mathbf{E} + \mathbf{E}^{\mathrm{T}}\mathbf{R}^{-1}\mathbf{H}] \\
&= \mathbf{A} - \epsilon\mathbf{B}
\end{aligned} \tag{14.4.10}$$

where $\mathbf{A} = \mathbf{H}^{\mathrm{T}}\mathbf{R}^{-1}\mathbf{H}$ and $\mathbf{B} = -(\mathbf{H}^{\mathrm{T}}\mathbf{R}^{-1}\mathbf{E} + \mathbf{E}^{\mathrm{T}}\mathbf{R}^{-1}\mathbf{H})$.

Then

$$\begin{aligned}
(\mathbf{A} - \epsilon\mathbf{B})^{-1} &= [\mathbf{A}(\mathbf{I} - \epsilon\mathbf{A}^{-1}\mathbf{B})]^{-1} \\
&= (\mathbf{I} - \epsilon\mathbf{A}^{-1}\mathbf{B})^{-1}\mathbf{A}^{-1}.
\end{aligned} \tag{14.4.11}$$

Recall from Appendix B that if $\mathbf{C}$ is a matrix such that $\|\mathbf{C}\| < 1$, then we can write

$$(\mathbf{I} - \mathbf{C})^{-1} = \mathbf{I} + \mathbf{C} + \mathbf{C}^2 + \mathbf{C}^3 + \cdots \tag{14.4.12}$$

which is a power series expansion quite akin to the well-known geometric series, In applying this result to the r.h.s. of (14.4.11), first assume that the perturbation

$\epsilon\mathbf{E}$ is such that $\|\epsilon\mathbf{A}^{-1}\mathbf{B}\| < 1$. Under this condition, using (14.4.12) and using only the first degree term in ϵ, we obtain

$$(\mathbf{A} - \epsilon\mathbf{B})^{-1} = (\mathbf{I} - \epsilon\mathbf{A}^{-1}\mathbf{B})^{-1}\mathbf{A}^{-1} = \mathbf{A}^{-1} + \epsilon\mathbf{A}^{-1}\mathbf{B}\mathbf{A}^{-1}. \tag{14.4.13}$$

Now consider

$$\begin{aligned}(\overline{\mathbf{H}}^{T}\mathbf{R}^{-1}\overline{\mathbf{H}})^{-1}\overline{\mathbf{H}}^{\mathrm{T}}\mathbf{R}^{-1} &= [\mathbf{A}^{-1} + \epsilon\mathbf{A}^{-1}\mathbf{B}\mathbf{A}^{-1}](\mathbf{H} + \epsilon\mathbf{E})^{\mathrm{T}}\mathbf{R}^{-1}\\ &= \mathbf{A}^{-1}\mathbf{H}^{\mathrm{T}}\mathbf{R}^{-1} + \epsilon[\mathbf{A}^{-1}\mathbf{E}^{\mathrm{T}}\mathbf{R}^{-1}\\ &\quad + \mathbf{A}^{-1}\mathbf{B}\mathbf{A}^{-1}\mathbf{H}^{\mathrm{T}}\mathbf{R}^{-1}].\end{aligned} \tag{14.4.14}$$

Combining (14.4.14) with (14.4.9) we get (substituting for $\mathbf{B}$)

$$\hat{\mathbf{x}} = \hat{\mathbf{x}}_{\mathrm{LS}} + \epsilon\mathbf{A}^{-1}\mathbf{E}^{\mathrm{T}}\mathbf{R}^{-1}(\mathbf{z} - \mathbf{H}\,\hat{\mathbf{x}}_{\mathrm{LS}}) - \epsilon\mathbf{A}^{-1}\mathbf{H}^{\mathrm{T}}\mathbf{R}^{-1}\mathbf{E}\,\hat{\mathbf{x}}_{\mathrm{LS}}. \tag{14.4.15}$$

Since we are unaware of the change, substituting $\mathbf{z} = \mathbf{H}\mathbf{x} + \mathbf{v}$ and taking expectations we get

$$E(\hat{\mathbf{x}}) = \mathbf{x} - \epsilon\mathbf{A}^{-1}\mathbf{H}^{\mathrm{T}}\mathbf{R}^{-1}\mathbf{E}\mathbf{x} \tag{14.4.16}$$

where the second term on the r.h.s. denotes the bias. In the scalar case, that is, when $\mathbf{H} = h$, $\mathbf{R} = 1$ and $\mathbf{E} = 1$, this relation (14.4.16) reduces to (14.1.6). We leave it as an exercise to compute the effect of this perturbation on the covariance of $\hat{\mathbf{x}}$.

Exercises

14.1 Let $\mathbf{A}$, $\mathbf{B}$, $\mathbf{C}$ be three 2×2 real matrices. By explicit multiplication verify the identity $\mathrm{tr}(\mathbf{ABC}) = \mathrm{tr}(\mathbf{CAB}) = \mathrm{tr}(\mathbf{BCA})$.

14.2 Let $\mathbf{v} = (v_1, v_2)^{\mathrm{T}}$ and let $\mathbf{A} = \begin{bmatrix} a & b \\ b & c \end{bmatrix}$ be a symmetric matrix. Verify the following

$$\mathrm{tr}(\mathbf{v}\mathbf{v}^{\mathrm{T}}\mathbf{A}) = av_1^2 + cv_2^2 + 2bv_1v_2.$$

14.3 If $\mathbf{v} = (v_1, v_2)^{\mathrm{T}}$ is such that $E(\mathbf{v}) = 0$ and $E(\mathbf{v}\mathbf{v}^{\mathrm{T}}) = \sigma^2\mathbf{I}$, that is, $E(v_1v_2) = 0$, then using the result in (Exercise 14.1) verify

$$E[\mathrm{tr}(\mathbf{v}\mathbf{v}^{\mathrm{T}}\mathbf{A})] = (a + c)\sigma^2 = \sigma^2\mathrm{tr}(\mathbf{A}).$$

14.4 Let $\mathbf{P} = \mathbf{H}(\mathbf{H}^{\mathrm{T}}\mathbf{H})^{-1}\mathbf{H}^{\mathrm{T}}$ be an orthogonal projection matrix then verify that

$$\mathrm{tr}(\mathbf{H}(\mathbf{H}^{\mathrm{T}}\mathbf{H})^{-1}\mathbf{H}^{\mathrm{T}}) = \mathrm{tr}[(\mathbf{H}^{\mathrm{T}}\mathbf{H})(\mathbf{H}^{\mathrm{T}}\mathbf{H})^{-1}] = \mathrm{tr}(\mathbf{I}_n) = n.$$

Note: It should be interesting to note that the trace of an orthogonal projection matrix is also its rank.

14.5 Compute the effect of the perturbation in (14.4.8) on the covariance of $\hat{\mathbf{x}}$ given in (14.4.9).

Notes and references

The material covered in this chapter is rather standard in the literature – Melsa and Cohn (1978) and Sage and Melsa (1971). For a discussion of the Gauss–Markov theorem refer to Rao (1973) and Brammer and Siffling (1989).

15

Maximum likelihood method

In this chapter, we provide an introduction to the basic principles of point estimation using the Fisher's Framework where the unknown parameter $\mathbf{x} \in \mathbb{R}^n$ to be estimated is treated as constant and the conditional probability distribution $p(\mathbf{z}|\mathbf{x})$ is assumed to be known. This conditional distribution when considered as a function of $\mathbf{x}$ is known as the **likelihood function** $L(\mathbf{x}|\mathbf{z}) = p(\mathbf{z}|\mathbf{x})$. The basic idea of the maximum likelihood method introduced by Fisher in the early 1920s is quite simple in that given a set of observations $\mathbf{z}$, it seeks to find a value of $\mathbf{x}$ that maximizes the probability or likelihood of observing this sample $\mathbf{z}$.

Section 15.1 describes the basic framework of this method. Many of the salient properties of the maximum likelihood estimates are contained in Section 15.2. A discussion of the nonlinear case – when the observations are a nonlinear function of the unknown parameter – is contained in Section 15.3.

15.1 The maximum likelihood method

Let $\mathbf{z} \in \mathbb{R}^m$, $\mathbf{x} \in \mathbb{R}^n$ and $h : \mathbb{R}^n \to \mathbb{R}^m$. It is assumed that

$$\mathbf{z} = h(\mathbf{x}) + \mathbf{v} \tag{15.1.1}$$

where $\mathbf{v}$ is the additive measurement noise. It is assumed that the functional form of the multivariate distribution $p(\mathbf{v})$ of $\mathbf{v}$ is known but it may contain certain unknown parameters. It is also assumed that

$$\left.\begin{array}{l} E(\mathbf{v}) = 0; \quad \mathrm{Cov}(\mathbf{v}) = E(\mathbf{v}\mathbf{v}^{\mathrm{T}}) = \mathbf{R} \\ E(\mathbf{v}\mathbf{x}^{\mathrm{T}}) = 0 \end{array}\right\} \tag{15.1.2}$$

that is, $\mathbf{v}$ and $\mathbf{x}$ are **uncorrelated**. Knowing $p(\mathbf{v})$ and using (15.1.1), we can derive the functional form of the multivariate distribution of $\mathbf{z}$ conditional on the unknown, namely, $p(\mathbf{z}|\mathbf{x})$. Since $\mathbf{x}$ is an unknown constant, so is $h(\mathbf{x})$ and hence $p(\mathbf{z}|\mathbf{x})$ is essentially a translation of $p(\mathbf{v})$. As an example, if $p(\mathbf{v}) = N(0, \Sigma)$, a multivariate normal with mean zero and covariance matrix Σ, then $p(\mathbf{z}|\mathbf{x}) = N(h(x), \Sigma)$, a multivariate normal with mean $h(\mathbf{x})$ and covariance matrix Σ.

The method of maximum likelihood is predicated on the assumption that the functional form of $p(\mathbf{z}|\mathbf{x})$ is known. While this function $p(\mathbf{z}|\mathbf{x})$ is a probability distribution function of $\mathbf{z}$ given $\mathbf{x}$, it has a **dual interpretation** of a likelihood function when considered as a function of $\mathbf{x}$ for a given sample of observation $\mathbf{z}$. Thus, define the likelihood function

$$L(\mathbf{x}|\mathbf{z}) = p(\mathbf{z}|\mathbf{x}). \tag{15.1.3}$$

By definition, maximum likelihood estimate $\mathbf{x}_{\text{ML}}$ is one that maximizes $L(\mathbf{x}|\mathbf{z})$, that is

$$L(\mathbf{x}_{\text{ML}}|\mathbf{z}) \geq L(\hat{\mathbf{x}}|\mathbf{z}) \tag{15.1.4}$$

for any other estimate $\hat{\mathbf{x}}$ of $\mathbf{x}$. In other words, $\hat{\mathbf{x}}_{\text{ML}}$ is one that maximizes the probability of observing the given sample of observations $\mathbf{z}$.

Computationally, it is often convenient to work with the logarithm of the likelihood function. In this case $\hat{\mathbf{x}}_{\text{ML}}$ is defined by

$$\ln L(\hat{\mathbf{x}}_{\text{ML}}|\mathbf{z}) \geq \ln L(\hat{\mathbf{x}}|\mathbf{z}) \tag{15.1.5}$$

for any other estimate $\hat{\mathbf{x}}$. Thus, a necessary condition for the maximum (Appendix D) is that

$$\nabla_{\mathbf{x}}[\ln L(\mathbf{x}|\mathbf{z})] = \frac{1}{L(\mathbf{x}|\mathbf{z})} \nabla_{\mathbf{x}} L(\mathbf{x}|\mathbf{z}) = 0. \tag{15.1.6}$$

The gradient of $\ln L(\mathbf{x}|\mathbf{z})$ with respect to $\mathbf{x}$ is often known as the **score** which is a vector of size n.

Remark 15.1.1 It is assumed in (15.1.1) that the noise corrupting the observation is additive in nature. We want to emphasize that this vector $\mathbf{v}$ is **not** observable. Based on the properties of the measurement system, one could pre-compute statistical properties of $\mathbf{v}$ and hence the assumption that $E(\mathbf{v}) = 0$ and $E(\mathbf{v}\mathbf{v}^{\text{T}}) = \mathbf{R}$. If $E(\mathbf{v}) \neq 0$, then the measurement system introduces a bias in the observation and this would call for re-calibration of the system. If the second-order property ($\mathbf{R}$) is **not** known, then we could include this unknown as a part of the unknown parameter $\mathbf{x}$ to be estimated.

The following is an illustration of this method.

Example 15.1.1 Consider a special case of a linear model

$$\mathbf{z} = \mathbf{H}\mu + \mathbf{v} \tag{15.1.7}$$

where $\mathbf{z} = (z_1, z_2, \ldots, z_m)^{\mathrm{T}}, \mathbf{H} = (1, 1, \ldots, 1)^{\mathrm{T}}, \mu \in \mathbb{R}$ and $\mathbf{v} = (v_1, v_2, \ldots, v_m)^{\mathrm{T}}$. Thus, $z_i = \mu + v_i$ for $i = 1$ to m. Let $\mathbf{v}$ be such that v_i are independent and identically distributed normal variables with mean zero and unknown variance σ^2, that is, $v_i \sim iid\ N(0, \sigma^2)$.

Hence

$$E(\mathbf{v}\mathbf{v}^{\mathrm{T}}) = \mathbf{R} = \mathrm{Diag}(\sigma^2, \sigma^2, \ldots, \sigma^2)$$

is an $m \times m$ diagonal matrix with σ^2 along the diagonal. Thus, $\mathbf{v} \sim N(0, \sigma^2\mathbf{I})$ and $\mathbf{z} \sim N(\mathbf{H}\mu, \sigma^2\mathbf{I})$.

The likelihood function

$$\begin{aligned} L(\mathbf{x}|\mathbf{z}) &= p(\mathbf{z}|\mathbf{x}) \\ &= (2\pi)^{-\frac{m}{2}}(\sigma^2)^{-\frac{m}{2}} \exp[-\tfrac{1}{2\sigma^2}(\mathbf{z} - \mathbf{H}\mu)^{\mathrm{T}}(\mathbf{z} - \mathbf{H}\mu)]. \end{aligned}$$

The log likelihood function becomes

$$l(\mathbf{x}|\mathbf{z}) = \ln L(\mathbf{x}|\mathbf{z}) = -\frac{m}{2}\ln 2\pi - \frac{m}{2}\ln(\sigma^2) - \frac{1}{2\sigma^2}\sum_{i=1}^{m}(z_i - \mu)^2. \tag{15.1.8}$$

The necessary condition for the maximum is given by

$$\begin{aligned} 0 = \nabla_{\mathbf{x}} \ln L(\mathbf{x}|\mathbf{z}) &= \begin{bmatrix} \dfrac{\partial \ln L(\mathbf{x}|\mathbf{z})}{\partial \mu} \\ \dfrac{\partial \ln L(\mathbf{x}|\mathbf{z})}{\partial \sigma^2} \end{bmatrix} \\ &= \begin{bmatrix} \dfrac{1}{\sigma^2}\displaystyle\sum_{i=1}^{m}(z_i - \mu) \\ -\dfrac{m}{2}\left(\dfrac{1}{\sigma^2}\right) + \dfrac{1}{2\sigma^4}\displaystyle\sum_{i=1}^{m}(z_i - \mu)^2 \end{bmatrix} \end{aligned} \tag{15.1.9}$$

from which we obtain the maximum likelihood estimates as

$$\hat{\mu}_{\mathrm{ML}} = \frac{1}{m}\sum_{i=1}^{m} z_i = \bar{z} \tag{15.1.10}$$

and

$$\hat{\sigma}^2_{\mathrm{ML}} = \frac{1}{m}\sum_{i=1}^{m}(z_i - \bar{z})^2. \tag{15.1.11}$$

Comparing these with the results in Examples(13.2.2) and (13.2.3), it follows that

(a) $\hat{\mu}_{\mathrm{ML}}$ is an **unbiased** and an **efficient** estimate
(b) $\hat{\sigma}^2_{\mathrm{ML}}$ is a **biased** estimate of σ^2.

To verify that (15.1.10) and (15.1.11) correspond to the maximum, let us compute the Hessian of $l(\mathbf{x}|\mathbf{z})$ in (15.1.8). It can be verified from (15.1.9) that

$$\frac{\partial^2 l(\mathbf{x}|\mathbf{z})}{\partial \mu^2} = -\frac{m}{\sigma^2}$$

$$\frac{\partial^2 l(\mathbf{x}|\mathbf{z})}{\partial (\sigma^2)^2} = \frac{m}{2\sigma^4} - \frac{1}{\sigma^6}\sum_{i=1}^{m}(z_i - \mu)^2 \tag{15.1.12}$$

and

$$\frac{\partial^2 l(\mathbf{x}|\mathbf{z})}{\partial \mu \, \partial \sigma^2} = -\frac{1}{\sigma^4}\sum_{i=1}^{m}(z_i - \mu).$$

Using (15.1.10), it can be verified that

$$\sum_{i=1}^{m}(z_i - \hat{\mu}_{\text{ML}}) = \sum_{i=1}^{m} z_i - m\hat{\mu}_{\text{ML}} = 0$$

and (15.1.13)

$$\sum_{i=1}^{m}(z_i - \bar{z})^2 = \sum_{i=1}^{m}(z_i - \hat{\mu}_{\text{ML}})^2 = m\,\hat{\sigma}^2_{\text{ML}}.$$

Now combining (15.1.12) and (15.1.13), it can be verified (Exercise 15.1) that the Hessian of $l(\mathbf{x}|\mathbf{z})$ evaluated at $(\hat{\mu}_{\text{ML}}, \hat{\sigma}^2_{\text{ML}})$ is given by

$$\nabla^2 l(\mathbf{x}|\mathbf{z}) = \begin{bmatrix} -\frac{m}{\sigma^2} & 0 \\ 0 & -\frac{m}{2\sigma^4} \end{bmatrix} \tag{15.1.14}$$

which is a diagonal matrix with negative entries along the diagonal. Hence this Hessian is negative definite and hence the estimates in (15.1.10)–(15.1.11) correspond to the maximum of $l(\mathbf{x}|\mathbf{z})$.

Remark 15.1.2 when σ^2 is unknown, maximizing $\ln L(\mathbf{x}|\mathbf{z})$ in (15.1.8) is the same as minimizing

$$(\mathbf{z} - \mathbf{H}\mu)^{\mathrm{T}}(\mathbf{z} - \mathbf{H}\mu)$$

which is, in fact, the least squares criterion used in Chapter 14. In other words, when the underlying distribution of $\mathbf{v}$ is a Gaussian with unknown covariance, then the maximum likelihood method reduces to the statistical least squares method discussed in Chapter 14.

15.2 Properties of maximum likelihood estimates

In this section, we catalog many of the key properties of the maximum likelihood estimates without proof. For details refer to many of the excellent texts included in the references.

(M1) **Consistency** $\hat{\mathbf{x}}_{\text{ML}}$ is a consistent estimate. That is, the multivariate (sampling) distribution of the random variable $\hat{\mathbf{x}}_{\text{ML}}$ is increasingly clustered around the unknown true value of $\mathbf{x}$. Stated formally, for any $\epsilon > 0$,

$$\text{Prob}[\ \|\mathbf{x} - \hat{\mathbf{x}}_{\text{ML}}\| > \epsilon\ ] \to 0.$$

as the number m of samples grows without bound.

(M2) **Asymptotic normality** The actual distribution of $\hat{\mathbf{x}}_{\text{ML}}$ can be approximated by a multivariate normal distribution,

$$\hat{\mathbf{x}}_{\text{ML}} \sim N(\mathbf{x}, \mathbf{I}^{-1}(\mathbf{x}))$$

where

$$\mathbf{I}(\mathbf{x}) = -E[\nabla_{\mathbf{x}}^2 \ln L(\mathbf{x}|\mathbf{z})]$$

which is the negative of the expected value of the Hessian of $\ln L(\mathbf{x}|\mathbf{z})$.

(M3) **Asymptotic efficiency** The estimate $\hat{\mathbf{x}}_{\text{ML}}$ tends to its minimum value $\mathbf{I}^{-1}(\mathbf{x})$ dictated by the well known Cramer–Rao inequality given in Section 13.2. In other words, the maximum likelihood estimate becomes an efficient estimate as m, the sample size, grows without bound.

(M4) **Invariance** Let $\hat{\mathbf{x}}_{\text{ML}}$ be the maximum likelihood estimate of $\mathbf{x}$ and let $g(\mathbf{x})$ be a function of $\mathbf{x}$. Then $g(\hat{\mathbf{x}}_{\text{ML}})$ is the maximum likelihood estimate of $g(\mathbf{x})$.

We now illustrate these properties using a general linear model.

Example 15.2.1 Let $\mathbf{z} = \mathbf{H}\mathbf{x} + \mathbf{v}$ where $\mathbf{H} \in \mathbb{R}^{m\times n}$, $\mathbf{z}, \mathbf{v} \in \mathbb{R}^m$ and $\mathbf{x} \in \mathbb{R}^n$. Let $\mathbf{v} \sim N(0, \sigma^2\mathbf{I})$. Consider now the problem of estimating $(\mathbf{x}^{\text{T}}, \sigma^2)^{\text{T}}$. Then

$$l(\mathbf{x}|\mathbf{z}) = \ln L(\mathbf{x}|\mathbf{z}) = -\frac{m}{2}\ln(2\pi) - \frac{m}{2}\ln(\sigma^2) - \frac{1}{2\sigma^2}((\mathbf{z} - \mathbf{H}\mathbf{x}))^{\text{T}}((\mathbf{z} - \mathbf{H}\mathbf{x})) \tag{15.2.1}$$

then

$$\nabla_{\mathbf{x}}\, l(\mathbf{x}|\mathbf{z}) = -\frac{1}{\sigma^2}[\mathbf{H}^{\text{T}}\mathbf{H}\mathbf{x} - \mathbf{H}^{\text{T}}\mathbf{z}] = -\frac{1}{\sigma^2}\mathbf{H}^{\text{T}}\mathbf{v}$$
$$\frac{\partial l(\mathbf{x}|\mathbf{z})}{\partial \sigma^2} = -\frac{m}{2\sigma^2} + \frac{1}{2\sigma^4}((\mathbf{z} - \mathbf{H}\mathbf{x}))^{\text{T}}((\mathbf{z} - \mathbf{H}\mathbf{x})).$$

Setting these to zero and solving, we obtain

$$\hat{\mathbf{x}}_{\text{ML}} = (\mathbf{H}^{\text{T}}\mathbf{H})^{-1}\mathbf{H}^{\text{T}}\mathbf{z} \tag{15.2.2}$$

which is the same as the least squares estimate (refer to Section 14.1) and

$$\hat{\sigma}^2_{\text{ML}} = \frac{1}{m}\hat{\mathbf{r}}^{\text{T}}\hat{\mathbf{r}} \tag{15.2.3}$$

where $\hat{\mathbf{r}}$ denotes the model residual given by

$$\hat{\mathbf{r}} = \mathbf{z} - \mathbf{H}\hat{\mathbf{x}}_{\text{ML}}. \tag{15.2.4}$$

We now compute the Hessian of $l(\mathbf{x}|\mathbf{z})$:

$$\nabla^2_{\mathbf{x}} l(\mathbf{x}|\mathbf{z}) = -\frac{\mathbf{H}^{\mathrm{T}}\mathbf{H}}{\sigma^2}; \qquad -E(\nabla^2_{\mathbf{x}}\, l(\mathbf{x}|\mathbf{z})) = \frac{\mathbf{H}^{\mathrm{T}}\mathbf{H}}{\sigma^2}$$

$$\frac{\partial^2 l(\mathbf{x}|\mathbf{z})}{\partial(\sigma^2)^2} = \frac{m}{2\sigma^4} - \frac{1}{\sigma^6}\mathbf{v}^{\mathrm{T}}\mathbf{v}; \; -E\left(\frac{\partial^2 l(\mathbf{x}|\mathbf{z})}{\partial(\sigma^2)^2}\right) = \frac{m}{2\sigma^4}$$

since $E(\mathbf{v}^{\mathrm{T}}\mathbf{v}) = m\sigma^2$.

$$\frac{\partial^2 l(\mathbf{x}|\mathbf{z})}{\partial \mathbf{x}\, \partial\sigma^2} = \frac{\mathbf{H}^{\mathrm{T}}\mathbf{v}}{\sigma^4}; \quad -E\left(\frac{\partial^2 l(\mathbf{x}|\mathbf{z})}{\partial \mathbf{x}\, \partial\sigma^2}\right) = 0.$$

Hence

$$\mathbf{I}(\mathbf{x}) = -E\begin{bmatrix} \nabla^2_{\mathbf{x}} \ln(\mathbf{x}|\mathbf{z}) & \dfrac{\partial^2 l(\mathbf{x}|\mathbf{z})}{\partial \mathbf{x}\, \partial\sigma^2} \\ \dfrac{\partial^2 \ln(\mathbf{x}|\mathbf{z})}{\partial \mathbf{x}\, \partial\sigma^2} & \dfrac{\partial^2 l(\mathbf{x}|\mathbf{z})}{\partial(\sigma^2)^2} \end{bmatrix} = \begin{bmatrix} \frac{\mathbf{H}^{\mathrm{T}}\mathbf{H}}{\sigma^2} & 0 \\ 0 & \frac{m}{2\sigma^4} \end{bmatrix}.$$

and

$$\mathbf{I}^{-1}(\mathbf{x}) = \begin{bmatrix} \sigma^2(\mathbf{H}^{\mathrm{T}}\mathbf{H})^{-1} & 0 \\ 0 & \frac{2\sigma^4}{m} \end{bmatrix}.$$

Substituting (15.2.2) and (15.2.3) in (15.2.1), we obtain the maximum value $\hat{l}$ of $l(\mathbf{x}|\mathbf{z})$:

$$\hat{l} = \frac{m}{2}\ln 2\pi - \frac{m}{2}\ln\left(\frac{\hat{\mathbf{r}}^{\mathrm{T}}\hat{\mathbf{r}}}{m}\right) - \frac{m}{2}.$$

Taking exponential on both sides

$$\begin{aligned} \hat{L} &= (2\pi)^{\frac{m}{2}} \mathrm{e}^{-\frac{m}{2}} (\hat{\mathbf{r}}^{\mathrm{T}}\hat{\mathbf{r}})^{-\frac{m}{2}} \\ &= (\tfrac{2\pi}{\mathrm{e}})^{\frac{m}{2}} (\hat{\mathbf{r}}^{\mathrm{T}}\hat{\mathbf{r}})^{-\frac{m}{2}}. \end{aligned} \tag{15.2.5}$$

15.3 Nonlinear case

For completeness, in this section we indicate the major steps involved in obtaining the maximum likelihood estimate when $\mathbf{z} = h(\mathbf{x}) + \mathbf{v}$ where $h(\mathbf{x})$ in general is a nonlinear function. Assume $\mathbf{v} \sim N(0, \sigma^2\mathbf{I})$. Then, $\mathbf{z} \sim N(h(\mathbf{x}), \sigma^2\mathbf{I})$ and the log likelihood function becomes

$$l(\mathbf{x}|\mathbf{z}) = -\frac{m}{2}\ln(2\pi) - \frac{m}{2}\ln(\sigma^2) - \frac{1}{2\sigma^2}(\mathbf{z} - h(\mathbf{x}))^{\mathrm{T}}(\mathbf{z} - h(\mathbf{x})). \tag{15.3.1}$$

Clearly, this is a nonlinear optimization problem and is solved iteratively using the process illustrated in Chapter 7.

In particular, this can be done by using either the first-order method that relies on a linear approximation to $l(\mathbf{x}|\mathbf{z})$ or the second-order method that uses a quadratic approximation to $l(\mathbf{x}|\mathbf{z})$ at a given operating point. These ideas are well developed in Chapter 7 in Part II and to save space, they are **not** repeated here.

Exercises

15.1 Verify (15.1.14).
Hint: Substitute the values of $\hat{\mu}_{\text{ML}}$ and $\hat{\sigma}^2_{\text{ML}}$ from (15.1.10) and (15.1.11) respectively for μ and σ^2 in (15.1.12) and simplify.

15.2 Let $z = \log x + v$ where x, v and hence z are scalars. Let the distribution of v, namely $p(v)$, be unimodal with mode at $v = 0$, that is $p\,(0) > p\,(a)$ for any $a \neq 0$. Then verify that $\hat{x}_{\text{ML}} = \mathrm{e}^z$.

15.3 Let $z_i = x_1 + x_2 v_i$ for $i = 1$ to m and $v_i \sim iid\ N(0, 1)$. Find the maximum likelihood estimates for x_1 and x_2.

Notes and references

Maximum likelihood method has been the standard workhorse in estimation theory for several decades and many mathematical software packages have ready-to-use routines for conducting parameter estimation using this approach. Rao (1973) and Melsa and Cohn (1978) contain a very good introduction to this method.

16
Bayesian estimation method

This chapter provides an overview of the classical **Bayesian** method for point estimation. The main point of departure of this method from other methods is that it considers the unknown $\mathbf{x}$ as a random variable. All the prior knowledge about this unknown is summarized in the form of a known prior distribution $p(\mathbf{x})$ of $\mathbf{x}$. If $\mathbf{z}$ is the set of observations that contains information about the unknown $\mathbf{x}$, this distribution is often given in the form of a conditional distribution $p(\mathbf{z}|\mathbf{x})$. The basic idea is to combine these two pieces of information to obtain an optimal estimate of $\mathbf{x}$, called the Bayes estimate.

The Bayesian framework is developed in Section 16.1. Special classes of Bayesian estimators – Bayes least squares estimate leading to the **conditional mean** (which is also the **minimum variance estimate**), **conditional mode**, and **conditional median** estimates are derived in Section 16.2.

16.1 The Bayesian framework

Let $\mathbf{x} \in \mathbb{R}^n$ be the **unknown** to be estimated and $\mathbf{z} \in \mathbb{R}^m$ be the observations that contain information about the unknown $\mathbf{x}$ to be estimated. The distinguishing feature of the Bayes framework is that it also treats the unknown $\mathbf{x}$ as a **random variable**. It is assumed that a **prior distribution** $p(\mathbf{x})$ is known. This distribution summarizes our initial belief about the unknown. It is assumed that nature picks a value of $\mathbf{x}$ from the distribution $p(\mathbf{x})$ but decides to tease us by not disclosing her choice, thereby defining a **game**. In this game, we are only allowed to observe $\mathbf{z}$ whose conditional distribution $p(\mathbf{z}|\mathbf{x})$ is known. The idea is to combine these two pieces of information – the prior distribution $p(\mathbf{x})$ and the conditional distribution $p(\mathbf{z}|\mathbf{x})$ along with the sample $\mathbf{z}$ drawn from it in an "**optimal**" fashion to obtain the "**best**" estimate of $\mathbf{x}$. This optimization problem has come to be known as the **game against nature** or **one-person game**.

Let $\hat{\mathbf{x}} = \hat{\mathbf{x}}(\mathbf{z})$ denote the estimate of $\mathbf{x}$ based on the given sample observation $\mathbf{z}$. Define

$$\tilde{\mathbf{x}} = \mathbf{x} - \hat{\mathbf{x}} \tag{16.1.1}$$

the **error** in the estimate $\hat{\mathbf{x}}$. The "best" estimate is one that makes this error **small** in some acceptable measure. Our first task then is to quantify the size of this error. To this end, we define a **cost function** $c : \mathbb{R}^n \to \mathbb{R}$ satisfying the following conditions:

(1) $c(0) = 0$
(2) $c(\cdot)$ is a non-decreasing function of the norm of its argument. That is, for any two vectors $\mathbf{a}$, $\mathbf{b}$ in $\mathbb{R}^n$

$$c(\mathbf{a}) \leq c(\mathbf{b}) \quad \text{if} \quad \|\mathbf{a}\| \leq \|\mathbf{b}\|$$

The idea is to use this cost function to **size up** the error $\tilde{\mathbf{x}}$. We now give some examples of useful cost functions.

(a) **Weighted sum of squared error** Let $\mathbf{W} \in \mathbb{R}^{n \times n}$ be a symmetric and positive definite matrix. Then

$$\begin{aligned} c(\tilde{\mathbf{x}}) = \tilde{\mathbf{x}}^{\mathrm{T}} \mathbf{W} \tilde{\mathbf{x}} &= (\mathbf{x} - \hat{\mathbf{x}})^{\mathrm{T}} \mathbf{W} (\mathbf{x} - \hat{\mathbf{x}}) \\ &= \|(\mathbf{x} - \hat{\mathbf{x}})\|_{\mathbf{W}}^2 \end{aligned} \tag{16.1.2}$$

denotes the weighted sum of the squared error. When $\mathbf{W} = \mathbf{I}$, the identity matrix, this reduces to the popular sum of the squared error.

(b) **Uniform cost function** Let $\epsilon > 0$ be a small, fixed real number. Define

$$c(\tilde{\mathbf{x}}) = \begin{cases} 0, & \text{if } \|\mathbf{x}\| \leq \epsilon \\ 1, & \text{otherwise} \end{cases} \tag{16.1.3}$$

(c) **Absolute error** For the special case when the unknown is a scalar, we can use

$$c(\tilde{x}) = |(x - \hat{x})| \tag{16.1.4}$$

namely, the absolute value of the error as the measure of the size of the error.

(d) **Symmetric and convex cost function** If

$$c(\tilde{\mathbf{x}}) = c(-\tilde{\mathbf{x}}) \tag{16.1.5a}$$

then $c(\cdot)$ is a **symmetric** function. In addition, if for any two $\mathbf{x}$ and $\mathbf{y}$ in $\mathbb{R}^n$

$$c(a\mathbf{x} + (1 - a)\mathbf{y}) \leq ac(\mathbf{x}) + (1 - a)c(\mathbf{y}) \tag{16.1.5b}$$

for any a, $0 \leq a \leq 1$, then $c(\cdot)$ is called a **convex function**.

These are by no means exhaustive and are meant to provide examples of the choice of the cost function.

Statement of the Problem Given $p(\mathbf{x})$, $p(\mathbf{z}|\mathbf{x})$, the sample $\mathbf{z}$, and the choice of the cost function $c(\cdot)$, our goal is to find an estimate $\hat{\mathbf{x}}$ that minimizes the expected cost $B(\hat{\mathbf{x}}) = E[c(\tilde{\mathbf{x}})]$, called the **Bayes' cost function**.

We now move on to deriving an explicit expression for the Bayes' cost function $B(\hat{\mathbf{x}})$:

$$B(\hat{\mathbf{x}}) = E[c(\tilde{\mathbf{x}})] = \int_{\mathbb{R}^m} \int_{\mathbb{R}^n} c(\mathbf{x} - \hat{\mathbf{x}})\, p(\mathbf{x}, \mathbf{z}) \mathrm{d}\mathbf{x} \mathrm{d}\mathbf{z} \tag{16.1.6}$$

where $p(\mathbf{x}, \mathbf{z})$ is the **joint distribution** of $\mathbf{x}$ and $\mathbf{z}$ and the integrals (16.1.6) are multidimensional integrals over the product space $(\mathbb{R}^n \times \mathbb{R}^m)$. Since

$$p(\mathbf{x}, \mathbf{z}) = p(\mathbf{z}|\mathbf{x})\, p(\mathbf{x}) = p(\mathbf{x}|\mathbf{z})\, p(\mathbf{z}), \tag{16.1.7}$$

we obtain the well-known **Bayes' formula**

$$p(\mathbf{x}|\mathbf{z}) = \frac{p(\mathbf{z}|\mathbf{x}) p(\mathbf{x})}{p(\mathbf{z})} \tag{16.1.8}$$

where

$$p(\mathbf{z}) = \int_{\mathbb{R}^n} p(\mathbf{x}, \mathbf{z}) \mathrm{d}\mathbf{x} = \int_{\mathbb{R}^n} p(\mathbf{z}|\mathbf{x}) p(\mathbf{x}) \mathrm{d}\mathbf{x} \tag{16.1.9}$$

is the **marginal distribution** of $\mathbf{z}$. This conditional distribution $p(\mathbf{x}|\mathbf{z})$ is known as the **posterior distribution** of $\mathbf{x}$ given the observation $\mathbf{z}$. It follows from (16.1.8) that this posterior distribution combines the information that is contained in the prior $p(\mathbf{x})$ and the conditional distribution $p(\mathbf{z}|\mathbf{x})$ in a very natural way. A little reflection would immediately indicate that any meaningful estimation scheme must exploit the combined information in $p(\mathbf{x}|\mathbf{z})$. Combining (16.1.7) and (16.1.8) with (16.1.6), we can rewrite the latter as

$$B(\hat{\mathbf{x}}) = \int_{\mathbb{R}^m} B(\hat{\mathbf{x}}|\mathbf{z}) p(\mathbf{z}) \mathrm{d}\mathbf{z} \tag{16.1.10}$$

where

$$B(\hat{\mathbf{x}}|\mathbf{z}) = \int_{\mathbb{R}^n} c(\mathbf{x} - \hat{\mathbf{x}}) p(\mathbf{x}|\mathbf{z}) \mathrm{d}\mathbf{x}. \tag{16.1.11}$$

Since $p(\mathbf{z}) \geq 0$, it follows that minimizing $B(\hat{\mathbf{x}}|\mathbf{z})$ would indeed minimize $B(\hat{\mathbf{x}})$. In other words, we could either minimize $B(\hat{\mathbf{x}})$ directly or else minimize $B(\hat{\mathbf{x}}|\mathbf{z})$, to obtain the optimal estimate.

16.2 Special classes of Bayesian estimates

Given the above framework, we now define several families of Bayesian estimators by varying the choice of the cost function in (16.1.9).

(A) **Bayes' least squares estimator** This class of estimator is obtained by choosing the cost function to be the weighted sum of squared error given in (16.1.2). As a first step in this derivation, define

$$\mu = E[\mathbf{x}|\mathbf{z}] = \int_{\mathbb{R}^n} \mathbf{x}\, p(\mathbf{x}|\mathbf{z}) \mathrm{d}\mathbf{x} \tag{16.2.1}$$

which is the mean of the posterior distribution in (16.1.7). By the property of the conditional expectation (Appendix F), this conditional mean $\mu \in \mathbb{R}^n$ is a function of the observation $\mathbf{z}$. Then

$$\begin{aligned}
B(\hat{\mathbf{x}}) &= E[c(\tilde{\mathbf{x}})] \\
&= E[(\mathbf{x} - \hat{\mathbf{x}})^{\mathrm{T}}\mathbf{W}(\mathbf{x} - \hat{\mathbf{x}})] \\
&= E[(\mathbf{x} - \mu + \mu - \hat{\mathbf{x}})^{\mathrm{T}}\mathbf{W}(\mathbf{x} - \mu + \mu - \hat{\mathbf{x}})] \\
&= E[(\mathbf{x} - \mu)^{\mathrm{T}}\mathbf{W}(\mathbf{x} - \mu)] + E[(\mu - \hat{\mathbf{x}})^{\mathrm{T}}\mathbf{W}(\mu - \hat{\mathbf{x}})] \\
&\quad + 2E[(\mathbf{x} - \mu)^{\mathrm{T}}\mathbf{W}(\mu - \hat{\mathbf{x}})].
\end{aligned} \tag{16.2.2}$$

Now, using the **iterated law** of conditional expectation (Appendix F), we can rewrite the third term on the r.h.s. of (16.2.2) as

$$E[(\mathbf{x} - \mu)^{\mathrm{T}}\mathbf{W}(\mu - \hat{\mathbf{x}})] = E\{E[(\mathbf{x} - \mu)^{\mathrm{T}}\mathbf{W}(\mu - \hat{\mathbf{x}})|\mathbf{z}]\}.$$

Since both μ and $\hat{\mathbf{x}}$ are functions of $\mathbf{z}$, we obtain

$$\begin{aligned}
E[(\mathbf{x} - \mu)^{\mathrm{T}}\mathbf{W}(\mu - \hat{\mathbf{x}})|\mathbf{z}] &= (\mu - \hat{\mathbf{x}})^{\mathrm{T}}\mathbf{W}E[(\mathbf{x} - \mu)|\mathbf{z}] \\
&= (\mu - \hat{\mathbf{x}})^{\mathrm{T}}\mathbf{W}(E(\mathbf{x}|\mathbf{z}) - \mu) \\
&= 0 \ \text{ by (16.2.1)}.
\end{aligned}$$

Thus, the third term on the r.h.s. of (16.2.2) vanishes, leaving behind

$$B(\hat{\mathbf{x}}) = E[(\mathbf{x} - \mu)^{\mathrm{T}}\mathbf{W}(\mathbf{x} - \mu)] + E[(\mu - \hat{\mathbf{x}})^{\mathrm{T}}\mathbf{W}(\mu - \hat{\mathbf{x}})]. \tag{16.2.3}$$

Recall that the only **control** we have is the choice of the estimate $\hat{\mathbf{x}}$ which in turn affects only the second term in (16.2.3) but not the first. Since both the terms on the r.h.s. of (16.2.3) are non-negative, setting $\hat{\mathbf{x}}_{\mathrm{MS}} = \mu$ would minimize $B(\hat{\mathbf{x}})$. Stated in other words, the conditional mean μ in (16.2.1) minimizes the Bayes' cost function (16.2.2) and is called the **Bayes' least squares estimate for x**. An explicit expression for $\hat{\mathbf{x}}_{\mathrm{MS}}$ is given by

$$\begin{aligned}
\hat{\mathbf{x}}_{\mathrm{MS}} &= E[\mathbf{x}|\mathbf{z}] \\
&= \int_{\mathbb{R}^n} \mathbf{x}\left(\frac{p(\mathbf{z}|\mathbf{x})p(\mathbf{x})}{p(\mathbf{z})}\right)\mathrm{d}x \\
&= \frac{\displaystyle\int_{\mathbb{R}^n} \mathbf{x}p(\mathbf{z}|\mathbf{x})p(\mathbf{x})\mathrm{d}x}{\displaystyle\int_{\mathbb{R}^n} p(\mathbf{z}|\mathbf{x})p(\mathbf{x})\mathrm{d}x}.
\end{aligned} \tag{16.2.4}$$

Remark 16.2.1 Another look at the derivation We now illustrate the derivation of the above Bayes' least squares estimate by minimizing $B(\hat{\mathbf{x}}|\mathbf{z})$ in (16.1.10):

$$B(\hat{\mathbf{x}}|\mathbf{z}) = \int_{\mathbb{R}^n} (\mathbf{x} - \hat{\mathbf{x}})^{\mathrm{T}}\mathbf{W}(\mathbf{x} - \hat{\mathbf{x}})p(\mathbf{x}|\mathbf{z})\mathrm{d}\mathbf{x}. \tag{16.2.5}$$

By setting the gradient of $B(\hat{\mathbf{x}}|\mathbf{z})$ w.r. to $\hat{\mathbf{x}}$ to zero we obtain

$$\begin{aligned} 0 &= \nabla_{\hat{\mathbf{x}}} B(\hat{\mathbf{x}}|\mathbf{z}) \\ &= -2\mathbf{W} \int_{\mathbb{R}^m} (\mathbf{x} - \hat{\mathbf{x}}) p(\mathbf{x}|\mathbf{z}) \mathrm{d}\mathbf{z} \end{aligned}$$

from which we obtain

$$\begin{aligned} \int_{\mathbb{R}^m} \mathbf{x} p(\mathbf{x}|\mathbf{z}) \mathrm{d}\mathbf{z} &= \int_{\mathbb{R}^n} \hat{\mathbf{x}}_{\mathrm{MS}} p(\mathbf{x}|\mathbf{z}) \mathrm{d}\mathbf{x} \\ &= \hat{\mathbf{x}}_{\mathrm{MS}} \int_{\mathbb{R}^n} p(\mathbf{x}|\mathbf{z}) \mathrm{d}\mathbf{x} = \hat{\mathbf{x}}_{\mathrm{MS}} \end{aligned} \tag{16.2.6}$$

since $\hat{\mathbf{x}}_{\mathrm{MS}}$ depends only on $\mathbf{z}$ and not on $\mathbf{x}$. That is, once again we obtain $\hat{\mathbf{x}}_{\mathrm{MS}}$ as the conditional expectation of $\mathbf{x}$.

Remark 16.2.2 A generalization It turns out that the conditional expectation of $\mathbf{x}$ is an optimal estimate for a wide range of choices of the cost function. It can be shown that if the cost function $c(\cdot)$ is **symmetric** and **convex** and if the conditional distribution $p(\mathbf{x}|\mathbf{z})$ is **unimodal** then the conditional expectation is in fact an optimal estimate for $\mathbf{x}$.

We now state several important properties of the Bayes' least squares estimate $\hat{\mathbf{x}}_{\mathrm{MS}}$ in (16.2.4).

(a) **Unbiasedness** Consider

$$\begin{aligned} E[\mathbf{x} - \hat{\mathbf{x}}_{\mathrm{MS}}] &= E\{E[\mathbf{x} - \hat{\mathbf{x}}_{\mathrm{MS}}|\mathbf{z}]\} \\ &= E\{E[\mathbf{x}|\mathbf{z}] - \hat{\mathbf{x}}_{\mathrm{MS}}\} \\ &= 0, \end{aligned} \tag{16.2.7}$$

since $\hat{\mathbf{x}}_{\mathrm{MS}}$ is a function of $\mathbf{z}$.

That is, $E(\hat{\mathbf{x}}_{\mathrm{MS}}) = E(\mathbf{x})$ and hence $\hat{\mathbf{x}}_{\mathrm{MS}}$ is an **unbiased estimate**. Further, from

$$E[\tilde{\mathbf{x}}] = E[\mathbf{x} - \hat{\mathbf{x}}_{\mathrm{MS}}] = 0 \tag{16.2.8}$$

it follows that the mean of the estimation error is zero.

(b) **Minimum (error) variance property** Setting $\mathbf{W} = \mathbf{I}$ in the expression for $B(\hat{\mathbf{x}}|\mathbf{z})$ in (16.2.5), since $\hat{\mathbf{x}}_{\mathrm{MS}}$ is an unbiased estimate, we get

$$B(\hat{\mathbf{x}}_{\mathrm{MS}}|\mathbf{z}) = \int_{\mathbb{R}^n} (\hat{\mathbf{x}} - \hat{\mathbf{x}}_{\mathrm{MS}})^{\mathrm{T}} (\hat{\mathbf{x}} - \hat{\mathbf{x}}_{\mathrm{MS}}) p(\mathbf{x}|\mathbf{z}) \mathrm{d}\mathbf{x} \tag{16.2.9}$$

which has the natural interpretation of the (total) **variance** of the error $\tilde{\mathbf{x}}$. Since $\hat{\mathbf{x}}_{\mathrm{MS}}$ also minimizes $B(\hat{\mathbf{x}}|\mathbf{z})$, the Bayes' least squares estimate is also the **minimum (error) variance estimate**

Example 16.2.1 Consider the problem of estimating an unknown scalar x using the observation z where $z = x + v$ and v is such $v \sim N(0, \sigma_v^2)$ and $x \sim N(\mathbf{m}_x, \sigma_x^2)$, where, recall $N(a, b^2)$ denotes the normal distribution with mean a and variance b^2. Further it is assumed that the unknown x and the observation noise v are **uncorrelated**. It can be shown[†] that $z \sim N(\mathbf{m}_x, \sigma^2)$ where $\sigma^2 = \sigma_v^2 + \sigma_x^2$.

Let us begin by computing the conditional density of z given x. Since $z = x + v$, it can be verified that

$$p(z|x) = N(x, \sigma_v^2).$$

Using all this information, we now compute the posterior distribution $p(\mathbf{x}|\mathbf{z})$ as

$$\begin{aligned} p(\mathbf{x}|\mathbf{z}) &= \frac{p(z|x)p(x)}{p(z)} \\ &= \frac{N(x, \sigma_v^2)\, N(\mathbf{m}_x, \sigma_x^2)}{N(\mathbf{m}_x, \sigma^2)} \\ &= \beta \, \exp\{-\frac{1}{2}[\frac{(z-x)^2}{\sigma_v^2} + \frac{(x-\mathbf{m}_x)^2}{\sigma_x^2} - \frac{(z-\mathbf{m}_x)^2}{\sigma^2}]\} \end{aligned} \quad (16.2.10)$$

where β is a constant. (Exercise 16.1)

Simplifying the term in the square brackets, we obtain

$$\begin{aligned} \frac{(z-x)^2}{\sigma_v^2} + \frac{(x-\mathbf{m}_x)^2}{\sigma_x^2} - \frac{(z-\mathbf{m}_x)^2}{\sigma^2} &= x^2[\frac{1}{\sigma_v^2} + \frac{1}{\sigma_x^2}] - 2x[\frac{z}{\sigma_v^2} + \frac{\mathbf{m}_x}{\sigma_x^2}] \\ &\quad + [\frac{z^2}{\sigma_v^2} + \frac{\mathbf{m}_x^2}{\sigma_x^2} - \frac{(z-\mathbf{m}_x)^2}{\sigma^2}]. \end{aligned} \quad (16.2.11)$$

Define

$$\frac{1}{\sigma_e^2} = \frac{1}{\sigma_v^2} + \frac{1}{\sigma_x^2} = \frac{\sigma_v^2 + \sigma_x^2}{\sigma_v^2 \sigma_x^2} \quad (16.2.12)$$

and

$$\frac{\hat{\mathbf{x}}_{\text{MS}}}{\sigma_e^2} = \frac{z}{\sigma_v^2} + \frac{\mathbf{m}_x}{\sigma_x^2}. \quad (16.2.13)$$

It can be verified (Exercise 16.2) that

$$\frac{\hat{\mathbf{x}}_{\text{MS}}^2}{\sigma_e^2} = \frac{z^2}{\sigma_v^2} + \frac{\mathbf{m}_x^2}{\sigma_x^2} - \frac{(z-\mathbf{m}_x)^2}{\sigma^2}. \quad (16.2.14)$$

Now combining (16.2.12)–(16.2.14), we can rewrite the r.h.s. of (16.2.11) as

$$\tfrac{1}{\sigma_e^2}[x^2 - 2x\hat{\mathbf{x}}_{\text{MS}} + \hat{\mathbf{x}}_{\text{MS}}^2] = \tfrac{1}{\sigma_e^2}(x - \hat{\mathbf{x}}_{\text{MS}})^2. \quad (16.2.15)$$

[†] Since x and v are both normal variates, that they are uncorrelated implies that they are also independent. It is well known that the distribution of the sum of two independent random variables is given by the **convolution** of their distribution. It can be verified that the convolution of two normal distributions is again a normal distribution.

Now combining all these with (16.2.9), we readily obtain

$$p(\mathbf{x}|\mathbf{z}) = \alpha \ \exp[-\frac{1}{2}\frac{(x-\hat{\mathbf{x}}_{\text{MS}})^2}{\sigma_e^2}] \tag{16.2.16}$$

where σ_e^2 and $\hat{\mathbf{x}}_{\text{MS}}$ are defined in (16.2.12) and (16.2.13). The least squares estimate given by $E[\mathbf{x}|\mathbf{z}]$ is

$$\begin{aligned}\hat{\mathbf{x}}_{\text{MS}} &= (\frac{\sigma_e^2}{\sigma_x^2})\mathbf{m}_x + (\frac{\sigma_e^2}{\sigma_v^2})z\\ &= (\frac{\sigma_v^2}{\sigma_x^2+\sigma_v^2})\mathbf{m}_x + (\frac{\sigma_x^2}{\sigma_x^2+\sigma_v^2})z\\ &= \alpha\mathbf{m}_x + (1-\alpha)z \end{aligned} \tag{16.2.17}$$

where $\alpha = \sigma_v^2/(\sigma_x^2+\sigma_v^2) > 0$. That is, $\hat{\mathbf{x}}_{\text{MS}}$ is a convex combination of $\mathbf{m}_x$ and z and lies in the line segment joining $\mathbf{m}_x$ and z. Thus, if $\sigma_x^2 > \sigma_v^2$, then α is closer to zero and the estimate $\hat{\mathbf{x}}_{\text{MS}}$ is dominated by the observation $\mathbf{z}$. On the other hand, if $\sigma_v^2 > \sigma_x^2$, then α is closer to unity and $\mathbf{m}_x$ is the dominant component in $\hat{x}_{\text{MS}}$. The variance of this estimate is given by

$$\sigma_e^2 = \frac{\sigma_x^2\sigma_v^2}{\sigma_x^2+\sigma_v^2} \tag{16.2.18}$$

Interest in this estimate is largely a result of this adaptive character of (16.2.17).

Example 16.2.2 For later use, we now provide a multivariate extension of the above example. Consider the problem of estimating an unknown $\mathbf{x} \in \mathbb{R}^n$ using the observations $\mathbf{z} \in \mathbb{R}^m$ where $\mathbf{z} = \mathbf{Hx} + \mathbf{v}$, $\mathbf{H} \in \mathbb{R}^{m\times n}$ is of full rank, and $\mathbf{v} \in \mathbb{R}^m$. Assume the following:

(1) $\mathbf{v} \sim N(0, \Sigma_{\mathbf{v}})$, $\Sigma_{\mathbf{v}} \in \mathbb{R}^{m\times m}$ symmetric and positive definite
(2) $\mathbf{x} \in N(\mathbf{m}_x, \Sigma_x)$ where $\mathbf{m}_x \in \mathbb{R}^n$ and $\Sigma_x \in \mathbb{R}^{n\times n}$ symmetric and positive definite
(3) $\mathbf{v}$ and $\mathbf{x}$ are uncorrelated

From this it follows that

$$\mathbf{Hx} \sim N(\mathbf{Hm}_x, \mathbf{H}\Sigma_{\mathbf{x}}\mathbf{H}^{\mathrm{T}}).$$

Since $\mathbf{x}$ and $\mathbf{v}$ are uncorrelated so are $\mathbf{Hx}$ and $\mathbf{v}$. Given that $\mathbf{Hx}$ and $\mathbf{v}$ are both normal, it follows that $\mathbf{z} = \mathbf{Hx} + \mathbf{v}$ is also normal. Hence the distribution of $\mathbf{z}$ depends only on its mean and covariance matrix, which we now compute

$$\begin{aligned} E(\mathbf{z}) &= E(\mathbf{Hx}+\mathbf{v}) = \mathbf{Hm}_x.\\ \text{Cov}(\mathbf{z}) &= E[(\mathbf{z}-\mathbf{Hm}_x)(\mathbf{z}-\mathbf{Hm}_x)^{\mathrm{T}}]\\ &= E[(\mathbf{H}(\mathbf{x}-\mathbf{m}_x)+\mathbf{v})(\mathbf{H}(\mathbf{x}-\mathbf{m}_x)+\mathbf{v})^{\mathrm{T}}]\\ &= \mathbf{H}E[(\mathbf{x}-\mathbf{m}_x)(\mathbf{x}-\mathbf{m}_x)^{\mathrm{T}}]\mathbf{H}^{\mathrm{T}} + E(\mathbf{vv}^{\mathrm{T}})\\ &= \mathbf{H}\Sigma_x\mathbf{H}^{\mathrm{T}} + \Sigma_v. \end{aligned}$$

Thus

$$p(\mathbf{z}) = N(\mathbf{Hm}_x, \Sigma) \tag{16.2.19}$$

where

$$\Sigma = (\mathbf{H}\Sigma_x\mathbf{H}^{\mathrm{T}} + \Sigma_v). \tag{16.2.20}$$

We now compute the conditional distribution of $\mathbf{z}$ given $\mathbf{x}$, which is again normal.

$$\begin{aligned} E[\mathbf{z}|\mathbf{x}] &= E[\mathbf{Hx} + \mathbf{v}|\mathbf{x}] \\ &= \mathbf{Hx} + E(\mathbf{v}) = \mathbf{Hx} \\ E[(\mathbf{z} - \mathbf{Hx})(\mathbf{z} - \mathbf{Hx})^{\mathrm{T}}|\mathbf{x}] &= E[\mathbf{v}\mathbf{v}^{\mathrm{T}}|\mathbf{x}] \\ &= \Sigma_v \end{aligned}$$

and

$$p(\mathbf{z}|\mathbf{x}) = N(\mathbf{Hx}, \Sigma_v). \tag{16.2.21}$$

Using Bayes' rule, we obtain the posterior distribution as

$$\begin{aligned} p(\mathbf{x}|\mathbf{z}) &= \frac{p(\mathbf{z}|\mathbf{x})p(\mathbf{x})}{p(\mathbf{z})} \\ &= \frac{N(\mathbf{Hx}, \Sigma_v)N(m_x, \Sigma_x)}{N(\mathbf{Hm}_x, \Sigma)} \\ &= \alpha \exp\{-\tfrac{1}{2}[(\mathbf{z} - \mathbf{Hx})^{\mathrm{T}}\Sigma_v^{-1}(\mathbf{z} - \mathbf{Hx}) + (\mathbf{x} - \mathbf{m}_x)^{\mathrm{T}}\Sigma_x^{-1}(\mathbf{x} - \mathbf{m}_x) \\ &\quad - (\mathbf{z} - \mathbf{Hm}_x)^{\mathrm{T}}\Sigma^{-1}(\mathbf{z} - \mathbf{Hm}_x)]\}. \end{aligned} \tag{16.2.22}$$

The terms inside the square brackets in the exponent of the r.h.s. of (16.2.22) after simplification become

$$\begin{aligned} &\mathbf{x}^{\mathrm{T}}[\mathbf{H}^{\mathrm{T}}\Sigma_v^{-1}\mathbf{H} + \Sigma_x^{-1}]\mathbf{x} - 2[\mathbf{H}^{\mathrm{T}}\Sigma_v^{-1}\mathbf{z} + \Sigma_x^{-1}\mathbf{m}_x]^{\mathrm{T}}\mathbf{x} \\ &+ \mathbf{z}^{\mathrm{T}}\Sigma_v^{-1}\mathbf{z} + \mathbf{m}_x^{\mathrm{T}}\Sigma_x^{-1}\mathbf{m}_x - (\mathbf{z} - \mathbf{Hm}_x)^{\mathrm{T}}\Sigma^{-1}(\mathbf{z} - \mathbf{Hm}_x). \end{aligned} \tag{16.2.23}$$

Equating (16.2.23) with

$$\begin{aligned} (\mathbf{x} - \hat{\mathbf{x}}_{\mathrm{MS}})^{\mathrm{T}}\Sigma_e^{-1}(\mathbf{x} - \hat{\mathbf{x}}_{\mathrm{MS}}) &= \mathbf{x}^{\mathrm{T}}\Sigma_e^{-1}\mathbf{x} - 2\hat{\mathbf{x}}_{\mathrm{MS}}^{\mathrm{T}}\Sigma_e^{-1}\mathbf{x} \\ &\quad + \hat{\mathbf{x}}_{\mathrm{MS}}^{\mathrm{T}}\Sigma_e^{-1}\hat{\mathbf{x}}_{\mathrm{MS}} \end{aligned} \tag{16.2.24}$$

we obtain

$$\Sigma_e = (\mathbf{H}^{\mathrm{T}}\Sigma_v^{-1}\mathbf{H} + \Sigma_x^{-1})^{-1} \tag{16.2.25}$$

and

$$\hat{\mathbf{x}}_{\mathrm{MS}} = \Sigma_e\,[\mathbf{H}^{\mathrm{T}}\Sigma_v^{-1}\mathbf{z} + \Sigma_x^{-1}\mathbf{m}_x]. \tag{16.2.26}$$

Thus, the least squares estimate is given by

$$\hat{\mathbf{x}}_{\mathrm{MS}} = (\mathbf{H}^{\mathrm{T}}\Sigma_v^{-1}\mathbf{H} + \Sigma_x^{-1})^{-1}[\mathbf{H}^{\mathrm{T}}\Sigma_v^{-1}\mathbf{z} + \Sigma_x^{-1}\mathbf{m}_x] \tag{16.2.27}$$

and the covariance matrix of this estimate is given by Σ_e in (16.2.25).

(B) **Maximum posterior estimate** This class of estimates is obtained by using the uniform cost function in (16.1.3) to define $B(\mathbf{x}|\mathbf{z})$ in (16.1.11). First define

$$S_\epsilon = \{\mathbf{x} \in \mathbb{R}^n | \|(\mathbf{x} - \hat{\mathbf{x}})\| > \epsilon\}$$

and

$$\mathcal{S}_\epsilon^c = \mathbb{R}^n - \mathcal{S}_\epsilon = \{\mathbf{x} \in \mathbb{R}^n |\ \|(\mathbf{x} - \hat{\mathbf{x}})\| \le \epsilon\}.$$

It can be verified that the volume of this ϵ-cube in $\mathbb{R}^n$ is given by

$$\mathrm{VOLUME}(\mathcal{S}_\epsilon^c) = (2\epsilon)^n$$

Now substituting (16.1.3) in (16.1.11), we obtain (where the subscript U denotes the uniform cost)

$$\begin{aligned} B_U(\hat{\mathbf{x}}|\mathbf{z}) &= \int_{\mathcal{S}_\epsilon} p(\mathbf{x}|\mathbf{z})\mathrm{d}x \\ &= 1 - \int_{\mathcal{S}_\epsilon^c} p(\mathbf{x}|\mathbf{z})\mathrm{d}x \\ &= 1 - (2\epsilon)^n p(\hat{\mathbf{x}}|\mathbf{z}) \end{aligned} \tag{16.2.28}$$

where the last line is obtained by applying the standard **mean value theorem** to the integral over $\mathcal{S}_\epsilon^c$. The expression on the r.h.s. of (16.2.29) is minimum when $p(\hat{\mathbf{x}}|\mathbf{z})$ is a maximum. Accordingly, we define the estimator $\hat{\mathbf{x}}_{\mathrm{U}}$ that minimizes $B_U(\hat{\mathbf{x}}|\mathbf{z})$ as the one that maximizes the posterior distribution $p(\mathbf{z}|\mathbf{x})$, that is,

$$p(\hat{\mathbf{x}}_U|\mathbf{z}) \ge p(\hat{\mathbf{x}}|\mathbf{z}). \tag{16.2.29}$$

Consequently, $\hat{\mathbf{x}}_U$ is called **maximum a posteriori estimate** (MAP) which is also known as the **conditional mode** estimate.

An equivalent characterization of $\hat{\mathbf{x}}_{\mathrm{U}}$ is given in terms of the vanishing of the gradient of $p(\mathbf{x}|\mathbf{z})$ as

$$0 = \nabla_x p(\mathbf{x}|\mathbf{z}) = \frac{1}{p(\mathbf{z})}\nabla_x[p(\mathbf{z}|\mathbf{x})p(\mathbf{x})] \tag{16.2.30}$$

since $p(\mathbf{z})$ is independent of $\mathbf{x}$. Sometimes, it is convenient to express this relation in terms of the gradient of the logarithm of $p(\mathbf{x}|\mathbf{z})$ as follows:

$$0 = \frac{1}{p(\mathbf{z})}[\nabla_x \ln p(\mathbf{z}|\mathbf{x}) + \nabla_x \ln p(\mathbf{x})]. \tag{16.2.31}$$

This latter formulation is very helpful in cases when the distribution is normal.

(C) **Conditional median estimate** This class of estimators is derived by using the absolute value of the error as the cost function as in (16.1.4). While the use of this criterion is restricted to the scalar case, that is, the unknown $x \in \mathbb{R}$ $(n = 1)$, it has a very natural interpretation. Substituting (16.1.4) into $B(\hat{x}|z)$

in (16.1.11), the latter becomes (the subscript A denotes absolute value cost function)

$$\begin{aligned} B_A &= \int_{-\infty}^{\infty} |(x - \hat{x})|\ p(x|z)\mathrm{d}x \\ &= -\int_{-\infty}^{\hat{x}} (x - \hat{x})\ p(x|z)\mathrm{d}x + \int_{\hat{x}}^{\infty} (x - \hat{x})\ p(x|z)\mathrm{d}x. \end{aligned} \tag{16.2.32}$$

Now taking the derivatives of both sides w.r. to $\hat{x}$ and equating to zero, we obtain

$$0 = \frac{\mathrm{d}B_A(\hat{x}|z)}{\mathrm{d}\,\hat{x}} = \int_{-\infty}^{\hat{x}_A} p(x|z)\mathrm{d}x - \int_{\hat{x}_A}^{\infty} p(x|z)\mathrm{d}z,$$

from which we get

$$\int_{-\infty}^{\hat{x}_A} p(x|z)\mathrm{d}x = \int_{\hat{x}_A}^{\infty} p(x|z)\mathrm{d}x = \frac{1}{2}. \tag{16.2.33}$$

That is, $\hat{x}_A$ satisfying (16.2.33) is the **median** of the posterior distribution $p(x|z)$.

Exercises

16.1 Compute the value of the constant β in (16.2.10).
16.2 Verify the correctness of the relation (16.2.14).

Notes and references

The material covered in this chapter is rather standard in the literature – refer to Melsa and Cohn (1978) and Jazwinski (1970). Also refer to the survey paper by Cohn (1997).

17

From Gauss to Kalman: sequential, linear minimum variance estimation

In all of the Chapters 14 through 16, we have concentrated on the basic optimality of the estimators derived using different philosophies – **least sum of squared errors, minimum variance estimates** (Chapter 14), **maximum likelihood estimates** (Chapter 15), and optimality using several key parameters of the posterior distribution including the **conditional mean, mode and median** (Chapter 16). In this concluding chapter of Part IV, we turn to analyzing the structure of certain class of optimal estimates. For example, we only know that the conditional mean of the posterior distribution is a minimum variance estimate. But this mean, in general, could be a nonlinear function of the observations $\mathbf{z}$. This observation brings us to the following structural question: when is a linear function of the observations optimal? Understanding the structural properties of an estimator is extremely important and is a major determinant in evaluating the computational feasibility of these estimates.

In Section 17.1 we derive conditions under which a linear function of the observations defines a minimum variance estimate. We then extend this analysis in Section 17.2 to the sequential framework where it is assumed that we have two pieces of information about the unknown, (a) an a priori estimate $\mathbf{x}^-$ and its associated covariance matrix Σ_- and (b) a new observation $\mathbf{z}$ and its covariance matrix Σ_v. We derive conditions under which a linear function of $\mathbf{x}^-$ and $\mathbf{z}$ will lead to a minimum variance estimate $\mathbf{x}^+$ of the unknown $\mathbf{x}$. This development embodies the essence of the celebrated Kalman filtering technique which is the basis for the **sequential** or **on-line linear minimum variance** estimation. Over the past four decades this latter method has become a workhorse in countless practical estimation problems in many branches of engineering, atmospheric and geophysical sciences, finance and economics.

17.1 Linear minimum variance estimation

Let

$$\mathbf{z} = \mathbf{Hx} + \mathbf{v} \tag{17.1.1}$$

be the observations about the unknown $\mathbf{x} \in \mathbb{R}^n$ where $\mathbf{z} \in \mathbb{R}^m$ and $\mathbf{H} \in \mathbb{R}^{m \times n}$ and $\mathbf{v} \in \mathbb{R}^m$ is the (unobservable) observation noise. Since $\mathbf{z}$ is a linear function of the observation, it is tempting to ask the question: when is a **linear function** of $\mathbf{z}$ a **minimum variance** estimate for $\mathbf{x}$? In this section, we answer this question by deriving the optimality properties of a general class of linear estimators – a confluence of optimality and linear structure which is computationally very appealing.

It is assumed that $\mathbf{x}$ is, in general, random and the method relies on the following assumptions relating only to the **second-order** properties of both $\mathbf{x}$ and $\mathbf{v}$.

(a) $E(\mathbf{v}) = 0$, $\text{Cov}(\mathbf{v}) = E(\mathbf{v}\mathbf{v}^\text{T}) = \Sigma_v$.
(b) $E[\mathbf{x}] = \mathbf{m}$ and $\text{Cov}(\mathbf{x}) = E[(\mathbf{x} - \mathbf{m})(\mathbf{x} - \mathbf{m})^\text{T}] = \Sigma_x$.
(c) Both Σ_v and Σ_x are symmetric and positive definite.
(d) $\mathbf{v}$ and $\mathbf{x}$ are uncorrelated.

We seek an estimate $\hat{\mathbf{x}}$ of the form

$$\hat{\mathbf{x}} = \mathbf{b} + \mathbf{A}\mathbf{z} \tag{17.1.2}$$

where $\mathbf{b} \in \mathbb{R}^n$ and $\mathbf{A} \in \mathbb{R}^{n \times m}$. Let $\tilde{\mathbf{x}} = \mathbf{x} - \hat{\mathbf{x}}$ denote the **error** in the linear estimate $\hat{\mathbf{x}}$ in (17.1.2). Our goal is to find $\mathbf{b}$ and $\mathbf{A}$ such that the sum of the expected value of the squares of the components of $\tilde{\mathbf{x}}$ is a minimum. That is, we seek to minimize

$$\begin{aligned} E[\tilde{\mathbf{x}}^\text{T}\tilde{\mathbf{x}}] &= E[(\mathbf{x} - \hat{\mathbf{x}})^\text{T}(\mathbf{x} - \hat{\mathbf{x}})] \\ &= E[\text{tr}[(\mathbf{x} - \hat{\mathbf{x}})^\text{T}(\mathbf{x} - \hat{\mathbf{x}})] \quad \text{(trace of a scalar is the scalar)} \\ &= E[\text{tr}[(\mathbf{x} - \hat{\mathbf{x}})(\mathbf{x} - \hat{\mathbf{x}})^\text{T}] \quad [\text{tr}(\mathbf{AB}) = \text{tr}(\mathbf{BA})] \\ &= \text{tr}[E(\mathbf{x} - \hat{\mathbf{x}})(\mathbf{x} - \hat{\mathbf{x}})^\text{T}] \quad \text{(E is a linear operator)} \\ &= \text{tr}(\mathbf{P}) \end{aligned} \tag{17.1.3}$$

where

$$\mathbf{P} = E[(\mathbf{x} - \hat{\mathbf{x}})(\mathbf{x} - \hat{\mathbf{x}})^\text{T}]. \tag{17.1.4}$$

The first and a rather obvious condition comes from requiring that $\hat{\mathbf{x}}$ in (17.1.2) is an **unbiased** estimate. For, unless $E[\hat{\mathbf{x}}] = E[\mathbf{x}]$, (17.1.3) will **not** correspond to the variance of $\hat{\mathbf{x}}$. From

$$\begin{aligned} \mathbf{m} = E[\hat{\mathbf{x}}] &= E[\mathbf{b} + \mathbf{A}\mathbf{z}] \\ &= \mathbf{b} + \mathbf{A}\mathbf{H}E[\mathbf{x}] = \mathbf{b} + \mathbf{A}\mathbf{H}\mathbf{m} \text{ (since } E(\mathbf{v}) = 0), \end{aligned}$$

we obtain

$$\mathbf{b} = (\mathbf{I} - \mathbf{A}\mathbf{H})\mathbf{m} \tag{17.1.5}$$

as the condition for unbiasedness of $\hat{\mathbf{x}}$. Hence $\hat{\mathbf{x}}$ in (17.1.2) becomes

$$\hat{\mathbf{x}} = \mathbf{m} + \mathbf{A}(\mathbf{z} - \mathbf{H}\mathbf{m}). \tag{17.1.6}$$

Substituting (17.1.6) into (17.1.4), the latter becomes

$$\begin{aligned}\mathbf{P} &= E\{[(\mathbf{x}-\mathbf{m}) - \mathbf{A}(\mathbf{z}-\mathbf{Hm})][(\mathbf{x}-\mathbf{m}) - \mathbf{A}(\mathbf{z}-\mathbf{Hm})]^{\mathrm{T}}\} \\ &= E[(\mathbf{x}-\mathbf{m})(\mathbf{x}-\mathbf{m})^{\mathrm{T}}] + \mathbf{A}E[(\mathbf{z}-\mathbf{Hm})(\mathbf{z}-\mathbf{Hm})^{\mathrm{T}}]\mathbf{A}^{\mathrm{T}} \\ &\quad - \mathbf{A}E[(\mathbf{z}-\mathbf{Hm})(\mathbf{x}-\mathbf{m})^{\mathrm{T}}] - E[(\mathbf{x}-\mathbf{m})(\mathbf{z}-\mathbf{Hm})^{\mathrm{T}}]\mathbf{A}^{\mathrm{T}}. \end{aligned} \quad (17.1.7)$$

But from (17.1.1), it follows that

$$(\mathbf{z}-\mathbf{Hm}) = \mathbf{H}(\mathbf{x}-\mathbf{m}) + \mathbf{v}. \quad (17.1.8)$$

Since $(\mathbf{x}-\mathbf{m})$ and $\mathbf{v}$ are uncorrelated, substituting (17.1.8) into (17.1.7) and simplifying, we obtain

$$\mathbf{P} = \Sigma_x + \mathbf{ADA}^{\mathrm{T}} - \mathbf{AH}\Sigma_x - \Sigma_x\mathbf{H}^{\mathrm{T}}\mathbf{A}^{\mathrm{T}} \quad (17.1.9)$$

where

$$\mathbf{D} = (\mathbf{H}\Sigma_x\mathbf{H}^{\mathrm{T}} + \Sigma_v). \quad (17.1.10)$$

$\mathbf{D}$ is a symmetric matrix. Since Σ_x and Σ_v are both positive definite, so is $\mathbf{D}$, which we assume to hold as a basic requirement.

Our goal can now be restated as follows. Find the matrix $\mathbf{A}$ that minimizes the trace of the covariance matrix $\mathbf{P}$ of $\hat{\mathbf{x}}$ in (17.1.9) that is a quadratic function of the matrix $\mathbf{A}$. Since the tr($\mathbf{P}$) is the sum of the diagonal elements of $\mathbf{P}$, we can achieve this goal by minimizing each of the diagonal elements of $\mathbf{P}$. To this end, consider the ith diagonal element of $\mathbf{P}$:

$$\mathbf{P}_{ii} = (\Sigma_x)_{ii} + \mathbf{A}_{i*}\mathbf{DA}_{i*}^{\mathrm{T}} - \mathbf{A}_{i*}\mathbf{b}_{i*}^{\mathrm{T}} - \mathbf{b}_{i*}\mathbf{A}_{i*}^{\mathrm{T}} \quad (17.1.11)$$

where

$(\Sigma_x)_{ii}$ is the ith diagonal element of Σ_x.
$\mathbf{A}_{i*}$ is the ith row of $\mathbf{A}$, and
$\mathbf{b}_{i*}$ is the ith row of the $n \times m$ matrix $\Sigma_x\mathbf{H}^{\mathrm{T}}$.

Since $\mathbf{A}_{i*}\mathbf{b}_{i*}^{\mathrm{T}} = \mathbf{b}_{i*}\mathbf{A}_{i*}^{\mathrm{T}}$, we can rewrite (17.1.11) as

$$\mathbf{P}_{ii} = \mathbf{A}_{i*}\mathbf{DA}_{i*}^{\mathrm{T}} - 2\mathbf{b}_{i*}\mathbf{A}_{i*}^{\mathrm{T}} + (\Sigma_x)_{ii} \quad (17.1.12)$$

which can be rewritten in the standard quadratic form as

$$\mathbf{P}_{ii} = \mathbf{y}^{\mathrm{T}}\mathbf{Dy} - 2\mathbf{b}^{\mathrm{T}}\mathbf{y} + \mathbf{c}$$

where $\mathbf{y} = \mathbf{A}_{i*}^{\mathrm{T}}$, $\mathbf{b} = \mathbf{b}_{i*}^{\mathrm{T}}$ and $\mathbf{c} = (\Sigma_x)_{ii}$. From

$$\nabla_y\mathbf{P}_{ii} = 2(\mathbf{Dy} - \mathbf{b}) = 0,$$

we obtain a condition for the ith row of $\mathbf{A}$ as $\mathbf{y} = \mathbf{D}^{-1}\mathbf{b}$ or

$$\mathbf{A}_{i*}^{\mathrm{T}} = \mathbf{D}^{-1}\mathbf{b}_{i*}^{\mathrm{T}}. \quad (17.1.13)$$

Since $\nabla_y^2 \mathbf{P}_{ii} = \mathbf{D}$, positive definite, it follows that (17.1.13) is a minimum. Combining the solutions on (17.1.13) for each $i = 1, 2, \ldots, n$, we readily obtain the condition for the minimum of the tr($\mathbf{P}$) as

$$\left[\mathbf{A}_{1*}^{\mathrm{T}}\, \mathbf{A}_{2*}^{\mathrm{T}}\, \cdots\, \mathbf{A}_{m*}^{\mathrm{T}}\right] = \mathbf{D}^{-1}\left[\mathbf{b}_{1*}^{\mathrm{T}}\, \mathbf{b}_{2*}^{\mathrm{T}}\, \cdots\, \mathbf{b}_{m*}^{\mathrm{T}}\right]$$

which can be succinctly written as

$$\begin{aligned} \mathbf{A}^{\mathrm{T}} = \mathbf{D}^{-1}\mathbf{H}\Sigma_x \quad \text{or} \quad \mathbf{A} &= \Sigma_x \mathbf{H}^{\mathrm{T}}\mathbf{D}^{-1} \\ &= \Sigma_x \mathbf{H}^{\mathrm{T}}[\mathbf{H}\Sigma_x\mathbf{H}^{\mathrm{T}} + \Sigma_v]^{-1}. \end{aligned} \tag{17.1.14}$$

Combining with (17.1.6) and (17.1.9), **the linear minimum variance** estimate is

$$\hat{\mathbf{x}} = \mathbf{m} + \Sigma_x \mathbf{H}^{\mathrm{T}}[\mathbf{H}\Sigma_x\mathbf{H}^{\mathrm{T}} + \Sigma_v]^{-1}[\mathbf{z} - \mathbf{H}\mathbf{m}] \tag{17.1.15}$$

and its **covariance matrix** is

$$\mathbf{P} = \Sigma_x - \Sigma_x \mathbf{H}^{\mathrm{T}}[\mathbf{H}\Sigma_x\mathbf{H}^{\mathrm{T}} + \Sigma_v]^{-1}\mathbf{H}\Sigma_x. \tag{17.1.16}$$

The methodological significance of the above derivation lies in the fact that we managed to convert a minimization w.r. to the matrix to a collection of standard quadratic minimization problems each of which is then solved independently. There are at least three more (different) ways of approaching this minimization problem. Since each of these ideas is interesting and very useful, we provide a quick overview of these methods in the following.

Remark 17.1.1 Perturbation Method Let $\mathbf{B}$ be the (optimal) matrix, we are seeking and the matrix $\mathbf{A}$ in (17.1.2) be of the form

$$\mathbf{A} = \mathbf{B} + \epsilon\mathbf{E} \tag{17.1.17}$$

where ϵ is a small real number and $\mathbf{E}$ is an arbitrary matrix. This form for $\mathbf{A}$ in (17.1.17) is obtained by adding a perturbation $\epsilon\mathbf{E}$ to $\mathbf{B}$. Substituting (17.1.17) in (17.1.9) we obtain a matrix which is a function of ϵ as

$$\begin{aligned} g(\epsilon) = {} & [\Sigma_x + \mathbf{B}\mathbf{D}\mathbf{B}^{\mathrm{T}} - \mathbf{B}\mathbf{H}\Sigma_x - \Sigma_x\mathbf{H}^{\mathrm{T}}\mathbf{B}^{\mathrm{T}}] \\ & + \epsilon[\mathbf{E}\mathbf{D}\mathbf{B}^{\mathrm{T}} + \mathbf{B}\mathbf{D}\mathbf{E}^{\mathrm{T}} - \mathbf{E}\mathbf{H}\Sigma_x - \Sigma_x\mathbf{H}^{\mathrm{T}}\mathbf{E}^{\mathrm{T}}] \\ & + \epsilon^2[\mathbf{E}\mathbf{D}\mathbf{E}^{\mathrm{T}}]. \end{aligned} \tag{17.1.18}$$

Since $g(\epsilon)$ is a quadric function of ϵ, a little reflection would reveal that the optimality condition is given by

$$\frac{\mathrm{d}g(\epsilon)}{\mathrm{d}\epsilon}\Big|_{\epsilon=0} = 0 \tag{17.1.19}$$

for any matrix $\mathbf{E}$. Applying this condition to (17.1.18) we obtain

$$\begin{aligned} \frac{\mathrm{d}g(\epsilon)}{\mathrm{d}\epsilon}\Big|_{\epsilon=0} &= \mathbf{E}[\mathbf{D}\mathbf{B}^{\mathrm{T}} - \mathbf{H}\Sigma_x] + [\mathbf{B}\mathbf{D} - \Sigma_x\mathbf{H}^{\mathrm{T}}]\mathbf{E}^{\mathrm{T}} \\ &= 0 \end{aligned}$$

for all $\mathbf{E}$ which is true exactly when

$$\mathbf{BD} = \Sigma_x \mathbf{H}^T \quad \text{or} \quad \mathbf{B} = \Sigma_x \mathbf{H}^T \mathbf{D}^{-1}. \tag{17.1.20}$$

Indeed, not surprisingly this expression for the optimal matrix is the same as in (17.1.14). This type of perturbation method is deeply rooted in calculus of variation.

Remark 17.1.2 Completing the perfect square: We now present an algebraic technique which is a matrix analog of the the well-known method of completing the perfect square. Consider

$$\mathbf{P}(\mathbf{A}) = \Sigma_x + \mathbf{ADA}^T - \mathbf{AH}\Sigma_x - \Sigma_x \mathbf{H}^T \mathbf{A}^T. \tag{17.1.21}$$

Now, add and subtract $\Sigma_x(\mathbf{H}^T\mathbf{D}^{-1}\mathbf{H})\Sigma_x$ to the r.h.s. of (17.1.21) which on simplification becomes

$$\begin{aligned}\mathbf{P}(\mathbf{A}) &= [\mathbf{A} - \Sigma_x \mathbf{H}^T \mathbf{D}^{-1}]\mathbf{D}[\mathbf{A} - \Sigma_x \mathbf{H}^T \mathbf{D}^{-1}]^T \\ &\quad + \Sigma_x - \Sigma_x(\mathbf{H}^T\mathbf{D}^{-1}\mathbf{H})\Sigma_x.\end{aligned} \tag{17.1.22}$$

On rewriting, we get

$$\begin{aligned}&\mathbf{P}(\mathbf{A}) - \{\Sigma_x - \Sigma_x(\mathbf{H}^T\mathbf{D}^{-1}\mathbf{H})\Sigma_x\} \\ &= [\mathbf{A} - \Sigma_x \mathbf{H}^T \mathbf{D}^{-1}]\mathbf{D}[\mathbf{A} - \Sigma_x \mathbf{H}^T \mathbf{D}^{-1}]^T \\ &\geq 0\end{aligned} \tag{17.1.23}$$

where the r.h.s. in general, is a non-null, positive semi-definite matrix. Hence $\mathbf{P}(\mathbf{A})$ takes the minimum value when

$$\mathbf{P} = \Sigma_x - \Sigma_x(\mathbf{H}^T\mathbf{D}^{-1}\mathbf{H})\Sigma_x$$

or exactly when

$$\mathbf{A} = \Sigma_x \mathbf{H}^T \mathbf{D}^{-1}$$

which is the optimal choice of $\mathbf{A}$ again confirming (17.1.14).

Remark 17.1.3 Differentiation of Trace From (17.1.21) we obtain

$$\text{tr}[\mathbf{P}(\mathbf{A})] = \text{tr}[\Sigma_x] + \text{tr}[\mathbf{ADA}^T] - \text{tr}[\mathbf{AH}\Sigma_x] - \text{tr}[\Sigma_x \mathbf{H}^T \mathbf{A}^T].$$

Using the results relating to the differentiation of the trace of a matrix w.r. to a matrix in Appendix C, it follows that

$$\frac{\partial \text{tr}[\mathbf{P}(\mathbf{A})]}{\partial \mathbf{A}} = 2\mathbf{AD} - \Sigma_x \mathbf{H}^T - \Sigma_x \mathbf{H}^T = 0, \tag{17.1.24}$$

from which we obtain

$$\mathbf{AD} = \Sigma_x \mathbf{H}^T \quad \text{or} \quad \mathbf{A} = \Sigma_x \mathbf{H}^T \mathbf{D}^{-1} \tag{17.1.25}$$

as the optimal choice.

Relation to Bayes' minimum variance estimate Recall from Section 16.2 that the Bayes' least squares estimate is the conditional mean of the posterior distribution. Since this conditional mean is also unbiased, it turns out that the conditional mean as the least squares estimate is also a minimum variance estimate. In this section, we have now derived the linear, unbiased minimum variance estimate. In the following, we examine the relation between these two versions of the minimum variance estimates, in particular the one that is given in Example 16.2 especially (16.2.25)–(16.2.26) and the one in (17.1.15)–(17.1.16). Thanks to the Sherman–Morrison–Woodbury matrix inversion lemma (Appendix B). It turns out that despite the apparent differences in their form, the estimate $\hat{\mathbf{x}}_{\mathrm{MS}}$ in (16.2.25) and the estimate $\hat{\mathbf{x}}$ in (17.1.15) are in fact one and the same. We begin by applying the above said matrix inversion result to the inverse of $\mathbf{D}$ defined in (17.1.10). From Appendix B it can be verified that

$$\begin{aligned}\mathbf{D}^{-1} &= [\mathbf{H}\Sigma_x\mathbf{H}^{\mathrm{T}} + \Sigma_v]^{-1}\\ &= \Sigma_v^{-1} - \Sigma_v^{-1}\mathbf{H}[\mathbf{H}^{\mathrm{T}}\Sigma_v^{-1}\mathbf{H} + \Sigma_x^{-1}]^{-1}\mathbf{H}^{\mathrm{T}}\Sigma_v^{-1}.\end{aligned} \tag{17.1.26}$$

Multiplying both sides on the left by $\Sigma_x\mathbf{H}^{\mathrm{T}}$ we obtain

$$\begin{aligned}&\Sigma_x\mathbf{H}^{\mathrm{T}}[\mathbf{H}\Sigma_x\mathbf{H}^{\mathrm{T}} + \Sigma_v]^{-1}\\ &\quad= \Sigma_x\mathbf{H}^{\mathrm{T}}\Sigma_v^{-1} - \Sigma_x\mathbf{H}^{\mathrm{T}}\Sigma_v^{-1}\mathbf{H}[\mathbf{H}^{\mathrm{T}}\Sigma_v^{-1}\mathbf{H} + \Sigma_x^{-1}]^{-1}\mathbf{H}^{\mathrm{T}}\Sigma_v^{-1}\\ &\quad= \{\Sigma_x - \Sigma_x\mathbf{H}^{\mathrm{T}}\Sigma_v^{-1}\mathbf{H}[\mathbf{H}^{\mathrm{T}}\Sigma_v^{-1}\mathbf{H} + \Sigma_x^{-1}]^{-1}\}\mathbf{H}^{\mathrm{T}}\Sigma_v^{-1}\\ &\quad= \{\Sigma_x[\mathbf{H}^{\mathrm{T}}\Sigma_v^{-1}\mathbf{H} + \Sigma_x^{-1}] - \Sigma_x\mathbf{H}^{\mathrm{T}}\Sigma_v^{-1}\mathbf{H}\}[\mathbf{H}^{\mathrm{T}}\Sigma_v^{-1}\mathbf{H} + \Sigma_x^{-1}]^{-1}\mathbf{H}^{\mathrm{T}}\Sigma_v^{-1}\\ &\quad= [\mathbf{H}^{\mathrm{T}}\Sigma_v^{-1}\mathbf{H} + \Sigma_x^{-1}]^{-1}\mathbf{H}^{\mathrm{T}}\Sigma_v^{-1}\end{aligned} \tag{17.1.27}$$

where the second line on the r.h.s. is obtained by taking the common factor $\mathbf{H}^{\mathrm{T}}\Sigma_v^{-1}$ on the right and the third line is obtained by again taking the factor $[\mathbf{H}^{\mathrm{T}}\Sigma_v^{-1}\mathbf{H} + \Sigma_x^{-1}]^{-1}$ on the right and the fourth line is rather obvious.

Now, substituting (17.1.27) into the r.h.s. of (17.1.15), the latter becomes

$$\begin{aligned}\hat{\mathbf{x}} &= \mathbf{m} + [\mathbf{H}^{\mathrm{T}}\Sigma_v^{-1}\mathbf{H} + \Sigma_x^{-1}]^{-1}\mathbf{H}^{\mathrm{T}}\Sigma_v^{-1}[\mathbf{z} - \mathbf{H}\mathbf{m}]\\ &= [\mathbf{H}^{\mathrm{T}}\Sigma_v^{-1}\mathbf{H} + \Sigma_x^{-1}]^{-1}\mathbf{H}^{\mathrm{T}}\Sigma_v^{-1}\mathbf{z}\\ &\quad + \{\mathbf{I} - [\mathbf{H}^{\mathrm{T}}\Sigma_v^{-1}\mathbf{H} + \Sigma_x^{-1}]^{-1}\mathbf{H}^{\mathrm{T}}\Sigma_v^{-1}\mathbf{H}\}\mathbf{m}.\end{aligned} \tag{17.1.28}$$

But the second term on the r.h.s. on the second line of (17.1.28) can be rewritten as

$$\begin{aligned}&[\mathbf{H}^{\mathrm{T}}\Sigma_v^{-1}\mathbf{H} + \Sigma_x^{-1}]^{-1}\{[\mathbf{H}^{\mathrm{T}}\Sigma_v^{-1}\mathbf{H} + \Sigma_x^{-1}] - \mathbf{H}^{\mathrm{T}}\Sigma_v^{-1}\mathbf{H}\}\mathbf{m}\\ &\quad= [\mathbf{H}^{\mathrm{T}}\Sigma_v^{-1}\mathbf{H} + \Sigma_x^{-1}]^{-1}[\Sigma_x^{-1}\mathbf{m}].\end{aligned} \tag{17.1.29}$$

Combining (17.1.29) with (17.1.28), we obtain

$$\hat{\mathbf{x}} = [\mathbf{H}^{\mathrm{T}}\Sigma_v^{-1}\mathbf{H} + \Sigma_x^{-1}]^{-1}[\mathbf{H}^{\mathrm{T}}\Sigma_v^{-1}\mathbf{z} + \Sigma_x^{-1}\mathbf{m}] \tag{17.1.30}$$

which is exactly $\hat{\mathbf{x}}_{\mathrm{MS}}$ in (16.2.26).

Table 17.1.1 *Duality in minimum variance estimation* $\mathbf{z} = \mathbf{Hx} + \mathbf{v}$, $\mathbf{m} = E\mathbf{x}$

Bayesian estimate	Linear minimum variance estimate
$\hat{\mathbf{x}}_{MS} = [\mathbf{H}^T\Sigma_v^{-1}\mathbf{H} + \Sigma_x^{-1}]^{-1} \cdot [\mathbf{H}^T\Sigma_v^{-1}\mathbf{z} + \Sigma_x^{-1}\mathbf{m}]$ – (16.2.26)	$\hat{\mathbf{x}} = \mathbf{m} + \Sigma_x\mathbf{H}^T[\mathbf{H}\Sigma_x\mathbf{H}^T + \Sigma_v]^{-1}[\mathbf{z} - \mathbf{Hm}]$ – (17.1.15)
$\Sigma_e = [\mathbf{H}^T\Sigma_v^{-1}\mathbf{H} + \Sigma_x^{-1}]^{-1}$– (16.2.25)	$\mathbf{P} = \Sigma_x - \Sigma_x\mathbf{H}^T[\mathbf{H}\Sigma_x\mathbf{H}^T + \Sigma_v]^{-1}\mathbf{H}\Sigma_x$ – (17.1.16)
$[\mathbf{H}^T\Sigma_v^{-1}\mathbf{H} + \Sigma_x^{-1}] \in \mathbb{R}^{n\times n}$	$[\mathbf{H}\Sigma_x\mathbf{H}^T + \Sigma_v] \in \mathbb{R}^{m\times m}$
State space formulation	Observation space formulation
Preferred when $n < m$	Preferred when $m < n$

Again applying the matrix inversion lemma to Σ_e in (16.2.25), we obtain

$$\begin{aligned}\Sigma_e &= (\mathbf{H}^T\Sigma_v^{-1}\mathbf{H} + \Sigma_x^{-1})^{-1}\\ &= \Sigma_x - \Sigma_x\mathbf{H}^T[\mathbf{H}\Sigma_x\mathbf{H}^T + \Sigma_v]^{-1}\mathbf{H}\Sigma_x\\ &= \mathbf{P} \quad \text{in (17.1.16)}\end{aligned}$$

which establishes the equivalence between (16.2.25)–(16.2.26) and (17.1.15)–(17.1.16). There is a natural duality between these formulations of the minimum variance estimation problem as shown in Table 17.1.1. From this table, it is clear that the Bayesian estimate calls for inverting $n \times n$ matrices and the linear minimum variance estimate needs the inverse of $m \times m$ matrices. Since $\mathbf{z} \in \mathbb{R}^m$ and $\mathbf{x} \in \mathbb{R}^n$, the linear minimum variance approach is also called the **observation space** formulation and the Bayes' approach is called the **state space formulation**. From the computational point of view, given the equivalence between these formulations, state space approach is to be preferred when $n < m$ (under-determined case) and observation space approach is to be preferred when $n > m$ (over-determined or the inconsistent case).

17.2 Kalman filtering: a first look

Let $\mathbf{x} \in \mathbb{R}^n$ be the unknown constant to be estimated. It is assumed that we have the luxury of knowing an unbiased **prior estimate** $\mathbf{x}^-$ of $\mathbf{x}$ with a **known covariance matrix** Σ_-. In the absence of any other information we would use $\mathbf{x}^-$ in place of $\mathbf{x}$. But then, a new observation $\mathbf{z} \in \mathbb{R}^m$ containing information about the unknown $\mathbf{x}$ arrives on the scene. The question now is: how to mix these two pieces of information – the prior estimate $\mathbf{x}^-$ and the new observation $\mathbf{z}$ in an optimal fashion to arrive at a new posterior estimate $\mathbf{x}^+$. To fix the ideas, it is assumed that

$$\mathbf{z} = \mathbf{Hx} + \mathbf{v} \tag{17.2.1}$$

where $\mathbf{H} \in \mathbb{R}^{m\times n}$ and the (unobservable) observation noise is such that $E(\mathbf{v}) = 0$, $E(\mathbf{v}\mathbf{v}^T) = \Sigma_v$ and both $\mathbf{x}$ and $\mathbf{x}^-$ are uncorrelated with $\mathbf{v}$.

An astute reader can readily decipher the undercurrent of the Bayesian influence (Chapter 16). Our inquiry in here is, however, motivated by our quest to examine the structured aspects of minimum variance estimation when we have two pieces of information. Thus, the developments in this section are conceptually very similar to those given in Section 17.1.

Let $\mathbf{L} \in \mathbb{R}^{n\times n}$ and $\mathbf{K} \in \mathbb{R}^{n\times m}$ be two matrices. Define the new (posterior) estimate $\mathbf{x}^+$ as

$$\mathbf{x}^+ = \mathbf{L}\mathbf{x}^- + \mathbf{K}\mathbf{z} \tag{17.2.2}$$

which is a **linear function** of $\mathbf{x}^-$ and $\mathbf{z}$. Our goal is to find $\mathbf{L}$ and $\mathbf{K}$ such that $\mathbf{x}^+$ is an unbiased linear minimum variance estimate for $\mathbf{x}$.

The unbiasedness condition requires that

$$\begin{aligned}\mathbf{x} &= E[\mathbf{x}^+]\\ &= E[\mathbf{L}\mathbf{x}^- + \mathbf{K}\mathbf{z}]\\ &= E[\mathbf{L}\mathbf{x}^- + \mathbf{K}(\mathbf{H}\mathbf{x} + \mathbf{v})]\\ &= (\mathbf{L} + \mathbf{K}\mathbf{H})\mathbf{x}.\end{aligned} \tag{17.2.3}$$

Since $\mathbf{x}^-$ is an unbiased estimate for $\mathbf{x}$ and $E(\mathbf{v}) = 0$, we get

$$\mathbf{L} + \mathbf{K}\mathbf{H} = \mathbf{I} \quad \text{or} \quad \mathbf{L} = \mathbf{I} - \mathbf{K}\mathbf{H}. \tag{17.2.4}$$

Combining this with (17.2.2), we get

$$\mathbf{x}^+ = \mathbf{x}^- + \mathbf{K}[\mathbf{z} - \mathbf{H}\mathbf{x}^-] \tag{17.2.5a}$$

$$= (\mathbf{I} - \mathbf{K}\mathbf{H})\mathbf{x}^- + \mathbf{K}\mathbf{z}. \tag{17.2.5b}$$

The term $(\mathbf{z} - \mathbf{H}\mathbf{x}^-)$ is often called the **innovation** or the new information contained in the observation $\mathbf{z}$. Since the expression (17.2.5b) is quite similar to (17.1.6), the derivations to follow are quite similar to those in Section 17.1. This estimate $\mathbf{x}^+$ is random since $\mathbf{z}$ is. The sum of the variance of the components of $\mathbf{x}^+$ is given by

$$\begin{aligned}\mathrm{Var}(\mathbf{x}^+) &= E[(\mathbf{x}^+ - \mathbf{x})^{\mathrm{T}}(\mathbf{x}^+ - \mathbf{x})]\\ &= E\{\mathrm{tr}[(\mathbf{x}^+ - \mathbf{x})^{\mathrm{T}}(\mathbf{x}^+ - \mathbf{x})]\} && \text{[trace of a scalar]}\\ &= E\{\mathrm{tr}[(\mathbf{x}^+ - \mathbf{x})(\mathbf{x}^+ - \mathbf{x})^{\mathrm{T}}]\} && [\mathrm{tr}(\mathbf{AB}) = \mathrm{tr}(\mathbf{BA})]\\ &= \mathrm{tr}\{E[(\mathbf{x}^+ - \mathbf{x})(\mathbf{x}^+ - \mathbf{x})^{\mathrm{T}}]\} && (E \text{ is a linear operator})\\ &= \mathrm{tr}[\Sigma^+]\end{aligned} \tag{17.2.6}$$

where

$$\Sigma^+ = E(\mathbf{x}^+ - \mathbf{x})(\mathbf{x}^+ - \mathbf{x})^{\mathrm{T}} \tag{17.2.7}$$

is the (posterior) covariance of the new estimate $\mathbf{x}^+$. Using (17.2.5) in (17.2.7) we obtain an explicit expression for Σ^+.

$$
\begin{aligned}
\Sigma^+ &= E\{[(\mathbf{I}-\mathbf{KH})(\mathbf{x}^- - \mathbf{x}) + \mathbf{Kv}][(\mathbf{I}-\mathbf{KH})(\mathbf{x}^- - \mathbf{x}) + \mathbf{Kv}]^{\mathrm{T}}\} \\
&= (\mathbf{I}-\mathbf{KH})[E(\mathbf{x}^- - \mathbf{x})(\mathbf{x}^- - \mathbf{x})^{\mathrm{T}}](\mathbf{I}-\mathbf{KH})^{\mathrm{T}} \\
&\qquad + \mathbf{K}\,E[\mathbf{v}\mathbf{v}^{\mathrm{T}}]\mathbf{K}^{\mathrm{T}} \qquad (\mathbf{x}^- \text{ and } \mathbf{v} \text{ are uncorrelated}) \\
&= (\mathbf{I}-\mathbf{KH})\Sigma_-(\mathbf{I}-\mathbf{KH})^{\mathrm{T}} + \mathbf{K}\Sigma_v\mathbf{K}^{\mathrm{T}} \\
&= \Sigma_- + \mathbf{KDK}^{\mathrm{T}} - \mathbf{KH}\Sigma_- - \Sigma_-\mathbf{H}^{\mathrm{T}}\mathbf{K}^{\mathrm{T}}
\end{aligned} \tag{17.2.8}
$$

where

$$
\mathbf{D} = (\mathbf{H}\Sigma_-\mathbf{H}^{\mathrm{T}} + \Sigma_v). \tag{17.2.9}
$$

By exploiting the similarity between (17.1.9)–(17.1.10) and (17.2.8)–(17.2.9), we readily obtain the value of $\mathbf{K}$ that minimizes the $\mathrm{tr}(\Sigma^+)$ as

$$
\begin{aligned}
\mathbf{K} &= \Sigma_-\mathbf{H}^{\mathrm{T}}\mathbf{D}^{-1} \\
&= \Sigma_-\mathbf{H}^{\mathrm{T}}[\mathbf{H}\Sigma_-\mathbf{H}^{\mathrm{T}} + \Sigma_x]^{-1}.
\end{aligned} \tag{17.2.10}
$$

Combining (17.2.9)–(17.2.10) with (17.2.5) we obtain the linear, unbiased, minimum variance estimate

$$
\mathbf{x}^+ = \mathbf{x}^- + \Sigma_-\mathbf{H}^{\mathrm{T}}[\mathbf{H}\Sigma_-\mathbf{H}^{\mathrm{T}} + \Sigma_v]^{-1}[\mathbf{z} - \mathbf{Hx}^-]. \tag{17.2.11}
$$

The covariance matrix of this minimum variance estimate is obtained by using (17.2.9)–(17.2.10) in (17.2.8) which on simplification becomes

$$
\Sigma^+ = \Sigma_- - \Sigma_-\mathbf{H}^{\mathrm{T}}[\mathbf{H}\Sigma_-\mathbf{H}^{\mathrm{T}} + \Sigma_v]^{-1}\mathbf{H}\Sigma_-. \tag{17.2.12}
$$

Several comments are in order.

(a) The matrix $\mathbf{K}$ in (17.2.10) is called the **Kalman gain matrix** in honor of Kalman who first developed this method. Notice that the above derivation uses only the second-order information – mean and variance.

(b) The expression for the (posterior) covariance matrix depends only on the measurement strategy and **not** on the actual observations $\mathbf{z}$ and hence can be computed even before obtaining the first observation. This is one of the advantages of this method since it enables the analyst/designer to evaluate competing measurement strategies and pick the "right" strategy.

(c) In the above derivation, consistent with the goals of part IV, it was assumed that the unknown $\mathbf{x}$ is a **fixed** vector. This method, however, directly carries over to the case when $\mathbf{x}$ denotes the state of a dynamical system driven by noise. This analysis is pursued in Part VII. In fact Kalman's original contribution was couched in a dynamical context (Kalman 1960).

(d) The expression for the estimate $\mathbf{x}^+$ in (17.2.11) and its covariance Σ^+ in (17.2.12) are exactly of the same form as those of $\hat{\mathbf{x}}$ (17.1.15) and $\mathbf{P}$ in (17.1.16), respectively. Referring to the Table (17.1.1), since (17.1.15) and (17.1.16) are the same Bayesian estimates derived in Section 16.2, it readily follows that the expressions (17.2.11) and (17.2.12) are also of the same form as the Bayesian estimates given in Table 17.1.1. This similarity between the Kalman's derivation and Bayesian derivation should not be surprising given that Kalman's derivation is based on the prior estimate $\mathbf{x}^-$ and the new observation $\mathbf{z}$.

We conclude this section with an interesting remark relating to the role and impact of the prior information compared to the observations in obtaining the new estimate $\mathbf{x}^+$. It turns out, not surprisingly, that the prior information can, after all, be treated as an additional observation. This connection is established by invoking an important matrix factorization result from Appendix B.

Since Σ_- is symmetric and positive definite so is Σ_-^{-1} and it is well known that there exists a non-singular matrix $\mathbf{R}$ (called **square root** matrix) such that

$$\Sigma_-^{-1} = \mathbf{R}^{\mathrm{T}}\mathbf{R} \quad \text{or} \quad \Sigma_- = \mathbf{R}^{-1}\mathbf{R}^{-\mathrm{T}}. \tag{17.2.13}$$

Given $\mathbf{R}$, now define $\mathbf{z}^-$ which is an artificial observation induced by the prior estimate $\mathbf{x}^-$ as

$$\mathbf{z}^- = \mathbf{R}\mathbf{x}^- = \mathbf{R}\mathbf{x} + \mathbf{v}^-. \tag{17.2.14}$$

Notice that the square matrix $\mathbf{R}$ now plays the role of the (artificial) measurement system. Since $\mathbf{x}^-$ is an unbiased (prior) estimate $\mathbf{x}$, it follows that

$$E[\mathbf{z}^-] = \mathbf{R}\mathbf{x} \quad \text{and} \quad E(\mathbf{v}^-) = \mathbf{O} \tag{17.2.15}$$

and

$$\begin{aligned} E[\mathbf{v}^-(\mathbf{v}^-)^{\mathrm{T}}] &= E[(\mathbf{z}^- - \mathbf{R}\mathbf{x})(\mathbf{z}^- - \mathbf{R}\mathbf{x})^{\mathrm{T}}] \\ &= E[\mathbf{R}(\mathbf{x}^- - \mathbf{x})(\mathbf{x}^- - \mathbf{x})^{\mathrm{T}}\mathbf{R}^{\mathrm{T}}] \\ &= \mathbf{R}\Sigma_-\mathbf{R}^{\mathrm{T}} \\ &= \mathbf{I} \quad \text{(using 17.2.13)}. \end{aligned}$$

Now combine $\mathbf{z}^-$ with $\mathbf{z}$ to get a new extended observation vector

$$\begin{bmatrix} \mathbf{z}^- \\ \mathbf{z} \end{bmatrix} = \begin{bmatrix} \mathbf{R} \\ \mathbf{H} \end{bmatrix} \mathbf{x} + \begin{bmatrix} \mathbf{v}^- \\ \mathbf{v} \end{bmatrix}$$

or

$$\bar{\mathbf{z}} = \bar{\mathbf{H}}\mathbf{x} + \bar{\mathbf{v}} \tag{17.2.16}$$

where

$$\bar{\mathbf{z}} = \begin{bmatrix} \bar{\mathbf{z}} \\ \mathbf{z} \end{bmatrix} \in \mathbb{R}^{(n+m)}, \quad \bar{\mathbf{v}} = \begin{bmatrix} \mathbf{v}^- \\ \mathbf{v} \end{bmatrix} \in \mathbb{R}^{(n+m)} \quad \text{and} \quad \begin{bmatrix} \mathbf{R} \\ \mathbf{H} \end{bmatrix} \in \mathbb{R}^{(n+m)\times n}.$$

Let

$$\mathbf{W} = \begin{bmatrix} \mathbf{I} & 0 \\ 0 & \Sigma_v^{-1} \end{bmatrix} \tag{17.2.17}$$

be the $(n+m) \times (n+m)$ weight matrix. Now define the weighted norm of the residual in (17.2.16) as

$$\begin{aligned} f(\mathbf{x}) &= (\bar{\mathbf{z}} - \bar{\mathbf{H}}\mathbf{x})^{\mathrm{T}}\mathbf{W}(\bar{\mathbf{z}} - \bar{\mathbf{H}}\mathbf{x}) \\ &= [(\mathbf{z}^- - \mathbf{R}\mathbf{x})^{\mathrm{T}}(\mathbf{z} - \mathbf{H}\mathbf{x})^{\mathrm{T}}] \begin{bmatrix} \mathbf{I} & 0 \\ 0 & \Sigma_v^{-1} \end{bmatrix} \begin{bmatrix} (\mathbf{z}^- - \mathbf{R}\mathbf{x}) \\ (\mathbf{z} - \mathbf{H}\mathbf{x}) \end{bmatrix} \\ &= (\mathbf{z}^- - \mathbf{R}\mathbf{x})^{\mathrm{T}}(\mathbf{z}^- - \mathbf{R}\mathbf{x}) + (\mathbf{z} - \mathbf{H}\mathbf{x})^{\mathrm{T}}\Sigma_v^{-1}(\mathbf{z} - \mathbf{H}\mathbf{x}) \\ &= (\mathbf{x}^- - \mathbf{x})^{\mathrm{T}}\Sigma_-^{-1}(\mathbf{x}^- - \mathbf{x}) + (\mathbf{z} - \mathbf{H}\mathbf{x})^{\mathrm{T}}\Sigma_v^{-1}(\mathbf{z} - \mathbf{H}\mathbf{x}) \end{aligned} \tag{17.2.18}$$

where the last line is obtained by substituting $\mathbf{z}^- = \mathbf{R}\mathbf{x}^-$ and simplifying.

Computing the gradient of $f(\mathbf{x})$ and setting it to zero we obtain the optimal least squares estimate as the solution of (Exercise 17.1 and 17.2)

$$[\Sigma_-^{-1} + \mathbf{H}^{\mathrm{T}}\Sigma_v^{-1}\mathbf{H}]\mathbf{x} = [\mathbf{H}^{\mathrm{T}}\Sigma_v\mathbf{z} + \Sigma_-^{-1}\mathbf{x}^-]. \tag{17.2.19}$$

By comparing this with the Bayes' estimate in Table 17.1.1, it follows that it naturally leads to the same estimate.

Exercises

17.1 Compute the gradient of $f(\mathbf{x})$ in (17.2.18) and verify that the minimizing $\mathbf{x}$ is given by the solution of (17.2.19).

17.2 Compute the Hessian of $f(\mathbf{x})$ in (17.2.18) and verify that it is positive definite.

Notes and references

The idea of linear recursive minimum variance estimation, thanks to Kalman (1960), has virtually revolutionized the way estimation theory is applied to solve problems from a wide variety of disciplines – communication and control (Jazwinski (1970), Sage and Melsa (1971), Maybeck (1979), Catlin (1989), Brammer and Siffling (1989), Sorenson (1966)), aerospace applications (Gelb (1974), Bucy and Joseph (1968)), geophysical data assimilation (Cohn (1997), Ghil et al. (1981), (1991) and (1997)), and econometric and financial domain (Harvey (1989) and Hamilton (1994)), to name a few. Kailath (1974) provides an authoritative and critical overview of the development of linear estimation theory. Also refer to Sorenson (1980) for a broad overview of estimation theory. Schweppe (1973) provides a very balanced view of various approaches to estimation in a dynamical context.

Algorithmic and computational aspects of Kalman filtering is covered in Bierman (1977). Ghil and Malanotte-Rizzoli (1991) covers the meteorological applications of Kalman filtering.

PART V

Data assimilation: stochastic/static models

18

Data assimilation – static models: concepts and formulation

In this opening chapter of Part V we develop the basic concepts leading to the formulation of the so-called data assimilation problem for static models. This problem arises in a wide variety of application domains and accordingly it goes with different terminologies that are unique to an application domain. For example, in oceanography and geological exploration, it is known as the **inverse** problem. In meteorology, this is known as the **retrieval** problem, **objective analysis**, **three dimensional variational assimilation (3DVAR)** problem, to mention a few. Henceforth, we use the term static data assimilation problem, retrieval problem, and inverse problem interchangeably. Despite these differences in the origin and the peculiarities of the labels, there is a common mathematical structure – **a unity in diversity** – that underlie all of these problems. The primary aim of this chapter is to develop this common framework. In Part VI we develop the data assimilation for dynamic models.

In Section 18.1, we describe the basic building blocks leading to the statement of the data assimilation problem for the static model. It turns out that this problem is intrinsically **under-determined** (where the number n of unknown variables is larger than the number, m of equations) which in turn implies that the solution space has a **large degree of freedom** (equal to n – m) leading to **infinitely many** solutions. Any attempt to induce uniqueness of the solution calls for the reduction of the dimensionality of the solution space. This is achieved by a general technique that is called regularization. A comprehensive review of the various regularization techniques that are commonly used is given in Section 18.2.

18.1 The static data assimilation problem: a first look

We begin by describing the basic building blocks leading to the formulation of this problem.

(a) **Space–Time domain**

Most of the static data assimilation problems of interest involve two or three dimensions of the physical space we live in. Since our aim is to focus on the mathematical formulation of this problem, we first illustrate the key ideas using a two-dimensional space domain. Extension to three dimensions is obvious and will be pursued when appropriate. Accordingly, assume that we are given a finite rectangular domain, D, whose boundaries are parallel to the standard coordinate x and y axes and is defined by $a \leq x \leq b$ and $c \leq y \leq d$ for some $a < b$ and $c < d$. Refer to Figure 18.1.1 for an illustration.

The **time** interval over which the observations of interest are obtained or available for this data assimilation problem is assumed to be so small that for practical purposes, we consider the time to be **fixed**.

(b) **Computational grid**

Given the domain of interest, our first task is to embed a computational grid in this domain. Let n_x and n_y denote the number of grid points along the x and y boundaries that define the domain. Refer to Figure 18.1.1 for an illustration. Let $h_x = (b - a)/(n_x - 1)$ and $h_y = (d - c)/(n_y - 1)$ denote the included grid spacings in the x and y directions and $\Delta = h_x h_y$ denote the **unit** area. Thus, the domain is divided into $(b - a)(d - c)/\Delta = (n_x - 1)(n_y - 1)$ units.

There are two equivalent ways to number the grid points. First, is the usual **double index** ij notation where $1 \leq i \leq n_x$ and $1 \leq j \leq n_y$, quite similar to the coordinate representation of a point in two dimensions. Thus, a point labelled ij is at a distance $(i - 1)h_x$ along the x-direction and $(j - 1)h_y$ along the y-direction where all the distances are measured w.r.t. the south-west corner of the origin. The second is using a **single index** k for the point ij using an invertible mapping where

$$k = (j - 1)n_x + i. \tag{18.1.1}$$

It can be verified that this single indexing corresponds to a numbering of the grid points in the **row major order** (left to right and bottom up) as illustrated in Figure 18.1.1.

(c) **The (unknown) true state**

At each grid point i, $1 \leq i \leq n$ is defined a state variable x_i called the **true state** of the nature which is **unknown**. Each of these x_i's could be a scalar or itself be a vector with say, L components: $\mathbf{x}_i = (x_{i1}, x_{i2}, \ldots, x_{iL})^{\mathrm{T}}$ for some integer $L \geq 1$. For example, $L = 5$ and $\mathbf{x}_i$ may have the following five components:

x_{i1} – temperature
x_{i2} – pressure
x_{i3} – specific humidity
x_{i4} – magnitude of the east-west wind
x_{i5} – magnitude of the north-south wind

all measured at the space location indexed by i.

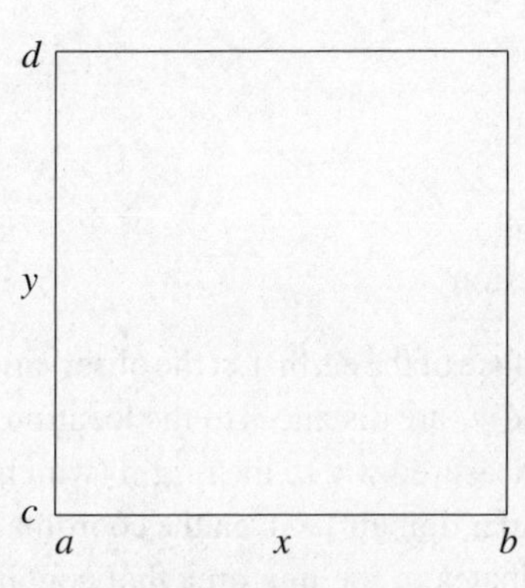

(*a*) The given domain D.

15 25 35 45
14 24 34 44
13 23 33 43
12 22 32 42
11 21 31 41

(*b*) A 4×5 computational grid indexed using ij-notation with $1 \le i \le 4$ and $1 \le j \le 5$. $n_x = 4$ and $n_y = 5$ and $n = 20$. The point 11 is taken as the origin of the domain.

17 18 19 20
z_4 z_5
13 14 15 16
z_2 z_3
9 10 11 12
z_1
5 6 7 8
1 2 3 4

(*c*) A 4×5 grid indexed in the row major order where $k = (j-1)n_x + i$. Thus, the node with 23 is also labelled 7. Grid point 1 is taken as the origin.

Fig. 18.1.1 A view of the domain D with n computational grid points and m observation stations.

Let $\mathbf{x} = (x_1, x_2, \ldots, x_n)^{\mathrm{T}}$ denote the vector of state variable where $\mathbf{x} \in \mathbb{R}^n$ where each x_i may denote a block of size $L \ge 1$. The goal of the retrieval problem is to obtain a good estimate of this unknown state vector.

(d) **Observations**

The value of the (unknown) true state is to be estimated using a set of m observations $\mathbf{z} = (z_1, z_2, \ldots, z_m)^{\mathrm{T}}$ where each $\mathbf{z}_j = (z_{j1}, z_{j2}, \ldots, z_{jM})^{\mathrm{T}}$ may contain $M(\ge 1)$ observables.

For example, when $M = 4$

$$\begin{aligned} z_{j1} &- \text{temperature} \\ z_{j2} &- \text{pressure} \\ z_{j3} &- \text{wind speed} \\ z_{j4} &- \text{wind direction} \end{aligned}$$

all measured say, at two meters above the surface of the earth. Let the observation z_j be located at a point (x_j, y_j) where x_j and y_j are distances to the location z_j along the x and y directions, respectively measured w.r. to the origin (which is the south-west corner of the chosen rectangular domain). Given the coordinates (x_j, y_j) we can readily compute the coordinates of the unit area that contains z_j as follows:

$$i = \lceil \frac{x_j}{h_x} \rceil \quad \text{and} \quad j = \lceil \frac{y_j}{h_y} \rceil, \tag{18.1.2}$$

where $\lceil x \rceil$ called the ceiling of x is the smallest integer greater than or equal to x. Then, z_j lies in the unit area whose south-west corner has coordinates ij. It is often the case that the distribution of these m observations in the domain may **not** be uniform and that the location of the observation may **not** coincide with that of the grid point. This disparity between the location of the grid points and the observation stations calls for an interpolation scheme between the uniform network of grid points and the non-uniform network of observation stations.

(e) **Relation between observations and the state variables**

Recall that the components of the observation vector $\mathbf{z}$ denote a set of observables such as temperature, pressure, humidity, wind speed and direction, an x-ray image, etc. The state vector $\mathbf{x}$ also denotes a set of physical quantities of direct interest in the model. Let $h : \mathbb{R}^n \longrightarrow \mathbb{R}^m$ given by

$$\mathbf{z} = h(\mathbf{x}) + \nu \tag{18.1.3}$$

where $h(\mathbf{x}) = (h_1(\mathbf{x}), h_2(\mathbf{x}), \ldots, h_m(\mathbf{x}))^{\mathrm{T}}$ with each $h_i : \mathbb{R}^n \longrightarrow \mathbb{R}$ and $\mathbf{z} = (z_1, z_2, \ldots, z_m)^{\mathrm{T}}$. This function h that relates the observables to the state variable is also known as the **forward operator** or the **mathematical model** for the observations. Accordingly, $\mathbb{R}^n$ is known as the **model space** and $\mathbb{R}^m$ as the **observation space** and the function $h(.)$ denotes the **static model** of interest in this data assimilation problem. The vector $\nu \in \mathbb{R}^m$ is the observation noise which characterizes the property of the measuring instruments. It is assumed that

$$E(\nu) = 0 \quad \text{and} \quad \mathrm{Cov}(\nu) = E(\nu\nu^{\mathrm{T}}) = \mathbf{R} \tag{18.1.4}$$

where $\mathbf{R} \in \mathbb{R}^{m \times m}$ is a known real, symmetric and positive definite matrix that represents the covariance/correlation structure of the measurement noise.

Given the reality that the observation stations and computational grid points are not the same, it is necessary to **decompose** the function $h(.)$ into two components. To this end, define $h^\circ : \mathbb{R}^m \longrightarrow \mathbb{R}^m$ and a $h^{\mathrm{I}} : \mathbb{R}^n \longrightarrow \mathbb{R}^m$ such that

$$\mathbf{z} = h^\circ(\mathbf{x}^\circ) \quad \text{and} \quad \mathbf{x}^\circ = h^{\mathrm{I}}(\mathbf{x}) \tag{18.1.5}$$

where $\mathbf{x}^\circ = (x_1^\circ, x_2^\circ, \ldots, x_m^\circ)$ and

$$\mathbf{z} = h^\circ(h^{\mathrm{I}}(\mathbf{x})) = (h^\circ \circ h^{\mathrm{I}})(\mathbf{x}) = h(\mathbf{x}). \tag{18.1.6}$$

Thus, $h^\circ(.)$ **converts** the set of m observations into a set of m state variables $\mathbf{x}^\circ$ at the m observation locations and the **interpolation** function $h^{\mathrm{I}}(.)$ then relates the m-vector $\mathbf{x}^\circ$ to the n-vector $\mathbf{x}$ onto the computational grid. Notice that if the observation network and the computational grid are the same, then there is no need for interpolation and $h(\mathbf{x}) = h^\circ(\mathbf{x})$, otherwise $h(\mathbf{x})$ is the composite $(h^\circ \circ h^{\mathrm{I}})(\mathbf{x}) = h^\circ(h^{\mathrm{I}}(\mathbf{x}))$ of the two mappings h° and h^{I}.

The choice of $h^\circ(.)$ depends on the problem on hand, state variables of interest in the analysis and the nature and type of quantities that are observable. Observations of interest may come from balloons, radars, satellites, or ground stations. Generally, $h^\circ(.)$ represents the **physical laws** that relate the state variables and the observables such as for example **Stefan's law** of radiation, **Planck's law** of black body radiation, **Faraday's laws** relating to generation of electricity, laws from **fluid dynamics** and/or **thermodynamics** or else may depend on the **empirical laws** that relate rain drop size to reflectivity, to mention a few. Accordingly, this part h° of the forward operator can be non-linear function.

There is a wide variety of choices for the interpolation including both the linear and nonlinear schemes. To fix the ideas we now describe an example of the **linear interpolation** scheme where h^{I} takes the form of an $m \times n$ matrix.

(f) **The linear interpolation**

As a first step, let us describe the basic ideas of the linear interpolation in one dimension. Referring to the figure 18.1.2, let the jth observation station be located in the grid spacing enclosed by the grid points i and $i + 1$. Let a be the fraction of the distance (measured in units of the grid spacing $h_{\mathbf{x}}$ of the location of z_j from the right grid point $(i + 1)$. Let x_j° be the value of the state variable at this observation station that is recovered from $\mathbf{z}$ using $h^\circ(.)$. Then the linear interpolation that relates x_j° to x_i and x_{i+1} is given by

$$\frac{x_j^\circ - x_i}{1 - a} = \frac{x_{i+1} - x_j^\circ}{a} \tag{18.1.7}$$

or

$$a x_i + (1 - a) x_{i+1} = x_j^\circ. \tag{18.1.8}$$

Now consider the one-dimensional grid in Figure 18.1.2. Let a_j be the fraction of the distance of z_j from the east boundary (right end) of the grid

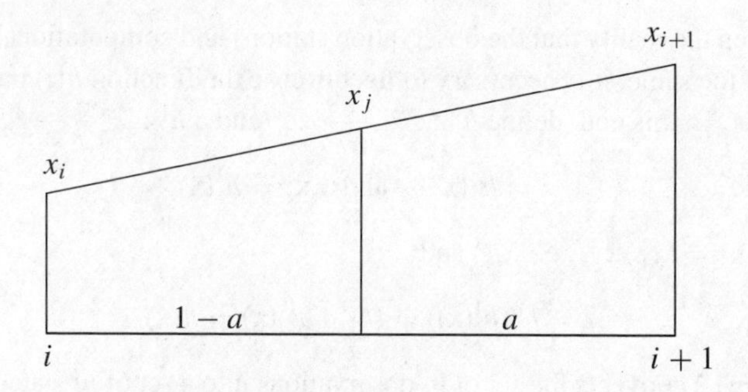

(*a*) The linear interpolation – an illustration.

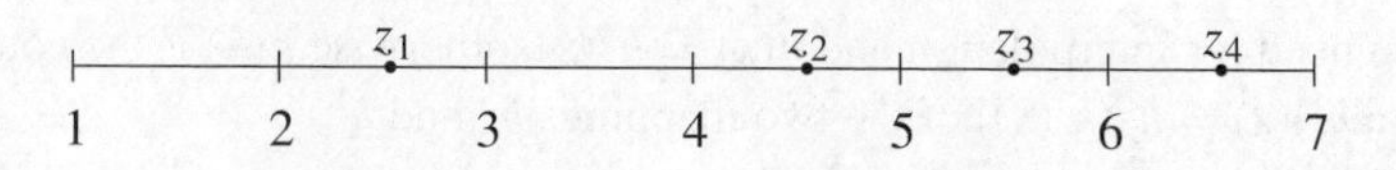

(*b*) One-dimensional example – computational grid with 7 points and observation network with 4 stations.

Fig. 18.1.2 Interpolation in one dimension.

spacing that contains z_j, and let x_j° be the value of the state variable recovered from z_j using $h^\circ(.)$. Given x_j° and a_j for $j = 1, 2, 3$, and 4, we can apply (18.1.8) repeatedly to each of these locations to obtain the following matrix relation between $\mathbf{x}^\circ = (x_1^\circ, x_2^\circ, x_3^\circ, x_4^\circ)^{\mathrm{T}}$ and $\mathbf{x} = (x_1, x_2, \ldots, x_7)^{\mathrm{T}}$:

$$\mathbf{x}^\circ = \mathbf{H}\mathbf{x} \tag{18.1.9}$$

and $\mathbf{H}$ is a 4×7 matrix given by

$$\mathbf{H} = \begin{bmatrix} 0 & a_1 & \acute{a}_1 & 0 & 0 & 0 & 0 \\ 0 & 0 & 0 & a_2 & \acute{a}_2 & 0 & 0 \\ 0 & 0 & 0 & 0 & a_3 & \acute{a}_3 & 0 \\ 0 & 0 & 0 & 0 & 0 & a_4 & \acute{a}_4 \end{bmatrix} \tag{18.1.10}$$

where $\acute{a}_j = 1 - a_j$ for simplicity in notation. It can be verified that there is a maximum of two non-zero elements in each row of $\mathbf{H}$ and that the sum of the elements in each row of $\mathbf{H}$ is one. The rows of $\mathbf{H}$ are linearly independent and hence $\mathbf{H}$ is a maximal rank, equal to four (= number of observations), which is the number of rows in $\mathbf{H}$ (Exercise 18.1).

In extending this idea to two dimensions, consider the observation z_j located in the unit grid area enclosed by the four grid points $\{k, k+1, k+n_x, k+n_x+1\}$ as shown in Figure 18.1.3. Let a_j and b_j denote the fraction of the distances (measured in units of $h_{\mathbf{x}}$ and $h_{\mathbf{y}}$, respectively) of z_j from the north-east boundary point $(k + n_x + 1)$. Applying (18.1.8) first

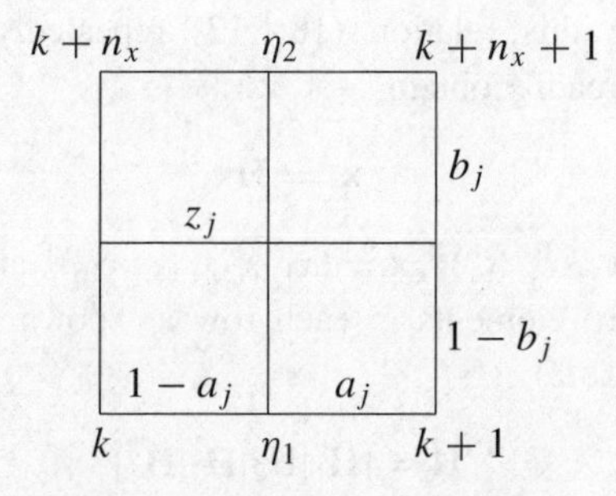

(*a*) Linear interpolation – an illustration.

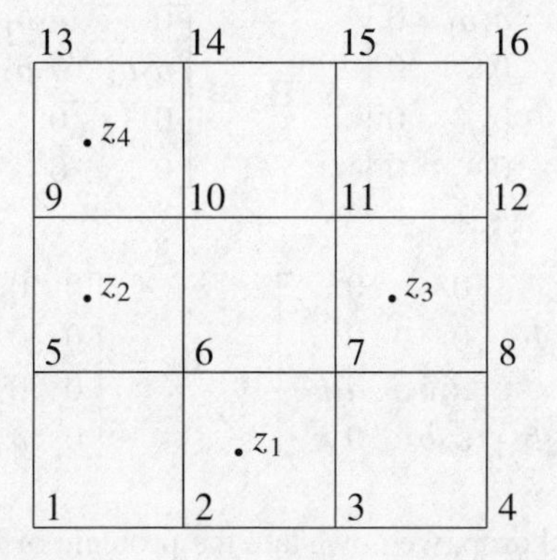

(*b*) Two-dimensional example – computational grid with 16 points and observation network with 4 stations.

Fig. 18.1.3 Interpolation in two dimensions.

along the vertical direction, we obtain

$$b_j \eta_1 + (1 - b_j)\eta_2 = x_i^\circ. \tag{18.1.11}$$

Now applying (18.1.8) to each of η_1 and η_2 along the horizontal direction leads to

$$\left.\begin{aligned} a_j x_k + (1 - a_j)x_{k+1} &= \eta_1 \\ a_j x_{k+n_x} + (1 - a_j)x_{k+n_x+1} &= \eta_2 \end{aligned}\right\} \tag{18.1.12}$$

Combining these, we obtain

$$a_j b_j x_k + \acute{a}_j b_j x_{k+1} + a_j \acute{b}_j x_{k+n_x} + \acute{a}_j \acute{b}_j x_{k+n_x+1} = x_j^\circ \tag{18.1.13}$$

where $\acute{a}_j = 1 - a_j$ and $\acute{b}_j = 1 - b_j$.

Now applying this relation (18.1.13) repeatedly to each of the four observations, we readily obtain

$$\mathbf{x}^\circ = \mathbf{H}\mathbf{x} \tag{18.1.14}$$

where $\mathbf{x}^\circ = (x_1^\circ, x_2^\circ, x_3^\circ, x_4^\circ)^T$, $\mathbf{x} = (x_1, x_2, \ldots, x_{16})^T$ and $\mathbf{H}$ is a 4×16 matrix with four non-zero elements in each row as shown in the partitioned form below, (Exercise 18.2).

$$\mathbf{H} = [\mathbf{H}_1 | \mathbf{H}_2 | \mathbf{H}_3 | \mathbf{H}_4] \tag{18.1.15}$$

where

$$\mathbf{H}_1 = \begin{bmatrix} 0 & a_1 b_1 & \acute{a}_1 b_1 & 0 \\ 0 & 0 & 0 & 0 \\ 0 & 0 & 0 & 0 \\ 0 & 0 & 0 & 0 \end{bmatrix}, \quad \mathbf{H}_2 = \begin{bmatrix} 0 & a_1 \acute{b}_1 & \acute{a}_1 \acute{b}_1 & 0 \\ a_2 b_2 & \acute{a}_2 b_2 & 0 & 0 \\ 0 & 0 & a_3 b_3 & \acute{a}_3 b_3 \\ 0 & 0 & 0 & 0 \end{bmatrix}$$

$$\mathbf{H}_3 = \begin{bmatrix} 0 & 0 & 0 & 0 \\ a_2 \acute{b}_2 & \acute{a}_2 \acute{b}_2 & 0 & 0 \\ 0 & 0 & a_3 \acute{b}_3 & \acute{a}_3 \acute{b}_3 \\ 0 & a_4 b_4 & \acute{a}_4 b_4 & 0 \end{bmatrix}, \quad \mathbf{H}_4 = \begin{bmatrix} 0 & 0 & 0 & 0 \\ 0 & 0 & 0 & 0 \\ 0 & 0 & 0 & 0 \\ 0 & a_4 \acute{b}_4 & \acute{a}_4 \acute{b}_4 & 0 \end{bmatrix}.$$

.

Against this backdrop, we now state the problem of interest to us.

The Static Data Assimilation Problem Given two positive integers n and m with $n > m$,

(a) a rectangular domain
(b) the computational grid with n points
(c) the location and the value of m observations $\mathbf{z} = (z_1, z_2, \ldots, z_m)^T$
(d) the function $h^\circ : \mathbb{R}^m \longrightarrow \mathbb{R}^m$ and the **interpolation** scheme $h^I : \mathbb{R}^n \longrightarrow \mathbb{R}^m$ where $h(\mathbf{x}) = h^\circ(h^I(\mathbf{x}))$ where $\mathbf{x} \in \mathbb{R}^n$, find the vector $\mathbf{x}$ such that $h(x)$ "best" fits the observation $\mathbf{z}$.

The vector $\mathbf{x}$ that is obtained as the solution of the above problem is called **the analysis** and is often denoted by $\mathbf{x}_a$.

18.2 A classification of strategies for solution

The data assimilation or the retrieval problem as stated above is an underdetermined problem since $n > m$. Thus, the solution space has $(n - m)$ degrees of freedom giving rise to infinitely many solutions. In such a situation, uniqueness of the solution is obtained by imposing additional constraints. We now examine some of the available strategies for guaranteeing uniqueness of the solution to the retrieval problem.

To simplify discussion, it is assumed that the function $h(\mathbf{x})$ denoting the forward operator is linear. That is, there exists a matrix $\mathbf{H} \in \mathbb{R}^{m\times n}$ such that

$$\mathbf{z} = \mathbf{H}\mathbf{x} + \nu \tag{18.2.1}$$

where ν is the observation noise with the known second-order properties as stated in (18.1.3)

Strategy I

Following the developments in Section 5.3, we may choose the path of formulating this problem as follows: Given $\mathbf{z}$, among all the vectors $\mathbf{x}$ that satisfy $\mathbf{z} = \mathbf{H}\mathbf{x}$, find the one with a minimum norm. This is accomplished by defining the Lagrangian

$$L(\mathbf{x}, \lambda) = \frac{1}{2}\mathbf{x}^{\mathrm{T}}\mathbf{x} + \lambda^{\mathrm{T}}(\mathbf{z} - \mathbf{H}\mathbf{x}) \tag{18.2.2}$$

where $\lambda = (\lambda_1, \lambda_2, \ldots, \lambda_m)^{\mathrm{T}}$ is the so-called undetermined Lagrangian multiplier.

The vector $\mathbf{x}_a$ that minimizes $L(\mathbf{x}, \lambda)$ is obtained by solving

$$\left.\begin{aligned} \nabla_{\mathbf{x}} L(\mathbf{x}, \lambda) &= \mathbf{x} - \mathbf{H}^T\lambda = 0 \\ \text{and} \qquad & \\ \nabla_{\lambda} L(\mathbf{x}, \lambda) &= \mathbf{z} - \mathbf{H}\mathbf{x} \quad = 0 \end{aligned}\right\} \tag{18.2.3}$$

from which we obtain

$$\mathbf{x}_a = \mathbf{H}^{+}\mathbf{z} \tag{18.2.4}$$

where $\mathbf{H}^{+} = \mathbf{H}^{\mathrm{T}}(\mathbf{H}\mathbf{H}^{\mathrm{T}})^{-1}$ is the Moore–Penrose **generalized inverse** (Appendix B) of H.

Since the problem is underdetermined, this idea of strictly vanishing residual $\mathbf{r}(\mathbf{x}) = \mathbf{z} - \mathbf{H}\mathbf{x}$ may seem natural at first sight. However, this may be acceptable only when there is no noise corrupting the observation or when we do not have any recourse to obtaining the properties of the measurement noise. In the retrieval problem of interest to us, it is assumed that there is observation noise with known, second-order properties. If we take the presence of this noise into account and rework the above analysis, we will obtain

$$\mathbf{x}_a = \mathbf{H}^{+}(\mathbf{z} - \mathbf{v}). \tag{18.2.5}$$

However, the noise vector $\mathbf{v}$ must **not** be observable and hence this formula for $\mathbf{x}_a$ is all but useless. Stated in other words, this strategy of strict enforcement of the vanishing of the residual is **not** as desirable as it may look at first sight.

In search of a more realistic and philosophically appealing approach we now seek to weaken the requirements of the vanishing of the residual, $\mathbf{r}(\mathbf{x}) = (\mathbf{z} - \mathbf{H}\mathbf{x})$.

Strategy II

Consider a least squares approach of minimizing the weighted norm of $\mathbf{r}(\mathbf{x})$. Recall from Chapter 6 that the norm of the residual when weighted by the inverse of the known covariance matrix $\mathbf{R}$ of the observation errors is **invariant** under linear transformation of the observation space. Accordingly, consider a new objective function

$$J_0(\mathbf{x}) = \frac{1}{2}(\mathbf{z} - \mathbf{H}\mathbf{x})^{\mathrm{T}}\mathbf{R}^{-1}(\mathbf{z} - \mathbf{H}\mathbf{x}). \tag{18.2.6}$$

We now seek $\hat{\mathbf{x}}_{\mathrm{LS}}$ that minimizes this $J_0(\mathbf{x})$. It can be verified that the gradient and the Hessian are given by

$$\nabla J_0(\mathbf{x}) = -\mathbf{H}^{\mathrm{T}}\mathbf{R}^{-1}\mathbf{z} + (\mathbf{H}^{\mathrm{T}}\mathbf{R}^{-1}\mathbf{H})\mathbf{x} \tag{18.2.7}$$

and

$$\nabla^2 J_0(\mathbf{x}) = \mathbf{H}^{\mathrm{T}}\mathbf{R}^{-1}\mathbf{H}. \tag{18.2.8}$$

Setting the gradient to zero, we obtain $\hat{\mathbf{x}}_{\mathrm{LS}}$ as the solution of

$$(\mathbf{H}^{\mathrm{T}}\mathbf{R}^{-1}\mathbf{H})\mathbf{x} = \mathbf{H}^{\mathrm{T}}\mathbf{R}^{-1}\mathbf{z}. \tag{18.2.9}$$

Several observations are in order.

(a) **The matrix $\mathbf{H}^{\mathrm{T}}\mathbf{R}^{-1}\mathbf{H}$ is singular**
Recall that $\mathbf{H} \in \mathbb{R}^{m\times n}$ with Rank($\mathbf{H}$) $= \min(n, m) = m$ by assumption and $\mathbf{R}^{-1} \in \mathbb{R}^{m\times m}$ is a symmetric and positive definite matrix. Hence $\mathbf{H}^{\mathrm{T}}\mathbf{R}^{-1}\mathbf{H}$ is an $n \times n$ symmetric matrix such that the

$$\begin{aligned}\mathrm{Rank}(\mathbf{H}^{\mathrm{T}}\mathbf{R}^{-1}\mathbf{H}) &\le \min\{\mathrm{Rank}(\mathbf{H}), \mathrm{Rank}(\mathbf{R}^{-1})\}\\ &= m\\ &< n.\end{aligned}$$

In other words, the matrix on the l.h.s. of (18.2.9) is singular and the retrieval problem becomes **ill-posed**.

(b) **The matrix $\mathbf{H}^{\mathrm{T}}\mathbf{R}^{-1}\mathbf{H}$ is positive semi-definite** For any $\mathbf{y} \in \mathbb{R}^n$

$$\mathbf{y}^{\mathrm{T}}(\mathbf{H}^{\mathrm{T}}\mathbf{R}^{-1}\mathbf{H})\mathbf{y} = (\mathbf{H}\mathbf{y})^{\mathrm{T}}\mathbf{R}^{-1}(\mathbf{H}\mathbf{y}) \ge 0 \tag{18.2.10}$$

since $\mathbf{R}^{-1}$ is positive definite. However, since the Rank($\mathbf{H}$) $= m$, only m of its n columns are linearly independent and the null space of $\mathbf{H}$

$$N(\mathbf{H}) = \{\mathbf{y}|\mathbf{H}\mathbf{y} = 0\}$$

is of dimension $(n - m)$. Hence, for all $0 \neq \mathbf{y} \in N(\mathbf{H})$ $\mathbf{H}\mathbf{y} = 0$ and equality holds good in (18.2.10) for all $\mathbf{y} \in N(\mathbf{H})$. Hence, $\mathbf{H}^{\mathrm{T}}\mathbf{R}^{-1}\mathbf{H}$ is positive semidefinite.

(c) If the observation network is such that there is no correlation between the measurement error between two distinct stations, and if all the instruments in these

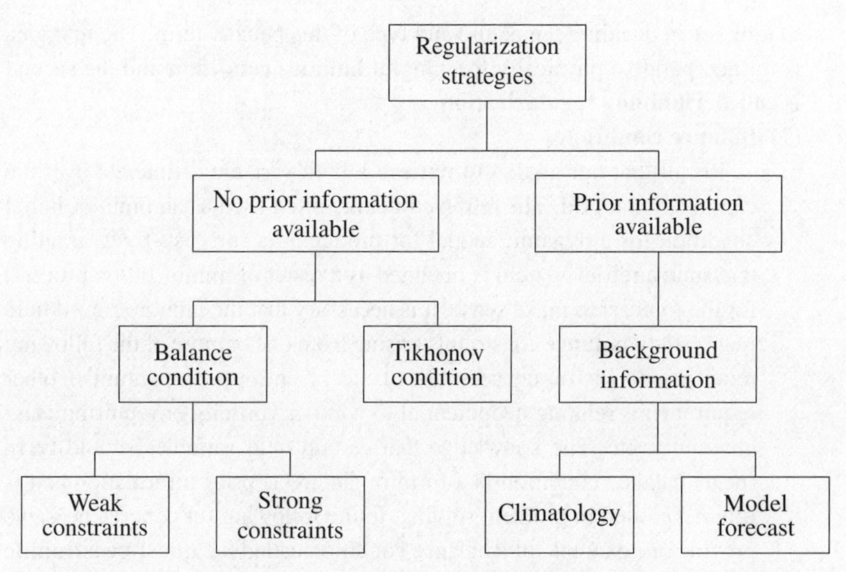

Fig. 18.2.1 A classification of regularization strategies.

stations are identical, then $\mathbf{R} \in \mathbb{R}^{m\times m}$ becomes a diagonal matrix with a common diagonal entry, $\mathbf{R} = \text{Diag}(\sigma^2, \sigma^2, \ldots, \sigma^2)$ where σ^2 denotes the common variance of these identical instruments. In this case, $\mathbf{R}^{-1} = \text{Diag}(\alpha, \alpha, \ldots, \alpha)$ with $\alpha = \sigma^{-2}$. Substituting this in (18.2.6) and simplifying, it can be verified that

$$J_0(\mathbf{x}) = \frac{\alpha}{2}(\mathbf{z} - \mathbf{Hx})^{\mathrm{T}}(\mathbf{z} - \mathbf{Hx}). \tag{18.2.11}$$

This form of $J_0(\mathbf{x})$ is known as the **penalty term**. In other words, the weighted norm of the residual includes the penalty term as a special case.

Despite the fact that the second strategy leads to an ill-posed problem, since the underlying idea of using the weighted norm of the residual is quite appealing from both the computational and philosophical point of view, we now look for ways to converting an ill-posed to a well-posed problem. The process of converting an ill-posed to a well-posed problem is called **regularization**. Roughly speaking, regularization relates to the process of adding constraints (natural or artificial) so as to reduce the dimensionality of the solution space in question. We now present an overview and a classification of the regularization strategies that are routinely used in the geophysical domain. Referring to the Figure 18.2.1, at the highest level these strategies fall into two groups depending on the nature and type of prior information available.

(A) **No prior information is available** When no direct credible prior information is available, regularization is achieved by adding a smoothing term which is often designed as a quadratic penalty term. There are essentially two directions to look for in deciding the nature and type of this penalty term. The first idea is to incorporate a physically meaningful **balance condition** and the second is called **Tikhonov regularization**.

(1) **Balance conditions**

Recall that our goal is to retrieve a field variable of interest over the computational grid. The retrieved field is often used as an input or initial condition for a dynamic model for producing a forecast. Even granting that such a retrieved field is obtained as a result of minimization process, for the forecast to make sense, it is necessary that the input retrieved field must satisfy **balance** constraints arising from one or more of the following requirements such as conservation of energy, entropy, momentum, or other requirements relating geopotential to wind or vorticity or requiring mass continuity, etc. The knowledge that certain field variables should be in (near) balance condition is a form of (indirect) prior information and is often exploited in problem solving. In the following for concreteness, we provide one example of a balance condition called the **quasi-geostrophic balance** (See 3.7.2 but without friction). In this case, the geopotential is related to the wind as follows:

$$\frac{\partial \phi}{\partial \mathbf{x}} = f_0 \mathbf{v} \quad \text{and} \quad \frac{\partial \phi}{\partial \mathbf{y}} = -f_0 \mathbf{u} \tag{18.2.12}$$

where $\phi = \phi(\mathbf{x}, \mathbf{y})$ is the geopotential and $\mathbf{u} = \mathbf{u}(\mathbf{x}, \mathbf{y})$ and $\mathbf{v} = \mathbf{v}(\mathbf{x}, \mathbf{y})$ are the east-west and north-south wind components, respectively, and f_0 is the Coriolis parameter (Holton (1972), Daley (1991)). Thus, if we are retrieving the geopotential from its observations, it would be useful to measure the u and v wind components as well so that we enforce the balance condition (18.2.12) on the retrieved geopotential field.

These constraints are discretized on the computational grid using suitable discretization schemes and expressed as algebraic expressions relating the geopotential and the observed wind components.

Let $\phi = (\phi_1, \phi_2, \phi_3, \ldots, \phi_n)^{\mathrm{T}}$ be the discretized form ϕ over the grid (Exercise 18.3). Then the resulting algebraic expression can be expressed as a function

$$g(\phi) = 0 \tag{18.2.13}$$

where $g(\phi) = (g_1(\phi), g_2(\phi), \ldots, g_p(\phi))^{\mathrm{T}}$. Since the geostrophic constraint (18.2.12) is linear in ϕ, for this case $g(\phi)$ in (18.2.13) will be a linear in ϕ. In general, the balance constraints can be nonlinear in which case its discretized counterpart will be a nonlinear algebraic equation in the retrieved field variable.

Given the balance constraint, the question is: how best to integrate it into the basic retrieval problem? This can be done in two ways – using it as a **strong constraint** or as a **weak constraint** (Sasaki (1958), (1969), and (1970)).

(a) **Strong constraint** In this approach, strict enforcement of the balance constraints is required. This is accomplished by defining the Lagrangian

$$L(\mathbf{x}, \lambda) = \frac{1}{2}(\mathbf{z} - \mathbf{H}\mathbf{x})^{\mathrm{T}}\mathbf{R}^{-1}(\mathbf{z} - \mathbf{H}\mathbf{x}) + \lambda^{\mathrm{T}} g(\mathbf{x}) \tag{18.2.14}$$

where the generic state variable is used in place of ϕ in (18.2.13), and $\lambda \in \mathbb{R}^p$ is the **undetermined** Lagrangian multiplier.

(b) **Weak constraint** Since the balance constraints themselves are approximations to reality, in this approach balance constraints are included as a quadratic penalty term as

$$J(\mathbf{x}) = \frac{1}{2}(\mathbf{z} - \mathbf{H}\mathbf{x})^{\mathrm{T}}\mathbf{R}^{-1}(\mathbf{z} - \mathbf{H}\mathbf{x}) + \frac{\alpha}{2}\, \|g(\mathbf{x})\|^2 \tag{18.2.15}$$

for some $\alpha > 0$. The idea is, if α is large, minimizing $J(\mathbf{x})$ will force $\mathbf{x}$ in such a way so as to keep the value of the norm of $g(\mathbf{x})$ small so that the balance condition (18.2.13) is very nearly satisfied. Early algorithms based on **polynomial approximation** used balance constraints effectively (Daley (1991) and Thiébaux and Pedder (1987)).

(2) **Tikhonov regularization**

When there is no prior information nor any other known balance condition required of the field variable in question, an artificial smoothing or penalty term is added as follows:

$$J(\mathbf{x}) = \frac{1}{2}(\mathbf{z} - \mathbf{H}\mathbf{x})^{\mathrm{T}}\mathbf{R}^{-1}(\mathbf{z} - \mathbf{H}\mathbf{x}) + \frac{\alpha}{2}\, \|\mathbf{x}\|^2 \tag{18.2.16}$$

which is very similar in spirit to the idea of weak constraint in (18.2.15) (Chapter 5). Most of the approaches in tomography, geological exploration routinely use Tikhonov type regularization (Chapter 19).

(B) **Prior Information Available**

When the prior information is available, it could appear in various forms. We now examine the nature and type of prior information that may be available at our disposal.

(1) **Knowledge of the background state $\mathbf{x}_{\mathrm{B}}$** Let $\mathbf{x}$ be the field variable that is being retrieved. It is often the case that we have prior information about $\mathbf{x}$. This information may come from two sources – **climatology** and **previous forecast** in the meteorological application. Climatological background gives valuable information, but it is not specific to a given weather situation. Rather, it gives bounds on the weather gleaned from typically large data sets. For example, if you are planning on a trip to Sydney, Australia in January, we know that it must be early summer and the maximum

temperature will be in the range of 35–40°C. Forecasts typically provide information specific to a given weather regime and accordingly is likely to highlight or focus on a subrange of the climatology. This information – be it from climatology or forecast – is called the **background state** and is denoted by $\mathbf{x}_B$.

(2) **Knowledge of the covariance/correlation structure of $\mathbf{x}_B$**
Since the forecast or climatology provide probable information, to be able to use this information in a statistically sensible manner, we need to have detailed information about its probabilistic characteristics. At the least we should have information about the spatial correlations of the probable values of the background field. This correlation can be computed from the climatological data or from the forecast data. For example, using the climatological data base covering North America, we can compute the spatial correlogram of the temperature anomaly (which is the difference between the long term average and the actual maximum temperature) for each of the twelve months in a year or each of the four seasons, etc. Likewise, we can compute the forecast error covariance by systematically archiving the forecast values and comparing them to "truth" – e.g., accurate analyses in data rich regions.

At the highest level one may be able to assume detailed information about the multivariate distribution of the background error. For example, we can often assume that observational errors are normally distributed. In such cases, we have the luxury of assuming a multivariate normal distribution for the background errors.

Given this prior information in the form of $\mathbf{x}_B$ and the associated statistical properties, we can combine it with the observation $\mathbf{z}$ and its statistical characteristics in a number of ways – **statistical least squares** (Chapter 14), **minimum variance estimation** (Chapter 17), **maximum likelihood** (Chapter 15), and **Bayesian estimation** (Chapter 16).

Retrieval methods that exploit the prior information are one form; the others are: successive correction methods, optimal interpolation methods, and the Bayesian approach that includes minimum variance Kalman filtering techniques. A comprehensive review of these algorithms and their properties are described in Chapters 19 and 20.

Exercises

18.1 Consider the one-dimensional grid with $n = 7$ points and $m = 4$ observations given in Figure 18.1.2.

(a) Assuming unit grid length, generate four random numbers from a uniform distribution in [0,1] to decide the actual locations of the four observations in the grid as shown in this figure.

(b) Using this information, numerically assemble the 4×7 matrix interpolation matrix $\mathbf{H}$.
(c) Compute $\mathbf{H}\mathbf{H}^{\mathrm{T}}$ and $\mathbf{H}^{\mathrm{T}}\mathbf{H}$.

18.2 Repeat the above exercise on the two-dimensional grid with $n = 164$ points and $m = 4$ observations in Figure 18.1.3.

18.3 Discretize the quasi-geostrophic balance constraint given in (18.2.11) on the 4×4 grid given in Figure 18.1.3.

Notes and references

In the mid-1950s, Yoshi Sasaki wrote a dissertation at the University of Tokyo that introduced the meteorological community to the variational calculus approach to dynamics (Sasaki 1955). Carl Eckart (Eckart 1960) performed the same service for the oceanographic community. Both of these efforts rested on Clebsch's work in the mid-nineteenth century (Clebsch 1857; Bateman 1932). Sasaki (1958) then viewed data assimilation from this variational viewpoint – essentially the Gaussian approach applied to continuous media. In this work of the late 1950s, Sasaki considered static conditions such as the wind laws, but by the late 1960s he expanded his view to include dynamical laws (Sasaki 1969, 1970). The work was the foundation for variational methods of assimilation applied to operational data analysis in Norway [Odd Haag (unpublished)] and in the United States [Lewis and Grayson (1972), Lewis (1972)]. A succinct review of the various formulations including strong and weak constraints is found in LeDimet and Talagrand (1986). The formulation of the problem in Section 18.1 is rather standard and we refer the reader to many excellent textbooks on this topic. The books by Menke (1984), Bennett (1992) and (2002), Parker (1994), and Tarantola (1987) are tuned to audiences in geophysics and oceanography and those by Daley (1991), Thiébaux and Pedder (1987), Gandin (1963), Bengtsson et al. (1981) are written with the meteorological interest in mind.

19

Classical algorithms for data assimilation

With the advent of numerical weather prediction (NWP) in the post-WWII period of the 1950s, it became clear that numerical weather map analysis was a necessary adjunct to the prediction. Whereas a subjective hand analysis of weather data was the standard through the mid-twentieth century, the time constraints of operational NWP made it advisable to analyze the variables on a network of grid points in an objective fashion. It is worth mentioning, however, that the limited availability of computers in the early 1950s led the Norwegian weather service to prepare initial data for the one-level barotropic forecast by hand. In fact, the forecast (a 24-hour prediction) was also prepared graphically by methods that were akin to those discussed in Chapter 2 (Section 2.6). These forecasts compared favorably with the computer-generated results in the USA and Sweden. Nevertheless, by the mid-1950s, objective data assimilation via computer was the rule. We review several of these early methods of data assimilation.

In section 19.1 we provide a brief review of the **polynomial approximation** method with a discussion of two ways of handling the **balance constraints** – as **strong** and **weak constraints**. An algorithm based on Tikhonov regularization is described in section 19.2. A class of iterative techniques known as **successive correction methods** (SCM) is reviewed in section 19.3 along with convergence properties of these schemes. The concluding section 19.4 derives the mechanics of the **optimum interpolation** method and its relation to SCM.

19.1 Polynomial approximation method

In conjunction with the first successful (NWP) research at Princeton University in the late 1940s, Hans Panofsky (1949) developed a method of objectively analyzing the meteorological variables on a two-dimensional surface ($x - y$ coordinate system). This method rested upon the least squares fit of a general third-order polynomial (in $x - y$) to a set of observations. Since the general third-order polynomial has ten coefficients, the minimum requisite set of observations to find a solution is exactly ten. It is generally advisable to include more observations than the minimum set since the observations contain error and some degree of smoothing is desirable. Further, since the polynomial cannot be expected to represent the variation of the meteorological variable over spatial dimension that is large compared to important smaller-scale structure, the domain of interest (continental USA, e.g.) is divided into sub-domains where separate polynomial fits are found. It then became a problem to guarantee some measure of continuity in the field across those boundaries.

Once the coefficients are optimally found by minimizing the squared departure between the polynomial and the observations, the value of the variable is determined as a function of $x - y$ in the sub-domain where observations are analyzed. Gilchrist and Cressman (1954) essentially followed this line of attack; but rather than defining sub-domains, they found a separate polynomial fit (in their case, a second-degree polynomial) for each grid point in the domain of interest.

Local polynomial method is a special case of fitting functions to irregularly spaced data in one, two, and three dimensions – a topic of widespread interest in numerical analysis, especially the theory of splines (de Boor (1978) and Bartels et al. (1987)).

We now illustrate the basic ideas using a simple example.

Example 19.1.1 (See Exercise 19.1) Consider a square domain whose side is ten units long. Embed a 10×10 uniform grid in this domain. Thus, $n = 100$. Randomly generate a set of forty locations ($m = 40$) using a uniform probability density function (pdf). Let $\mathbf{x}_j = (x_{j1}, x_{j2})^{\mathrm{T}} \in \mathbb{R}^2$ be the location of the jth observation station for $j = 1$ to 40. Let z_j denote the temperature at the station j. Randomly generate z_j from a uniform pdf in the interval [85°F, 95°F]. Let

$$p(\mathbf{x}) = (x_1, x_2)^{\mathrm{T}} \begin{bmatrix} a_1 & a_2 \\ a_2 & a_3 \end{bmatrix} \begin{bmatrix} x_1 \\ x_2 \end{bmatrix} + (a_4, a_5) \begin{bmatrix} x_1 \\ x_2 \end{bmatrix} + a_6 \tag{19.1.1}$$

be the second-degree polynomial in $x_1\, x_2$. Define $\mathbf{y} = (a_1, a_2, a_3, a_4, a_5, a_6)^{\mathrm{T}} \in \mathbb{R}^6$, the vector of unknown coefficients. Let $S_i(d)$ denote the circular region centered at the grid point i and of radius $d (> 0)$, called the **radius of influence**. For each grid point i, now define

$$J_i(\mathbf{y}) = \sum_{x_j \in S_i(d)} W_{ij} [p(\mathbf{x}_j) - z_j]^2 \tag{19.1.2}$$

where W_{ij} is the (empirical) weight which generally has an inverse relation to the distance between the ith grid point and the jth observation station and the summation is restricted to only those stations that lie in the region of influence.

Clearly, $J_i(\mathbf{y})$ is a quadratic function in $\mathbf{y}$, and the method seeks to find the $\mathbf{y}$ that minimizes (19.1.2). This minimizer is obtained by setting the gradient $\nabla_{\mathbf{y}} J_i(\mathbf{y})$ to zero and checking to see if the Hessian at the minimizer is positive definite. Since $\mathbf{y}$ has six coefficients, this results in a set of six linear equations in six unknowns.

By solving this system repeatedly at each grid point, we obtain the retrieved field over the grid (the method of Gilchrist and Cressman (1954)).

We encourage the reader to complete Exercise 19.1 at this time.

Several observations are in order.

(a) Since the weights W_{ij} are inversely proportional to the distance between the ith grid point and the jth observation location, observations that are closer have a larger influence than those that are farther away. Thus, when combined with the nonuniform observation density around the grid point, this leads to a retrieved field of nonuniform quality.

(b) When the number of coefficients is greater than the number of observations in the region of influence, this leads to an under-determined problem. **Balance conditions** are used to induce uniqueness. In closing this section, we illustrate the two standard ways of using the balance conditions.

Example 19.1.2 Let $\mathbf{x} = (x_1, x_2, x_3)^{\mathrm{T}}$ and consider

$$J(\mathbf{x}) = \mathbf{x}^{\mathrm{T}}\mathbf{B}\mathbf{x} + \mathbf{d}^{\mathrm{T}}\mathbf{x} \tag{19.1.3}$$

where

$$\mathbf{B} = \begin{bmatrix} 1 & 0 & 0 \\ 0 & 1 & 0 \\ 0 & 0 & 1 \end{bmatrix} \quad \text{and} \quad \mathbf{d} = \begin{pmatrix} -2 \\ 0 \\ 4 \end{pmatrix}.$$

This $J(\mathbf{x})$ plays the role of the $J_i(\mathbf{y})$ in (19.1.2) where we, for simplicity, have replaced $\mathbf{y} \in \mathbb{R}^6$ with $\mathbf{x} \in \mathbb{R}^3$.

Let $g : \mathbb{R}^3 \longrightarrow \mathbb{R}$ and the functional form of the balance constraint be given by

$$g(\mathbf{x}) = 0. \tag{19.1.4}$$

In general, there could be more than one such constraint and each of these constraints could be linear or nonlinear in $\mathbf{x}$. For definiteness, it is assumed that $g(\mathbf{x})$ is linear and is given by,

$$g(\mathbf{x}) = \mathbf{a}^{\mathrm{T}}\mathbf{x} - 2 = 0 \tag{19.1.5}$$

where $\mathbf{a} = (1, -1, 2)^{\mathrm{T}}$. Given the objective function $J(\mathbf{x})$ in (19.1.2), and the linear constraint in (19.1.4), there are two ways to combine them.

Strong constraint formulation Let λ be the undetermined multiplier and define the Lagrangian

$$L(\mathbf{x}, \lambda) = J(\mathbf{x}) + \lambda g(\mathbf{x}). \tag{19.1.6}$$

In minimizing $L(\mathbf{x}, \lambda)$ set

$$\nabla_{\mathbf{x}} L(\mathbf{x}, \lambda) = \nabla J(\mathbf{x}) + \lambda \mathbf{a} = 0 \tag{19.1.7}$$

and

$$\nabla_\lambda L(\mathbf{x}, \lambda) = \mathbf{a}^\mathrm{T}\mathbf{x} - 2 = 0. \tag{19.1.8}$$

Substituting

$$\nabla J(\mathbf{x}) = 2\mathbf{Bx} + \mathbf{d}$$

into the first equation and solving it we obtain $x_1 = 1 - \lambda/2$ and $x_2 = \lambda/2$ and $x_3 = -2 - \lambda$.

Substituting these into the second equation, we obtain $\lambda = -5/3$. Combining these obtain the minimizer as

$$x_1^* = \frac{11}{6}, \; x_2^* = -\frac{5}{6}, \quad \text{and} \quad x_3^* = -\frac{1}{3}, \tag{19.1.9}$$

and this is called the **strong solution**.

Weak constraint formulation Define

$$\begin{aligned} J_\alpha(\mathbf{x}) &= J(\mathbf{x}) + \alpha \, \|g(\mathbf{x})\|^2 \\ &= \mathbf{x}^\mathrm{T}\mathbf{Bx} + \mathbf{dx} + \alpha(\mathbf{a}^\mathrm{T}\mathbf{x} - 2)^2 \end{aligned}$$

where $\alpha(> 0)$ is called the **penalty** parameter. Then, setting

$$\nabla J_\alpha(\mathbf{x}) = 2\mathbf{Bx} + \mathbf{d} + 2\alpha\mathbf{a}[\mathbf{a}^\mathrm{T}\mathbf{x} - 2] = 0$$

we obtain the minimizer as the solution of

$$[\mathbf{B} + \alpha\mathbf{a}\mathbf{a}^\mathrm{T}]\mathbf{x} = 2\alpha\mathbf{a} - \frac{\mathbf{d}}{2}. \tag{19.1.10}$$

Substituting for **B**, **a** and **d**, this linear system can be solved for **x** as an explicit function of α, and $\mathbf{x}(\alpha)$ is called the **weak solution**. (Exercise 19.2)

Dividing both sides of (19.1.10) by α and letting α grow without bound, in the limit it becomes

$$[\mathbf{a}\mathbf{a}^\mathrm{T}]\mathbf{x} = 2\mathbf{a} \tag{19.1.11}$$

that is,

$$\begin{bmatrix} 1 & -1 & 2 \\ -1 & 1 & -2 \\ 2 & -2 & 4 \end{bmatrix} \begin{bmatrix} x_1 \\ x_2 \\ x_3 \end{bmatrix} = \begin{bmatrix} 2 \\ -2 \\ 4 \end{bmatrix}. \tag{19.1.12}$$

The matrix $[\mathbf{a}\mathbf{a}^\mathrm{T}]$ on the l.h.s. of (19.1.11) is of rank one and hence is singular. However, it can be verified that the r.h.s. vector $2\mathbf{a}$ lies in the **range space** (**span** of the columns) of $[\mathbf{a}\mathbf{a}^\mathrm{T}]$. Hence it has a consistent solution. Indeed, it can be verified that the strong solution (19.1.9) also satisfies (19.1.11). Stated in other words, the weak solution converges to the strong solution as the penalty parameter α grows without bound.

19.2 Tikhonov regularization method

When no explicit prior information is available about the unknown, this method calls for augmenting $J_0(\mathbf{x})$ in (18.2.6) by adding a **quadratic penalty** or **regularization** term as follows.

Pick a real constant $\alpha > 0$ and define

$$J(\mathbf{x}) = J_0(\mathbf{x}) + J_p(\mathbf{x}) \tag{19.2.1}$$

where the **observation** term

$$J_0(\mathbf{x}) = \frac{1}{2}(\mathbf{z} - \mathbf{H}\mathbf{x})^{\mathrm{T}}\mathbf{R}^{-1}(\mathbf{z} - \mathbf{H}\mathbf{x}) \tag{19.2.2}$$

and the **penalty** term

$$J_p(\mathbf{x}) = \frac{\alpha}{2}\mathbf{x}^{\mathrm{T}}\mathbf{x}. \tag{19.2.3}$$

The retrieval problem is then stated as follows: given $\mathbf{z}$, $\mathbf{H}$, $\mathbf{R}$, and α, find the $\mathbf{x}$ that minimizes $J(\mathbf{x})$ in (19.2.1). The gradient and the Hessian of $J(\mathbf{x})$ are given by

$$\nabla J(\mathbf{x}) = -\mathbf{H}^{\mathrm{T}}\mathbf{R}^{-1}\mathbf{z} + (\mathbf{H}^{\mathrm{T}}\mathbf{R}^{-1}\mathbf{H} + \alpha\mathbf{I})\mathbf{x} \tag{19.2.4}$$

and

$$\nabla^2 J(\mathbf{x}) = (\mathbf{H}^{\mathrm{T}}\mathbf{R}^{-1}\mathbf{H} + \alpha\mathbf{I}). \tag{19.2.5}$$

The minimizer of $J(\mathbf{x})$ is then obtained by setting the gradient to zero and is given by the solution of the linear system

$$(\mathbf{H}^{\mathrm{T}}\mathbf{R}^{-1}\mathbf{H} + \alpha\mathbf{I})\mathbf{x} = \mathbf{H}^{\mathrm{T}}\mathbf{R}^{-1}\mathbf{z}. \tag{19.2.6}$$

Since $\alpha > 0$, it follows that

$$\begin{aligned} \mathbf{y}^{\mathrm{T}}[\mathbf{H}^{\mathrm{T}}\mathbf{R}^{-1}\mathbf{H} + \alpha\mathbf{I}]\mathbf{y} &= (\mathbf{H}\mathbf{y})^{\mathrm{T}}\mathbf{R}^{-1}(\mathbf{H}\mathbf{y}) + \alpha\mathbf{y}^{\mathrm{T}}\mathbf{y} \\ &\geq 0 \end{aligned} \tag{19.2.7}$$

for any $\mathbf{y} \in \mathbb{R}^n$ with equality holding good only when $\mathbf{y} = 0$. That is, the matrix $(\mathbf{H}^{\mathrm{T}}\mathbf{R}^{-1}\mathbf{H} + \alpha\mathbf{I})$ is positive definite and hence is non-singular and the solution of (19.2.6) is indeed the minimizer of $J(\mathbf{x})$. In other words, adding the penalty term $J_p(\mathbf{x})$ does the trick of transforming an ill-posed problem to a well-posed problem.

Like everything else in life, this idea of regularization comes with its own set of advantages and new challenges. The advantages are: it preserves the practical aspects of the least squares approach and helps alleviate the mathematical difficulty of ill-posedness and provides a unique solution. However, the hidden challenge relates to deciding the best value of α. From (19.2.6) it follows that α plays the role of a **trade-off** parameter. If α is small, then $J_0(\mathbf{x})$ term dominates and if α is large then the penalty term dominates. In practice there is no clear-cut rationale for the choice of α.

To understand the impact of the uniform term on computation, let λ_i for $i = 1$ to n be the n eigenvalues of $\mathbf{H}^{\mathrm{T}}\mathbf{R}^{-1}\mathbf{H}$. Since this matrix is of rank m, without loss of generality, let

$$\left.\begin{array}{c}\lambda_1 \geq \lambda_2 \geq \cdots \geq \lambda_m > 0 \\ \text{and} \\ \lambda_{m+1} = \lambda_{m+2} = \cdots = \lambda_n = 0\end{array}\right\}$$

If μ_i for $i = 1$ to n are eigenvalues of $(\mathbf{H}^{\mathrm{T}}\mathbf{R}^{-1}\mathbf{H} + \alpha\mathbf{I})$ then it can be verified (Appendix B) that $\mu_i = \lambda_i + \alpha$. Hence,

$$\kappa_2(H^{\mathrm{T}}R^{-1}H + \alpha I) = \frac{\mu_1}{\mu_n} = \frac{\lambda_1 + \alpha}{\alpha} = 1 + \frac{\lambda_1}{\alpha} \ll \kappa_2(\mathbf{H}^{\mathrm{T}}\mathbf{R}^{-1}\mathbf{H}) \qquad (19.2.8)$$

where $\kappa_2(\mathbf{A})$ is called the **spectral condition** number of $\mathbf{A}$ (Appendix B). Thus, computationally larger α is better. In view of the fact α helps to reduce the spectral condition number, it is also called the **damping factor** and the solution of (19.2.6) is called a **damped solution**.

19.3 Structure functions

In the mid-1950s, the Swedish Hydrological Service began operational weather prediction (Wiin-Nielsen 1991). In support of the prediction, a numerical weather map analysis of the 500 mb geopotential height was developed. The scheme was developed by two of Carl Rossby's proteges, Páll Bergthorsson and Bo Döös (Bergthorsson and Döös 1955). The following statement from their paper identifies the weakness of polynomial fitting:

> *In our investigation of this problem at the University of Stockholm, we reached the conclusion that quite often it is not possible to get a reasonable analysis only by means of interpolation between synoptic observations. It is quite clear that the distance between the observations must be small compared with the size of the systems to be analyzed. This is certainly not the case in many areas as over the oceans. In such cases any interpolation method will fail, independent of whether it is linear, quadratic, or cubic . If however, some observations were available in the area 12 hours ago, a twelve hour forecast is probably a better approximation than the interpolated analysis.*

This viewpoint set meteorological data assimilation on a pathway from which it has never veered. Although details differ at the various operational centers worldwide, the idea of using a forecast to help determine the final analysis remains a centerpiece of meteorological data assimilation. Bergthorsson had been a forecaster in Iceland before Rossby called him to Sweden to work on this project. The computerized map analysis followed the same pattern used by the practical forecaster – namely, augment the limited data at a given time by extrapolation of historical data to the desired region at the present time. And the best extrapolation is typically via the dynamical prediction model.

Although the method of Bergthorsson and Döös has important details that we leave to a reading of their paper, the essence of their scheme rests on weighting the analysis increment (the difference between forecast and observation at the observation station) and the forecast. In short, if the observed height is greater than the forecast at a set of observation locations surrounding the grid point, then the estimate at the grid point is a weighted sum of these increments added to the forecast at the grid point – in this case, an increase in the forecasted value. The spatial correlation of forecast errors were used to determine the weights. These structure functions, based on the assiduous work of comparing forecasts with observations, have proved to be representative of those used today. (These will be discussed further in conjunction with optimal interpolation methodology).

19.4 Iterative methods

As the power of computers increased in the 1950s–1960s, iterative methods of data assimilation became feasible. And, indeed, iteration remains justified for several reasons: (1) the scales of motion in weather regimes span the dimension of the globe down to local circulations at land/sea interfaces or in complex terrain, and (2) the governing constraints are typically nonlinear and solutions to optimization problems cannot be achieved in a single step. These issues of nonlinearity are addressed later. Building on the concept introduced by Bergthorsson and Döös (1955), the NWP component of the US Weather Bureau introduced an iterative method of data assimilation that has come to be called Successive Correction Method (SCM) or Cressman's method after its originator George Cressman (Cressman 1959). As in the case of the Bergthorsson/Döös scheme, the observation increment or difference between observation and forecast at the station became the central variable in the scheme.

Let us first discuss the iterative data assimilation problem generally and then we specifically describe Cressman's method as well as others.

Let $\mathbf{x} \in \mathbb{R}^n$ denote the unknown field variable to be retrieved over a computational grid with n grid points embedded in a two-dimensional domain. Refer to Section 18.1 for details. Let $\mathbf{x}_B \in \mathbb{R}^n$ denote the **known prior** information about the unknown. This $\mathbf{x}_B$ may be the result of a previous forecast or it could have been derived from the climatology, where, recall that both $\mathbf{x}$ and $\mathbf{x}_B$ are defined over the same computational grid.

Let $\mathbf{z} = (z_1, z_2, \ldots, z_m)^T \in \mathbb{R}^m$ be the set of m observations from m observation stations distributed in the domain of interest. To simplify the discussion below, it is assumed that both $\mathbf{z}$ and $\mathbf{x}$ refer to the same physical entity such as temperature, pressure, etc. That is, referring to (18.1.3), the function $h^\circ(.)$ is an identity function and that $\mathbf{x}^\circ = \mathbf{z}$. Thus, $h(\mathbf{x})$ reduces to the interpolation function $h^I(.)$. Again to simplify the notation, it is assumed that $h^I(.)$ is a linear interpolation function and we denote forward model as $\mathbf{z} = \mathbf{H}\mathbf{x}$ where $\mathbf{H} \in \mathbb{R}^{m \times n}$.

To begin with we have two pieces of information: the given **background field** $\mathbf{x}_\mathrm{B}$ on the computational grid and the observation $\mathbf{z} = \mathbf{x}^\circ$ on the observation network which can be mapped to the computational grid using the interpolation matrix $\mathbf{H}$ as $\mathbf{Hz}$. The essential idea is to combine these two pieces of information using an iterative scheme. To this end, let $\mathbf{x}_k$ denote the kth approximation to the unknown $\mathbf{x}$. Initially, $\mathbf{x}_0 = \mathbf{x}_\mathrm{B}$, the known prior. Then $\mathbf{x}_{k+1}$ for $k \geq 0$ is defined as

$$\mathbf{x}_{k+1} = \mathbf{x}_k + \mathbf{QW}[\mathbf{z} - \mathbf{H}\mathbf{x}_k] \tag{19.4.1}$$

where $\mathbf{W} \in \mathbb{R}^{n\times m}$ is the **weight matrix** and $\mathbf{Q} \in \mathbb{R}^{n\times n}$ is a (diagonal) matrix with normalizing constants across its diagonal. Just about every known iterative method for solving the retrieval problem can be derived from (19.4.1) by specializing the choices for the matrices $\mathbf{W}$ and $\mathbf{Q}$.

(a) **Cressman's Method** This scheme assumes that the desired scalar field (variable such as geopotential height or temperature) at a grid point is the sum of the background and a weighted sum of the increments inside the radius of influence surrounding the grid point. The weights are given by

$$W_{ij} = \begin{cases} \frac{d^2 - r_{ij}^2}{d + r_{ij}^2}, & \text{if } r_{ij} \leq d \\ 0, & \text{otherwise} \end{cases} \tag{19.4.2}$$

where d and r_{ij} are the radius of influence and distance between grid point and observation, respectively, as defined earlier. Again, $\mathbf{Q}$ is a diagonal matrix with

$$Q_{ii}^{-1} = \left[\sum_{j=1}^{m} W_{ij} + \frac{\sigma_0^2}{\sigma_\mathrm{B}^2} \right] \tag{19.4.3}$$

where σ_0^2 is the known common variance of the observational error and σ_B^2 is the known common variance of the background error.

Remark 19.4.1 The assumption that the observational error has some variance σ_0^2 across all the m stations is plausible only when all the stations use the same type of instrument for their measurement. But the assumption of common variance for the background error may not hold all the time. However, it must be recognized that computing the statistical properties of the background error is a thorny problem at best that has received much attention and remains challenging – see references at end of chapter.

(b) **Barnes (1964) scheme** This scheme defines the weights as

$$W_{ij} = \begin{cases} \exp[-\frac{r_{ij}^2}{d^2}], & \text{if } r_{ij} \leq d \\ 0, & \text{otherwise} \end{cases} \tag{19.4.4}$$

Barnes (1964) suggested an adaptive version where d is varied iteratively as

$$d_{k+1} = \gamma d_k \tag{19.4.5}$$

for some real constant $0 < \gamma < 1$ where k is the iteration number given in (19.4.1).

This weighting scheme is patterned after the normal or Gaussian distribution and is often used when there is no prior information (See Exercise 19.4).

Example 19.4.1 (Exercise 19.3) Consider the retrieval problem stated in Example 19.1.1 having 100 grid points and 40 observation stations located randomly in a 10 unit $\times$ 10 unit square domain. Generate the background $\mathbf{x}_B$ as follows. The value of $\mathbf{x}_B$ at the ith grid point

$$x_{B,i} = 90 + \epsilon_i$$

where ϵ_i is a zero mean random variable with uniform distribution in the range $[-1, 1]$. Using this $\mathbf{x}_B$ and the observation $\mathbf{z}$ generated in Example 19.1.1, implement the Cressman's method using the weights in (19.3.4). Iterate and examine if $\mathbf{x}_k$ converges to any limit. Notice that this implementation requires the knowledge of the interpolation matrix $\mathbf{H}$. This 40×100 matrix can be readily obtained using the techniques described in Section 18.1 once we know the location of the observation stations relative to the computational grid.

The weighted sum of increments is called the "correction", i.e., the value to be added to the background to obtain the new estimate (generally different from the background which is the zeroth-order estimate). With this estimate serving as an improved background, the radius of influence is decreased and a new set of increments is found (via interpolation from grid points to observation point). With the revised weight function, a weighted sum of increments provides another correction. This iterative approach continues until the radius of influence is representative of the smallest scales that can be resolved by the observations. In Cressman's early work with mesh size the order of 200 km ($= \Delta x$), four scans were used where the radius of influence decreased from 4.75 Δx to 1.80 Δx.

(c) **Convergence of iterative schemes** Once the properties of a basic iterative scheme are well understood, attention soon shifts to the analysis of its long-term behavior. The question such as "does the iterative scheme converge and if it does what is the limit?" becomes important. In the following we develop the basic ideas relating to the convergence of the iterative scheme described in (19.4.1).

To simplify the notation, define $\eta_k = \mathbf{H}\mathbf{x}_k$, that is, $\eta_k \in \mathbb{R}^m$ is the interpolated version of the iterate $\mathbf{x}_k$ at the observation locations. Since $\mathbf{x}_0 = \mathbf{x}_B$, $\eta_0 = \mathbf{H}\mathbf{x}_B$ is the interpolated background field. Substituting this in (19.4.1), we get

$$\mathbf{x}_{k+1} = \mathbf{x}_k + \mathbf{QW}[\mathbf{z} - \eta_k]. \tag{19.4.6}$$

Iterating this, as $\mathbf{x}_0 = \mathbf{x}_B$, we obtain

$$\mathbf{x}_k = \mathbf{x}_B + \mathbf{QW}\sum_{j=0}^{k-1}[\mathbf{z} - \eta_j]. \tag{19.4.7}$$

Thus, the long-term behavior of $\mathbf{x}_k$ depends on the summation in the second term of the r.h.s. of (19.4.7). A necessary condition for convergence of this sum is that for large j, $(\mathbf{z} - \eta_j)$ must vanish. We now determine when this condition is satisfied for large j.

To this end, pick a non-singular matrix $\mathbf{T} \in \mathbb{R}^{m\times m}$ and define a related iteration

$$\eta_{k+1} = \eta_k + \mathbf{T}[\mathbf{z} - \eta_k]. \tag{19.4.8}$$

Notice that while the iteration (19.4.1) is defined on the computational grid, this new iterative scheme defining η_k is defined on the observation network. Now subtracting $\mathbf{z}$ from both sides of (19.4.8), it becomes

$$\begin{aligned}(\eta_{k+1} - \mathbf{z}) &= (\eta_k - \mathbf{z}) + \mathbf{T}(\mathbf{z} - \eta_k)\\ &= (\mathbf{I} - \mathbf{T})(\eta_k - \mathbf{z})\end{aligned}$$

which on iteration becomes

$$\eta_k - \mathbf{z} = (\mathbf{I} - \mathbf{T})^k(\eta_0 - \mathbf{z}) \tag{19.4.9}$$

where $\mathbf{I} \in \mathbb{R}^{m\times m}$ is the identity matrix. Substituting (19.4.9) in (19.4.7), the latter becomes

$$\mathbf{x}_k = \mathbf{x}_B + \mathbf{QW}\sum_{j=0}^{k-1}(\mathrm{I} - \mathrm{T})^j(\mathbf{z} - \eta_0). \tag{19.4.10}$$

Thus, the necessary condition – vanishing of $(\mathbf{z} - \eta_j)$ translates into vanishing of $(\mathbf{I} - \mathbf{T})^j(\mathbf{z} - \eta_0)$. Since $(\mathbf{z} - \eta_0)$ is fixed, this can happen exactly when $(\mathbf{I} - \mathbf{T})^j$ goes to zero as j becomes unbounded.

Since $(\mathbf{I} - \mathbf{T})^j \longrightarrow 0$ as $j \longrightarrow \infty$ exactly when the spectral radius of $(\mathbf{I} - \mathbf{T})$ is less than unity (Appendix B), we immediately obtain the following necessary and sufficient condition for the convergence of (19.4.1):

$$\rho(\mathbf{I} - \mathbf{T}) < 1 \tag{19.4.11}$$

where $\rho(\mathbf{A})$ is the spectral radius of the matrix $\mathbf{A} = \mathbf{I} - \mathbf{T}$. Under this condition, we obtain (Appendix B)

$$\begin{aligned}(\mathbf{I} - \mathbf{A})^{-1} = \textstyle\sum_{j=0}^{\infty}\mathbf{A}^j &= \textstyle\sum_{j=0}^{k-1}\mathbf{A}^j + \mathbf{A}^k\sum_{j=0}^{\infty}\mathbf{A}^j\\ &= \textstyle\sum_{i=0}^{k-1}\mathbf{A}^i + \mathbf{A}^k(\mathbf{I} - \mathbf{A})^{-1}\end{aligned}$$

or

$$\sum_{j=0}^{k-1} \mathbf{A}^j = [\mathbf{I} - \mathbf{A}^k][\mathbf{I} - \mathbf{A}]^{-1}$$

that is,

$$\sum_{j=0}^{k-1} (\mathbf{I} - \mathbf{T})^j = [\mathbf{I} - (\mathbf{I} - \mathbf{T})^k]\mathbf{T}^{-1}. \tag{19.4.12}$$

Substituting (19.4.12) into (19.4.10), we have

$$\mathbf{x}_k = \mathbf{x}_{\mathrm{B}} + \mathbf{QW}[\mathbf{I} - (\mathbf{I} - \mathbf{T})^k]\mathbf{T}^{-1}(\mathbf{z} - \eta_0). \tag{19.4.13}$$

Since $(\mathbf{I} - \mathbf{T})^k \rightarrow 0$ as $k \rightarrow \infty$, when (19.4.11) is true, we obtain the analysis

$$\mathbf{x}_a = \lim_{k \longrightarrow \infty} \mathbf{x}_k = \mathbf{x}_{\mathrm{B}} + \mathbf{QWT}^{-1}(\mathbf{z} - \eta_0)$$

or

$$\mathbf{x}_a - \mathbf{x}_{\mathrm{B}} = \mathbf{QWT}^{-1}(\mathbf{z} - \eta_0). \tag{19.4.14}$$

The vector $(\mathbf{x}_a - \mathbf{x}_{\mathrm{B}})$ is called the **analysis increment** defined on the computational grid and $(\mathbf{z} - \eta_0)$ is the **observation increment** defined on the observation network.

Define $\mathbf{K} \in \mathbb{R}^{m \times n}$ where

$$\mathbf{QWT}^{-1} = \mathbf{K} \quad \text{or} \quad \mathbf{KT} = \mathbf{QW}. \tag{19.4.15}$$

Given $\mathbf{T}$ such that $\rho(\mathbf{I} - \mathbf{T}) < 1$ along with $\mathbf{W}$ and $\mathbf{Q}$, we can obtain the rows of $\mathbf{K}$ by solving the collection of linear systems $\mathbf{KT} = \mathbf{QW}$. In this case, we can rewrite (19.4.14) as

$$\mathbf{x}_a - \mathbf{x}_{\mathrm{B}} = \mathbf{K}(\mathbf{z} - \eta_0). \tag{19.4.16}$$

The remaining issue is : how to pick $\mathbf{T}$ to satisfy (19.4.11). To this end recall from Appendix B that if λ and μ are the eigenvalues of $\mathbf{T}$ and $(\mathbf{I} - \mathbf{T})$ respectively, then $\mu = 1 - \lambda$. Hence, by (19.4.11) we have

$$|\mu| = |1 - \lambda| < 1. \tag{19.4.17}$$

That is, the eigenvalues of $\mathbf{T}$ must lie in the unit disk in the complex plane centered at $(1, 0)$.

Here is a simple recipe for picking the matrix $\mathbf{T}$ satisfying the above condition. Start with a symmetric positive definite matrix $\mathbf{C}$ whose eigenvalues are known to lie in the real interval (a, b) for some $0 < a < b < \infty$. That is, $\mathbf{C}$ could be a covariance matrix. Since the spectrum of a matrix can be squeezed by preconditioning, let $\mathbf{M}$ be a diagonal matrix such that

$$\mathbf{T} = \mathbf{CM}. \tag{19.4.18}$$

The idea here is to choose **M** such that it transforms the spectrum of **C** that lies in the interval (a, b) to that of **T** in the required interval (0, 2). Given **C**, there are numerous choices for the preconditioner **M**. In other words, there are infinitely many choices of **T** in (19.4.18) satisfying the convergence condition (19.4.11).

From (19.4.16) it follows that the resulting analysis increment $(\mathbf{x}_a - \mathbf{x}_B)$ for a given observation increment $(\mathbf{z} - \eta_0)$ depends on **K** which in turn depends on the choice of the matrix **T**. This raises an interesting question. Can we force the analysis increment obtained by the iterative scheme to the same as that obtained by an optimal scheme (such as optimal interpolation) starting from the same observation increment? The answer is indeed YES and is obtained by a clever choice of the matrix **T**, and we will indicate one such choice in Section 19.5. This possibility of being able to force the behavior of an iterative scheme to match that of an optimal scheme is quite appealing both philosophically and computationally.

This simple but elegant framework for proving convergence of iterative schemes was developed independently by Bratseth (1986) in Norway and Franke and Gordon (1983). Also refer to Franke (1988) and Seaman (1988) for experimental verification of these theoretical claims.

19.5 Optimal interpolation method

Successive correction methods were routinely used by operational weather centers worldwide in 1960s–1970s – Sweden, USA, Japan, to name a few. In the former Soviet Union, a technique called **optimal interpolation** (OI) was championed by Lev Gandin. This technique belongs to a class of methods developed earlier and independently by Norbert Wiener (1949) in USA and Kolmogorov (1941) in the former Soviet Union. A review of the work by these two celebrated world-class mathematicians is found in the introduction of Yaglom's treatise on stochastic process (Yaglom (1962)). In the following we provide a synopsis of this method of OI.

Consider a computational grid with n grid points embedded in a two-dimensional domain. Let the m observation stations also be embedded in this same domain. See Section 18.1 for details. Consider a field variable of interest over this domain – such as temperature, pressure, etc. at a given time.

19.5.1 Perfect observations

Let z_j denote the observation of this field variable at the jth observation station, for $j = i$ to m, and $\mathbf{z} = (z_i, z_2, \ldots, z_m)^{\mathrm{T}}$ denote the vector of these m observations at the chosen instant in time.

As an example, let the field variable of interest denote the surface temperature. Let z_j denote the temperature at the International Airport in Chicago at 12:00 noon on Christmas Day. A plot of the time series of temperature observations at this station on this day over the past several years would reveal that z_j is indeed a random variable. In the following, it is assumed that the time series of the past observations at each of the m observation stations are **stationary** (See Appendix F for a definition of stationarity).

Let $\bar{z}_j$ denote the long-term time average of z_j obtained from the time series of past observations at the station j. Since this time average $\bar{z}_j$ is known, it constitutes the **prior** or **background** information about z_j. Define

$$\tilde{z}_j = z_j - \bar{z}_j \tag{19.5.1}$$

which represents the **anomaly** or the **observation increment** at this station j. Let $\bar{\mathbf{z}}_j = (\bar{z}_1, \bar{z}_2, \ldots, \bar{z}_m)^{\mathrm{T}}$ and $\tilde{\mathbf{z}}_j = (\tilde{z}_1, \tilde{z}_2, \ldots, \tilde{z}_m)^{\mathrm{T}}$.

It is assumed that the observations are made using a **perfect instrument** and that there is **no measurement error**.

It can be verified that

$$E[\tilde{\mathbf{z}}] = E[\mathbf{z} - \bar{\mathbf{z}}] = 0 \tag{19.5.2}$$

and

$$\begin{aligned} \mathbf{C} = \mathrm{Cov}(\mathbf{z}) &= E[(\mathbf{z} - \bar{\mathbf{z}})(\mathbf{z} - \bar{\mathbf{z}})^{\mathrm{T}}] \\ &= E[\tilde{\mathbf{z}}\tilde{\mathbf{z}}^{\mathrm{T}}] \end{aligned} \tag{19.5.3}$$

where

$$c_{jk} = E[\tilde{z}_j \tilde{z}_k] \tag{19.5.4}$$

denotes the covariance between the observations at the stations j and k where $i \leq j$, $k \leq m$. While $\mathbf{C} \in \mathbb{R}^{m \times m}$ is **symmetric**, it is also assumed that it is a **positive definite** matrix. This covariance matrix $\mathbf{C}$ describes the **inherent spatial covariance structure** of the field variable in question and it can be readily computed from the historical time series data from the observation stations. Hence, in the following it is assumed that the matrix $\mathbf{C}$ is **known**.

Covariance between a grid point and observations Our aim is to estimate the value of the same field variable at the grid point i. To avoid confusion, we denote this quantity as x_i, for $i = 1$ to n. This x_i is also called the **analysis**.

Let $\bar{x}_i$ denote the average value of x_i obtained from the historical time series of observations of the field variable at this grid point i. Recall that this computation is no different from that of obtaining $\bar{z}_j$ described above. Clearly, this $\bar{x}_i$ for $i = 1, 2, \ldots, n$ constitutes the **prior** or the **background** information about the x_i we are seeking. Let

$$\tilde{x}_i = x_i - \bar{x}_i \tag{19.5.5}$$

denote the **anomaly** or the **analysis increment**. It can be verified that

$$E[\tilde{x}_i] = 0 \quad \text{and} \quad d_{ji} = E[\tilde{z}_j\tilde{x}_i] = E[\tilde{x}_i\tilde{z}_j] \tag{19.5.6}$$

denote the covariance of the field variance between the jth observation station and the ith grid point. Let $\mathbf{d}_{*i} = (d_{1i}, d_{2i}, \ldots, d_{mi})^{\mathrm{T}}$ denote the vector covariance between the m observation stations and the chosen grid point.

This covariance d_{ji} can be computed from the past times of observations at the station j and the grid point i. Henceforth, it is assumed that the vector $\mathbf{d}_{*i}$ of covariances is **known** for each $i = 1, 2, \ldots, n$.

Statement of problem The essence of the OI data assimilation problem is to express the analysis increment $\tilde{x}_i$ as a linear combination of the observation increments. Namely,

$$\tilde{x}_i = \sum_{j=1}^{m} W_j\tilde{z}_j = \mathbf{W}^{\mathrm{T}}\tilde{\mathbf{z}} \tag{19.5.7}$$

where $\mathbf{W} = (W_1, W_2, \ldots, W_m)^{\mathrm{T}}$ is the **unknown weight vector** to be determined.

Against this backdrop we now state the problem: Given (*a*) the vector of increments $\tilde{\mathbf{z}}$ and its symmetric positive definite covariance matrix $\mathbf{C}$, and (*b*) the background or prior information $\bar{x}_i$ at the grid point i and the vector $\mathbf{d}_{*i}$ of covariance of the observations (covariance between the observation stations and the grid point i), find the weight vector $\mathbf{W}$ that minimizes the expected value of the residual in (19.5.7). That is, find a $\mathbf{W} \in \mathbb{R}^m$ that minimizes the mean square error given by

$$f(\mathbf{W}) = E[\tilde{x}_i - \mathbf{W}^{\mathrm{T}}\tilde{\mathbf{z}}]^2. \tag{19.5.8}$$

Expanding the r.h.s. of (19.5.8) and using a series of algebraic manipulations, we get

$$\begin{aligned} f(\mathbf{W}) &= E[\tilde{x}_i^2 - 2\tilde{x}_i(\mathbf{W}^{\mathrm{T}}\tilde{\mathbf{z}}) + (\mathbf{W}^{\mathrm{T}}\tilde{\mathbf{z}})^2] \\ &= E[\tilde{x}_i^2 - 2(\tilde{x}_i\tilde{\mathbf{z}}^{\mathrm{T}})\mathbf{W} + \mathbf{W}^{\mathrm{T}}\tilde{\mathbf{z}}\tilde{\mathbf{z}}^{\mathrm{T}}\mathbf{W}] \\ &= E(\tilde{x}_i^2) - 2E[\tilde{x}_i\tilde{\mathbf{z}}^{\mathrm{T}}]\mathbf{W} + \mathbf{W}^{\mathrm{T}}E[\tilde{\mathbf{z}}\tilde{\mathbf{z}}^{\mathrm{T}}]\mathbf{W} \\ &= \mathrm{Var}(\tilde{x}_i) - 2\mathbf{d}_{*i}^{\mathrm{T}}\mathbf{W} + \mathbf{W}^{\mathrm{T}}\mathbf{C}\mathbf{W} \end{aligned} \tag{19.5.9}$$

where $\mathbf{d}_{*i}$ and $\mathbf{C}$ are defined in (19.5.6) and (19.5.3), respectively. The gradient and the Hessian of $f(\mathbf{W})$ are given by

$$\nabla f(\mathbf{W}) = -2\mathbf{d}_{*i} + 2\mathbf{C}\mathbf{W} \quad \text{and} \quad \nabla^2 f(\mathbf{W}) = 2\mathbf{C}. \tag{19.5.10}$$

Since $\mathbf{C}$ is positive definite, by setting the gradient to zero, we obtain the minimizer of $f(\mathbf{W})$ as the solution of the linear system

$$\mathbf{C}\mathbf{W} = \mathbf{d}_{*i}. \tag{19.5.11}$$

Solving this equation for $\mathbf{W}$ and using it in (19.5.7) we obtain the optimal value of the analysis increment. When the increment is added to the prior value $\bar{x}_i$, we obtain the optimal analysis.

19.5.2 Noisy observations

In practice, the observations are rarely noise free. Let us now attack the more realistic case where observations contain measurement noise.

$$\mathbf{z} = \bar{\mathbf{z}} + \tilde{\mathbf{z}} + \mathbf{v} \tag{19.5.12}$$

where the new term $\mathbf{v} \in \mathbb{R}^m$ is the observation noise. It is assumed that

$$E(\mathbf{v}) = 0 \quad \text{and} \quad \text{Cov}(\mathbf{v}) = E(\mathbf{v}\mathbf{v}^{\mathrm{T}}) = \mathbf{R} \tag{19.5.13}$$

where $\mathbf{R}$ is the known symmetric and positive definite matrix. Further, it is natural to assume that the intrinsic and physically based variation in $\tilde{\mathbf{z}}$ is uncorrelated with the observation noise $\mathbf{v}$, that is

$$E(\mathbf{v}^{\mathrm{T}}\tilde{\mathbf{z}}) = 0 = E(\tilde{\mathbf{z}}\mathbf{v}^{\mathrm{T}}). \tag{19.5.14}$$

With these assumptions and (19.5.2), we obtain

$$E(\tilde{\mathbf{z}}) = 0$$

and

$$\begin{aligned} \text{Cov}(\tilde{\mathbf{z}}) &= E\{[(\mathbf{z} - \bar{\mathbf{z}}) - \mathbf{v}][(\mathbf{z} - \bar{\mathbf{z}}) - \mathbf{v}]^{\mathrm{T}}\} \\ &= E[\tilde{\mathbf{z}}\tilde{\mathbf{z}}^{\mathrm{T}}] + E[\mathbf{v}\mathbf{v}^{\mathrm{T}}] \\ &= \mathbf{C} + \mathbf{R}. \end{aligned} \tag{19.5.15}$$

Thus, in the presence of observation noise, the effective covariance matrix of the observation increments is the sum of its intrinsic covariance matrix and the observational covariance matrix.

Turning our attention to the ith grid point, recall that we are seeking to estimate the analysis increment $\tilde{x}_i = x_i - \bar{x}_i$. Since there is no measurement involved, we need not account for observation noise. Further, it is reasonable to assume that this analysis increment $\tilde{x}_i$ is **uncorrelated** with the observation noise component v_j for all $j = 1$ to m. That is,

$$E[\tilde{x}_i v_j] = 0 \text{ for all } 1 \le j \le m. \tag{19.5.16}$$

Following (19.5.6) we have

$$d_{ji} = E[(\tilde{z}_j - v_j)\tilde{x}_i] \tag{19.5.17}$$

$$= E[\tilde{z}_j \tilde{x}_i]. \tag{19.5.18}$$

By repeating the above analysis, we now obtain the following analog of (19.5.11) that defines the **optimal weight** as (Exercise 19.5).

$$(\mathbf{C}+\mathbf{R})\mathbf{W} = \mathbf{d}_{*i}. \tag{19.5.19}$$

In view of the similarity between (19.5.11) and (19.5.19) in the following discussion we will only use the relation (19.5.11). All the comments carry over to (19.5.19) with $\mathbf{C}$ replaced by $(\mathbf{C}+\mathbf{R})$.

Several observations are in order:

(a) **Minimum value of the mean square error** $f(\mathbf{W})$ From (19.5.11), the minimizing weight vector is given by

$$\mathbf{W}_* = \mathbf{C}^{-1}\mathbf{d}_{*i}. \tag{19.5.20}$$

Substituting this value into (19.5.8) and simplifying, the minimum value of the mean square error is given by

$$f(\mathbf{W}_*) = \mathrm{Var}(\tilde{x}_i) - \mathbf{d}_{*i}^{\mathrm{T}}\,\mathbf{C}\,\mathbf{d}_{*i}. \tag{19.5.21}$$

The first term on the r.h.s is the variance of the a priori or background value of the analysis increment. The second term is the direct result of using the linear combination of the observation increments. Since $\mathbf{C}$ is positive definite, so is $\mathbf{C}^{-1}$ and hence this second term is also positive. Consequently, the net effect of using the observation increments is to reduce the value of the mean square error in the estimate of the analysis increments. This method has come to be known as the **optimal interpolation** method.

(b) **Computational cost** As we change the grid point i from 1 through n, the r.h.s. of (19.5.11) changes but the matrix $\mathbf{C}$ on the l.h.s. remains the same. Given that $\mathbf{C}$ is symmetric and positive definite, first obtain the **Cholesky decomposition**, (Chapter 9) $\mathbf{C} = \mathbf{G}\mathbf{G}^{\mathrm{T}}$ where $\mathbf{G}$ is a lower triangular matrix. This step requires $O(m^3)$ flops. We can then use this Cholesky factor $\mathbf{G}$ repeatedly to solve (19.5.11) in $O(m^3 + nm^2)$ flops.

(c) **Data selection and statistical interpolation**

In actual practice, $m \approx 10^6$ and $n \approx 10^7$ and solving (19.5.11) for these values of m and n can be a daunting task. To overcome this difficulty, it is a common practice to use only a smaller subset, of say $n_i (\ll m)$ observations that lie in a chosen **region of influence** around the grid point i. That is,

$$\tilde{x}_i - \sum_{j \in S_i} W_j \tilde{z}_j \tag{19.5.22}$$

where S_i is the pre-specified sub-domain surrounding the grid point containing n_i observations. The net effect of this **data selection** strategy is that, while we are still required to solve linear systems of the type (19.5.11) to determine the optimal weights in (19.5.22), since the size n_i of these systems are much smaller than m, it results in a considerable saving in computational time.

This modification of the optimal interpolation resulting from the data select strategy is called **statistical interpolation**. Refer to Lorenc (1981) and Daley (1991) for details.

(d) **Relation to time series analysis** A special case of the linear system of the type (19.5.11) routinely arises in modelling **stationary time series** using **autoregressive** models (Box and Jenkins (1970)) and goes by the name **Yule–Walker** equation. In this special case, the matrix **C** on the l.h.s. of (19.5.11) in addition to being symmetric and positive definite is also a **Toeplitz** matrix.

By definition a Toeplitz structure restricts the elements along each diagonal of a matrix to be the same. As an example a 4×4 Toeplitz matrix **A** is given by

$$\mathbf{A} = \begin{bmatrix} a_0 & a_1 & a_2 & a_3 \\ b_1 & a_0 & a_1 & a_2 \\ b_2 & b_1 & a_0 & a_1 \\ b_3 & b_2 & b_1 & a_0 \end{bmatrix}. \tag{19.5.23}$$

This general Toeplitz matrix is **not** symmetric in the usual sense of the definition (w.r. to the **main** or **principal diagonal** consisting of the elements (a_0, a_0, a_0, a_0) that runs from the north-west to the south-east corner of the matrix). But it possesses another form of symmetry called **persymmetry** which is the symmetry w.r. to the **main anti-diagonal** consisting of (a_3, a_1, b_1, b_3) that runs from the north-east to the south-west corner. It can be verified that every Toeplitz matrix is persymmetric but not vice versa. For example, the following matrix.

$$\mathbf{B} = \begin{bmatrix} a_0 & a_1 & a_3 \\ b_2 & b_1 & a_1 \\ b_3 & b_2 & a_0 \end{bmatrix}$$

is persymmetric and not a Toeplitz matrix.

Returning to the matrix **C** in (19.5.11), recall that C_{ij} denotes the intrinsic covariance of the field variable between the observation stations i and j. In addition, let **C** be such that C_{ij} depends on $|i - j|$ and **not** on i and j. This new requirement when combined with its symmetry properly forces **C** to be of the type

$$\mathbf{C} = \begin{bmatrix} c_0 & c_1 & c_2 & \cdots & c_{m-1} \\ c_1 & c_0 & c_1 & \cdots & c_{m-2} \\ c_2 & c_1 & c_0 & \cdots & c_{m-3} \\ \vdots & \vdots & \vdots & \vdots & \vdots \\ c_{m-1} & c_{m-2} & c_{m-3} & \cdots & c_0 \end{bmatrix} \tag{19.5.24}$$

which is **symmetric**, **Toeplitz**, and **positive definite** matrix. The linear system (19.5.11) when the matrix **C** is of the type (19.5.24) is called the **Yule–Walker** system of equations. Levinson (1947a and 1947b) has developed a special class of algorithms requiring only $\Theta(m^2)$ flops to solve this Yule–Walker system. A thorough discussion of this and other related algorithms for solving Toeplitz system is given in Golub and Van Loan (1989) [Exercise 19.6].

(e) **Stationary random function in two dimensions**

The above approach is predicated on the assumption that the field variable of interest such as the **temperature, pressure** is a random function of the space variables. Wiener and Kolmogorov were the first to independently develop the theory of least squares for stochastic processes which is the study of random functions.

Motivated by the military applications during the early years of WWII, Wiener in 1941 developed a method using some of the sophisticated mathematical techniques from the theory of Fourier transforms and integral equations he had earlier developed in his work. High level of mathematical sophistication combined with the classified nature of the report containing these results contributed to the lack of widespread dissemination it truly deserved. Recognizing the fundamental nature of this work, Norman Levinson (1947a and 1947b) re-derived Wiener's results in discrete time using simple heuristic arguments. Levinson's papers were included as Appendices B and C in Wiener's book that was later published in 1949. Since then **Wiener filtering** technique, as it has come to be called, is routinely used and is a part of the folklore in the vast arena of digital signal processing.

When Wiener concentrated on the spectral or frequency domain approach using the Fourier transform theory of which he was a master, Kolmogorov (1941) using the Hilbert space formulation studied the same class of problems in the arena of discrete time domain.

We close this observation by citing some of the many applications of the Wiener filtering theory.

Meteorology In the mid 1950s, L. Gandin championed the application of Wiener's approach to the problem of **objective analysis** of meteorological field variables. Gandin called the resulting approach an optimal interpolation and it soon became one of the standard approaches in meteorology. A succinct summary of his many faceted contributions is contained in Gandin (1963). As a member of the World Meteorological Organization's Commission in 1956–7, Arnt Eliassen made a study of objective analysis in conjunction with the design of observational networks. In this study, he introduced the idea of using the covariances of data at the stations and came tantalizingly close to introducing OI to meteorology. As he said in his oral history interview, "The method was developed further by several other people, in particular Lev Gandin" (Eliassen 1990, p. 23).

Mining and Geology In the early 1950s, D. G. Krige (1951) in South Africa applied Wiener's ideas to solve the spatial estimation problem of interest in mining. The noted French Geostatistician G. Matheron (1963) expanded on Krige's work and coined the term **Kriging** to denote the class of minimum variance estimation methods. For a detailed account of Kriging and many of its variations, refer to the paper by Journel (1977) and a recent book by Chiles and Delfiner (1999).

Forestry Early applications of Wiener's method to statistical estimation problems of interest in forestry in Sweden are reported in Matérn (1960).

Refer to Yaglom (1962) for a general introduction to the theory of random functions.

Exercises

19.1 (a) Solve the polynomial approximation described in Example 19.1.1 using d = three grid lengths.

(b) Draw the contour plots of the temperature at the observation locations.

(c) Draw the contour plots of the retrieved temperature field on the computational grid. Compare these two contours and comment.

19.2 (a) Solve the linear system (19.1.10) with

$$\mathbf{B} = \begin{bmatrix} 1 & 0 & 0 \\ 0 & 1 & 0 \\ 0 & 0 & 1 \end{bmatrix} \mathbf{d} = \begin{pmatrix} -2 \\ 0 \\ 4 \end{pmatrix} \quad \text{and} \quad \mathbf{a} = \begin{pmatrix} 1 \\ -1 \\ 2 \end{pmatrix}$$

explicitly for the weak solution $x(\alpha)$.

(b) Compute $\lim_{\alpha \longrightarrow \infty} x(\alpha)$ and verify that it is equal to the strong solution given in (19.1.9).

19.3 Perform the Cressman-type iterative scheme on the sample problem described in Example 19.3.1.

19.4 Implement the Barnes' scheme on the model problem described in Example 19.1.1. Plot the contours of the retrieved field resulting from this exercise.

19.5 By replacing $\mathbf{C}$ by $(\mathbf{C} + \mathbf{R})$ in (19.5.9), verify that (19.5.19) is the minimizer for this modified function. Compute the minimum value of the corresponding mean square error.

19.6 Consider a stationary and homogeneous random field variable in two dimensions. Explore if there exists an arrangement of m observation stations such that the intrinsic covariance of the field variable in question is such that $C_{ij} = C_{|i-j|}$.

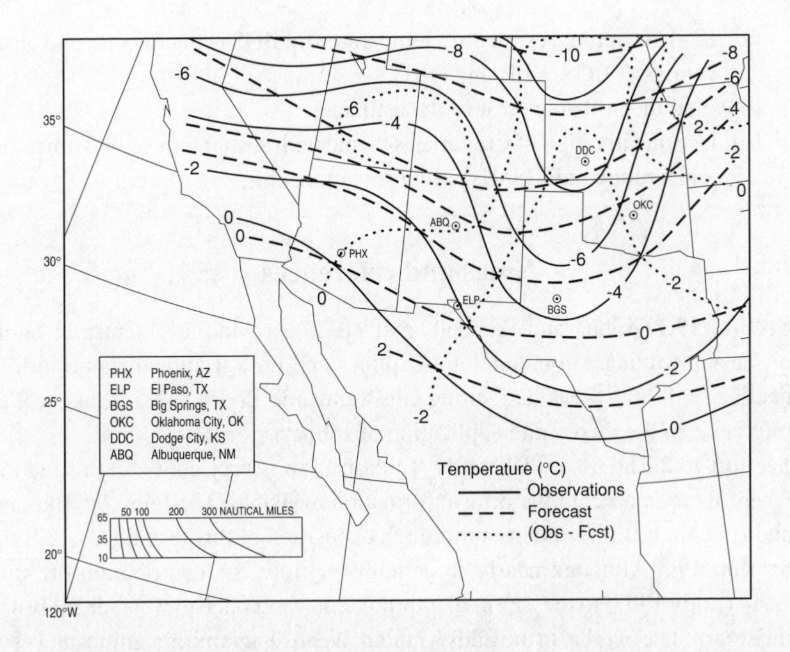

19.7 On the accompanying synoptic map over the southwest and southcentral USA, you notice three sets of contours for tropospheric temperature (lower troposphere). One set is the forecast (background), one set is the observations, and the other set is the difference between forecast and observation. Several observing stations are shown:

City/State	Identifier
Albuquerque, NM	ABQ
El Paso, TX	ELP
Oklahoma City, OK	OKC
Dodge City, KS	DDC
Phoenix, AZ	PHX
Big Springs, TX	BGS
Fort Worth, TX	FTW

Estimate (visually) the observation, forecast, and difference (observation minus forecast) at each of these stations.

Assume the following error characteristics:

Forecast error variance: $(1.0\,°C)^2$ $(=\sigma_{\mathrm{F}}^2)$

Observation error variance: $(0.5\,°C)^2$ $(=\sigma_{\mathrm{o}}^2)$

Spatial Error Covariance of Forecast: $\sigma_{\mathrm{F}}^2 \exp\left\{-0.5\left(\frac{d_{ij}}{L}\right)^2\right\}$

where d_{ij} is distance between stations and $L = 300$ nm (naut. mile).

Using only these stations (BGS, ELP, and OKC):

(a) Find the optimal estimate of temperature at BGS using forecast and observations at BGS, ELP, and OKC.
(b) Find the probable error of the estimate.
(c) Use one iteration of Cressman scheme with scan radius of 400 nm to find temperature at BGS. Use all observation sites.

Notes and references

Section 19.1 Wahba and Wendelberger (1980) provide an interesting family of ideas involving a combination of splines and the variational framework for dealing with both weak and strong constraints. de Boor (1978) is an excellent reference on the theory and application of splines.

Section 19.2 The use of Tikhonov regularization is very common in situations where there is no credible prior information available. The book by Tikhonov and Arsenin (1977) contains a thorough exposition of this technique.

Section 19.3 Although nearly forgotten over time, the contribution of Amos Eddy (Eddy 1967) to the "optimal interpolation" approach to weather analysis is important. The overlap in the Eddy/Gandin "Venn Diagram" is significant, yet the approaches are different, not unlike the overlap in the Wiener/Kolmogorov and Kalman works discussed earlier. A valuable exercise results in comparing Eddy (1967) and Gandin (1963). Ian Rutherford was among the first meteorologists to exhibit the value of OI in operations (Rutherford (1972)). Seaman (1977) provides an especially valuable introduction to OI.

Section 19.4 Franke (1988) and Seaman (1977) contain a good comparison of the results of applying successive correction and statistical interpolation (refer to section 15.4) methods to real data. A recent tutorial by Trapp and Doswell (2000) provides a comprehensive evaluation of many of the classical methods for objective analysis of radar data.

Section 19.5 For a thorough examination of OI as used in meteorology, the reader is referred to Lorenc (1981). This paper describes the methodology required to extend Gandin's approach to a multi-variate approach that includes observations of wind along with geopotential and temperature – it is a 3-d multivariate OI ("Statistical Interpolation") where quality control is an added valuable element of the analysis. A more general approach to statistical interpolation that includes the 1981 paper as a subset is found in Lorenc (1986), (1988), (1992), (1995), (1997) and (1998). Also refer to Bergman (1979) for an interesting application of OI. Courtier et al. (1998) contains a detailed account of the 3DVAR implementation. The paper entitled "The Anatomy of Inverse Problems" by Scales and Snieder (2000) provides a succinct summary of challenges associated with this problem.

An elaboration on the various forms of the balance conditions used in meteorology is found in Holton (1972). The use of these constraints in data assimilation was first formulated by Sasaki (1958). This paper followed in the pattern of Gauss's original work, i.e., there was no account for the error structure in the data/observations. The work by Gandin (1963) took account of the error structure but did not enforce the constraints in the "strong constraint" sense. Gandin's original work used climatology as background. A succinct review of the various formulations including strong and weak constraints as well as the Tikhonov regularization is found in LeDimet and Talagrand (1986) (linked to "adjoint method").

20

3DVAR: a Bayesian formulation

This chapter develops the solution to the retrieval problem (stated in Chapter 18) using the Bayesian framework and draws heavily from Part III (especially Chapters 16 and 17). The method based on this framework has also come to be known as the three dimensional variational method – 3DVAR for short. This class of methods is becoming the industry standard for use at the operational weather prediction centers around the world (Parrish and Derber (1992), Lorenc (1995), Gauthier et al. (1996), Cohn et al. (1998), Rabier et al. (1998). Andersson et al. (1998)). This global method does not require any form of **data selection** strategy which local methods depend on.

From the algorithmic perspective, there are two ways of approach for this problem – use of the **model space** ($\mathbb{R}^n$) or use of the **observation space** ($\mathbb{R}^m$) (refer to Chapter 17). While these two approaches are logically equivalent, there is a computational advantage to model space when $n < m$, whereas the advantage goes to observation space when $n > m$.

In Section 20.1, we derive the Bayesian framework for the problem. The straightforward solution for the special case when the forward operator is linear is covered in Section 20.2. The following Section 20.3 brings out the duality between the model space and the observation space formulations using the notion of preconditioning. The general case of nonlinear method is treated in the next two sections with the second-order method in Section 20.4 and the first-order method in Section 20.5. The first-order method for the nonlinear case closely resembles the linear formulation.

20.1 The Bayesian formulation

We begin by describing the basic components that underlie this formulation.

(A) **The true state**

Let $\mathbf{x} \in \mathbb{R}^n$ denote the **unknown** true state of a field variable (such as pressure, temperature, etc.) over the computational grid, and $\mathbb{R}^n$ is called the **model space** or the **grid space**. Our goal is to estimate this unknown state.

(B) **The prior or background information**

While $\mathbf{x}$ is not known, we may know quite a bit about it in the form of a **prior** or **background** information. This prior knowledge may be derived from the long term climatological data or from a previous forecast, etc. Let $\mathbf{x}_b \in \mathbb{R}^n$ denote this background field. Let

$$\tilde{\mathbf{x}}_b = \mathbf{x} - \mathbf{x}_b \tag{20.1.1}$$

denote the background error. It is assumed that

$$E(\tilde{\mathbf{x}}_b) = 0 \tag{20.1.2}$$

and that the (spatial) covariance structure of $\mathbf{x}_b$ is given by

$$\mathbf{B} = E[\tilde{\mathbf{x}}_b \tilde{\mathbf{x}}_b^{\mathrm{T}}] \tag{20.1.3}$$

where $\mathbf{B} \in \mathbb{R}^{n\times n}$ is a symmetric and positive definite matrix.

Assumption 1 The unknown (true) state has a multivariate normal distribution with known mean $\mathbf{x}_b$ and covariance matrix $\mathbf{B}$ which is positive definite. That is, $\mathbf{x} \sim N(\mathbf{x}_b, \mathbf{B})$ or

$$p(\mathbf{x}) = \frac{1}{(2\pi)^{\frac{n}{2}} |\mathbf{B}|^{\frac{1}{2}}} \exp[-J_b(\mathbf{x})] \tag{20.1.4}$$

where $|\mathbf{B}|$ is the determinant of $\mathbf{B}$ and

$$J_b(\mathbf{x}) = \frac{1}{2}(\mathbf{x} - \mathbf{x}_b)^{\mathrm{T}} \mathbf{B}^{-1} (\mathbf{x} - \mathbf{x}_b). \tag{20.1.5}$$

(C) **Observation**

Let $\mathbf{z} \in \mathbb{R}^m$ be the observation about the unknown and $\mathbb{R}^m$ is called the **observation space**. Let

$$\mathbf{z} = h(\mathbf{x}) + \mathbf{v} \tag{20.1.6}$$

where $h : \mathbb{R}^n \longrightarrow \mathbb{R}^m$ is denoted by $h(\mathbf{x}) = (h_1(\mathbf{x}), h_2(\mathbf{x}), \ldots, h_m(\mathbf{x}))^{\mathrm{T}}$ is a vector-valued function of the vector $\mathbf{x}$ and $\mathbf{v} \in \mathbb{R}^m$ is the unobservable measurement noise vector. It is assumed that

$$E(\mathbf{v}) = 0 \quad \text{and} \quad \mathbf{R} = \mathrm{Cov}(\mathbf{v}) = E(\mathbf{v}\mathbf{v}^{\mathrm{T}}) \tag{20.1.7}$$

where $\mathbf{R} \in \mathbb{R}^{m\times n}$ is a known symmetric and positive definite matrix. Further, this observation noise $\mathbf{v}$ is not correlated to $\mathbf{x}$, the true state, nor the background state $\mathbf{x}_b$. Hence,

$$E[\mathbf{v}\tilde{\mathbf{x}}_b^{\mathrm{T}}] = E[\tilde{\mathbf{x}}_b \mathbf{v}^{\mathrm{T}}] = 0. \tag{20.1.8}$$

Assumption 2 The observation noise $\mathbf{v}$ has a multivariate normal distribution, $\mathbf{v} \sim N(0, \mathbf{R})$, and $\mathbf{R}$ is positive definite. Hence, from (20.1.6) it follows that $\mathbf{z} \sim N(h(\mathbf{x}), \mathbf{R})$ or

$$p(\mathbf{z}|\mathbf{x}) = \frac{1}{(2\pi)^{\frac{m}{2}} |\mathbf{R}|^{\frac{1}{2}}} \exp[-J_o(\mathbf{x})] \tag{20.1.9}$$

and

$$J_o(\mathbf{x}) = \frac{1}{2}(\mathbf{z} - h(\mathbf{x}))^{\mathrm{T}} \mathbf{R}^{-1} (\mathbf{z} - \mathbf{H}(\mathbf{x})). \tag{20.1.10}$$

(D) **Bayes' rule**

Our goal is to compute the posterior distribution $p(\mathbf{x}|\mathbf{z})$. Referring to Appendix F and Chapter 16, this is done using the well-known Bayes' rule:

$$p(\mathbf{x}|\mathbf{z}) = \frac{p(\mathbf{x}, \mathbf{z})}{p(\mathbf{z})} = \frac{p(\mathbf{z}|\mathbf{x})p(\mathbf{x})}{p(\mathbf{z})}. \tag{20.1.11}$$

The denominator term $p(\mathbf{z})$ is given by

$$p(\mathbf{z}) = \int_{\mathbb{R}^n} p(\mathbf{z}|\mathbf{x})p(\mathbf{x})\mathrm{d}\mathbf{x} \tag{20.1.12}$$

where $\mathrm{d}\mathbf{x} = \mathrm{d}x_1 \mathrm{d}x_2 \ldots \mathrm{d}x_n$ is the n-dimensional infinitesimal volume, is independent of $\mathbf{x}$ and plays the role of a normalizing constant. Denoting $C = [p(\mathbf{z})]^{-1}$, we rewrite (20.1.11) as

$$p(\mathbf{x}|\mathbf{z}) = Cp(\mathbf{z}|\mathbf{x})p(\mathbf{x}). \tag{20.1.13}$$

Substituting (20.1.4) and (20.1.9) into the above, the latter becomes

$$p(\mathbf{z}|\mathbf{x}) = \frac{C}{(2\pi)^{\frac{m}{2}+\frac{n}{2}} |\mathbf{R}|^{\frac{1}{2}} |\mathbf{B}|^{\frac{1}{2}}} \exp[-(J_o(\mathbf{x}) + J_b(\mathbf{x}))]. \tag{20.1.14}$$

The maximum posterior estimate (MAP) (Chapter 16) is obtained by finding the value of $\mathbf{x}$ that maximizes the r.h.s of (20.1.14). This can also be accomplished by taking the natural logarithms of both sides and maximizing the resulting function of $\mathbf{x}$. since

$$\ln p(\mathbf{x}|\mathbf{z}) = \ln C_1 - [J_o(\mathbf{x}) + J_b(\mathbf{x})] \tag{20.1.15}$$

where C_1 denotes the constant term in the r.h.s.of (20.1.14). Clearly, (20.1.15) is a maximum exactly when

$$\begin{aligned} J(\mathbf{x}) &= J_o(\mathbf{x}) + J_b(\mathbf{x}) \\ &= \tfrac{1}{2}(\mathbf{z} - h(\mathbf{x}))^{\mathrm{T}} \mathbf{R}^{-1} (\mathbf{z} - h(\mathbf{x})) + \tfrac{1}{2}(\mathbf{x} - \mathbf{x}_b)^{\mathrm{T}} \mathbf{B}^{-1} (\mathbf{x} - \mathbf{x}_b) \end{aligned} \tag{20.1.16}$$

is a **minimum.**

Several comments are in order.

(a) **Relation of least squares**

Comparing the above formulation with those given in Chapters 16, 17, and 18, it follows that when the observations are linear functions of the state and if all the distributions involved are normal, then Bayes MAP estimate is equivalent to the least squares estimate.

(b) **Sampling and interpolation errors**

The field variables that are being observed (such as temperature, pressure, wind, etc.) are continuous (random) functions of the space variables. This field variable, in general, is a composite of signals of various scales. Using the tools from the Fourier transform theory (Appendix G), we can in fact express the field as a linear combination of signal components of various wavelengths. It is well known that the total energy in the field variable is the sum of the energy associated with each component signal.

In meteorological practice, we, however, **sample** this continuous function either using the computational grid or using the network of observational stations and choose to represent the continuous function by a vector with finite number of components such as $\mathbf{x} \in \mathbb{R}^n$ or $\mathbf{z} \in \mathbb{R}^m$. This sampling unless done correctly will always result in an error that goes with the name **representation error**. A well-known result in **sampling theory** may be stated as follows: **unless the sampling interval is less than half the wavelength of the signal with the smallest wavelength that is present in the field, we cannot reconstruct the field from its samples**. Since grid spacings and/or spacings between the observation stations are fixed, when measuring the field variables that exhibit large variations (such as wind, for example) will always suffer from this error, denoted by $\mathbf{e}_s \in \mathbb{R}^m$.

There is also another source of error. Recall that $h(\mathbf{x}) = h^{\circ}(h^{\mathrm{I}}(\mathbf{x}))$. Even granting that the physical and/or empirical laws represented by $h^{\circ}(.)$ are perfect, the interpolation function $h^{\mathrm{I}}(\mathbf{x})$ is always associated with an error. Let $\mathbf{e}_{\mathrm{I}}$ denote the interpolation error. It is assumed that these two errors are additive and $\mathbf{f} = \mathbf{e}_s + \mathbf{e}_{\mathrm{I}}$ is known as the **representative** error.

While these errors are difficult to quantify, it is a standard practice to treat it as random. Accordingly, it is assumed that $\mathbf{f} \sim N(0, \mathbf{F})$ and that $\mathbf{f}$ is **not** correlated to $\mathbf{x}$, $\mathbf{x}_b$, and $\mathbf{v}$. If we explicitly include this error, then (20.1.6) becomes

$$\mathbf{z} = h(\mathbf{x}) + \mathbf{v} + \mathbf{f} \tag{20.1.17}$$

and $\mathbf{z} \sim N(h(\mathbf{x}), \mathbf{R} + \mathbf{F})$. Thus, the effect of including $\mathbf{f}$ is to change the matrix $\mathbf{R}$ into $(\mathbf{R} + \mathbf{F})$. In the light of this argument, to simplify the notation, we tacitly assume that $\mathbf{R}$ contains information about $\mathbf{F}$.

(D) **Statement of the problem**

Returning to the minimization of $J(\mathbf{x})$ in (20.1.16), the minimizer of $J(\mathbf{x})$ is obtained by solving the following nonlinear algebraic equation:

$$0 = \nabla J(\mathbf{x}) = \mathbf{B}^{-1}(\mathbf{x} - \mathbf{x}_b) - \mathbf{D}_h^{\mathrm{T}}(\mathbf{x})\mathbf{R}^{-1}[\mathbf{z} - h(\mathbf{x})] \tag{20.1.18}$$

where $\mathbf{D}_h(\mathbf{x})$ is the $m \times n$ Jacobian matrix of $h(\mathbf{x})$ (Appendix C).

There is a well-established body of literature devoted to solving nonlinear algebraic equations – Ortega and Rhienboldt (1970) and Dennis and Schnabel (1996). These methods are based on the classical Newton's iterative methods (Chapter 11) and in general are computationally demanding especially when the dimensions are large.

The rest of the chapter is devoted to solving this minimization problem. In the interest of pedagogy, we first discuss the simple case when $h(\mathbf{x})$ is a linear function of $\mathbf{x}$ before tackling the nonlinear case.

20.2 The linear case

In this section, we analyze the minimization of $J(\mathbf{x})$ in (20.1.16) when $h(\mathbf{x}) = \mathbf{Hx}$, a nonlinear function of $\mathbf{x}$. Substituting this, we get

$$J(\mathbf{x}) = \frac{1}{2}(\mathbf{x} - \mathbf{x}_b)^{\mathrm{T}}\mathbf{B}^{-1}(\mathbf{x} - \mathbf{x}_b) + \frac{1}{2}(\mathbf{z} - \mathbf{Hx})^{\mathrm{T}}\mathbf{R}^{-1}(\mathbf{z} - \mathbf{Hx}). \tag{20.2.1}$$

This combined quadratic form has been analyzed from various points of view each of which has its own implication on the computational efforts needed in finding its minimum. In this and in the following sections, we systematically categorize these ideas.

(A) **The basic form of solution in model space**

The gradient and the Hessian of $J(\mathbf{x})$ in (20.2.1) are given by

$$\begin{aligned} \nabla J(\mathbf{x}) &= \mathbf{B}^{-1}(\mathbf{x} - \mathbf{x}_b) + (\mathbf{H}^{\mathrm{T}}\mathbf{R}^{-1}\mathbf{H})\mathbf{x} - \mathbf{H}^{\mathrm{T}}\mathbf{R}^{-1}\mathbf{z} \\ &= (\mathbf{B}^{-1} + \mathbf{H}^{\mathrm{T}}\mathbf{R}^{-1}\mathbf{H})\mathbf{x} - (\mathbf{B}^{-1}\mathbf{x}_b + \mathbf{H}^{\mathrm{T}}\mathbf{R}^{-1}\mathbf{z}) \end{aligned} \tag{20.2.2}$$

and

$$\nabla^2 J(\mathbf{x}) = (\mathbf{B}^{-1} + \mathbf{H}^{\mathrm{T}}\mathbf{R}^{-1}\mathbf{H}). \tag{20.2.3}$$

Since $\mathbf{B}$ and $\mathbf{R}$ are assumed to be positive definite, so are $\mathbf{B}^{-1}$ and $\mathbf{R}^{-1}$ and hence the above Hessian is positive definite irrespective of whether $\mathbf{H}$ is of full rank or not. The minimizer is obtained by setting the gradient to zero and is given by the solution of the linear system

$$(\mathbf{B}^{-1} + \mathbf{H}^{\mathrm{T}}\mathbf{R}^{-1}\mathbf{H})\mathbf{x} = (\mathbf{B}^{-1}\mathbf{x}_b + \mathbf{H}^{\mathrm{T}}\mathbf{R}^{-1}\mathbf{z}). \tag{20.2.4}$$

The matrix on the l.h.s is an $n \times n$ symmetric and positive definite matrix, and this has come to be known as the **model space approach**. We can readily apply the conjugate gradient algorithms (Chapter 11) to solve the above linear system.

(B) **Incremental form in model space**

It is useful to recast the above problem in another equivalent form called the **incremental form**. To arrive at this form, add and subtract $\mathbf{Hx}_b$ to the second term on the r.h.s.of (20.2.1). Denoting $\delta\mathbf{x} = \mathbf{x} - \mathbf{x}_b$, $J(\mathbf{x})$ becomes

$$J(\delta\mathbf{x}) = \frac{1}{2}(\delta\mathbf{x})^{\mathrm{T}}\mathbf{B}^{-1}(\delta\mathbf{x}) + \frac{1}{2}(\mathbf{d} - \mathbf{H}\delta\mathbf{x})^{\mathrm{T}}\mathbf{R}^{-1}(\mathbf{d} - \mathbf{H}\delta\mathbf{x}) \tag{20.2.5}$$

where $\mathbf{d} = \mathbf{z} - \mathbf{Hx}_b$. Setting the gradient

$$\nabla J(\delta\mathbf{x}) = \mathbf{B}^{-1}\delta\mathbf{x} + (\mathbf{H}^{\mathrm{T}}\mathbf{R}^{-1}\mathbf{H})\delta\mathbf{x} - \mathbf{H}^{\mathrm{T}}\mathbf{R}^{-1}\mathbf{d} \tag{20.2.6}$$

to zero, we obtain the minimizing increment as the solution of the linear system

$$(\mathbf{B}^{-1} + \mathbf{H}^{\mathrm{T}}\mathbf{R}^{-1}\mathbf{H})\delta\mathbf{x} = \mathbf{H}^{\mathrm{T}}\mathbf{R}^{-1}[\mathbf{z} - \mathbf{Hx}_b]. \tag{20.2.7}$$

Again, this form requires the solution of this $n \times n$ system in the model space.

(C) **Basic form or solution in observation space**

By invoking the **Sherman–Morrison–Woodbury** inversion formula (Appendix B), it can be verified that

$$(\mathbf{B}^{-1} + \mathbf{H}^{\mathrm{T}}\mathbf{R}^{-1}\mathbf{H})^{-1} = \mathbf{B} - \mathbf{BH}^{\mathrm{T}}[\mathbf{R} + \mathbf{HBH}^{\mathrm{T}}]^{-1}\mathbf{HB}. \tag{20.2.8}$$

Then, from (20.2.7) and (20.2.8) we have

$$\begin{aligned}
\delta\mathbf{x} &= (\mathbf{B}^{-1} + \mathbf{H}^{\mathrm{T}}\mathbf{R}^{-1}\mathbf{H})^{-1}\mathbf{H}^{\mathrm{T}}\mathbf{R}^{-1}(\mathbf{z} - \mathbf{Hx}_b) \\
&= \mathbf{BH}^{\mathrm{T}}\mathbf{R}^{-1}(\mathbf{z} - \mathbf{Hx}_b) - \mathbf{BH}^{\mathrm{T}}[\mathbf{R} + \mathbf{HBH}^{\mathrm{T}}]^{-1}(\mathbf{HBH}^{\mathrm{T}})\mathbf{R}^{-1}(\mathbf{z} - \mathbf{Hx}_b) \\
&= \mathbf{BH}^{\mathrm{T}}[\mathbf{I} - (\mathbf{R} + \mathbf{HBH}^{\mathrm{T}})^{-1}(\mathbf{HBH}^{\mathrm{T}})]\mathbf{R}^{-1}(\mathbf{z} - \mathbf{Hx}_b) \\
&= \mathbf{BH}^{\mathrm{T}}[(\mathbf{R} + \mathbf{HBH}^{\mathrm{T}})^{-1}[(\mathbf{R} + \mathbf{HBH}^{\mathrm{T}}) - \mathbf{HBH}^{\mathrm{T}}]]\mathbf{R}^{-1}(\mathbf{z} - \mathbf{Hx}_b) \\
&= \mathbf{BH}^{\mathrm{T}}(\mathbf{R} + \mathbf{HBH}^{\mathrm{T}})^{-1}(\mathbf{z} - \mathbf{Hx}_b).
\end{aligned} \tag{20.2.9}$$

It is convenient to compute this solution in two steps: first solve

$$(\mathbf{R} + \mathbf{HBH}^{\mathrm{T}})\mathbf{w} = (\mathbf{z} - \mathbf{Hx}_b) \tag{20.2.10}$$

for $\mathbf{w}$ and then compute

$$\delta\mathbf{x} = \mathbf{BH}^{\mathrm{T}}\mathbf{w}. \tag{20.2.11}$$

The matrix on the l.h.s of (20.2.10) is an $m \times m$ symmetric and positive definite and this formulation is known as the observation space approach. Again, notice that we can solve (20.2.10) using the classical conjugate gradient method.

Several comments are in order.

(a) **Minimum Variance Solution** In deriving the basic form of the solution (20.2.10)–(20.2.11) in the observation space, we used an indirect approach by invoking the **matrix inversion** formula. These equations can be derived rather directly by reformulating it as the linear **minimum variance estimation problem** as was done in Cohn et al. (1998) as well as in Chapter 17.

Refer to Table 17.1.1 for an exposé of the duality between the model space and the observation space formulation. At the NASA Data Assimilation Office, Goddard Space Flight Center, Pfaendtner et al. (1995) developed the so-called Physical Space Statistical Analysis System (PSAS) which is essentially the basic observation space approach given in (20.2.10)–(20.2.11). Cohn et al. (1998) provide an excellent summary of this minimum variance method and its relation to other approaches based on the model space formulation.

(b) **Computational aspects of the solution**

To understand the computational aspects of solving (20.2.7) and (20.2.10), we first need to establish the properties of the matrices on the l.h.s of these two equations. To this end, first recall that

$$\left.\begin{array}{l} \text{Rank}(\mathbf{B}) = n,\ \text{Rank}(\mathbf{R}) = m \\ \text{and assume that} \\ \text{Rank}(\mathbf{H}) = \text{Rank}(\mathbf{H}^{\mathrm{T}}) = \min(m, n) \end{array}\right\} \tag{20.2.12}$$

that is, **H** is of full rank.

It is convenient to consider two cases.

Case (A) **Overdetermined system**: $m > n$ In this case,

$$\begin{aligned} \text{Rank}(\mathbf{H}^{\mathrm{T}}\mathbf{R}^{-1}\mathbf{H}) &= \min\{\text{Rank}(\mathbf{R}), \text{Rank}(\mathbf{H})\} \\ &= n \end{aligned} \tag{20.2.13}$$

and the $n \times n$ symmetric matrix $\mathbf{A}_{\mathrm{M}} = (\mathbf{B}^{-1} + \mathbf{H}^{\mathrm{T}}\mathbf{R}^{-1}\mathbf{H})$ is of full rank n (hence nonsingular) and positive definite. Let

$$\lambda_1 \geq \lambda_2 \geq \lambda_3 \geq \cdots \geq \lambda_n > 0 \tag{20.2.14}$$

be the n (positive) eigenvalues of $\mathbf{A}_{\mathrm{M}}$. Then while its **spectral condition number** (Appendix B)

$$\kappa_2(\mathbf{A}_{\mathrm{M}}) = \frac{\lambda_1}{\lambda_n} \tag{20.2.15}$$

is guaranteed to be finite, since the smallest eigenvalue λ_n can be very small, this condition number could be very large.

Similarly, it can be verified that the Rank$(\mathbf{HBH}^{\mathrm{T}}) = n$ and the $m \times m$ symmetric matrix $\mathbf{A}_0 = (\mathbf{R} + \mathbf{HBH}^{\mathrm{T}})$ is of rank $n < m$. That is, $\mathbf{A}_0$ is **rank deficient** (hence singular) and positive semi-definite. If

$$\mu_1 \geq \mu_2 \geq \cdots \geq \mu_m = 0 \tag{20.2.16}$$

are the m eigenvalues of $\mathbf{A}_0$, then its spectral condition number

$$\kappa_2(\mathbf{A}_0) = \frac{\mu_1}{\mu_n} = \infty. \tag{20.2.17}$$

Consequently, when the estimation problem is overdetermined, the model space formulation leads to a well-posed problem and the observation space formulation gives rise to an ill-posed problem.

Case (B) Underdetermined case: $m < n$

By repeating the above argument, it can be verified that $\text{Rank}(\mathbf{H}^{\mathrm{T}}\mathbf{R}^{-1}\mathbf{H}) = m$ and the $n \times n$ symmetric matrix $\mathbf{A}_M$ is rank deficient (hence singular) and positive semi-definite and $\kappa_2(\mathbf{A}_M) = \infty$.

Similarly, $\text{Rank}(\mathbf{HBH}^{\mathrm{T}}) = m$ and the $m \times m$ symmetric matrix $\mathbf{A}_0$ is of full rank (hence nonsingular) and positive definite and $\kappa_2(\mathbf{A}_0)$ is finite.

In this case we obtain the dual result namely, the observation space formulation is well-posed and the model space formulation is ill-posed.

The conjugate gradient algorithm (Chapter 11) is the method of choice for solving linear systems with symmetric positive definite matrices. Because of round-off errors resulting from the finite precision arithmetic, its convergence properties critically depend on the condition number of the matrices $(\mathbf{B}^{-1} + \mathbf{H}^{\mathrm{T}}\mathbf{R}^{-1}\mathbf{H})$ when $m > n$ and that of $(\mathbf{R} + \mathbf{HBH}^{\mathrm{T}})$ when $m < n$. In these cases as observed above, while these condition numbers are guaranteed to be finite, they could take large values leading to unduly slow convergence, especially in large-scale problems of interest in meteorology.

The above analysis naturally leads us to looking for ways to accelerate the convergence of conjugate gradient algorithms by **taming** the condition number of the matrices involved. This is often accomplished by using a strategy called preconditioning (Chapter 11).

20.3 Pre-conditioning and duality

In this section we develop two preconditioning strategies – one for the model space and the other for the observation space.

(A) **Preconditioned incremental form in model space**

Since the matrix $\mathbf{B}$ is symmetric and positive definite, we can decompose $\mathbf{B}$ (Appendix B) as the product of its square root, namely,

$$\mathbf{B} = \mathbf{B}^{\frac{1}{2}}\mathbf{B}^{\frac{1}{2}}. \tag{20.3.1}$$

Define a new variable $\mathbf{u}$ using the linear transformation

$$\delta\mathbf{x} = \mathbf{B}^{\frac{1}{2}}\mathbf{u}. \tag{20.3.2}$$

Substituting (20.3.2) into (20.2.5), we get

$$J(\mathbf{u}) = \frac{1}{2}\mathbf{u}^{\mathrm{T}}\mathbf{u} + \frac{1}{2}(\mathbf{d} - \mathbf{HB}^{\frac{1}{2}}\mathbf{u})\mathbf{R}^{-1}(\mathbf{d} - \mathbf{HB}^{\frac{1}{2}}\mathbf{u}). \tag{20.3.3}$$

Hence,

$$\nabla J(\mathbf{u}) = [\mathbf{I} + \mathbf{B}^{\frac{1}{2}}(\mathbf{H}^{\mathrm{T}}\mathbf{R}^{-1}\mathbf{H})\mathbf{B}^{\frac{1}{2}}]\mathbf{u} - \mathbf{B}^{\frac{1}{2}}\mathbf{H}^{\mathrm{T}}\mathbf{R}^{-1}\mathbf{d} \tag{20.3.4}$$

and

$$\nabla^2 J(\mathbf{u}) = [\mathbf{I} + \mathbf{B}^{\frac{1}{2}}(\mathbf{H}^{\mathrm{T}}\mathbf{R}^{-1}\mathbf{H})\mathbf{B}^{\frac{1}{2}}]. \tag{20.3.5}$$

The minimizer is obtained by solving the $n \times n$ linear system

$$(\mathbf{I} + \mathbf{A}_{\mathrm{M}})\mathbf{u} = \mathbf{B}^{\frac{1}{2}}\mathbf{H}^{\mathrm{T}}\mathbf{R}^{-1}\mathbf{d} \tag{20.3.6}$$

where

$$\mathbf{A}_{\mathrm{M}} = \mathbf{B}^{\frac{1}{2}}(\mathbf{H}^{\mathrm{T}}\mathbf{R}^{-1}\mathbf{H})\mathbf{B}^{\frac{1}{2}}. \tag{20.3.7}$$

(B) **Preconditioned incremental form in observation space**

It can be verified that the solution of the linear system (20.2.10) is the minimizer of the associated quadratic form given by

$$f(\mathbf{w}) = \frac{1}{2}\mathbf{w}^{\mathrm{T}}[\mathbf{R} + \mathbf{H}\mathbf{B}\mathbf{H}^{\mathrm{T}}]\mathbf{w} - \mathbf{w}^{\mathrm{T}}\mathbf{d}. \tag{20.3.8}$$

Just as we did with the $\mathbf{B}$ matrix, since $\mathbf{R}$ is also symmetric and positive definite, we can decompose it as

$$\mathbf{R} = \mathbf{R}^{\frac{1}{2}}\mathbf{R}^{\frac{1}{2}} \tag{20.3.9}$$

and define a new variable $\mathbf{y}$ as

$$\mathbf{w} = \mathbf{R}^{-\frac{1}{2}}\mathbf{y}. \tag{20.3.10}$$

Substituting this into (20.3.8), the latter becomes

$$f(\mathbf{y}) = \frac{1}{2}\mathbf{y}^{\mathrm{T}}[\mathbf{I} + \mathbf{R}^{-\frac{1}{2}}(\mathbf{H}\mathbf{B}\mathbf{H}^{\mathrm{T}})\mathbf{R}^{-\frac{1}{2}}]\mathbf{y} - \mathbf{y}^{\mathrm{T}}\mathbf{R}^{-\frac{1}{2}}\mathbf{d}. \tag{20.3.11}$$

Hence

$$\nabla f(\mathbf{y}) = [\mathbf{I} + \mathbf{R}^{-\frac{1}{2}}(\mathbf{H}\mathbf{B}\mathbf{H}^{\mathrm{T}})\mathbf{R}^{-\frac{1}{2}}]\mathbf{y} - \mathbf{R}^{-\frac{1}{2}}\mathbf{d} \tag{20.3.12}$$

and

$$\nabla^2 f(\mathbf{y}) = [\mathbf{I} + \mathbf{R}^{-\frac{1}{2}}(\mathbf{H}\mathbf{B}\mathbf{H}^{\mathrm{T}})\mathbf{R}^{-\frac{1}{2}}]. \tag{20.3.13}$$

The minimizer is obtained by solving the $m \times m$ linear system

$$(\mathbf{I} + \mathbf{A}_0)\mathbf{y} = \mathbf{R}^{-\frac{1}{2}}\mathbf{d} \tag{20.3.14}$$

where

$$\mathbf{A}_0 = \mathbf{R}^{-\frac{1}{2}}(\mathbf{H}\mathbf{B}\mathbf{H}^{\mathrm{T}})\mathbf{R}^{-\frac{1}{2}}. \tag{20.3.15}$$

A note on the nomenclature is in order. While we have tried to associate each of the above forms of solution with the two basic spaces – the model and the observation space, in the literature various authors have used different descriptors. At the National Center for Environmental Prediction (NCEP) in Washington DC, preconditioned incremental form of the spectral model is used in (20.3.6) under the label **spectral statistical interpolation (SSI) analysis system** (Parrish and Derber 1992).

Courtier (1997) has very effectively used the preconditioned incremental form in both the model and observation space to bring out the similarity between these two formulations. In the following, we provide an exposé of the analysis contained in Courtier (1997).

We begin by establishing that the $n \times n$ matrix $\mathbf{A}_M$ in (20.3.7) and the $m \times m$ matrix $\mathbf{A}_0$ in (20.3.15) share the same set of non-zero eigenvalues. To this end, let λ and ξ be the eigenvalues and the eigenvector pair for the matrix $\mathbf{A}_0$. Then, using a series of obvious manipulations, we get

$$\begin{aligned}
\mathbf{R}^{-\frac{1}{2}}(\mathbf{HBH}^{\mathrm{T}})\mathbf{R}^{-\frac{1}{2}}\xi &= \lambda\xi \\
(\mathbf{B}^{\frac{1}{2}}\mathbf{H}^{\mathrm{T}}\mathbf{R}^{-\frac{1}{2}})[\mathbf{R}^{-\frac{1}{2}}(\mathbf{HBH}^{\mathrm{T}})\mathbf{R}^{-\frac{1}{2}}]\xi &= \lambda(\mathbf{B}^{\frac{1}{2}}\mathbf{H}^{\mathrm{T}}\mathbf{R}^{-\frac{1}{2}})\xi \\
(\mathbf{B}^{\frac{1}{2}}(\mathbf{H}^{\mathrm{T}}\mathbf{R}^{-1}\mathbf{H})\mathbf{B}^{\frac{1}{2}})[\mathbf{B}^{\frac{1}{2}}\mathbf{H}^{\mathrm{T}}\mathbf{R}^{-\frac{1}{2}}]\xi &= \lambda(\mathbf{B}^{\frac{1}{2}}\mathbf{H}^{\mathrm{T}}\mathbf{R}^{-\frac{1}{2}})\xi \\
(\mathbf{B}^{\frac{1}{2}}(\mathbf{H}^{\mathrm{T}}\mathbf{R}^{-1}\mathbf{H})\mathbf{B}^{\frac{1}{2}})\eta &= \lambda\eta
\end{aligned} \tag{20.3.16}$$

from which it follows that λ and $\eta = (\mathbf{B}^{\frac{1}{2}}\mathbf{H}^{\mathrm{T}}\mathbf{R}^{-\frac{1}{2}})\xi$ are the eigenvalue and the corresponding eigenvector of $\mathbf{A}_{\mathrm{M}} = \mathbf{B}^{\frac{1}{2}}(\mathbf{H}^{\mathrm{T}}\mathbf{R}^{-1}\mathbf{H})\mathbf{B}^{\frac{1}{2}}$.

For definiteness, let

$$\lambda_1 \geq \lambda_2 \geq \lambda_3 \geq \cdots \geq \lambda_n \tag{20.3.17}$$

be the eigenvalues of $\mathbf{A}_{\mathrm{M}}$ and

$$\mu_1 \geq \mu_2 \geq \mu_3 \geq \cdots \geq \mu_m \tag{20.3.18}$$

be those of $\mathbf{A}_0$. Here again, we consider two cases.

Case A. Overdetermined System: $m > n$ It can be verified that

$$\begin{aligned}
\mathrm{Rank}(\mathbf{A}_{\mathrm{M}}) &= \min\{\mathrm{Rank}(\mathbf{R}), \mathrm{Rank}(\mathbf{B}), \mathrm{Rank}(\mathbf{H})\} \\
&= n = \mathrm{Rank}(\mathbf{A}_0).
\end{aligned} \tag{20.3.19}$$

In this case, the $n \times n$ symmetric matrix $\mathbf{A}_{\mathrm{M}}$ is of full rank (hence non-singular) and positive definite but the $m \times m$ symmetric matrix $\mathbf{A}_0$ is rank deficient (hence singular) and positive semi-definite. This fact when combined with the equality of eigenvalues of $\mathbf{A}_{\mathrm{M}}$ and $\mathbf{A}_0$ proved above leads to the inescapable conclusion that

$$\left.\begin{aligned}
&\mu_i = \lambda_i > 0 \text{ for } i = 1 \text{ to } n \\
\text{and}& \\
&\mu_j = 0 \text{ for } j = n+1, n+2, \ldots, m
\end{aligned}\right\} \tag{20.3.20}$$

However, the eigenvalues of $(\mathbf{I} + \mathbf{A}_\mathrm{M})$ are $(1 + \lambda_i)$ for $i = 1$ to n and those of $(\mathbf{I} + \mathbf{A}_0)$ are $((1 + \mu_i)$ for $i = 1$ to m. Hence, both of these matrices, $(\mathbf{I} + \mathbf{A}_\mathrm{M})$ and $(\mathbf{I} + \mathbf{A}_0)$ are non-singular and their spectral condition numbers are given by

$$\kappa_2(\mathbf{I} + \mathbf{A}_\mathrm{M}) = \frac{1 + \lambda_1}{1 + \lambda_n} < 1 + \mu_1 = \mathcal{K}(\mathbf{I} + \mathbf{A}_0) < \infty. \tag{20.3.21}$$

Case B. Underdetermined System: $m < n$ In this case, it can be verified that

$$\mathrm{Rank}(\mathbf{A}_\mathrm{M}) = m = \mathrm{Rank}(\mathbf{A}_0)$$

and $\mathbf{A}_\mathrm{M}$ is rank deficient (hence singular) and positive semi-definite and $\mathbf{A}_0$ is of full rank (hence non-singular) and positive definite. Consequently,

$$\left.\begin{array}{ll} & \lambda_i = \mu_i > 0 \text{ for } i = 1 \text{ to } m \\ \text{and} & \\ & \lambda_j = 0 \text{ for } j = m+1, m+2, \ldots, n \end{array}\right\} \tag{20.3.22}$$

Here again, both $(\mathbf{I} + \mathbf{A}_\mathrm{M})$ and $(\mathbf{I} + \mathbf{A}_0)$ are non-singular and their condition numbers are given by

$$\kappa_2(\mathbf{I} + \mathbf{A}_0) = \frac{1 + \mu_1}{1 + \mu_n} < 1 + \lambda_1 = \kappa_2(\mathbf{I} + \mathbf{A}_\mathrm{M}). \tag{20.3.23}$$

Thus, in contrast with the basic formulation described in Section 20.2 wherein one of the forms leads to an ill-posed problem depending only on the values of m and n, the primary import of preconditioning is that it forces both the formulations to be well-posed for all values of m and n. This in turn implies that we truly have a choice of the formulation dictated by the total computation efforts needed in arriving at the solution which depends on the values of m and n and other structural properties of the matrices involved.

The downside of this preconditioning is that it requires the decomposition of the matrix **B** or **R**.

20.4 The nonlinear case: second-order method

Following the deeply rooted traditions in the contemporary literature on nonlinear optimization theory (Nash and Sofer (1996) and Dennis and Schnabel (1996)), in this section we describe a framework for minimizing the nonlinear objective function $J(\mathbf{x})$ in (20.1.16) which can be rewritten as

$$J(\mathbf{x}) = J_\mathrm{b}(\mathbf{x}) + J_\mathrm{o}(\mathbf{x}) \tag{20.4.1}$$

where

$$\left.\begin{aligned} J_b(\mathbf{x}) &= \tfrac{1}{2}(\mathbf{x}-\mathbf{x}_b)^T\mathbf{B}^{-1}(\mathbf{x}-\mathbf{x}_b) \\ \text{and} \qquad & \\ J_o(\mathbf{x}) &= \tfrac{1}{2}(\mathbf{z}-h(\mathbf{x}))^T\mathbf{R}^{-1}(\mathbf{z}-h(\mathbf{x})) \end{aligned}\right\} \tag{20.4.2}$$

This framework is based on developing the **full quadratic** approximation and is called the **second-order method** (See Chapter 7 for details).

Let $\mathbf{x}_*$ be the minimizer of $J(\mathbf{x})$ where $J : \mathbb{R}^n \to \mathbb{R}$ is the given nonlinear functional. Let $\mathbf{x}_c$ be the current operating point which denotes the initial approximation to $\mathbf{x}_*$. The idea rests upon replacement of $J(\mathbf{x})$ by a **quadratic approximation**, say $Q(\mathbf{y})$ around $\mathbf{x}_c$ where $\mathbf{x} = \mathbf{x}_c + \mathbf{y}$. Then,

$$\begin{aligned} Q(\mathbf{y}) &= J(\mathbf{x}_c + \mathbf{y}) \\ &= J(\mathbf{x}_c) + [\nabla J(\mathbf{x}_c)]^T\mathbf{y} + \tfrac{1}{2}\mathbf{y}^T[\nabla^2 J(\mathbf{x}_c)]\mathbf{y} \end{aligned} \tag{20.4.3}$$

which is the **second-order Taylor series** representation of $J(\mathbf{x})$ around $\mathbf{x}_c$ (Appendix C). Then the solution $\mathbf{y}_*$ of

$$\nabla Q(\mathbf{y}) = \nabla J(\mathbf{x}_c) + \nabla^2 J(\mathbf{x}_c)\mathbf{y} = 0 \tag{20.4.4}$$

or

$$\nabla^2 J(\mathbf{x}_c)\mathbf{y} = -\nabla J(\mathbf{x}_c) \tag{20.4.5}$$

is the minimizer of $Q(\mathbf{y})$. This equation is the basis for Newton's method for minimization. The idea is, once $\mathbf{y}_*$ is found, then $(\mathbf{x}_c + \mathbf{y}_*)$ becomes the new operating point and the entire procedure is repeated until convergence.

Applying this framework to $J(\mathbf{x})$ in (20.4.1), we obtain

$$\left.\begin{aligned} \nabla J(\mathbf{x}) &= \nabla J_b(\mathbf{x}) + \nabla J_o(\mathbf{x}) \\ \nabla^2 J(\mathbf{x}) &= \nabla^2 J_b(\mathbf{x}) + \nabla^2 J_o(\mathbf{x}) \end{aligned}\right\} \tag{20.4.6}$$

with

$$\nabla J_b(\mathbf{x}) = \mathbf{B}^{-1}(\mathbf{x}-\mathbf{x}_b),\ \nabla^2 J_b(\mathbf{x}) = \mathbf{B}^{-1} \tag{20.4.7}$$

and

$$\nabla J_o(\mathbf{x}) = -\mathbf{D}_h^T(\mathbf{x})\mathbf{R}^{-1}[\mathbf{z} - h(\mathbf{x})]. \tag{20.4.8}$$

Since (20.4.7) involves product of matrices and vectors that are functions of the state vector $\mathbf{x}$, computation of $\nabla^2 J_o(\mathbf{x})$ requires care and deliberation. Recall that the kth column [also the kth row since $\nabla^2 J(\mathbf{x})$ is symmetric] of $\nabla^2 J_o(\mathbf{x})$ is the gradient of the kth element of the vector $\partial \nabla J_o(\mathbf{x})/\partial x_k$. We first identify this element in (20.4.8). To this end, let

$$(h(\mathbf{x}) - \mathbf{z}) = g(\mathbf{x}) = (g_1(\mathbf{x}), g_2(\mathbf{x}), \ldots, g_m(\mathbf{x}))^T \tag{20.4.9}$$

where

$$g_j(\mathbf{x}) = [h_j(\mathbf{x}) - z_j]. \tag{20.4.10}$$

Also, recall that the transpose of the Jacobian of $h(\mathbf{x})$ is a $n \times m$ matrix (Appendix C).

$$\mathbf{D}_h^{\mathrm{T}}(\mathbf{x}) = \left[\frac{\partial h_j}{\partial x_i} \right] \tag{20.4.11}$$

where $1 \le i \le n$ is the row index and $1 \le j \le m$ is the column index. Let $e(\mathbf{x}) = (e_1(\mathbf{x}), e_2(\mathbf{x}), \ldots, e_m(\mathbf{x}))^{\mathrm{T}}$ denote the m-vector corresponding to the kth row of $\mathbf{D}_h^{\mathrm{T}}(\mathbf{x})$. That is,

$$e(\mathbf{x}) = \left(\frac{\partial h_1}{\partial x_k}, \frac{\partial h_2}{\partial x_k}, \ldots, \frac{\partial h_m}{\partial x_k} \right)^{\mathrm{T}} \tag{20.4.12}$$

and

$$e_i(\mathbf{x}) = \frac{\partial h_i}{\partial x_k} \text{ for } 1 \le i \le m. \tag{20.4.13}$$

In terms of this simplified notation, the kth element of $\nabla J_o(\mathbf{x})$ is given by

$$\frac{\partial J_o(\mathbf{x})}{\partial x_k} = e^{\mathrm{T}}(\mathbf{x})\mathbf{R}^{-1}g(\mathbf{x}). \tag{20.4.14}$$

Then, the kth column of the Hessian, $\nabla^2 J_o(\mathbf{x})$, is the sum of the two vectors (Appendix C):

$$\nabla \left(\frac{\partial J_o(\mathbf{x})}{\partial x_k} \right) = \mathbf{D}_e^{\mathrm{T}}(\mathbf{x})\mathbf{R}^{-1}g(\mathbf{x}) + \mathbf{D}_g^{\mathrm{T}}(\mathbf{x})\mathbf{R}^{-1}e(\mathbf{x}) \tag{20.4.15}$$

where $\mathbf{D}_e(\mathbf{x})$ and $\mathbf{D}_g(\mathbf{x})$ are the Jacobians of $e(\mathbf{x})$ in (20.4.12) and $g(\mathbf{x})$ in (20.4.9), respectively. Further, simplifying our notation, we label the first and second terms on the r.h.s.of (20.4.15) as vector1 and vector2, respectively. From this, we see that the required Hessian of $J_o(\mathbf{x})$ can be computed as the sum of two matrices, call them $\mathbf{E}_1$ and $\mathbf{E}_2$, and the mathematical expression for this Hessian is written as:

$$\nabla^2 J_o(\mathbf{x}) = \mathbf{E}_1 + \mathbf{E}_2 \tag{20.4.16}$$

where the kth column of $\mathbf{E}_1$ is given by vector1 and that of $\mathbf{E}_2$ by vector2.

Computation of $\mathbf{E}_1$ from Vector1

Let

$$\mathbf{R}^{-1}g(\mathbf{x}) = b(\mathbf{x}) = (b_1(\mathbf{x}), b_2(\mathbf{x}), \ldots, b_m(\mathbf{x}))^{\mathrm{T}} \tag{20.4.17}$$

Using (20.4.13), we have

$$\mathbf{D}_e^{\mathrm{T}}(\mathbf{x}) = \left[\frac{\partial e_j}{\partial x_i} \right] = \left[\frac{\partial^2 h_j}{\partial x_i \partial x_k} \right] \tag{20.4.18}$$

where $1 \leq i \leq n$ is the row index and $1 \leq j \leq m$ is the column index. Thus, the kth column of $\mathbf{E}_1$ is given by the matrix vector product $\mathbf{D}_h(\mathbf{x})b(\mathbf{x})$. That is, kth column of $\mathbf{E}_1$ = Vector1 = $\mathbf{D}_e(\mathbf{x})b(\mathbf{x})$

$$\begin{bmatrix} \sum_{j=1}^{m} \left(\frac{\partial^2 h_j}{\partial x_1 \partial x_k}\right) b_j(\mathbf{x}) \\ \sum_{j=1}^{m} \left(\frac{\partial^2 h_j}{\partial x_2 \partial x_k}\right) b_j(\mathbf{x}) \\ \vdots \\ \sum_{j=1}^{m} \left(\frac{\partial^2 h_j}{\partial x_n \partial x_k}\right) b_j(\mathbf{x}) \end{bmatrix}. \tag{20.4.19}$$

We can now recover each of the n columns of $\mathbf{E}_1$ by simply varying k from 1 through n. Now, since each element of the matrix is the sum of exactly m terms, we can express $\mathbf{E}_1$ as the sum of exactly m matrices as follows:

$$\mathbf{E}_1 = \nabla^2 h_1(\mathbf{x}) b_1(\mathbf{x}) + \nabla^2 h_2(\mathbf{x}) b_2(\mathbf{x}) + \cdots + \nabla^2 h_m(\mathbf{x}) b_m(\mathbf{x}) \tag{20.4.20}$$

where $\nabla^2 h_i(\mathbf{x})$ is the Hessian of the ith component of $h_i(\mathbf{x})$ of $h(\mathbf{x})$.

Computation of $\mathbf{E}_2$ from Vector2

Recall that the kth column of $\mathbf{E}_2$ is given by

$$\text{Vector2} = \mathbf{D}_g^{\mathrm{T}}(\mathbf{x})\mathbf{R}^{-1} e(\mathbf{x}) \tag{20.4.21}$$

where from (20.4.12), we know that $e(\mathbf{x})$ is a column vector that is the kth row of $\mathbf{D}_h^{\mathrm{T}}(\mathbf{x})$ (which is the same as the kth column vector of $\mathbf{D}_h(\mathbf{x})$). Further, from (20.4.10), since $g(\mathbf{x}) = (h(\mathbf{x}) - \mathbf{z})$, where $\mathbf{y}$ is a known constant vector, it immediately follows from the definition that

$$\mathbf{D}_g(\mathbf{x}) = \mathbf{D}_h(\mathbf{x}). \tag{20.4.22}$$

Combining these arguments with (20.4.21), we have

$$k\text{th column of } \mathbf{E}_2 = \text{Vector2} = \mathbf{D}_h^{\mathrm{T}}(\mathbf{x})\mathbf{R}^{-1}\{k\text{th column of } \mathbf{D}_h(\mathbf{x})\}$$

and hence

$$\mathbf{E}_2 = \mathbf{D}_h^{\mathrm{T}}(\mathbf{x})\mathbf{R}^{-1}\mathbf{D}_h(\mathbf{x}). \tag{20.4.23}$$

We are now ready to assemble the Hessian $\nabla^2 J(\mathbf{x})$. Combining (20.4.6), (20.4.7), (20.4.16), (20.4.20), and (20.4.23), it follows that

$$\nabla^2 J(\mathbf{x}) = \left[\mathbf{B}^{-1} + \mathbf{D}_h^{\mathrm{T}}(\mathbf{x})\mathbf{R}^{-1}\mathbf{D}_h(\mathbf{x}) + \sum_{m}^{i=1} \nabla^2 h_i(\mathbf{x}) b_i(\mathbf{x}) \right] \tag{20.4.24}$$

where $b(\mathbf{x}) = \mathbf{R}^{-1}[h(\mathbf{x}) - \mathbf{z}]$. We now return to Newton's equation (20.4.5) that defines the second-order method. By combining (20.4.24) with (20.4.5), (20.4.7), and (20.4.8), and evaluating all the new quantities at $\mathbf{x}_c$ (since $\mathbf{x}_c$ is the operating point), from (20.4.5), we get

$$[\mathbf{B}^{-1} + \mathbf{D}_h^{\mathrm{T}}(\mathbf{x}_c)\mathbf{R}^{-1}\mathbf{D}_h(\mathbf{x}_c) + \textstyle\sum_{i=1}^{m} \nabla^2 h_i(\mathbf{x}_c) b_i(\mathbf{x}_c)]\mathbf{y}$$
$$= \mathbf{D}_h^{\mathrm{T}}(\mathbf{x}_c)\mathbf{R}^{-1}[\mathbf{z} - h(\mathbf{x}_c)] - \mathbf{B}^{-1}(\mathbf{x}_c - \mathbf{x}_b) \tag{20.4.25}$$

where $b(\mathbf{x}) = \mathbf{R}^{-1}[h(\mathbf{x}_c) - \mathbf{z}]$.

The solution to (20.4.25) yields $\mathbf{y}(= \mathbf{x} - \mathbf{x}_c)$, the so-called second-order analysis increment at $\mathbf{x}_c$.

20.5 Special Case: first-order method

In deriving the first-order method as a special case, let us look at the second-order method from an alternative point of view. To this end, expand $h(\mathbf{x})$ around $\mathbf{x}_c$ using the second-order approximation as shown below.

Recall from Appendix C that (with $\mathbf{x} = \mathbf{x}_c + \mathbf{y}$)

$$h(\mathbf{x}) = h(\mathbf{x}_c) + \mathbf{D}_h(\mathbf{x}_c)\mathbf{y} + \psi(\mathbf{y}) \tag{20.5.1}$$

where for simplicity in notation, we have

$$\psi(\mathbf{y}) = \frac{1}{2}\mathbf{y}^{\mathrm{T}}\mathbf{D}_h^2(\mathbf{x}_c)\mathbf{y}, \tag{20.5.2}$$

a vector with $\psi(\mathbf{y}) = (\psi_i(\mathbf{y}), \psi_2(\mathbf{y}), \ldots, \psi_n(\mathbf{y}))^{\mathrm{T}}$, $\psi_1(\mathbf{y}) = \mathbf{y}^{\mathrm{T}}\mathbf{B}_i\mathbf{y}$ and $\mathbf{B}_i = \nabla^2 h_i(\mathbf{x}_c)$. (These Hessian matrices $\mathbf{B}_i$ are not to be confused with the background error covariance matrix $\mathbf{B}$). Thus, each component of $\psi(\mathbf{y})$ is a quadratic form in $\mathbf{y}$. Now, substituting (20.5.1) and (20.5.2) into (20.4.1) and using the notation in Section 20.4, we get

$$J(\mathbf{x}) = \tfrac{1}{2}\,[-g(\mathbf{x}_c) - \mathbf{D}_h(\mathbf{x}_c)\mathbf{y} - \psi(\mathbf{y})]^{\mathrm{T}}\,\mathbf{R}^{-1}\,[-g(\mathbf{x}_c) - \mathbf{D}_h(\mathbf{x}_c)\mathbf{y} - \psi(\mathbf{y})]$$
$$+ \tfrac{1}{2}(\mathbf{y} + \mathbf{x}_c - \mathbf{x}_b)^{\mathrm{T}}\mathbf{B}^{-1}(\mathbf{y} + \mathbf{x}_c - \mathbf{x}_b). \tag{20.5.3}$$

It can be verified that the r.h.s.of (20.5.3) is a fourth-degree polynomial in $\mathbf{y}$. To obtain a quadratic approximation, we neglect the third- and fourth-degree terms in (20.5.3). This quadratic approximation is given by,

$$Q'(\mathbf{y}) = (\frac{1}{2})g^{\mathrm{T}}(\mathbf{x}_c)\mathbf{R}^{-1}g(\mathbf{x}_c) + g^{\mathrm{T}}(\mathbf{x}_c)\mathbf{R}^{-1}\mathbf{D}_h(\mathbf{x}_c)\mathbf{y}$$
$$+ (\frac{1}{2})\mathbf{y}^{\mathrm{T}}\left[\mathbf{B}^{-1} + \mathbf{D}_h^{\mathrm{T}}(\mathbf{x}_c)\mathbf{R}^{-1}\mathbf{D}_h(\mathbf{x}_c)\right]\mathbf{y}$$
$$+ g^{\mathrm{T}}(\mathbf{x}_c)\mathbf{R}^{-1}\psi(\mathbf{y}) + \frac{1}{2}(\mathbf{y} + \mathbf{x}_c - \mathbf{x}_b)^{\mathrm{T}}\mathbf{B}^{-1}(\mathbf{y} + \mathbf{x}_c - \mathbf{x}_b). \tag{20.5.4}$$

It can be verified that $Q'(\mathbf{y})$ as given in (20.5.4) is indeed the same as $Q(\mathbf{y})$ in (20.4.3).

As preparation for computation of the gradient of (20.5.4), let us first compute the gradient of the next to the last term in (20.5.4). Using (20.4.17), we have

$$
\begin{aligned}
\nabla\left[g^{\mathrm{T}}(\mathbf{x}_c)\mathbf{R}^{-1}\psi(\mathbf{y})\right] &= \nabla\left[b^{\mathrm{T}}(\mathbf{x}_c)\psi(\mathbf{y})\right] \\
&= \left(\frac{1}{2}\right)\nabla\left[\sum_{i=1}^{n} b_i(\mathbf{x}_c)(\mathbf{y}^{\mathrm{T}}\mathbf{B}_i\mathbf{y})\right] \\
&= \sum_{i=1}^{n} b_i(\mathbf{x}_c)(\mathbf{B}_i\mathbf{y}) \\
&= \sum_{i=1}^{n} b_i(\mathbf{x}_c)\left[\nabla^2 h_i(\mathbf{x}_c)\mathbf{y}\right].
\end{aligned}
\tag{20.5.5}
$$

Using (20.5.5), it can be verified that

$$
\begin{aligned}
\nabla Q'(\mathbf{y}) = {} & \mathbf{D}_h^{\mathrm{T}}(\mathbf{x}_c)\mathbf{R}^{-1}g(\mathbf{x}_c) + \left[\mathbf{B}^{-1} + \mathbf{D}_h^{\mathrm{T}}(\mathbf{x}_c)\mathbf{R}^{-1}\mathbf{D}_h(\mathbf{x}_c)\right]\mathbf{y} \\
& + \sum_{i=1}^{n} b_i(\mathbf{x}_c)\left[\nabla^2 h_i(\mathbf{x}_c)\right]\mathbf{y} + \mathbf{B}^{-1}(\mathbf{x}_c - x_{\mathrm{b}}).
\end{aligned}
\tag{20.5.6}
$$

Setting (20.5.6) to zero, we find the optimum $\mathbf{y}$ as the solution of the following equation

$$
\begin{aligned}
&[\mathbf{B}^{-1} + \mathbf{D}_h^{\mathrm{T}}(\mathbf{x}_c)\mathbf{R}^{-1}\mathbf{D}_h(\mathbf{x}_c) + \sum_{i=1}^{n} b_i(\mathbf{x}_c)\nabla^2 h_i(\mathbf{x}_c)]\mathbf{y} \\
&= \mathbf{D}_h^{\mathrm{T}}(\mathbf{x}_c)\mathbf{R}^{-1}[\mathbf{z} - h(\mathbf{x}_c)] - \mathbf{B}^{-1}(\mathbf{x}_c - \mathbf{x}_{\mathrm{b}})
\end{aligned}
\tag{20.5.7}
$$

which is identical to (20.4.25).

And as we might expect, by setting $D_h^2 = 0$ in (20.5.2), (20.5.7) reduces to

$$
\left[\mathbf{B}^{-1} + \mathbf{D}_h^{\mathrm{T}}(\mathbf{x}_c)\mathbf{R}^{-1}\mathbf{D}_h(\mathbf{x}_c)\right]\mathbf{y} = \mathbf{D}_h^{\mathrm{T}}(\mathbf{x}_c)\mathbf{R}^{-1}[\mathbf{z} - h(\mathbf{x}_c)] - \mathbf{B}^{-1}(\mathbf{x}_c - \mathbf{x}_{\mathrm{b}}) \tag{20.5.8}
$$

which is the first-order method used in Daley and Barker (2001), Lorenc (1986), and Tarontola (1987). In short, the standard first-order method achieves a quadratic form of J (second-degree polynomial in $\mathbf{y}$) by assuming that h can be approximated by a Taylor expansion up to the first-degree term in $\mathbf{y}$. However, if we assume h is approximated by the expansion out to the second-degree term, then J will be a fourth-degree polynomial in $\mathbf{y}$. When we examine this form of J up to the second-degree polynomial in $\mathbf{y}$ ("quadratic J"), we find that there are terms involving the Hessian of $h(\mathbf{x})$ that are unaccounted for in the first-order method. We thus say that the first-order method is a "partial" quadratic approximation, whereas the second-order method is a "full" quadratic approximation to J.

The Hessians of $Q(\mathbf{y})$ in (20.4.3) and $Q'(\mathbf{y})$ in (20.5.4) are identical and given by

$$\nabla^2 Q(\mathbf{y}) = \nabla^2 Q'(\mathbf{y}) = \left[\mathbf{B}^{-1} + \mathbf{D}_h^{\mathrm{T}}(\mathbf{x}_c)\mathbf{R}^{-1}\mathbf{D}_h(\mathbf{x}_c) + \sum_{i=1}^{n} b_i(\mathbf{x}_c)\nabla^2 h_i(\mathbf{x}_c) \right]. \tag{20.5.9}$$

Recall that $[\mathbf{B}^{-1} + \mathbf{D}_h^{\mathrm{T}}(\mathbf{x}_c)\mathbf{R}^{-1}\mathbf{D}_h(\mathbf{x}_c)]$ is positive definite by assumption. However, the positive definiteness of the matrix on the r.h.s.of (20.5.9) depends on the second-order properties of the nonlinear forward operator $h(\mathbf{x})$ and the scaled or normalized observation innovation $b(\mathbf{x}_c) = -\mathbf{R}^{-1}[\mathbf{z} - h(\mathbf{x}_c)]$.

Exercises

20.1 (Courtier (1997)) Verify that the matrices $\mathbf{R}^{-\frac{1}{2}}(\mathbf{HBH}^{\mathrm{T}})\mathbf{R}^{-\frac{1}{2}}$ and $(\mathbf{HBH}^{\mathrm{T}})\mathbf{R}^{-1}(\mathbf{HBH}^{\mathrm{T}})^{\frac{1}{2}}$ have the same set of (non-zero) eigenvalues.

20.2 (Courtier (1997)) Verify that $(\mathbf{HBH}^{\mathrm{T}})\mathbf{R}^{-1}(\mathbf{HBH}^{\mathrm{T}})^{\frac{1}{2}}$ and $\mathbf{B}^{\frac{1}{2}}(\mathbf{HR}^{-1}\mathbf{H})\mathbf{B}^{-\frac{1}{2}}$ have the same set of (non-zero) eigenvalues.

20.3 (Courtier (1997)) Express $f(\mathbf{w})$ in (20.3.8) as a new quadratic form in $\mathbf{s}$ where $\mathbf{s} = (\mathbf{HBH}^{\mathrm{T}})^{\frac{1}{2}}\mathbf{w}$, assuming $(\mathbf{HBH}^{\mathrm{T}})^{\frac{1}{2}}$ is non singular. Compute the gradient and the Hessian of the new resulting quadratic form.

Notes and references

As mentioned in the introduction, the 3DVAR-based approach has now replaced all the earlier ideas using the optimal interpolation, successive correction, etc. and is now routinely used by weather centers around the world. The review paper by Lorenc (1986) offers a comprehensive review of 3DVAR as practiced in meteorology. Determining estimates of $\mathbf{B}$, the background error covariance matrix, is an especially challenging component of data assimilation since we never know the true state in atmospheric application. Various strategies have been put forward including those discussed in Hollingsworth and Lönnberg (1986), Fisher and Courtier (1995), Parrish and Derber (1992), Courtier, et al. (1998), Parrish et al. (1997), and Weaver and Courtier (2001).

Section 20.1 For an exposition of the Bayesian approach, refer to Purser (1984), Lorenc (1986),(1988), and Tarantola (1987).

Sections 20.2–20.3 Duality is treated in Cohn et al. (1998) and Courtier (1997). This latter reference provides a good summary of the preconditioned approach. Preconditioned formulation is used in Parrish and Derber (1992).

Sections 20.4–20.5 The derivation of the second-order method is taken from Lakshmivarahan, Honda and Lewis (2003). First-order method is routinely used in practice – Lorenc and Hammon (1998), Daley and Barker (2001), Huang (2000).

Nonlinear forward operators also arise in integrating the data quality control as a part of the retrieval methodology. Refer to Lorenc and Hammon (1988) Ingleby and Lorenc (1993) and Andersson and Järvinen (1999).

21
Spatial digital filters

In this chapter we provide an overview of the role and use of **spatial** digital filters in solving the retrieval problem of interest in this part V. This chapter begins with a classification of filters in Section 21.1. Nonrecursive filters are covered in Section 21.2 and a detailed account of the recursive filters and their use is covered in Section 21.3.

21.1 Filters: A Classification

The word filter in **spatial digital filter** is used in the same (functional) sense as used in coffee filter, water filter, filter lenses, to name a few, that is, to prevent the passage of some unwanted items – the coffee bean sediments from the concoction, impurities in the drinking water, a light of particular wavelength or color from passing through. Spatial filters are designed to prevent the passage of signal components of a specified frequency or wavelength. For example, the **low-pass** filter is designed to suppress or filter out high frequency or smaller wavelength signals. There is a vast corpus of literature dealing with filters in general. In this section we provide a useful classification of these filters.

Filters can be classified along at least five different dimensions depending on the type of signals and the properties of the filter. Refer to Figure 21.1.1. Signals to be filtered can be in **analog** or **digital** form and signals can be a function of **time** and/or **space**. For example, in **time series modelling** we deal with digital signals in discrete time and **analog computers** use continuous time signals. In meteorology one is often interested in the spatial features of a disturbance affecting the weather system. Filters can be classified using **structure, functionality** and **causality**. Structurally, a filter can be classified into two categories – **recursive** or **non-recursive** type. Recursive filters have **infinite memory** whereas non-recursive filters have only **finite memory**. In the functional classification, we have **low-pass**, **high-pass** and **band-pass** filters. Causality is a constraint imposed by on-line/real-time applications where the current actions/decisions are to be based only on the information available from the past without having to anticipate the future. In the

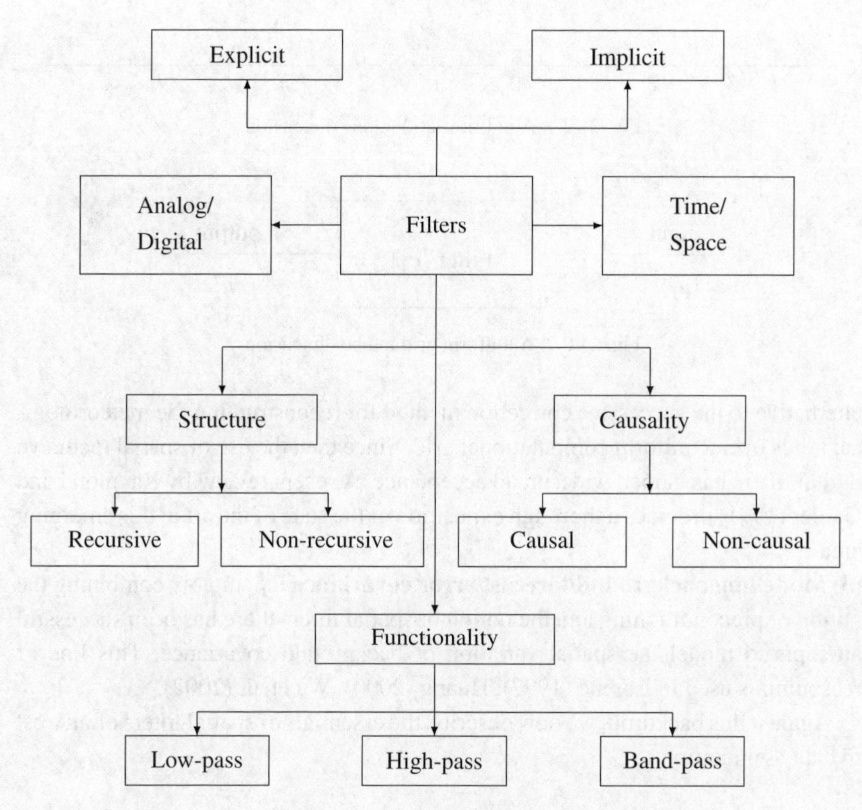

Fig. 21.1.1 A classification of filters.

off-line model where all the signals to be processed are available, a non-causal approach is justified. Filters can also be divided into **explicit** or **implicit** filters. While explicit filters directly compute the filtered output, implicit filters need matrix inversion. Explicit filters are used in real-time signal processing and implicit filters are often used in meteorology. Accordingly, we can have digital, time domain low-pass recursive filters; digital spatial low-pass, non-recursive filters to name a few of the possibilities.

The use of filters in meteorology can be classified into three groups.

(1) Statistical numerical schemes The use of non-recursive spatial filters in meteorology dates back to the 1950s when Shuman (1957) demonstrated their use in stabilizing the numerical solution of the balance equation. Robert (1966) used similar filters for integrating the general circulation primitive equation (spectral) model with central differences for controlling the instability due to the friction term. Asselin (1972) uses spatial filters in conjunction with the leap-frog semi-implicit and fully-implicit schemes. Also see Orszag (1971) for related ideas.

(2) Solution to the retrieval problem Purser and McQuigg (1982) in an unpublished but widely known report demonstrated the use of recursive spatial filters as an

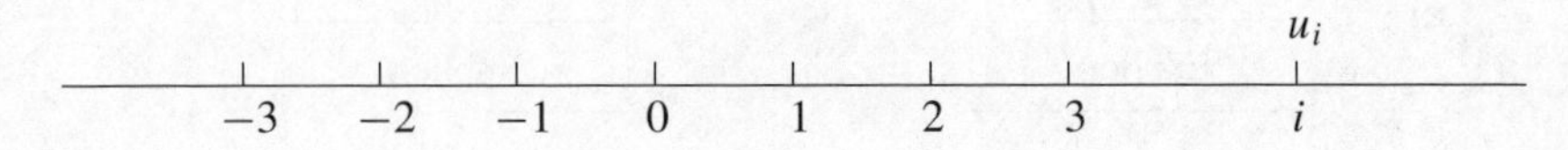

Fig. 21.2.1 A uniform grid in one dimension.

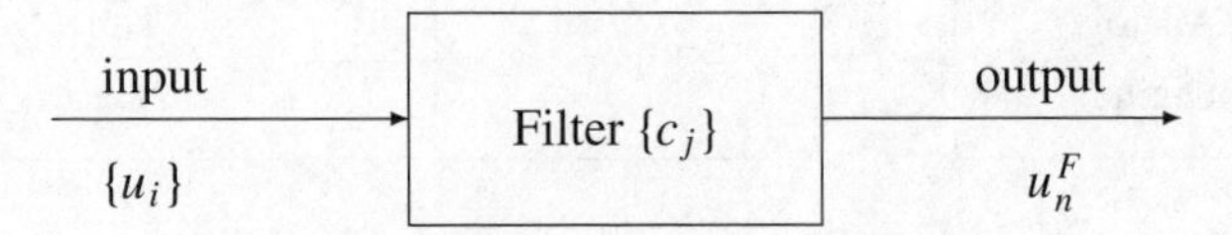

Fig. 21.2.2 A uniform grid in one dimension.

alternative to the successive correction method for reconstructing the meteorological fields over a uniform computational grid. Since then the use of spatial recursive digital filters has gained widespread acceptance. A recent review by Raymond and Garder (1991) provides a thorough exposition of the state of the art in this emerging area.

(3) Modelling background/forecast error covariance By suitably combining the notion of preconditioning and the notion of spatial filters there has been successful attempts to model the spatial variation of background covariance. This line of reasoning is used in Lorenc (1997), Huang (2000), Wu et al. (2002).

Against this backdrop, we now describe the essentials of spatial filters of interest to data assimilation.

21.2 Non-recursive filters

Let $u : \mathbb{R} \to \mathbb{R}$ where $u(x)$ denotes the scalar field variable as a function of the scalar (space) variable x. For example, $u(x)$ may denote the temperature or pressure at the point x. For purposes of our analysis, embed a two-way infinite uniform grid with Δx as the grid spacing. Refer to Figure 21.2.1. Let $u_i = u(i\,\Delta x)$ be the discretization or the sampling of the continuous field variable $u(x)$ at the grid points $i = 0, \pm 1, \pm 2, \pm 3, \ldots$ Let u_i^{F} be the **filtered version** of the field variable. A non-recursive filter is specified by a system of weights $\{\ldots, c_{-3}, c_{-2}, c_{-1}, c_0, c_1, c_2, c_3, \ldots\}$. Given the input field $\{u_i\}$ and the system of weights $\{c_j\}$, the filtered output u_n^{F} is defined by (Refer to Figure 21.2.2)

$$u_n^{\mathrm{F}} = \sum_{i=-\infty}^{\infty} c_i\, u_{n-i} \tag{21.2.1}$$

for $n = 0, \pm 1, \pm 2, \ldots$ By changing the indices, it can be verified that

$$u_n^{\mathrm{F}} = \sum_{i=-\infty}^{\infty} c_i\, u_{n-i} = \sum_{i=-\infty}^{\infty} c_{n-i}\, u_i. \tag{21.2.2}$$

This operation of obtaining $\{u_n^F\}$ from $\{u_i\}$ and $\{c_j\}$ is quite basic and is called the **convolution** of $\{u_i\}$ and $\{c_j\}$ and is succinctly denoted by

$$u^F = c * u. \tag{21.2.3}$$

This is also an example of a **non-causal** filter since u_n^F in (21.2.1) depends on values of u_i for both $i \le n$ (past) and $i > n$ (future). In practice, since the infinite summation is not feasible, we often define a (symmetric) window of weights c_i for $i = -N$ to N and use the following finite version

$$u_n^F = \sum_{i=-N}^{N} c_i\, u_{n-i}. \tag{21.2.4}$$

Clearly, this window of length $(2N + 1)$ is defined by the stencil of weights represented by

$$[c_{-N},\, c_{-N+1}, \ldots,\, c_{-2},\, c_{-1},\, c_0,\, c_1,\, c_2, \ldots,\, c_{N-1},\, c_N].$$

The properties of the non-recursive moving window filter is uniquely defined by the length $(2N + 1)$ of the window and the distribution of the weights within the window.

For purposes of illustration, consider first the case where the individual observations of the field variable u_i are corrupted by an additive white noise v_i where

$$E(v_i) = 0, \quad \mathrm{Var}(v_i) = \sigma^2 \quad \text{and} \quad E(v_i\, v_j) = 0 \quad \text{for } i \ne j. \tag{21.2.5}$$

Then, the filtered field is given by

$$u_n^F = \sum_{i=-N}^{N} c_i\, (u_{n-i} + v_{n-i}). \tag{21.2.6}$$

Hence

$$E(u_n^F) = \sum_{i=-N}^{N} c_i\, u_{n-i} \tag{21.2.7}$$

and

$$\begin{aligned}\mathrm{Var}(u_n^F) &= E[u_n^F - E(u_n^F)]^2 \\ &= \sigma^2 \sum_{i=-N}^{N} c_i^2.\end{aligned} \tag{21.2.8}$$

Thus, the filter either amplifies or dampens the input variance depending on

$$\sum_{i=-N}^{N} c_i^2 \gtrless 1. \tag{21.2.9}$$

Since noise is associated with high frequency, a low-pass non-recursive moving window filter by definition must be designed to dampen the high frequency noise components. That is, the weights c_i must be such that

$$\sum_{i=-N}^{N} c_i^2 < 1. \tag{21.2.10}$$

Also recall that one standard technique for removing the effect of noise is to use the **smoothing** or the **averaging** process. This in turn suggests that c_i's must satisfy

$$\sum_{i=-N}^{N} c_i = 1. \tag{21.2.11}$$

Against this backdrop, we now state the low-pass filter design problem: Find the weights

$$c = \{c_i \mid i = 0, \pm 1, \pm 2, \ldots, \pm N\}$$

such that the variance of u_n^{F} in (21.2.8) is a minimum when c_i's are required to satisfy (21.2.11). This standard constrained minimization problem is solved by using the **Lagrangian multiplier** method (Appendix C). Let

$$L(\lambda, c) = \sigma^2 \sum_{i=-N}^{N} c_i^2 + \lambda \left(\sum_{i=-N}^{N} c_i - 1 \right). \tag{21.2.12}$$

By setting the first derivative of L w.r.t. c_i for $i = -N$ to N and λ to zero and solving the resulting equations (Exercise 21.1) it can be verified that the minimizing values of c_i are given by

$$c_i = \frac{1}{2N+1}, \quad \text{for all } i.$$

The minimum value of the variance of u_n^{F} is then given by

$$\sigma^2 \sum_{i=-N}^{N} \frac{1}{(2N+1)^2} = \frac{\sigma^2}{(2N+1)} < \sigma^2 \tag{21.2.13}$$

which in turn guarantees the low-pass nature of the filter. Thus, the $(2N+1)$ window, low-pass, symmetric moving window of minimum variance is given by the stencil

$$\frac{1}{2N+1}[1, 1, \ldots, 1, 1, 1, \ldots, 1, 1]$$

and

$$u_n^{\mathrm{F}} = \frac{1}{2N+1} \sum_{i=-N}^{N} u_{n-i}. \tag{21.2.14}$$

We now describe some of the examples of the low-pass filters used repeatedly in meteorological applications.

Shuman (1957) filter This is a three-point symmetric moving average filter defined by the stencil

$$\left[\frac{c}{2}, (1-c), \frac{c}{2}\right]$$

for some $0 < c < 1$. Hence

$$\begin{aligned} u_n^{\mathrm{F}} &= \frac{c}{2} u_{n-1} + (1-c)\, u_n + \frac{c}{2} u_{n+1} \\ &= u_n + \frac{c}{2}[u_{n-1} - 2\, u_n + u_{n+1}]. \end{aligned} \tag{21.2.15}$$

It can be easily verified that r.h.s.of (21.2.15) is a finite difference approximation to the second-order linear differential operator $\eta = \left(1 + \frac{c}{2}\frac{\mathrm{d}^2}{\mathrm{d}x^2}\right)$ and hence

$$u_n^{\mathrm{F}} \approx \left(1 + \frac{c}{2}\frac{\mathrm{d}^2}{\mathrm{d}x^2}\right) u(x)|_{x=n\,\Delta x}. \tag{21.2.16}$$

Spectral analysis of Shuman filter

The actual filtering properties of the Shuman or any other filter can be best understood by performing analysis in the **spectral** or **frequency** domain. Assuming that the field variable $u(x)$ is periodic in x, we can express $u(x)$ approximately as a finite sum of **Fourier components** (Appendix G) of the form

$$u(x) = A_0 + \sum_{i=1}^{k} \mathbf{A}_i \cos\left(\frac{2\pi}{L_i}\right) x$$

where L_i is the **wavelength** of the ith component. Recall that $f_i = 1/L_i$ is the **frequency** and $k_i = 2\pi/L_i$ is the **wave number** of the ith component. Since the operation associated with the filter in (21.2.15) is linear, without loss of generality in the following analysis it is assumed that $k = 1$ and

$$u(x) = A_0 + A \cos kx. \tag{21.2.17}$$

Sampling this $u(x)$ at the grid points in Figure 21.2.1 we get

$$\left.\begin{aligned} u_n &= u(x_n) = A_0 + A \cos kn\,\Delta x \\ u_{n\pm1} &= u(x_{n\pm1}) = A_0 + A \cos k(n \pm 1)\,\Delta x \end{aligned}\right\} \tag{21.2.18}$$

Substituting (21.2.18) into (21.2.15) and simplifying (Exercise 21.2) we get

$$u_n^{\mathrm{F}} = A_0 + A^{\mathrm{F}} \cos kn\,\Delta x \tag{21.2.19}$$

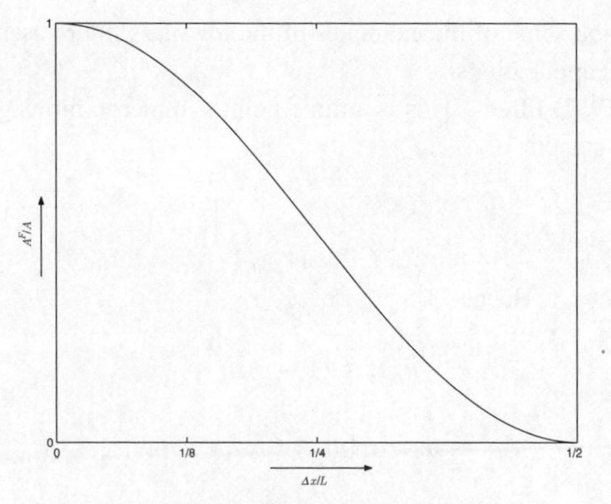

Fig. 21.2.3 Variation of A^F.

where

$$\frac{A^{\mathrm{F}}}{A} = \{1 - c\,[1 - \cos(k\,\Delta x)]\}$$

$$= \left[1 - 2c\sin^2\left(\frac{\pi\,\Delta x}{L}\right)\right]. \qquad (21.2.20)$$

Several comments are in order.

(1) **The phase of the filtered output** The filtered component has the same phase as the input component, that is, there is no phase shift resulting from this filtering. This is an intrinsic property of symmetric filters such as the Shuman filter.

(2) **The amplitude of the filtered output** From (21.2.20) it follows that the amplitude A^{F} of the filtered output is a function of the **stencil parameter** c, the **input wave number** k, and the **grid spacing** Δx. It can be verified that $A^{\mathrm{F}} = 0$ when $c = 1/2$ and for the wavelength $L = 2\Delta x$. That is, signals whose wavelength is twice the grid spacing are totally eliminated or filtered out. All signals with wavelength $L > 2\Delta x$ are dampened. Signals of wavelength $L < 2\Delta x$ cannot be resolved by this grid and would appear as signals with longer wavelength due to **aliasing**. These latter signals of longer wavelengths are also dampened by this filter. The variation of (A^{F}/A) as a function of $(\Delta x/L)$ is given in Figure 21.2.3, from which the low-pass property of this filter becomes very evident.

(3) **Shuman filter in two dimensions** The above analysis of the 1-d filter can be readily extended to multiple dimensions. We illustrate the major steps in this extension by deriving the Shuman filters in two dimensions.

Consider a two-way infinite 2-d grid with Δx and Δy as grid spacings in the x and y directions. Let $u(x, y)$ be the scalar field variable of interest. The

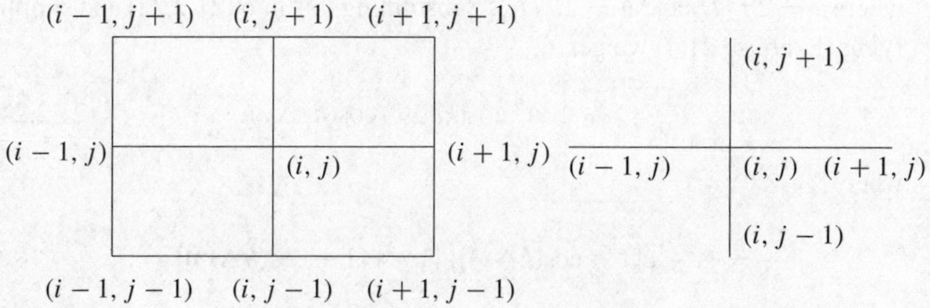

$(i-1, j+1)$ $(i, j+1)$ $(i+1, j+1)$

$(i-1, j)$ (i, j) $(i+1, j)$

$(i-1, j-1)$ $(i, j-1)$ $(i+1, j-1)$

(*a*) Nine-point stencil

$(i, j+1)$

$(i-1, j)$ (i, j) $(i+1, j)$

$(i, j-1)$

(*b*) Five-point stencil

Fig. 21.2.4 Two forms of the stencils for 2-d filters.

sampled value of $u(x, y)$ at the grid point (x, y) is given by

$$u_{ij} = u(i\,\Delta x,\ j\,\Delta y).$$

In analogy with the 1-d analysis, define operators

$$\left.\begin{aligned} \eta_i(u_{ij}) &= u_{ij} + \tfrac{c}{2}[u_{i-1,j} - 2u_{ij} + u_{i+1,j}] \\ &\approx \left(1 + \tfrac{c}{2}\tfrac{\partial^2}{\partial x^2}\right) u(x, y)|_{x=i\,\Delta x,\ y=j\,\Delta y} \\ \text{and} \qquad & \\ \eta_j(u_{ij}) &= u_{ij} + \tfrac{c}{2}[u_{i,j-1} - 2u_{ij} + u_{i,j+1}] \\ &\approx \left(1 + \tfrac{c}{2}\tfrac{\partial^2}{\partial y^2}\right) u(x, y)|_{x=i\,\Delta x,\ y=j\,\Delta y} \end{aligned}\right\} \tag{21.2.21}$$

which are the Shuman operators along the x and y directions. Then, u^{F}_{ij} can be defined by

$$\begin{aligned} u^{\mathrm{F}}_{ij} &= \eta_j\,[\eta_i\,(u_{ij})] \\ &= u_{ij} + \frac{c(1-c)}{2}\,[u_{i-1,j} + u_{i,j+1} + u_{i+1,j} + u_{i,j-1} - 4u_{ij}] \\ &\quad + \frac{c^2}{4}\,[u_{i-1,j+1} + u_{i+1,j+1} + u_{i+1,j-1} + u_{i-1,j-1}] \end{aligned} \tag{21.2.22}$$

which is an operator defined on the nine-point stencil (Exercise 21.3). The typical nine-point stencil is given in Figure 21.2.4.

Let

$$u(x, y) = A_0 + A\cos\left(\frac{2\pi}{L_x}\right) x \cos\left(\frac{2\pi}{L_y}\right) y.$$

Then

$$u_{ij} = A_0 + A\cos(ki\,\Delta x)\cos(hj\,\Delta y) \tag{21.2.23}$$

where $k = 2\pi/L_x$ and $h = 2\pi/L_y$. Substituting this into (21.2.22) and simplifying (Exercise 21.4) we get

$$u_{ij}^{\mathrm{F}} = A_0 + A^{\mathrm{F}} \cos(ki\Delta x)\cos(hj\Delta y) \tag{21.2.24}$$

where

$$\begin{aligned}\frac{A^{\mathrm{F}}}{A} &= \{1 - c[1 - \cos(k\Delta x)]\}\{1 - c[1 - \cos(h\Delta y)]\} \\ &= \left[1 - 2c\sin^2\left(\frac{k\Delta x}{2}\right)\right]\left[1 - 2c\sin^2\left(\frac{h\Delta y}{2}\right)\right].\end{aligned} \tag{21.2.25}$$

Stated in other words, the effect of this nine-point stencil is equivalent to applying the three-point stencil twice – first in the x-direction and then along the y-direction.

Alternatively, we can define

$$\begin{aligned}u_{ij}^{\mathrm{F}} &= \frac{1}{2}\,[\eta_i(u_{ij}) + \eta_j(u_{ij})] \\ &= u_{ij} + \frac{c}{4}[u_{i-1,j} + u_{i+1,j} + u_{i,j-1} + u_{i,j+1} - 4u_{ij}]\end{aligned} \tag{21.2.26}$$

leading to a simple five-point stencil (Figure 21.2.4). It can be verified (Exercise 21.5) that for u_{ij} in (21.2.23) we get

$$u_{ij}^{\mathrm{F}} = A_0 + A^{\mathrm{F}} \cos(ki\Delta x)\cos(hj\Delta x)$$

where

$$\begin{aligned}\frac{A^{\mathrm{F}}}{A} &= 1 - c\left\{\left[1 - \frac{1}{2}[\cos(k\Delta x) + \cos(h\Delta y)\right]\right\} \\ &= 1 - c\left[\sin^2\left(\frac{k\Delta x}{2}\right) + \sin^2\left(\frac{h\Delta y}{2}\right)\right].\end{aligned} \tag{21.2.27}$$

(4) **Extensions** Shapiro (1975) analyzed the properties of several extensions of the Shuman filters. These and other extensions are pursued in Exercises (21.6)–(21.7).

21.3 Recursive filters

Consider the uniform 1-d grid as in Figure 21.2.1. The general r**th order**, **non-causal**, **recursive** filter is given by

$$u_n^{\mathrm{F}} = \sum_{i=1}^{r} \alpha_i\, u_{n-i}^{\mathrm{F}} + \sum_{j=-k}^{k} \beta_j\, u_{n-j} \tag{21.3.1}$$

where the constants r, k, $\alpha_1, \alpha_2, \ldots, \alpha_r, \beta_0, \beta_{\pm 1}, \beta_{\pm 2}, \ldots, \beta_{\pm k}$ define the characteristics of this filter. This is called the rth **order** filter since u_n^{F} depends on the past r filtered values $u_{n-1}^{\mathrm{F}}, u_{n-2}^{\mathrm{F}}, \ldots, u_{n-r}^{F}$ and **non-causal** since u_n^{F} depends on the present and the past unfiltered inputs $u_n, u_{n-1}, \ldots, u_{n-k}$ and the future unfiltered inputs $u_{n+1}, u_{n+2}, \ldots, u_{n+k}$. The rth order filter, in addition also needs r initial values $u_0^{\mathrm{F}}, u_{-1}^{\mathrm{F}}, u_{-2}^{\mathrm{F}}, \ldots, u_{-r+1}^{\mathrm{F}}$. A **causal** version of the rth order recursive filter is given by

$$u_n^{\mathrm{F}} = \sum_{i=1}^{r} \alpha_i \, u_{n-i}^{\mathrm{F}} + \sum_{j=0}^{k} \beta_j \, u_{n-j}. \tag{21.3.2}$$

In the following we only consider the causal versions of the recursive filters.

(A) The first-order forward filter

To get a good handle on the behavior of the recursive filters, consider a **first-order** filter given by

$$u_n^{\mathrm{A}} = \alpha u_{n-1}^{\mathrm{A}} + (1-\alpha) u_n \tag{21.3.3}$$

for some $0 < \alpha < 1$. This is called the **forward** filter since the index n increases from left ($-\infty$) to right ($+\infty$). By expanding (21.3.3) (Exercise 21.9) it can be verified that

$$u_n^{\mathrm{A}} = \alpha^n u_0^{\mathrm{A}} + (1-\alpha) \sum_{k=0}^{n-1} \alpha^k u_{n-k} \tag{21.3.4}$$

where u_0^{A} is the initial value for u_n^{A}. Since $\alpha < 1$ for larger n, we can represent u_n^{A} as

$$u_n^{\mathrm{A}} = \sum_{k=0}^{n-1} \mathbf{g}_{k+1} u_{n-k} \tag{21.3.5}$$

where

$$\mathbf{g}_{k+1} = \begin{cases} (1-\alpha)\,\alpha^k & , \quad \text{for } k \geq 0 \\ 0 & , \quad \text{for } k < 0 \end{cases} \tag{21.3.6}$$

Basic properties of this exponential weight sequence are explored in Exercise 21.10. This operation that defines the filtered output sequence $u^{\mathrm{F}} = \{u_n^{\mathrm{F}}\}$ as a linear combination of input sequence $u = \{u_n\}$ with the exponential weighting sequence $g^+ = \{\mathbf{g}_{k+1}\}$ is called the **discrete convolution** (Appendix G) and is denoted by

$$u^{\mathrm{A}} = g^+ * u. \tag{21.3.7}$$

By combining the fact that $g^+ = 0$ for $k < 0$ with the causality constraint that u_n^{F} does not depend on the input u_j for $j > n$, it can be verified that the definition of convolution in (21.2.3) reduces to (21.3.7).

Spectral or frequency characterization of the recursive filters can be easily obtained by invoking the theory of discrete time Fourier transforms as described in

Appendix G. Let $u^{\mathrm{A}}(f)$, $u(f)$, and $G^{+}(f)$ denote the discrete time Fourier transforms (DFT) of $\{u_n^{\mathrm{F}}\}$, $\{u_n\}$, and $\{\mathbf{g}_{k+1}\}$, respectively. By taking the DFT on both sides of (21.3.7) we get

$$u^{\mathrm{A}}(f) = G^{+}(f)\,u(f). \tag{21.3.8}$$

That is, the DFT of the output is the product of the DFT of the filter with that of the input.

Let

$$\begin{aligned} u^{\mathrm{A}}(f) &= A_{0}(f)\,\mathrm{e}^{\mathrm{i}\,\theta_0(f)} \\ u(f) &= A_{\mathrm{I}}(f)\,\mathrm{e}^{\mathrm{i}\,\theta_{\mathrm{I}}(f)} \\ G^{+}(f) &= A_{g}(f)\,\mathrm{e}^{\mathrm{i}\,\theta_g(f)} \end{aligned} \tag{21.3.9}$$

be the standard polar representation of these Fourier transforms where $A(f)$ is the amplitude and $\theta(f)$ is the phase and $\mathrm{i} = \sqrt{-1}$. Substituting (21.3.9) into (21.3.8) it follows that

$$\begin{aligned} A_0(f) &= A_g(f)\,A_{\mathrm{I}}(f) \\ &\text{and} \\ \theta_0(f) &= \theta_g(f) + \theta_{\mathrm{I}}(f). \end{aligned} \tag{21.3.10}$$

Thus, for a given input sequence $\{u_n\}$, the properties of the output are uniquely determined by the properties of the filter $G^{+}(f)$. For the $\{g^{+}\}$ sequence defined in (21.3.6), it can be verified that

$$\left.\begin{aligned} G^{+}(f) &= \tfrac{(1-\alpha)}{1-\alpha \mathrm{e}^{-\mathrm{i}2\pi f}} \\ &= A_g^{+}(f)\,\mathrm{e}^{\mathrm{i}\theta_g^{+}(f)} \end{aligned}\right\} \tag{21.3.11}$$

where

$$|\,A_g^{+}(f)\,|^2 = \frac{(1-\alpha)^2}{(1-\alpha)^2 + 2\alpha\,[1 - \cos 2\pi f]} \tag{21.3.12}$$

and

$$\tan[\theta_g^{+}(f)] = -\frac{\alpha\,\sin 2\pi f}{1 - \alpha\,\cos 2\pi f}. \tag{21.3.13}$$

Thus, the first-order forward filter introduces a phase shift between the input and the output.

Recall that a Shuman-type non-recursive low-pass filter does not introduce any phase shift. Our goal is to design low-pass recursive filters that do not introduce any phase shift. A little reflection immediately suggests that a "**backward**" filter suitably designed can introduce a phase shift equal in magnitude but opposite in sign. Then, by combining one sweep of the forward filter followed by that of the backward filter we can obtain a recursive filter that potentially has no phase shift.

Motivated by this intuition, we now move on to describing the first-order backward filter.

(B) The first-order backward filter

Using the output u^{A} of the forward filter as the input, we now define a new sequence $u^{\mathrm{F}} = \{u_n^{\mathrm{F}}\}$ using

$$u_n^{\mathrm{F}} = \alpha\, u_{n+1}^{\mathrm{F}} + (1-\alpha)\, u_n^{\mathrm{A}}. \tag{21.3.14}$$

Notice that u_n^{F} is related to u_n^{A} in the same way as u_n^{A} is related to u_n except that they operate in opposite directions. In view of this similarity, the analysis of this backward filter is quite similar to that of (21.3.4). Iterating (21.3.14), it can be verified that

$$u_n^{\mathrm{F}} = \alpha u_{2n}^{\mathrm{F}} + (1-\alpha) \sum_{j=0}^{n-1} \alpha^j u_{n+j}^{\mathrm{A}}.$$

Since $0 < \alpha < 1$, for large n, we have

$$\begin{aligned} u_n^{\mathrm{F}} &= (1-\alpha) \sum_{j=0}^{n-1} \alpha^j u_{n+j}^{\mathrm{A}} \\ &= (1-\alpha) \sum_{i=0}^{-(n-1)} \alpha^{-i} u_{n-i}^{\mathrm{A}} \quad \text{(by change of indices)} \\ &= \sum_{i=0}^{-(n-1)} g_{\mathrm{I}}^{-} u_{n-i}^{\mathrm{F}} \end{aligned} \tag{21.3.15}$$

where

$$g_{\mathrm{I}}^{-} = \begin{cases} (1-\alpha)\,\alpha^{-i} &, \quad \text{for } i \le 0 \\ 0 &, \quad \text{for } i > 0 \end{cases} \tag{21.3.16}$$

The expression on the r.h.s.of (21.3.15) is the discrete convolution between the new weight sequence $g^{-} = \{g_{\mathrm{I}}^{-}\}$ and $\{u_n^{\mathrm{A}}\}$ which can be succinctly denoted by

$$u^{\mathrm{F}} = g^{-} * u^{\mathrm{A}}. \tag{21.3.17}$$

Now taking the DFT on both sides, we get

$$u^{\mathrm{F}}(f) = G^{-}(f) u^{\mathrm{A}}(f). \tag{21.3.18}$$

where $G^{-}(f)$, the DFT of $\{g_{\mathrm{I}}^{-}\}$ in (21.3.16) is given by

$$G^{-}(f) = \frac{(1-\alpha)}{1 - \alpha\, \mathrm{e}^{\mathrm{i}2\pi f}}. \tag{21.3.19}$$

Comparing this expression with $G^{+}(f)$ in (21.3.11), it follows that $G^{-}(f)$ is the complex conjugate of $G^{+}(f)$. Hence they have the same amplitude but are of opposite phase as desired. Hence, if

$$G^-(f) = A_g^-(f)\,e^{i\theta_g^-(f)} \tag{21.3.20}$$

then (Exercise 21.12)

$$A_g^-(f) = A_g^+(f) \quad \text{and} \quad \theta_g^-(f) = -\theta_g^+(f). \tag{21.3.21}$$

We now combine the forward and the backward filter to obtain the following:

(C) A recursive low-pass filter

Given $\{u_n\}$, for any $0 < \alpha < 1$, define two exponential weight sequences g^+ and g^- given in (21.3.6) and (21.3.16) respectively. Then one sweep of the forward filter using g^+ followed by one sweep of the backward filter using g^- gives

$$u^{\mathrm{F}} = g^- * u^{\mathrm{A}} \quad \text{and} \quad u^{\mathrm{A}} = g^+ * u \tag{21.3.22}$$

or equivalently, using the DFT we get

$$u^{\mathrm{F}}(f) = G^-(f)\,u^{\mathrm{A}}(f) \quad \text{and} \quad u^{\mathrm{A}}(f) = G^+(f)\,u(f).$$

Combining these, we get

$$u^{\mathrm{F}}(f) = [G^-(f)\,G^+(f)]\,u(f) = S(f)\,u(f). \tag{21.3.23}$$

Using (21.3.12)–(21.2.13) and (21.3.20)–(21.3.21) it follows that (see Example G.3.3 in Appendix G)

$$\begin{aligned} S(f) = G^-(f)\,G^+(f) &= \frac{(1-\alpha)^2}{(1-\alpha)^2 + 2\alpha[1 - \cos 2\pi f]} \\ &= \frac{(1-\alpha)^2}{1 + \frac{\alpha}{(1-\alpha)^2}[2\sin \pi f]^2}. \end{aligned} \tag{21.3.24}$$

That is, the attenuation factor $[G^-(f)\,G^+(f)]$ as a function of f is real, positive and is less than 1 for all f. Hence the combination defines a recursive low-pass filter.

(D) Choice of the filter parameter

One rationale for the choice of the filter parameter α is to match the spatial variance of the scalar field that is being filtered with the variance of the weights of this combined filter. This is done by combining the two relations in (21.3.22) to get

$$u^{\mathrm{F}} = g^- * (g^+ * u) = (g^- * g^+) * u = s * u \tag{21.3.25}$$

where $s = (g^- * g^+)$. It can be verified that, from Appendix G, $s = \{s_n\}$ with

$$s_{\pm n} = \left(\frac{1-\alpha}{1+\alpha}\right)\alpha^{|n|}. \tag{21.3.26}$$

The basic properties of this sequence are explored in Exercise (21.13), where it is shown that the mean of $s = \{s_n\}$ is zero and its variance is $2\alpha/(1-\alpha)^2$. Hence, if the variance of the input field is R^2, then we require that

$$R^2 = \frac{2\alpha\,(\Delta x)^2}{(1-\alpha)^2} \tag{21.3.27}$$

where $2\alpha/(1-\alpha)^2$ is the variance of $s = \{s_n\}$ (Exercise 21.13). Thus, given R^2, one can readily solve the quadratic equation

$$R^2(1-\alpha)^2 = 2\alpha(\Delta x)^2 \tag{21.3.28}$$

to obtain α.

Properties of the convolution of s with itself are examined in Exercise 21.14.

(E) Gaussian filter as the limit of recursive filters

From (21.3.22), we can relate $u^{\mathrm{F}}(f)$ and $u(f)$ using

$$u^{\mathrm{F}}(f) = S(f)\,u(f)$$

or equivalently, in the spatial domain using (21.3.25) as

$$u^{\mathrm{F}} = s * u$$

where $s(f)$ is the DFT of the sequence s. Let $o(f)$ be the output of an n-fold cascade of the filter $s(f)$, namely

$$o(f) = [s(f)]^n v(f)$$

or equivalently

$$\begin{aligned} O = \{o_n\} &= (s * s * \cdots * s) * u \\ &= (s^{*n}) * u. \end{aligned} \tag{21.3.29}$$

From the Example G.2.4 and the basic result relating to the repeated convolution in Appendix G, it follows that

$$(s^{*n}) \longrightarrow \text{Gaussian filter,}$$

whose variance is n times the variance of s. Hence, if R^2 is the measured variance of the input field, then the parameter α for the recursive filter is obtained by solving

$$R^2 = \frac{2\alpha n(\Delta x)^2}{(1-\alpha)^2}. \tag{21.3.30}$$

21.4 Higher-order recursive filters

In this section we first derive an implicit representation of the filter $s(f)$ described in Section 21.3. This representation easily lends itself to the design of higher-order recursive filters. First rewrite (21.3.3) and (21.3.14) as

$$u_n = \frac{1}{1-\alpha}\left[u_n^{\mathrm{A}} - \alpha u_{n-1}^{\mathrm{A}}\right] \tag{21.4.1}$$

and

$$u_n^{\mathrm{A}} = \frac{1}{1-\alpha}\left[u_n^{\mathrm{F}} - \alpha u_{n+1}^{\mathrm{F}}\right]. \tag{21.4.2}$$

Substituting the latter on the r.h.s.of (21.4.1) and simplifying, we get

$$u_n = u_{n-1}^{\mathrm{F}} - \frac{\alpha}{(1-\alpha)^2}\left[u_n^{\mathrm{F}} - 2u_n^{\mathrm{F}} + u_{n+1}^{\mathrm{F}}\right] \tag{21.4.3}$$

$$\approx [1 - aD^2]\, u_n^{\mathrm{F}}(x)|_{x=n\Delta x} \tag{21.4.4}$$

where $D^2 = \mathrm{d}^2/\mathrm{d}x^2$ and (by Exercise 2.13)

$$a = \frac{\alpha}{(1-\alpha)^2} = \frac{1}{2}\,\mathrm{Var}(s_n) \tag{21.4.5}$$

where $\mathrm{Var}(s_n)$ denotes the centered spatial second moment of $\{s_n\}$. Now, using (21.4.3), define a symmetric, diagonally dominant, tridiagonal matrix $\mathbf{M}$ (of infinite size) whose ith row is given by

$$\mathbf{M}_{i*} = [\cdots - a \quad 1 + 2a \quad - a \cdots].$$

Using this matrix, we can now represent u^{F} **implicitly** as

$$u = Mu^{\mathrm{F}} \tag{21.4.6}$$

where u and u^{F} are infinite vectors given by

$$u = (\ldots u_{-2},\, u_{-1},\, u_0,\, u_1,\, u_2, \ldots)$$

and

$$u^{\mathrm{F}} = (\ldots u_{-2}^{\mathrm{F}},\, u_{-1}^{\mathrm{F}},\, u_0^{\mathrm{F}},\, u_1^{\mathrm{F}},\, u_2^{\mathrm{F}}, \ldots).$$

Thus, one application of the filter is equivalent to computing

$$u^{\mathrm{F}} = \mathbf{M}^{-1} u. \tag{21.4.7}$$

Higher-order low-pass implicit filters are defined using functions of the second-order differential operator (21.4.4) and the Shuman-type averaging operator-defined by

$$A^2 = \frac{1}{4}[1 \quad 2 \quad 1]. \tag{21.4.8}$$

Following Raymond and Garder (1991) we can define several families of implicit filters as follows: for any integer $p \geq 1$ and a real number $\epsilon > 0$, define

Table 21.4.1 *Stencil for the differential operator* D^{2p}

p	stencil for D^{2p}						
1	[1	−2	1]				
	$n-1$	n	$n+1$				
2	[1	−4	6	−4	1]		
	$n-2$	$n-1$	n	$n+1$	$n+2$		
3	[1	−6	15	−20	15	−6	1]
	$n-3$	$n-2$	$n-1$	n	$n-1$	$n-2$	$n-3$

Table 21.4.2 *Stencil for the averaging operator* A^{2p}

p	stencil for A^{2p}
1	$\frac{1}{4}$[1 2 1]
2	$\frac{1}{16}$[1 4 6 4 1]
3	$\frac{1}{64}$[1 6 15 20 15 6 1]

(a) **Sine filter**

$$u_n = [1 + (-1)^p \epsilon D^{2p}] u_n^{\mathrm{F}} \tag{21.4.9}$$

(b) **Tangent filter**

$$[A^{2p}] u_n = [A^{2p} + (-1)^p \epsilon D^{2p}] u_n^{\mathrm{F}} \tag{21.4.10}$$

(c) **Cosine complement filter**

$$[A^{2p}] u_n = [A^{2p} + \epsilon] u_n^{\mathrm{F}} \tag{21.4.11}$$

The stencils for D^{2p} and A^{2p} for $p = 1, 2, 3$ are given in Table 21.4.1 and 21.4.2. Derivation and analysis of the spectral representation for these filters is pursued in Exercise (21.15).

21.5 Variational analysis using spatial filters

The standard approach to the 3DVAR seeks to minimize (Chapter 20)

$$J(\mathbf{x}) = J_{\mathrm{b}}(\mathbf{x}) + J_{\mathrm{O}}(\mathbf{x}) \tag{21.5.1}$$

where

$$J_{\mathrm{b}}(\mathbf{x}) = \frac{1}{2}(\mathbf{x} - \bar{\mathbf{x}})^{\mathrm{T}} \mathbf{B}^{-1} (\mathbf{x} - \bar{\mathbf{x}}) \tag{21.5.2}$$

and

$$J_0(\mathbf{x}) = \frac{1}{2}(\mathbf{z} - \mathbf{Hx})^{\mathrm{T}}\mathbf{R}^{-1}(\mathbf{z} - \mathbf{Hx}) \tag{21.5.3}$$

where $\mathbf{x}$, $\bar{\mathbf{x}} \in \mathbb{R}^n$, $\mathbf{z} \in \mathbb{R}^m$, $\mathbf{B} \in \mathbb{R}^{n\times n}$ and $\mathbf{H} \in \mathbb{R}^{m\times n}$. The vector $\bar{\mathbf{x}}$ is called the **background** which is the prior information and $\mathbf{B}$ is the covariance of the error in $\bar{\mathbf{x}}$. The vector $\mathbf{z}$ is the observation vector and $\mathbf{R}$ denotes the covariance of the observational errors. The minimizing $\mathbf{x}$ is given by the solution of the linear system (Chapter 20)

$$(\mathbf{B}^{-1} + \mathbf{H}^{\mathrm{T}}\mathbf{R}^{-1}\mathbf{H})(\mathbf{x} - \bar{\mathbf{x}}) = \mathbf{H}^{\mathrm{T}}\mathbf{R}^{-1}\,[\mathbf{z} - \mathbf{H}\bar{\mathbf{x}}] \tag{21.5.4}$$

where $(\mathbf{x} - \bar{\mathbf{x}})$ is called the **analysis increment** and $(\mathbf{z} - \mathbf{H}\bar{\mathbf{x}})$ is called the **innovation**. Much of the challenge associated with solving this linear system relates to the properties of the matrix on the l.h.s. of 21.5.4 – in particular, the knowledge of the background error covariance matrix $\mathbf{B}$ and the spectral condition number of the matrix $(\mathbf{B}^{-1} + \mathbf{H}^{\mathrm{T}}\mathbf{R}^{-1}\mathbf{H})$.

(A) Models for the background error covariance matrix B

One of the most useful and elegant assumption about the spatial correlation is that it has a homogeneous and isotropic spatial structure specified by the Gaussian structure. Given $\mathbf{B}$, define

$$\mathbf{E} = \mathrm{Diag}(\mathbf{B}_{11}^{1/2}, \mathbf{B}_{22}^{1/2}, \ldots, \mathbf{B}_{nn}^{1/2}) \tag{21.5.5}$$

the diagonal matrix of the standard deviations. Then

$$\mathbf{E}^{-1/2}\mathbf{B}\mathbf{E}^{-1/2} \tag{21.5.6}$$

is the correlation matrix corresponding to the covariance matrix $\mathbf{B}$. The idea is to require that this correlation has a Gaussian structure.

Recall from Section 21.3 that the n-fold cascade of the filter $s(f)$ has the Gaussian structure. This in turn implies that we could hope to realize the Gaussian spatial correlation implied by the matrix $\mathbf{B}$ by repeated application of the recursive spatial filters.

(B) Conditioning of the matrix $(\mathbf{B}^{-1} + \mathbf{H}^{\mathrm{T}}\mathbf{R}^{-1}\mathbf{H})$

The sensitivity and hence the quality of the solution of (21.5.4) is directly related to the spectral condition number of $(\mathbf{B}^{-1} + \mathbf{H}^{\mathrm{T}}\mathbf{R}^{-1}\mathbf{H})$. Recall that this condition number amplifies the small errors resulting from the finite precision arithmetic (Appendix B). A standard method for **taming** the condition number is to use the concept of preconditioning (Chapter 12). It turns out that we can indeed exploit the recursive-filter-based realization of $\mathbf{B}$ as a tool for preconditioning the matrix $(\mathbf{B}^{-1} + \mathbf{H}^{\mathrm{T}}\mathbf{R}^{-1}\mathbf{H})$. The actual link between the choice of precondition and the recursive filter is provided by designing the filter to match the square root of the matrix $\mathbf{B}$.

Stated in other words, the versatility of the recursive spatial filters relate to their ability **to kill two birds in one shot** – realize the Gaussian spatial correlation as well as to help tame the condition number. We describe this design in two stages.

(a) **Preconditioning** By way of motivating the need for preconditioning, we begin by analyzing the properties of the system matrix in (21.5.4). Recall that while the matrix $(\mathbf{B}^{-1} + \mathbf{H}^{\mathrm{T}}\mathbf{R}^{-1}\mathbf{H})$ is positive definite, we do not have any control over its spectrum. Its maximum eigenvalue can be very large and/or its minimum eigenvalue while remaining positive can be very small. Thus, the spectral condition number which is the ratio of the maximum to the minimum eigenvalues can indeed be very large. One standard method to **squeeze** the spectrum of $(\mathbf{B}^{-1} + \mathbf{H}^{\mathrm{T}}\mathbf{R}^{-1}\mathbf{H})$ is to use a form of preconditioning. This is accomplished by a suitable coordinate transformation which is described below.

Recall from Chapter 9 that any symmetric matrix $\mathbf{B}$ can be factored multiplicatively as

$$\mathbf{B} = \mathbf{C}\mathbf{C}^{\mathrm{T}}. \tag{21.5.7}$$

Using this factor matrix $\mathbf{C}$, define a linear transformation of the variables in (21.5.1) as

$$(\mathbf{x} - \bar{\mathbf{x}}) = \mathbf{C}\mathbf{w}. \tag{21.5.8}$$

Substituting (21.5.8) into (21.5.1), we get a new representation of the functional $J(\cdot)$ in the new coordinate system:

$$J(\mathbf{w}) = \frac{1}{2}\mathbf{w}^{\mathrm{T}}\mathbf{w} + \frac{1}{2}(\hat{\mathbf{z}} - \mathbf{H}\mathbf{C}\mathbf{w})^{\mathrm{T}}\mathbf{R}^{-1}(\hat{\mathbf{z}} - \mathbf{H}\mathbf{C}\mathbf{w}) \tag{21.5.9}$$

where $\hat{\mathbf{z}} = \mathbf{z} - \mathbf{H}\,\bar{\mathbf{x}}$. Then, the minimizer of $J(\mathbf{w})$ is given by the solution of the linear system

$$[\mathbf{I} + \mathbf{C}^{\mathrm{T}}\mathbf{H}^{\mathrm{T}}\mathbf{R}^{-1}\mathbf{H}\mathbf{C}]\mathbf{w} = \mathbf{C}^{\mathrm{T}}\mathbf{H}^{\mathrm{T}}\mathbf{R}^{-1}\hat{\mathbf{z}}. \tag{21.5.10}$$

We now examine the spectral properties of the new matrix on the l.h.s. of (21.5.10). It can be verified that $\mathbf{C}^{\mathrm{T}}\mathbf{H}^{\mathrm{T}}\mathbf{R}^{-1}\mathbf{H}\mathbf{C} \in \mathbb{R}^{n\times n}$, $\mathrm{Rank}(\mathbf{C}) = n$, and $\mathrm{Rank}(\mathbf{C}^{\mathrm{T}}\mathbf{H}^{\mathrm{T}}\mathbf{R}^{-1}\mathbf{H}\mathbf{C}) = m$. Hence, the eigenvalues of $(\mathbf{C}^{\mathrm{T}}\mathbf{H}^{\mathrm{T}}\mathbf{R}^{-1}\mathbf{H}\mathbf{C})$ are such that

$$\lambda_1 \geq \lambda_2 \geq \cdots \geq \lambda_m > \lambda_{m+1} = \lambda_{m+2} = \cdots = \lambda_n = 0.$$

Thus, the condition number of $\mathbf{C}^{\mathrm{T}}\mathbf{H}^{\mathrm{T}}\mathbf{R}^{-1}\mathbf{H}\mathbf{C}$ is infinite. Recall that if λ is an eigenvalue of $\mathbf{A}$, then $(1 + \lambda)$ is the corresponding eigenvalue of $(\mathbf{I} + \mathbf{A})$. Accordingly the eigenvalues μ_i of $(\mathbf{I} + \mathbf{C}^{\mathrm{T}}\mathbf{H}^{\mathrm{T}}\mathbf{R}^{-1}\mathbf{H}\mathbf{C})$ are such that $\mu_{\mathrm{I}} = 1 + \lambda_i$ and

$$\mu_1 \geq \mu_2 \geq \cdots \geq \mu_m \geq \mu_{m+1} = \mu_{m+2} = \cdots = \mu_n = 1.$$

Herein lies the impact of the preconditioning – the smallest eigenvalue is bounded below by unity. Hence μ_1 is indeed the condition number of the matrix on the l.h.s of (21.5.10).

(b) A Filter-based Implementation of C
To get a grip on the basic ideas, consider the problem in one space dimension. If **B** is the given covariance matrix, then its corresponding correlation structure is given by the matrix in (21.5.6). Then, in the recursive filter based approach, this correlation is realized as follows:

$$\mathbf{E}^{-1/2}\mathbf{B}\mathbf{E}^{-1/2} = [(\mathbf{D}_x\mathbf{C}_x)(\mathbf{D}_x\mathbf{C}_x)^{\mathrm{T}}]^n \tag{21.5.11}$$

where $\mathbf{D}_x$ is a normalizing diagonal matrix and $\mathbf{C}_x$ is a matrix representation of the filter defined by the differential operator in (21.4.4), namely

$$\mathbf{C}_x^{-1} = [1 - a\mathbf{D}_x^2]. \tag{21.5.12}$$

Using the standard three-point stencil

$$\mathbf{D}_x^2 = \frac{\partial^2}{\partial x^2} \approx \left[\begin{array}{ccc} 1 & -2 & 1 \\ i-1 & i & i+1 \end{array} \right]$$

we get

$$(\mathbf{C}_x^{-1} g)_i = -ag_{i-1} + (1+2a)g_i - ag_{i+1}.$$

Choosing appropriate boundary conditions (Hayden and Purser 1995), it can be verified that $\mathbf{C}_x^{-1}$ has the following tridiagonal structure:

$$\mathbf{C}_x^{-1} = \begin{bmatrix} c_1 & b_1 & 0 & 0 & 0 & 0 \\ b_1 & c_2 & b_1 & 0 & 0 & 0 \\ 0 & b_1 & c_2 & b_1 & 0 & 0 \\ 0 & 0 & b_1 & c_2 & b_1 & 0 \\ 0 & 0 & 0 & b_1 & c_2 & b_1 \\ 0 & 0 & 0 & 0 & b_1 & c_1 \end{bmatrix}$$

where $b_1 = -a$, $c_2 = 1 + 2a$, and $c_1 = 1 + a$. The normalizing diagonal matrix is computed as follows: the ith diagonal element D_{ii} of $\mathbf{D}_x$ is given by

$$D_{ii} = R_{ii}^{1/2} \tag{21.5.13}$$

where R_{ii} is the ith diagonal entry of the product $\mathbf{C}_x\mathbf{C}_x^{\mathrm{T}}$. Since the l.h.s of (21.5.11) is a correlation matrix, the diagonal matrix $\mathbf{D}_x$ so defined normalizes the product $\mathbf{C}_x\mathbf{C}_x^{\mathrm{T}}$ so that the r.h.s.of (21.5.11) is also a correlation matrix. The minimization of $J(\mathbf{w})$ in (21.5.9) can be achieved by using the standard gradient algorithm (Chapter 10) using

$$\mathbf{w}_{k+1} = \mathbf{w}_k - \alpha\,\nabla J(\mathbf{w}_k) \tag{21.5.14}$$

or by using the conjugate gradient algorithm in Chapter 11. In either case, the idea is whenever multiplication of a vector by the matrix **C** is encountered, this operation is replaced by the recursive filter operation on the vector.

(C) An Alternate Transformation Huang (2000) used another transformation using **B** instead of its square root **C** to obtain a form suitable for the application of spatial filters. Let

$$(\mathbf{x} - \bar{\mathbf{x}}) = \mathbf{B}\mathbf{V}. \tag{21.5.15}$$

Substituting this in (21.5.1), the latter becomes

$$J(\mathbf{V}) = \frac{1}{2}\mathbf{V}^{\mathrm{T}}\mathbf{B}\mathbf{V} + \frac{1}{2}[\hat{\mathbf{z}} - \mathbf{H}\mathbf{B}\mathbf{V}]^{\mathrm{T}}\mathbf{R}^{-1}[\hat{\mathbf{z}} - \mathbf{H}\mathbf{B}\mathbf{V}]. \tag{21.5.16}$$

Then

$$\nabla J(\mathbf{V}) = \mathbf{B}[\mathbf{V} + \mathbf{H}^{\mathrm{T}}\mathbf{R}^{-1}(\mathbf{B}\mathbf{V} - \hat{\mathbf{z}})]. \tag{21.5.17}$$

The minimum of $J(\mathbf{V})$ can be achieved using the gradient algorithm:

$$\mathbf{V}_{k+1} = \mathbf{V}_k - \alpha\,\nabla J(\mathbf{V})$$

where the multiplication of the vector **B** is replaced by the application of the recursive filter on that vector. By setting $V_0 = 0$, there is no need for inverting the matrix **B**. Alternatively, we could also apply the conjugate gradient method for minimizing $J(\mathbf{V})$.

Exercises

21.1 For the $L(\lambda, c)$ in (21.2.12), compute the derivatives $\partial L/\partial c_i$ and $\partial L/\partial \lambda$. By setting these derivatives to zero, verify that the minimizing value of $c_i = \frac{1}{2N+1}$.

21.2 Verify the correctness of (21.2.19)–(21.2.20).

21.3 Using the definition in (21.2.21), verify the expression for u_{ij}^{F} in (21.2.22).

21.4 Using u_{ij} in (21.2.23), verify the expression for A^{F}/A in (21.2.25) using the nine-point operator in (21.2.22).

21.5 Verify the expression for A^{F}/A for the five-point stencil given in (21.2.25).

21.6 **Shapiro (1975) filters** Define the half grid length difference operator

$$\delta\,u_i = u_{i+1/2} - u_{i-1/2}.$$

Then $\delta^2\,u_i = \delta(\delta(u_i)) = u_{i-1} - 2u_i + u_{i+1}$.

(1) Verify that

$$\left(1 + \frac{\delta^2}{4}\right)u_i = \frac{1}{4}[u_{i-1} + 2u_i + u_{i+1}]$$

is the Shuman operator in (21.2.15) with $c = 1/2$.

(2) Compute the stencils for the following operators:

(*a*) $\left(1 - \frac{\delta^2}{4}\right)\left(1 + \frac{\delta^2}{4}\right)u_i$

(*b*) $\left(1 + \frac{\delta^4}{16}\right)\left(1 - \frac{\delta^4}{16}\right)u_i$

(3) Prove or disprove

$$\left(1-\frac{\delta^2}{4}\right)\left(1+\frac{\delta^2}{4}\right)=\left(1+\frac{\delta^2}{4}\right)\left(1-\frac{\delta^2}{4}\right)$$

and

$$\left(1-\frac{\delta^2}{16}\right)\left(1+\frac{\delta^2}{16}\right)=\left(1+\frac{\delta^2}{16}\right)\left(1-\frac{\delta^2}{16}\right).$$

21.7 **Shapiro filters (1975)** Recall that we can expand

$$\left[1+\left(\frac{\delta}{2}\right)^2\right]^{-1}=1-\left(\frac{\delta}{2}\right)^2+\left(\frac{\delta}{2}\right)^4-\left(\frac{\delta}{2}\right)^6+\cdots$$

For $p = 1, 2, 3, \ldots$, we can define a family of higher-order filters using

$$\left[1+\left(\frac{\delta}{2}\right)^2\right]\left[1-\left(\frac{\delta}{2}\right)^2+\left(\frac{\delta}{2}\right)^4-\cdots(-1)^p\left(\frac{\delta}{2}\right)^{2p}\right]u_i.$$

Compute the stencils for $p = 1, 2, 3$, and 4.

21.8 Consider the moving average filter

$$u_n^{\mathrm{F}} = u_n + \frac{c}{2}[u_{n-1} - 2u_n + u_{n+1}].$$

(*a*) Verify that

$$\frac{1}{2N+1}\sum_{i=-N}^{N} u_n^{\mathrm{F}} = \frac{1}{2N+1}\sum_{i=-N}^{N} u_n + \frac{c}{4N+2}[u_{-N-1} - u_{-N} - u_p + u_{p+1}].$$

(*b*) Verify that the average of the filtered quantity on the l.h.s. converges to the average of the unfiltered quantity given by the first term on the r.h.s. as $N \to \infty$.

21.9 Using the recurrence (21.3.3), derive the formula in (21.3.4) by successive substitution.

21.10 **Properties of the sequence g^+ in (21.3.6)** Let $g_k^+ = (1-\alpha)\alpha^k$ for $k \geq 0$ and $g_k^+ = 0$ for $k < 0$.

(*a*) Verify that $\sum_{k=0}^{\infty} g_k^+ = 1$. That is, g_k^+ defines the **discrete exponential probability distribution** on the integers $\{0, 1, 2, 3, \ldots\}$ where the integer is associated with the probability g_k^+.

(*b*) Show that the **mean** of this distribution is given by

$$M_1 = \sum_{k=0}^{\infty} k g_k^+ = \frac{\alpha}{1-\alpha}.$$

Hint: Recall $\sum_{i=0}^{\infty}\alpha^i = \frac{1}{1-\alpha}$. Differentiating both sides we get $\sum_{i=0}^{\infty} i\alpha^{i-1} = \frac{1}{(1-\alpha)^2}$.

(*c*) Show that the **second moment** of this distribution is given by

$$M_2 = \sum_{k=0}^{\infty} k^2 g_k^+ = \frac{\alpha(1+\alpha)}{(1-\alpha)^2}.$$

(*d*) Verify that the variance of this distribution is given by

$$M_2 - M_1^2 = \frac{\alpha}{(1-\alpha)^2}.$$

(*e*) Repeat the computations in (*a*) through (*d*) for the dual sequence g^- where $g_k^- = (1-\alpha)\,\alpha^{-k}$ for $k \le 0$ and $g_k^- = 0$ for $k > 0$.

21.11 **Continuous version of the sequence** g^+
Define

$$g^+(x) = \begin{cases} 1/\lambda \exp(-x/\lambda) & , \quad \text{for } x \ge 0 \\ 0 & , \quad \text{for } x < 0 \end{cases}$$

(*a*) Verify that $\int_0^\infty g^+(x)\,\mathrm{d}x = 1$.
(*b*) Verify that the first moment

$$M_1 = \int_0^\infty x g^+(x)\,\mathrm{d}x = \lambda.$$

(*c*) Verify that the second moment

$$M_2 = \int_0^\infty x^2 g^+(x)\,\mathrm{d}x = 2\lambda^2.$$

(*d*) Verify that the variance $\sigma^2 = M_2 - M_1^2 = \lambda^2$.
(*e*) Repeat the computations in (*a*) through (*d*) for the dual function $g^-(x) = \frac{1}{\lambda}\exp(x/\lambda)$ for $x \le 0$ and $g^-(x) = 0$ for $x > 0$.

21.12 Using the expression for $G^+(f)$ in (21.3.11) and $G^-(f)$ in (21.3.19), verify the claim in (21.3.21).

21.13 **Properties of the sequence** $s_{\pm n}$ **in (21.3.26)**
(1) Verify that $\sum_{n=-\infty}^{\infty} s_n = 1$.
(2) Show that the mean $M_1 = \sum_{n=-\infty}^{\infty} n s_n = 0$.
(3) Show that the second moment

$$M_2 = \sum_{n=-\infty}^{\infty} n^2 s_n = \frac{2\alpha}{(1-\alpha)^2}.$$

(4) Since the mean $M_1 = 0$, $\mathrm{Var}(s_n) = M_2$.
Remark If x^+ and x^- are two independent random variables with distributions g^+ and g^- respectively, then the distribution of $x = x^+ + x^-$ is given by s which is the **convolution** of g^+ and g^- (Appendix F). Clearly the variance of x is the sum of the variances of x^+ and x^-.

21.14 Let s be the sequence given in (21.3.25). Compute $s^{*2} = s * s$, the convolution of s with itself.

(*a*) Verify that the mean of the this sequence s^{*2} is zero and its variance is twice the variance of s.

(*b*) If s^{*n} is the n-fold convolution of s with itself, then verify that the mean of s^{*n} is zero and its variance is n times the variance of s.

21.15 **Higher-order implicit filters** [Raymond and Garder (1991)]

(*a*) Let $u_n(k) = A\,\mathrm{e}^{\mathrm{i}kn\Delta x}$. Compute $u_n^{\mathrm{F}}(k)$ for the three filters in (21.4.9)–(21.4.11).

(*b*) Compute the ratio $H(k) = u_n^{\mathrm{F}}(k)/u_n(k)$ for each of these filters.

(*c*) Plot the variation of the amplitude of $H(k)$ similar to the plot in Figure 21.2.3.

Notes and references

Section 21.1 For a comprehensive discussion of the classification, design and application of digital filters refer to Oppenheim and Schaffer (1975), Hamming (1989).

Section 21.2 Shuman (1957), Shapiro (1970) and (1975), Whittlesey (1964), Assselin (1972), Robert (1966) and Orszag (1971) contain a thorough discussion of the analysis and application of non-recursive filters. Also refer to Hamming (1989) and Oppenheim and Schaffer (1975).

Section 21.3 Application of recursive filters to data smoothing problem in meteorology began with the work of Purser and McQuigg (1985). Since then it has been extended in several directions – refer to Hayden and Purser (1988) and (1995), Purser (1987), Lorenc (1986), (1992) and (1997), Devenyi and Benjamin (1998), Wu, Purser and Parrish (2002). The papers by Raymond (1988), Raymond and Garder (1988) and Raymond and Garder (1991) contain a thorough discussion and a review of the literature in this area. The concept of Gaussian filters and their use in meteorology began with the paper by Barnes (1964).

Section 21.4 Raymond and Garder (1988) and Raymond and Garder (1991) contain an excellent discussion of the design and applications of higher-order filters.

Section 21.5 Application of recursive filters to the problem of 3-d variational analysis problem began with Purser and McQuigg in 1982. See Hayden and Purser (1988) and (1995), Lorenc (1997), Huang (2000) and Wu, Purser and Parrish (2002) for further details. Development of recursive filters has been an extremely active area of research. Refer to Purser et al. (2003a) and (2003b) for a discussion of recursive filters to model spatially inhomogeneous and anisotropic covariance functions.

Thiébaux and Pedder (1987), Schlatter (1975), and Daley (1991) has a good discussion on modelling spatial correlation.

PART VI

Data assimilation: deterministic/dynamic models

22

Dynamic data assimilation: the straight line problem

In this opening chapter of Part VI, we introduce the basic principles of data assimilation using the now classical Lagrangian framework. This is done using a very simple dynamical system representing a particle moving in a straight line at a constant velocity, and hence the title "straight line problem".

In Section 22.1 the statement of the problem is given. First by reformulating this problem as one of fitting data to a straight line, we compute the required solution in closed form in Section 22.2. This solution is used as a benchmark against which the basic iterative scheme for data assimilation is compared. The first introduction to the iterative algorithm (which has come to be known as the adjoint method) for data assimilation is derived in Section 22.3. Section 22.4 describes a practical method for experimental analysis of this class of algorithms based on the notion of Monte Carlo type twin experiments.

22.1 A statement of the inverse problem

We begin by describing a common physical phenomenon of interest in everyday life. A particle is observed to be moving in a **straight line** at a constant velocity, say, $\alpha > 0$. The problem is to model the dynamics of motion of this particle and then **predict** its position at a future instant in time, say $t > 0$.

Let $x(t) \in \mathbb{R}$ denote the state representing the position of the particle at time $t \geq 0$, where $x(t_0) = x_0$ is the **initial state**. Refer to Figure 22.1.1.

In this setup, the dynamics of motion of the particle can be adequately described by the following.

Model equation

$$\frac{\mathrm{d}x}{\mathrm{d}t} = \alpha, \quad \text{where } x(0) = x_0. \tag{22.1.1}$$

Clearly, the model solution

$$x(t) = x_0 + \alpha t \tag{22.1.2}$$

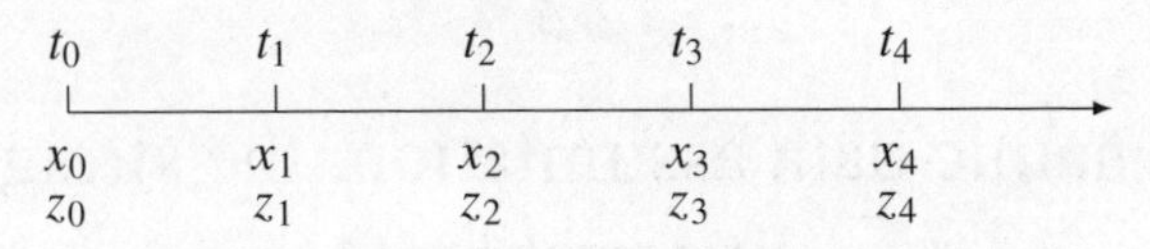

Fig. 22.1.1 A pictorial view.

denotes the position of the moving particle at time t. Since x_0 and α together "control" the position of the particle, the vector $\mathbf{c} = (x_0, \alpha)^T \in \mathbb{R}^2$ in meteorological circles is often called the **control vector** or simply **control**.

Assumption 22.1.1 It is assumed that both x_0 and α are **not** known a priori.

Indeed, if the control is known, then using (22.1.2) we know the position of the particle for all times. This is called the **direct** problem. It is this assumption about the lack of a priori knowledge of the control that makes the assimilation and prediction problem non-trivial and interesting. Thus, in the absence of a priori knowledge about the control $\mathbf{c}$, to make a meaningful prediction of the position of the particle, we must first concentrate on **estimating** the control, which has come to be known as the **inverse** problem. Once a reliable estimate of the control is obtained, we can **predict** the position at a future time as required, by using (22.1.2).

To this end, we **measure** the state (position) of the particle at prescribed time instances,

$$0 = t_0 < t_1 < t_2 < \cdots < t_N. \tag{22.1.3}$$

Observations Let

$$z_0 < z_1 < z_2 < \cdots < z_N \tag{22.1.4}$$

denote the **observed** positions of the particle where z_i is the observation of the position at time epoch t_i for $i = 0, 1, 2, \ldots, N$. Since it is almost a curse that we rarely measure without error, we once again model the observation as the sum of true position plus a random component as follows:

$$z_i = x(t_i) + v_i \tag{22.1.5}$$

where v_i denotes the **measurement noise**. For convenience we make the following assumptions on the noise.

Assumption 22.1.2
(1) v_i's are random variables which are mutually independent and identically distributed.
(2) $E(v_i) = 0$, $E(v_i^2) = \sigma^2$, $E(v_i v_j) = 0$ for all $i \neq j$.

In (22.1.5), it is assumed that the noise corrupting the observation is **additive**. There are other ways in which noise can enter, for example in a **multiplicative** fashion. Additive noise is easier to handle than their multiplicativej counterparts.

Fundamental to any estimation problem is a choice of a criterion or optimality condition. There are basically two rules that govern the choice of these criteria. First, it must have **physical significance** and second it must be **mathematically tractable**. The least squares criterion introduced by Gauss and Legendre two centuries ago remains the standard. For the problem on hand, this criterion can be expressed as follows.

Criterion

$$J(\mathbf{c}) = \sum_{i=0}^{N} \frac{1}{\sigma^2}(x(t_i) - z_i)^2 . \tag{22.1.6}$$

Notice that $J(\mathbf{c})$ denotes the sum of the squares of the difference between the state of the model as seen through the model (22.1.1) and the observations in (22.1.5).

In (22.1.6) while z_i's are the known observations, $x(t_i)$, the state at time t_i is **not** a free variable. In fact, $x(t_i)$'s depend indirectly on the control **c** through the model dynamics in (22.1.1). Thus, $J(\mathbf{c})$ in (22.1.6) is **not** an explicit function of **c**. In fact, much of the difficulty and hence the challenge in data assimilation is due to the fact that the criterion is **not** an explicitly known function of the control.

With all the basic ingredients in place we now state a prototype of the dynamic data assimilation problem.

Statement of the inverse problem Given the **deterministic model** (22.1.1) and a set of **noisy observations** $\{z_i \mid 0 \leq i \leq N\}$, the problem is to estimate the control **c** such that it minimizes $J(\mathbf{c})$ in (22.1.6) when the $x(t_i)$'s are constrained to be the solution of the model equation (22.1.1).

Thus, a typical assimilation problem is recast as a **minimization** of a cost functional subject to an **equality constraint** defined by the model dynamics.

The basic tools for characterizing the solution of this class of constrained minimization problem and the algorithms for finding the minimizing solutions are carefully developed in Chapters 10–12 in Part III, and in solving the above assimilation problem, we heavily draw upon the methodology of these chapters. A word of caution is in order, however. In these chapters on optimization, it is often assumed that the functional to be minimized is known **explicitly** as the function of the variables with respect to which the minimization is sought. A typical example is as follows: minimize $\phi(x_1, x_2) = -x_1 x_2$ when $x_1 + x_2 = 1$. In the dynamic data assimilation problem of interest in meteorology we often seek to minimize a functional $J(\mathbf{c})$ when it is **not** known explicitly. The functional $J(\mathbf{c})$ depends on **c** through the states that are defined by the dynamical equations.

Since the model solution (22.1.2) defines a straight line with slope $\alpha > 0$ and intercept x_0, the above problem is often called the **straight line problem**. From this angle, this problem reduces to a familiar problem of "**fitting a straight line**" to a set of observations.

While the straight line problem is fairly simple and is usually introduced in a first course on statistics and numerical analysis, our interest in this problem

Table 22.1.1 *Observations of a moving particle*

	$i = 0$	1	2	3
t_i	0.0	1.0	2.0	3.0
z_i	1.0	3.0	2.0	3.0

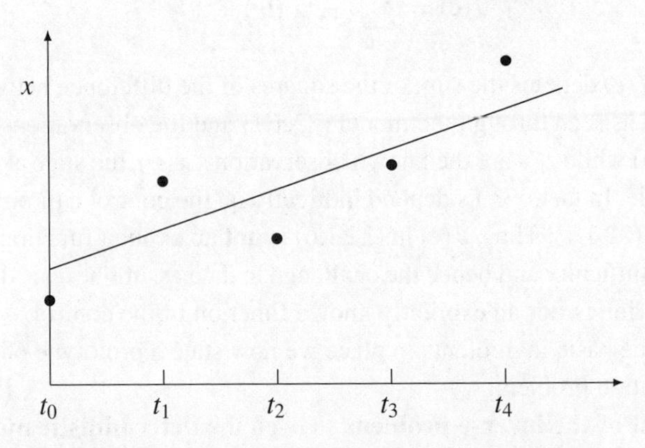

Fig. 22.1.2 Another look at the straight line problem.

stems from different directions. First, since the estimates of the control can be computed in closed form (see Section 22.2 below), this enables us to compare the goodness of iterative optimization schemes for data assimilation. Second, the control is a two-dimensional vector which enables us to plot the contours of $J(\mathbf{c})$ in analyzing the progress of the iterates of the minimization process. Third, the two components of the control vector, x_0 and α, are not dimensionally similar – x_0 is a distance and α is the velocity. This would enable us to demonstrate the difficulties associated with estimating the control with dimensionally heterogeneous components by appropriately scaling the variables.

We conclude this section by providing a numerical example of this problem.

Example 22.1.1 Let (z_i, t_i) be given as in Table 22.1.

Assuming $\sigma^2 = 1$, we can readily verify, using $x(t_i) = x_0 + \alpha t_i$ and $t_i = i$ for i = 0,1,2, and 3, that

$$\begin{aligned} J(\mathbf{c}) &= (x(t_0) - z_0)^2 + (x(t_1) - z_1)^2 + (x(t_2) - z_2)^2 + (x(t_3) - z_3)^2 \\ &= 4x_0^2 + 12x_0\alpha + 14\alpha^2 - 2x_0\,(z_0 + z_1 + z_2 + z_3) \\ &\quad - 2\alpha\,(z_1 + 2z_2 + 3z_3) + \left(z_0^2 + z_1^2 + z_2^2 + z_3^2\right). \end{aligned}$$

Letting $\mathbf{c} = (x_0, \alpha)^{\mathrm{T}}$, $J(\mathbf{c})$ can be written using the vector matrix notation as follows:

$$J(\mathbf{c}) = \frac{1}{2}\mathbf{c}^{\mathrm{T}}\mathbf{A}\mathbf{c} + \mathbf{b}^{\mathrm{T}}\mathbf{c} + d \qquad (22.1.7)$$

where

$$\begin{aligned}\mathbf{A} &= \begin{bmatrix} 8 & 12 \\ 12 & 28 \end{bmatrix} \\ \mathbf{b} &= \begin{bmatrix} -2\,(z_0 + z_1 + z_2 + z_3) \\ -2\,(z_1 + 2z_2 + 3z_3) \end{bmatrix} \\ d &= \left(z_0^2 + z_1^2 + z_2^2 + z_3^2\right),\end{aligned}$$

substituting the values of z_i from Table 22.1 we have $\mathbf{b} = (-18, -32)^{\mathrm{T}}$ and $d = 23$.

22.2 A closed form solution

The idea is to convert $J(\mathbf{c})$ in (22.1.6) into an explicit function of $\mathbf{c}$ by substituting $x(t_i) = x_0 + \alpha t_i$ using (22.1.2). Thus, $J(\mathbf{c})$ becomes

$$J(\mathbf{c}) = \frac{1}{\sigma^2}\sum_{i=0}^{N}(x(t_i) - z_i)^2 = \frac{1}{\sigma^2}\sum_{i=0}^{N}(x_0 + \alpha t_i - z_i)^2 \qquad (22.2.1)$$

which is clearly a quadratic polynomial in x_0 and α.

Let

$$\nabla J(\mathbf{c}) = \left(\frac{\partial J}{\partial x_0}, \frac{\partial J}{\partial \alpha}\right)^{\mathrm{T}} \qquad (22.2.2)$$

denote the **gradient** vector of J with respect to the components x_0 and α of $\mathbf{c}$. From the basic principles of minimization (refer to Chapters 10–12), it follows that the minimizing value of $\mathbf{c}$ is obtained by solving

$$\nabla J(\mathbf{c}) = 0. \qquad (22.2.3)$$

Differentiating (22.2.1) with respect to x_0 and α in turn, (22.2.3) in component form can be expressed as follows (Exercise 22.1)

$$(N+1)\,x_0 + S_t\alpha = S_z, \quad S_t x_0 + S_{t^2}\alpha = S_{tz} \qquad (22.2.4)$$

where it can be verified that

$$S_t = \sum_{i=0}^{N} t_i, \quad S_z = \sum_{i=0}^{N} z_i, \quad S_{t^2} = \sum_{i=0}^{N} t_i^2, \quad S_{tz} = \sum_{i=0}^{N} t_i\, z_i. \qquad (22.2.5)$$

Solving (22.2.4), the minimizing $\mathbf{c}^* = (x_0^*, \alpha^*)^{\mathrm{T}}$ is given by (Exercise 22.2)

$$\alpha^* = \frac{S_{tz} - \frac{S_t S_z}{N+1}}{S_{t^2} - \frac{S_t^2}{N+1}} = \frac{\sum_{i=0}^{N} (z_i - \bar{z})\,(t_i - \bar{t})}{\sum_{i=0}^{N} (t_i - \bar{t})^2} = \frac{\sigma_{zt}^2}{\sigma_t^2} \tag{22.2.6}$$

where

$$\bar{t} = \frac{S_t}{N+1} \quad \text{and} \quad \bar{z} = \frac{S_z}{N+1} \tag{22.2.7}$$

and

$$x_0^* = \frac{S_z}{N+1} - \frac{S_t}{N+1}\,\alpha^* = \bar{z} - \bar{t}\alpha^*. \tag{22.2.8}$$

Example 22.2.1 We illustrate these computations using the data in Table 22.1.1. From Table 22.1, it can be verified that

$$\begin{aligned} S_t &= 6.0 & S_z &= 9.0 \\ S_{t^2} &= 14.0 & S_{tz} &= 16.0. \end{aligned}$$

Thus, (22.2.4) becomes

$$\begin{aligned} 4.0x_0 + 6.0\alpha &= 9.0 \\ 6.0x_0 + 14.0\alpha &= 16.0. \end{aligned}$$

Solving these we obtain

$$x_0^* = 1.5 \quad \text{and} \quad \alpha^* = 0.5.$$

Thus, the regression line is given by $x(t) = 1.5 + 0.5t$.

Remark 22.2.1 An astute reader may have already noticed the fact that we have converted a constrained minimization problem into an unconstrained problem. This was possible by substituting $x(t_i)$ in terms x_0 and α using the model equation. While in principle this is always possible, except in simple cases such as the straight line problem, this substitution becomes infeasible. The use of standard packages for symbolic manipulation such as the MAPLE, MATHEMATICA could mitigate this difficulty (Exercise 22.3). Since the model equation (22.2.1) is linear, in this case $J(\mathbf{c})$ is a quadratic polynomial in the components of $\mathbf{c}$. In general, when the model is a non-linear model, understandably, $J(\mathbf{c})$ will be a higher-order polynomial in the components of $\mathbf{c}$.

Remark 22.2.2 The linearity of the model has another important consequence. As observed above, $J(\mathbf{c})$ is a quadratic polynomial in x_0 and α. Hence, it can be verified that $J(\mathbf{c})$ is unimodal, that is, it has a unique minimum. However, non-linearity of the model renders $J(\mathbf{c})$ a higher-order polynomial in $\mathbf{c}$ and, hence $J(\mathbf{c})$ can have multiple minima. Data assimilation problems that lead to the possibility of multimodality present formidable challenges. (Exercise 22.4) .

Once $\mathbf{c}^* = (x_0^*, \alpha^*)$ is known, then we can compute the minimum value of the sum of the squared-error (*SSE*) as follows :

$$\text{SSE} = \sum_{i=0}^{N} \left[\left(x_0^* + \alpha^* t_i \right) - z_i \right]^2. \tag{22.2.9}$$

We now consider several special cases:

CASE A Only one observation z_i at time t_i is available, (that is $N = 0$)

In this case, the linear system (22.2.4) reduces to

$$x_0 + \alpha t_i = z_i, \quad x_0 t_i + \alpha t_i^2 = z_i t_i. \tag{22.2.10}$$

Since the second equation is a constant multiple of the first, there is one equation in two unknowns and the system (22.2.10) is singular. Physically, there are an infinite number of lines that can pass through a single point (observation). Mathematically, there are an infinite number of (x_0, α) that pass through the point. And as expected, SSE = 0 in this case.

CASE B Two observations z_1 and z_2 at time instants t_1 and t_2 are available.

In this case, the linear system (22.2.4) becomes

$$\begin{aligned} x_0 + \alpha\left(\tfrac{t_1+t_2}{2}\right) &= \left(\tfrac{z_1+z_2}{2}\right) \\ x_0(t_1 + t_2) + \alpha\left(t_1^2 + t_2^2\right) &= (t_1 z_1 + t_2 z_2). \end{aligned} \tag{22.2.11}$$

This system is non-singular and the unique solution of (22.2.11) is given by

$$x_0^* = \frac{z_1 t_2 - z_2 t_1}{t_2 - t_1}, \quad \alpha^* = \frac{z_2 - z_1}{t_2 - t_1}. \tag{22.2.12}$$

Substituting this in (22.2.9) it can be verified that in this case SSE $= 0$ as well (Exercise 22.5). In essence, there is one line that can pass through the given points (observations). Although SSE $= 0$, the estimated state can suffer serious error since the observations are generally erroneous.

CASE C Three or more observations. In this general case, the general analysis leading to (22.2.6) and (22.2.8) holds good and it can be verified that in this case SSE > 0 (Exercise 22.6).

Remark 22.2.3 According to the fundamental theorem in numerical analysis, there exists an nth degree polynomial that exactly fits a collection of $(n + 1)$ points. Case B considered above is the special case of this fundamental result when $n = 1$. So, it is not surprising that SSE $= 0$ in this case. However, when you fit a straight line to a set of three or more points, it is generally the case that SSE > 0.

Remark 22.2.4 We now show that the estimates x_0^* and α^* given in (22.2.8) and (22.2.6) are indeed **unbiased** (Chapter 13) estimates. To this end, consider

$$z_i = x_0 + \alpha t_i + v_i.$$

Summing both sides from 0 to N and dividing by $(N+1)$ we obtain (using the notation in 22.2.7)

$$\bar{z} = x_0 + \alpha\,\bar{t} + \bar{v} \tag{22.2.13}$$

where analogously

$$\bar{v} = \frac{1}{N+1}\sum_{i=0}^{N} v_i .$$

Hence, we obtain

$$[z_i - \bar{z}] = \alpha\,[t_i - \bar{t}] + [\,v_i - \bar{v}\,]. \tag{22.2.14}$$

Substituting (22.2.13) into (22.2.6) we obtain

$$\alpha^* = \frac{\sum_{i=0}^{N}(z_i - \bar{z})(t_i - \bar{t})}{\sum_{i=0}^{N}(t_i - \bar{t})^2} = \frac{\sum_{i=0}^{N}\alpha\,(t_i - \bar{t})^2 + \sum_{i=0}^{N}(t_i - \bar{t})(v_i - \bar{v})}{\sum_{i=0}^{N}(t_i - \bar{t})^2}$$

$$= \alpha + \frac{\sum_{i=0}^{N}(t_i - \bar{t})(v_i - \bar{v})}{\sum_{i=0}^{N}(t_i - \bar{t})^2} .$$

Recall that t_i's are not random and are fixed a priori, and $E[v_i] = 0$ implies $E\,[\,\bar{v}\,] = 0$. Thus we obtain (since x_0 and α are not random)

$$E[\,\alpha^*] = E\,[\alpha] + \frac{\sum_{i=0}^{N}(t_i - \bar{t})\,E\,[v_i - \bar{v}]}{\sum_{i=0}^{N}(t_i - \bar{t})^2} = E\,[\alpha] = \alpha \tag{22.2.15}$$

that is, α^* is an unbiased estimate of α. Now, combining (22.2.8) and (22.2.13), we have

$$x_0^* = \bar{z} - \bar{t}\alpha^* = x_0 + \bar{t}(\alpha - \alpha^*) + \bar{v}.$$

Taking expectations

$$E[\,x_0^*\,] = E[\,x_0\,] + \bar{t}E[\,\alpha - \alpha^*\,] + E\,[\,\bar{v}\,].$$

From (22.2.15) and the properties of v_i we immediately obtain

$$E[\,x_0^*\,] = E[\,x_0\,] = x_0.$$

That is, x_0^* is an **unbiased** estimate of x_0 .

Remark 22.2.5 The optimal regression line is given by (using 22.2.8)

$$x(t_i) = x_0^* + \alpha^* t_i = \bar{z} + \alpha^*(t_i - \bar{t}).$$

Clearly, when $t_i = \bar{t}$, we obtain $x(t_i) = \bar{z}$, that is, the optimal regression line passes through the point $(\bar{t}, \bar{z})$, where $\bar{z}$ is the **centroid** of the observations and $\bar{t}$ is the **centroid** of the time instances at which observations are obtained .

Remark 22.2.6 Define α_i to be the slope of the line segment joining (t_i, z_i) and $(\bar{t}, \bar{z})$, that is,

$$\alpha_i = \frac{z_i - \bar{z}}{t_i - \bar{t}} \quad \text{for} \quad i = 0, 1, 2, \ldots, N.$$

Define a system of weights w_i as follows:

$$w_i = (t_i - \bar{t}\,)^2.$$

The optimal α^*given by (22.2.6) can be rewritten as

$$\alpha^* = \frac{\sum_{i=0}^{N} (z_i - \bar{z})(t_i - \bar{t})}{\sum_{i=0}^{N} (t_i - \bar{t})^2} = \frac{\sum_{i=0}^{N} w_i \alpha_i}{\sum_{i=0}^{N} w_i} = \sum_{i=0}^{N} a_i \alpha_i$$

where $a_i = w_i / \sum_{i=0}^{N} w_i > 0$ and $\sum_{i=0}^{N} a_i = 1$.

Thus, the optimal value of the slope of the regression line is a **convex** combination of the slopes of the line segments joining (t_i, z_i) and $(\bar{t}, \bar{z})$.

22.3 The Lagrangian approach: discrete time formulation

In this approach we first discretize the model equations using a scheme that preserves the **fidelity** of the model. For the model in question, a simple Euler scheme for discretization will be adequate (i.e., no truncation error). Using the standard one-sided approximation to the time derivative, the model equation (22.1.1) can be written in the **discrete form** as

$$\frac{x_k - x_{k-1}}{\Delta t} = \alpha \quad \text{or} \quad x_k - x_{k-1} = \alpha \Delta t \tag{22.3.1}$$

where $x_k = x(k\Delta t)$ for some small but fixed $\Delta t > 0$ and for all $k \geq 1$. Rewrite (22.3.1) as

$$-x_{k-1} + x_k = \alpha \Delta t \tag{22.3.2}$$

and define $\mathbf{x} = (x_1, x_2, \ldots, x_N)^{\mathrm{T}}$ to be the vector whose ith component represents the state of the system at time i where recall that x_0 is the initial position. In the parlance of dynamical systems the sequence of states that constitutes the components of the vector $\mathbf{x}$ is often called the **orbit** of $\mathbf{c}$, the control vector. In meteorological circles, this sequence of states is also called the **forward model solution** starting from the control $\mathbf{c}$.

It can be verified that (22.3.2) can be succinctly written in the matrix-vector form as follows:

$$\mathbf{F}\mathbf{x} = \mathbf{b} \tag{22.3.3}$$

where $\mathbf{F}$ is an $N \times N$ **lower bidiagonal** matrix

$$\begin{bmatrix} 1 & 0 & \cdots & 0 & 0 \\ -1 & 1 & \cdots & 0 & 0 \\ \vdots & \vdots & \vdots & \vdots & \vdots \\ 0 & 0 & \cdots & 1 & 0 \\ 0 & 0 & \cdots & -1 & 1 \end{bmatrix} \tag{22.3.4}$$

and $\mathbf{b}$ is an N-vector given by

$$\mathbf{b} = (\alpha\Delta t + x_0,\ \alpha\Delta t,\ \alpha\Delta t,\ \ldots,\ \alpha\Delta t)^{\mathrm{T}}. \tag{22.3.5}$$

Notice that the linear equation (22.3.3) is the discrete analog of the continuous time model in (22.1.1).

Let

$$z_k = x_k + v_k \tag{22.3.6}$$

be the observation of the state of the system at time k, for $0 \le k \le N$. The criterion $J(\mathbf{c})$ in (22.1.6) now takes the form

$$J(\mathbf{c}) = \frac{1}{\sigma^2} \sum_{i=0}^{N} (x_i - z_i)^2. \tag{22.3.7}$$

The problem is to minimize $J(\mathbf{c})$ when x_i's are given to be the solution of the system (22.3.3). We now convert this constrained minimization problem into an unconstrained minimization problem by defining the associated Lagrangian as follows (refer to Appendix D for details).

$$L(\mathbf{c},\ \mathbf{x},\ \lambda) = J(\mathbf{c}) + \sum_{i=1}^{N} \lambda_i [x_i - x_{i-1} - \alpha\Delta t\,] \tag{22.3.8}$$

where $\lambda = (\lambda_1, \lambda_2, \ldots, \lambda_N)^{\mathrm{T}}$ is an N-vector of **undetermined** Lagrangian multipliers. Notice that λ_i is associated with the state transition from time instant $(i-1)$ to i.

Remark 22.3.1 The original problem of minimizing $J(\mathbf{c})$ in (22.3.7) subject to the linear equality constraints in (22.3.1) is said to have **two** degrees of freedom since $\mathbf{c}$ is a vector of dimension two. This constrained problem is now converted into an unconstrained problem of minimizing $L(\mathbf{c},\ \mathbf{x},\ \lambda)$ in (22.3.8). This latter problem has $(2N + 2)$ degrees of freedom which is equal to the total number of distinct variables in $L(\mathbf{c},\ \mathbf{x},\ \lambda)$. Thus, the difficulty of dealing with the constraints is **traded** for increased dimensionality of the problem .

Now differentiating $L(\mathbf{c}, \mathbf{x}, \lambda)$ with respect to the variables in $\mathbf{c}$, $\mathbf{x}$ and λ, we obtain the following:

$$\nabla_c L(\mathbf{c}, \mathbf{x}, \lambda) = \begin{pmatrix} \dfrac{\partial L}{\partial x_0} \\ \\ \dfrac{\partial L}{\partial \alpha} \end{pmatrix} = \begin{pmatrix} \dfrac{2}{\sigma^2}(x_0 - z_0) - \lambda_1 \\ -\Delta t \displaystyle\sum_{i=1}^{N} \lambda_i \end{pmatrix}, \tag{22.3.9}$$

$$\nabla_x L = \left(\frac{\partial L}{\partial x_1}, \frac{\partial L}{\partial x_2}, \ldots, \frac{\partial L}{\partial x_n} \right)^{\mathrm{T}} \tag{22.3.10}$$

where

$$\frac{\partial L}{\partial x_i} = \frac{2}{\sigma^2}(x_i - z_i) + \lambda_i - \lambda_{i+1}, \quad 1 \le i \le N \tag{22.3.11}$$

with $\lambda_{N+1} = 0$, and

$$\nabla_\lambda L = \left(\frac{\partial L}{\partial \lambda_1}, \frac{\partial L}{\partial \lambda_2}, \ldots, \frac{\partial L}{\partial \lambda_N} \right)^{\mathrm{T}} \tag{22.3.12}$$

where

$$\frac{\partial L}{\partial \lambda_i} = x_i - x_{i-1} - \alpha \Delta t, \quad 1 \le i \le N. \tag{22.3.13}$$

For a given set of observations, z_i, $0 \le i \le N$, the values of the variables that minimize $L(\mathbf{c}, \mathbf{x}, \lambda)$ are given by the solution of

$$\nabla_c L = 0, \quad \nabla_x L = 0, \quad \nabla_\lambda L = 0. \tag{22.3.14}$$

Notice that $\nabla_\lambda L = 0$ is indeed the N constraints given by the model equation (22.3.1) which are trivially true. From (22.3.11), the N equations in $\nabla_x L = 0$ can be rewritten as

$$\lambda_i - \lambda_{i+1} = -\tfrac{2}{\sigma^2}(x_i - z_i), \quad 1 \le i \le N$$

or equivalently in matrix-vector form as

$$\mathbf{B}\lambda = \mathbf{e} \tag{22.3.15}$$

where $\mathbf{B}$ is an $N \times N$ **upper bidiagonal** matrix

$$\begin{bmatrix} 1 & -1 & \cdots & 0 & 0 \\ 0 & 1 & \cdots & 0 & 0 \\ \vdots & \vdots & \vdots & \vdots & \vdots \\ 0 & 0 & \cdots & 1 & -1 \\ 0 & 0 & \cdots & 0 & 1 \end{bmatrix}$$

and $\mathbf{e}$ is an N-vector given by

$$\mathbf{e} = -\frac{2}{\sigma^2}(x_1 - z_1, x_2 - z_2, \ldots, x_N - z_N)^{\mathrm{T}}. \tag{22.3.16}$$

Remark 22.3.2 Since $x_i - z_i$ denotes the error between the model solution and the observation, the vector **e** is often called the **error vector**.

Similarly, from (22.3.9) $\nabla_c L = 0$ leads to

$$\lambda_1 = \frac{2}{\sigma^2}(x_0 - z_0), \quad \sum_{i=1}^{N} \lambda_i = 0. \tag{22.3.17}$$

That is, the problem of solving (22.3.14) reduces to one of solving simultaneously (22.3.15) and (22.3.17) and we now turn our attention to solving this latter problem.

Suppose we pick a $\mathbf{c} = (x_0, \alpha)$ out of the blue sky and compute its orbit using (22.3.3). Then compute **e** in (22.3.16) using the calculated orbit and the given observations and solve for λ using (22.3.15) as required by $\nabla_x L = 0$. If the values of λ so obtained also satisfy the equation $\nabla_c L = 0$, then we are done. The chosen value of $\mathbf{c} = (x_0, \alpha)$ is indeed the optimum value we are looking for. However, it is highly unlikely that a randomly chosen **c** will simultaneously satisfy $\nabla_x L = 0$ and $\nabla_c L = 0$. Instead of searching for the optimal **c** randomly, we suggest the following iterative algorithm for minimizing $L(\mathbf{c}, \mathbf{x}, \lambda)$ in (22.3.8).

An algorithm for minimization

Step 1 Pick a starting vector, say $\mathbf{c}_{\text{old}} = (x_0, \alpha)$.

Step 2 Compute the orbit $x = (x_1, x_2, \ldots, x_N)$ using the forward model equation (22.3.3).

Step 3 Using the given set of observations z_i, $1 \le i \le N$ and the orbit computed in Step 2, calculate the error vector **e** from (22.3.16).

Step 4 Solve (22.3.15) for λ.

Step 5 If the resulting value of $\nabla_c L$ is zero then we are done. Else go to Step 6.

Step 6 Compute $\mathbf{c}_{\text{new}}$ using $\mathbf{c}_{\text{old}}$ and $\nabla_c L$ in one of the several variations of the minimization algorithms in Chapters 10–12. The gradient algorithm, for example, defines

$$\mathbf{c}_{\text{new}} = \mathbf{c}_{\text{old}} - \beta \nabla_c L$$

where β is the step size obtained by the standard one-dimensional search in Chapter 10.

Step 7 Set $\mathbf{c}_{\text{old}} \leftarrow \mathbf{c}_{\text{new}}$ and go to Step 2.

Several comments are in order.

Remark 22.3.3 First notice that the matrix **B** in (22.3.15) is the **transpose** or the **adjoint** of the matrix **F** in (22.3.3). In view of this relationship, systems (22.3.3) and (22.3.15) namely $\mathbf{Fx} = \mathbf{b}$ and $\mathbf{B}\lambda = \mathbf{F}^{\mathrm{T}}\lambda = \mathbf{e}$ are called **adjoint systems**, and the above minimization algorithm is often known as the **adjoint** algorithm.

Recall that **A** is a lower bidiagonal matrix and **B** is an upper bidiagonal matrix. In solving the systems (22.3.3) and (22.3.15), it is often useful to think of recovering the components of **x** from x_1 to x_N but those of λ from λ_N to λ_1. Since solving

$\mathbf{Fx} = \mathbf{b}$ is equivalent to **forward** integration of the model equation (22.1.1), solving $\mathbf{B}\lambda = \mathbf{e}$ is often termed as **backward** integration.

Remark 22.3.4 In implementing this algorithm in Step 5 we must use our appropriate stopping criterion. One possibility is to check for the norm of $\nabla_c L$ namely stop if $\| \nabla_c L \|_2 < \varepsilon$ where $\varepsilon = 10^{-6}$, typically the machine precision. Another indication of convergence could be based on the values of the criterion function. If you do not see any appreciable change in the values of the criterion, it could signal convergence. In practice one must use as many indicators as possible. However, one must also weigh in the cost of computing these indicators since it could easily add up to the overall cost.

Remark 22.3.5 In the above derivation it was tacitly assumed that observations of the state of the system are available at each of the grid points used in discretizing the model equation. Often this may not be the case. In the meteorological domain the location of observations is decided quite independently of the type of grid that is used in solving the problem. In other words, it is more realistic to assume that the observations may be available at points other than the grid points. In this case, in formulating the data assimilation problem we have one of two options: either interpolate the observations across the grid and use the interpolated values in defining the criterion or else one can interpolate the state of the model equation and compute the interpolated value of the state where observations are available .

22.4 Monte Carlo via twin experiments

Once an adequate model of a physical phenomenon of interest has been chosen, the next logical step is to go out and observe the phenomenon and measure appropriate quantities of interest. For the problem of the moving particle in a straight line considered in this chapter, the positions of the particle at prescribed time epochs constitute an appropriate set. But to obtain actual data is invariably expensive and very time consuming. Before one makes such an investment on data collection, one needs to understand how the model and the method interact. This is often done using a broad umbrella of techniques that has come to be known as the **twin experiment** which basically involves the following steps:

Step 1 Identify the control variable $\mathbf{c}$ for the given model and assign values to the elements of $\mathbf{c}$.

Step 2 Starting with these values, perform a forward execution of the model and compute the state variables $x(t_i)$, $i = 0, 1, 2, \ldots, N$.

Step 3 Generate a sequence $v_i = 0, 1, 2, \ldots, N$ of uncorrelated Gaussian random variates from a population with mean zero and known variance.

Step 4 Generate the observation z_i as the sum of $x(t_i)$ computed in Step 2 and v_i generated in Step 3. that is, $z_i = x(t_i) + v_i$, $i = 0, 1, \ldots, N$.

Table 22.4.1

t_i	0	1	2	3
$x(t_i)$	2.5	2.0	2.5	3.0
v_i	−0.5	1.0	−0.5	0
z_i	1.0	3.0	2.0	3.0

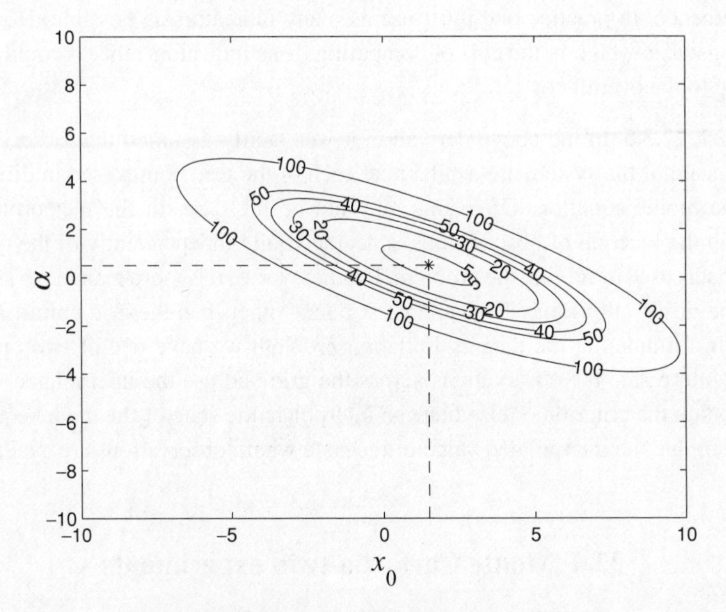

Fig. 22.4.1 Contours of $J(\mathbf{c})$ in the x_0–α plane.

Step 5 Now combine the observations z_i and the model and estimate $\mathbf{c}^*$, the optimal value of the control using the method described in Section 22.3.

In the following we illustrate this process using the Example 22.1.1 . Let the model be $x(t_i) = x_0 + \alpha t_i$, with $\mathbf{c} = (x_0, \alpha)^{\mathrm{T}}$. Choose $\mathbf{c} = (1.5, 0.5)^{\mathrm{T}}$ and generate $x(t_i)$ shown in Table 22.4.1.

It can be verified that (refer to Example 22.1.1)

$$J(\mathbf{c}) = \sum_{i=0}^{3} (x(t_i) - z_i)^2 = \sum_{i=0}^{3} (x_0 + \alpha t_i - z_i)^2 = \frac{1}{2}\, \mathbf{c}^{\mathrm{T}}\mathbf{A}\mathbf{c} + \mathbf{b}^{\mathrm{T}}\mathbf{c} + d \qquad (22.4.1)$$

where

$$\mathbf{A} = \begin{bmatrix} 8 & 12 \\ 12 & 28 \end{bmatrix} \quad \mathbf{b} = \begin{pmatrix} -18 \\ -32 \end{pmatrix} \quad \text{and} \quad d = 23.$$

The contours of $J(\mathbf{c})$ are plotted in the x_0–α plane in Figure 22.4.1. The eigenvalues of the matrix $\mathbf{A}$ are given by $\lambda_1 = 33.6205$ and $\lambda_2 = 2.3795$. Recall that the

Table 22.4.2 *Performance of Gradient Algorithm*

Iteration Number k	Value of $J(\mathbf{c}_k)$ for various starting points $\mathbf{c}(0)$			
	$\mathbf{c}(0) = (3, 7)$	$\mathbf{c}(0) = (3, -7)$	$\mathbf{c}(0) = (-3, 7)$	$\mathbf{c}(0) = (-3, -7)$
0	719	663	323	1275
1	3.50952	22.66601	49.40266	2.32102
2	1.50562	2.17724	8.63737	1.50052
3	1.50001	1.52166	2.56345	1.50000
4	1.50000	1.50069	1.65845	
5		1.50002	1.52360	
6		1.50000	1.50351	
7			1.50052	
8			1.50007	
9			1.50001	
10			1.50000	

Table 22.4.3 *Performance of Conjugate Gradient Method*

Iteration Number k	Value of $J(\mathbf{c}_k)$ for various starting points $\mathbf{c}(0)$			
	$\mathbf{c}(0) = (3, 7)$	$\mathbf{c}(0) = (3, -7)$	$\mathbf{c}(0) = (-3, 7)$	$\mathbf{c}(0) = (-3, -7)$
0	719	663	323	1275
1	3.50952	22.66601	49.40266	2.32102
2	1.50000	1.49999	1.49999	1.50000

reciprocals of the square root of these eigenvalues are the semi-axes of the standard ellipse corresponding to the quadratic form in (22.4.1). Also recall that the optimal $\mathbf{c}^* = (1.5, 0.5)^{\mathrm{T}}$ is already known to us in this twin experiment.

Tables 22.4.2 and 22.4.3 provide the comparative performance of the gradient and the conjugate gradient algorithms respectively.

Exercises

22.1 Verify the correctness of (22.2.4).

22.2 Verify that $\mathbf{c}^*$ in (22.2.6)–(22.2.8) is indeed the solution of (22.2.4).

22.3 Using your favorite symbolic manipulation system (such as MAPLE, MATHEMATICA, etc.) express $J(\mathbf{c})$ explicitly as a polynomial in the components of $\mathbf{c}$.

22.4 Consider the following functions:

(*a*) $(x-1)(x-2)$ (*e*) $(x-1)^2(x-2)$
(*b*) $(x-1)(x-2)(x-3)$ (*f*) $(x-1)^2(x-2)^2$
(*c*) $(x-1)(x-2)(x-3)(x-4)$ (*g*) $(x-1)^3(x-2)$
(*d*) $(x-1)^2$ (*h*) $(x-1)(x-2)^2(x-4)$

(1) By qualitative analysis first predict the number of minima in the above functions.
(2) Actually plot and verify your predictions.
(3) Is there a generalization of the pattern to higher-order polynomials?

22.5 (a) Verify that x_0^* and α^*given in (22.2.12) is the solution to (22.2.11) for the special case of two observations.
(b) Again verify that SSE = 0 for this case.

22.6 Pick any subset of k observations for k = 1,2,3, and 4. Compute x_0^* and α^* and the corresponding SSE. Plot SSE vs. k.

22.7 Solve the system $\mathrm{B}\lambda = \mathrm{e}$ in (22.3.15) explicitly in closed form and verify that the solution λ_i is given by

$$\lambda_i = -\frac{2}{\sigma^2}\sum_{j=i}^{N}(x_j - z_j).$$

22.8 Using the value of λ_i computed in (Exercise 22.7), verify that

$$\frac{\partial L}{\partial x_0} = \frac{2}{\sigma^2}\sum_{i=0}^{N}(x_i - z_i)$$

and

$$\begin{aligned}\frac{\partial L}{\partial \alpha} &= -\Delta t \sum_{i=1}^{N}\lambda_i \\ &= \frac{2\Delta t}{\sigma^2}\sum_{i=1}^{N} i(x_i - z_i) = \frac{2\Delta t}{\sigma^2}\sum_{i=0}^{N} i(x_i - z_i)\end{aligned}$$

since $t_0 = 0$.

22.9 Using (22.2.1) compute $\partial J/\partial x_0$ and $\partial J/\partial \alpha$ and verify using the results of Exercise 22.8 that

$$\frac{\partial L}{\partial x_0} = \frac{\partial J}{\partial x_0} \quad \text{and} \quad \frac{\partial L}{\partial \alpha} = \frac{\partial J}{\partial \alpha}$$

or stated succinctly, $\nabla_c L = \nabla_c J$.

Notes and references

Section 22.1 The standard problem of fitting a straight line to a collection of data is stated in a dynamic context. This section demonstrates how the problem of fitting data to a dynamic model gives rise to a minimization problem under equality constraints. A mathematical framework for solving minimization problem under equality constraints is described in Appendix D. This so-called straight line problem is analyzed in detail by Lewis (1990).

Section 22.2 This section provides the link between the dynamic data assimilation and the classical regression analysis. For a treatment of regression analysis, refer to Draper and Smith (1966).

Section 22.3 The Lagrangian approach to dynamic data assimilation is now classic. Thacker and Long (1988) contains a good introduction to fitting dynamical models to data. Lanczos's book on variational methods explains the undetermined Lagrange multipliers in a most pedagogical manner (Lanczos 1970).

Section 22.4 The technique of using twin experiments to evaluate the goodness of a class of assimilation methods is standard in the meteorology literature. They have come to be called OSSE's (Observation System Simulation Experiments) where model output is used to create observations (with additive noise), and then the consequences of employing these observations in data assimilation experiments are analyzed.

23

First-order adjoint method: linear dynamics

In the opening chapter of Part VI we considered a very special dynamical model for pedagogical reasons. Having gained some working knowledge of the methodology for solving the inverse problem using the Lagrangian framework, we now consider the general linear dynamical system. Once we understand the underpinnings of this methodology in the context of a general linear dynamical system, its applicability to a wide variety of linear models is possible.

When compared to Chapter 22, the contents of this chapter are a generalization in one sense and a specialization in another. The generalization comes from the fact that we consider the generic linear dynamical system where the state variables are vectors instead of scalars. The specialty, on the other hand, comes from the fact that we only consider the problem of estimating the **initial condition** instead of an initial condition and a parameter ($\mathbf{x}_0$ and α in the straight line problem).

It could be argued that since few models of interest in real world applications are linear, this chapter's value is essentially academic. While this argument carries some weight, it should be recognized that linear analysis has a fundamental role to play in development of adjoint method for non-linear dynamical systems. For example, one standard approach to non-linear system analysis is using the so-called **perturbation method**. In this method the non-linear problem is reduced to a local analysis of an associated linear system. Next we want to demonstrate that the data assimilation problem is intrinsically challenging, even when the system is controlled by linear dynamics and observations are linear functions of the state variables.

This chapter is organized as follows. The statement of the inverse problem is given in Section 23.1. Conditions for observability and closed form solution are given in Section 23.2. Section 23.3 describes the Lagrangian method. An algorithmic framework for minimization is given in Section 23.4. The adjoint method for solving the inverse problem is described in Section 23.5. An alternate method for finding the adjoint – a discrete counterpart to integration by parts – is found in Section 23.6.

23.1 A statement of the inverse problem

Let $t \in [0, \infty]$ denote the time variable and let $\mathbf{x} \in \mathbb{R}^n$ with $\mathbf{x} = (x_1, x_2, \ldots, x_n)^{\mathrm{T}}$ denote the state of a system where T denotes the transpose. Let $\mathbf{A} \in \mathbb{R}^{n\times n}$ be a real matrix that is assumed to be **non-singular**.

Model Consider the linear system of the type

$$\frac{\mathrm{d}\mathbf{x}}{\mathrm{d}t} = \mathbf{A}\mathbf{x} \tag{23.1.1}$$

with $\mathbf{x}(0) = \mathbf{x}_0 = \mathbf{c}$ being the initial condition.

The solution of (23.1.1) (refer to Chapter 32) is given by

$$\mathbf{x}(t) = \mathrm{e}^{\mathbf{A}t}\mathbf{x}_0. \tag{23.1.2}$$

The vector of derivatives given by $\mathbf{A}\mathbf{x}$ in (23.1.1) defines the **field** and the collection $\mathbf{x}(t)$ in (23.1.2) for all initial conditions $\mathbf{x}_0 \in \mathbb{R}^n$ is called the **flow** of the system. When the matrix $\mathbf{A}$ is a constant matrix, that is, independent of time, then (23.1.1) is called a **constant coefficient** linear system or simply **autonomous** system. When $\mathbf{A}$ changes with time, it represents a **variable coefficient** or a **nonautonomous system**. The matrix $\mathrm{e}^{\mathbf{A}t}$ relates the state $\mathbf{x}_0$ of the system at time $t = 0$ to state $\mathbf{x}(t)$ at any time t and hence is called the **state transition matrix**, and is often denoted by $\mathbf{L}(t) = \mathrm{e}^{\mathbf{A}t}$. In analogy with the scalar exponential function, it can be easily verified that

$$\begin{aligned} &\text{(a) } \mathbf{L}(0) &&= \mathbf{I}, \text{ the identity matrix} \\ &\text{(b) } \mathbf{L}(t_1 + t_2) &&= \mathbf{L}(t_1)\mathbf{L}(t_2). \end{aligned}$$

Pick a small real number $\Delta t > 0$, and define $\mathbf{x}_k = \mathbf{x}(k\Delta t)$. Now discretizing the equation (23.1.1), using the standard Euler scheme, we obtain

$$\mathbf{x}_{k+1} = \mathbf{M}\mathbf{x}_k \tag{23.1.3}$$

where the $n \times n$ matrix $\mathbf{M}$ is given by

$$\mathbf{M} = (\mathbf{I} + \Delta t\mathbf{A}). \tag{23.1.4}$$

Iterating (23.1.3), we obtain

$$\mathbf{x}_k = \mathbf{M}^k\mathbf{c} = \mathbf{L}_k\mathbf{c} \tag{23.1.5}$$

where $\mathbf{L}_k = \mathbf{M}^k$ denotes the k-step transition matrix (Exercise 23.1).

Observations It is assumed that the observation vector $\mathbf{z} \in \mathbb{R}^m$ is a **linear function** of the state vector $\mathbf{x} \in \mathbb{R}^n$ corrupted by a mean zero additive noise $\mathbf{v} \in \mathbb{R}^m$ with known variance and which is assumed to be **temporally uncorrelated**.

Let $\mathbf{H} \in \mathbb{R}^{m\times n}$. The observations are then modeled by the following relation:

$$\mathbf{z}_k = \mathbf{H}\mathbf{x}_k + \mathbf{v}_k \tag{23.1.6}$$

where $\mathbf{v}_k$, the **noise** vector, is such that

$$\mathbf{E}[\mathbf{v}_k] = 0 \quad \text{and} \quad \mathbf{E}[\mathbf{v}_{k_1}^{\mathrm{T}}\mathbf{v}_{k_2}] = \mathbf{R}\delta_{k_1k_2} \tag{23.1.7}$$

where $\delta_{k_1k_2}$ is the Kronecker delta. That is, $\mathbf{v}_k$ is a **white noise** sequence.

Criterion Let $\mathbf{W} \in \mathbb{R}^{m\times m}$ be a symmetric, positive definite matrix called the weight matrix. Given $\mathbf{W}$, we define

$$J(\mathbf{c}) = \frac{1}{2}\sum_{k=0}^{N} \langle\, [\mathbf{z}_k - \mathbf{H}\mathbf{x}_k]\,,\ \mathbf{W}[\mathbf{z}_k - \mathbf{H}\mathbf{x}_k]\,\rangle \tag{23.1.8}$$

$$= \frac{1}{2}\sum_{k=0}^{N} [\mathbf{z}_k - \mathbf{H}\mathbf{x}_k]^{\mathrm{T}}\ \mathbf{W}[\mathbf{z}_k - \mathbf{H}\mathbf{x}_k]\,. \tag{23.1.9}$$

A meaningful choice for $\mathbf{W}$ is $\mathbf{W} = \mathbf{R}^{-1}$, the inverse of the covariance matrix $\mathbf{R}$ of the observation noise.

Statement of problem Given the set of noisy observations $\{\mathbf{z}_k|\ 0 \le k \le N\}$, the problem is to estimate the control $\mathbf{c}$ that minimizes $J(\mathbf{c})$ in (23.1.8) when $\mathbf{x}_k$ is subjected to the equality constraints defined by (23.1.3).

23.2 Observability and a closed form solution

Given $\{\mathbf{z}_0, \mathbf{z}_1, \ldots, \mathbf{z}_N\}$, under what conditions can we uniquely determine the initial state x_0 of the model (23.1.3)? In other words, how useful are observations in determining the initial state? In answering this question, recall that

$$\begin{aligned} \mathbf{z}_k &= \mathbf{H}\mathbf{x}_k + \mathbf{v}_k \\ &= \mathbf{H}\mathbf{M}^k\mathbf{c} + \mathbf{v}_k\,. \end{aligned} \tag{23.2.1}$$

Thus,

$$\begin{bmatrix} \mathbf{z}_0 \\ \mathbf{z}_1 \\ \vdots \\ \mathbf{z}_N \end{bmatrix} = \begin{bmatrix} \mathbf{H} \\ \mathbf{H}\mathbf{M} \\ \vdots \\ \mathbf{H}\mathbf{M}^N \end{bmatrix} \mathbf{c} + \begin{bmatrix} \mathbf{v}_0 \\ \mathbf{v}_1 \\ \vdots \\ \mathbf{v}_N \end{bmatrix} \tag{23.2.2}$$

$$\mathbf{Z} = \mathcal{H}\mathbf{c} + \mathbf{V} \tag{23.2.3}$$

where

$$\mathbf{Z} = \begin{bmatrix} \mathbf{z}_0 \\ \mathbf{z}_1 \\ \vdots \\ \mathbf{z}_N \end{bmatrix}, \quad \mathcal{H} = \begin{bmatrix} \mathbf{H} \\ \mathbf{H}\mathbf{M} \\ \vdots \\ \mathbf{H}\mathbf{M}^N \end{bmatrix} \quad \text{and} \quad \mathbf{V} = \begin{bmatrix} \mathbf{v}_0 \\ \mathbf{v}_1 \\ \vdots \\ \mathbf{v}_N \end{bmatrix}$$

with $\mathbf{Z} \in \mathbb{R}^{(N+1)\times m}$, $\mathbf{V} \in \mathbb{R}^{(N+1)\times m}$, $\mathcal{H} \in \mathbb{R}^{(N+1)m\times n}$ and $\mathbf{c} \in \mathbb{R}^n$.

The standard method for solving this over-determined system is by the method of generalized least squares (Chapter 5).

Let

$$\mathbf{W} = \text{Diag}(\mathbf{R}^{-1}, \mathbf{R}^{-1}, \ldots, \mathbf{R}^{-1}) \in \mathbb{R}^{(N+1)m\times(N+1)m}$$

where $\mathbf{R} \in \mathbb{R}^{m\times m}$ is the covariance of $\mathbf{v}_i$ defined in (23.1.7). The least squares solution that minimizes (Chapter 5)

$$\begin{aligned} f(\mathbf{c}) &= (\mathbf{Z} - \mathcal{H}\mathbf{c})^{\mathrm{T}}\mathbf{W}(\mathbf{Z} - \mathcal{H}\mathbf{c}) \\ &= \sum_{i=0}^{N}(\mathbf{z}_i - \mathbf{H}\mathbf{M}^i\mathbf{c})^{\mathrm{T}}\mathbf{R}^{-1}(\mathbf{z}_i - \mathbf{H}\mathbf{M}^i\mathbf{c}) \end{aligned} \tag{23.2.4}$$

is given by

$$\mathbf{c}^* = (\mathcal{H}^{\mathrm{T}}\mathbf{W}\mathcal{H})^{-1}\mathcal{H}\mathbf{W}\mathbf{Z}. \tag{23.2.5}$$

Clearly this solution exists if and only if the **observability matrix**

$$\mathcal{H}^{\mathrm{T}}\mathbf{W}\mathcal{H} = \sum_{i=0}^{N}(\mathbf{M}^{i-1})^{\mathrm{T}}\mathbf{H}^{\mathrm{T}}\mathbf{R}^{-1}\mathbf{H}\mathbf{M}^{i-1} \tag{23.2.6}$$

is non-singular or of full rank. For more discussion on observability refer to Chapter 26.

In this linear case since the functional form of the trajectory of (23.1.3) as function of the control $\mathbf{c}$ is known explicitly as in (23.1.5), we can in fact convert the constrained minimization problem into an unconstrained form by substituting $\mathbf{x}_k = \mathbf{L}_k\mathbf{c}$ in $J(\mathbf{c})$. Indeed, we obtain

$$J(\mathbf{c}) = \frac{1}{2}\sum_{k=0}^{N}(\mathbf{z}_k - \mathbf{H}\mathbf{L}_k\mathbf{c})^{\mathrm{T}}\,\mathbf{R}^{-1}(\mathbf{z}_k - \mathbf{H}\mathbf{L}_k\mathbf{c})\,. \tag{23.2.7}$$

Consider the typical term

$$\begin{aligned} g_k &= (\mathbf{z}_k - \mathbf{H}\mathbf{L}_k\mathbf{c})^{\mathrm{T}}\,\mathbf{R}^{-1}(\mathbf{z}_k - \mathbf{H}\mathbf{L}_k\mathbf{c}) \\ &= \mathbf{c}^{\mathrm{T}}\mathbf{B}_k\mathbf{c} - 2\mathbf{b}_k^{\mathrm{T}}\mathbf{c} + d_k \end{aligned} \tag{23.2.8}$$

where

$$\left.\begin{aligned} \mathbf{B}_k &= \mathbf{L}_k^{\mathrm{T}}\mathbf{H}^{\mathrm{T}}\mathbf{R}^{-1}\mathbf{H}\mathbf{L}_k \\ \mathbf{b}_k &= \mathbf{L}_k^{\mathrm{T}}\mathbf{H}^{\mathrm{T}}\mathbf{R}^{-1}\mathbf{z}_k \\ d_k &= \mathbf{z}_k^{\mathrm{T}}\mathbf{R}^{-1}\mathbf{z}_k\,. \end{aligned}\right\} \tag{23.2.9}$$

Substituting (23.2.8) into (23.2.7), we obtain

$$J(\mathbf{c}) = \frac{1}{2}\mathbf{c}^{\mathrm{T}}\mathbf{B}\mathbf{c} - \mathbf{b}^{\mathrm{T}}\mathbf{c} + \frac{d}{2} \tag{23.2.10}$$

where

$$\mathbf{B} = \sum_{k=0}^{N} \mathbf{B}_k, \quad \mathbf{b} = \sum_{k=0}^{N} \mathbf{b}_k \quad \text{and} \quad d = \sum_{k=0}^{N} d_k. \tag{23.2.11}$$

From (23.2.10) it immediately follows that $J(\mathbf{c})$ is **unimodal**. That is, when the model has the linear dynamics and observations are a linear function of the state, then the mean square criterion gives rise to a unimodal objective function.

It can be verified that $\nabla_{\mathbf{c}} J$ is given by

$$\nabla_{\mathbf{c}} J(\mathbf{c}) = \mathbf{B}\mathbf{c} - \mathbf{b}. \tag{23.2.12}$$

Hence, the optimum value of $\mathbf{c}$ is given by

$$\mathbf{B}\mathbf{c} = \mathbf{b}. \tag{23.2.13}$$

Substituting for $\mathbf{B}$ and $\mathbf{b}$, this equation becomes

$$\left[\sum_{k=0}^{N} \mathbf{L}_k^{\mathrm{T}} \mathbf{H}^{\mathrm{T}} \mathbf{R}^{-1} \mathbf{H} \mathbf{L}_k\right] \mathbf{c} = \left[\sum_{k=0}^{N} \mathbf{L}_k^{\mathrm{T}} \mathbf{H}^{\mathrm{T}} \mathbf{R}^{-1} \mathbf{z}_k\right]. \tag{23.2.14}$$

While this is indeed a closed form solution, this form is far from being practical. In fact, it is computationally very expensive and is prone to round-off errors (Exercise 23.2).

Despite the difficulty of using this closed form solution, the point of this exercise is to demonstrate two important facts. First, $J(\mathbf{c})$ is **unimodal**. Second, data assimilation problems for this case of linear dynamics and observations that are linear functions of the state remain challenging and computationally demanding (Exercise 23.2). This is the primary reason for seeking clever iterative schemes to optimally estimate the initial conditions. Such an iterative scheme is developed in the following section.

In closing, we encourage the reader to pursue Exercise 23.3 which represents an extension of the development in this section to the case when the matrix $\mathbf{A}$ of the model (23.1.1) and the matrix $\mathbf{H}$ of the observation system in (23.1.6) both vary as a function of the time index k.

23.3 A method for finding the gradient: Lagrangian approach

Let us begin by computing the orbit of the linear discrete time system in (23.1.3). To this end, we rewrite it as follows: for $k \geq 1$

$$-\mathbf{M}\mathbf{x}_{k-1} + \mathbf{x}_k = 0 \quad \text{with } \mathbf{x}_0 = \mathbf{c}. \tag{23.3.1}$$

Define a block-partitioned vector $\mathbf{X}$ as

$$\mathbf{X} = (\mathbf{x}_1, \mathbf{x}_2, \ldots, \mathbf{x}_N)^{\mathrm{T}} \tag{23.3.2}$$

of block dimension N with each block of size n, that is, $\mathbf{x}_k = (\mathbf{x}_{1k}, \mathbf{x}_{2k}, \ldots, \mathbf{x}_{nk})^{\mathrm{T}}$. Clearly, the elements of the vector $\mathbf{X}$ constitute the orbit of (23.1.3). Using $\mathbf{X}$, the system (23.3.1) can be succinctly written as

$$\mathbf{F}\mathbf{X} = \mathbf{b} \tag{23.3.3}$$

where $\mathbf{F}$ is an $N \times N$ lower block bi-diagonal matrix of the form

$$\mathbf{F} = \begin{bmatrix} \mathbf{I} & 0 & 0 & \cdots & 0 & 0 & 0 \\ -\mathbf{M} & \mathbf{I} & 0 & \cdots & 0 & 0 & 0 \\ 0 & -\mathbf{M} & \mathbf{I} & \cdots & 0 & 0 & 0 \\ \vdots & \vdots & \vdots & \cdots & \vdots & \vdots & \vdots \\ 0 & 0 & 0 & \cdots & -\mathbf{M} & \mathbf{I} & 0 \\ 0 & 0 & 0 & \cdots & 0 & -\mathbf{M} & \mathbf{I} \end{bmatrix} \tag{23.3.4}$$

and $\mathbf{b}$ is a block-partitioned vector in conformity with $\mathbf{x}$ and is given by

$$\mathbf{b} = (\mathbf{M}\mathbf{c}, 0, 0, \ldots, 0)^{\mathrm{T}}. \tag{23.3.5}$$

Since the orbit of $\mathbf{X}$ represents the forward solution obtained from the initial condition $\mathbf{x}_0 = \mathbf{c}$, the system (23.3.3) has come to be known as the **forward system**.

Now define a vector $\boldsymbol{\lambda}_k$ as

$$\boldsymbol{\lambda}_k = (\lambda_{1k}, \lambda_{2k}, \ldots, \lambda_{nk})^{\mathrm{T}}$$

and let

$$\boldsymbol{\lambda} = (\boldsymbol{\lambda}_1, \boldsymbol{\lambda}_2, \ldots, \boldsymbol{\lambda}_N)^{\mathrm{T}}$$

be a block-partitioned vector of block dimension N where each block is of size n. Using the $J(\mathbf{c})$ defined in (23.1.8) and the dynamics in (23.1.3), we now introduce the Lagrangian

$$L(\mathbf{c}, \mathbf{X}, \boldsymbol{\lambda}) = J(\mathbf{c}) + \sum_{k=1}^{N} \boldsymbol{\lambda}_k^{\mathrm{T}} [-\mathbf{M}\mathbf{x}_{k-1} + \mathbf{x}_k] \tag{23.3.6}$$

where each term of the summation on the right-hand side is an inner product of $\boldsymbol{\lambda}_k$, the kth **Lagrangian multiplier vector** and the vector $[-\mathbf{M}\mathbf{x}_{k-1} + \mathbf{x}_k]$ representing the state transition from time instant $(k-1)$ to k. Clearly, L is a function of $(2N+1)$ vector variables each of which is of size n. Thus, there are a total of $(2N+1)n$ variables in L. Substituting for $J(\mathbf{c})$ in (23.3.6), the latter becomes

$$\begin{aligned} L(\mathbf{c}, \mathbf{X}, \boldsymbol{\lambda}) = &\frac{1}{2} \sum_{k=0}^{N} [\mathbf{z}_k - \mathbf{H}\mathbf{x}_k]^{\mathrm{T}} \mathbf{R}^{-1} [\mathbf{z}_k - \mathbf{H}\mathbf{x}_k] \\ &+ \sum_{k=1}^{N} \boldsymbol{\lambda}_k^{\mathrm{T}} [-\mathbf{M}\mathbf{x}_{k-1} + \mathbf{x}_k]. \end{aligned} \tag{23.3.7}$$

By differentiating L with respect to $\mathbf{c}$, and the components of $\mathbf{X}$ and $\boldsymbol{\lambda}$, we obtain

$$\nabla_{\mathbf{c}} L = \mathbf{H}^{\mathrm{T}}\mathbf{R}^{-1}[\mathbf{H}\mathbf{c} - \mathbf{z}_0] - \mathbf{M}^{\mathrm{T}}\boldsymbol{\lambda}_1 \tag{23.3.8}$$

$$\nabla_{\mathbf{x}_k} L = \mathbf{H}^{\mathrm{T}}\mathbf{R}^{-1}[\mathbf{H}\mathbf{x}_k - \mathbf{z}_k] + \boldsymbol{\lambda}_k - \mathbf{M}^{\mathrm{T}}\boldsymbol{\lambda}_{k+1} \tag{23.3.9}$$

$$\nabla_{\boldsymbol{\lambda}_k} L = -\mathbf{M}\mathbf{x}_{k-1} + \mathbf{x}_k \tag{23.3.10}$$

for $1 \le k \le N$ where $\boldsymbol{\lambda}_{N+1} = 0$.

Since the right-hand side of (23.3.10) represents the dynamic constraints, we already have

$$\nabla_{\boldsymbol{\lambda}_k} L = 0 \quad \text{for} \quad 1 \le k \le N$$

as required.

Clearly, $\nabla_{\mathbf{x}_k} L$ vanishes when

$$\boldsymbol{\lambda}_k - \mathbf{M}^{\mathrm{T}}\boldsymbol{\lambda}_{k+1} = -\mathbf{H}^{\mathrm{T}}\mathbf{W}[\mathbf{H}\mathbf{x}_k - \mathbf{z}_k]. \tag{23.3.11}$$

Define an n vector

$$\mathbf{e}_k = \mathbf{H}^{\mathrm{T}}\mathbf{W}[\mathbf{z}_k - \mathbf{H}\mathbf{x}_k] \tag{23.3.12}$$

and a block-partitioned vector

$$\mathbf{e} = (\mathbf{e}_1, \mathbf{e}_2, \ldots, \mathbf{e}_N)^{\mathrm{T}} \tag{23.3.13}$$

of block dimension N. Since $\boldsymbol{\lambda}_{N+1} = 0$, we can rewrite (23.3.11) succinctly as

$$\mathbf{B}\boldsymbol{\lambda} = \mathbf{e} \tag{23.3.14}$$

where

$$\mathbf{B} = \begin{bmatrix} \mathbf{I} & -\mathbf{M}^{\mathrm{T}} & 0 & \cdots & 0 & 0 \\ 0 & \mathbf{I} & -\mathbf{M}^{\mathrm{T}} & \cdots & 0 & 0 \\ \vdots & \vdots & \vdots & \cdots & \vdots & \vdots \\ 0 & 0 & 0 & \cdots & \mathbf{I} & -\mathbf{M}^{\mathrm{T}} \\ 0 & 0 & 0 & \cdots & 0 & \mathbf{I}. \end{bmatrix}. \tag{23.3.15}$$

Several observations are in order.

Remark 23.3.1 Notice that the matrix B in (23.3.15) is the **transpose** or the **adjoint** of the matrix F in (23.3.4). This structural relationship between the equation defining the Lagrangian multiplier $\boldsymbol{\lambda}$'s in (23.3.15) and the orbit $\mathbf{X}$ in (23.3.3) is quite intrinsic (also refer to Remark 23.3.3) to data assimilation problems. Since $\boldsymbol{\lambda}$'s occur in the first degree in $L(\mathbf{c}, \mathbf{X}, \boldsymbol{\lambda})$, it turns out that the equations defining $\boldsymbol{\lambda}$'s through $\nabla_{\mathbf{x}_k} L = 0$ are always linear in $\boldsymbol{\lambda}$'s, irrespective of whether the model dynamics is linear or nonlinear. However, in the case of nonlinear models, it turns out (as will be seen in Chapter 24) that the linear equations defining $\boldsymbol{\lambda}$'s will be the adjoint of a linearized version of the nonlinear dynamics obtained by the standard first-order perturbation technique .

As **B** is a simple, sparse, structured matrix, we can compute the solution of (23.3.14) by using the standard **back substitution** method. Indeed, for $k = N, N-2, \ldots, 2, 1$,

$$\boldsymbol{\lambda}_k = \sum_{j=k}^{N} (\mathbf{M}^{\mathrm{T}})^{j-k}\, \mathbf{e}_j. \tag{23.3.16}$$

Remark 23.3.2 Since the subsystem (23.3.15) is solved in the decreasing order of indices – from $\boldsymbol{\lambda}_N$ to $\boldsymbol{\lambda}_1$ this system is often called the **backward system** as opposed to the **forward system** (23.3.3). where the **x**'s are solved in the increasing order of indices – from $\mathbf{x}_1$ to $\mathbf{x}_N$.

Now substituting for $\boldsymbol{\lambda}_1$ in (23.3.8), the latter becomes

$$\begin{aligned}\nabla_{\mathbf{c}} L &= -\mathbf{e}_0 - \mathbf{M}^{\mathrm{T}} \boldsymbol{\lambda}_1 \\ &= -\mathbf{e}_0 - \sum_{j=1}^{N} (\mathbf{M}^{\mathrm{T}})^j \mathbf{e}_j \\ &= -\sum_{j=0}^{N} (\mathbf{M}^{\mathrm{T}})^j\, \mathbf{e}_j. \end{aligned} \tag{23.3.17}$$

Comparing this with (23.2.12) it can be verified that (Exercise 23.6)

$$\nabla_{\mathbf{c}} L = \nabla_{\mathbf{c}} J.$$

Thus, the gradient of $J(\mathbf{c})$ at the point **c** is computed by the following procedure:

Step 1 Starting with **c**, compute the orbit **X** using the **forward equation FX = b** in (23.3.3)).

Step 2 Using the observations $\left\{ \mathbf{z}_j \middle|\, 1 \le j \le N \right\}$ and the orbit **X** computed in Step 1, compute the vector **e**.

Step 3 Solve the **backward system B$\boldsymbol{\lambda}$ = e** in (23.3.15).

Step 4 Substitute for $\boldsymbol{\lambda}_1$ in (23.3.8) and obtain $\nabla_{\mathbf{c}} L$.

Computationally Steps 1 and 3 are the most demanding and these two steps involve solution of large, sparse, structured linear systems. Except for the nature of indexing – Step 1 recovers $\mathbf{x}_i$'s in the increasing order of indices, and Step 3 recovers $\boldsymbol{\lambda}_i$'s in the decreasing order of indices – these computations are essentially similar. Steps 2 and 4, on the other hand, involve routine evaluation of expressions.

In the contemporary literature on vector/parallel processing, several classes of algorithms are currently available for solving large, sparse, structured systems. Depending on the availability of parallel and vector processors, one can draw upon this body of algorithms and accelerate the computation of the gradient of $J(\mathbf{c})$ at any given point **c** (Exercise 23.7).

23.4 An algorithm for finding the optimal estimate

To motivate the rationale for an iterative algorithm, first compute $\nabla_{\mathbf{c}} J(\mathbf{c}) = 0$ for a chosen value of **c** using the method in Section 23.3. If $\nabla_{\mathbf{c}} J(\mathbf{c}) = 0$, then the value of **c** is the optimal estimate. However, it is highly improbable that a randomly chosen value of **c** will indeed be the optimal value. One could randomly keep picking **c** and testing if $\nabla_{\mathbf{c}} J(\mathbf{c})$ is zero until the optimal value is found. In lieu of such a **brute-force** algorithm, we need a systematic method for seeking the value of **c** at which $J(\mathbf{c})$ is a minimum. Using the norm of the gradient as a discriminant, the following iterative procedure provides a very natural framework for finding the optimum.

Step 1 Choose a vector **c** and call it $\mathbf{c}_{\text{old}}$.

Step 2 Using the method described in Section 23.3, compute $\nabla_{\mathbf{c}} J(\mathbf{c}_{\text{old}})$.

Step 3 If $\|\nabla_{\mathbf{c}} J(\mathbf{c}_{\text{old}})\| < \varepsilon$ for some prespecified $\varepsilon > 0$, then stop; otherwise, go to Step 4.

Step 4 Compute

$$\mathbf{c}_{\text{new}} = \mathbf{c}_{\text{old}} - \beta \nabla_{\mathbf{c}} J(\mathbf{c}_{\text{old}}) \tag{23.4.1}$$

for some step length parameter $\beta > 0$. Set $\mathbf{c}_{\text{old}} \leftarrow \mathbf{c}_{\text{new}}$ and go to Step 2.

The iterative scheme embodied in (23.4.1) is called the **steepest descent** approach to minimization. (See Chapter 10)

We encourage the reader to conduct the computer-based Monte Carlo type twin experiment described in Exercise (23.10).

Remark 23.4.1 There are basically two ingredients in any minimization algorithm – the **direction** of search and the **step length** in that direction. All the iterative schemes for minimization differ in the way in which these two factors are chosen. The steepest descent algorithm as given in Chapter 10 and the **conjugate gradient** algorithm described in Chapter 11 are two basic methods. Most of the FORTRAN subroutine libraries such as IMSL have several well-tested packages for minimization. We encourage the reader to become familiar with these standard packages .

Remark 23.4.2 There is another family of minimization algorithms called the **Newton** and **quasi-Newton** algorithms. These algorithms in addition to the gradient $\nabla_{\mathbf{c}} J$ need information on the **Hessian** $\nabla_{\mathbf{c}}^2 J(\mathbf{c})$ of $J(\mathbf{c})$. While estimating the Hessian, in principle, is computationally intense, there is a welcome trade-off. As a rule Newton-like algorithms using some form of information on the Hessian have very good (e.g., quadratic) convergence rates. Motivated by this scenario, there have been several attempts at computing the Hessian of $J(\mathbf{c})$. These have led to an emerging theory of the so-called second-order adjoint method. This latter method is described in Chapter 25.

23.5 A second method for computing the gradient: the adjoint operator approach

In this section we describe an alternate approach based on the principle and properties of the **adjoint** of a linear operator and the **inner product** for computing the gradient of J. We begin by recalling the following definition of the **adjoint** A^{T}, of a linear operator A.

Let $\mathbf{x}, \mathbf{y} \in \mathbb{R}^n$ and $\mathbf{A} \in \mathbb{R}^{n \times n}$. Then

$$< \mathbf{x}, \mathbf{A}\mathbf{y} > = < \mathbf{A}^{\mathrm{T}}\mathbf{x}, \mathbf{y} > \tag{23.5.1}$$

where $< \cdot, \ \cdot >$ denotes the usual inner product. Let $\mathbf{a}, \mathbf{b} \in \mathbb{R}^n$. Then, if

$$< \mathbf{a}, \mathbf{x} > = < \mathbf{b}, \mathbf{x} > \tag{23.5.2}$$

for all $\mathbf{x}$, then $\mathbf{a} = \mathbf{b}$.

There are two other basic facts that are fundamental to this approach. First relates to the definition of the **directional derivative** of J (Appendix C). Let

$$\delta \mathbf{c} = (\delta c_1, \delta c_2, \ldots, \delta c_n)^{\mathrm{T}} \tag{23.5.3}$$

be an **increment** in or a perturbation of $\mathbf{c}$. Let δJ be the **first variation** in J induced by $\delta \mathbf{c}$. From the definition of the first variation, it follows that

$$\delta J = < \delta \mathbf{c}, \nabla_{\mathbf{c}} J(\mathbf{c}) > \tag{23.5.4}$$

that is, the first variation δJ is **linearly** related to the increment $\delta \mathbf{c}$ where $\nabla_{\mathbf{c}} J(\mathbf{c})$ is the gradient we are seeking. The second and the last component needed in this approach consists in deriving an expression for the rate of growth of the initial perturbation $\delta \mathbf{c}$ as seen through the linear dynamical equation (23.1.3) which is repeated here for convenience:

$$\mathbf{x}_{k+1} = \mathbf{M}\mathbf{x}_k. \tag{23.5.5}$$

To this end, first choose an initial condition $\mathbf{c}$ and compute the orbit of (23.5.5) by solving the forward system in (23.3.3). The resulting sequence of states is often distinguished by a special name, **base state** or **nominal trajectory** and is denoted by

$$\bar{\mathbf{x}} = (\bar{\mathbf{x}}_0, \ \bar{\mathbf{x}}_1, \ \bar{\mathbf{x}}_2, \ldots, \bar{\mathbf{x}}_N).$$

Refer to Figure 23.5.1. Now, let $\delta \mathbf{c}$ in (23.5.3) denote an increment in $\mathbf{c}$. Let

$$\mathbf{x} = (\mathbf{x}_0, \mathbf{x}_1, \mathbf{x}_2, \ldots, \mathbf{x}_N)$$

be the orbit resulting from this initial perturbed state. From (23.5.5) we have

$$\bar{\mathbf{x}}_{k+1} = \mathbf{M}\bar{\mathbf{x}}_k \quad \text{with} \quad \bar{\mathbf{x}}_0 = \mathbf{c}$$

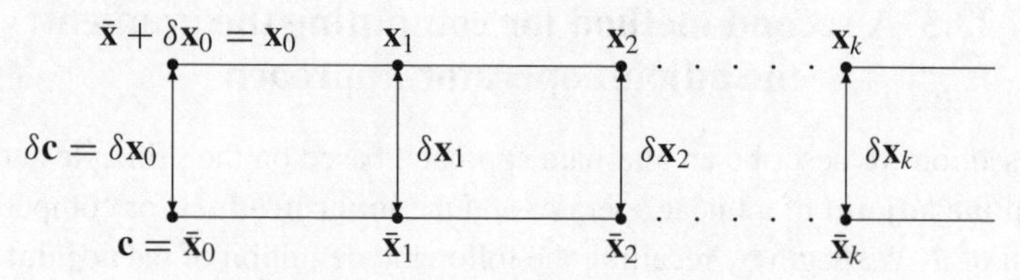

Fig. 23.5.1 Evolution of perturbation.

and

$$\mathbf{x}_{k+1} = \mathbf{M}\mathbf{x}_k \quad \text{with} \quad \mathbf{x}_0 = \mathbf{c} + \delta\mathbf{c}.$$

If $\delta\mathbf{x}_k = \mathbf{x}_k - \bar{\mathbf{x}}_k$, then it can be verified that

$$\delta\mathbf{x}_{k+1} = \mathbf{M}\,\delta\mathbf{x}_k \quad \text{with} \quad \delta\,\mathbf{x}_0 = \delta\mathbf{c}. \tag{23.5.6}$$

That is, the perturbation $\delta\mathbf{x}_k$ in the state $\mathbf{x}_k$ at time k resulting from the initial perturbation $\delta\mathbf{c}$ evolves according to the same linear dynamics (23.5.5). Solving (23.5.6), we readily obtain

$$\delta\mathbf{x}_k = \mathbf{M}^k\delta\mathbf{c}. \tag{23.5.7}$$

Remark 23.5.1 Based on the principles of Lyapunov Stability (Part VIII), we can readily conclude that the growth in the initial error as manifested in (23.5.7) is related to the distribution of the eigenvalues of the matrix $\mathbf{M}$. Let

$$\boldsymbol{\lambda}(\mathbf{M}) = \{\lambda_1, \lambda_2, \ldots, \lambda_n\}$$

denote the **spectrum** which is the set of eigenvalues of $\mathbf{M}$. Then, it can be verified that

$$\delta\mathbf{x}_k \to 0$$

if and only if the **spectral radius**, $\rho(\mathbf{M})$ of $\mathbf{M}$ is less than 1, that is

$$\rho(\mathbf{M}) = \max_{1\le i\le n} \{\,|\lambda_i|\,\} < 1.$$

Otherwise, there exists at least one eigenvalue of magnitude larger than unity and the perturbation will grow **exponentially**. It will become evident later that this spectral analysis of $\mathbf{M}$ constitutes the backbone of the **predictability** theory as used in contemporary meteorology today.

Once we derive the dynamics of the evolution of the initial perturbation, we are now ready to compute δJ rather explicitly. To this end, recall that

$$J(\mathbf{c}) = \frac{1}{2}\sum_{k=0}^{N} \langle(\mathbf{z}_k - \mathbf{H}\mathbf{x}_k)\ ,\ \ \mathbf{W}(\mathbf{z}_k - \mathbf{H}\mathbf{x}_k)\rangle \tag{23.5.8}$$

with $\mathbf{W} = \mathbf{R}^{-1}$. From first principles, it can be verified (Appendix B) that

$$\delta J = \sum_{k=0}^{N} \langle \mathbf{W}(\mathbf{H}\mathbf{x}_k - \mathbf{z}_k), \quad \mathbf{H}\delta\mathbf{x}_k \rangle. \tag{23.5.9}$$

Now substituting for $\delta\mathbf{x}_k$ from (23.5.7) and using the adjoint property (23.5.1) repeatedly we obtain

$$\begin{aligned}
\delta J &= \sum_{k=0}^{N} \langle \mathbf{W}(\mathbf{H}\mathbf{x}_k - \mathbf{z}_k), \quad \mathbf{H}\mathbf{M}^k \delta\mathbf{c} \rangle \\
&= \sum_{k=0}^{N} \langle \mathbf{H}^{\mathrm{T}}\mathbf{W}(\mathbf{H}\mathbf{x}_k - \mathbf{z}_k), \quad \mathbf{M}^k \delta\mathbf{c} \rangle \qquad (23.5.10) \\
&= \sum_{k=0}^{N} \langle (\mathbf{M}^{\mathrm{T}})^k \, \mathbf{H}^{\mathrm{T}}\mathbf{W}(\mathbf{H}\mathbf{x}_k - \mathbf{z}_k), \quad \delta\,\mathbf{c} \rangle \qquad (23.5.11)
\end{aligned}$$

where $(\mathbf{M}^k)^{\mathrm{T}} = (\mathbf{M}^{\mathrm{T}})^k$ in view of Exercise 23.5.

Since

$$\langle \mathbf{x}_1 + \mathbf{x}_2, \ \mathbf{y} \rangle = \langle \mathbf{x}_1, \mathbf{y} \rangle + \langle \mathbf{x}_2, \mathbf{y} \rangle$$

we can rewrite (23.5.11) as

$$\delta J = \left\langle \sum_{k=0}^{N} (\mathbf{M}^{\mathrm{T}})^k \, \mathbf{H}^{\mathrm{T}}\mathbf{W}(\mathbf{H}\mathbf{x}_k - \mathbf{z}_k), \ \delta\,\mathbf{c} \right\rangle. \tag{23.5.12}$$

Comparing (23.5.12) with (23.5.4) (since $< \mathbf{x}, \mathbf{y} > = < \mathbf{y}, \mathbf{x} >$) we obtain (Bravo!)

$$\nabla_{\mathbf{c}} J(\mathbf{c}) = \sum_{k=0}^{N} (\mathbf{M}^{\mathrm{T}})^k \mathbf{H}^{\mathrm{T}}\mathbf{W}\,(\mathbf{H}\mathbf{x}_k - \mathbf{z}_k). \tag{23.5.13}$$

We encourage the reader to compare this expression with (23.2.12) and (23.3.17).

As observed in Section 23.2, from a computational perspective this is a wild monster too difficult to tame. Hence, in the following we describe a simple-minded recursive algorithm for computing the expression on the right-hand side of (23.5.13).

To this end, we make use of the sequence of vectors $\mathbf{f}_k \in \mathbb{R}^n$ defined by

$$\mathbf{f}_k = \mathbf{H}^{\mathrm{T}}\mathbf{W}(\mathbf{H}\mathbf{x}_k - \mathbf{z}_k). \tag{23.5.14}$$

Let $\bar{\boldsymbol{\lambda}}_k \in \mathbb{R}^n$ be a sequence defined for $k = N, N-1, \ldots, 2, 1, 0$ by the recurrence relation:

$$\left.\begin{aligned}
\bar{\boldsymbol{\lambda}}_N &= \mathbf{f}_N \\
\bar{\boldsymbol{\lambda}}_k &= \mathbf{M}^{\mathrm{T}}\bar{\boldsymbol{\lambda}}_{k+1} + \mathbf{f}_k
\end{aligned}\right\} \tag{23.5.15}$$

Solving this latter recurrence, it can be verified that

$$\bar{\boldsymbol{\lambda}}_0 = \sum_{k=0}^{N} (\mathbf{M}^{\mathrm{T}})^k \, \mathbf{f}_k. \tag{23.5.16}$$

Table 23.5.1

k	0	1	2	3	4	5	6	7	8	9
z_k	0.0010	0.3055	−0.1104	0.7613	0.0224	0.0962	0.3925	0.0693	0.0161	−0.2208

Table 23.5.2

Iteration Number k	Values of $J(\mathbf{c}_k)$ starting point $\mathbf{c}(0) = (1, 19)$	
	gradient method	conjugate gradient method
0	68.2765	68.2765
1	0.8195	0.8195
pt 2	0.3316	0.3280
3	0.3280	

Comparing this expression with (23.5.13), we can immediately conclude that $\bar{\boldsymbol{\lambda}}_0$ is indeed the gradient we are seeking. Notice, the difference, however, is that as opposed to computing the expression (23.5.14), $\bar{\boldsymbol{\lambda}}_0$ can be incrementally computed by simple **back substitution** implied by the recurrence (23.5.15).

We illustrate the above development using the following:

Example 23.5.1 Let $n = 2$ and $m = 1$ with $\mathbf{x}_{k+1} = \mathbf{M}\mathbf{x}_k$ and $z_k = \mathbf{H}\mathbf{x}_k + v_k$ where

$$\mathbf{M} = \begin{bmatrix} 0.7 & 0.9 \\ 0.1 & 0.5 \end{bmatrix}, \quad \mathbf{H} = \begin{bmatrix} 0.2 \; 0.1 \end{bmatrix}, \quad v_k \sim N(0, \sigma^2)$$

with $\sigma^2 = 0.1$. The eigenvalues of $\mathbf{M}$ are given by $\lambda_1 = 0.9162$ and $\lambda_2 = 0.2838$ and the model is stable. Using $\mathbf{x}_0 = (0.2, 0.2)^{\mathrm{T}}$ we created a set of $N = 10$ observations given in Table 23.5.1.

The functional form of $J(\mathbf{c})$ computed using (23.2.10) is given by

$$J(\mathbf{c}) = \frac{1}{2}\mathbf{c}^{\mathrm{T}}\mathbf{B}\,\mathbf{c} - \mathbf{b}^{\mathrm{T}}\mathbf{c} + \frac{d}{2}$$

where

$$\mathbf{B} = \begin{bmatrix} 0.1358 & 0.2084 \\ 0.2084 & 0.3866 \end{bmatrix}, \quad \mathbf{b} = \begin{bmatrix} 0.1568 \\ 0.3068 \end{bmatrix}, \quad d = 0.9028.$$

The plot of the contours of $J(\mathbf{c})$ is given in Figure 23.5.2. The eigenvalues of $\mathbf{B}$ are given by $\mu_1 = 0.0180$ and $\mu_2 = 0.5045$ which in turn confirms the closed ellipses that form the contours. The optimal value is given by $\mathbf{c}^* = \mathbf{B}^{-1}\mathbf{b} =$

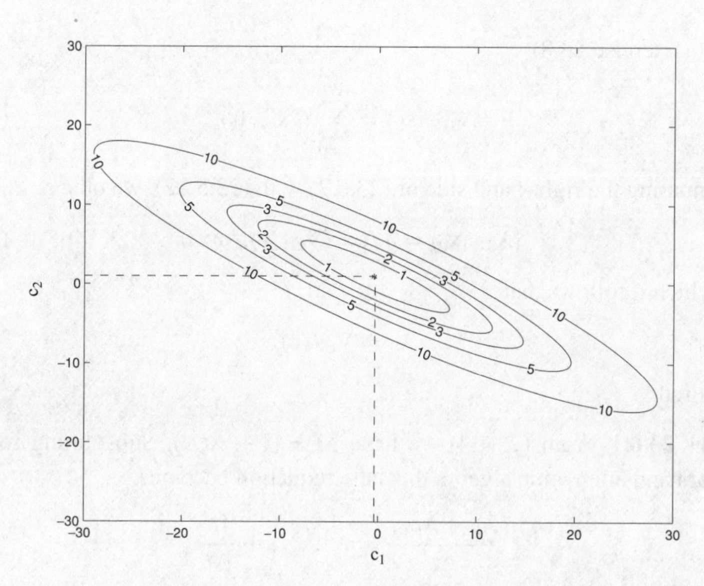

Fig. 23.5.2 Contours of $J(\mathbf{c})$ in the c_1–c_2 plane.

$(-0.3698, 0.9930)^{\mathrm{T}}$ and the optimal value of $J(\mathbf{c}^*) = 0.3280$. Results of the iterative minimization using the gradient and conjugate gradient methods are listed in Table 23.5.2.

23.6 Method of integration by parts

There is also an alternate method for verifying that $\bar{\boldsymbol{\lambda}}_0$ is the gradient of $J(\mathbf{c})$. The importance of this method is that it is very general and is the discrete analog of the method based on the **integration by parts** which is widely used in the study of adjoint operators in a continuous domain.

Take the inner product of $(\delta\mathbf{x}_k - \mathbf{M}\delta\mathbf{x}_{k-1})$ which is the dynamics of error in (23.5.6) and the vector $\bar{\boldsymbol{\lambda}}_k$ defined in (23.5.15) and add the resulting expression from $k = 1$ to N. We obtain

$$
\begin{aligned}
0 &= \sum_{k=1}^{N} \left\langle \bar{\boldsymbol{\lambda}}_k,\ (\delta\mathbf{x}_k - \mathbf{M}\delta\mathbf{x}_{k-1}) \right\rangle \\
&= \sum_{k=1}^{N} \left\{ \left\langle \bar{\boldsymbol{\lambda}}_k,\ \delta\mathbf{x}_k \right\rangle - \left\langle \mathbf{M}^{\mathrm{T}}\bar{\boldsymbol{\lambda}}_k,\ \delta\mathbf{x}_{k-1} \right\rangle \right\} \\
&= \sum_{k=1}^{N} \left\{ \left\langle \bar{\boldsymbol{\lambda}}_k,\ \delta\mathbf{x}_k \right\rangle - \left\langle \left(\bar{\boldsymbol{\lambda}}_{k-1} - \mathbf{f}_k \right),\ \delta\mathbf{x}_{k-1} \right\rangle \right\} \\
&= -\left\langle \bar{\boldsymbol{\lambda}}_0,\ \delta\mathbf{x}_0 \right\rangle + \sum_{k=0}^{N} \langle \delta\mathbf{x}_k,\ \mathbf{f}_k \rangle.
\end{aligned}
\tag{23.6.1}
$$

That is, (Exercise 23.8)

$$\langle \bar{\boldsymbol{\lambda}}_0,\ \delta \mathbf{x}_0 \rangle = \sum_{k=0}^{N} \langle \delta \mathbf{x}_k,\ \mathbf{f}_k \rangle. \tag{23.6.2}$$

By comparing the right-hand side of (23.6.2) with (23.5.12), we obtain

$$\langle \bar{\boldsymbol{\lambda}}_0, \delta \mathbf{x}_0 \rangle = \delta J = \langle \nabla_{\mathbf{c}} J(\mathbf{c}),\ \ \delta \mathbf{x}_0 \rangle \tag{23.6.3}$$

from which it follows that

$$\bar{\boldsymbol{\lambda}}_0 = \nabla_{\mathbf{c}} J(\mathbf{c})$$

as required.

Remark 23.6.1 From (23.1.4) we have $\mathbf{M} = (\mathbf{I} + \Delta t \mathbf{A})$. Substituting for $\mathbf{M}$ in (23.5.15) and after some algebra this latter equation becomes

$$\frac{\bar{\boldsymbol{\lambda}}_k - \bar{\boldsymbol{\lambda}}_{k+1}}{\Delta t} = \mathbf{A}^{\mathrm{T}} \bar{\boldsymbol{\lambda}}_{k+1} + \frac{\mathbf{f}_k}{\Delta t}. \tag{23.6.4}$$

In the limit as $\Delta \mathrm{t} \rightarrow 0$, this can be written as

$$-\frac{\mathrm{d}\bar{\boldsymbol{\lambda}}}{\mathrm{d}t} = \mathbf{A}^{\mathrm{T}} \bar{\boldsymbol{\lambda}}\,(t) + \mathbf{f}(t) \tag{23.6.5}$$

with $\boldsymbol{\lambda}(t_f) = \mathbf{f}(t_f)$ where $t_f = \Delta t N$. It can be verified that the model dynamics

$$\frac{\mathrm{d}\mathbf{x}}{\mathrm{d}t} = \mathbf{A}\mathbf{x}\,(t)$$

and the **unforced** version (23.6.5), namely

$$-\frac{\mathrm{d}\bar{\boldsymbol{\lambda}}}{\mathrm{d}t} = \mathbf{A}^{\mathrm{T}} \bar{\boldsymbol{\lambda}}\,(t)$$

are the **adjoint** pair of equations. In view of this relation, in meteorological parlance the recurrence relation defining $\bar{\boldsymbol{\lambda}}_k$ in (23.5.15) has come to be known as the **adjoint equation**, and the method of computing $\nabla J(\mathbf{c})$ as $\bar{\boldsymbol{\lambda}}_0$ by solving this adjoint equation is called the **adjoint method**.

Remark 23.6.2 An astute reader may already have noticed the similarity between the backward system (23.5.15) in $\boldsymbol{\lambda}_k$ and the adjoint equation (23.3.14) in $\bar{\boldsymbol{\lambda}}_k$. (Exercise 23.9). In view of this similarity, the method for computing the gradient using the Lagrangian framework has also been referred to as the **adjoint method** in the literature.

Exercises

23.1 True or False: when $\mathbf{A}$ is singular (non-singular) then so is $\mathbf{M} = (\mathbf{I} + \Delta t \mathbf{A})$ for some, small $\Delta t > 0$.

23.2 Recall that the multiplication of two $n \times n$ real matrices requires n^3 real multiplications and $n^2(n-1)$ real additions, and that the addition of two $n \times n$ real matrices requires n^2 real additions.

(a) Compute the number of real multiplications and real additions needed to compute the matrix $\mathbf{B}$ and the vector $\mathbf{b}$ using (23.2.11) and (23.2.9).

23.3 Consider the following linear time varying system

$$\frac{d\mathbf{x}}{dt} = \mathbf{A}(t)\,\mathbf{x} \quad \text{with} \quad \mathbf{x}(0) = \mathbf{x}_0 = \mathbf{c}.$$

Discretizing using the Euler scheme we obtain

$$\mathbf{x}_{k+1} = \mathbf{M}_k \mathbf{x}_k$$

where

$$\mathbf{M}_k = [I + \Delta t \mathbf{A}_k] \quad \text{and} \quad \mathbf{A}_k = \mathbf{A}(k\Delta t)$$

for some, small $\Delta t > 0$. Iterating this we obtain

$$\mathbf{x}_k = \mathbf{R}_k \mathbf{c} \quad \text{where} \quad \mathbf{R}_k = \mathbf{M}_{k-1}\mathbf{M}_{k-2} \cdots \mathbf{M}_1 \mathbf{M}_0.$$

Let $\mathbf{z}_k = \mathbf{H}_k \mathbf{x}_k + \mathbf{v}_k$ and consider

$$J(\mathbf{c}) = \frac{1}{2} \sum_{k=0}^{N} (\mathbf{z}_k - \mathbf{H}\mathbf{x}_k)^{\mathrm{T}}\, \mathbf{W}(\mathbf{z}_k - \mathbf{H}\mathbf{x}_k).$$

(a) Express $J(\mathbf{c})$ explicitly as a function of $\mathbf{c}$.
(b) Verify that $J(\mathbf{c})$ is unimodal.
(c) Derive an expression for optimal $\mathbf{c}$ in the form $\mathbf{B}\mathbf{c} = \mathbf{d}$.
(d) Compute the number of real additions and multiplications required in obtaining the matrix $\mathbf{B}$ and vector $\mathbf{d}$.
(e) Compare this result with that obtained in Exercise (23.2).

23.4 Prove that the matrix $\mathbf{B}$ in (23.3.15) is non-singular.

23.5 Verify that $(\mathbf{M}^k)^{\mathrm{T}} = (\mathbf{M}^{\mathrm{T}})^k$, i.e., the transpose of the kth power of $\mathbf{M}$ is the kth power of the transpose of $\mathbf{M}$

23.6 Compare (23.2.12) and (23.3.17) and verify that $\nabla_{\mathbf{c}} L = \nabla_{\mathbf{c}} J$.

23.7 A **brute-force method** for computing $\nabla_{\mathbf{c}} J(\mathbf{c})$ at a given $\mathbf{c}$ may be stated as follows: Let $\mathbf{e}_i = (0, 0, \ldots, 1, \ldots, 0)^{\mathrm{T}}$ denote the ith unit vector in $\mathbb{R}^n$ and let $h > 0$ denote a small real number. Then, ith partial derivative of J denoted by $\partial J / \partial c_i$ can be approximated as

$$\frac{\partial J}{\partial c_i} = \frac{J(\mathbf{c} + \mathbf{e}_i h) - J(\mathbf{c} - \mathbf{e}_i h)}{2h}.$$

Thus

$$\nabla_{\mathbf{c}} J(\mathbf{c}) = \left(\frac{\partial J}{\partial c_1}, \frac{\partial J}{\partial c_2}, \ldots \frac{\partial J}{\partial c_n} \right)^{\mathrm{T}}$$

can be obtained. Computation of $\partial J/\partial c_i$ needs the following steps.

Step A Starting from an initial state $\mathbf{x}_0 = (\mathbf{c} + \mathbf{e}_i h)$ first compute the orbit by solving the forward system (23.3.3). Using the $\mathbf{x}_i$'s computed and the given set of observations $\{\mathbf{z}_i \mid 0 \le i \le N\}$ in (23.1.6), compute the value of $J(\mathbf{c} + \mathbf{e}_i h)$.

Step B Repeat Step A by starting again from $\mathbf{x}_0 = (\mathbf{c} - \mathbf{e}_i h)$ and compute $J(\mathbf{c} - \mathbf{e}_i h)$.

Step C Compute an approximation to $\partial J/\partial c_i$ using the two-sided approximation given above.

Notice that this three-step procedure is to be repeated n times, one for each direction.

(a) Compute the total number of times the forward equation (23.3.3) needs to be solved in computing $\nabla_{\mathbf{c}} J(\mathbf{c})$.

(b) Compare the amount of work required by this brute-force method with the Lagrangian approach described in Section 23.3.

23.8 Verify all the steps leading to (23.6.2) from (23.6.1).

23.9 Rewrite (23.5.15) in the matrix-vector form and verify that it corresponds to an upper block bi-diagonal system of the type similar to (23.3.15).

23.10 **Monte Carlo type twin experiment**

(a) Let $\mathbf{x} = (x_1, x_2)^{\mathrm{T}}$ and consider the linear system

$$\mathbf{x}_{k+1} = \mathbf{M}\mathbf{x}_k$$

where

$$\mathbf{M} = \begin{bmatrix} \mathrm{a} & \mathrm{b} \\ \mathrm{c} & \mathrm{d} \end{bmatrix} \quad \text{and} \quad \mathbf{x}_0 = \mathbf{c} = \begin{pmatrix} c_1 \\ c_2 \end{pmatrix}.$$

(b) Let

$$\mathbf{z}_k = \mathbf{H}\mathbf{x}_k + \mathbf{v}_k$$

be the scalar observations where $\mathbf{H} = [h_1, h_2] \in \mathbb{R}^{1\times 2}$ and $\mathbf{v}_k$ is the white Gaussian noise with mean, $E(\mathbf{v}_k) = 0$ and variance $E(\mathbf{v}_k^2) = \sigma^2$.

(c) Using the method described in Section 23.3, develop a program to compute $\nabla_{\mathbf{c}} J$.

(d) Conduct a Monte Carlo type twin experiment analogous to the one described in Section 22.4.

Step 0 First select the elements of the matrices $\mathbf{M}$ and $\mathbf{H}$ randomly and keep them fixed.

Step 1 Pick a vector $\mathbf{x}_0 = \mathbf{c}$ randomly and compute the trajectory $\mathbf{x}_0, \mathbf{x}_1, \mathbf{x}_2, \mathbf{x}_3, \ldots, \mathbf{x}_N$.

Step 2 Generate a sequence of Gaussian random variables $\mathbf{v}_k$ from $N(0, \sigma^2)$ and generate the observations $\mathbf{z}_k$, $k = 0, 1, 2, \ldots, N$.

Step 3 Compute the objective function $J(\mathbf{c})$ and plot the contour of $J(\mathbf{c})$. (See Part (e))

Step 4 Minimize $J(\mathbf{c})$ using the steepest descent algorithm (Chapter 10) and the Conjugate Gradient algorithm (Chapter 11).

(e) Compute the matrix of the quadratic form $J(\mathbf{c})$ explicitly and evaluate its eigenvalues. Relate the shape of the contours to the eigenvalues.

(f) Examine the effect of changing the following parameters.

(1) Variance σ^2 of the observation noise.

(2) Number of observations.

(3) Location of the observations.

(4) Change the observation matrix $\mathbf{H}$: try $\mathbf{H} = [0, h_2]$ and $\mathbf{H} = [h_1, 0]$.

(5) Vary the system matrix $\mathbf{M}$.

23.11 Problem based on discussion in Section 3.4. Measurements of the current and depth of a section of river are as follows:

$$\tilde{v} \text{ (current): } 5\,\mathrm{m\,s^{-1}}$$
$$\tilde{D} \text{ (Depth): } 10\,\mathrm{m}$$

The instruments, current meter and depth finder, exhibit the following error variances:

$$\sigma_v^2 = (0.5\,\mathrm{m\,s^{-1}})^2$$
$$\sigma_D^2 = (0.5\,\mathrm{m})^2$$

A stationary gravity wave of length L is generated and an eye-ball estimate of its length is

$$\tilde{L} = 14\,\mathrm{m}$$

where the error variance of this estimate is

$$\sigma_L^2 = (2\,\mathrm{m})^2\,.$$

Obtain estimates of the current, depth, and stationary wavelength (denoted by v, D, and L) such that the functional

$$J = \frac{1}{\sigma_v^2}(v - \tilde{v})^2 + \frac{1}{\sigma_D^2}(D - \tilde{D})^2 + \frac{1}{\sigma_L^2}(L - \tilde{L})^2$$

is minimized subject to the constraint

$$\tanh\left(\frac{2\pi D}{L}\right) = \frac{v^2}{gD}\left(\frac{2\pi D}{L}\right) = \frac{v^2 \cdot 2\pi}{gL}.$$

Notes and references

Section 23.1 Refer to Ghil and Malanotte-Rizzoli (1991) for the statement of the assimilation problem of interest in the geophysical domain. The data assimilation problem is very closely related to the so-called inverse problem. For details refer to Tarantola (1987).

Section 23.2 The closed form solution given in this section is largely of theoretical interest and does not lead to any useful solution.

Section 23.3 For a discussion of the Lagrangian multiplier approach refer to Appendix D as well as Thacker and Long (1988) and Lanczos (1970).

Section 23.4 Refer to Part III for a succinct discussion of the optimization methods. Refer to Dennis and Schnabel (1996) for more details.

Section 23.5 In the early 1980s, Francois LeDimet (1982) introduced the meteorological community to optimization strategies that stemmed from the work of control theorists in France (notably Lyon). The adjoint operator approach began with the work of LeDimet and Talagrand (1986). Also refer to Lewis and Derber (1985), Talagrand and Courtier (1987) and (1993) and Courtier and Talagrand (1990). An interesting account and a commentary on the adjoint method is given in Errico (1997).

Section 23.6 The method of integration by parts is often used in the continuous time formulation. This method is also used in Chapter 25 in the context of the second-order adjoint method.

24

First-order adjoint method: nonlinear dynamics

In this chapter we extend the first-order adjoint method developed in the context of linear dynamical systems in Chapter 23 to the case of general nonlinear dynamical systems. Since most of the models of interest in research and operations are nonlinear, the contents of this chapter are especially applicable to real world problems.

In Chapter 23 we have brought out the similarities and differences between the two methods – the Lagrangian approach and the adjoint operator theoretic approach – for computing the gradient of the functional representing the desired criterion. Since we have featured the Lagrangian method twice – in Chapters 22 and 23 – we use the adjoint operator theoretic approach in this chapter. Our decision to use alternate approaches in different chapters is driven by two goals. First it provides variety and is hopefully more stimulating to the reader. Second, and more importantly, it enables the reader to acquire dexterity with the spectrum of tools that can be applied to real world problems.

The statement of the inverse problem is contained in Section 24.1. The first-order perturbation method for quantifying the growth of error is described in Section 24.2. Computation of the gradient using the adjoint method and an algorithm for minimization are given in Section 24.3 and 24.4 respectively. In Section 24.5, we introduce the reader to computation of model output sensitivity via adjoint modeling, i.e., the change of model output with respect to elements of the control vector.

24.1 Statement of the inverse problem

Let $k \in \{0, 1, 2, 3, \ldots\}$ denote the discrete time variables respectively. Let $\mathbf{x} \in \mathbb{R}^n$ with $\mathbf{x} = (x_1, x_2, \ldots, x_n)^{\mathrm{T}}$ and $\boldsymbol{\alpha} \in \mathbb{R}^p$ with $\boldsymbol{\alpha} = (\alpha_1, \alpha_2, \ldots, \alpha_p)^{\mathrm{T}}$. Let $\mathbf{F} : \mathbb{R}^n \times \mathbb{R}^p \to \mathbb{R}^n$ such that $\mathbf{F} = (F_1, F_2, \ldots, F_n)^{\mathrm{T}}$ where $F_i = F_i(\mathbf{x}, \boldsymbol{\alpha})$ for $1 \le i \le n$.

Model Consider the following nonlinear system of the type

$$\frac{\partial \mathbf{x}}{\partial t} = \mathbf{F}(\mathbf{x}, \boldsymbol{\alpha}) \tag{24.1.1}$$

with $\mathbf{x}(0) = \mathbf{c}$, being the initial condition. The vector $\mathbf{x}(t)$ denotes the **state** of the system at time t, $\boldsymbol{\alpha}$ is a set of physical or empirical parameters, and $\mathbf{F}(\mathbf{x}, \boldsymbol{\alpha})$ represents the field at the point $\mathbf{x}$. It is assumed that each component F_i of $\mathbf{F}$ is "sufficiently smooth" in both $\mathbf{x}$ and $\boldsymbol{\alpha}$ so that the solution of (24.1.1) exists and is unique.

Remark 24.1.1 In general, the function $\mathbf{F}$ could depend on the state variable $\mathbf{x}$ as well as its derivatives with respect to the standard space variables x, y, and z. For example, the Burger–Bateman equation can be written as

$$\frac{\partial u}{\partial t} = -u\frac{\partial u}{\partial x} + \varepsilon\frac{\partial^2 u}{\partial x^2} = f(u, \varepsilon). \tag{24.1.2}$$

Here u is the state variable, x is the space variable, and ε is a parameter called the **diffusion constant** or **diffusivity**. Notice that the right-hand side depends on u, $\partial u/\partial x$, $\partial^2 u/\partial x^2$, and ε. There should be no confusion between the generic state variable $\mathbf{x}$ and the conventional space variables x, y, z.

The equation (24.1.1) can be discretized using a variety of schemes (see Richtmyer 1957) and the resulting discrete version of the dynamics can be represented as

$$\mathbf{x}_{k+1} = \mathbf{M}(\mathbf{x}_k, \boldsymbol{\alpha}) \tag{24.1.3}$$

where $\mathbf{M} : \mathbb{R}^n \times \mathbb{R}^p \to \mathbb{R}^n$ with $\mathbf{M} = (M_1, M_2, \ldots, M_n)^{\mathrm{T}}$, $M_i = M_i(\mathbf{x}_k, \boldsymbol{\alpha})$ for $1 \le i \le n$, and $\mathbf{x}_0 = \mathbf{c}$ is the initial condition.

Remark 24.1.2 When the equation (24.1.1) is discretized, the size of the resulting state vector $\mathbf{x}_k$ in (24.1.3) and the nature of the mapping $\mathbf{M}$ that defines the field critically depend on several factors: the **number** of **space variables** involved in the original dynamics (24.1.1), the **size** of the grid in each space variable, the nature of the **stencil** used in arriving at a discrete approximation, etc. For the example in (24.1.2), if we discretize $u(x, t)$ using 100 subintervals in the x direction and 40 subintervals in the t direction, the dynamical equations take the following form

$$\mathbf{u}(k+1) = \mathbf{f}(\mathbf{u}(k), \varepsilon), \quad 0 \le k \le 40 \tag{24.1.4}$$

where $\mathbf{u}(k) = (\mathbf{u}(0, k\Delta t), \mathbf{u}(\Delta x, k\Delta t), \mathbf{u}(2\Delta x, k\Delta t), \ldots, \mathbf{u}(100\Delta x, k\Delta t))^{\mathrm{T}}$, that is, $\mathbf{u}(k) \in \mathbb{R}^{101}$ and $\mathbf{f} : \mathbb{R}^{101} \times \mathbb{R} \to \mathbb{R}^{101}$. In essence, when we say $\mathbf{x}(t)$ in (24.1.1) is a vector of size n, and $\mathbf{x}_k$ in (24.1.3) is also a vector of size n, obviously the value of n is **not** necessarily the same in both the cases. Much like $\mathbf{x}$ is used as a generic variable for the state, n is the generic size of a state variable, and this should **not** cause any confusion.

We now move on to characterizing the multi-step state transition map induced by the discrete dynamical equation (24.1.3). To this end, first define inductively

$$\mathbf{M}^{(1)}(\mathbf{x}, \boldsymbol{\alpha}) = \mathbf{M}(\mathbf{x}, \boldsymbol{\alpha})$$

and the ***k*-fold iterate**

$$\mathbf{M}^{(k)}(\mathbf{x}, \alpha) = \mathbf{M}\big(\mathbf{M}^{(k-1)}(\mathbf{x}, \alpha), \alpha\big). \tag{24.1.5}$$

A little reflection will immediately lead to the following:

$$\begin{aligned}\mathbf{x}_k &= \mathbf{M}(\mathbf{x}_{k-1}, \alpha) = \mathbf{M}(\mathbf{M}(\mathbf{x}_{k-2}, \alpha), \alpha) = \mathbf{M}^{(2)}(\mathbf{x}_{k-2}, \alpha)\\ &= \cdots = \mathbf{M}^{(k)}(\mathbf{x}_0, \alpha)\end{aligned} \tag{24.1.6}$$

that is, $\mathbf{M}^{(k)}(\mathbf{x}, \alpha)$ indeed denotes the required k-step state transition mapping.

Remark 24.1.3 For the special case when (24.1.1) is a linear system, that is, when

$$\mathbf{x}_{k+1} = \mathbf{M}\mathbf{x}_k$$

where $\mathbf{M} \in \mathbb{R}^{n \times n}$ is a matrix, then clearly

$$\mathbf{x}_k = \mathbf{M}^k \mathbf{x}_0.$$

Thus, the k-fold iterate becomes the product of the kth power of $\mathbf{M}$ and $\mathbf{x}_0$. Much of the difficulty associated with the analysis of nonlinear dynamics is a direct consequence of the difficulty of computing the k-fold iterate $\mathbf{M}^{(k)}(\mathbf{x}, \alpha)$ of the map $\mathbf{M}(\mathbf{x}, \alpha)$ that defines the field.

Observations Let $\mathbf{h} : \mathbb{R}^n \to \mathbb{R}^m$ and $\mathbf{v} \in \mathbb{R}^m$. Let

$$\mathbf{z}_k = \mathbf{h}(\mathbf{x}_k) + \mathbf{v}_k \tag{24.1.7}$$

denote the observations of the state $\mathbf{x}_k$ at time k where $\mathbf{h} = (h_1, h_2, \ldots h_m)^{\mathrm{T}}$ with $h_i = h_i(\mathbf{x}_k)$. Here it is assumed that the observation is a nonlinear function of the state and is corrupted by an additive noise vector $\mathbf{v}_k$ with the properties

$$E[\mathbf{v}_k] = 0, \quad \mathrm{Cov}(\mathbf{v}_k) = \mathbf{R}, \quad \text{and} \quad E\left[\mathbf{v}_{k_1}\mathbf{v}_{k_2}^{\mathrm{T}}\right] = 0 \quad \text{for} \quad k_1 \neq k_2. \tag{24.1.8}$$

Criterion Let $\mathbf{W} \in \mathbb{R}^{m \times m}$ be a symmetric, positive, definite matrix. Define

$$J(\mathbf{c}) = \sum \langle (\mathbf{z}_k - \mathbf{h}(\mathbf{x}_k)), \mathbf{W}(\mathbf{z}_k - \mathbf{h}(\mathbf{x}_k)) \rangle \tag{24.1.9}$$

with $\mathbf{x}_0 = \mathbf{c}$. An obvious choice for $\mathbf{W}$ is $\mathbf{R}^{-1}$.

We now state a version of the problem of interest to us.

Statement of the Inverse Problem Assume that the parameter α is known. Given the set of observations $\{\mathbf{z}_k \,|\, 0 \leq k \leq N\}$, the problem is to estimate the control $\mathbf{c}$ that minimizes $J(\mathbf{c})$ in (24.1.9) where $\mathbf{x}_k$ is constrained by the model dynamics (24.1.3).

Remark 24.1.4 The problem of estimating the parameters α is mathematically no different from that of estimating the initial conditions. We assume α is known at this juncture to facilitate our discussion of similarities and differences in assimilation under linear and nonlinear constraints.

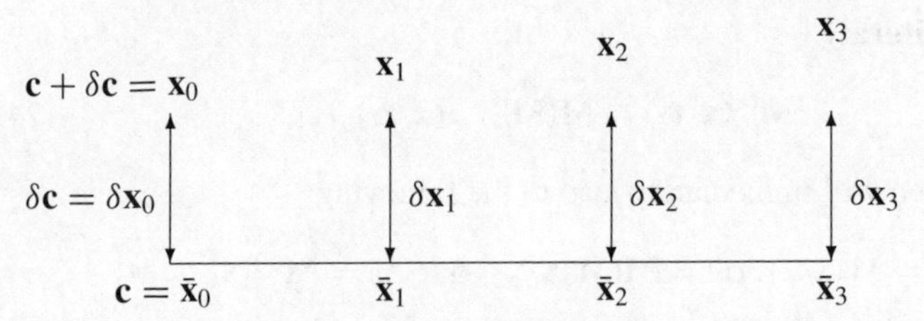

Fig. 24.2.1 An illustration of base and perturbed state.

24.2 First-order perturbation analysis

Recall from Section 23.5 that the adjoint operator theoretic method for computing the gradient of $\mathbf{J}$ critically depends on the evolution of the perturbation in the initial condition. We begin by quantifying the dynamics of perturbation: Choose $\mathbf{c} \in \mathbb{R}^n$ and let $\bar{\mathbf{x}}_1, \bar{\mathbf{x}}_2, \bar{\mathbf{x}}_3, \ldots$ be the **base state trajectory** computed from $\bar{\mathbf{x}}_0 = \mathbf{c}$. The trajectory $\mathbf{x}_1, \mathbf{x}_2, \mathbf{x}_3, \ldots$ computed from $\bar{\mathbf{x}}_0 = \mathbf{c} + \delta\mathbf{c}$ is called the **perturbed trajectory**. Refer to Figure 24.2.1 for an illustration.

Clearly, the actual evolution of the perturbation is given by

$$\mathbf{x}_k - \bar{\mathbf{x}}_k = \mathbf{M}^{(k)}(\mathbf{c} + \delta\mathbf{c}, \alpha) - \mathbf{M}^{(k)}(\mathbf{c}, \alpha), \tag{24.2.1}$$

the difference between the perturbed nonlinear evolutions where the first term represents the perturbed initial condition and the second term the base state initial condition. It is often advantageous to approximate this evolution with a linear model. This approximate characterization of the evolution of the initial perturbation is obtained by using the first-order Taylor Series expansion of the map $\mathbf{M}(\mathbf{x}, \alpha)$ that defines the field. To this end, we first introduce the Jacobian $\mathbf{D}_M(\mathbf{x})$ of $\mathbf{M}(\mathbf{x}, \alpha)$:

$$\mathbf{D}_M(\mathbf{x}) = \begin{bmatrix} \frac{\partial M_1}{\partial x_1} & \frac{\partial M_1}{\partial x_2} & \frac{\partial M_1}{\partial x_3} & \cdots & \frac{\partial M_1}{\partial x_n} \\ \frac{\partial M_2}{\partial x_1} & \frac{\partial M_2}{\partial x_2} & \frac{\partial M_2}{\partial x_3} & \cdots & \frac{\partial M_2}{\partial x_n} \\ \vdots & \vdots & \vdots & \cdots & \vdots \\ \frac{\partial M_n}{\partial x_1} & \frac{\partial M_n}{\partial x_2} & \frac{\partial M_n}{\partial x_3} & \cdots & \frac{\partial M_n}{\partial x_n} \end{bmatrix}. \tag{24.2.2}$$

Using the first-order Taylor series expansion we obtain

$$\mathbf{M}(\mathbf{x}_0, \alpha) = \mathbf{M}(\bar{\mathbf{x}}_0 + \delta\mathbf{c}, \alpha) \approx \mathbf{M}(\bar{\mathbf{x}}_0, \alpha) + \mathbf{D}_M(\bar{\mathbf{x}}_0)\delta\mathbf{c}. \tag{24.2.3}$$

Defining (recall $\delta\mathbf{x}_0 = \delta\mathbf{c}$)

$$\delta\mathbf{x}_1 = \mathbf{D}_M(\bar{\mathbf{x}}_0)\delta\mathbf{c}$$

we see that $\bar{\mathbf{x}}_1 + \delta\mathbf{x}_1$ is an **approximation** to $\mathbf{x}_1$ to **first-order** accuracy. From

$$\mathbf{M}(\bar{\mathbf{x}}_1 + \delta\mathbf{x}_1, \boldsymbol{\alpha}) \approx \mathbf{M}(\bar{\mathbf{x}}_1, \boldsymbol{\alpha}) + \mathbf{D}_M(\bar{\mathbf{x}}_1)\delta\mathbf{x}_1$$

and denoting

$$\delta\mathbf{x}_2 = \mathbf{D}_M(\mathbf{x}_1)\delta\mathbf{x}_1$$

we get $\bar{\mathbf{x}}_2 + \delta\mathbf{x}_2$ to be an approximation to $\mathbf{x}_2$. Continuing this argument, we can inductively define

$$\delta\mathbf{x}_{k+1} = \mathbf{D}_M(\bar{\mathbf{x}}_k)\,\delta\mathbf{x}_k \tag{24.2.4}$$

where $\bar{\mathbf{x}}_k + \delta\mathbf{x}_k$ is an approximation to $\mathbf{x}_k$. Notice (24.2.4) is a non-autonomous **linear dynamical system** where the one-step state transition matrix $\mathbf{D}_M(\bar{\mathbf{x}}_k)$ is evaluated along the base state. In meteorological circles, this system (24.2.4) has come to be known as the **tangent linear system** (TLS). (Exercise 24.2)

Since this system is used repeatedly, in the following we simplify the notation as

$$\mathbf{D}_M(k) = \mathbf{D}_M(\bar{\mathbf{x}}_k) \tag{24.2.5}$$

Rewriting (24.2.4) as

$$\delta\mathbf{x}_{k+1} = \mathbf{D}_M(k)\delta\mathbf{x}_k \tag{24.2.6}$$

and iterating the latter we obtain

$$\delta\mathbf{x}_k = \mathbf{D}_M(k-1)\mathbf{D}_M(k-2)\ldots\mathbf{D}_M(1)\mathbf{D}_M(0)\delta\mathbf{c}. \tag{24.2.7}$$

For $i \leq j$, we define

$$\mathbf{D}_M(j:i) = \mathbf{D}_M(j)\mathbf{D}_M(j-1)\ldots\mathbf{D}_M(i) \tag{24.2.8}$$

we can rewrite (24.2.7) as

$$\delta\mathbf{x}_k = \mathbf{D}_M(k-1:0)\delta\mathbf{c} \tag{24.2.9}$$

Stated in other words $\mathbf{D}_M(k-1:0)$ denotes the k-step transition matrix from time step 0 to k.

The iterative scheme in (24.2.6) that defines the TLS can also be written in a matrix-vector form. Let $\delta\mathbf{x} = (\delta\mathbf{x}_1, \delta\mathbf{x}_2, \ldots, \delta\mathbf{x}_N)^{\mathrm{T}}$. Then (24.2.6) becomes

$$\mathbf{F}\delta\mathbf{x} = \mathbf{b} \tag{24.2.10}$$

where $\mathbf{F}$ is an $N \times N$ block-partitioned matrix given by

$$\mathbf{F} = \begin{bmatrix} \mathbf{I} & 0 & 0 & \cdots & 0 & 0 \\ -\mathbf{D}_M(1) & \mathbf{I} & 0 & \cdots & 0 & 0 \\ 0 & -\mathbf{D}_M(2) & \mathbf{I} & \cdots & 0 & 0 \\ \vdots & \vdots & \vdots & \cdots & \vdots & \vdots \\ 0 & 0 & 0 & \cdots & \mathbf{I} & 0 \\ 0 & 0 & 0 & \cdots & -\mathbf{D}_M(N-1) & \mathbf{I} \end{bmatrix} \tag{24.2.11}$$

and $\mathbf{b}$ is a block-partitioned vector given by

$$\mathbf{b} = (\mathbf{D}_M(0)\delta\mathbf{c}, 0, 0 \ldots 0)^{\mathrm{T}}.$$

Several remarks are in order.

Remark 24.2.1 Define a quantity

$$\begin{aligned} r_1(k) &= \frac{\|\mathbf{x}_k - \bar{\mathbf{x}}_k\|}{\|\delta\mathbf{c}\|} \\ &= \frac{\left\|\mathbf{M}^{(k)}(\mathbf{c} + \delta\mathbf{c}, \boldsymbol{\alpha}) - \mathbf{M}^{(k)}(\mathbf{c}, \boldsymbol{\alpha})\right\|}{\|\delta\mathbf{c}\|} \end{aligned} \tag{24.2.12}$$

which is the ratio of the norm of the actual perturbation as seen through the nonlinear dynamics (24.1.3) at time k to the norm of the initial perturbation. Notice that $r_1(k)$ is an implicit function of c. If $r_1(k)$ is greater than 1, then it implies that the system (24.2.6) **magnifies** or **amplifies** the initial error. Thus, values of $r_1(k) > 1$ would indicate that the system is **unstable** leading to growth of initial perturbation (Part VIII). On the other hand, if $r_1(k) \leq 1$, then the system does not magnify the perturbation and the system is **stable** (Exercise 24.3).

Remark 24.2.2 Define a related quantity

$$r_2(k) = \frac{\|\delta\mathbf{x}_k\|}{\|\delta\mathbf{c}\|} \tag{24.2.13}$$

which is the ratio of the norm of the approximate perturbation $\delta\mathbf{x}_k$ as seen through the tangent linear system (24.2.6) to the norm of the initial perturbation. Again, it can be verified that $r_2(k)$ is also an implicit function of $\mathbf{c}$. Using (24.2.9) we can rewrite (24.2.13) as

$$\begin{aligned} r_2(k) &= \frac{\|\mathbf{D}_M(k-1:0)\delta\mathbf{c}\|}{\|\delta\mathbf{c}\|} \\ &= \frac{(\delta\mathbf{c})^{\mathrm{T}}[\mathbf{D}_M^{\mathrm{T}}(k-1:0)\mathbf{D}_M(k-1:0)](\delta\mathbf{c})}{(\delta\mathbf{c})^{\mathrm{T}}(\delta\mathbf{c})} \end{aligned} \tag{24.2.14}$$

If $r_2(k) > 1$, then the TLS magnifies the error which is indicative of the fact that the TLS is **unstable**. However, if $r_2(k) \leq 1$, then the TLS is **stable** (Exercise 24.4).

Remark 24.2.3 The sequences $\{r_1(k)\}_{k\geq 1}$ and $\{r_2(k)\}_{k\geq 1}$ are **indices** much like the **consumer price index**, **Dow Jones Industrial Average**, **S&P 500 Index** and the like. They relate the magnitude of the perturbation at time k to that at the initial time. Notice that $r_1(k)$ can be computed by running the nonlinear model (24.1.3) forward twice – starting from $\mathbf{c}$ and $\mathbf{c} + \delta\mathbf{c}$. Likewise, $r_2(k)$ can be computed by running the tangent linear model (24.2.6) forward once. But this involves first the evaluation of the Jacobian $\mathbf{D}_M(x)$ along the base orbit and then running the model (24.2.6).

A comparison of the plot of $r_1(k)$ and $r_2(k)$ would reveal the goodness of the linear approximation in quantifying the propagation of initial perturbation.

Remark 24.2.4 It is primarily through the computation of the ratio $r_1(k)$ that Edward Lorenz in 1963 discovered the existence of **deterministic chaos** as we know it today.

Example 24.2.1 Consider the model (Lorenz (1960))

$$\left.\begin{aligned}\frac{dx_1}{dt} &= \alpha_1 x_2 x_3\\ \frac{dx_2}{dt} &= \alpha_2 x_1 x_3\\ \frac{dx_3}{dt} &= \alpha_3 x_1 x_2\end{aligned}\right\} \tag{24.2.15}$$

Discretizing it using the standard forward Euler scheme, we obtain

$$\mathbf{x}_{k+1} = \mathbf{M}(\mathbf{x}_k) \tag{24.2.16}$$

where $\mathbf{x}_k = (x_{1k}, x_{2k}, x_{3k})^T$, $\mathbf{M}(\mathbf{x}) = (M_1(\mathbf{x}), M_2(\mathbf{x}), M_3(\mathbf{x}))^T$ where

$$\left.\begin{aligned}M_1(\mathbf{x}_k) &= x_{1k} + (\alpha_1 \Delta t)x_{2k}x_{3k}\\ M_2(\mathbf{x}_k) &= x_{2k} + (\alpha_2 \Delta t)x_{1k}x_{3k}\\ M_3(\mathbf{x}_k) &= x_{3k} + (\alpha_3 \Delta t)x_{1k}x_{2k}\end{aligned}\right\} \tag{24.2.17}$$

It can be verified that

$$\mathbf{D_M}(\mathbf{x}) = \begin{bmatrix} 1 & (\alpha_1 \Delta t)x_3 & (\alpha_1 \Delta t)x_2 \\ (\alpha_2 \Delta t)x_3 & 1 & (\alpha_2 \Delta t)x_1 \\ (\alpha_3 \Delta t)x_2 & (\alpha_3 \Delta t)x_1 & 1 \end{bmatrix}. \tag{24.2.18}$$

For $\alpha_1 = -0.553$, $\alpha_2 = 0.451$, $\alpha_3 = 0.051$, and $\Delta t = 0.1$, a plot of the base trajectory of the model in (24.2.17) starting from $\bar{\mathbf{x}}_0 = (1.0, 0.1, 0.0)^T$ and the perturbed trajectory starting from $\mathbf{x}_0 = \bar{\mathbf{x}}_0 + \varepsilon_0$ where $\varepsilon_0 = (0.1, 0.1, 0.1)^T$ are shown in Figure 24.2.2. Notice that there is a considerable difference between these two solutions which in turn indicates the sensitive dependence of this model on the initial condition. (Refer to Chapter 32 for details). Plots of the evolution of $r_1(k)$ and $r_2(k)$ for these two initial conditions are given in Figure 24.2.3. Figure 24.2.4 provides a plot of the ensemble of the components of $\mathbf{x}_k$ and $r_1(k)$ and $r_2(k)$ using N = 1000 samples of the initial error ε_0 drawn from a normal distribution $N(\bar{\mathbf{x}}_0, \mathbf{P}_0)$ where $\bar{\mathbf{x}}_0 = (1.0, 0.1, 0.0)^T$ and $\mathbf{P}_0 = \text{Diag}(\sigma_1^2, \sigma_2^2, \sigma_3^2)$ where $\sigma_1 = 0.1$ and $\sigma_2 = \sigma_3 = 0.01$.

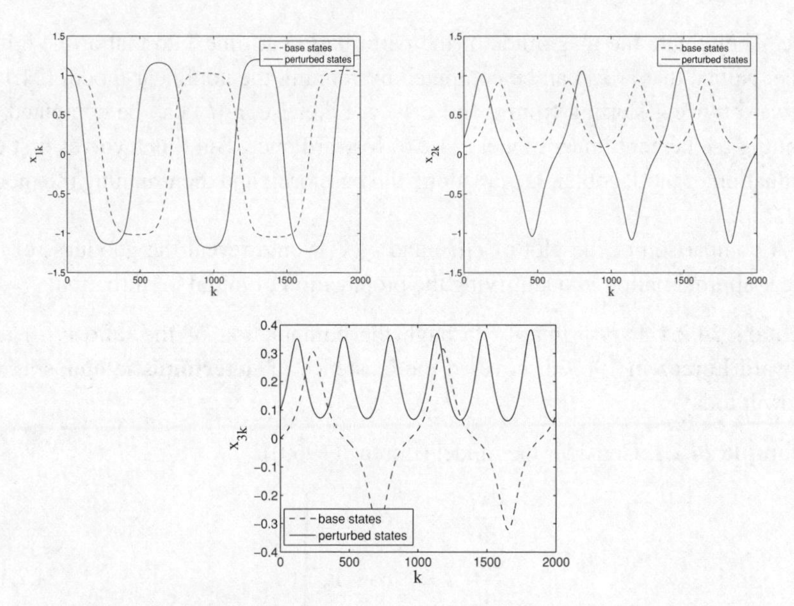

Fig. 24.2.2 Plot of the trajectories of (24.2.15) starting from the base state $\mathbf{x}_0 = (1.0, 0.1, 0.0)^{\mathrm{T}}$ and the perturbed state $\mathbf{x}_0 = (1.1, 0.2, 0.1)^{\mathrm{T}}$.

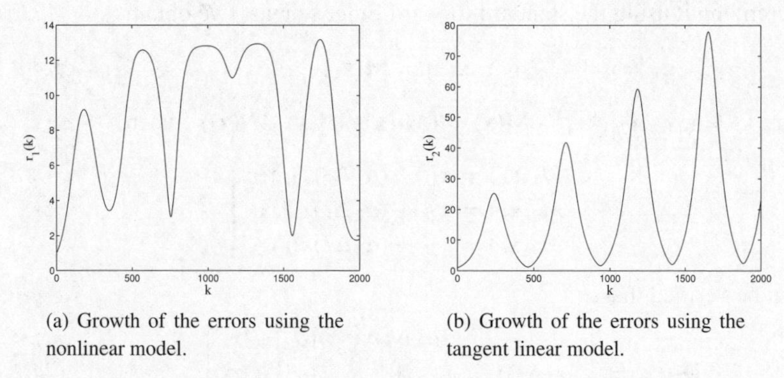

(a) Growth of the errors using the nonlinear model.

(b) Growth of the errors using the tangent linear model.

Fig. 24.2.3 Plot of $r_1(k)$ and $r_2(k)$ for trajectories starting from $\bar{\mathbf{x}}_0 = (1.0, 0.1, 0.0)^{\mathrm{T}}$ and $\mathbf{x}_0 = (1.1, 0.2, 0.1)$.

24.3 Computation of the gradient of $J(\mathbf{c})$

We are now ready to take up the main task of computing the gradient of $J(\mathbf{c})$ using the **adjoint operator** approach used in Section 23.5 (Exercise 24.7). From (24.1.9) recall that

$$J(\mathbf{c}) = \frac{1}{2} \sum_{k=0}^{N} \langle (\mathbf{z}_k - \mathbf{h}(\mathbf{x}_k)) \, , \mathbf{W} \, (\mathbf{z}_k - \mathbf{h}(\mathbf{x}_k)) \rangle. \tag{24.3.1}$$

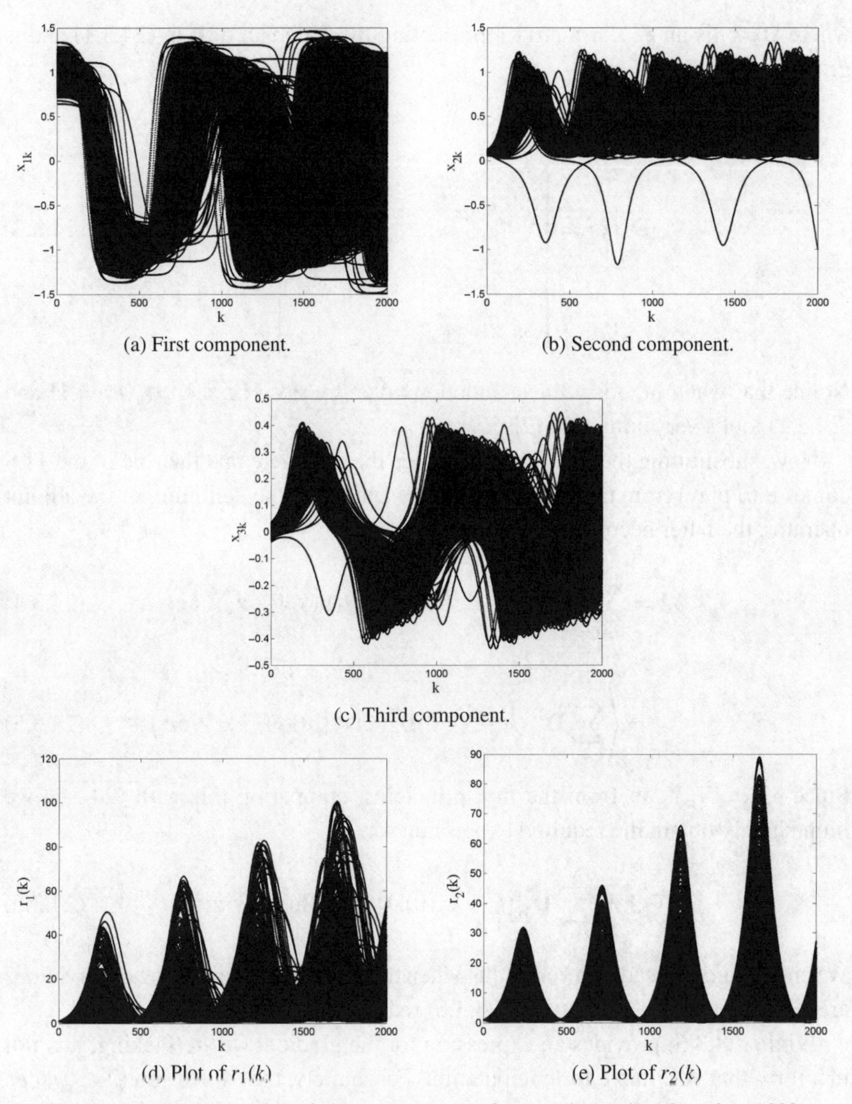

(a) First component. (b) Second component.

(c) Third component.

(d) Plot of $r_1(k)$ (e) Plot of $r_2(k)$

Fig. 24.2.4 Ensemble plot of the components of $\mathbf{x}_k$, $r_1(k)$, and $r_2(k)$ using 1000 samples of initial errors drawn from $N(\bar{\mathbf{x}}_0, \mathbf{P}_0)$ where $\bar{\mathbf{x}}_0 = (1.0, 0.1, 0.0)^{\mathrm{T}}$ and $\mathbf{P}_0 = \mathrm{Diag}(\sigma_1^2, \sigma_2^2, \sigma_3^2)$, $\sigma_1 = 0.1$, $\sigma_2 = \sigma_3 = 0.01$.

Let $\delta\mathbf{J}$ denote the first variation in $J(\mathbf{c})$ induced by the initial perturbation $\delta\mathbf{c}$ in $\mathbf{x}_0 = \mathbf{c}$. From first principles it can be verified that

$$\delta\mathbf{J} = \sum_{k=0}^{N} \langle \mathbf{W}(\mathbf{h}(\mathbf{x}_k) - \mathbf{z}_k)\, , \mathbf{D}_h(k)\delta\mathbf{x}_k \rangle \tag{24.3.2}$$

where $\mathbf{D}_h(k)$ is an $m \times n$ matrix representing the Jacobian of $\mathbf{h}$ in (24.3.1) and is given by

$$\mathbf{D}_h(k) = \begin{bmatrix} \frac{\partial h_1}{\partial x_1} & \frac{\partial h_1}{\partial x_2} & \cdots & \frac{\partial h_1}{\partial x_n} \\ \frac{\partial h_2}{\partial x_1} & \frac{\partial h_2}{\partial x_2} & \cdots & \frac{\partial h_2}{\partial x_n} \\ \vdots & \vdots & \cdots & \vdots \\ \frac{\partial h_m}{\partial x_1} & \frac{\partial h_m}{\partial x_2} & \cdots & \frac{\partial h_m}{\partial x_n} \end{bmatrix}_{x=x(k)} . \tag{24.3.3}$$

Notice that when $\mathbf{h}(x)$ is a linear function given by say, $\mathbf{Hx}$, then $\mathbf{D}_h(k) = \mathbf{H}$ and (24.3.2) looks very similar to (23.5.9).

Now, substituting the value of $\delta\mathbf{x}_k$ (notice that it is here that the role of the TLS comes into play) from (24.2.9) into (24.3.2) and using the definition of the adjoint operator, the latter becomes

$$\delta\mathbf{J} = \sum_{k=0}^{N} \left\langle \mathbf{D}_M^{\mathrm{T}}(k-1:0)\mathbf{D}_h^{\mathrm{T}}(k)\mathbf{W}(\mathbf{h}(\mathbf{x}_k) - \mathbf{z}_k)\,,\,\delta\mathbf{c} \right\rangle \tag{24.3.4}$$

$$= \left\langle \sum_{k=0}^{N} \mathbf{D}_M^{\mathrm{T}}(k-1:0)\mathbf{D}_h^{\mathrm{T}}(k)\mathbf{W}(\mathbf{h}(\mathbf{x}_k) - \mathbf{z}_k)\,,\,\delta\mathbf{c} \right\rangle . \tag{24.3.5}$$

Since $\delta\mathbf{J} = \langle \nabla_{\mathbf{c}}\mathbf{J}, \delta\mathbf{c} \rangle$ from the first principles, comparing this with (24.3.5) we immediately obtain the required expression for

$$\nabla_{\mathbf{c}}\mathbf{J} = \sum_{k=0}^{N} \mathbf{D}_M^{\mathrm{T}}(k-1:0)\mathbf{D}_h^{\mathrm{T}}(k)\mathbf{W}(\mathbf{h}(\mathbf{x}_k) - \mathbf{z}_k). \tag{24.3.6}$$

We encourage the reader to verify that when the model is linear and the observations are a linear function of the state, (24.3.6) reduces to (23.5.13).

While (24.3.6) provides an expression for the gradient we are seeking, it is **not** in a form that is suitable for computation. Fortunately, the expression (24.3.6) can be computed as a solution of a recurrence relation as shown below.

Let $\mathbf{f}_k \in \mathbb{R}^n$ be a sequence of vectors defined by

$$\mathbf{f}_k = \mathbf{D}_h^{\mathrm{T}}(k)\mathbf{W}(\mathbf{h}(\mathbf{x}_k) - \mathbf{z}_k)\,. \tag{24.3.7}$$

Let $\bar{\lambda}_k \in \mathbb{R}^n$ be a sequence defined for $k = N, N-1, \ldots, 2, 1, 0$ by

$$\bar{\lambda}_k = \mathbf{D}_M^{\mathrm{T}}(k)\bar{\lambda}_{k+1} + \mathbf{f}_k, \qquad \bar{\lambda}_N = \mathbf{f}_N. \tag{24.3.8}$$

This latter recurrence can be succinctly written in the matrix-vector form as follows. Let

$$\bar{\lambda} = \left(\bar{\lambda}_0, \bar{\lambda}_1, \bar{\lambda}_2, \ldots, \bar{\lambda}_N\right)^{\mathrm{T}} \tag{24.3.9}$$

and

$$\mathbf{f} = (\mathbf{f}_0, \mathbf{f}_1, \mathbf{f}_2, \ldots, \mathbf{f}_N)^{\mathrm{T}} \tag{24.3.10}$$

be two block-partitioned vectors. Then (24.3.8) can be recast as

$$\mathbf{B}\bar{\lambda} = \mathbf{f} \tag{24.3.11}$$

where $\mathbf{B}$ is the $(N+1) \times (N+1)$ upper block bidiagonal matrix

$$\mathbf{B} = \begin{bmatrix} \mathbf{I} & -\mathbf{D}_M^{\mathrm{T}}(0) & 0 & 0 & \cdots & 0 & 0 \\ 0 & \mathbf{I} & -\mathbf{D}_M^{\mathrm{T}}(1) & 0 & \cdots & 0 & 0 \\ 0 & 0 & \mathbf{I} & -\mathbf{D}_M^{\mathrm{T}}(2) & \cdots & 0 & 0 \\ \vdots & \vdots & \vdots & \vdots & \ddots & \vdots & \vdots \\ 0 & 0 & 0 & 0 & \cdots & \mathbf{I} & -\mathbf{D}_M^{\mathrm{T}}(N-1) \\ 0 & 0 & 0 & 0 & \cdots & 0 & \mathbf{I} \end{bmatrix} \tag{24.3.12}$$

Solving this recurrence by iterating it backward from $k = N$ to $k = 0$, we readily obtain that

$$\nabla_{\mathbf{c}} J = \lambda_0 = \sum_{k=0}^{N} \mathbf{D}_M^{\mathrm{T}}(k-1:0)\mathbf{f}_k. \tag{24.3.13}$$

The equation (24.3.8) has come to be known as the **First-order adjoint** equation.

The above development readily leads to the following algorithm for computing $\nabla_{\mathbf{c}} J$.

Step 1 Starting with a $\mathbf{c}$, compute the orbit using (24.1.3).

Step 2 Using the observations $\{\mathbf{z}_j \mid 0 \le j \le N\}$ and the orbit computed in Step 1 compute the vector $\mathbf{f}$ in (24.3.6).

Step 3 Solve the backward or the adjoint equation (24.3.8), and solve for $\bar{\lambda}$. Clearly, $\nabla_{\mathbf{c}} J(\mathbf{c}) = \bar{\lambda}_0$.

24.4 An algorithm for finding the optimal estimate

Once $\nabla_{\mathbf{c}} J(\mathbf{c})$ is computed we can use it in an algorithm quite analogous to the developments in Section 23.4. For purposes of completeness we merely state the algorithm.

Step 1 Choose a vector $\mathbf{c}$ and call it $\mathbf{c}_{\text{old}}$.

Step 2 Using the method in Section 23.4 compute $\nabla_{\mathbf{c}} J(\mathbf{c})$.

Step 3 If $\|\nabla_{\mathbf{c}} J(\mathbf{c})\| < \varepsilon$, for some prespecified $\varepsilon > 0$, stop. Else go to Step 4.

Step 4 Compute

$$\mathbf{c}_{\text{new}} = \mathbf{c}_{\text{old}} - \beta \nabla_{\mathbf{c}} J(\mathbf{c}) \tag{24.4.1}$$

for some step length parameter $\beta > 0$. Set $\mathbf{c}_{\text{old}} \leftarrow \mathbf{c}_{\text{new}}$ and go to Step 2.

We encourage the reader to perform the Monte Carlo type twin experiment using the Lorenz's model as described in Exercise 24.9.

The following example taken from Chapter 3 illustrates this methodology.

Example 24.4.1 Consider a nonlinear dynamics in two dimensions given by

$$\dot{x_1} = ax_1x_2 \quad \text{and} \quad \dot{x_2} = bx_1^2. \tag{24.4.2}$$

It can be verified that

$$\frac{\mathrm{d}}{\mathrm{d}t}(x_1^2 + x_2^2) = 2(x_1\dot{x_1} + x_2\dot{x_2}) = 2x_1^2x_2(a+b).$$

This system is conservative when $a + b = 0$ and non-conservative otherwise. When $a = 1/2$ and $b = -1/2$, (24.4.2) reduces to the first two equations resulting from the spectral expansion of the Burgers' equation described in Section 3.3.

Let $\mathbf{z} = \mathbf{h}(\mathbf{x}) + \mathbf{v}$ denote the observations, where $\mathbf{z} \in \mathbb{R}^2$ and

$$h_1(\mathbf{x}) = ax_1x_2 \quad \text{and} \quad h_2(\mathbf{x}) = bx_1^2 \tag{24.4.3}$$

and $\mathbf{v} \sim N(0, \mathbf{R})$ with $\mathbf{R} = \text{Diag}(\sigma_1^2, \sigma_2^2)$. Notice that $\mathbf{h}(\mathbf{x})$ is the same as the r.h.s.of the model in (24.4.2).

Discretizing the model in (24.4.2) we get $\mathbf{x}_k = \mathbf{M}(\mathbf{x}_k)$ where $\mathbf{M}(\mathbf{x}) = (M_1(\mathbf{x}), M_2(\mathbf{x}))^{\mathrm{T}}$, $M_1(\mathbf{x}) = x_1 + a\Delta t x_1 x_2$ and $M_2(\mathbf{x}) = x_2 + b\Delta t x_1^2$. The Jacobian of $\mathbf{M}(\mathbf{x})$ and $\mathbf{h}(\mathbf{x})$ are given by

$$\mathbf{D_M}(\mathbf{x}) = \begin{bmatrix} 1 + a\Delta t x_2 & a\Delta t x_1 \\ 2b\Delta x_1 & 1 \end{bmatrix}$$

and

$$\mathbf{D_h}(\mathbf{x}) = \begin{bmatrix} ax_2 & ax_1 \\ 2bx_1 & 0 \end{bmatrix}.$$

Let $a = 1/2$, $b = -1/2$, and $\sigma_1^2 = \sigma_2^2 = 0.01$.

Generate observations Starting the discrete time model from the base initial state $\bar{\mathbf{x}}_0 = (1, 1)$, compute the base trajectory $\bar{\mathbf{x}}_k$ for $k = 0$ to 10. Generate a set of eleven random vectors $\mathbf{v}_k \sim N(0, \mathbf{R})$. Let $\mathbf{z}_k = \mathbf{h}(\bar{\mathbf{x}}_k) + \mathbf{v}_k$, for $k = 0$ to 10 be the set of all observations.

Analysis of the criterion $J(\mathbf{c})$ Using the above set of observations, a plot of the contour of $J(\mathbf{c})$ in (24.3.1) is given in Figure 24.4.1. It turns out that this $J(\mathbf{c})$ has two minima. A cross section of $J(\mathbf{c})$ along the diagonal line, $x_1 = x_2$ given in Figure 24.4.2 provides another view of this multi-minima.

Minimization Trajectories of the gradient algorithm starting from $(2, 2)^{\mathrm{T}}$ and $(2, -2)^{\mathrm{T}}$ are shown in Figure 24.4.1. Table 24.4.1 gives the values of $J(\mathbf{c})$ along these trajectories.

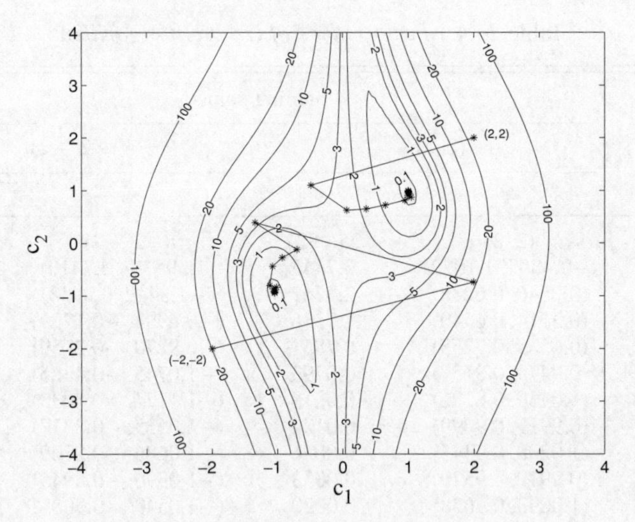

Fig. 24.4.1 Contours of $J(\mathbf{c})$ for the Example 24.4.1 .

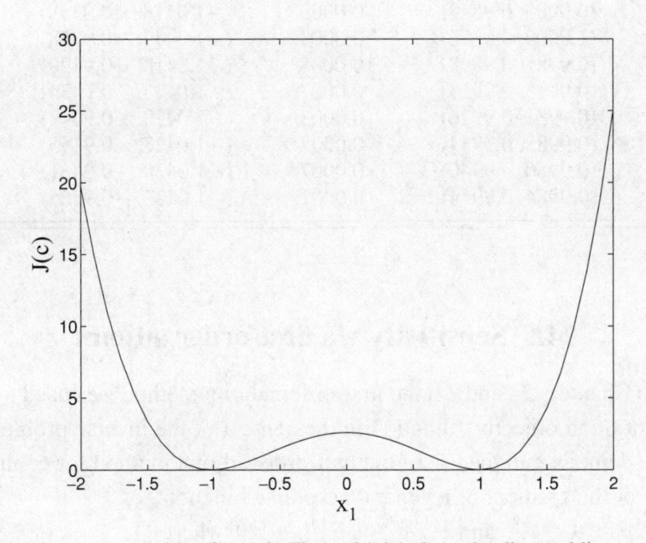

Fig. 24.4.2 A cross section of $J(\mathbf{c})$ in Figure 24.4.1 along the diagonal line $x_1 = x_2$.

The aim of this example is to bring out the challenges that underlie the multidimensional nonlinear minimization which is the basis for the 4DVAR methods. In this example, by our design we know that the true minimum is the one in the first quadrant that is close to $(1, 1)^{\mathrm{T}}$. In large-scale problems of interest in geophysical domain, there is no way of knowing how many minima are there, let alone deciding which one is the right one.

We invite the reader to examine the cases when $(a + b) > 0$ and $(a + b) < 0$.

Table 24.4.1 *Performance of Gradient Algorithm*

Iteration Number k	Starting points (2, 2) $\mathbf{c}_k$	$J(\mathbf{c}_k)$	(−2, −2) $\mathbf{c}_k$	$J(\mathbf{c}_k)$
0	(2, 2)	25.0522	(−2, −2)	19.7650
1	(−0.4855, 1.1022)	3.7445	(1.9953, −0.7410)	18.3859
2	(0.0646, 0.6288)	2.4762	(−1.3442, 0.3953)	3.5031
3	(0.3614, 0.6540)	1.7386	(−0.6994, −0.1229)	1.3967
4	(0.6535, 0.7255)	0.7851	(−0.9270, −0.2759)	0.6979
5	(0.9475, 0.8157)	0.0793	(−1.0745, −0.4385)	0.3248
6	(1.0449, 0.8722)	0.0218	(−1.1174, −0.8033)	0.0471
7	(0.9919, 0.9099)	0.0122	(−1.0135, −0.8377)	0.0320
8	(1.0208, 0.9345)	0.0056	(−1.0686, −0.8700)	0.0134
9	(0.9979, 0.9520)	0.0033	(−1.0390, −0.8945)	0.0097
10	(1.0093, 0.9638)	0.0020	(−1.0540, −0.9083)	0.0076
11	(0.9981, 0.9737)	0.0014	(−1.0399, −0.9241)	0.0066
12	(1.0038, 0.9796)	0.0010	(−1.0477, −0.9304)	0.0062
13	(0.9983, 0.9842)	0.0009	(−1.0418, −0.9358)	0.0060
14	(1.0009, 0.9872)	0.0008	(−1.0447, −0.9388)	0.0059
15	(0.9981, 0.9898)	0.0008	(−1.0418, −0.9420)	0.0059
16	(0.9995, 0.9913)	0.0007	(−1.0433, −0.9434)	0.0058
17	(0.9981, 0.9926)	0.0007	(−1.0420, −0.9448)	0.0058
18	(0.9988, 0.9934)	0.0007	(−1.0427, −0.9455)	0.0058
19	(0.9980, 0.9940)	0.0007	(−1.0420, −0.9461)	0.0058
20	(0.9984, 0.9944)	0.0007	(−1.0423, −0.9465)	0.0058

24.5 Sensitivity via first-order adjoint

Thus far in Chapters 23 and 24, the first-order adjoint method was used to compute the gradient of an objective function in the context of the inverse problem. In this section we demonstrate the use of the first-order adjoint method for computing the sensitivity or the gradient of a general **response** function.

Let $\mathbf{y} \in \mathbb{R}^N$, $\mathbf{u} \in \mathbb{R}^M$ and $\mathbf{F} : \mathbb{R}^N \times \mathbb{R}^M \to \mathbb{R}^N$ where

$$\mathbf{F}(\mathbf{y}, \mathbf{u}) = 0 \tag{24.5.1}$$

denotes the **model** equation. Here $\mathbf{y}$ is called the **state** of the system and $\mathbf{u}$ is the **control** vector which denotes initial/boundary conditions or parameters in the system. Let $G : \mathbb{R}^N \times \mathbb{R}^M \to \mathbb{R}$ where $G(\mathbf{y}, \mathbf{u})$ denotes the response function. Our goal is to compute the gradient $\nabla_{\mathbf{u}} G$ which is a measure of the sensitivity of G w.r.to $\mathbf{u}$.

There are at least three different methods to compute this gradient.

Direct method This method calls for first solving (24.5.1) for $\mathbf{y}$ as an explicit function of $\mathbf{u}$, say $\mathbf{y} = \mathbf{y}(\mathbf{u})$. Substituting this $\mathbf{y}$ in G, obtain $G(\mathbf{y}(\mathbf{u}), \mathbf{u})$ from which the required gradient can then be computed explicitly

This method, while conceptually simple, is often difficult to implement especially when the model equations are nonlinear in which case we may not be able to compute the solution $\mathbf{y} = \mathbf{y}(\mathbf{u})$ of (24.5.1) explicitly.

Finite difference method If $\mathbf{u} = (u_1, u_2, \ldots, u_M)^{\mathrm{T}}$, then for $i = 1$ to M, we can approximate the ith component of the gradient using

$$\frac{G(\mathbf{y}, \mathbf{u} + \alpha_i \mathbf{e}_i) - G(\mathbf{y}, \mathbf{u} - \alpha_i \mathbf{e}_i)}{2\alpha_i} \tag{24.5.2}$$

for small values of α_i where $\mathbf{e}_i$ is the standard ith unit vector. However, this calls for computing the model solution $\mathbf{y}$ for two values of $\mathbf{u}$ namely $(\mathbf{u} \pm \alpha_i \mathbf{e}_i)$, which in turn requires solving the model equation (24.5.1) a total of $2M$ times. Obviously, this method can be computationally expensive.

First-order adjoint method Recall that the first variation of G is given by

$$\delta G = \left\langle \delta \mathbf{y}, \frac{\partial G}{\partial \mathbf{y}} \right\rangle + \left\langle \delta \mathbf{u}, \frac{\partial G}{\partial \mathbf{u}} \right\rangle \tag{24.5.3}$$

where $\partial G/\partial \mathbf{y}$ and $\partial G/\partial \mathbf{u}$ are the vectors of partial derivatives of G w.r.to $\mathbf{y}$ and $\mathbf{u}$, respectively. Similarly, the first variation in $\mathbf{F}(\mathbf{y}, \mathbf{u})$ is given by

$$\delta F = (\mathbf{D}_{\mathbf{F}}^{\mathbf{y}})\delta \mathbf{y} + (\mathbf{D}_{\mathbf{F}}^{\mathbf{u}})\delta \mathbf{u} \tag{24.5.4}$$

where

$$\mathbf{D}_{\mathbf{F}}^{\mathbf{y}} = \left[\frac{\partial F_i}{\partial y_j} \right] \quad 1 \le i \le N, \quad 1 \le j \le N$$

and

$$\mathbf{D}_{\mathbf{F}}^{\mathbf{u}} = \left[\frac{\partial F_i}{\partial u_j} \right] \quad 1 \le i \le N, \quad 1 \le j \le M$$

are the Jacobians of $\mathbf{F}$ w.r.to $\mathbf{y}$ and $\mathbf{u}$ respectively.

Let $\mathbf{p} \in \mathbb{R}^N$ be an arbitrary vector called the **adjoint** variable. Taking the inner product of both sides of (24.5.4) with $\mathbf{p}$ gives

$$\langle (\mathbf{D}_{\mathbf{F}}^{\mathbf{y}})\delta \mathbf{y}, \mathbf{p} \rangle + \langle (\mathbf{D}_{\mathbf{F}}^{\mathbf{u}})\delta \mathbf{u}, \mathbf{p} \rangle = 0 \tag{24.5.5}$$

which using the adjoint property (refer to 23.5.1) becomes

$$\langle \delta \mathbf{y}, (\mathbf{D}_{\mathbf{F}}^{\mathbf{y}})^{\mathrm{T}} \mathbf{p} \rangle = -\langle \delta \mathbf{u}, (\mathbf{D}_{\mathbf{F}}^{\mathbf{u}})^{\mathrm{T}} \mathbf{p} \rangle. \tag{24.5.6}$$

Now define $\mathbf{p}$ by setting

$$(\mathbf{D}_{\mathbf{F}}^{\mathbf{y}})^{\mathrm{T}} \mathbf{p} = \frac{\partial G}{\partial \mathbf{y}}. \tag{24.5.7}$$

Step 1 Given $\mathbf{F}(\mathbf{y}, \mathbf{u})$, compute the Jacobians $\mathbf{D}_{\mathbf{F}}^{\mathbf{y}}$ and $\mathbf{D}_{\mathbf{F}}^{\mathbf{u}}$

Step 2 Given $G(\mathbf{y}, \mathbf{u})$, compute

$$\frac{\partial G}{\partial \mathbf{y}} \quad \text{and} \quad \frac{\partial G}{\partial \mathbf{u}}$$

Step 3 Define $\mathbf{p}$ as the solution of

$$(\mathbf{D}_{\mathbf{F}}^{\mathbf{y}})^{\mathrm{T}}\mathbf{p} = \frac{\partial G}{\partial \mathbf{y}}$$

Step 4 The required gradient is

$$\nabla_{\mathbf{u}} G = -(\mathbf{D}_{\mathbf{F}}^{\mathbf{u}})^{\mathrm{T}}\mathbf{p} + \frac{\partial G}{\partial \mathbf{u}}$$

Fig. 24.5.1 First-order adjoint sensitivity computation.

This is the analog of the backward tangent linear system that defines the adjoint variable (Refer to (24.3.7)). Combining (24.5.7) with (24.5.3) and using (24.5.6) we get

$$\begin{aligned}\delta G &= \langle \delta\mathbf{y}, (\mathbf{D}_{\mathbf{F}}^{\mathbf{y}})^{\mathrm{T}}\mathbf{p}\rangle + \left\langle \delta\mathbf{u}, \frac{\partial G}{\partial \mathbf{u}}\right\rangle \\ &= -\langle \delta\mathbf{u}, (\mathbf{D}_{\mathbf{F}}^{\mathbf{u}})^{\mathrm{T}}\mathbf{p}\rangle + \left\langle \delta\mathbf{u}, \frac{\partial G}{\partial \mathbf{u}}\right\rangle \\ &= \left\langle \delta\mathbf{u}, -(\mathbf{D}_{\mathbf{F}}^{\mathbf{u}})^{\mathrm{T}}\mathbf{p} + \frac{\partial G}{\partial \mathbf{u}}\right\rangle. \end{aligned} \tag{24.5.8}$$

From first principles, since $\delta G = \langle \delta\mathbf{u}, \nabla_{\mathbf{u}} G\rangle$, combining this with (24.5.8) we obtain the required expression for the gradient as

$$\nabla_{\mathbf{u}} G = -(\mathbf{D}_{\mathbf{F}}^{\mathbf{u}})^{\mathrm{T}}\mathbf{p} + \frac{\partial G}{\partial \mathbf{u}}. \tag{24.5.9}$$

This method is summarized in the form of an algorithm in Figure 24.5.1.

In the special case when G is a function of $\mathbf{y}$ and does not depend on $\mathbf{u}$ explicitly, then $\partial G/\partial \mathbf{u} = 0$ in (24.5.9).

Example 24.5.1 Consider the scalar dynamics with $x_{k+1} = ax_k$ with x_0 as the initial condition. Let $\mathbf{y} = (x_1, x_2, x_3)$ and $\mathbf{u} = (x_0, a)$, that is, $N = 3$ and $M = 2$. Then $\mathbf{F}(\mathbf{y}, \mathbf{u}) = (F_1(\mathbf{y}, \mathbf{u}), F_2(\mathbf{y}, \mathbf{u}), F_3(\mathbf{y}, \mathbf{u}))^{\mathrm{T}}$ where

$$\begin{aligned} F_1(\mathbf{y}, \mathbf{u}) &= x_1 - ax_0 \\ F_2(\mathbf{y}, \mathbf{u}) &= x_2 - ax_1 \\ F_3(\mathbf{y}, \mathbf{u}) &= x_3 - ax_2. \end{aligned}$$

Let $G(\mathbf{y}, \mathbf{u}) = (x_3 - z)^2$ for some constant z. Then

$$\frac{\partial G}{\partial \mathbf{y}} = (0, 0, 2(x_3 - z))^{\mathrm{T}} \quad \text{and} \quad \frac{\partial G}{\partial \mathbf{u}} = 0.$$

Also

$$\mathbf{D}_{\mathbf{F}}^{\mathbf{y}} = \begin{bmatrix} 1 & 0 & 0 \\ -a & 1 & 0 \\ 0 & -a & 1 \end{bmatrix} \quad \text{and} \quad \mathbf{D}_{\mathbf{F}}^{\mathbf{u}} = \begin{bmatrix} -a & -x_0 \\ 0 & -x_1 \\ 0 & -x_2 \end{bmatrix}.$$

Equation (24.5.7) becomes

$$\begin{bmatrix} 1 & -a & 0 \\ 0 & 1 & -a \\ 0 & 0 & 1 \end{bmatrix} \begin{bmatrix} p_1 \\ p_2 \\ p_3 \end{bmatrix} = \begin{bmatrix} 0 \\ 0 \\ 2(x_3 - z) \end{bmatrix}.$$

Solving this we obtain

$$\mathbf{p} = 2(x_3 - z)(a^2 \quad a \quad 1)^{\mathrm{T}}.$$

Hence

$$\begin{aligned} \nabla_{\mathbf{u}} G = -(\mathbf{D}_{\mathbf{F}}^{\mathbf{u}})^{\mathrm{T}} \mathbf{p} &= 2(x_3 - z) \begin{bmatrix} a^3 \\ a^2 x_0 + a x_1 + x_2 \end{bmatrix} \\ &= 2(x_3 - z) \begin{bmatrix} a^3 \\ 3a^2 x_0 \end{bmatrix}. \end{aligned}$$

Since $x_3 = a^3 x_0$, we can in fact verify directly that

$$\frac{\partial G}{\partial x_0} = 2(x_3 - z)\frac{\partial x_3}{\partial x_0} = 2a^3(x_3 - z)$$

and

$$\frac{\partial G}{\partial a} = 2(x_3 - z)\frac{\partial x_3}{\partial a} = 6a^2(x_3 - z)x_0.$$

Exercises

24.1 Let $0 \le b \le 4$ and $x \in [0, 1]$. Consider the nonlinear dynamics

$$\mathbf{x}_{k+1} = f(\mathbf{x}_k, b) = b\mathbf{x}_k(1 - \mathbf{x}_k)$$

(a) Compute $f^{(2)}$ and $f^{(3)}$ and plot $f^{(2)}$ and $f^{(3)}$ as a function of $x \in [0, 1]$.
(b) Analyze the shape of $f^{(2)}$ and $f^{(3)}$ for various values of b in the range $[0, 4]$.

24.2 Compute the tangent linear systems for the dynamics in Exercise 24.1.

24.3 Compute and plot $r_1(k)$ vs. k for various values of $0 \le x_0 \le 1$ and $0 < b \le 4$ for the model in Exercise 24.1.

24.4 Compute and plot $r_2(k)$ vs. k for various values of $0 \le x_0 \le 1$ and $0 < b \le 4$ for the model in Exercise 24.1.

24.5 Compare $\{r_1(k)\}_{k\ge1}$ and $\{r_2(k)\}_{k\ge1}$ and comment on the performance of TLS in handling the perturbation.

24.6 Compute and plot the ratio $r_1(k)$ and $r_2(k)$ for the following dynamical systems.

(a) **Two species population model**

$$\frac{\mathrm{d}x}{\mathrm{d}t} = x + y - x(x^2 + y^2)$$
$$\frac{\mathrm{d}y}{\mathrm{d}t} = -x + y - y(x^2 + y^2)$$

(b) **Lorenz's model** $a = 10$, $b = 8/3$, $0 < r < 30$

$$\left.\begin{aligned}\frac{\mathrm{d}x}{\mathrm{d}t} &= -ax + ay\\ \frac{\mathrm{d}y}{\mathrm{d}t} &= -xz + rx - y\\ \frac{\mathrm{d}z}{\mathrm{d}t} &= xy - bz\end{aligned}\right\}$$

(c) **Another "Burgers' equation"** Burgers sought to explore laminar and turbulent flow through the equations below. Here u represents the mean or laminar motion while v represents the turbulent flow. P is a constant pressure gradient force and α is the viscosity. (See Burgers 1939, Sect. 5).

$$\frac{\mathrm{d}u}{\mathrm{d}t} = P - \alpha u - v^2$$
$$\frac{\mathrm{d}v}{\mathrm{d}t} = uv - \alpha v$$

24.7 Following the developments in Section 23.3 reformulate the problem of computing the gradient of $J(\mathbf{c})$ using the Lagrangian framework.

24.8 Let $F(y, u) = \mathrm{e}^{yu} - u = 0$ and $G(y, u) = (y - B)^2$. Compute $\nabla_u G$ using the direct method and the adjoint method.

24.9 Monte Carlo type Twin Experiment using the **Lorenz's** model in Exercise (24.6(b)).

(a) **step 0** Discretize the Lorenz's model using the Euler discretization and express it as

$$\mathbf{x}_{k+1} = \mathbf{M}(\mathbf{x}_k). \qquad (*)$$

step 1 Compute the Jacobian $\mathbf{D_M}(\mathbf{x})$.
step 2 Pick an initial condition $\mathbf{x}_0 = \mathbf{c} = (c_1, c_2, c_3)^{\mathrm{T}}$ randomly and compute the trajectory $\mathbf{x}_0, \mathbf{x}_1, \mathbf{x}_2, \ldots, \mathbf{x}_N$ of $(*)$.
step 3 Evaluate the Jacobian $\mathbf{D_M}$ along this trajectory.
step 4 Generate observations $k = 0, 1, 2, \ldots, N$

$$\mathbf{z}_k = \mathbf{x}_k + \mathbf{v}_k \qquad (**)$$

where $\mathbf{v}_k$ is the Gaussian white noise with mean zero and covariance matrix $\mathbf{R} = \text{Diag}(\sigma_1^2, \sigma_2^2, \sigma_3^2)$.

step 5 Compute the standard least squares criterion $J(\mathbf{c})$ and identify the matrix of this quadratic form. Compute its eigenvalues.

step 6 Minimize $J(\mathbf{c})$ using the methods in Chapters 10–12.

(b) Examine the effect of the following

(1) Change the number and location of the observations.

(2) Change the variance of the observation noise.

(3) Change the value of the parameter $r = 1, 10, 20, 25, 28, 29$ and comment on the results for each of these choices.

24.10 Compute derivative of $G(x) = x^2$ w.r.to u when $F(x, u) = 3x^2 + ux - 1 = 0$ using the direct method and the first order disjoint method.

24.11 Compute the derivative of $J(x, z) = \frac{1}{2}(x - z)^2$ with respect to u when $F(x, u) = 2x^2 - u = 0$.

24.12 Compute the derivative of $J(x, z) = \frac{1}{2}(x - z)^2$ w.r.to u when $F(x, u) = e^{xu} - x = 0$.

24.13 Minimize $J(x, u) = x^2 + u^2$ when $xu = 1$.

24.14 Minimize $J(x) = \frac{1}{2}ax^2$ when $x + bu + c = 0$ where $a > 0$ and b and c are nonzero real constants.

24.15 Maximize $V(r, h) = \pi r^2 h$ when $2\pi r^2 + 2\pi r h = A_0$, a given fixed constant.

24.16 Minimize $J(\mathbf{x}, \mathbf{u}) = \frac{1}{2}\mathbf{x}^T\mathbf{Q}\mathbf{x} + \frac{1}{2}\mathbf{u}^T\mathbf{R}\mathbf{u}$ when $\mathbf{x}$ and $\mathbf{u}$ are constrained by

$$\mathbf{A}\mathbf{x} + \mathbf{B}\mathbf{u} + \mathbf{c} = 0$$

where $\mathbf{x} \in \mathbb{R}^n$, $\mathbf{u} \in \mathbb{R}^m$, $\mathbf{c} \in \mathbb{R}^n$, $\mathbf{Q} \in \mathbb{R}^{n\times n}$ and $\mathbf{R} \in \mathbb{R}^{m\times m}$ are symmetric and positive definite, $\mathbf{A} \in \mathbb{R}^{n\times n}$ is nonsingular and $\mathbf{B} \in \mathbb{R}^{n\times m}$.

24.17 The equation for a projectile motion with reduced gravity may be written as

$$\ddot{y} = -g(1 - e^{t/\theta}) \quad \text{and} \quad \ddot{x} = 0$$

where $x = x(t)$ and $y = y(t)$ are the horizontal and vertical positions of the projectile at time t, g is the acceleration due to gravity and θ is the time scale (constant) and $\theta \gg t$, the time of flight. Since t/θ is small, approximating $e^{-t/\theta} \approx 1 - t/\theta$ we get

$$\ddot{y} = -\frac{gt}{\theta} \quad \text{and} \quad \ddot{x} = 0. \qquad (*)$$

Sketch the trajectory and compare to normal gravity case.

(1) Assuming $x(0) = y(0) = 0$, $\dot{x}(0)$ and $\dot{y}(0)$ are given, verify that the solution of these equations are given by

$$x(t) = \dot{x}(0)t \quad \text{and} \quad y(t) = \dot{y}(0)t - \frac{gt^3}{6\theta}.$$

(2) Find the sensitivity of $y(t)$ w.r.t. θ.

(3) Using the central difference approximation, discretize (*) and verify that

$$y_{n+1} = -\frac{gn\tau}{\theta}\tau^2 + 2y_n - y_{n-1} \quad \text{for} n \geq 1 \qquad (**)$$

where $y_1 = \dot{y}(0)\tau$ and $y_0 = y(0) = 0$. Compute the sensitivity of y_3 w.r.t. θ.

(4) If z_1, z_2, z_3 are observations of the positions y_1, y_2, and y_3 respectively, find the sensitivity of J w.r.t. θ where

$$J(\theta) = (y_1 - z_1)^2 + (y_2 - z_2)^2 + (y_3 - z_3)^2$$

and y_2 and y_3 are defined in (**).

24.18 The following data from the US Census Bureau reflects the uncertainty in the estimate of population in the early decades of our country's history. From data in the World Almanac, 2002, pp376–377), we have:

index(i)	year	population ($\tilde{P}_i$)
0	1800	3, 929, 000
1	1850	23, 191, 876
2	1900	76, 212, 168
3	1950	151, 325, 798
4	2000	281, 421, 906

We often use the exponential growth equation to study population evolution. In discrete form,

$$P_{i+1} - P_i = kP_i, \quad k > 0, \quad i = 1, 2, 3, 4.$$

Find the parameter k and P_0 by minimizing

$$J = \sum_{i=0}^{4} \sigma_i (P_i - \tilde{P}_i)^2$$

subject to the four constraints and where $\sigma_i / \sigma_0 = 4$ for $i = 1, 2, 3$, and 4.

Notes and references

Section 24.1 The description of the inverse problem is standard in the literature.
Section 24.2 The techniques of perturbation analysis are rather standard in applied mathematics. Refer to Errico et al. (1993) for an assessment of the effectiveness of the perturbation analysis.

Section 24.3 This section follows the paradigm described in Section 23.5. Refer to LeDimet and Talagrand (1986). Lewis and Derber (1985) use the classical Lagrangian formulation used in Section 23.3. Thacker and Long (1988) were the first researchers to clarify the use of Lagrange multiplier method in geophysical data assimilation problems. Their paper deserves a careful reading by students. Also refer to Derber (1989), Derber and Bouttier (1999), Derber and Rosati (1989), Sun and Ficker et al. (1991) and Zupanski et al. (2000) for more details.
Section 24.4 Refer to Chapters 10–12 and Dennis and Schnabel (1996) for a description of many versions of first-order and second-order optimization algorithms.
Section 24.5 This section follows LeDimet, Navon and Descau (2002). The "other" Burgers equation in Exercise 24.6 is discussed in Burgers (1939) where this example describes the interplay between the laminar component u and the turbulent component v in the presence of a pressure force P. The reader is referred to the ambitious effort by Tomi Vukicevic and colleagues where the atmosphere's cloudiness is estimated by assimilating visible and infrared radiance data into a mesoscale prediction model (Vukicevic et al. (2004)).

25
Second-order adjoint method

In the variational approach, the dynamic data assimilation problem is recast as a minimization of the least squares performance criterion subject to the dynamic constraints. The first-order adjoint methods described in Chapters 22–24 enable us to compute the gradient of this objective function. Since the convergence of the gradient algorithm can be slow, especially in nonlinear problems of interest in geophysical applications, the gradient obtained using the first-order adjoint method is often used in conjunction with the quasi-Newton methods (Chapter 12) to obtain faster convergence. The strength of the quasi-Newton methods lies in their ability to extract the approximate Hessian of the objective function which in turn is used in a Newton-like algorithm. It is well known that minimization algorithms using the Hessian information perform better. Thus it behooves us to ponder the following question: in addition to the gradient, can we directly compute the Hessian related information, namely the Hessian-vector product? If this information can be obtained, we can then use it in conjunction with the conjugate gradient algorithm to obtain faster convergence. A framework for using the Hessian-vector product within the conjugate gradient algorithm framework is described in Section 12.3.

In this chapter we derive the so-called **second-order adjoint method** for computing simultaneously the gradient and the Hessian-vector product. The derivation for the scalar case is given in Section 25.1 and its extension to include the vector case is given in 25.2. Section 25.3 describes an application of the second-order adjoint method for computing the sensitivity of a response function. First-order adjoint sensitivity computations are given in Section 24.5.

25.1 Second-order adjoint method: scalar case

Let $M : \mathbb{R} \rightarrow \mathbb{R}$ and $h : \mathbb{R} \rightarrow \mathbb{R}$. Let $x_k \in \mathbb{R}$ for $k = 0, 1, 2, \ldots$ denote the state of a scalar nonlinear dynamical system whose evolution is given by

$$x_{k+1} = M(x_k) \tag{25.1.1}$$

where $x_0 = c$ is the unknown initial condition. Let

$$z_k = h(x_k) + v_k \tag{25.1.2}$$

denote the observation which is a nonlinear function of the state x_k subjected to the addition of scalar noise v_k where $E(v_k) = 0$, $\text{Var}(v_k) = R_k > 0$ and v_k is serially **uncorrelated**. That is, v_k is a **white noise** sequence.

Given a set of observations $z_k : k = 0, 1, \ldots, N$, our goal is to find the initial condition $x_0 = c$ that minimizes

$$J(c) = \frac{1}{2}\sum_{k=0}^{N}(h(x_k) - z_k)^2 R_k^{-1} \tag{25.1.3}$$

when the states evolve according to the dynamics in (25.1.1).

First and second variations of $J(c)$

As a first step towards computing the gradient and the Hessian of $J(c)$, we compute the first and second variations of $J(c)$ as follows. (Refer to Appendix C)

The first variation $\delta J(c)$ is given by

$$\begin{aligned}\delta J(c) &= \sum_{k=0}^{N} \frac{\partial h}{\partial x_k} R_k^{-1}[h(x_k) - z_k]\delta x_k \\ &= \sum_{k=0}^{N} f_k \delta x_k\end{aligned} \tag{25.1.4}$$

where

$$f_k = \frac{\partial h}{\partial x_k} R_k^{-1}[h(x_k) - z_k]. \tag{25.1.5}$$

Now taking the first variation of both sides of (25.1.4) we get (using the chain rule in Appendix C)

$$\begin{aligned}\delta^2 J(c) &= \sum_{k=0}^{N} \delta[f_k \delta x_k] \\ &= \sum_{k=0}^{N}[(\delta f_k)(\delta x_k) + f_k \delta^2 x_k]\end{aligned} \tag{25.1.6}$$

where $\delta^2 J(c)$ and $\delta^2 x_k$ denote the second variation of $J(c)$ and x_k respectively. Thus, computing $\delta J(c)$ and $\delta^2 J(c)$ reduces to one of computing f_k, δf_k, δx_k and $\delta^2 x_k$. Since f_k can be computed readily using (25.1.5), we now take up the computation of δf_k.

Now taking the first variation of both sides of (25.1.5) and using the chain rule, we obtain

$$\begin{aligned}\delta f_k &= \delta\left(\frac{\partial h}{\partial x_k}\right) R_k^{-1}[h(x_k) - z_k] + \frac{\partial h}{\partial x_k} R_k^{-1} \delta[h(x_k) - z_k] \\ &= \left(\frac{\partial^2 h}{\partial x_k^2}\right) R_k^{-1}[h(x_k) - z_k]\delta x_k + \left(\frac{\partial h}{\partial x_k}\right)^2 R_k^{-1} \delta x_k. \end{aligned} \tag{25.1.7}$$

Substituting (25.1.7) into (25.1.6), we get

$$\begin{aligned}\delta^2 J(c) &= \sum_{k=0}^{N} \left\{ \frac{\partial^2 h}{\partial x_k^2} R_k^{-1}[h(x_k) - z_k] + \left(\frac{\partial h}{\partial x_k}\right)^2 R_k^{-1} \delta x_k \right\} \delta x_k \\ &\quad + \sum_{k=0}^{N} f_k \delta^2 x_k. \end{aligned} \tag{25.1.8}$$

Dynamics of the first and second variation of x_k

We now move on to quantifying the dynamics of evolution of δx_k and $\delta^2 x_k$ using (25.1.1).

Taking the first variation of both sides of (25.1.1) we get

$$\delta x_{k+1} = \left(\frac{\partial M}{\partial x_k}\right) \delta x_k \tag{25.1.9}$$

where $\delta x_0 = \delta c$. This linear non-autonomous dynamical system is known as the **tangent linear system** (Chapter 24). Now taking the second variation of both sides of (25.1.9) and using the chain rule, we get

$$\begin{aligned}\delta^2 x_{k+1} &= \delta\left(\frac{\partial M}{\partial x_k}\right) \delta x_k + \left(\frac{\partial M}{\partial x_k}\right) \delta^2 x_k \\ &= \left(\frac{\partial M}{\partial x_k}\right) \delta^2 x_k + \left(\frac{\partial^2 M}{\partial x_k^2}\right) (\delta x_k)^2 \end{aligned} \tag{25.1.10}$$

where $\delta^2 x_0 = 0$. This is also a non-autonomous linear system quite similar to (25.1.9) but with an extra forcing term that depends on the Hessian of M.

First-order adjoint and gradient computation

Let $\lambda_k \in \mathbb{R}$ denote the sequence of first-order adjoint variables defined by the first-order adjoint equation (Chapter 24)

$$\left.\begin{aligned} \lambda_k &= \left(\tfrac{\partial M}{\partial x_k}\right) \lambda_{k+1} + f_k \\ \text{where} \qquad & \\ \lambda_N &= f_N \end{aligned}\right\} \tag{25.1.11}$$

Now rewrite (25.1.11) and (25.1.9) as

$$\lambda_k - \left(\frac{\partial M}{\partial x_k}\right)\lambda_{k+1} - f_k = 0 \tag{25.1.12}$$

and

$$\delta x_{k+1} - \left(\frac{\partial M}{\partial x_k}\right)\delta x_k = 0. \tag{25.1.13}$$

By way of eliminating the first derivative term in M, multiplying equation (25.1.12) by δx_k and equation (25.1.13) by $-\lambda_{k+1}$ and adding, it follows that

$$\lambda_k \delta x_k - \lambda_{k+1}\delta x_{k+1} - f_k \delta x_k = 0.$$

Now summing over k ranging from 0 to $N-1$, we get

$$\sum_{k=0}^{N-1}(\lambda_k \delta x_k - \lambda_{k+1}\delta x_{k+1}) = \sum_{k=0}^{N-1} f_k \delta x_k.$$

Cancelling out the terms in this telescoping sum on the l.h.s., and using $\lambda_N = f_N$ and $\delta x_0 = \delta c$ we get

$$\begin{aligned}\lambda_0 \delta c &= \sum_{k=0}^{N} f_k \delta x_k \\ &= \delta J(c) \quad \text{using (25.1.4).}\end{aligned} \tag{25.1.14}$$

From first principles

$$\delta J(c) = \frac{\partial J(c)}{\partial c}\delta c. \tag{25.1.15}$$

Comparing these two expressions for $\delta J(c)$, it follows that

$$\lambda_0 = \frac{\partial J(c)}{\partial c} \tag{25.1.16}$$

which is the required gradient. In other words, by solving the first-order adjoint equation (25.1.11) which is a non-autonomous, linear recurrence relation backward in time from $k = N$ to $k = 0$, we get $\lambda_0 = \frac{\partial J(c)}{\partial c}$.

The second-order adjoint and Hessian information

As a first step towards deriving the Hessian information, take the first variation of both sides of (25.1.11) leading to

$$y_k = \left(\frac{\partial M}{\partial x_k}\right) y_{k+1} + \left[\left(\frac{\partial^2 M}{\partial x_k^2}\right)\delta x_k\right]\lambda_{k+1} + \delta f_k \tag{25.1.17}$$

where $y_k = \delta\lambda_k$ for simplicity in notation and $y_N = \delta\lambda_N = \delta f_N$ with δf_k as given in (25.1.7). The variable y_k is called the second-order adjoint variable and equation (25.1.17) is called the **second-order adjoint equation**.

Rewrite (25.1.17) and (25.1.10) as

$$y_k - \left(\frac{\partial M}{\partial x_k}\right) y_{k+1} - \left[\left(\frac{\partial^2 M}{\partial x_k^2}\right) \delta x_k\right] \lambda_{k+1} - \delta f_k = 0 \tag{25.1.18}$$

and

$$\delta^2 x_{k+1} - \left(\frac{\partial M}{\partial x_k}\right) \delta^2 x_k - \left(\frac{\partial^2 M}{\partial x_k^2}\right) (\delta x_k)^2 = 0. \tag{25.1.19}$$

Multiply (25.1.18) by δx_k and (25.1.19) by $-\lambda_{k+1}$ and add the resulting two equations to get

$$\begin{aligned} \delta x_k y_k - \left(\frac{\partial M}{\partial x_k}\right) \delta x_k y_{k+1} - \delta f_k \delta x_k \\ -\lambda_{k+1} \delta^2 x_{k+1} + \left(\frac{\partial M}{\partial x_k}\right) \delta^2 x_k \lambda_{k+1} = 0. \end{aligned} \tag{25.1.20}$$

Now recall that

$$\left(\frac{\partial M}{\partial x_k}\right) \lambda_{k+1} = \lambda_k - f_k \quad \text{using (12.1.11)}$$

and

$$\left(\frac{\partial M}{\partial x_k}\right) \delta x_k = \delta x_{k+1} \quad \text{using (12.1.9).}$$

Substituting these into (25.1.20) and summing it over k ranging from 0 to $N-1$, we obtain

$$\begin{aligned} &\sum_{k=0}^{N-1} \{(\lambda_k - f_k)\delta^2 x_k - \lambda_{k+1} \delta^2 x_{k+1}\} \\ &= \sum_{k=0}^{N-1} \{-\delta x_k \delta \lambda_k + \delta x_{k+1} \delta \lambda_{k+1} + \delta f_k \delta x_k\}. \end{aligned} \tag{25.1.21}$$

By cancelling the terms in this telescoping sum and substituting $\delta x_0 = \delta c, \delta^2 x_0 = 0$, $\lambda_N = f_N$ and $y_N = \delta \lambda_N = \delta f_N$, we get

$$\left.\begin{aligned} y_0(\delta c) &= \textstyle\sum_{k=0}^{N} (\delta x_k \delta f_k + f_k \delta^2 x_k) \\ &= \delta^2 J(c) \quad \text{using (25.1.6)} \end{aligned}\right\} \tag{25.1.22}$$

From first principles, (since δc is fixed) it follows that

$$\begin{aligned} \delta^2 J(c) = \delta(\delta(J(c))) &= \delta \left[\frac{\partial J}{\partial c} \delta c\right] \\ &= \left(\frac{\partial^2 J}{\partial c^2}\right) (\delta c)^2. \end{aligned} \tag{25.1.23}$$

Model $x_{k+1} = M(x_k), \quad x_0 = c$

Observation $z_k = h(x_k) + v_k$

$$E(v_k) = 0 \text{ and } \mathrm{Var}(v_k) = R_k > 0$$

Tangent Linear System

$$\delta x_{k+1} = \left(\frac{\partial M}{\partial x_k}\right)\delta x_k, \quad \delta x_0 = \delta c$$

First-order adjoint equation

$$\lambda_k = \left(\frac{\partial M}{\partial x_k}\right)\lambda_{k+1} + f_k, \quad \lambda_N = f_N$$

$$f_k = \left(\frac{\partial h}{\partial x_k}\right)R_k^{-1}[h(x_k) - z_k]$$

Second-order adjoint equation

$$y_k = \left(\frac{\partial M}{\partial x_k}\right)y_{k+1} + \left(\frac{\partial^2 M}{\partial x_k^2}\right)\delta x_k \lambda_{k+1} + \delta f_k, \quad y_N = \delta f_N$$

$$\delta f_k = \left(\frac{\partial^2 h}{\partial x_k^2}\right)R_k^{-1}[h(x_k) - z_k]\delta x_k + \left(\frac{\partial h}{\partial x_k}\right)^2 R_k^{-1}\delta x_k$$

Gradient and Hessian information

$$\lambda_0 = \frac{\partial J}{\partial c}, \qquad y_0 = \left(\frac{\partial^2 J}{\partial c^2}\right)\delta c$$

Fig. 25.1.1 The second-order adjoint method: scalar case.

Comparing these two expressions for $\delta^2 J(c)$ we immediately obtain

$$y_0 = \left(\frac{\partial^2 J}{\partial c^2}\right)\delta c \tag{25.1.24}$$

which is the required Hessian information.

Stated in other words, by solving the second-order adjoint equation (25.1.17) which is a linear non-autonomous recurrence relation backward in time, we obtain that y_0 is the sought after Hessian information.

A summary of these equations is given in Figure 25.1.1

Special case Consider the case when $M(\cdot)$ and $h(\cdot)$ are linear functions, say

$$M(x) = ax \quad \text{and} \quad h(x) = bx \tag{25.1.25}$$

for some real constants a and b. Then

$$\frac{\partial M}{\partial x} = a, \quad \frac{\partial h}{\partial x} = b \quad \text{and} \quad \frac{\partial^2 M}{\partial x^2} = 0 = \frac{\partial^2 h}{\partial x^2}.$$

Hence

$$\delta f_k = \left(\frac{\partial h}{\partial x_k}\right)^2 R_k^{-1}\delta x_k \tag{25.1.26}$$

and the second-order adjoint equation reduces to

$$y_k = \left(\frac{\partial M}{\partial x_k} \right) y_{k+1} + \delta f_k \tag{25.1.27}$$

with $y_N = \delta f_N$.

25.2 Second-order adjoint method: vector case

Let $\mathbf{M} : \mathbb{R}^n \to \mathbb{R}^n$ with $\mathbf{M}(\mathbf{x}) = (M_1(\mathbf{x}), M_2(\mathbf{x}), \ldots, M_n(\mathbf{x}))^{\mathrm{T}}$ and $\mathbf{h} : \mathbb{R}^n \to \mathbb{R}^m$ with $\mathbf{h}(\mathbf{x}) = (h_1(\mathbf{x}), h_2(\mathbf{x}), \ldots, h_m(\mathbf{x}))^{\mathrm{T}}$. Let $\mathbf{x}_k$ denote the state of a nonlinear dynamical system whose time evolution is given by

$$\mathbf{x}_{k+1} = \mathbf{M}(\mathbf{x}_k) \tag{25.2.1}$$

with $\mathbf{x}_0 = \mathbf{c}$. Let

$$\mathbf{z}_k = \mathbf{h}(\mathbf{x}_k) + \mathbf{v}_k \tag{25.2.2}$$

denote the observations where $\mathbf{v}_k \in \mathbb{R}^m$ is a white noise sequence with

$$E(\mathbf{v}_k) = 0 \quad \text{and} \quad \mathrm{Cov}(\mathbf{v}_k) = \mathbf{R}_k \in \mathbb{R}^{m \times m}.$$

Given $\mathbf{z}_k : k = 0, 1, \ldots, N$, consider the problem of finding the initial condition $\mathbf{c} \in \mathbb{R}^n$ that minimizes

$$\left.\begin{aligned} J(\mathbf{c}) &= \textstyle\sum_{k=0}^N \mathbf{J}_k(\mathbf{c}) \\ \text{and} \qquad & \\ \mathbf{J}_k(\mathbf{c}) &= \tfrac{1}{2}\langle(\mathbf{h}(\mathbf{x}_k) - \mathbf{z}_k), \mathbf{R}_k^{-1}(\mathbf{h}(\mathbf{x}_k) - \mathbf{z}_k)\rangle \end{aligned}\right\} \tag{25.2.3}$$

when the states $\mathbf{x}_k$ are constrained by the dynamical equation (25.2.1).

Except for the complication of dealing with vectors and matrices, the following analysis parallels that in Section 25.1. To save space we only indicate the major steps leaving the details of algebra as an exercise to the reader.

We begin by computing the first two variations of $J(\mathbf{c})$.

First and second variation The first variation of $\mathbf{J}_k(\mathbf{c})$ is given by (Appendix C)

$$\begin{aligned} \delta \mathbf{J}_k(\mathbf{c}) &= \frac{1}{2} \delta[(\mathbf{h}(\mathbf{x}_k) - \mathbf{z}_k)^{\mathrm{T}} \mathbf{R}_k^{-1} (\mathbf{h}(\mathbf{x}_k) - \mathbf{z}_k)] \\ &= \langle \mathbf{f}_k, \delta \mathbf{x}_k \rangle \end{aligned} \tag{25.2.4}$$

where

$$\mathbf{f}_k = \mathbf{D}_{\mathbf{h}}^{\mathrm{T}}(\mathbf{x}_k) \mathbf{R}_k^{-1} [(\mathbf{h}(\mathbf{x}_k) - \mathbf{z}_k)] \tag{25.2.5}$$

and $\mathbf{D_h}(\mathbf{x}) \in \mathbb{R}^{m\times n}$ is the **Jacobian** of $\mathbf{h}(\mathbf{x})$. Hence

$$\delta J(\mathbf{c}) = \sum_{k=0}^{N} \langle \mathbf{f}_k, \delta\mathbf{x}_k\rangle. \tag{25.2.6}$$

By applying the chain rule to (25.2.4) we now obtain the second variation $\delta^2\mathbf{J}_k(\mathbf{c})$ as

$$\delta^2\mathbf{J}_k(\mathbf{c}) = \langle \delta\mathbf{f}_k, \delta\mathbf{x}_k\rangle + \langle \mathbf{f}_k, \delta^2\mathbf{x}_k\rangle \tag{25.2.7}$$

where (since $\mathbf{z}_k$ is a given constant vector)

$$\delta\mathbf{f}_k = \delta[\mathbf{D}_\mathbf{h}^\mathrm{T}(\mathbf{x}_k)]\mathbf{R}_k^{-1}[(\mathbf{h}(\mathbf{x}_k) - \mathbf{z}_k)] + \mathbf{D}_\mathbf{h}^\mathrm{T}(\mathbf{x}_k)\mathbf{R}_k^{-1}\delta[\mathbf{h}(\mathbf{x}_k)] \tag{25.2.8}$$

and

$$\delta[\mathbf{h}(\mathbf{x}_k)] = \mathbf{D_h}(\mathbf{x}_k)\delta\mathbf{x}_k. \tag{25.2.9}$$

To get a handle on computing $\delta[\mathbf{D}_\mathbf{h}^T(\mathbf{x})]$, first we consider the following example.

Example 25.2.1 Let $\mathbf{y} = (y_1, y_2)^\mathrm{T}$ and $\mathbf{g} : \mathbb{R}^2 \to \mathbb{R}^2$ with $\mathbf{g}(\mathbf{y}) = (g_1(\mathbf{y}), g_2(\mathbf{y}))^\mathrm{T}$. Then

$$\mathbf{D_g}(\mathbf{y}) = \begin{bmatrix} \frac{\partial g_1}{\partial y_1} & \frac{\partial g_1}{\partial y_2} \\ \frac{\partial g_2}{\partial y_1} & \frac{\partial g_2}{\partial y_2} \end{bmatrix}.$$

If $\mathbf{A} = [a_{ij}]$, then define $\delta(\mathbf{A}) = [\delta(a_{ij})]$. In computing $\delta[\mathbf{D_g}(\mathbf{y})]$, let ∇ denote the gradient operator with respect to $\mathbf{y}$. Then

$$\begin{aligned} \delta\left(\frac{\partial g_1}{\partial y_1}\right) &= \left\langle \nabla\left(\frac{\partial g_1}{\partial y_1}\right), \delta\mathbf{y}\right\rangle \\ &= \frac{\partial^2 g_1}{\partial y_1^2}\delta y_1 + \frac{\partial^2 g_1}{\partial y_1 \partial y_2}\delta y_2. \end{aligned}$$

Similarly

$$\delta\left(\frac{\partial g_1}{\partial y_2}\right) = \frac{\partial^2 g_1}{\partial y_1 \partial y_2}\delta y_1 + \frac{\partial^2 g_1}{\partial y_2^2}\delta y_2.$$

Combining these, the first row of $\delta[\mathbf{D_g}(\mathbf{y})]$ is given by

$$\begin{aligned} \left[\delta\left(\frac{\partial g_1}{\partial y_1}\right), \delta\left(\frac{\partial g_1}{\partial y_2}\right)\right] &= (\delta y_1, \delta y_2) \begin{bmatrix} \frac{\partial^2 g_1}{\partial y_1^2} & \frac{\partial^2 g_1}{\partial y_1 \partial y_2} \\ \frac{\partial^2 g_1}{\partial y_2 \partial y_1} & \frac{\partial^2 g_1}{\partial y_2^2} \end{bmatrix} \\ &= (\delta\mathbf{y})^\mathrm{T}\nabla^2 g_1(\mathbf{y}). \end{aligned}$$

$\nabla^2 g_1(\mathbf{x})$ is the Hessian (which is a symmetric) matrix of $g_1(\mathbf{y})$.

Similarly the second row of $\delta[\mathbf{D_g}(\mathbf{y})]$ is

$$\left[\delta\left(\frac{\partial g_2}{\partial y_1}\right), \delta\left(\frac{\partial g_2}{\partial y_2}\right)\right] = (\delta\mathbf{y})^{\mathrm{T}}\nabla^2 g_2(\mathbf{y}).$$

.

Hence, we define

$$\mathbf{D}^2_{\mathbf{g}}(\mathbf{y}, \delta\mathbf{y}) = \delta[\mathbf{D_g}(\mathbf{y})] = \begin{bmatrix} (\delta\mathbf{y})^{\mathrm{T}}\nabla^2 g_1(\mathbf{y}) \\ (\delta\mathbf{y})^{\mathrm{T}}\nabla^2 g_2(\mathbf{y}) \end{bmatrix}$$

and

$$\delta[\mathbf{D}^{\mathrm{T}}_{\mathbf{g}}(\mathbf{y})] = [\nabla^2 g_1(\mathbf{y})\delta\mathbf{y}, \nabla^2 g_2(\mathbf{y})\delta\mathbf{y}] = [\mathbf{D}^2_{\mathbf{g}}(\mathbf{y}, \delta\mathbf{y})]^{\mathrm{T}}.$$

By generalizing this example and applying to $\mathbf{h}(\mathbf{x})$, we readily see that

$$\begin{aligned}\delta[\mathbf{D}^{\mathrm{T}}_{\mathbf{h}}(\mathbf{x}_k)] &= [\nabla^2 h_1(\mathbf{x})\delta\mathbf{x}, \nabla^2 h_2(\mathbf{x})\delta\mathbf{x}, \ldots, \nabla^2 h_m(\mathbf{x})\delta\mathbf{x}] \\ &= [\mathbf{D}^2_{\mathbf{h}}(\mathbf{x}, \delta\mathbf{x})]^{\mathrm{T}}. \end{aligned} \tag{25.2.10}$$

which is an $n \times m$ matrix.

Now substituting (25.2.9) and (25.2.10) in (25.2.8) we obtain for later reference that

$$\begin{aligned}\delta\mathbf{f}_k = {} & [\mathbf{D}^2_{\mathbf{h}}(\mathbf{x}_k, \delta\mathbf{x}_k)]^{\mathrm{T}}\mathbf{R}_k^{-1}[(\mathbf{h}(\mathbf{x}_k) - \mathbf{z}_k)] \\ & + \mathbf{D}^{\mathrm{T}}_{\mathbf{h}}(\mathbf{x}_k)\mathbf{R}_k^{-1}\mathbf{D_h}(\mathbf{x}_k)\delta\mathbf{x}_k. \end{aligned} \tag{25.2.11}$$

Dynamics of the first and second variation of $\mathbf{x}_k$

Taking the first variation of (25.2.1) we get the so-called **tangent linear** system

$$\delta\mathbf{x}_{k+1} = \mathbf{D_M}(\mathbf{x}_k)\delta\mathbf{x}_k \tag{25.2.12}$$

where $\mathbf{D_M}(\mathbf{x}) \in \mathbb{R}^{n\times n}$ is the Jacobian of $\mathbf{M}(\mathbf{x})$ and $\delta\mathbf{x}_0 = \delta\mathbf{c}$. Now taking the second variation of both sides using the chain rule, we have

$$\delta^2\mathbf{x}_{k+1} = \delta[\mathbf{D_M}(\mathbf{x}_k)]\delta\mathbf{x}_k + \mathbf{D_M}(\mathbf{x}_k)\delta^2\mathbf{x}_k \tag{25.2.13}$$

where, by Example 25.2.1

$$\begin{aligned}\delta[\mathbf{D_M}(\mathbf{x}_k)] &= \begin{bmatrix} (\delta\mathbf{x}_k)^{\mathrm{T}}\nabla^2\mathbf{M}_1(\mathbf{x}_k) \\ (\delta\mathbf{x}_k)^{\mathrm{T}}\nabla^2\mathbf{M}_2(\mathbf{x}_k) \\ \vdots \\ (\delta\mathbf{x}_k)^{\mathrm{T}}\nabla^2\mathbf{M}_n(\mathbf{x}_k) \end{bmatrix} \\ &= \mathbf{D}^2_{\mathbf{M}}(\mathbf{x}_k, \delta\mathbf{x}_k) \end{aligned} \tag{25.2.14}$$

is an $n \times n$ matrix. Substituting (25.2.14) into (25.2.13) we get

$$\delta^2\mathbf{x}_{k+1} = \mathbf{D_M}(\mathbf{x}_k)\delta^2\mathbf{x}_k + \mathbf{D}^2_{\mathbf{M}}(\mathbf{x}_k, \delta\mathbf{x}_k)\delta\mathbf{x}_k \tag{25.2.15}$$

where $\delta^2\mathbf{x}_0 = 0$ and $\delta\mathbf{x}_0 = \delta\mathbf{c}$.

Notice that both (25.2.12) and (25.2.15) are linear non-autonomous systems.

First-order adjoint and gradient of $J(\mathbf{c})$

Let $\boldsymbol{\lambda}_k \in \mathbb{R}^n$. Then the first-order adjoint equation is given by

$$\boldsymbol{\lambda}_k = \mathbf{D}_{\mathbf{M}}^{\mathrm{T}}(\mathbf{x}_k)\boldsymbol{\lambda}_{k+1} + \mathbf{f}_k \tag{25.2.16}$$

with $\boldsymbol{\lambda}_N = \mathbf{f}_N$ where $\mathbf{f}_k$ is defined in (25.2.5).

Now take the inner product of both sides of (25.2.12) with $-\boldsymbol{\lambda}_{k+1}$ and inner product of both sides of (25.1.16) with $\delta\mathbf{x}_k$ and adding, we obtain

$$\langle -\boldsymbol{\lambda}_{k+1}, \delta\mathbf{x}_{k+1}\rangle + \langle \delta\mathbf{x}_k, \boldsymbol{\lambda}_k\rangle = \langle \mathbf{f}_k, \delta\mathbf{x}_k\rangle.$$

Now summing both sides over k ranging from 0 to $N-1$ and using the facts that $\boldsymbol{\lambda}_N = \mathbf{f}_N$, $\delta\mathbf{x}_0 = \delta\mathbf{c}$, after simplification, it becomes

$$\begin{aligned}\langle \boldsymbol{\lambda}_0, \delta\mathbf{c}\rangle &= \sum_{k=0}^{N}\langle \mathbf{f}_k, \delta\mathbf{x}_k\rangle \\ &= \delta J(\mathbf{c}) \qquad\qquad \text{using (25.2.6)} \\ &= \langle \nabla J(\mathbf{c}), \delta\mathbf{c}\rangle \qquad \text{(from first principles).}\end{aligned} \tag{25.2.17}$$

Hence $\boldsymbol{\lambda}_0 = \nabla J(\mathbf{c})$ which is obtained by computing the recurrence (25.2.16) backward in time.

Second-order adjoint and hessian vector product

Taking the first variation of both sides of (25.2.16) and representing $\delta\boldsymbol{\lambda}_k$ by $\mathbf{y}_k \in \mathbb{R}^n$ we get the second-order adjoint equation using (25.2.10)

$$\begin{aligned}\mathbf{y}_k &= \delta[\mathbf{D}_{\mathbf{M}}^{\mathrm{T}}(\mathbf{x}_k)]\boldsymbol{\lambda}_{k+1} + \mathbf{D}_{\mathbf{M}}^{\mathrm{T}}(\mathbf{x}_k)\mathbf{y}_{k+1} + \delta\mathbf{f}_k \\ &= [\mathbf{D}_{\mathbf{M}}^{2}(\mathbf{x}_k,\ \delta\mathbf{x}_k)]^{\mathrm{T}}\boldsymbol{\lambda}_{k+1} + \mathbf{D}_{\mathbf{M}}^{\mathrm{T}}(\mathbf{x}_k)\mathbf{y}_{k+1} + \delta\mathbf{f}_k\end{aligned} \tag{25.2.18}$$

where $\mathbf{y}_N = \delta\mathbf{f}_N$. Now take the inner product of both sides of (25.2.15) by $-\boldsymbol{\lambda}_{k+1}$ and the inner product of both sides of (25.2.18) by $\delta\mathbf{x}_k$ and adding (to eliminate the $\mathbf{D}_{\mathbf{M}}^{2}(\mathbf{x}_k,\ \delta\mathbf{x}_k)$ term) we obtain

$$\begin{aligned}&\langle -\boldsymbol{\lambda}_{k+1}, \delta^2\mathbf{x}_{k+1}\rangle + \langle \delta\mathbf{x}_k, \mathbf{y}_k\rangle \\ &= \langle -\boldsymbol{\lambda}_{k+1}, \mathbf{D}_{\mathbf{M}}(\mathbf{x}_k)\delta^2\mathbf{x}_k\rangle + \langle \delta\mathbf{x}_k, \mathbf{D}_{\mathbf{M}}^{\mathrm{T}}(\mathbf{x}_k)\mathbf{y}_{k+1}\rangle + \langle \delta\mathbf{x}_k, \delta\mathbf{f}_k\rangle \\ &= -\langle \mathbf{D}_{\mathbf{M}}^{\mathrm{T}}(\mathbf{x}_k)\boldsymbol{\lambda}_{k+1}, \delta^2\mathbf{x}_k\rangle + \langle \mathbf{D}_{\mathbf{M}}(\mathbf{x}_k)\delta\mathbf{x}_k, \mathbf{y}_{k+1}\rangle + \langle \delta\mathbf{x}_k, \delta\mathbf{f}_k\rangle\end{aligned} \tag{25.2.19}$$

where we have used the property $\langle \mathbf{x}, \mathbf{A}\mathbf{y}\rangle = \langle \mathbf{A}^{\mathrm{T}}\mathbf{x}, \mathbf{y}\rangle$. But from (25.2.12) and (25.2.16) we have

$$\mathbf{D}_{\mathbf{M}}(\mathbf{x}_k)\delta\mathbf{x}_k = \delta\mathbf{x}_{k+1} \quad \text{and} \quad \mathbf{D}_{\mathbf{M}}^{\mathrm{T}}(\mathbf{x}_k)\boldsymbol{\lambda}_{k+1} = \boldsymbol{\lambda}_k - \mathbf{f}_k.$$

Substituting these back into the r.h.s. of (25.2.19) and summing both sides of the resulting expression over k ranging from 0 to $N-1$, and using $\delta\mathbf{x}_0 = \delta\mathbf{c}$, $\delta^2\mathbf{x}_0 = 0$,

Model $\mathbf{x}_{k+1} = \mathbf{M}(\mathbf{x}_k), \quad \mathbf{x}_0 = \mathbf{c}$

Observation $\mathbf{z}_k = \mathbf{h}(\mathbf{x}_k) + \mathbf{v}_k$

$$E(\mathbf{v}_k) = 0 \text{ and } \mathrm{Cov}(\mathbf{v}_k) = \mathbf{R}_k$$

Tangent Linear System

$$\delta\mathbf{x}_{k+1} = \mathbf{D}_{\mathbf{M}}(\mathbf{x}_k)\delta\mathbf{x}_k, \quad \delta\mathbf{x}_0 = \delta\mathbf{c}$$

First-order adjoint equation

$$\boldsymbol{\lambda}_k = \mathbf{D}_{\mathbf{M}}^{\mathrm{T}}(\mathbf{x}_k)\boldsymbol{\lambda}_{k+1} + \mathbf{f}_k$$

$$\boldsymbol{\lambda}_N = \mathbf{f}_N$$

$$\mathbf{f}_k = \mathbf{D}_{\mathbf{h}}^{\mathrm{T}}(\mathbf{x}_k)\mathbf{R}_k^{-1}[(\mathbf{h}(\mathbf{x}_k) - \mathbf{z}_k)]$$

Second-order adjoint equation

$$\mathbf{y}_k = \mathbf{D}_{\mathbf{M}}^{\mathrm{T}}(\mathbf{x}_k)\mathbf{y}_{k+1} + [\mathbf{D}_{\mathbf{M}}^2(\mathbf{x}_k, \delta\mathbf{x}_k)]^{\mathrm{T}}\boldsymbol{\lambda}_{k+1} + \delta\mathbf{f}_k$$

$$\mathbf{y}_N = \delta\mathbf{f}_N$$

$$\delta\mathbf{f}_k = [\mathbf{D}_{\mathbf{h}}^2(\mathbf{x}_k, \delta\mathbf{x}_k)]^{\mathrm{T}}\mathbf{R}_k^{-1}[(\mathbf{h}(\mathbf{x}_k) - \mathbf{z}_k)]$$
$$+\mathbf{D}_{\mathbf{h}}^{\mathrm{T}}(\mathbf{x}_k)\mathbf{R}_k^{-1}\mathbf{D}_{\mathbf{h}}(\mathbf{x}_k)\delta\mathbf{x}_k$$

Gradient and Hessian information

$$\boldsymbol{\lambda}_0 = \nabla J(\mathbf{c}), \quad \text{and} \quad \mathbf{y}_0 = \nabla^2 J(\mathbf{c})\,\delta\mathbf{c}$$

Fig. 25.2.1 The second-order adjoint method: nonlinear system.

$\boldsymbol{\lambda}_N = \mathbf{f}_N$ and $\mathbf{y}_N = \delta\mathbf{f}_N$, we obtain after simplification

$$\begin{aligned}
\langle \mathbf{y}_0, \delta\mathbf{c}\rangle &= \sum_{k=0}^{N} \left[\langle \mathbf{f}_k, \delta^2\mathbf{x}_k\rangle + \langle \delta\mathbf{f}_k, \delta\mathbf{x}_k\rangle\right] \\
&= \delta^2 J(\mathbf{c}) \qquad \text{using (25.2.7)} \\
&= \langle \nabla^2 J(\mathbf{c})\delta\mathbf{c}, \delta\mathbf{c}\rangle \qquad \text{(from first principles).}
\end{aligned} \tag{25.2.20}$$

This in turn leads to

$$\mathbf{y}_0 = \nabla^2 J(\mathbf{c})\delta\mathbf{c}$$

which is the Hessian-vector product.

A summary of this method is given in Figure 25.2.1.

Model $\mathbf{x}_{k+1} = \mathbf{M}_k \mathbf{x}_k, \quad \mathbf{x}_0 = \mathbf{c}$

Observation $\mathbf{z}_k = \mathbf{H}_k \mathbf{x}_k + \mathbf{v}_k$

Tangent Linear System

$$\delta\mathbf{x}_{k+1} = \mathbf{M}_k \delta\mathbf{x}_k, \quad \delta\mathbf{x}_0 = \delta\mathbf{c}$$

First-order Adjoint

$$\boldsymbol{\lambda}_k = \mathbf{M}_k^{\mathrm{T}} \boldsymbol{\lambda}_{k+1} + \mathbf{f}_k$$

$$\boldsymbol{\lambda}_N = \mathbf{f}_N$$

$$\mathbf{f}_k = \mathbf{H}_k^{\mathrm{T}} \mathbf{R}_k^{-1} [\mathbf{H}_k \mathbf{x}_k - \mathbf{z}_k]$$

Second-order Adjoint

$$\mathbf{y}_k = \mathbf{M}_k^{\mathrm{T}} \mathbf{y}_{k+1} + \delta\mathbf{f}_k$$

$$\mathbf{y}_N = \delta\mathbf{f}_N$$

$$\delta\mathbf{f}_k = \mathbf{H}_k^{\mathrm{T}} \mathbf{R}_k^{-1} \mathbf{H}_k \delta\mathbf{x}_k$$

Gradient and Hessian information

$$\boldsymbol{\lambda}_0 = \nabla J(\mathbf{c}), \quad \text{and} \quad \mathbf{y}_0 = \nabla^2 J(\mathbf{c})\delta\mathbf{c}$$

Fig. 25.2.2 The second-order adjoint method: linear system.

Special case Consider when $\mathbf{M}(\mathbf{x})$ and $\mathbf{h}(\mathbf{x})$ are linear that is

$$\mathbf{M}(\mathbf{x}_k) = \mathbf{M}_k \mathbf{x} \quad \text{and} \quad \mathbf{h}(\mathbf{x}_k) = \mathbf{H}_k \mathbf{x}$$

where $\mathbf{M}_k \in \mathbb{R}^{n\times n}$ and $\mathbf{H}_k \in \mathbb{R}^{m\times n}$. Then

$$\mathbf{D_M}(\mathbf{x}) = \mathbf{M}_k \quad \text{and} \quad \mathbf{D_h}(\mathbf{x}) = \mathbf{H}_k$$

with

$$\mathbf{D}^2_{\mathbf{M}}(\mathbf{x}, \delta\mathbf{x}) = 0 \quad \text{and} \quad \mathbf{D}^2_{\mathbf{h}}(\mathbf{x}, \delta\mathbf{x}) = 0.$$

Substituting these we obtain the second-order method which is summarized in Figure 25.2.2.

25.3 Second-order adjoint sensitivity

In this section we provide an extension of the first-order adjoint sensitivity computations presented in Section 24.5.

Let $\mathbf{F} : \mathbb{R}^N \times \mathbb{R}^M \times \mathbb{R}^K \to \mathbb{R}^N$ and let

$$\mathbf{F}(\mathbf{y}, \mathbf{u}, \boldsymbol{\alpha}) = 0 \tag{25.3.1}$$

be the model equation where $\mathbf{y} \in \mathbb{R}^N$ denotes the **state**, $\mathbf{u} \in \mathbb{R}^M$, the variables including the **initial/boundary** conditions, and $\boldsymbol{\alpha} \in \mathbb{R}^K$, the **parameters** of the model. It is tacitly assumed that there exists a unique solution $\mathbf{y} = \mathbf{y}(\mathbf{u}, \boldsymbol{\alpha})$ of the model equation (25.3.1). It is also assumed that the model variable $\mathbf{u}$ is **not** known a priori. This unknown variable $\mathbf{u}$ is often estimated using a given set of observations $\mathbf{z} \in \mathbb{R}^L$ by invoking a data assimilation procedure described in this book. The optimal estimate $\hat{\mathbf{u}} = \hat{\mathbf{u}}(\boldsymbol{\alpha}, \mathbf{z})$ is obtained as the minimizer of an **objective function** $J(\mathbf{y}, \mathbf{u}, \mathbf{z})$ where $J : \mathbb{R}^N \times \mathbb{R}^M \times \mathbb{R}^L \to \mathbb{R}$. By initializing the model (25.3.1) with $\mathbf{u} = \hat{\mathbf{u}}(\boldsymbol{\alpha}, \mathbf{z})$, we obtain the **optimal state** $\hat{\mathbf{y}} = \hat{\mathbf{y}}(\boldsymbol{\alpha}, \mathbf{z})$ which is a function of the parameter $\boldsymbol{\alpha}$ and the observation $\mathbf{z}$. Our goal in this section is to compute the sensitivity of the given **response** function $G(\hat{\mathbf{y}})$ (which is $G(\mathbf{y})$ evaluated along the optimal state $\mathbf{y} = \hat{\mathbf{y}}$) with respect to the parameter $\boldsymbol{\alpha}$ and/or the observation $\mathbf{z}$ where $G : \mathbb{R}^N \to \mathbb{R}$.

The first step is to derive the necessary condition for the minimum of $J(\mathbf{y}, \mathbf{u}, \mathbf{z})$ with respect to $\mathbf{u}$ for a fixed $\boldsymbol{\alpha}$ when $\mathbf{y}$ and $\mathbf{u}$ are related by the model equation (25.3.1). Clearly, this step involves the computation of the gradient $\nabla_{\mathbf{u}} J = \nabla_{\mathbf{u}} J(\mathbf{y}, \mathbf{u}, \mathbf{z})$ under the constraint of the equation (25.3.1). A little reflection would reveal that this gradient computation can be performed by applying the first-order adjoint method of Section 24.5 using $J(\mathbf{y}, \mathbf{u}, \mathbf{z})$ in place of $G(\mathbf{y}, \mathbf{u})$.

Let $\delta\mathbf{u}$ be a perturbation or variation in $\mathbf{u}$ and let $\delta\mathbf{y}$ denote the induced variation in $\mathbf{y}$. Then recall (Appendix C) that the first variation δJ in J is given by

$$\delta J = \left\langle \delta\mathbf{y}, \frac{\partial J}{\partial \mathbf{y}} \right\rangle + \left\langle \delta\mathbf{u}, \frac{\partial J}{\partial \mathbf{u}} \right\rangle = \langle \delta\mathbf{u}, \nabla_{\mathbf{u}} J \rangle \tag{25.3.2}$$

where $\partial J / \partial \mathbf{y} \in \mathbb{R}^N$ and $\partial J / \partial \mathbf{u} \in \mathbb{R}^N$ are the partial derivatives of J with respect to $\mathbf{y}$ and $\mathbf{u}$, respectively.

By way of expressing $\delta\mathbf{y}$ in terms of $\delta\mathbf{u}$, we take the first variation of both sides of (25.3.1) for a fixed $\boldsymbol{\alpha}$ to obtain

$$\delta\mathbf{F} = (\mathbf{D}_{\mathbf{F}}^{\mathbf{y}})\delta\mathbf{y} + (\mathbf{D}_{\mathbf{F}}^{\mathbf{u}})\delta\mathbf{u} = 0 \tag{25.3.3}$$

where $\mathbf{D}_{\mathbf{F}}^{\mathbf{y}} \in \mathbb{R}^{N\times N}$ and $\mathbf{D}_{\mathbf{F}}^{\mathbf{u}} \in \mathbb{R}^{N\times M}$ are the Jacobians of $\mathbf{F}$ with respect to $\mathbf{y}$ and $\mathbf{u}$ respectively. Taking the inner product of both sides of (25.3.3) with $\mathbf{p} \in \mathbb{R}^N$ called the first-order adjoint variable and using the adjoint property (refer to (23.5.1)), we get

$$\langle \delta\mathbf{y}, (\mathbf{D}_{\mathbf{F}}^{\mathbf{y}})^{\mathrm{T}}\mathbf{p} \rangle = -\langle \delta\mathbf{u}, (\mathbf{D}_{\mathbf{F}}^{\mathbf{u}})^{\mathrm{T}}\mathbf{p} \rangle. \tag{25.3.4}$$

Now choosing $\mathbf{p}$ such that

$$\frac{\partial J}{\partial \mathbf{y}} = (\mathbf{D}_{\mathbf{F}}^{\mathbf{y}})^{\mathrm{T}}\mathbf{p} \tag{25.3.5}$$

in (25.3.4) and substituting this into (25.3.2) leads to

$$\delta J = \langle \delta \mathbf{y}, (\mathbf{D}_F^{\mathbf{y}})^{\mathrm{T}} \mathbf{p} \rangle + \left\langle \delta \mathbf{u}, \frac{\partial J}{\partial \mathbf{u}} \right\rangle = \left\langle \delta \mathbf{u}, -(\mathbf{D}_F^{\mathbf{u}})^{\mathrm{T}} \mathbf{p} + \frac{\partial J}{\partial \mathbf{u}} \right\rangle$$

from which we obtain the necessary condition for the optimality of J with respect to $\mathbf{u}$ as

$$\nabla_{\mathbf{u}} J = -(\mathbf{D}_F^{\mathbf{u}})^{\mathrm{T}} \mathbf{p} + \frac{\partial J}{\partial \mathbf{u}} = 0. \tag{25.3.6}$$

By combining these optimality conditions (25.3.5) and (25.3.6) with the original model equation (25.3.1), we restate our problem as follows: compute the sensitivity of $G(\mathbf{y})$ with respect to $\boldsymbol{\alpha}$ and/or $\mathbf{z}$ using the first-order adjoint method in Section 24.5 when the new set of model variables $(\mathbf{y}, \mathbf{p})$ are related to $\boldsymbol{\alpha}$ and $\mathbf{z}$ through the new **extended** model equations:

$$\left.\begin{aligned} \mathbf{F}(\mathbf{y}, \mathbf{u}, \boldsymbol{\alpha}) &= 0 \\ \frac{\partial J}{\partial \mathbf{y}} - (\mathbf{D}_F^{\mathbf{y}})^{\mathrm{T}} \mathbf{p} &= 0 \\ \frac{\partial J}{\partial \mathbf{u}} - (\mathbf{D}_F^{\mathbf{u}})^{\mathrm{T}} \mathbf{p} &= 0 \end{aligned}\right\} \tag{25.3.7}$$

Much of the remaining challenge is largely due to the two additional set of equations in (25.3.7) representing the necessary conditions for the optimality of $\mathbf{u}$.

We first introduce some useful notations. Let

$$\nabla_{\mathbf{y}}^2 J = \left[\frac{\partial^2 J}{\partial y_i \partial y_j} \right] \in \mathbb{R}^{N \times N} \quad \text{and} \quad \nabla_{\mathbf{u}}^2 J = \left[\frac{\partial^2 J}{\partial u_i \partial u_j} \right] \in \mathbb{R}^{M \times M} \tag{25.3.8}$$

be the Hessian of J w.r.to $\mathbf{y}$ and $\mathbf{u}$ respectively. Also, let

$$\nabla_{\mathbf{yu}}^2 J = \left[\frac{\partial^2 J}{\partial y_i \partial u_j} \right] \in \mathbb{R}^{N \times M} \quad \text{and} \quad \nabla_{\mathbf{uy}}^2 J = \left[\nabla_{\mathbf{yu}}^2 J \right]^{\mathrm{T}}. \tag{25.3.9}$$

Similarly, we can define $\nabla_{\mathbf{y}}^2 F_i$, $\nabla_{\mathbf{u}}^2 F_i$, $\nabla_{\boldsymbol{\alpha}}^2 F_i$, $\nabla_{\mathbf{yu}}^2 F_i$, $\nabla_{\mathbf{y}\boldsymbol{\alpha}}^2 F_i$, and $\nabla_{\mathbf{u}\boldsymbol{\alpha}}^2 F_i$ for $i = 1, 2, \ldots, N$.

For definiteness, in the following, we illustrate the computation of the sensitivity of $G(\mathbf{y})$ w.r.to $\boldsymbol{\alpha}$ for a fixed $\mathbf{z}$. Let $\delta\boldsymbol{\alpha}$ be the perturbation in $\boldsymbol{\alpha}$ and let $\delta\mathbf{u}$ and $\delta\mathbf{y}$ be the induced variations in $\mathbf{u}$ and $\mathbf{y}$ respectively. The first variation δG in G is given by

$$\delta G = \left\langle \delta \mathbf{y}, \frac{\partial G}{\partial \mathbf{y}} \right\rangle = \langle \delta \boldsymbol{\alpha}, \nabla_{\boldsymbol{\alpha}} G \rangle \tag{25.3.10}$$

where $\partial G / \partial \mathbf{y} \in \mathbb{R}^N$ is the partial derivative of G w.r.to $\mathbf{y}$ and $\nabla_{\boldsymbol{\alpha}} G$, the gradient of G w.r.to $\boldsymbol{\alpha}$ is the required sensitivity of G.

We now compute this sensitivity in the following steps.

Step 1 The first variation of the first equation in (25.3.7), which is a relation in $\mathbb{R}^N$, is given by

$$\delta \mathbf{F} = (\mathbf{D}_\mathbf{F}^\mathbf{y})\delta \mathbf{y} + (\mathbf{D}_\mathbf{F}^\mathbf{u})\delta \mathbf{u} + (\mathbf{D}_\mathbf{F}^\alpha)\delta \boldsymbol{\alpha} = 0 \tag{25.3.11}$$

where $\mathbf{D}_\mathbf{F}^\alpha \in \mathbb{R}^{N\times K}$ is the Jacobian of $\mathbf{F}$ w.r.to $\boldsymbol{\alpha}$.

Step 2 Let δ denote the generic first variation operator. Taking the first variation of the second equation in (25.3.7) we get

$$\delta \left[\frac{\partial J}{\partial \mathbf{y}}\right] - \delta \left[(\mathbf{D}_\mathbf{F}^\mathbf{y})^\mathrm{T}\mathbf{p}\right] = 0. \tag{25.3.12}$$

Consider the first term on the l.h.s of (25.3.12):

$$\delta \left[\frac{\partial J}{\partial \mathbf{y}}\right] = \delta_\mathbf{y}\left[\frac{\partial J}{\partial \mathbf{y}}\right] + \delta_\mathbf{u}\left[\frac{\partial J}{\partial \mathbf{y}}\right] + \delta_\alpha\left[\frac{\partial J}{\partial \mathbf{y}}\right] \tag{25.3.13}$$

where δ_α, $\delta_\mathbf{u}$, and $\delta_\mathbf{y}$ are the first variation operators w.r.to $\boldsymbol{\alpha}$, $\mathbf{u}$, and $\mathbf{y}$ respectively. From first principles, we readily obtain

$$\delta_\mathbf{y}\left[\frac{\partial J}{\partial \mathbf{y}}\right] = \delta_\mathbf{y}\begin{bmatrix} \frac{\partial J}{\partial y_1} \\ \frac{\partial J}{\partial y_2} \\ \vdots \\ \frac{\partial J}{\partial y_N} \end{bmatrix} = \begin{bmatrix} (\delta \mathbf{y})^\mathrm{T}\nabla_\mathbf{y}(\frac{\partial J}{\partial y_1}) \\ (\delta \mathbf{y})^\mathrm{T}\nabla_\mathbf{y}(\frac{\partial J}{\partial y_2}) \\ \vdots \\ (\delta \mathbf{y})^\mathrm{T}\nabla_\mathbf{y}(\frac{\partial J}{\partial y_N}) \end{bmatrix} = \left[\nabla_\mathbf{y}^2 J\right]\delta \mathbf{y}. \tag{25.3.14}$$

Similarly

$$\delta_\mathbf{u}\left[\frac{\partial J}{\partial \mathbf{y}}\right] = \left[\nabla_{\mathbf{yu}}^2 J\right]\delta \mathbf{u}. \tag{25.3.15}$$

Since J does not depend on $\boldsymbol{\alpha}$ explicitly, we have

$$\delta_\alpha\left[\frac{\partial J}{\partial \mathbf{y}}\right] = 0. \tag{25.3.16}$$

Consider the second term on the l.h.s of (25.3.12):

$$\delta \left[(\mathbf{D}_\mathbf{F}^\mathbf{y})^\mathrm{T}\mathbf{p}\right] = \delta\left[(\mathbf{D}_\mathbf{F}^\mathbf{y})^\mathrm{T}\right]\mathbf{p} + (\mathbf{D}_\mathbf{F}^\mathbf{y})^\mathrm{T}\delta \mathbf{p}. \tag{25.3.17}$$

In view of the fact

$$\delta\left[(\mathbf{D}_\mathbf{F}^\mathbf{y})^\mathrm{T}\right] = \left[\delta(\mathbf{D}_\mathbf{F}^\mathbf{y})\right]^\mathrm{T}$$

we first compute

$$\delta\left[\mathbf{D}_\mathbf{F}^\mathbf{y}\right] = \delta_\mathbf{y}\left[\mathbf{D}_\mathbf{F}^\mathbf{y}\right] + \delta_\mathbf{u}\left[\mathbf{D}_\mathbf{F}^\mathbf{y}\right] + \delta_\alpha\left[\mathbf{D}_\mathbf{F}^\mathbf{y}\right]. \tag{25.3.18}$$

Again, from first principles we get

$$\delta_{\mathbf{y}}\left[\mathbf{D}_{\mathbf{F}}^{\mathbf{y}}\right] = \delta_{\mathbf{y}} \begin{bmatrix} (\nabla_{\mathbf{y}} F_1)^{\mathrm{T}} \\ (\nabla_{\mathbf{y}} F_2)^{\mathrm{T}} \\ \vdots \\ (\nabla_{\mathbf{y}} F_N)^{\mathrm{T}} \end{bmatrix} = \begin{bmatrix} (\delta\mathbf{y})^{\mathrm{T}} \nabla_{\mathbf{y}}^2 F_1 \\ (\delta\mathbf{y})^{\mathrm{T}} \nabla_{\mathbf{y}}^2 F_2 \\ \vdots \\ (\delta\mathbf{y})^{\mathrm{T}} \nabla_{\mathbf{y}}^2 F_N \end{bmatrix}$$
$$= \left[\delta\mathbf{y},\, \nabla_{\mathbf{y}}^2 \mathbf{F}\right] \in \mathbb{R}^{N \times N}. \tag{25.3.19}$$

Similarly,

$$\delta_{\mathbf{u}}\left[\mathbf{D}_{\mathbf{F}}^{\mathbf{y}}\right] = \delta_{\mathbf{u}} \begin{bmatrix} (\nabla_{\mathbf{y}} F_1)^{\mathrm{T}} \\ (\nabla_{\mathbf{y}} F_2)^{\mathrm{T}} \\ \vdots \\ (\nabla_{\mathbf{y}} F_N)^{\mathrm{T}} \end{bmatrix} = \begin{bmatrix} (\delta\mathbf{u})^{\mathrm{T}} \nabla_{\mathbf{uy}}^2 F_1 \\ (\delta\mathbf{u})^{\mathrm{T}} \nabla_{\mathbf{uy}}^2 F_2 \\ \vdots \\ (\delta\mathbf{u})^{\mathrm{T}} \nabla_{\mathbf{uy}}^2 F_N \end{bmatrix}$$
$$= \left[\delta\mathbf{u},\, \nabla_{\mathbf{uy}}^2 \mathbf{F}\right] \in \mathbb{R}^{N \times N} \tag{25.3.20}$$

and

$$\delta_{\alpha}\left[\mathbf{D}_{\mathbf{F}}^{\mathbf{y}}\right] = \delta_{\alpha} \begin{bmatrix} (\nabla_{\mathbf{y}} F_1)^{\mathrm{T}} \\ (\nabla_{\mathbf{y}} F_2)^{\mathrm{T}} \\ \vdots \\ (\nabla_{\mathbf{y}} F_N)^{\mathrm{T}} \end{bmatrix} = \begin{bmatrix} (\delta\boldsymbol{\alpha})^{\mathrm{T}} \nabla_{\alpha\mathbf{y}}^2 F_1 \\ (\delta\boldsymbol{\alpha})^{\mathrm{T}} \nabla_{\alpha\mathbf{y}}^2 F_2 \\ \vdots \\ (\delta\boldsymbol{\alpha})^{\mathrm{T}} \nabla_{\alpha\mathbf{y}}^2 F_N \end{bmatrix}$$
$$= \left[\delta\mathbf{u},\, \nabla_{\alpha\mathbf{y}}^2 \mathbf{F}\right] \in \mathbb{R}^{N \times N} \tag{25.3.21}$$

where $\nabla_{\mathbf{uy}}^2 F_i \in \mathbb{R}^{M \times N}$ and $\nabla_{\alpha\mathbf{y}}^2 F_i \in \mathbb{R}^{K \times N}$ for $i = 1, 2, \ldots, N$. Substituting (25.3.14)–(25.3.16) into (25.3.13); (25.3.19)–(25.3.21) into (25.3.17) and in turn substituting the resulting expressions in (25.3.13) and (25.3.17) into (25.3.12) we obtain the following relation in $\mathbb{R}^N$:

$$\left[\nabla_{\mathbf{y}}^2 J\right] \delta\mathbf{y} + \left[\nabla_{\mathbf{yu}}^2 J\right] \delta\mathbf{u} - \left[\delta\mathbf{y},\, \nabla_{\mathbf{y}}^2 \mathbf{F}\right]^{\mathrm{T}} \mathbf{p} - \left[\delta\mathbf{u},\, \nabla_{\mathbf{uy}}^2 \mathbf{F}\right]^{\mathrm{T}} \mathbf{p}$$
$$- \left[\delta\boldsymbol{\alpha},\, \nabla_{\alpha\mathbf{y}}^2 \mathbf{F}\right]^{\mathrm{T}} \mathbf{p} - (\mathbf{D}_{\mathbf{F}}^{\mathbf{y}})^{\mathrm{T}} \delta\mathbf{p} = 0. \tag{25.3.22}$$

Step 3 We now turn to computing the first variation of the third equation in (25.3.7). Since this equation is structurally similar to the second equation considered in Step 2, we only indicate the major steps leaving the verification as an exercise

(Exercise 25.2). Clearly, we obtain the following relation in $\mathbb{R}^M$:

$$
\begin{aligned}
&\delta\left[\frac{\partial J}{\partial \mathbf{u}}\right] - \delta\left[(\mathbf{D}_\mathbf{F}^\mathbf{u})^\mathrm{T}\mathbf{p}\right] \\
&= \delta_\mathbf{y}\left[\frac{\partial J}{\partial \mathbf{u}}\right] + \delta_\mathbf{u}\left[\frac{\partial J}{\partial \mathbf{u}}\right] - \delta_\mathbf{y}[\mathbf{D}_\mathbf{F}^\mathbf{u}]^\mathrm{T}\mathbf{p} - \delta_\mathbf{u}[\mathbf{D}_\mathbf{F}^\mathbf{u}]^\mathrm{T}\mathbf{p} - \delta_\alpha[\mathbf{D}_\mathbf{F}^\mathbf{u}]^\mathrm{T}\mathbf{p} - (\mathbf{D}_\mathbf{F}^\mathbf{u})^\mathrm{T}\delta_\mathbf{p} \\
&= [\nabla^2_{\mathbf{uy}}J]\delta\mathbf{y} + [\nabla^2_\mathbf{u}J]\delta\mathbf{u} - [\delta\mathbf{y}, \nabla^2_{\mathbf{yu}}\mathbf{F}]^\mathrm{T}\mathbf{p} - [\delta\mathbf{u}, \nabla^2_\mathbf{u}\mathbf{F}]^\mathrm{T}\mathbf{p} \\
&\quad -[\delta\boldsymbol{\alpha}, \nabla^2_{\alpha\mathbf{u}}\mathbf{F}]^\mathrm{T}\mathbf{p} - (\mathbf{D}_\mathbf{F}^\mathbf{u})^\mathrm{T}\delta\mathbf{p} \\
&= 0
\end{aligned}
\tag{25.3.23}
$$

where $[\delta\mathbf{y}, \nabla^2_{\mathbf{yu}}\mathbf{F}]$, $[\delta\mathbf{u}, \nabla^2_\mathbf{u}\mathbf{F}]$, and $[\delta\boldsymbol{\alpha}, \nabla^2_{\alpha\mathbf{u}}\mathbf{F}]$ are all matrices in $\mathbb{R}^{N\times M}$.

Step 4: Let $\mathbf{q} \in \mathbb{R}^N$ and $\mathbf{r} \in \mathbb{R}^M$ be the two second-order adjoint variables. Taking the inner product of (25.3.11) and (25.3.22) with $\mathbf{q}$ and that of (25.3.23) with $\mathbf{r}$ and adding all the resulting expressions we get

$$
\begin{aligned}
&\langle\mathbf{q}, (\mathbf{D}_\mathbf{F}^\mathbf{y})\delta\mathbf{y}\rangle + \langle\mathbf{q}, (\mathbf{D}_\mathbf{F}^\mathbf{u})\delta\mathbf{u}\rangle + \langle\mathbf{q}, (\mathbf{D}_\mathbf{F}^\alpha)\delta\boldsymbol{\alpha}\rangle + \langle\mathbf{q}, (\nabla^2_\mathbf{y}J)\delta\mathbf{y}\rangle \\
&\langle\mathbf{q}, [\nabla^2_{\mathbf{yu}}J]\delta\mathbf{u}\rangle - \langle\mathbf{q}, [\delta\mathbf{y}, \nabla^2_\mathbf{y}\mathbf{F}]^\mathrm{T}\mathbf{p}\rangle - \langle\mathbf{q}, [\delta\mathbf{u}, \nabla^2_{\mathbf{uy}}\mathbf{F}]^\mathrm{T}\mathbf{p}\rangle \\
&-\langle\mathbf{q}, [\delta\boldsymbol{\alpha}, \nabla^2_{\alpha\mathbf{y}}\mathbf{F}]^\mathrm{T}\mathbf{p}\rangle - \langle\mathbf{q}, (\mathbf{D}_\mathbf{F}^\mathbf{y})^\mathrm{T}\delta\mathbf{p}\rangle + \langle\mathbf{r}, [\nabla^2_{\mathbf{uy}}J]\delta\mathbf{y}\rangle + \langle\mathbf{r}, [\nabla^2_\mathbf{u}J]\delta\mathbf{u}\rangle \\
&-\langle\mathbf{r}, [\delta\mathbf{y}, \nabla^2_{\mathbf{yu}}\mathbf{F}]^\mathrm{T}\mathbf{p}\rangle - \langle\mathbf{r}, [\delta\mathbf{u}, \nabla^2_\mathbf{u}\mathbf{F}]^\mathrm{T}\mathbf{p}\rangle - \langle\mathbf{r}, [\delta\boldsymbol{\alpha}, \nabla^2_{\alpha\mathbf{u}}\mathbf{F}]^\mathrm{T}\mathbf{p}\rangle \\
&-\langle\mathbf{r}, (\mathbf{D}_\mathbf{F}^\mathbf{u})^\mathrm{T}\delta\mathbf{p}\rangle \\
&= 0.
\end{aligned}
\tag{25.3.24}
$$

By way of simplifying this expression, we invoke the adjoint property to get

$$
\langle\mathbf{q}, [\delta\mathbf{y}, \nabla^2_\mathbf{y}\mathbf{F}]^\mathrm{T}\mathbf{p}\rangle = \langle[\delta\mathbf{y}, \nabla^2_\mathbf{y}\mathbf{F}]\mathbf{q}, \mathbf{p}\rangle. \tag{25.3.25}
$$

But, from the definition we have

$$
\begin{aligned}
[\delta\mathbf{y}, \nabla^2_\mathbf{y}\mathbf{F}]\mathbf{q} &= \begin{bmatrix} (\delta\mathbf{y})^\mathrm{T}[\nabla^2_\mathbf{y}F_1]\mathbf{q} \\ (\delta\mathbf{y})^\mathrm{T}[\nabla^2_\mathbf{y}F_2]\mathbf{q} \\ \vdots \\ (\delta\mathbf{y})^\mathrm{T}[\nabla^2_\mathbf{y}F_N]\mathbf{q} \end{bmatrix} = \begin{bmatrix} \mathbf{q}^\mathrm{T}[\nabla^2_\mathbf{y}F_1]\delta\mathbf{y} \\ \mathbf{q}^\mathrm{T}[\nabla^2_\mathbf{y}F_2]\delta\mathbf{y} \\ \vdots \\ \mathbf{q}^\mathrm{T}[\nabla^2_\mathbf{y}F_N]\delta\mathbf{y} \end{bmatrix} \\
&= [\mathbf{q}, \nabla^2_\mathbf{y}\mathbf{F}]\delta\mathbf{y}.
\end{aligned}
\tag{25.3.26}
$$

Combining this with (25.3.25), in view of the adjoint property we obtain

$$
\langle\mathbf{q}, [\delta\mathbf{y}, \nabla^2_\mathbf{y}\mathbf{F}]^\mathrm{T}\mathbf{p}\rangle = \langle[\mathbf{q}, \nabla^2_\mathbf{y}\mathbf{F}]\delta\mathbf{y}, \mathbf{p}\rangle = \langle\delta\mathbf{y}, [\mathbf{q}, \nabla^2_\mathbf{y}\mathbf{F}]^\mathrm{T}\mathbf{p}\rangle. \tag{25.3.27}
$$

Similarly, we have

$$\left.\begin{aligned}\langle \mathbf{q}, [\delta\mathbf{u}, \nabla^2_{\mathbf{uy}}\mathbf{F}]^{\mathrm{T}}\mathbf{p}\rangle &= \langle \delta\mathbf{u}, [\mathbf{q}, \nabla^2_{\mathbf{uy}}\mathbf{F}]^{\mathrm{T}}\mathbf{p}\rangle \\ \langle \mathbf{q}, [\delta\boldsymbol{\alpha}, \nabla^2_{\boldsymbol{\alpha}\mathbf{y}}\mathbf{F}]^{\mathrm{T}}\mathbf{p}\rangle &= \langle \delta\boldsymbol{\alpha}, [\mathbf{q}, \nabla^2_{\boldsymbol{\alpha}\mathbf{y}}\mathbf{F}]^{\mathrm{T}}\mathbf{p}\rangle \\ \langle \mathbf{r}, [\delta\mathbf{y}, \nabla^2_{\mathbf{yu}}\mathbf{F}]^{\mathrm{T}}\mathbf{p}\rangle &= \langle \delta\mathbf{y}, [\mathbf{r}, \nabla^2_{\mathbf{yu}}\mathbf{F}]^{\mathrm{T}}\mathbf{p}\rangle \\ \langle \mathbf{r}, [\delta\mathbf{u}, \nabla^2_{\mathbf{u}}\mathbf{F}]^{\mathrm{T}}\mathbf{p}\rangle &= \langle \delta\mathbf{u}, [\mathbf{r}, \nabla^2_{\mathbf{u}}\mathbf{F}]^{\mathrm{T}}\mathbf{p}\rangle \\ \langle \mathbf{r}, [\delta\boldsymbol{\alpha}, \nabla^2_{\boldsymbol{\alpha}\mathbf{u}}\mathbf{F}]^{\mathrm{T}}\mathbf{p}\rangle &= \langle \delta\boldsymbol{\alpha}, [\mathbf{r}, \nabla^2_{\boldsymbol{\alpha}\mathbf{u}}\mathbf{F}]^{\mathrm{T}}\mathbf{p}\rangle \end{aligned}\right\} \tag{25.3.28}$$

Substituting (25.3.28) into (25.3.24), using the adjoint property and collecting the like terms, we get

$$\langle \delta\mathbf{y}, \mathbf{A}\rangle + \langle \delta\mathbf{u}, \mathbf{B}\rangle + \langle \delta\boldsymbol{\alpha}, \mathbf{C}\rangle - \langle \delta\mathbf{p}, \mathbf{D}\rangle = 0 \tag{25.3.29}$$

where the vectors $\mathbf{A}, \mathbf{D} \in \mathbb{R}^N$, $\mathbf{B} \in \mathbb{R}^M$ and $\mathbf{C} \in \mathbb{R}^K$ are given by

$$\left.\begin{aligned}\mathbf{D} &= (\mathbf{D}^{\mathbf{y}}_{\mathbf{F}})\mathbf{q} + (\mathbf{D}^{\mathbf{u}}_{\mathbf{F}})\mathbf{r} \\ \mathbf{C} &= (\mathbf{D}^{\boldsymbol{\alpha}}_{\mathbf{F}})^{\mathrm{T}}\mathbf{q} - [\mathbf{q}, \nabla^2_{\boldsymbol{\alpha}\mathbf{y}}\mathbf{F}]^{\mathrm{T}}\mathbf{p} - [\mathbf{r}, \nabla^2_{\boldsymbol{\alpha}\mathbf{u}}\mathbf{F}]^{\mathrm{T}}\mathbf{p} \\ \mathbf{B} &= (\mathbf{D}^{\mathbf{u}}_{\mathbf{F}})^{\mathrm{T}}\mathbf{q} + [\nabla^2_{\mathbf{uy}}J]\mathbf{q} - [\mathbf{q}, \nabla^2_{\mathbf{uy}}\mathbf{F}]^{\mathrm{T}}\mathbf{p} + [\nabla^2_{\mathbf{u}}J]\mathbf{r} - [\mathbf{r}, \nabla^2_{\mathbf{u}}\mathbf{F}]^{\mathrm{T}}\mathbf{p} \\ \mathbf{A} &= (\mathbf{D}^{\mathbf{y}}_{\mathbf{F}})^{\mathrm{T}}\mathbf{q} + [\nabla^2_{\mathbf{y}}J]\mathbf{q} - [\mathbf{q}, \nabla^2_{\mathbf{y}}\mathbf{F}]^{\mathrm{T}}\mathbf{p} + [\nabla^2_{\mathbf{uy}}J]\mathbf{r} - [\mathbf{r}, \nabla^2_{\mathbf{yu}}\mathbf{F}]^{\mathrm{T}}\mathbf{p}\end{aligned}\right\} \tag{25.3.30}$$

Step 5 Setting

$$\mathbf{D} = 0 \tag{25.3.31}$$

$$\mathbf{A} = \frac{\partial G}{\partial \mathbf{y}} \tag{25.3.32}$$

we get two linear equations in two unknown adjoint variables $\mathbf{q}$ and $\mathbf{r}$. Solving these, we obtain the value of $\mathbf{q}$ and $\mathbf{r}$.

Step 6 Substituting (25.3.31)–(25.3.32) into (25.3.29) and combining the resulting expression with (25.3.10), it follows that

$$\delta G = \left\langle \delta\mathbf{y}, \frac{\partial G}{\partial \mathbf{y}}\right\rangle = \langle \delta\mathbf{y}, \mathbf{A}\rangle = -\langle \delta\mathbf{u}, \mathbf{B}\rangle - \langle \delta\boldsymbol{\alpha}, \mathbf{C}\rangle. \tag{25.3.33}$$

But, recall that the optimal $\mathbf{u} = \mathbf{u}(\boldsymbol{\alpha}, \mathbf{z})$ is obtained by solving the extended model equations in (25.3.7) using which we obtain

$$\delta\mathbf{u} = (\mathbf{D}^{\boldsymbol{\alpha}}_{\mathbf{u}})\delta\boldsymbol{\alpha} \tag{25.3.34}$$

where $\mathbf{D}^{\boldsymbol{\alpha}}_{\mathbf{u}} \in \mathbb{R}^{M\times K}$ is the Jacobian of $\mathbf{u}$ w.r.to $\boldsymbol{\alpha}$. Substituting (25.3.34) in (25.3.33), in view of the adjoint property we get

$$\delta G = \langle \delta\boldsymbol{\alpha}, -[\mathbf{C} + (\mathbf{D}^{\boldsymbol{\alpha}}_{\mathbf{u}})^{\mathrm{T}}\mathbf{B}]\rangle. \tag{25.3.35}$$

Step 1 Set up the extended model equations given in (25.3.7)
and solve for $\mathbf{y}$, $\mathbf{p}$ and $\mathbf{u}$ as a function of α and $\mathbf{z}$.

Step 2 Set up and solve (25.3.31)–(25.3.32) for the second-order
adjoint variables $\mathbf{q}$ and $\mathbf{r}$.

Step 3 Compute the sensitivity of G w.r.to α using (25.3.36).

Fig. 25.3.1 Algorithm for second-order adjoint sensitivity.

Hence, the required sensitivity of G w.r. to α is given by

$$\nabla_\alpha G = -[\mathbf{C} + (\mathbf{D}_\mathbf{u}^\alpha)^\mathrm{T}\mathbf{B}]. \tag{25.3.36}$$

We summarize this methodology in the form of an algorithm in Figure 25.3.1. The following comments are in order:

(1) Sensitivity w.r. to observations Since observations are often prone to errors, one might want to assess the sensitivity of a chosen response function $G(\mathbf{y})$ with respect to the perturbation $\delta\mathbf{z}$ in the observation $\mathbf{z}$. To this end, recall that the optimal solutions $\mathbf{y}$, $\mathbf{u}$, and $\mathbf{p}$ of the extended model equation (25.3.7) depend on α and $\mathbf{z}$. It stands to reason to expect that the model parameter α and the observation $\mathbf{z}$ are independent of each other. Hence, there is an inherent **duality** in the dependence of the optimal solution $\mathbf{y}$, $\mathbf{u}$, and $\mathbf{p}$ on α and $\mathbf{z}$. By exploiting this duality we can readily derive expressions for the sensitivity of $G(\mathbf{y})$ w.r. to $\mathbf{z}$. Thus, by keeping α fixed and repeating the Steps 1 through 6 of the above derivation, we can readily obtain expressions for the sensitivity of $G(\mathbf{y})$ w.r. to $\mathbf{z}$ (Exercise 25.3).

(2) Combined sensitivity We can in fact combine the sensitivity of G w.r.to α and $\mathbf{z}$ to obtain

$$\delta G = \langle \delta\alpha, \nabla_\mathbf{u} G\rangle + \langle \delta\mathbf{z}, \nabla_\mathbf{z} G\rangle$$

using the same methodology. We encourage the reader to derive expressions for this combined sensitivity (Exercise 25.4).

We illustrate this methodology using the following:

Example 25.3.1 Consider a scalar model equation ($N = M = K = 1$) with two ($L = 2$) observations where

$$\left.\begin{aligned} F(y, u, \alpha) &= y - \alpha u = 0 \\ J(y, u, \mathbf{z}) &= \tfrac{1}{2}(u - z_0)^2 + \tfrac{1}{2}(y - z_1)^2 \\ G(y) &= \tfrac{1}{2}y^2 \end{aligned}\right\} \tag{25.3.37}$$

We first compute all the required derivatives:

$$\left.\begin{aligned}
&\tfrac{\partial J}{\partial y} = (y - z_1),\ \tfrac{\partial J}{\partial u} = (u - z_0),\ \tfrac{\partial^2 J}{\partial y^2} = 1 = \tfrac{\partial^2 J}{\partial u^2},\ \tfrac{\partial^2 J}{\partial u \partial y} = 0\\
&\tfrac{\partial F}{\partial y} = 1,\ \tfrac{\partial F}{\partial u} = -\alpha,\ \tfrac{\partial F}{\partial \alpha} = -u\\
&\tfrac{\partial^2 F}{\partial y^2} = 0,\ \tfrac{\partial^2 F}{\partial u^2} = 0,\ \tfrac{\partial^2 F}{\partial \alpha^2} = 0\\
&\tfrac{\partial^2 F}{\partial y \partial u} = 0,\ \tfrac{\partial^2 F}{\partial u \partial \alpha} = -1,\ \tfrac{\partial^2 F}{\partial y \partial \alpha} = 0
\end{aligned}\right\} \tag{25.3.38}$$

Using these the extended model equations become

$$\left.\begin{aligned}
F(y, u, \alpha) = y - \alpha u = 0\\
\tfrac{\partial J}{\partial y} - (\tfrac{\partial F}{\partial y})p = (y - z_1) - p = 0\\
\tfrac{\partial J}{\partial u} - (\tfrac{\partial F}{\partial u})p = (u - z_0) + \alpha p = 0
\end{aligned}\right\} \tag{25.3.39}$$

Solving these three equations, the optimal values of u, y, and p as a function of α and $(z_0, z_1)^{\mathrm{T}}$ are given by

$$u = \frac{z_0 + \alpha z_1}{1 + \alpha^2},\ y = \alpha u \quad \text{and} \quad p = y - z_1. \tag{25.3.40}$$

The next step is to set up and solve (25.3.31)–(25.3.32) for q and r:

$$\left.\begin{aligned}
D &= (\tfrac{\partial F}{\partial y})q + (\tfrac{\partial F}{\partial u})r = q - \alpha r = 0\\
A &= (\tfrac{\partial F}{\partial y})q + (\tfrac{\partial^2 J}{\partial y^2})q - (\tfrac{\partial^2 F}{\partial y^2})pq + (\tfrac{\partial^2 J}{\partial u \partial y})r - (\tfrac{\partial^2 F}{\partial y \partial u})pr\\
&= 2q = \tfrac{\partial G}{\partial y} = y
\end{aligned}\right\} \tag{25.3.41}$$

Solving these, we get

$$q = \frac{y}{2} \quad \text{and} \quad r = \frac{y}{2\alpha}. \tag{25.3.42}$$

The values of B and C are given by (using (25.3.42))

$$\begin{aligned}
B &= \left(\frac{\partial F}{\partial u}\right) q + \left(\frac{\partial^2 J}{\partial u \partial y}\right) q - \left(\frac{\partial^2 F}{\partial u \partial y}\right) qp + \left(\frac{\partial^2 J}{\partial u^2}\right) r - \left(\frac{\partial^2 F}{\partial u^2}\right) pr\\
&= -q\alpha + r = \frac{y}{2\alpha}(1 - \alpha^2)
\end{aligned} \tag{25.3.43}$$

and

$$\begin{aligned}
C &= \left(\frac{\partial F}{\partial \alpha}\right) q - \left(\frac{\partial^2 F}{\partial \alpha \partial y}\right) pq - \left(\frac{\partial^2 F}{\partial \alpha \partial u}\right) pr\\
&= -uq + pr = -\frac{y z_1}{2\alpha}.
\end{aligned} \tag{25.3.44}$$

Now, from

$$u = \frac{z_0 + \alpha z_1}{1 + \alpha^2}$$

we get

$$\frac{\partial u}{\partial \alpha} = \frac{z_1(1-\alpha^2) - 2\alpha z_0}{(1+\alpha^2)^2}.$$

The sensitivity of G w.r.to α is then given by (Exercise 25.5)

$$\begin{aligned}\nabla_\alpha G = \frac{\partial G}{\partial \alpha} &= -[C + \frac{\partial u}{\partial \alpha} B] \\ &= \frac{y z_1}{2\alpha} - \frac{y(1-\alpha^2)}{2\alpha}\left[\frac{z_1(1-\alpha^2) - 2\alpha z_0}{(1+\alpha^2)^2}\right] \\ &= \frac{\alpha(z_0 + \alpha z_1)}{(1+\alpha^2)^3}\left[z_0(1-\alpha^2) + 2\alpha z_1\right]. \end{aligned} \tag{25.3.45}$$

Using the optimal values given in (25.3.40), it can be verified by direct computation that

$$G(y) = \frac{1}{2}y^2 = \frac{1}{2}\alpha^2 u^2 = \frac{\alpha^2}{2(1+\alpha^2)^2}[z_0 + \alpha z_1]^2. \tag{25.3.46}$$

By differentiating this expression w.r.to α, it can be verified that the sensitivity of G w.r.to α is again given by (25.3.45) (Exercise 25.6).

Exercises

25.1 Consider the dynamical system in three variables $\mathbf{x} = (x_1, x_2, x_3)^{\mathrm{T}}$

$$\begin{aligned} \frac{\mathrm{d}x_1}{\mathrm{d}t} &= -\left(\frac{1}{k^2} - \frac{1}{k^2 + l^2}\right) k l x_2 x_3 = M_1(\mathbf{x}) \\ \frac{\mathrm{d}x_2}{\mathrm{d}t} &= \left(\frac{1}{l^2} - \frac{1}{k^2 + l^2}\right) k l x_1 x_3 = M_2(\mathbf{x}) \\ \frac{\mathrm{d}x_3}{\mathrm{d}t} &= -\frac{1}{2}\left(\frac{1}{l^2} - \frac{1}{k^2}\right) k l x_2 x_3 = M_3(\mathbf{x}) \end{aligned}$$

which is known as the "maximum simplification" or "minimum" equation of Lorenz (1960) when $2\pi/l$ is the distance between successive zonal maxima and $2\pi/l$ is the wavelength of the disturbances (Refer to Chapter 3 for details). Let

$$\mathbf{z} = \mathbf{h}(\mathbf{x}) + \mathbf{v}$$

be the observation where $\mathbf{h}(\mathbf{x}) = (h_1(\mathbf{x}), h_2(\mathbf{x}), h_3(\mathbf{x}))^{\mathrm{T}}$, $h_i(\mathbf{x}) = M_i(\mathbf{x})$ for $i = 1, 2, 3$, and $\mathbf{v} = (v_1, v_2, v_3)^{\mathrm{T}} \sim \mathbf{v}(0, R)$.

(*a*) Derive the recurrence relations for the first-order and second-order adjoint methods.

(*b*) Compute the gradient and the Hessian vector product numerically by discretizing the above dynamics using the Euler scheme and using the value $k/l = 0.95$, initial condition $\mathbf{x}_0 = (1.0, 1.0, 0.0)^{\mathrm{T}}$ and $\mathbf{R} = \mathrm{Diag}(0.1, 0.1, 0.1)$.

25.2 Verify that the first variation of the third equation in (25.3.7) is given by (25.3.23).

25.3 By keeping α fixed and perturbing $\mathbf{z}$, derive an expression for the sensitivity of $G(\mathbf{y})$ w.r. to $\mathbf{z}$.

Hint: $\mathbf{F}(\mathbf{y}, \mathbf{u}, \alpha) = 0$ does not depend on $\mathbf{z}$ explicitly and $J(\mathbf{y}, \mathbf{u}, \mathbf{z})$ does not depend on α explicitly.

25.4 By simultaneously perturbing α and $\mathbf{z}$, derive an expression for the combined sensitivity of $G(\mathbf{y})$ w.r. to α and $\mathbf{z}$.

25.5 Using (25.3.40) verify the correctness of (25.3.45).

25.6 Compute the derivative of the r.h.s. of (25.3.46) w.r. to α.

25.7 The exact dynamical law is Burgers' equation with diffusion, namely,

$$\frac{\partial u}{\partial t} + u\frac{\partial u}{\partial x} = -\sigma^2 u, \quad \sigma^2 = 0.1$$

where $u = \sin x, 0 \le x \le 2\pi$, at $t = 0$, and where periodicity in x is assumed, i.e.,

$$u(x \pm 2\pi, t) = u(x).$$

In the case where we believe the dynamics to be Burgers' equation without diffusion,

$$\frac{\partial u}{\partial t} + u\frac{\partial u}{\partial x} = 0,$$

with the same initial condition and the assumed periodicity, our forecast will contain a systematic error – the amplitudes of the waves will be systematically too large.

The observations, surrogates of the truth, are produced by adding random noise to the truth, assumed to be the numerical solution to $u_t + uu_x + \sigma^2 u = 0$. When the numerical solution of $u_t + uu_x = 0$ is compared to the observations, it becomes clear that the assumed dynamics are systematically in error. To empirically account for this error, the dynamical law is changed by adding a time-independent function, i.e.,

$$u_t + uu_x = \phi(x).$$

The data assimilation problem then becomes determination of $\phi(x)$ such that the forecast better fits the observations. To be more specific, discretize the space into 16 equal spaced intervals in the x-domain,

$$16\Delta x = 2\pi,$$

and use a leapfrog integration scheme where $\Delta t = \Delta x = 2\pi/16$, i.e.,

$$u_i^{n+1} = u_i^{n-1} - u_i^n(u_{i+1}^n - u_{i-1}^n) + \phi_i .$$

Assume the initial condition is known exactly, then determine ϕ_i that minimizes

$$J = \sum_{i,n} (u_i^n - z_i^n)^2$$

where z_i^n are the observations. Experiment with varying number of observations. Discuss solution $\phi(x)$ in terms of the differences between truth and the assumed dynamics.

Notes and references

Section 25.1 and 25.2 The use of the second-order or Hessian information to accelerate the convergence of optimization methods took a strong hold in the 1960s leading to the development of a whole host of methods summarized in Chapter 12. The value of the Hessian in oceanographic data assimilation has been illuminated by Thacker (1989). To our knowledge Wang et al. (1992) was the first to use the second-order adjoint methods in dynamic data assimilation. Further extensions of their basic approach was then reported in Wang et al. (1995) and Wang et al. (1997). A recent review by LeDimet et al. (2002) provides a comprehensive summary of the principles and applications of this second-order method in the geophysical domain. Refer to Cacuci (2003) for further analysis of sensitivity.

This chapter contains the discrete version of the second-order methods described in these above-mentioned publications. A second-order method for the 3DVAR problem is given in Lakshmivarahan et al. (2003).

Section 25.3 This section follows the developments in LeDimet et al. (2002).

26

The 4DVAR problem: a statistical and a recursive view

In Chapters 22–25 we have discussed the solution to the **off-line**, 4DVAR problem of assimilating a given set of observations in deterministic/dynamic models using the classical least squares (Part II) method. In this framework, the adjoint method facilitates the computation of the gradient of the least squares objective function, which when used in conjunction with the minimization methods described in Part III, leads to the optimal initial conditions for the dynamic model. Even in the ideal case of a perfect dynamic model (error free model), the computed values of the optimal initial condition are noisy in response to erroneous observations. The deterministic approach in Chapter 22–25 are predicated on the assumption that the statistical properties of the noise corrupting the observations are **not** known a priori . The question is: if we are given additional information, say the second-order properties (mean and covariance) of the noisy observations, how can we use this information to derive the second-order properties of the optimal initial conditions? This can only be achieved by reliance on the statistical least squares method described in Chapter 14.

The goal of this chapter is two fold. The first is to apply the statistical least squares method of Chapter 14. More specifically, we derive explicit expressions for the **unbiased, optimal** (least squares) **estimate** of the initial condition and its covariance when the model is linear and the observations are a linear function of the state. Starting from this initial estimate and its covariance we then can predict the evolution of the state that best fits the data as well as its covariance as a function of time.

The second goal is to derive an equivalent **online** or **recursive** method for computing the estimate of the state as new observations arrive on the scene. The difference, however, is that instead of finding the optimal initial state, we seek to compute the optimal estimate $\widehat{\mathbf{x}}_N$ of the state $\mathbf{x}_N$ given that there are N observations, $N = 1, 2, 3, \ldots$ In particular, we seek to compute $\widehat{\mathbf{x}}_{N+1}$ based on $\widehat{\mathbf{x}}_N$ and the new observation $\mathbf{z}_{N+1}$.

This is the counterpart to Chapter 8 (in Part II) where we outlined the off-line solution to the deterministic least squares problem.

In addition, this chapter also provides a natural transition between the deterministic method and its statistical counterpart. More specifically, this chapter classifies the similarities and differences between off-line deterministic 4DVAR (Chapters 22–25) and the online or recursive statistical estimation method such as the Kalman filtering (Part VII).

26.1 A statistical analysis of the 4DVAR problem

Let $\mathbf{M}_k \in \mathbb{R}^{n\times n}$ and $\mathbf{H}_k \in \mathbb{R}^{m\times n}$ for $k = 0, 1, 2, \ldots$ Let $\mathbf{x}_k \in \mathbb{R}^n$ denote the state of a linear dynamical system

$$\mathbf{x}_{k+1} = \mathbf{M}_k \mathbf{x}_k \tag{26.1.1}$$

where the **initial condition** $\mathbf{x}_0 = \mathbf{c}$ is **not known**. We are given a set of observations for $k = 1, 2, \ldots, N$.

$$\mathbf{z}_k = \mathbf{H}_k \mathbf{x}_k + \mathbf{v}_k \tag{26.1.2}$$

with $\mathbf{v}_k \in \mathbb{R}^m$ is the **white noise** sequence with $E(\mathbf{v}_k) = 0$ and $\mathrm{Cov}(\mathbf{v}_k) = \mathbf{R}_k \in \mathbb{R}^{m\times m}$, a **known** symmetric and positive definite matrix.

The 4DVAR problem is: given $\{(\mathbf{z}_k, \mathbf{R}_k) : k = 1, 2, \ldots, N\}$, find the initial condition $\mathbf{x}_0 = \mathbf{c}$ that minimizes

$$\left.\begin{aligned} J(\mathbf{c}) &= \textstyle\sum_{k=1}^{N} \mathbf{J}_k(\mathbf{c}) \\ \mathbf{J}_k(\mathbf{c}) &= \tfrac{1}{2}(\mathbf{H}_k\mathbf{x}_k - \mathbf{z}_k)^{\mathrm{T}}\mathbf{R}_k^{-1}(\mathbf{H}_k\mathbf{x}_k - \mathbf{z}_k) \end{aligned}\right\} \tag{26.1.3}$$

when the states $\mathbf{x}_k$ are constrained by (26.1.1). The minimizing $\mathbf{c}$ is clearly a function of the observations and hence is **random**. Our goal is two fold; namely, find the minimizing $\mathbf{c}$ and its covariance. To this end, first define a chain of matrix products as

$$\mathbf{M}(j:i) = \begin{cases} \mathbf{M}_j\mathbf{M}_{j-1}\cdots\mathbf{M}_{i+1}\mathbf{M}_i, & \text{if } j \ge i \\ \mathbf{I}, & \text{if } j < i \end{cases} \tag{26.1.4}$$

By iterating (26.1.1) it can be verified that

$$\mathbf{x}_k = \mathbf{M}(k-1:0)\mathbf{c}. \tag{26.1.5}$$

Now substituting this into (26.1.3) we express $\mathbf{J}_k(\mathbf{c})$ explicitly as a function of $\mathbf{c}$:

$$\mathbf{J}_k(\mathbf{c}) = \frac{1}{2}(\mathbf{H}_k\mathbf{M}(k-1:0)\mathbf{c} - \mathbf{z}_k)^{\mathrm{T}}\mathbf{R}_k^{-1}(\mathbf{H}_k\mathbf{M}(k-1:0)\mathbf{c} - \mathbf{z}_k). \tag{26.1.6}$$

The gradient and the Hessian of $\mathbf{J}_k(\mathbf{c})$ are given by (Exercise 26.1)

$$\left.\begin{aligned} \nabla\mathbf{J}_k(\mathbf{c}) &= \mathbf{A}_k\mathbf{c} - \mathbf{B}_k\mathbf{z}_k \\ \nabla^2\mathbf{J}_k(\mathbf{c}) &= \mathbf{A}_k \end{aligned}\right\} \tag{26.1.7}$$

where

$$\left.\begin{aligned} \mathbf{A}_k &= \mathbf{M}^{\mathrm{T}}(k-1:0)\mathbf{H}_k^{\mathrm{T}}\mathbf{R}_k^{-1}\mathbf{H}_k\mathbf{M}(k-1:0) \\ \mathbf{B}_k &= \mathbf{M}^{\mathrm{T}}(k-1:0)\mathbf{H}_k^{\mathrm{T}}\mathbf{R}_k^{-1} \end{aligned}\right\} \tag{26.1.8}$$

The minimizing value of $\mathbf{c}$ is obtained as the solution $\widehat{\mathbf{c}}$ of the linear system

$$\begin{aligned} 0 = \nabla J(\mathbf{c}) &= \sum_{k=1}^{N} \nabla \mathbf{J}_k(\mathbf{c}) \\ &= \mathbf{Ac} - \sum_{k=1}^{N} \mathbf{B}_k\mathbf{z}_k \end{aligned} \tag{26.1.9}$$

where

$$\mathbf{A} = \sum_{k=1}^{N} \mathbf{A}_k. \tag{26.1.10}$$

That is,

$$\widehat{\mathbf{c}} = \mathbf{A}^{-1}\left(\sum_{k=1}^{N} \mathbf{B}_k\mathbf{z}_k\right) \tag{26.1.11}$$

and the Hessian of $J(\mathbf{c})$ evaluated at $c = \widehat{\mathbf{c}}$ is given by

$$\nabla^2\mathbf{J}(\widehat{\mathbf{c}}) = \sum_{k=1}^{N} \mathbf{A}_k = \mathbf{A}. \tag{26.1.12}$$

Since $\mathbf{R}_k$ is positive definite, $\mathbf{H}_k$ is of full rank and $\mathbf{M}_k$ is non-singular, it can be verified that $\mathbf{A}$ is symmetric and positive definite. Hence $\widehat{\mathbf{c}}$ is the unique minimizer of $J(\mathbf{c})$. Also notice that this unique minimizer is a **linear function** of the observation. Hence, it is known as the best linear estimate (Chapter 13).

We now establish some of the key statistical properties of $\widehat{\mathbf{c}}$ of interest to us.

$\widehat{\mathbf{c}}$ is an unbiased estimate

Substituting (26.1.2) for $\mathbf{z}_k$ and (26.1.5) for $\mathbf{x}_k$ in (26.1.11), the latter becomes

$$\begin{aligned} \widehat{\mathbf{c}} &= \mathbf{A}^{-1}\sum_{k=1}^{N} \mathbf{B}_k(\mathbf{H}_k\mathbf{x}_k + \mathbf{v}_k) \\ &= \mathbf{A}^{-1}\left(\sum_{k=1}^{N} \mathbf{B}_k\mathbf{H}_k\mathbf{M}(k-1:0)\right)\mathbf{c} + \mathbf{A}^{-1}\sum_{k=1}^{N} \mathbf{B}_k\mathbf{v}_k \\ &= \mathbf{A}^{-1}\left(\sum_{k=1}^{N} \mathbf{A}_k\right)\mathbf{c} + \mathbf{A}^{-1}\sum_{k=1}^{N} \mathbf{B}_k\mathbf{v}_k \quad \text{(using (26.1.8))} \\ &= \mathbf{c} + \mathbf{A}^{-1}\sum_{k=1}^{N} \mathbf{B}_k\mathbf{v}_k. \end{aligned} \tag{26.1.13}$$

Hence

$$
\begin{aligned}
E(\widehat{\mathbf{c}} - \mathbf{c}) &= \mathbf{A}^{-1} E\left(\sum_{k=1}^{N} \mathbf{B}_k \mathbf{v}_k\right) \\
&= \mathbf{A}^{-1} \sum_{k=1}^{N} \mathbf{B}_k E(\mathbf{v}_k) \\
&= 0.
\end{aligned} \tag{26.1.14}
$$

Hence, $\widehat{\mathbf{c}}$ is an **unbiased** least squares estimate (Chapter 13) of $\mathbf{c}$.

Variance of $\widehat{\mathbf{c}}$ It follows from (26.1.13) that

$$
(\widehat{\mathbf{c}} - \mathbf{c}) = \mathbf{A}^{-1} \sum_{k=1}^{N} \mathbf{B}_k \mathbf{v}_k.
$$

Hence, using the fact that $\mathbf{v}_k$ is serially uncorrelated, we have

$$
\begin{aligned}
\mathbf{P}_0 = \mathrm{Cov}(\widehat{\mathbf{c}}) &= E[(\widehat{\mathbf{c}} - \mathbf{c})(\widehat{\mathbf{c}} - \mathbf{c})^{\mathrm{T}}] \\
&= \mathbf{A}^{-1}\left[E\left\{\left(\sum_{k=1}^{N} \mathbf{B}_k \mathbf{v}_k\right)\left(\sum_{j=1}^{N} \mathbf{B}_j \mathbf{v}_j\right)^{\mathrm{T}}\right\}\right]\mathbf{A}^{-1} \\
&= \mathbf{A}^{-1}\left[\sum_{k=1}^{N}\sum_{j=1}^{N} \mathbf{B}_k E(\mathbf{v}_k \mathbf{v}_j^{\mathrm{T}}) \mathbf{B}_j^{\mathrm{T}}\right]\mathbf{A}^{-1} \\
&= \mathbf{A}^{-1}\left[\sum_{k=1}^{N} \mathbf{B}_k \mathbf{R}_k \mathbf{B}_k^{\mathrm{T}}\right]\mathbf{A}^{-1} \\
&= \mathbf{A}^{-1} = \left[\nabla^2 \mathbf{J}(\widehat{\mathbf{c}})\right]^{-1}
\end{aligned} \tag{26.1.15}
$$

since

$$
\begin{aligned}
\sum_{k=1}^{N} \mathbf{B}_k \mathbf{R}_k \mathbf{B}_k^{\mathrm{T}} &= \sum_{k=1}^{N} \mathbf{M}^{\mathrm{T}}(k-1:0)\mathbf{H}_k^{\mathrm{T}} \mathbf{R}_k^{-1} \mathbf{R}_k \mathbf{R}_k^{-1} \mathbf{H}_k \mathbf{M}(k-1:0) \\
&= \sum_{k=1}^{N} \mathbf{A}_k = \mathbf{A}.
\end{aligned}
$$

Combining these, it follows that $\widehat{\mathbf{c}}$ given in (26.1.11) is the **best linear unbiased estimate** (BLUE) whose covariance $\mathbf{P}_0$ is given in (26.1.15) which is the inverse of the Hessian of $J(\mathbf{c})$ at $\widehat{\mathbf{c}}$.

Starting from $\mathbf{x}_0 = \widehat{\mathbf{c}}$ and $\mathbf{P}_0$, and using the dynamics we can now predict the optimal trajectory and its associated covariance.

Prediction of optimal trajectory and its covariance

Clearly, the optimal trajectory is given by

$$
\widehat{\mathbf{x}}_k = \mathbf{M}(k-1:0)\widehat{\mathbf{c}}. \tag{26.1.16}
$$

Let $\mathbf{P}_k$ denote the covariance of $\widehat{\mathbf{x}}_k$. Then

$$
\begin{aligned}
\mathbf{P}_k &= \mathrm{Cov}(\widehat{\mathbf{x}}_k) \\
&= \mathrm{Cov}(\mathbf{M}(k-1:0)\widehat{\mathbf{c}}) \\
&= \mathbf{M}(k-1:0)\mathrm{Cov}(\widehat{\mathbf{c}})\mathbf{M}^{\mathrm{T}}(k-1:0) \\
&= \mathbf{M}(k-1:0)\mathbf{P}_0\mathbf{M}^{\mathrm{T}}(k-1:0).
\end{aligned} \tag{26.1.17}
$$

Now substituting $\mathbf{P}_0 = \mathbf{A}^{-1}$ and using (26.1.10) we obtain

$$
\begin{aligned}
\mathbf{P}_k &= \mathbf{M}(k-1:0)\left[\sum_{i=1}^{N}\mathbf{A}_i\right]^{-1}\mathbf{M}^{\mathrm{T}}(k-1:0) \\
&= \left[\mathbf{M}^{-\mathrm{T}}(k-1:0)\left(\sum_{i=1}^{N}\mathbf{A}_i\right)\mathbf{M}^{-1}(k-1:0)\right]^{-1} \\
&= \left[\sum_{i=1}^{N}\mathbf{M}^{-\mathrm{T}}(k-1:0)\mathbf{A}_i\mathbf{M}^{-1}(k-1:0)\right]^{-1}.
\end{aligned} \tag{26.1.18}
$$

Hence, substituting for $\mathbf{A}_i$ from (26.1.8) and simplifying

$$
\begin{aligned}
\mathbf{P}_k^{-1} &= \sum_{i-1}^{N}\mathbf{M}^{-\mathrm{T}}(k-1:0)\mathbf{M}^{\mathrm{T}}(i-1:0)\mathbf{H}_i^{\mathrm{T}}\mathbf{R}_i^{-1} \\
&\quad\cdot\mathbf{H}_i\mathbf{M}(i-1:0)\mathbf{M}^{-1}(k-1:0) \\
&= \sum_{i=1}^{k-1}\mathbf{M}^{-\mathrm{T}}(k-1:i)\mathbf{H}_i^{\mathrm{T}}\mathbf{R}_i^{-1}\mathbf{H}_i\mathbf{M}^{-1}(k-1:i) \\
&\quad+\mathbf{H}_k^{\mathrm{T}}\mathbf{R}_k^{-1}\mathbf{H}_k + \sum_{i=k+1}^{N}\mathbf{M}^{\mathrm{T}}(i-1:k)\mathbf{H}_i^{\mathrm{T}}\mathbf{R}_i^{-1}\mathbf{H}_i\mathbf{M}(i-1:k)
\end{aligned} \tag{26.1.19}
$$

where we have used the following property (Exercise 26.3)

$$
\left.
\begin{aligned}
&\mathbf{M}(i-1:0)\mathbf{M}^{-1}(k-1:0) \\
&\quad= \mathbf{I} && \text{if}\quad i=k \\
&\quad= \mathbf{M}_i^{-1}\mathbf{M}_{i+1}^{-1}\cdots\mathbf{M}_{k-1}^{-1} = \mathbf{M}^{-1}(k-1:i) && \text{if}\quad k>i \\
&\quad= \mathbf{M}_{i-1}\mathbf{M}_{i-2}\cdots\mathbf{M}_k = \mathbf{M}(i-1:k) && \text{if}\quad k<i
\end{aligned}
\right\} \tag{26.1.20}
$$

Several observations are in order:

(a) According to the classification of the statistical estimation problem (Chapter 27), the 4DVAR problem stated in the beginning of this section is known as the (off-line) **smoothing problem**. Hence, the estimate $\widehat{\mathbf{c}}$ in (26.1.11) in addition to being a BLUE is also known as the **smoothed estimate**.

(b) The expression for $\mathbf{P}_k^{-1}$, the inverse of the covariance matrix of $\widehat{\mathbf{x}}_k$ given in (26.1.19) is the weighted sum of $\mathbf{R}_k^{-1}$, the inverse of the covariance of the observations $\mathbf{z}_k$'s where the weight matrices are directly related to the model dynamics and the observation matrices.

(c) An important property of $\widehat{\mathbf{x}}_k$ given in (26.1.16) is that its covariance $\mathbf{P}_k$ given in (26.1.17) is "less than" the corresponding covariance obtained using the sequential estimation as shown in Section 26.2. This fact should **not** be surprising since this off-line estimate is a function of all the information contained in all of the observations $\mathbf{z}_1, \mathbf{z}_2, \ldots, \mathbf{z}_N$, whereas the sequential estimate $\widehat{\mathbf{x}}_k$ is only based on the first k observations $\mathbf{z}_1, \mathbf{z}_2, \ldots, \mathbf{z}_k$.

26.2 A recursive least squares formulation of 4DVAR

Let $\mathbf{x}_k \in \mathbb{R}^n$ and $\mathbf{M}_k \in \mathbb{R}^{n\times n}$ be a non-singular matrix for $k = 0, 1, 2, \ldots$ Consider a linear, nonautonomous deterministic, dynamical system (with no model noise)

$$\mathbf{x}_{k+1} = \mathbf{M}_k\mathbf{x}_k \tag{26.2.1}$$

where the initial condition $\mathbf{x}_0$ is a random variable with the following known prior information:

$$E(\mathbf{x}_0) = \mathbf{m}_0 \qquad \text{and} \qquad \text{Cov}(\mathbf{x}_0) = \mathbf{P}_0 \tag{26.2.2}$$

with $\mathbf{P}_0$ being a positive definite matrix.

The observations $\mathbf{z}_k \in \mathbb{R}^m$ for $k = 1, 2, 3, \ldots$ are given by

$$\mathbf{z}_k = \mathbf{H}_k\mathbf{x}_k + \mathbf{v}_k \tag{26.2.3}$$

where $\mathbf{H}_k \in \mathbb{R}^{m\times n}$ is of full rank and $\mathbf{v}_k \in \mathbb{R}^m$ is the observation noise vector with the following known properties:

$$E(\mathbf{v}_k) = 0 \qquad \text{and} \qquad \text{Cov}(\mathbf{v}_k) = \mathbf{R}_k \tag{26.2.4}$$

where $\mathbf{R}_k$ is an $m \times m$ positive definite matrix.

Given a set $\{\mathbf{z}_k | k = 1, 2, \ldots, N\}$ of N observations, our goal is to find an estimate $\widehat{\mathbf{x}}_N$ of $\mathbf{x}_N$ for $N = 1, 2, 3, \ldots$ To this end, define an objective function $\mathbf{J}_N$ given $\mathbf{z}_1, \mathbf{z}_2 \ldots, \mathbf{z}_N$ as

$$\mathbf{J}_N = \mathbf{J}_N^{\text{p}} + \mathbf{J}_N^{\text{o}} \tag{26.2.5}$$

where

$$\left.\begin{aligned} \mathbf{J}_N^{\text{p}} &= \tfrac{1}{2}(\mathbf{m}_0 - \mathbf{x}_0)^{\text{T}}\mathbf{P}_0^{-1}(\mathbf{m}_0 - \mathbf{x}_0) \\ \mathbf{J}_N^{\text{o}} &= \tfrac{1}{2}\textstyle\sum_{k=1}^{N}(\mathbf{z}_k - \mathbf{H}_k\mathbf{x}_k)^{\text{T}}\mathbf{R}_k^{-1}(\mathbf{z}_k - \mathbf{H}_k\mathbf{x}_k) \end{aligned}\right\} \tag{26.2.6}$$

Since $\mathbf{z}_1, \mathbf{z}_2, \ldots, \mathbf{z}_N$ are given, clearly $\mathbf{J}_N$ is a function of the states $\mathbf{x}_0, \mathbf{x}_1, \mathbf{x}_2, \ldots, \mathbf{x}_N$.

In this context it is useful to recall that the inverse of the covariance matrix is called the **information** matrix. Thus, if the eigenvalues of the covariance matrix are large, then those of the inverse are small and hence carry less information. Using this interpretation, it can be seen that the term $\mathbf{J}_N^{\mathrm{p}}$ relates to the term containing the prior information and $\mathbf{J}_N^{\mathrm{o}}$ relates to the term containing the collective information from all of the observations.

Our goal is to find an optimal estimate $\widehat{\mathbf{x}}_N$ that minimizes $\mathbf{J}_N$, where the states $\mathbf{x}_0, \mathbf{x}_1, \mathbf{x}_2, \ldots, \mathbf{x}_N$ are **constrained** by the evolution of the given dynamical model in (26.2.1). The first step in achieving this goal is to express each $\mathbf{x}_k$ in terms of $\mathbf{x}_N$ (instead of expressing $\mathbf{x}_k$ in terms of $\mathbf{x}_0$ as was done in Section 26.1) using the model equation. To this end, define $\mathbf{B}_k = \mathbf{M}_k^{-1}$ and using (26.1.4) define

$$\mathbf{B}(j:i) = \begin{cases} \mathbf{M}^{-1}(j:i) = \mathbf{M}_i^{-1}\mathbf{M}_{i+1}^{-1}\cdots\mathbf{M}_{j-1}^{-1}\mathbf{M}_j^{-1} = \\ \mathbf{B}_i\mathbf{B}_{i+1}\cdots\mathbf{B}_j, & \text{for} \quad j \geq i \\ \mathbf{I}, & \text{for} \quad j < i \end{cases} \tag{26.2.7}$$

Hence, using (26.2.7) and (26.2.1) we have

$$\left.\begin{aligned} \mathbf{x}_N &= \mathbf{M}_{N-1}\mathbf{M}_{N-2}\cdots\mathbf{M}_k\mathbf{x}_k = \mathbf{M}(N-1:k)\mathbf{x}_k \\ \mathbf{x}_k &= \mathbf{M}^{-1}(N-1:k)\mathbf{x}_N = \mathbf{B}(N-1:k)\mathbf{x}_N \end{aligned}\right\} \tag{26.2.8}$$

Similarly, let

$$\begin{aligned} \mathbf{m}_{k+1} &= \mathbf{M}_k\mathbf{m}_k \\ &= \mathbf{M}(k:0)\mathbf{m}_0 \end{aligned} \tag{26.2.9}$$

denote the trajectory of the model starting from $\mathbf{m}_0$. Hence

$$\mathbf{m}_0 = \mathbf{B}(N-1:0)\mathbf{m}_N. \tag{26.2.10}$$

Substituting for $\mathbf{x}_k$ using (26.2.8) and $\mathbf{m}_0$ using (26.2.10) into $\mathbf{J}_N$ in (26.2.5), we obtain

$$\begin{aligned} \mathbf{J}_N(\mathbf{x}_N) &= \frac{1}{2}(\mathbf{m}_N - \mathbf{x}_N)^{\mathrm{T}}\left[\mathbf{B}^{\mathrm{T}}(N-1:0)\mathbf{P}_0^{-1}\mathbf{B}(N-1:0)\right](\mathbf{m}_N - \mathbf{x}_N) \\ &+ \frac{1}{2}\sum_{k=1}^{N}(\mathbf{z}_k - \mathbf{H}_k\mathbf{B}(N-1:k)\mathbf{x}_N)^{\mathrm{T}}\mathbf{R}_k^{-1}(\mathbf{z}_k - \mathbf{H}_k\mathbf{B}(N-1:k)\mathbf{x}_N). \end{aligned} \tag{26.2.11}$$

Differentiating $\mathbf{J}_N(\mathbf{x}_N)$ with respect to $\mathbf{x}_N$ twice, we get the **gradient**

$$\begin{aligned} \nabla\mathbf{J}_N(\mathbf{x}_N) &= \mathbf{B}^{\mathrm{T}}(N-1:0)\mathbf{P}_0^{-1}\mathbf{B}(N-1:0)(\mathbf{x}_N - \mathbf{m}_N) \\ &+ \sum_{k=1}^{N}\mathbf{B}^{\mathrm{T}}(N-1:k)\mathbf{H}_k^{\mathrm{T}}\mathbf{R}_k^{-1}[\mathbf{H}_k\mathbf{B}(N-1:k)\mathbf{x}_N - \mathbf{z}_k] \end{aligned} \tag{26.2.12}$$

and the **Hessian**

$$\nabla^2 \mathbf{J}_N(\mathbf{x}_N) = \mathbf{B}^{\mathrm{T}}(N-1:0)\mathbf{P}_0^{-1}\mathbf{B}(N-1:0) + \sum_{k=1}^{N} \mathbf{B}^{\mathrm{T}}(N-1:k)\mathbf{H}_k^{\mathrm{T}}\mathbf{R}_k^{-1}\mathbf{H}_k\mathbf{B}(N-1:k). \qquad (26.2.13)$$

Setting the gradient to zero, we obtain the minimizer $\widehat{\mathbf{x}}_N$ as the solution of the linear system

$$\left[\mathbf{B}^{\mathrm{T}}(N-1:0)\mathbf{P}_0^{-1}\mathbf{B}(N-1:0) + \sum_{k=1}^{N} \mathbf{B}^{\mathrm{T}}(N-1:k)\mathbf{H}_k^{\mathrm{T}}\mathbf{R}_k^{-1}\mathbf{H}_k\mathbf{B}(N-1:k)\right]\widehat{\mathbf{x}}_N = \left[\mathbf{B}^{\mathrm{T}}(N-1:0)\mathbf{P}_0^{-1}\mathbf{B}(N-1:0)\mathbf{m}_N + \sum_{k=1}^{N} \mathbf{B}^{\mathrm{T}}(N-1:k)\mathbf{H}_k^{\mathrm{T}}\mathbf{R}_k^{-1}\mathbf{z}_k\right]. \qquad (26.2.14)$$

To simplify the notation, define

$$\left.\begin{aligned} \mathbf{F}_N^{\mathrm{p}} &= \mathbf{B}^{\mathrm{T}}(N-1:0)\mathbf{P}_0^{-1}\mathbf{B}(N-1:0) \\ \mathbf{F}_N^{o} &= \textstyle\sum_{k=1}^{N} \mathbf{B}^{\mathrm{T}}(N-1:k)\mathbf{H}_k^{\mathrm{T}}\mathbf{R}_k^{-1}\mathbf{H}_k\mathbf{B}(N-1:k) \\ \mathbf{f}_N^{p} &= \mathbf{F}_N^{\mathrm{p}}\mathbf{m}_N \\ \mathbf{f}_N^{\mathrm{o}} &= \textstyle\sum_{k=1}^{N} \mathbf{B}^{\mathrm{T}}(N-1:k)\mathbf{H}_k^{\mathrm{T}}\mathbf{R}_k^{-1}\mathbf{z}_k \end{aligned}\right\} \qquad (26.2.15)$$

Then (26.2.14) becomes

$$(\mathbf{F}_N^{\mathrm{p}} + \mathbf{F}_N^{o})\widehat{\mathbf{x}}_N = (\mathbf{f}_N^{p} + \mathbf{f}_N^{\mathrm{o}}). \qquad (26.2.16)$$

Remark 26.2.1 It is interesting to note that the matrix on the l.h.s of (26.2.14) is indeed the Hessian $\nabla^2\mathbf{J}_N(\mathbf{x}_N)$ which is also known as the **information matrix** with two components, $\mathbf{F}_N^{\mathrm{p}}$ denoting the prior information about $\widehat{\mathbf{x}}_N$ and $\mathbf{F}_N^{o}$ is the information contained in all of the observations about $\widehat{\mathbf{x}}_N$.

By induction, the minimizer $\widehat{\mathbf{x}}_{N+1}$ of $\mathbf{J}_{N+1}(\mathbf{x}_{N+1})$ is given by

$$(\mathbf{F}_{N+1}^{\mathrm{p}} + \mathbf{F}_{N+1}^{\mathrm{o}})\widehat{\mathbf{x}}_{N+1} = (\mathbf{f}_{N+1}^{p} + \mathbf{f}_{N+1}^{o}). \qquad (26.2.17)$$

While $\widehat{\mathbf{x}}_{N+1}$ can be obtained by solving (26.2.17) explicitly, the goal of the recursive framework is to express $\widehat{\mathbf{x}}_{N+1}$ as a function of $\widehat{\mathbf{x}}_N$ and $\mathbf{z}_{N+1}$. This calls for expressing $\mathbf{F}_{N+1}^{\mathrm{p}}$, $\mathbf{F}_{N+1}^{\mathrm{o}}$, $\mathbf{f}_{N+1}^{p}$, and $\mathbf{f}_{N+1}^{o}$ in terms of $\mathbf{F}_N^{\mathrm{p}}$, $\mathbf{F}_N^{o}$, $\mathbf{f}_N^{p}$, and $\mathbf{f}_N^{\mathrm{o}}$. We achieve this end in the following two steps.

STEP 1 Recursive Expression for $(\mathbf{F}_{N+1}^{\mathrm{p}} + \mathbf{F}_{N+1}^{\mathrm{o}})$

From (26.2.15) and (26.2.7), it can be verified that (Exercise 26.4)

$$\left.\begin{aligned} \mathbf{F}^{\mathrm{p}}_{N+1} &= \mathbf{B}^{\mathrm{T}}_N \mathbf{F}^{\mathrm{p}}_N \mathbf{B}_N \\ \text{and} \qquad & \\ \mathbf{F}^{\mathrm{o}}_{N+1} &= \mathbf{B}^{\mathrm{T}}_N \mathbf{F}^{o}_N \mathbf{B}_N + \mathbf{H}^{\mathrm{T}}_{N+1}\mathbf{R}^{-1}_{N+1}\mathbf{H}_{N+1} \end{aligned}\right\} \tag{26.2.18}$$

Hence

$$(\mathbf{F}^{\mathrm{p}}_{N+1} + \mathbf{F}^{\mathrm{o}}_{N+1}) = \mathbf{B}^{\mathrm{T}}_N[\mathbf{F}^{\mathrm{p}}_N + \mathbf{F}^{o}_N]\mathbf{B}_N + \mathbf{H}^{\mathrm{T}}_{N+1}\mathbf{R}^{-1}_{N+1}\mathbf{H}_{N+1}$$

or

$$(\widehat{\mathbf{P}}_{N+1})^{-1} = \mathbf{B}^{\mathrm{T}}_N(\widehat{\mathbf{P}}_N)^{-1}\mathbf{B}_N + \mathbf{H}^{\mathrm{T}}_{N+1}\mathbf{R}^{-1}_{N+1}\mathbf{H}_{N+1} \tag{26.2.19}$$

where we define

$$(\widehat{\mathbf{P}}_N)^{-1} = \mathbf{F}^{\mathrm{p}}_N + \mathbf{F}^{o}_N. \tag{26.2.20}$$

STEP 2 Recursive Expression for $(\mathbf{f}^{p}_{N+1} + \mathbf{f}^{o}_{N+1})$

Again using (26.2.15) and (26.2.7) we get (Exercise 26.5)

$$\mathbf{f}^{o}_{N+1} = \mathbf{B}^{\mathrm{T}}_N\mathbf{f}^{\mathrm{o}}_N + \mathbf{H}^{\mathrm{T}}_{N+1}\mathbf{R}^{-1}_{N+1}\mathbf{z}_{N+1}. \tag{26.2.21}$$

Hence, using (26.2.18)

$$\begin{aligned} \mathbf{f}^{p}_{N+1} + \mathbf{f}^{o}_{N+1} &= \mathbf{F}^{\mathrm{p}}_{N+1}\mathbf{m}_{N+1} + \mathbf{B}^{\mathrm{T}}_N\mathbf{f}^{\mathrm{o}}_N + \mathbf{H}^{\mathrm{T}}_{N+1}\mathbf{R}^{-1}_{N+1}\mathbf{z}_{N+1} \\ &= \mathbf{B}^{\mathrm{T}}_N\mathbf{F}^{\mathrm{p}}_N\mathbf{B}_N\mathbf{m}_{N+1} + \mathbf{B}^{\mathrm{T}}_N\mathbf{f}^{\mathrm{o}}_N + \mathbf{H}^{\mathrm{T}}_{N+1}\mathbf{R}^{-1}_{N+1}\mathbf{z}_{N+1}. \end{aligned} \tag{26.2.22}$$

But, from the definition of $\mathbf{B}_N$,

$$\mathbf{B}_N\mathbf{m}_{N+1} = \mathbf{B}_N\mathbf{M}_N\mathbf{m}_N = \mathbf{m}_N.$$

Hence, using (26.2.15), (26.2.16), and (26.2.20)

$$\begin{aligned} \mathbf{f}^{p}_{N+1} + \mathbf{f}^{o}_{N+1} &= \mathbf{B}^{\mathrm{T}}_N[\mathbf{F}^{\mathrm{p}}_N\mathbf{m}_N + \mathbf{f}^{\mathrm{o}}_N] + \mathbf{H}^{\mathrm{T}}_{N+1}\mathbf{R}^{-1}_{N+1}\mathbf{z}_{N+1} \\ &= \mathbf{B}^{\mathrm{T}}_N[\mathbf{f}^{p}_N + \mathbf{f}^{\mathrm{o}}_N] + \mathbf{H}^{\mathrm{T}}_{N+1}\mathbf{R}^{-1}_{N+1}\mathbf{z}_{N+1} \\ &= \mathbf{B}^{\mathrm{T}}_N[\mathbf{F}^{\mathrm{p}}_N + \mathbf{F}^{o}_N]\widehat{\mathbf{x}}_N + \mathbf{H}^{\mathrm{T}}_{N+1}\mathbf{R}^{-1}_{N+1}\mathbf{z}_{N+1} \\ &= \mathbf{B}^{\mathrm{T}}_N(\widehat{\mathbf{P}}_N)^{-1}\widehat{\mathbf{x}}_N + \mathbf{H}^{\mathrm{T}}_{N+1}\mathbf{R}^{-1}_{N+1}\mathbf{z}_{N+1}. \end{aligned} \tag{26.2.23}$$

Now, define

$$\mathbf{x}^{\mathrm{f}}_{N+1} = \mathbf{M}_N\widehat{\mathbf{x}}_N \qquad \text{or} \qquad \widehat{\mathbf{x}}_N = \mathbf{B}_N\mathbf{x}^{\mathrm{f}}_{N+1}. \tag{26.2.24}$$

Using this, we obtain

$$\begin{aligned} \mathbf{f}^{o}_{N+1} + \mathbf{f}^{p}_{N+1} &= \mathbf{B}^{\mathrm{T}}_N(\widehat{\mathbf{P}}_N)^{-1}\widehat{\mathbf{x}}_N + \mathbf{H}^{\mathrm{T}}_{N+1}\mathbf{R}^{-1}_{N+1}\mathbf{z}_{N+1} \\ &= \mathbf{B}^{\mathrm{T}}_N(\widehat{\mathbf{P}}_N)^{-1}\mathbf{B}_N\mathbf{x}^{\mathrm{f}}_{N+1} + \mathbf{H}^{\mathrm{T}}_{N+1}\mathbf{R}^{-1}_{N+1}\mathbf{z}_{N+1}. \end{aligned} \tag{26.2.25}$$

Now assembling (26.2.19) and (26.2.25) with (26.2.17), the latter becomes

$$[\mathbf{B}_N^{\mathrm{T}}(\widehat{\mathbf{P}}_N)^{-1}\mathbf{B}_N + \mathbf{H}_{N+1}^{\mathrm{T}}\mathbf{R}_{N+1}^{-1}\mathbf{H}_{N+1}]\widehat{\mathbf{x}}_{N+1}$$
$$= [\mathbf{B}_N^{\mathrm{T}}(\widehat{\mathbf{P}}_N)^{-1}\mathbf{B}_N\mathbf{x}_{N+1}^{\mathrm{f}} + \mathbf{H}_{N+1}^{\mathrm{T}}\mathbf{R}_{N+1}^{-1}\mathbf{z}_{N+1}]. \tag{26.2.26}$$

Now, define

$$(\mathbf{P}_{N+1}^{\mathrm{f}})^{-1} = \mathbf{B}_N^{\mathrm{T}}(\widehat{\mathbf{P}}_N)^{-1}\mathbf{B}_N. \tag{26.2.27}$$

Using this we readily obtain

$$\widehat{\mathbf{x}}_{N+1} = [(\mathbf{P}_{N+1}^{\mathrm{f}})^{-1} + \mathbf{H}_{N+1}^{\mathrm{T}}\mathbf{R}_{N+1}^{-1}\mathbf{H}_{N+1}]^{-1}$$
$$\cdot [(\mathbf{P}_{N+1}^{\mathrm{f}})^{-1}\mathbf{x}_{N+1}^{\mathrm{f}} + \mathbf{H}_{N+1}^{\mathrm{T}}\mathbf{R}_{N+1}^{-1}\mathbf{z}_{N+1}]. \tag{26.2.28}$$

The r.h.s. of (26.2.28) is the sum of two terms, the first of which is given by

$$[(\mathbf{P}_{N+1}^{\mathrm{f}})^{-1} + \mathbf{H}_{N+1}^{\mathrm{T}}\mathbf{R}_{N+1}^{-1}\mathbf{H}_{N+1}]^{-1}[(\mathbf{P}_{N+1}^{\mathrm{f}})^{-1}\mathbf{x}_{N+1}^{\mathrm{f}}].$$

Now adding and subtracting $\mathbf{H}_{N+1}^{\mathrm{T}}\mathbf{R}_{N+1}^{-1}\mathbf{H}_{N+1}\mathbf{x}_{N+1}^{\mathrm{f}}$, this term is equal to

$$[\,(\mathbf{P}_{N+1}^{\mathrm{f}})^{-1} + \mathbf{H}_{N+1}^{\mathrm{T}}\mathbf{R}_{N+1}^{-1}\mathbf{H}_{N+1}]^{-1}$$
$$\cdot[(\mathbf{P}_{N+1}^{\mathrm{f}})^{-1} + \mathbf{H}_{N+1}^{\mathrm{T}}\mathbf{R}_{N+1}^{-1}\mathbf{H}_{N+1} - \mathbf{H}_{N+1}^{\mathrm{T}}\mathbf{R}_{N+1}^{-1}\mathbf{H}_{N+1}]\mathbf{x}_{N+1}^{\mathrm{f}}$$
$$= \mathbf{x}_{N+1}^{\mathrm{f}} - \mathbf{K}_{N+1}\mathbf{H}_{N+1}\mathbf{x}_{N+1}^{\mathrm{f}} \tag{26.2.29}$$

where

$$\mathbf{K}_{N+1} = [(\mathbf{P}_{N+1}^{\mathrm{f}})^{-1} + \mathbf{H}_{N+1}^{\mathrm{T}}\mathbf{R}_{N+1}^{-1}\mathbf{H}_{N+1}]^{-1}\mathbf{H}_{N+1}^{\mathrm{T}}\mathbf{R}_{N+1}^{-1} \tag{26.2.30}$$

is called the (**Kalman**) **gain** matrix. Combining this with (26.2.28), we get the desired recursive expression

$$\widehat{\mathbf{x}}_{N+1} = \mathbf{x}_{N+1}^{\mathrm{f}} + \mathbf{K}_{N+1}[\mathbf{z}_{N+1} - \mathbf{H}_{N+1}\mathbf{x}_{N+1}^{\mathrm{f}}]. \tag{26.2.31}$$

A summary of this derivation is given in Figure 26.2.1.

A number of observations are in order.

(a) $\mathbf{P}_{N+1}^{\mathrm{f}}$ has a natural interpretation of being the covariance of $\mathbf{x}_{N+1}^{\mathrm{f}}$. To see this

$$(\mathbf{P}_{N+1}^{\mathrm{f}})^{-1} = \mathbf{B}_N^{\mathrm{T}}(\widehat{\mathbf{P}}_N)^{-1}\mathbf{B}_N$$

or

$$\mathbf{P}_{N+1}^{\mathrm{f}} = \mathbf{M}_N\widehat{\mathbf{P}}_N\mathbf{M}_N^{\mathrm{T}} \tag{26.2.32}$$

which is the standard relation that relates the covariance of $\widehat{\mathbf{x}}_N$ and $\mathbf{x}_{N+1}^{\mathrm{f}} = \mathbf{M}_N\widehat{\mathbf{x}}_N$. Similarly, $\widehat{\mathbf{P}}_{N+1}$ is the covariance of $\widehat{\mathbf{x}}_{N+1}$ (See Exercise 26.2).

(b) The above derivation based on the least squares formulation uses the information matrices $(\mathbf{P}_N^{\mathrm{f}})^{-1}$ and $(\widehat{\mathbf{P}}_N)^{-1}$ instead of the covariance matrices $\mathbf{P}_N^{\mathrm{f}}$ and $\widehat{\mathbf{P}}_N$.

Model $\mathbf{x}_{k+1} = \mathbf{M}_k \mathbf{x}_k; \quad \mathbf{x}_k = \mathbf{B}_k \mathbf{x}_{k+1}, \quad \mathbf{B}_k = \mathbf{M}_k^{-1}$

$\mathbf{x}_0$ is random with $E(\mathbf{x}_0) = \mathbf{m}_0$ and $\mathrm{Cov}(\mathbf{x}_0) = \mathbf{P}_0$

Observation $\mathbf{z}_k = \mathbf{H}_k \mathbf{x}_k + \mathbf{v}_k$

$$E(\mathbf{v}_k) = 0, \quad \mathrm{Cov}(\mathbf{v}_k) = \mathbf{R}_k$$

Recursive relation for the estimate

$$\widehat{\mathbf{x}}_0 = \mathbf{m}_0, \quad (\widehat{\mathbf{P}}_0)^{-1} = \mathbf{P}_0^{-1}$$

$$\mathbf{x}^{\mathrm{f}}_{N+1} = \mathbf{M}_N \widehat{\mathbf{x}}_N$$

$$(\mathbf{P}^{\mathrm{f}}_{N+1})^{-1} = \mathbf{B}_N^{\mathrm{T}} (\widehat{\mathbf{P}}_N)^{-1} \mathbf{B}_N$$

$$\begin{aligned}\mathbf{K}_{N+1} &= [(\mathbf{P}^{\mathrm{f}}_{N+1})^{-1} + \mathbf{H}^{\mathrm{T}}_{N+1} \mathbf{R}^{-1}_{N+1} \mathbf{H}_{N+1}]^{-1} \mathbf{H}^{\mathrm{T}}_{N+1} \mathbf{R}^{-1}_{N+1} \\ &= \widehat{\mathbf{P}}_{N+1} \mathbf{H}^{\mathrm{T}}_{N+1} \mathbf{R}^{-1}_{N+1}\end{aligned}$$

$$\widehat{\mathbf{x}}_{N+1} = \mathbf{x}^{\mathrm{f}}_{N+1} + \mathbf{K}_{N+1}[\mathbf{z}_{N+1} - \mathbf{H}_{N+1} \mathbf{x}^{\mathrm{f}}_{N+1}]$$

$$(\widehat{\mathbf{P}}_{N+1})^{-1} = (\mathbf{P}^{\mathrm{f}}_{N+1})^{-1} + \mathbf{H}^{\mathrm{T}}_{N+1} \mathbf{R}^{-1}_{N+1} \mathbf{H}_{N+1}$$

Fig. 26.2.1 Recursive estimate without model noise: information form.

Hence the recursive form given in Figure 26.2.1 has come to be known as the **information form**. A **dual** of this is called the **covariance form** which is derived in Chapter 27.

(c) We have already encountered this information form in the context of Bayesian estimation in Chapter 17. Refer to Table 17.1.1. The bridge that connects the **information form** in Figure 26.2.1 and the covariance form in Figure 27.2.2 is the classical **matrix inversion lemma** called the **Sherman–Morris–Woodbury** formula in Appendix B, which has been repeatedly used in Chapters 8, 17, and 27.

(d) The above derivation of the recursive least squares is due to P. Swirling in 1959 and is considered as a precursor to the Kalman filtering algorithm. Kalman (1960) and Kalman and Bucy (1961) derived the covariance form of the filter equations (refer to Part VII) using the principle of orthogonal projections which is also intimately related to the notion of least squares (Chapter 6).

(e) **Comparison** We conclude this section with a comparison of the variance of $\mathbf{x}_k$ obtained by the **off-line**, smoothing algorithm in Section 26.1 and the **online** or **sequential** algorithm of this section. Referring to (26.2.19), the variance $\widehat{\mathbf{P}}_k$ is given by the recurrence

$$(\widehat{\mathbf{P}}_k)^{-1} = \mathbf{B}^{\mathrm{T}}_{k-1} (\widehat{\mathbf{P}}_{k-1})^{-1} \mathbf{B}_{k-1} + \mathbf{H}^{\mathrm{T}}_k \mathbf{R}^{-1}_k \mathbf{H}_k. \tag{26.2.33}$$

Iterating this we obtain, using (26.2.7)

$$\begin{aligned}(\widehat{\mathbf{P}}_k)^{-1} &= \mathbf{B}^{\mathrm{T}}(k-1:0)(\widehat{\mathbf{P}}_0)^{-1}\mathbf{B}(k-1:0)\\ &\quad+\sum_{i=1}^{k}\mathbf{B}^{\mathrm{T}}(k-1:i)\mathbf{H}_i^{\mathrm{T}}\mathbf{R}_i^{-1}\mathbf{H}_i\mathbf{B}(k-1:i)\\ &= \mathbf{M}^{-\mathrm{T}}(k-1:0)(\widehat{\mathbf{P}}_0)^{-1}\mathbf{M}^{-1}(k-1:0)\\ &\quad+\sum_{i=1}^{k}\mathbf{M}^{-\mathrm{T}}(k-1:i)\mathbf{H}_i^{\mathrm{T}}\mathbf{R}_i^{-1}\mathbf{H}_i\mathbf{M}^{-1}(k-1:i)\end{aligned}$$

Since no prior information was used in the derivation of the off-line method, for fairness and equity in comparison, we set $(\widehat{\mathbf{P}}_0)^{-1} = 0$ in the above expression which leads to

$$\begin{aligned}(\widehat{\mathbf{P}}_k)^{-1} &= \sum_{i=1}^{k-1}\mathbf{M}^{-\mathrm{T}}(k-1:i)\mathbf{H}_i^{\mathrm{T}}\mathbf{R}_i^{-1}\mathbf{H}_i\mathbf{M}^{-1}(k-1:i)\\ &\quad+\mathbf{H}_k^{\mathrm{T}}\mathbf{R}_k^{-1}\mathbf{H}_k.\end{aligned} \tag{26.2.34}$$

Comparing this with the expression for $(\mathbf{P}_k)^{-1}$ in (26.1.19) we obtain that

$$(\mathbf{P}_k)^{-1} - (\widehat{\mathbf{P}}_k)^{-1} = \sum_{i=k+1}^{N}\mathbf{M}^{\mathrm{T}}(i-1:k)\mathbf{H}_i^{\mathrm{T}}\mathbf{R}_i^{-1}\mathbf{H}_i\mathbf{M}(i-1:k) \tag{26.2.35}$$

where the right-hand side is a positive definite matrix, from which we readily obtain

$$\widehat{\mathbf{P}}_k > \mathbf{P}_k \tag{26.2.36}$$

for all $k = 1, 2, \ldots, N$. That is, the smoothed estimate $\mathbf{x}_k$ in (26.1.16) has a "smaller" variance compared to the "filtered" estimate $\widehat{\mathbf{x}}_k$ derived in this section.

26.3 Observability, information and covariance matrices

Recall from Chapters 1 and 22 that observability relates to the ability to recover the past states from future observations. Using the dynamics (26.2.1) and observations in (26.2.2) and (26.2.8), we get

$$\mathbf{z}_k = \mathbf{H}_k\mathbf{x}_k + \mathbf{v}_k = \mathbf{H}_k\mathbf{M}(k-1:0)\mathbf{x}_0 + \mathbf{v}_k. \tag{26.3.1}$$

Now stacking these expressions for $\mathbf{z}_k$, $k = 1, 2, \ldots, N$ and arranging them in a partitioned matrix-vector form, we obtain

$$\begin{bmatrix}\mathbf{z}_1\\ \mathbf{z}_2\\ \vdots\\ \mathbf{z}_N\end{bmatrix} = \begin{bmatrix}\mathbf{H}_1\mathbf{M}(0:0)\\ \mathbf{H}_2\mathbf{M}(1:0)\\ \vdots\\ \mathbf{H}_N\mathbf{M}(N-1;0)\end{bmatrix}\mathbf{x}_0 + \begin{bmatrix}\mathbf{v}_1\\ \mathbf{v}_2\\ \vdots\\ \mathbf{v}_N\end{bmatrix}$$

or more succinctly as

$$\mathbf{z}(1:N) = \mathcal{H}\mathbf{x}_0 + \mathbf{v}(1:N) \tag{26.3.2}$$

where $\mathbf{z}(1:N) \in \mathbb{R}^{Nm}$, $\mathbf{v}(1:N) \in \mathbb{R}^{Nm}$ and $\mathcal{H} \in \mathbb{R}^{Nm\times n}$. The goal is to determine $\mathbf{x}_0$ given $\mathbf{z}(1:N)$. A little reflection would reveal that this is the **standard (over-determined) statistical least squares** problem (Chapter 13). The solution is obtained by minimizing

$$f(\mathbf{x}_0) = \frac{1}{2}[\mathcal{H}\mathbf{x}_0 - \mathbf{z}(1:N)]^{\mathrm{T}}\mathbf{R}^{-1}[\mathcal{H}\mathbf{x}_0 - \mathbf{z}(1:N)] \tag{26.3.3}$$

where

$$\mathbf{R} = \mathrm{Diag}[\mathbf{R}_1, \mathbf{R}_2, \cdots, \mathbf{R}_N] \in \mathbb{R}^{Nm\times Nm}.$$

The gradient and the Hessian of $f(\mathbf{x}_0)$ are given by

$$\nabla f(\mathbf{x}_0) = (\mathcal{H}^{\mathrm{T}}\mathbf{R}^{-1}\mathcal{H})\mathbf{x}_0 - \mathcal{H}^{\mathrm{T}}\mathbf{R}^{-1}\mathbf{z}(1:N) \tag{26.3.4}$$

and

$$\nabla^2 f(\mathbf{x}_0) = \mathcal{H}^{\mathrm{T}}\mathbf{R}^{-1}\mathcal{H}. \tag{26.3.5}$$

By setting the gradient to zero, it follows that the minimizing value of $\mathbf{x}_0$ is the solution of the linear system

$$(\mathcal{H}^{\mathrm{T}}\mathbf{R}^{-1}\mathcal{H})\mathbf{x}_0 = \mathcal{H}^{\mathrm{T}}\mathbf{R}^{-1}\mathbf{z}(1:N). \tag{26.3.6}$$

Clearly, the solution exists and is unique exactly when

$$\begin{aligned}\mathbf{O}_N &= \mathcal{H}^{\mathrm{T}}\mathbf{R}^{-1}\mathcal{H}\\ &= \sum_{k=1}^{N}\mathbf{M}^{\mathrm{T}}(k-1:0)\mathbf{H}_k^{\mathrm{T}}\mathbf{R}_k^{-1}\mathbf{H}_k\mathbf{M}(k-1:0)\end{aligned} \tag{26.3.7}$$

called the **observability matrix** is non-singular. This happens when $\mathcal{H}$ is of full rank.

Now, setting $(\mathbf{P}_0)^{-1} = 0$ in (26.2.14) we obtain the linear system that defines the sequential estimate $\widehat{\mathbf{x}}_N$ as

$$\begin{aligned}&\left[\sum_{k=1}^{N}\mathbf{B}^{\mathrm{T}}(N-1:k)\mathbf{H}_k^{\mathrm{T}}\mathbf{R}_k^{-1}\mathbf{H}_k\mathbf{B}(N-1:k)\right]\widehat{\mathbf{x}}_N\\ &\quad= \sum_{k=1}^{N}\mathbf{B}^{\mathrm{T}}(N-1:k)\mathbf{H}_k^{\mathrm{T}}\mathbf{R}_k^{-1}\mathbf{z}_k\end{aligned} \tag{26.3.8}$$

where the matrix on the l.h.s. is called the **information matrix** and is denoted by $\mathbf{F}_N^o$ as in (26.2.15). Again by setting $(\mathbf{P}_0)^{-1} = 0$ in (26.2.19), we readily see that

(Exercise 26.6)

$$
\begin{aligned}
(\widehat{\mathbf{P}}_N)^{-1} &= \mathbf{F}_N^o \\
&= \sum_{k=1}^{N} \mathbf{B}^{\mathrm{T}}(N-1:k)\mathbf{H}_k^{\mathrm{T}}\mathbf{R}_k^{-1}\mathbf{H}_k\mathbf{B}(N-1:k) \\
&= \sum_{k=1}^{N} \mathbf{M}^{-\mathrm{T}}(N-1:k)\mathbf{H}_k^{\mathrm{T}}\mathbf{R}_k^{-1}\mathbf{H}_k\mathbf{M}^{-1}(N-1:k) \\
&= \mathbf{M}^{-\mathrm{T}}(N-1:0)\mathbf{O}_N\mathbf{M}^{-1}(N-1:0).
\end{aligned} \tag{26.3.9}
$$

This intimate relation between the **observability matrix** $\mathbf{O}_N$, **information matrix** $\mathbf{F}_N^o$ and the **covariance matrix** $\widehat{\mathbf{P}}_N$ further attests the similarities between the off-line 4DVAR and the sequential estimation methods.

26.4 An extension

In this section for purposes of later comparison we enlarge the scope of the derivation of the recursive equations to include model noise. Let

$$\mathbf{x}_{k+1} = \mathbf{M}_k\mathbf{x}_k + \mathbf{w}_{k+1} \tag{26.4.1}$$

be the model dynamics and

$$\mathbf{z}_k = \mathbf{H}_k\mathbf{x}_k + \mathbf{v}_k \tag{26.4.2}$$

be the observations where

(a) $\mathbf{x}_0$ is random with $E(\mathbf{x}_0) = \mathbf{m}_0$ and $\mathrm{Cov}(\mathbf{x}_0) = \mathbf{P}_0$ and $\mathbf{x}_0$ is uncorrelated with $\mathbf{w}_k$ and $\mathbf{v}_k$
(b) $\mathbf{w}_k$ is a white noise sequence with $E(\mathbf{w}_k) = 0$, $\mathrm{Cov}(\mathbf{w}_k) = \mathbf{Q}_k$ and $\mathbf{w}_k$ is uncorrelated with $\mathbf{v}_k$ and
(c) $\mathbf{v}_k$ is a white noise sequence with $E(\mathbf{v}_k) = 0$ and $\mathrm{Cov}(\mathbf{v}_k) = \mathbf{R}_k$.

The inclusion of $\mathbf{w}_k$ only changes the expression for $\mathbf{P}_{N+1}^{\mathrm{f}}$ in (26.2.32) as

$$\mathbf{P}_{N+1}^{\mathrm{f}} = \mathbf{M}_N\widehat{\mathbf{P}}_N\mathbf{M}_N^{\mathrm{T}} + \mathbf{Q}_N. \tag{26.4.3}$$

Since the expression for $(\widehat{\mathbf{P}}_{N+1})^{-1}$ directly involves $(\mathbf{P}_{N+1}^{\mathrm{f}})^{-1}$, we now examine the explicit form of the inverse of the r.h.s. of (26.4.3). To this end, we invoke the matrix inverse formula (Appendix B)

$$(\mathbf{A} + \mathbf{X}^{\mathrm{T}}\mathbf{Y})^{-1} = \mathbf{A}^{-1} - \mathbf{A}^{-1}\mathbf{X}^{\mathrm{T}}[\mathbf{I} + \mathbf{Y}\mathbf{A}^{-1}\mathbf{X}^{\mathrm{T}}]^{-1}\mathbf{Y}\mathbf{A}^{-1}. \tag{26.4.4}$$

Model $\mathbf{x}_{k+1} = \mathbf{M}_k \mathbf{x}_k + \mathbf{w}_{k+1}, \quad \mathbf{B}_k = \mathbf{M}_k^{-1}$

$\mathbf{x}_0$ is random $\quad E(\mathbf{x}_0) = \mathbf{m}_0, \quad \text{Cov}(\mathbf{x}_0) = \mathbf{P}_0$

$\mathbf{w}_k$ is white noise with $E(\mathbf{w}_k) = 0$ and $\text{Cov}(\mathbf{w}_k) = \mathbf{Q}_k$

Observation

$\mathbf{z}_k = \mathbf{H}_k \mathbf{x}_k + \mathbf{v}_k$

$\mathbf{v}_k$ is white noise with $E(\mathbf{v}_k) = 0$ and $\text{Cov}(\mathbf{v}_k) = \mathbf{R}_k$

Recursive relation for the estimate

$$\widehat{\mathbf{x}}_0 = \mathbf{m}_0, \quad (\widehat{\mathbf{P}}_0)^{-1} = \mathbf{P}_0^{-1}$$

$$\mathbf{x}_{N+1}^{\mathrm{f}} = \mathbf{M}_N \mathbf{x}_N$$

$$(\mathbf{P}_{N+1}^{\mathrm{f}})^{-1} = (\mathbf{I} - \mathbf{G}_N)\mathbf{A}_N^{-1}$$

$$\mathbf{G}_N = \mathbf{A}_N^{-1}[\mathbf{A}_N^{-1} + \mathbf{Q}_N^{-1}]^{-1}$$

$$\mathbf{A}_N^{-1} = (\mathbf{M}_N \widehat{\mathbf{P}}_N \mathbf{M}_N^{\mathrm{T}})^{-1} = \mathbf{B}_N^{\mathrm{T}} (\widehat{\mathbf{P}}_N)^{-1} \mathbf{B}_N$$

$$\widehat{\mathbf{x}}_{N+1} = \mathbf{x}_{N+1}^{\mathrm{f}} + \mathbf{K}_{N+1}[\mathbf{z}_{N+1} - \mathbf{H}_{N+1}\mathbf{x}_{N+1}^{\mathrm{f}}]$$

$$\mathbf{K}_{N+1} = [(\mathbf{P}_{N+1}^{\mathrm{f}})^{-1} + \mathbf{H}_{N+1}^{\mathrm{T}} \mathbf{R}_{N+1}^{-1} \mathbf{H}_{N+1}]^{-1} \mathbf{H}_{N+1}^{\mathrm{T}} \mathbf{R}_{N+1}^{-1}$$

$$(\widehat{\mathbf{P}}_{N+1})^{-1} = (\mathbf{P}_{N+1}^{\mathrm{f}})^{-1} + \mathbf{H}_{N+1}^{\mathrm{T}} \mathbf{R}_{N+1}^{-1} \mathbf{H}_{N+1}$$

Fig. 26.4.1 Recursive estimation: information form.

Now setting $\mathbf{A}_N = \mathbf{M}_N \widehat{\mathbf{P}}_N \mathbf{M}_N^{\mathrm{T}}$, $\mathbf{X}^{\mathrm{T}} = \mathbf{Q}_N$, and $\mathbf{Y} = \mathbf{I}$ we obtain

$$\begin{aligned}(\mathbf{P}_{N+1}^{\mathrm{f}})^{-1} &= (\mathbf{A}_N + \mathbf{Q}_N)^{-1} \\ &= \mathbf{A}_N^{-1} - \mathbf{A}_N^{-1}\mathbf{Q}_N[\mathbf{I} + \mathbf{A}_N^{-1}\mathbf{Q}_N]^{-1}\mathbf{A}_N^{-1} \\ &= \mathbf{A}_N^{-1} - \mathbf{A}_N^{-1}[\mathbf{A}_N^{-1} + \mathbf{Q}_N^{-1}]^{-1}\mathbf{A}_N^{-1}.\end{aligned} \tag{26.4.5}$$

Defining

$$\mathbf{G}_N = \mathbf{A}_N^{-1}[\mathbf{A}_N^{-1} + \mathbf{Q}_N^{-1}]^{-1}$$

we can rewrite (Exercise 26.7)

$$(\mathbf{P}_{N+1}^{\mathrm{f}})^{-1} = (\mathbf{I} - \mathbf{G}_N)\mathbf{A}_N^{-1} \tag{26.4.6}$$

$$= (\mathbf{I} - \mathbf{G}_N)\mathbf{A}_N^{-1}(\mathbf{I} - \mathbf{G}_N)^{\mathrm{T}} + \mathbf{G}_N \mathbf{Q}_N^{-1} \mathbf{G}_N^{\mathrm{T}}. \tag{26.4.7}$$

This latter form expresses $(\mathbf{P}_{N+1}^{\mathrm{f}})^{-1}$ as a quadratic in $\mathbf{G}_N$ and is known as the **Joseph's form** (Refer to Chapter 28).

A summary of this extended version of the recursive estimation in information form is given in Figure 26.4.1. A note of caution is in order here. This form is not applicable when $\mathbf{Q}_N = 0$ in which case $\mathbf{P}^{\mathrm{f}}_N$ is given by (26.2.27).

Exercises

26.1 Compute the gradient and Hessian of $\mathbf{J}_k(\mathbf{c})$ in (26.1.6) and verify (26.1.7).

26.2 If $\mathbf{P} \in \mathbb{R}^{n\times n}$ is the covariance of $\mathbf{x} \in \mathbb{R}^n$, then verify that $\mathbf{APA}^{\mathbf{T}}$ is the covariance of $\underline{A}x$ where $\mathbf{A} \in \mathbb{R}^{n\times n}$.

26.3 Verify the correctness of (26.1.20).

26.4 Verify the correctness of (26.2.18).

26.5 Derive (26.2.21) from the expression for $\mathbf{f}^{o}_{N+1}$.

26.6 Verify the correctness of the derivation in (26.3.9).

26.7 Verify the relation (26.4.6).

Notes and references

Section 26.1 The derivation in this section is a direct extension of the results in Chapter 13 for the static version of the statistical least squares method.

Section 26.2 The sequential approach to the dynamic least squares estimation method described in this section was originally due to Swirling (1959). As will be seen in the next chapter, it is remarkably close to the Kalman filter formulation. The difference lies in the absence of model noise. The inverse covariance form of this sequential algorithm as described in this section has become a standard algorithm in the literature and is best suited to handle those cases when there is no prior information. The covariance version of this sequential algorithm known as the Kalman–Bucy filter method, developed by Kalman (1960) and Kalman and Bucy (1961), is described in Part VII. Refer to the books by Maybeck (1979) and Jazwinski (1970) for detailed treatment of these topics.

Section 26.3 The relation between the observability information and covariance matrices as developed in this section brings out the underlying fundamental relation between the off-line smoothing (4DVAR) methods and the on-line or sequential approach.

Section 26.4 The information form of the algorithm given in Figure 26.4.1 is the dual of the covariance form of the Kalman filter given in Figure 27.2.2.

PART VII

Data assimilation: stochastic dynamic models

27

Linear filtering – part I: Kalman filter

In this opening chapter of Part VII, we first provide a classification of three very basic estimation problems – **filtering**, **smoothing**, and **prediction**; terms that stem from the pioneering work of Wiener and Kolmogorov. We then derive the classic equations for linear filtering problem that are known as the Kalman filter equations.

27.1 Filtering, smoothing and prediction – a classification

Let $\mathbf{x}_k \in \mathbb{R}^n$ denote the **true state** of a dynamic system at time k given by

$$\mathbf{x}_{k+1} = \mathbf{M}(\mathbf{x}_k) + \mathbf{w}_{k+1} \tag{27.1.1}$$

where $\mathbf{M} : \mathbb{R}^n \to \mathbb{R}^n$ and $\mathbf{w}_k \in \mathbb{R}^n$ denotes the **noise vector** associated with the model (i.e., **model error**). This vector $\mathbf{w}_k$ is not directly observable but we assume knowledge of its second-order properties (mean and covariance).

We further assume a sequence $\mathbf{z}_k \in \mathbb{R}^m$ of **observations** given by

$$\mathbf{z}_k = h(\mathbf{x}_k) + \mathbf{v}_k \tag{27.1.2}$$

where $h : \mathbb{R}^n \to \mathbb{R}^m$ and $\mathbf{v}_k$ is the **observation noise** with known second-order properties. Let

$$\mathcal{F}_k = \{\, \mathbf{z}_i \mid 1 \le i \le k \,\} \tag{27.1.3}$$

denote the set of k observations. Clearly, $\mathcal{F}_k$ is a family of sets, steadily increasing as k increases. Let $\widehat{\mathbf{x}}_k$ denote the estimate of $\mathbf{x}_k$ at time k. The nature and character of this estimation problem depends on the time instant at which the estimate is required and the amount of information (in terms of the number of observations available) used in the estimation. The problem of computing (a) $\widehat{\mathbf{x}}_k$ given $\mathcal{F}_k$ is called the **filtering** problem, (b) $\widehat{\mathbf{x}}_k$ given $\mathcal{F}_N$ for some $k < N$ is called the **smoothing** problem, and (c) $\widehat{\mathbf{x}}_{k+s}$ given $\mathcal{F}_k$ for some $s \ge 1$ is called the **prediction** problem. Thus, while filtering and prediction problems use only the past and present information, smoothing

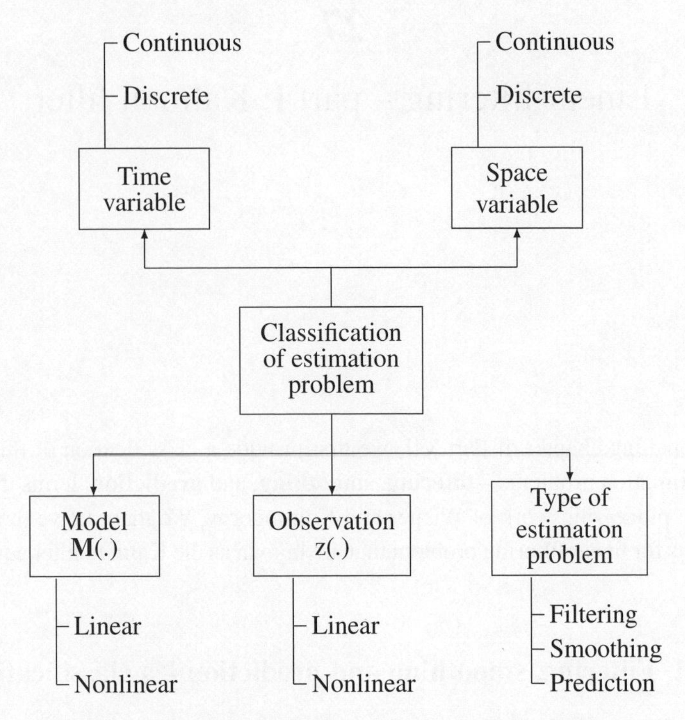

Fig. 27.1.1 A classification of the dynamic estimation problem.

uses all the past, present and the future information. Thus, smoothing is characteristically an **off-line** problem, but filtering and prediction can be recast as **online** problems.

We now combine these three classes of estimation problems with the properties of the mappings **M**(.) and h(.) in (27.1.1) and (27.1.2), respectively, to arrive at a broad spectrum of problems of interest in data assimilation. In this text, we tacitly assume that the time variable is **discrete** and the space variables such as $\mathbf{x}_k$ and $\mathbf{z}_k$ are **continuous**. Refer to Figure 27.1.1 for a general classification. The treatment of the continuous time version of this problem is more challenging and requires a good working knowledge of Ito's version of stochastic calculus. This is beyond our scope, but we refer the interested reader to Bucy and Joseph (1968). In this chapter, we consider the discrete time, continuous space filtering and prediction problems when the model is linear and the observations are linear functions of the state. Nonlinear versions of this problem are considered in Chapter 29.

Remark 27.1.1 From this broad-spectrum viewpoint, the dynamic data assimilation problem based on the variational approach (Part VI) is essentially an off-line smoothing problem. The intimate connection between Kalman filtering and smoothing and the variational solution has been addressed in Chapter 26.

27.2 Kalman filtering: linear dynamics

We begin by describing the basic building blocks.

(A) Dynamic model Consider a linear, non-autonomous dynamical system that evolves according to

$$\mathbf{x}_{k+1} = \mathbf{M}_k \mathbf{x}_k + \mathbf{w}_{k+1} \tag{27.2.1}$$

where $\mathbf{M}_k \in \mathbb{R}^{n \times n}$ is the (non-singular) system matrix that varies with time k, and $\mathbf{w}_k \in \mathbb{R}^n$ denotes the model error. It is assumed that $\mathbf{x}_0$ and $\mathbf{w}_k$ satisfy the following conditions:

(A1) $\mathbf{x}_0$ is random with **known mean** vector $E(\mathbf{x}_0) = \mathbf{m}_0$ and **known covariance** matrix $E[(\mathbf{x}_0 - \mathbf{m}_0)(\mathbf{x}_0 - \mathbf{m}_0)^{\mathrm{T}}] = P_0$,

(A2) The model error is **unbiased**, that is $E(\mathbf{w}_k) = 0$ for all k and is **temporally uncorrelated** (white noise), that is,

$$E[\mathbf{w}_k \mathbf{w}_j^{\mathrm{T}}] = \begin{cases} \mathbf{Q}_k & \text{if } j = k \\ 0 & \text{otherwise} \end{cases}$$

where $\mathbf{Q}_k \in \mathbb{R}^{n \times n}$ is symmetric and positive definite for all k, and

(A3) The model error $\mathbf{w}_k$ and the initial state $\mathbf{x}_0$ are **uncorrelated**: $E(\mathbf{w}_k \mathbf{x}_0^{\mathrm{T}}) = 0$ for all k.

(B) Observations Let $\mathbf{z}_k \in \mathbb{R}^m$ denote the observation at time k where $\mathbf{z}_k$ is related to $\mathbf{x}_k$ via

$$\mathbf{z}_k = \mathbf{H}_k \mathbf{x}_k + \mathbf{v}_k \tag{27.2.2}$$

where $\mathbf{H}_k \in \mathbb{R}^{m \times n}$ represents the time varying measurement system and $\mathbf{v}_k \in \mathbb{R}^m$ denotes the measurement noise with the following properties:

(B1) $\mathbf{v}_k$ has **mean zero** $E(\mathbf{v}_k) = 0$.

(B2) $\mathbf{v}_k$ is temporally uncorrelated:

$$E[\mathbf{v}_k \mathbf{v}_j^{\mathrm{T}}] = \begin{cases} \mathbf{R}_k & \text{if } j = k \\ 0 & \text{otherwise} \end{cases}$$

where $\mathbf{R}_k \in \mathbb{R}^{m \times m}$ is a symmetric and positive definite matrix, and

(B3) $\mathbf{v}_k$ is uncorrelated with the initial state $\mathbf{x}_0$ and the model error $\mathbf{w}_k$, that is,

$$E[\mathbf{x}_0 \mathbf{v}_k^{\mathrm{T}}] = 0 \quad \text{for all } k > 0$$

$$E[\mathbf{v}_k \mathbf{w}_j^{\mathrm{T}}] = 0 \quad \text{for all } k \text{ and } j$$

(C) Statement of the filtering problem Given that $\mathbf{x}_k$ evolves according to (27.2.1) and the set of observations

$$\mathcal{F}_k = \{\mathbf{z}_j \mid 1 \le j \le k\},$$

our goal is to find an estimate $\widehat{\mathbf{x}}_k$ of $\mathbf{x}_k$ that minimizes the mean squared error

$$E[(\mathbf{x}_k - \widehat{\mathbf{x}}_k)^{\mathrm{T}}(\mathbf{x}_k - \widehat{\mathbf{x}}_k)] = \operatorname{tr}\{E[(\mathbf{x}_k - \widehat{\mathbf{x}}_k)(\mathbf{x}_k - \widehat{\mathbf{x}}_k)^{\mathrm{T}}]\}. \tag{27.2.3}$$

If this estimate $\widehat{\mathbf{x}}_k$ is also **unbiased**, then the estimate we are seeking will be a minimum variance estimate.

Remark 27.2.1 The system matrix $\mathbf{M}_k$ in (27.2.1) is often obtained by discretization of a continuous time model which is normally specified by a system of ordinary or partial differential equations. In this context, the actual value of the unit of time interval (Δt) between k and $k+1$ in equation (27.2.1) is often decided by the **consistency** and **stability** of the discretization scheme. This interval Δt is often very small compared to the time interval at which successive sets of meteorological observations are available. As a typical example, while the observations may be available, say every three hours, the dynamic model is integrated in steps of 10 minutes. In this case, the model will undergo 18 steps of evolution between successive observation times.

Despite the mismatch in model time step and time interval between observation we will make the simplifying assumption that observations $\mathbf{z}_k$ are available at every tick of the model clock. An extension of methodology to cover the more general real-world situation will follow after we gain an understanding of the derivation in the special case.

Derivation of the Kalman filter

The derivation of this filter equation consists of two main steps: (*a*) the **forecast step** using the model and (*b*) the **data assimilation step**. In the first step, starting from an **optimal estimate** $\widehat{\mathbf{x}}_{k-1}$ at time $k-1$, use the model (27.2.1) to produce a forecast $\mathbf{x}_k^{\mathrm{f}}$ at time k. In the second step, we combine this **forecast** $\mathbf{x}_k^{\mathrm{f}}$ with the **observation** $\mathbf{z}_k$ to produce the **optimal estimate** $\widehat{\mathbf{x}}_k$. Thus, once $\widehat{\mathbf{x}}_0$, the initial optimal estimate is available, this process can be repeated as k increases.

(A) Model forecast step Recall that the initial state $\mathbf{x}_0$ is a random vector with known mean $E(\mathbf{x}_0)$ and known covariance matrix $\mathbf{P}_0$. Since there is no other information available at $k=0$, the unbiased estimate (Chapters 13–14) for $\mathbf{x}_0$ is its mean $E(\mathbf{x}_0)$. Accordingly, the initial value of the optimal estimate of the state vector is

$$\widehat{\mathbf{x}}_0 = E(\mathbf{x}_0) = \mathbf{m}_0. \tag{27.2.4}$$

Let $\widehat{\mathbf{e}}_0 = \mathbf{x}_0 - \widehat{\mathbf{x}}_0$ be the error in this estimate. Then the covariance of this error is given by

$$\widehat{\mathbf{P}}_0 = E[(\mathbf{x}_0 - \widehat{\mathbf{x}}_0)(\mathbf{x}_0 - \widehat{\mathbf{x}}_0)^{\mathrm{T}}] = \mathbf{P}_0. \tag{27.2.5}$$

Given $\widehat{\mathbf{x}}_0$, using the model (27.2.1) the predictable part of $\mathbf{x}_1$ is given by

$$\mathbf{x}_1^{\mathrm{f}} = E[\mathbf{x}_1 \,|\, \widehat{\mathbf{x}}_0] = E[\mathbf{M}_0\mathbf{x}_0 + \mathbf{w}_1 \,|\, \widehat{\mathbf{x}}_0] = \mathbf{M}_0\widehat{\mathbf{x}}_0 \tag{27.2.6}$$

since the model error $\mathbf{w}_1$ is not correlated with $\widehat{\mathbf{x}}_0$ and its mean is zero by assumption. This in turn implies that $\mathbf{x}_1^{\mathrm{f}}$ is **unbiased** and the error in this forecast is given by

$$\begin{aligned}\mathbf{e}_1^{\mathrm{f}} &= \mathbf{x}_1 - \mathbf{x}_1^{\mathrm{f}} \\ &= \mathbf{M}_0(\mathbf{x}_0 - \widehat{\mathbf{x}}_0) + \mathbf{w}_1 \\ &= \mathbf{M}_0\widehat{\mathbf{e}}_0 + \mathbf{w}_1. \end{aligned} \tag{27.2.7}$$

Hence, the covariance $\mathbf{P}_1^{\mathrm{f}}$ of the model forecast $\mathbf{x}_1^{\mathrm{f}}$ is

$$\begin{aligned}\mathbf{P}_1^{\mathrm{f}} &= E[\mathbf{e}_1^{\mathrm{f}}(\mathbf{e}_1^{\mathrm{f}})^{\mathrm{T}}] \\ &= E[(\mathbf{M}_0\widehat{\mathbf{e}}_0 + \mathbf{w}_1)(\mathbf{M}_0\widehat{\mathbf{e}}_0 + \mathbf{w}_1)^{\mathrm{T}}] \\ &= \mathbf{M}_0\widehat{\mathbf{P}}_0\mathbf{M}_0^{\mathrm{T}} + \mathbf{Q}_1 \end{aligned} \tag{27.2.8}$$

since $\mathbf{w}_1$ and $\mathbf{x}_0$ are uncorrelated.

Now, given $\mathbf{x}_1^{\mathrm{f}}$, we readily see that $\mathbf{H}_1\mathbf{x}_1^{\mathrm{f}}$ is the model counterpart to the observation $\mathbf{z}_1$ at time $k = 1$. Thus,

$$\begin{aligned}E[\mathbf{z}_1 \,|\, \mathbf{x}_1 = \mathbf{x}_1^{\mathrm{f}}] &= E[\mathbf{H}_1\mathbf{x}_1 + \mathbf{v}_1 \,|\, \mathbf{x}_1 = \mathbf{x}_1^{\mathrm{f}}] \\ &= \mathbf{H}_1\mathbf{x}_1^{\mathrm{f}} \end{aligned} \tag{27.2.9}$$

since $\mathbf{x}_1^{\mathrm{f}}$ and $\mathbf{v}_1$ are uncorrelated. Hence,

$$\begin{aligned}\mathrm{Cov}(\mathbf{z}_1 \,|\, \mathbf{x}_1^{\mathrm{f}}) &= \mathrm{Cov}(\mathbf{z}_1 - \mathbf{H}_1\mathbf{x}_1^{\mathrm{f}}) \\ &= \mathrm{Cov}(\mathbf{v}_1) \\ &= \mathbf{R}_1. \end{aligned} \tag{27.2.10}$$

Thus, at time $k = 1$, we have two pieces of information: (*a*) the forecast $\mathbf{x}_1^{\mathrm{f}}$ with covariance $\mathbf{P}_1^{\mathrm{f}}$ and (*b*) the observation $\mathbf{z}_1$ with its covariance $\mathbf{R}_1$. Our goal is to combine these two pieces of information to create an optimal estimate $\widehat{\mathbf{x}}_1$ with $\widehat{\mathbf{P}}_1$ as its covariance using the classical Bayesian framework described in Chapter 17.

Inductively, assume that we now have an optimal estimate $\widehat{\mathbf{x}}_{k-1}$ at time $k-1$, with $\widehat{\mathbf{P}}_{k-1}$ as its covariance. Refer to Figure 27.2.1. First compute the predictable part of $\mathbf{x}_k$ as

$$\mathbf{x}_k^{\mathrm{f}} = \mathbf{M}_{k-1}\widehat{\mathbf{x}}_{k-1}. \tag{27.2.11}$$

The error in this forecast is given by

$$\begin{aligned}\mathbf{e}_k^{\mathrm{f}} &= \mathbf{x}_k - \mathbf{x}_k^{\mathrm{f}} \\ &= \mathbf{M}_{k-1}(\mathbf{x}_{k-1} - \widehat{\mathbf{x}}_{k-1}) + \mathbf{w}_k \\ &= \mathbf{M}_{k-1}\widehat{\mathbf{e}}_{k-1} + \mathbf{w}_k. \end{aligned} \tag{27.2.12}$$

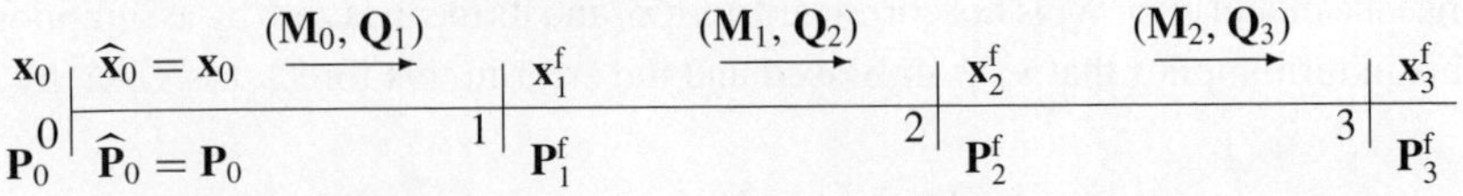

(a) Only the model is given and no observation

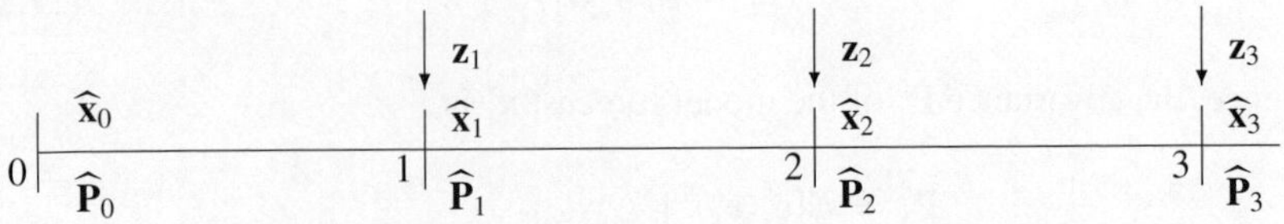

(b) No model, and only observations are given

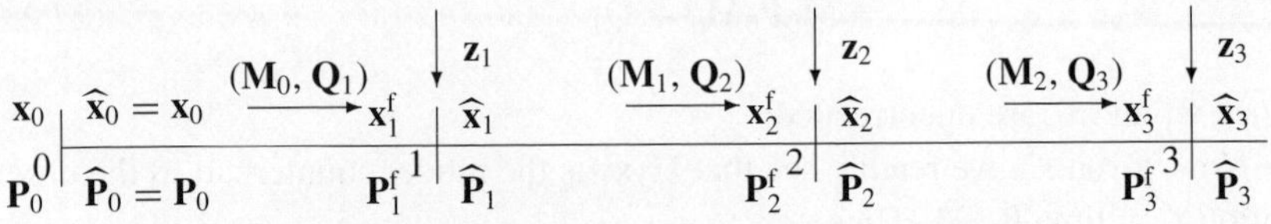

(c) Both the model and the observation are given

Fig. 27.2.1 A diagrammatic view of the role of observations and model in Kalman filtering.

Hence its covariance is given by

$$\begin{aligned}\mathbf{P}_k^{\mathrm{f}} &= E[\mathbf{e}_k^{\mathrm{f}}(\mathbf{e}_k^{\mathrm{f}})^{\mathrm{T}}] \\ &= E[(\mathbf{M}_{k-1}\widehat{\mathbf{e}}_{k-1} + \mathbf{w}_k)(\mathbf{M}_{k-1}\widehat{\mathbf{e}}_{k-1} + \mathbf{w}_k)^{\mathrm{T}}] \\ &= \mathbf{M}_{k-1}\widehat{\mathbf{P}}_{k-1}\mathbf{M}_{k-1}^{\mathrm{T}} + \mathbf{Q}_k. \end{aligned} \tag{27.2.13}$$

Given $\mathbf{x}_k^{\mathrm{f}}$, using the properties of $\mathbf{v}_k$, it follows that model counterpart of $\mathbf{z}_k$ is given by

$$\begin{aligned}E[\mathbf{z}_k \mid \mathbf{x}_k = \mathbf{x}_k^{\mathrm{f}}] &= E[\mathbf{H}_k\mathbf{x}_k + \mathbf{v}_k \mid \mathbf{x}_k^{\mathrm{f}}] \\ &= \mathbf{H}_k\mathbf{x}_k^{\mathrm{f}}.\end{aligned}$$

Hence

$$\mathrm{Cov}(\mathbf{z}_k \mid \mathbf{x}_k^{\mathrm{f}}) = \mathrm{Cov}(\mathbf{v}_k) = \mathbf{R}_k. \tag{27.2.14}$$

Thus, at time k, we have (*a*) the forecast $\mathbf{x}_k^{\mathrm{f}}$ with its covariance $\mathbf{P}_k^{\mathrm{f}}$ and (*b*) the observation $\mathbf{z}_k$ with $\mathbf{R}_k$ as its covariance. Our goal is to compute an optimal estimate $\widehat{\mathbf{x}}_k$ with $\widehat{\mathbf{P}}_k$ as its covariance by combining these two pieces of information to which we now turn our attention.

(B) Data Assimilation Step

Following the developments in Section 17.2 we now define an **unbiased** estimate $\widehat{\mathbf{x}}_k$ which is a linear function of $\mathbf{x}_k^{\mathrm{f}}$ and $\mathbf{z}_k$ as

$$\widehat{\mathbf{x}}_k = \mathbf{x}_k^{\mathrm{f}} + \mathbf{K}_k[\mathbf{z}_k - \mathbf{H}_k\mathbf{x}_k^{\mathrm{f}}] \tag{27.2.15}$$

where recall that $(\mathbf{z}_k - \mathbf{H}_k \mathbf{x}_k^f)$ is called the **innovation** which is obtained by purging the model counterpart of the observation $\mathbf{H}_k \mathbf{x}_k^f$ from $\mathbf{z}_k$. Note that this innovation is the counterpart to the analysis increment in the optimum interpolation method of data assimilation. The weighting matrix $\mathbf{K}_k \in \mathbb{R}^{n \times m}$ is called the **Kalman gain** matrix. The problem is to determine the matrix $\mathbf{K}_k$ such that it minimizes the variance in $\widehat{\mathbf{x}}_k$.

To this end, let us rewrite $\widehat{\mathbf{x}}_k$ in (27.2.15) in terms of quantities at time $(k-1)$ using the following relations:

$$\mathbf{x}_k^f = \mathbf{M}_{k-1}\widehat{\mathbf{x}}_{k-1}; \quad \mathbf{z}_k = \mathbf{H}_k \mathbf{x}_k + \mathbf{v}_k \quad \text{and} \quad \mathbf{x}_k = \mathbf{M}_{k-1}\mathbf{x}_{k-1} + \mathbf{w}_k.$$

Substituting these into (27.2.15) and simplifying, we obtain

$$\begin{aligned} \widehat{\mathbf{x}}_k &= \mathbf{x}_k^f + \mathbf{K}_k[\mathbf{H}_k(\mathbf{x}_k - \mathbf{x}_k^f) + \mathbf{v}_k] \\ &= \mathbf{M}_{k-1}\widehat{\mathbf{x}}_{k-1} + \mathbf{K}_k[\mathbf{H}_k\mathbf{M}_{k-1}(\mathbf{x}_{k-1} - \widehat{\mathbf{x}}_{k-1}) + \mathbf{H}_k\mathbf{w}_k + \mathbf{v}_k]. \end{aligned} \tag{27.2.16}$$

Then, the error $\widehat{\mathbf{e}}_k$ in $\widehat{\mathbf{x}}_k$ is given by

$$\begin{aligned} \widehat{\mathbf{e}}_k &= \mathbf{x}_k - \widehat{\mathbf{x}}_k \\ &= \mathbf{M}_{k-1}\widehat{\mathbf{e}}_{k-1} - \mathbf{K}_k\mathbf{H}_k\mathbf{M}_{k-1}\widehat{\mathbf{e}}_{k-1} + (\mathbf{I} - \mathbf{K}_k\mathbf{H}_k)\mathbf{w}_k - \mathbf{K}_k\mathbf{v}_k \\ &= (\mathbf{I} - \mathbf{K}_k\mathbf{H}_k)[\mathbf{M}_{k-1}\widehat{\mathbf{e}}_{k-1} + \mathbf{w}_k] - \mathbf{K}_k\mathbf{v}_k. \end{aligned} \tag{27.2.17}$$

The covariance of $\widehat{\mathbf{x}}_k$ is given by

$$\begin{aligned} \widehat{\mathbf{P}}_k &= E[\widehat{\mathbf{e}}_k(\widehat{\mathbf{e}}_k)^T] \\ &= (\mathbf{I} - \mathbf{K}_k\mathbf{H}_k)E[(\mathbf{M}_{k-1}\widehat{\mathbf{e}}_{k-1} + \mathbf{w}_k)(\mathbf{M}_{k-1}\widehat{\mathbf{e}}_{k-1} + \mathbf{w}_k)^T](\mathbf{I} - \mathbf{K}_k\mathbf{H}_k)^T \\ &\quad + \mathbf{K}_k E(\mathbf{v}_k\mathbf{v}_k^T)\mathbf{K}_k^T \end{aligned} \tag{27.2.18}$$

where the cross terms vanish since $\mathbf{v}_k$ is not correlated with $\mathbf{w}_j$ and $\mathbf{x}_j$.

Simplifying and using (27.2.13), we get

$$\begin{aligned} \widehat{\mathbf{P}}_k &= (\mathbf{I} - \mathbf{K}_k\mathbf{H}_k)[\mathbf{M}_{k-1}\widehat{\mathbf{P}}_{k-1}\mathbf{M}_{k-1}^T + \mathbf{Q}_k](\mathbf{I} - \mathbf{K}_k\mathbf{H}_k)^T + \mathbf{K}_k\mathbf{R}_k\mathbf{K}_k^T \\ &= (\mathbf{I} - \mathbf{K}_k\mathbf{H}_k)\mathbf{P}_k^f(\mathbf{I} - \mathbf{K}_k\mathbf{H}_k)^T + \mathbf{K}_k\mathbf{R}_k\mathbf{K}_k^T \\ &= \mathbf{P}_k^f - \mathbf{K}_k\mathbf{H}_k\mathbf{P}_k^f - \mathbf{P}_k^f\mathbf{H}_k^T\mathbf{K}_k^T + \mathbf{K}_k\mathbf{D}_k\mathbf{K}_k^T \end{aligned} \tag{27.2.19}$$

where

$$\mathbf{D}_k = [\mathbf{H}_k\mathbf{P}_k^f\mathbf{H}_k^T + \mathbf{R}_k]. \tag{27.2.20}$$

We now restate our problem: find the matrix $\mathbf{K}_k \in \mathbb{R}^{n \times m}$ that minimizes the $\text{tr}(\widehat{\mathbf{P}}_k)$.

Notice that the expression for $\widehat{\mathbf{P}}_k$ in (27.2.19) is structurally similar to (17.1.9) in Section 17.1 wherein we have solved this problem in **four** different ways. For completeness, we indicate only the key steps in solving this problem using the algebraic method of completing the perfect square. Adding and subtracting

$$\mathbf{P}_k^{\mathrm{f}}(\mathbf{H}_k^{\mathrm{T}}\mathbf{D}_k^{-1}\mathbf{H}_k)\mathbf{P}_k^{\mathrm{f}}$$

to the r.h.s of (27.2.19) and simplifying, we obtain

$$\widehat{\mathbf{P}}_k = \mathbf{P}_k^{\mathrm{f}} - \mathbf{P}_k^{\mathrm{f}}[\mathbf{H}_k^{\mathrm{T}}\mathbf{D}_k^{-1}\mathbf{H}_k]\mathbf{P}_k^{\mathrm{f}} + [\mathbf{K}_k - \mathbf{P}_k^{\mathrm{f}}\mathbf{H}_k^{\mathrm{T}}\mathbf{D}_k^{-1}]\mathbf{D}_k[\mathbf{K}_k - \mathbf{P}_k^{\mathrm{f}}\mathbf{H}_k^{\mathrm{T}}\mathbf{D}_k^{-1}]^{\mathrm{T}}. \tag{27.2.21}$$

The trace of the sum of matrices is the sum of their traces, and only the third term on the r.h.s of (27.2.21) depends on $\mathbf{K}_k$. Hence $\mathrm{tr}(\widehat{\mathbf{P}}_k)$ is minimum when

$$\begin{aligned}\mathbf{K}_k &= \mathbf{P}_k^{\mathrm{f}}\mathbf{H}_k^{\mathrm{T}}\mathbf{D}_k^{-1} \\ &= \mathbf{P}_k^{\mathrm{f}}\mathbf{H}_k^{\mathrm{T}}[\mathbf{H}_k\mathbf{P}_k^{\mathrm{f}}\mathbf{H}_k^{\mathrm{T}} + \mathbf{R}_k]^{-1}.\end{aligned} \tag{27.2.22}$$

Substituting this back into (27.2.21), we get

$$\begin{aligned}\widehat{\mathbf{P}}_k &= \mathbf{P}_k^{\mathrm{f}} - \mathbf{P}_k^{\mathrm{f}}\mathbf{H}_k^{\mathrm{T}}\mathbf{D}_k^{-1}\mathbf{H}_k\mathbf{P}_k^{\mathrm{f}} \\ &= \mathbf{P}_k^{\mathrm{f}} - \mathbf{P}_k^{\mathrm{f}}\mathbf{H}_k^{\mathrm{T}}[\mathbf{H}_k\mathbf{P}_k^{\mathrm{f}}\mathbf{H}_k^{\mathrm{T}} + \mathbf{R}_k]^{-1}\mathbf{H}_k\mathbf{P}_k^{\mathrm{f}} \\ &= \mathbf{P}_k^{\mathrm{f}} - \mathbf{K}_k\mathbf{H}_k\mathbf{P}_k^{\mathrm{f}} \\ &= (\mathbf{I} - \mathbf{K}_k\mathbf{H}_k)\mathbf{P}_k^{\mathrm{f}}.\end{aligned} \tag{27.2.23}$$

A summary of the Kalman filter equations is given in Figure 27.2.2.

A number of observations are in order.

(1) Minimum variance estimate Aside from Condition A1–A3 and B1–B3, we have **not** made any explicit assumptions regarding the probability distribution of $\mathbf{x}_0$, $\mathbf{w}_k$, and $\mathbf{v}_k$. The derivation only guarantees that the estimate is the best in the class of **linear**, **unbiased minimum variance** estimates. However, if we further assume that $\mathbf{x}_0$, $\mathbf{w}_k$, and $\mathbf{v}_k$ are **Gaussian**, that is $\mathbf{x}_0 \sim N(\mathbf{m}_0, \mathbf{P}_0)$, $\mathbf{w}_k \sim N(0, \mathbf{Q}_k)$ and $\mathbf{v}_k \sim N(0, \mathbf{R}_k)$, then the estimates obtained are the best including both linear and nonlinear unbiased and minimum variance estimates. In view of linearity for the present case, both $\mathbf{x}_k^{\mathrm{f}}$ and $\widehat{\mathbf{x}}_k$ are Gaussian, that is, $\mathbf{x}_k^{\mathrm{f}} \sim N(\mathbf{M}_{k-1}\widehat{\mathbf{x}}_{k-1}, \mathbf{P}_k^{\mathrm{f}})$ and $\widehat{\mathbf{x}}_k \sim N(\mathbf{x}_k^{\mathrm{f}} + \mathbf{K}_k(\mathbf{z}_k - \mathbf{H}_k\mathbf{x}_k^{\mathrm{f}}), \mathbf{P}_k^{\mathrm{f}})$.

(2) A simpler form for the Kalman gain matrix $\mathbf{K}_k$ By invoking the **Sherman–Morrison–Woodbury** formula for matrix inversion (Appendix B) and applying it to the second line (from the top) on the r.h.s of (27.2.23), we can write $\widehat{\mathbf{P}}_k$ as

$$\widehat{\mathbf{P}}_k = [(\mathbf{P}_k^{\mathrm{f}})^{-1} + \mathbf{H}_k^{\mathrm{T}}\mathbf{R}_k^{-1}\mathbf{H}_k]^{-1}. \tag{27.2.24}$$

Model $\mathbf{x}_{k+1} = \mathbf{M}_k \mathbf{x}_k + \mathbf{w}_{k+1}$

$E(\mathbf{w}_k) = 0 \quad \text{Cov}(\mathbf{w}_k) = \mathbf{Q}_k$

$\mathbf{x}_0$ is random with mean $\mathbf{m}_0$ and $\text{Cov}(\mathbf{x}_0) = \mathbf{P}_0$

Observation $\mathbf{z}_k = \mathbf{H}_k \mathbf{x}_k + \mathbf{v}_k$

$E(\mathbf{v}_k) = 0, \quad \text{Cov}(\mathbf{v}_k) = \mathbf{R}_k$

Model Forecast $\widehat{\mathbf{x}}_0 = E(\mathbf{x}_0), \quad \widehat{\mathbf{P}}_0 = \mathbf{P}_0$

$\mathbf{x}_k^{\text{f}} = \mathbf{M}_{k-1} \widehat{\mathbf{x}}_{k-1}$

$\mathbf{P}_k^{\text{f}} = \mathbf{M}_{k-1} \widehat{\mathbf{P}}_{k-1} \mathbf{M}_{k-1}^{\text{T}} + \mathbf{Q}_k$

Data Assimilation

$$
\begin{aligned}
\widehat{\mathbf{x}}_k &= \mathbf{x}_k^{\text{f}} + \mathbf{K}_k[\mathbf{z}_k - \mathbf{H}_k \mathbf{x}_k^{\text{f}}] \\
\mathbf{K}_k &= \mathbf{P}_k^{\text{f}} \mathbf{H}_k^{\text{T}} [\mathbf{H}_k \mathbf{P}_k^{\text{f}} \mathbf{H}_k^{\text{T}} + \mathbf{R}_k]^{-1} \\
&= \widehat{\mathbf{P}}_k \mathbf{H}_k^{\text{T}} \mathbf{D}_k^{-1} \\
\widehat{\mathbf{P}}_k &= \mathbf{P}_k^{\text{f}} - \mathbf{P}_k^{\text{f}} \mathbf{H}_k^{\text{T}} [\mathbf{H}_k \mathbf{P}_k^{\text{f}} \mathbf{H}_k^{\text{T}} + \mathbf{R}_k]^{-1} \mathbf{H}_k \mathbf{P}_k^{\text{f}} \\
&= [\mathbf{I} - \mathbf{K}_k \mathbf{H}_k] \mathbf{P}_k^{\text{f}}
\end{aligned}
$$

Fig. 27.2.2 A summary of Kalman filter: covariance form.

Now premultiplying the r.h.s. of (27.2.22) by $(\widehat{\mathbf{P}}_k \widehat{\mathbf{P}}_k^{-1})$ and using (27.2.24), we get

$$
\begin{aligned}
\mathbf{K}_k &= (\widehat{\mathbf{P}}_k \widehat{\mathbf{P}}_k^{-1}) \mathbf{P}_k^{\text{f}} \mathbf{H}_k^{\text{T}} [\mathbf{H}_k \mathbf{P}_k^{\text{f}} \mathbf{H}_k^{\text{T}} + \mathbf{R}_k]^{-1} \\
&= \widehat{\mathbf{P}}_k [(\mathbf{P}_k^{\text{f}})^{-1} + \mathbf{H}_k^{\text{T}} \mathbf{R}_k^{-1} \mathbf{H}_k] \mathbf{P}_k^{\text{f}} \mathbf{H}_k^{\text{T}} [\mathbf{H}_k \mathbf{P}_k^{\text{f}} \mathbf{H}_k^{\text{T}} + \mathbf{R}_k]^{-1} \\
&= \widehat{\mathbf{P}}_k [\mathbf{H}_k^{\text{T}} + \mathbf{H}_k^{\text{T}} \mathbf{R}_k^{-1} \mathbf{H}_k \mathbf{P}_k^{\text{f}} \mathbf{H}_k^{\text{T}}] [\mathbf{H}_k \mathbf{P}_k^{\text{f}} \mathbf{H}_k^{\text{T}} + \mathbf{R}_k]^{-1} \\
&= \widehat{\mathbf{P}}_k \mathbf{H}_k^{\text{T}} [\mathbf{I} + \mathbf{R}_k^{-1} \mathbf{H}_k \mathbf{P}_k^{\text{f}} \mathbf{H}_k^{\text{T}}] [\mathbf{H}_k \mathbf{P}_k^{\text{f}} \mathbf{H}_k^{\text{T}} + \mathbf{R}_k]^{-1} \\
&= \widehat{\mathbf{P}}_k \mathbf{H}_k^{\text{T}} \mathbf{R}_k^{-1} [\mathbf{R}_k + \mathbf{H}_k \mathbf{P}_k^{\text{f}} \mathbf{H}_k^{\text{T}}] [\mathbf{H}_k \mathbf{P}_k^{\text{f}} \mathbf{H}_k^{\text{T}} + \mathbf{R}_k]^{-1} \\
&= \widehat{\mathbf{P}}_k \mathbf{H}_k^{\text{T}} \mathbf{R}_k^{-1}
\end{aligned}
\tag{27.2.25}
$$

which is a much simpler form of $\mathbf{K}_k$.

(3) An interpretation of Kalman gain Consider the special case: $n = m, \mathbf{H}_k \equiv \mathbf{I}$ and $\mathbf{P}_k^{\text{f}}$ and $\mathbf{R}_k$ diagonal matrices given by

$$
\mathbf{P}_k^{\text{f}} = \text{Diag}(P_{11}^{\text{f}}, P_{22}^{\text{f}}, \ldots, P_{nn}^{\text{f}})
$$

and

$$\mathbf{R}_k = \text{Diag}(R_{11}, R_{22}, \ldots, R_{nn}).$$

Substituting these into (27.2.22), we obtain

$$\begin{aligned}\mathbf{K}_k &= \mathbf{P}^{\text{f}}_k[\mathbf{P}^{\text{f}}_k + \mathbf{R}_k]^{-1} \\ &= \text{Diag}\left(\frac{P^{\text{f}}_{11}}{P^{\text{f}}_{11} + R_{11}}, \frac{P^{\text{f}}_{22}}{P^{\text{f}}_{22} + R_{22}}, \ldots, \frac{P^{\text{f}}_{nn}}{P^{\text{f}}_{nn} + R_{nn}}\right).\end{aligned} \tag{27.2.26}$$

Now combine this with (27.2.15) to obtain

$$\widehat{\mathbf{x}}_k = (\mathbf{I} - \mathbf{K}_k)\mathbf{x}^{\text{f}}_k + \mathbf{K}_k\mathbf{z}_k$$

that is, the ith component of $\widehat{\mathbf{x}}_k$ is given by

$$\widehat{\mathbf{x}}_{i,k} = \left(\frac{R_{ii}}{P^{\text{f}}_{ii} + R_{ii}}\right)\mathbf{x}^{\text{f}}_{i,k} + \left(\frac{P^{\text{f}}_{ii}}{P^{\text{f}}_{ii} + R_{ii}}\right)\mathbf{z}_{i,k} \tag{27.2.27}$$

which is of the same form as in Example 16.2.1 in Chapter 16. Clearly, if P^{f}_{ii} is large, $\mathbf{K}_k$ assigns more weight to the observation and vice versa.

(4) $\widehat{\mathbf{P}}_k$ is independent of observations From the various forms on the r.h.s. of (27.2.27) it follows that the covariance $\widehat{\mathbf{P}}_k$ (of the optimal estimate $\widehat{\mathbf{x}}_k$) does **not** directly depend on $\mathbf{z}_k$ but only on its covariance $\mathbf{R}_k$ among others. This in turn implies that we can precompute $\widehat{\mathbf{P}}_k$ and analyze its long-term behavior even before the arrival of the first observation. This property of being able to compute and characterize the behavior of the covariance matrix of the optimal estimate is a unique and very desirable feature of this approach. This way one can examine and evaluate competing designs for the observation system.

(5) Special case: no observations In this case, there is no data assimilation step. Given $\widehat{\mathbf{x}}_0 = E(\mathbf{x}_0)$ and $\widehat{\mathbf{P}}_0 = \mathbf{P}_0$, we immediately get

$$\mathbf{x}^{\text{f}}_k = \widehat{\mathbf{x}}_k \quad \text{and} \quad \mathbf{P}^{\text{f}}_k = \widehat{\mathbf{P}}_k \quad \text{for all} \quad k \geq 0.$$

The evolution of the model forecast and its covariance are given by

$$\begin{aligned}\mathbf{x}^{\text{f}}_k &= \mathbf{M}_{k-1}\mathbf{x}^{\text{f}}_{k-1} \\ \mathbf{P}^{\text{f}}_k &= \mathbf{M}_{k-1}\mathbf{P}^{\text{f}}_{k-1}\mathbf{M}^{\text{T}}_{k-1} + \mathbf{Q}_k.\end{aligned} \tag{27.2.28}$$

Define a sequence of matrix products $\mathbf{M}(i:j)$ as

$$\mathbf{M}(i:j) = \begin{cases}\mathbf{M}_i\,\mathbf{M}_{i-1}\cdots\mathbf{M}_j & \text{if } i \geq j \\ \mathbf{I}, \text{ the identity matrix} & \text{if } i < j\end{cases} \tag{27.2.29}$$

and $\mathbf{M}^{\text{T}}(i:j)$ denotes the transpose of this product. By iterating (27.2.28) and using (27.2.29), we get

$$\mathbf{x}^{\text{f}}_k = \mathbf{M}(k-1:0)\,\mathbf{x}^{\text{f}}_0$$

and

$$\mathbf{P}^{\mathrm{f}}_k = \mathbf{M}(k-1:0)\mathbf{P}_0\mathbf{M}^{\mathrm{T}}(k-1:0)$$
$$+\sum_{j=0}^{k-1}\mathbf{M}(k-1:j+1)\mathbf{Q}_{j+1}\mathbf{M}^{\mathrm{T}}(k-1:j+1). \qquad (27.2.30)$$

In the special case when there is **no** model noise, that is, $\mathbf{Q}_j \equiv 0$, then (27.2.30) becomes

$$\mathbf{P}^{\mathrm{f}}_k = \mathbf{M}(k-1:0)\mathbf{P}_0\mathbf{M}^{\mathrm{T}}(k-1:0). \qquad (27.2.31)$$

This equation describes the evolution of the initial covariance as a function of time and constitutes the basis for the study of **stochastic dynamic systems**.

(6) Special case: no dynamics Consider the case when there is no dynamics, that is, $\mathbf{M}_k \equiv \mathbf{I}$, $\mathbf{w}_k \equiv 0$, and $\mathbf{Q}_k \equiv 0$. Then $\mathbf{x}_{k+1} = \mathbf{x}_k = \mathbf{x}$. The observations are given by

$$\mathbf{z}_k = \mathbf{H}_k\mathbf{x} + \mathbf{v}_k$$

with $E(\mathbf{v}_k) = 0$ and $\mathrm{Cov}(\mathbf{v}_k) = \mathbf{R}_k$. Referring to Figure 27.2.2, it follows that the forecast and its covariance are given by

$$\mathbf{x}^{\mathrm{f}}_k = \widehat{\mathbf{x}}_{k-1} \quad \text{with} \quad \widehat{\mathbf{x}}_0 = E(\mathbf{x}_0)$$
$$\mathbf{P}^{\mathrm{f}}_k = \widehat{\mathbf{P}}_{k-1} \quad \text{with} \quad \widehat{\mathbf{P}}_0 = \mathbf{P}_0.$$

The data assimilation step becomes

$$\mathbf{K}_k = \widehat{\mathbf{P}}_{k-1}\mathbf{H}^{\mathrm{T}}_k[\mathbf{H}_k\widehat{\mathbf{P}}_{k-1}\mathbf{H}^{\mathrm{T}}_k + \mathbf{R}_k]^{-1}$$
$$\widehat{\mathbf{x}}_k = \widehat{\mathbf{x}}_{k-1} + \mathbf{K}_k[\mathbf{z}_k - \mathbf{H}_k\widehat{\mathbf{x}}_{k-1}]$$
$$\widehat{\mathbf{P}}_k = \widehat{\mathbf{P}}_{k-1} - \widehat{\mathbf{P}}_{k-1}\mathbf{H}^{\mathrm{T}}_k[\mathbf{H}_k\widehat{\mathbf{P}}_{k-1}\mathbf{H}^{\mathrm{T}}_k + \mathbf{R}_k]^{-1}\mathbf{H}_k\widehat{\mathbf{P}}_{k-1}$$

which not surprisingly are the same as (17.2.11)–(17.2.12) for the **static Kalman filter**.

(7) Impact of perfect observations If the observations are perfect, then $\mathbf{R}_k \equiv 0$. Substituting this into the filter equations in Figure 27.2.2, we get

$$\mathbf{K}_k = \mathbf{P}^{\mathrm{f}}_k\mathbf{H}^{\mathrm{T}}_k[\mathbf{H}_k\mathbf{P}^{\mathrm{f}}_k\mathbf{H}^{\mathrm{T}}_k]^{+}$$

where $\mathbf{A}^{+}$ refers the **generalized inverse** of $\mathbf{A}$ (Refer to Chapter 5 and Appendix B) and

$$\widehat{\mathbf{P}}_k = (\mathbf{I} - \mathbf{K}_k\mathbf{H}_k)\mathbf{P}^{\mathrm{f}}_k$$
$$= \mathbf{P}^{\mathrm{f}}_k(\mathbf{I} - \mathbf{K}_k\mathbf{H}_k)^{\mathrm{T}} \qquad (\widehat{\mathbf{P}}_k \text{ is symmetric}).$$

From (27.2.19) we also see that

$$\widehat{\mathbf{P}}_k = (\mathbf{I} - \mathbf{K}_k\mathbf{H}_k)\mathbf{P}^{\mathrm{f}}_k(\mathbf{I} - \mathbf{K}_k\mathbf{H}_k)^{\mathrm{T}}$$
$$= (\mathbf{I} - \mathbf{K}_k\mathbf{H}_k)^2\mathbf{P}^{\mathrm{f}}_k.$$

Comparing these we obtain that $(\mathbf{I} - \mathbf{K}_k\mathbf{H}_k)$ is idempotent since

$$(\mathbf{I} - \mathbf{K}_k\mathbf{H}_k)^2 = (\mathbf{I} - \mathbf{K}_k\mathbf{H}_k).$$

Idempotent matrices are singular (Appendix B) and hence $\text{Rank}(\mathbf{I} - \mathbf{K}_k\mathbf{H}_k) \le n - 1$. Hence

$$\begin{aligned}\text{Rank}(\widehat{\mathbf{P}}_k) &\le \text{Min}\{\text{Rank}(\mathbf{I} - \mathbf{K}_k\mathbf{H}_k), \text{Rank}(\mathbf{P}^{\text{f}}_k)\}\\ &\le n - 1.\end{aligned}$$

Since $\widehat{\mathbf{P}}_k$ is a covariance matrix, this inequality implies that $\widehat{\mathbf{P}}_k$ must have at least one zero eigenvalue and hence **cannot** be positive definite. Thus if the observations are very nearly perfect then $\mathbf{R}_k$ is very small and this will cause **computational instability**.

(8) Residual checking The term

$$\mathbf{r}_k = (\mathbf{z}_k - \mathbf{H}_k\mathbf{x}^{\text{f}}_k)$$

appearing in the new estimate $\widehat{\mathbf{x}}_k$ given in (27.2.15) is called the **innovation** or **new information** or simply the **residual**. This term $\mathbf{r}_k \in \mathbb{R}^m$ is linearly transformed by $\mathbf{K}_k \in \mathbb{R}^{n\times m}$ and $\mathbf{K}_k\mathbf{r}_k$ is added to $\mathbf{x}^{\text{f}}_k$ to obtain $\widehat{\mathbf{x}}_k$. Rewriting $\mathbf{r}_k$ as

$$\mathbf{r}_k = \mathbf{H}_k(\mathbf{x}_k - \mathbf{x}^{\text{f}}_k) + \mathbf{v}_k = \mathbf{H}_k\mathbf{e}^{\text{f}}_k + \mathbf{v}_k$$

it follows that $E(\mathbf{r}_k) \equiv 0$ and

$$\begin{aligned}\text{Cov}(\mathbf{r}_k) &= E[\mathbf{r}_k\mathbf{r}^{\text{T}}_k]\\ &= \mathbf{H}_k\mathbf{P}^{\text{f}}_k\mathbf{H}^{\text{T}}_k + \mathbf{R}_k.\end{aligned}$$

The term $\mathbf{r}_k$ is routinely calculated, we can compute the first two moments of $\mathbf{r}_k$ and check them against these theoretical values to guarantee that the filter is working as it should. Any disparity between the computed moments of $\mathbf{r}_k$ and the theoretical values would point to the inadequacy of the model to explain the observations.

(9) Duality covariance vs. information forms The standard Kalman filter equations in Figure 27.2.2 is called the **covariance form** of the filter since it involves recurrence relation that directly updates the covariance matrices $\mathbf{P}^{\text{f}}_k$ and $\widehat{\mathbf{P}}_k$. By reformulating the statistical least squares method from a recursive point of view, we derived a dual form of the same filter called the **information form** in Figure 26.4.1. This latter form involves recurrence relation that updates the inverse of the covariance matrices $(\mathbf{P}^{\text{f}}_k)^{-1}$ and $(\widehat{\mathbf{P}}_k)^{-1}$. This inverse form of the filter is useful when there is no prior information about the initial state $\mathbf{x}_0$ in which case we can easily set $(\widehat{\mathbf{P}}_0)^{-1} = 0$ instead of $\widehat{\mathbf{P}}_0 = \infty$ (Refer to Chapter 16).

(10) Computational cost We now quantify the amount of work in terms of the number of **floating-point operations** (flops) that are needed to perform one iteration of the Kalman filter equations in Figure 27.2.2. To this end, recall from Appendix B that to multiply two matrices $\mathbf{A} \in \mathbb{R}^{n\times m}$ and $\mathbf{B} \in \mathbb{R}^{m\times r}$ it takes $2mnr$ flops – mnr multiplications and mnr additions. While it is true that in general

Table 27.2.1 *Estimation of the computational cost*

Item	Operation	Type of Computation	Cost
$\mathbf{x}^f_{k+1}$	$\mathbf{M}_k\widehat{\mathbf{x}}_k$	Matrix-vector Multiply	$2n^2$
$\mathbf{P}^f_{k+1}$	$\mathbf{H}_k\widehat{\mathbf{P}}_k\mathbf{H}_k^T + \mathbf{Q}_k$	Two matrix-matrix multiply + a matrix add	$4n^3 + n^2$
$\mathbf{K}_{k+1}$	$(\mathbf{H}_k\mathbf{P}^f_k\mathbf{H}_k^T + \mathbf{R}_k)$	Two matrix-matrix multiply + a matrix add	$4n^2m + m^2$
	$(\mathbf{H}_k\mathbf{P}^f_k\mathbf{H}_k^T + \mathbf{R}_k)^{-1}$	Inverse of a symmetric positive definite matrix	$\frac{1}{3}m^3$
	$\mathbf{P}^f_k\mathbf{H}_k^T(\mathbf{H}_k\mathbf{P}^f_k\mathbf{H}_k^T + \mathbf{R}_k)^{-1}$	Two matrix-matrix multiply	$2nm^2 + 2n^2m$
	Total cost of $\mathbf{K}_{k+1}$		$6n^2m + 2nm^2 + \frac{1}{3}m^3 + m^2$
$\widehat{\mathbf{P}}_{k+1}$	$[\mathbf{I} - \mathbf{K}_k\mathbf{H}_k]$	One matrix-matrix multiply and add identity matrix	$2n^2m + n$
	$(\mathbf{I} - \mathbf{K}_k\mathbf{H}_k)\mathbf{P}^f_{k+1}$	Matrix-matrix multiply	$2n^3$
	Total cost of $\widehat{\mathbf{P}}_{k+1}$		$2n^3 + 2n^2m + n^2$
$\widehat{\mathbf{x}}_{k+1}$	$(\mathbf{z}_k - \mathbf{H}_k\mathbf{x}^f_k)$	Matrix-vector multiply and a vector add	$2nm + m$
	$\mathbf{K}_k[\mathbf{z}_k - \mathbf{H}_k\mathbf{x}^f_k]$	Matrix-vector multiply	$2nm$
	$\mathbf{x}^f_k + \mathbf{K}_k[\mathbf{z}_k - \mathbf{H}_k\mathbf{x}^f_k]$	Vector add	n
	Total cost of $\widehat{\mathbf{x}}_{k+1}$		$4nm + n + m$

multiplication takes more time than addition, to simplify the process of estimating the cost, it is useful to assume a **unit cost model** where the unit of cost (measured in time) is equal to the maximum of the cost of performing a single operation – add, subtract, multiply, and divide. Using this convention, we now quantify the total cost in terms of the number of flops as a function of the **size** of the problem.

An itemized list of the cost for various steps of the Kalman filter is given in Table 27.2.1. It is evident from this table that the computation of the covariance matrices $\mathbf{P}^f_{k+1}$ and $\widehat{\mathbf{P}}_{k+1}$ is the most time-consuming part since in many of the applications $n \gg m$.

The following examples illustrate several key properties of the Kalman filter.

Example 27.2.1 Scalar dynamics with no observation
Consider a scalar, **first-order autoregressive (AR(1))** model

$$x_k = ax_{k-1} + w_k \tag{27.2.32}$$

where $a > 0$ and x_0 is random with **mean** m_0 and $\mathrm{Var}(x_0) = p_0$. The term w_k is temporally uncorrelated with $E(w_k) = 0$ and $\mathrm{Var}(w_k) = q > 0$. In addition x_0 and

w_k are uncorrelated. Iterating (27.2.32) we get

$$x_k = a^k x_0 + \sum_{j=1}^{k} a^{k-j} w_j. \tag{27.2.33}$$

Hence

$$E(x_k) = a^k E(x_0) = a^k m_0. \tag{27.2.34}$$

If p_k is the variance of x_k, then from (27.2.32) it follows that

$$p_k = \text{Var}(x_k) = \text{Var}(ax_{k-1} + w_k) = a^2 p_{k-1} + q. \tag{27.2.35}$$

Iterating (27.2.35), we get

$$p_k = a^{2k} p_0 + q \frac{(a^{2k} - 1)}{(a^2 - 1)}. \tag{27.2.36}$$

Thus, for a given m_0, p_0, and q, the behavior of x_k and its first two moments critically depends on the model parameter a. Depending on the range of values of a, three cases arise.

Case A: stable mode. $0 < a < 1$
From (27.2.34) and (27.2.36) it follows that

$$\lim_{k\to\infty} E(x_k) = 0 \quad \text{and} \quad \lim_{k\to\infty} p_k = \frac{q}{1 - a^2}$$

exponentially in time. Thus, x_k in the (mean square) limit tends to a random variable x^* with mean zero and variance equal to $q/(1 - a^2)$.

Case B: unstable mode. $1 < a < \infty$
In this case it can be verified that

$$\lim_{k\to\infty} E(x_k) = \infty \quad \text{and} \quad \lim_{k\to\infty} p_k = \infty$$

both increasing **exponentially** with time.

Case C: random walk. $a = 1$
In this case

$$x_k = x_0 + \sum_{i=1}^{k} w_i$$

with

$$E(x_k) = x_0$$

and

$$p_k = p_0 + kq.$$

Notice that in this case the variance increases **linearly** with time.

Example 27.2.2 Kalman filter: convergence of covariance
Consider the case of scalar dynamics and scalar observations

$$\begin{aligned} x_{k+1} &= a x_k + w_{k+1} \\ z_k &= h x_k + v_k \end{aligned} \tag{27.2.37}$$

where a and h are non-zero constants. It is assumed that $E(w_k) = 0 = E(v_k)$ and $\text{Var}(w_k) = q > 0$ and $\text{Var}(v_k) = r > 0$. The initial condition x_0 is a random variable with mean m_0 and $\text{Var}(x_0) = p_0$. It is further assumed that x_0, $\{w_k\}$, and $\{v_k\}$ are uncorrelated. The Kalman filter equations in Figure 27.2.2 when specialized to this scalar case becomes

$$\begin{aligned} x^{\text{f}}_{k+1} &= a\widehat{x}_k \\ p^{\text{f}}_{k+1} &= a^2 \widehat{p}_k + q \\ \widehat{x}_k &= x^{\text{f}}_k + K_k[z_k - h x^{\text{f}}_k] \\ K_k &= p^{\text{f}}_k h[h^2 p^{\text{f}}_k + r]^{-1} = \widehat{p}_k h r^{-1} \\ \widehat{p}_k &= p^{\text{f}}_k - (p^{\text{f}}_k)^2 h^2 [h^2 p^{\text{f}}_k + r]^{-1} \\ &= p^{\text{f}}_k r [h^2 p^{\text{f}}_k + r]^{-1}. \end{aligned} \tag{27.2.38}$$

Substituting for $\widehat{x}_k$ and x^{f}_k, we obtain the following linear first-order recurrence relations:

$$\begin{aligned} x^{\text{f}}_{k+1} &= a(1 - K_k h) x^{\text{f}}_k + a K_k z_k \\ \widehat{x}_{k+1} &= a(1 - K_{k+1} h)\widehat{x}_k + K_{k+1} h z_{k+1} \end{aligned} \tag{27.2.39}$$

or equivalently

$$\begin{aligned} e^{\text{f}}_{k+1} &= a(1 - K_k h) e^{\text{f}}_k + a K_k v_k + w_{k+1} \\ \widehat{e}_{k+1} &= a(1 - K_{k+1} h)\widehat{e}_k + (1 - K_{k+1} h) w_{k+1} - K_{k+1}\, v_{k+1}. \end{aligned} \tag{27.2.40}$$

These recurrence relations play a key role in the analysis of stability of the filter dynamics – see Example 27.2.3 and Exercise 27.2.

Similarly, substituting for $\widehat{p}_k$ in p^{f}_{k+1}, we obtain a first-order nonlinear recurrence

$$p^{\text{f}}_{k+1} = \frac{a^2\, p^{\text{f}}_k\, r}{h^2\, p^{\text{f}}_k + r} + q. \tag{27.2.41}$$

Dividing both sides by r, this reduces to

$$p_{k+1} = \frac{a^2 p_k}{h^2 p_k + 1} + \alpha \tag{27.2.42}$$

where $\alpha = q/r > 0$ and $p_k = p^{\text{f}}_k / r$. Notice that p_k in (27.2.42) depends only on the ratio α and **not** individually on q or r (Exercise 27.3). Equation (27.2.42) is called the **Riccati equation** which is a first-order, scalar, nonlinear recurrence relation.

Table 27.2.2 *Variation of* p^*

	p^*				
	α				
a	0.01	0.5	1.0	1.5	2.0
0.01	0.01	0.50	1.00	1.50	2.00
0.5	0.01	0.59	1.13	1.66	2.17
1.0	0.11	1.00	1.62	2.19	2.73
1.5	1.27	2.00	2.63	3.22	3.78
2.0	3.01	3.64	4.23	4.81	5.37

In the following we characterize the asymptotic properties of the solution of this Riccati equation.

To this end we first compute the **equilibrium points** of (27.2.42). Define (assuming $h = 1$ henceforth)

$$\left.\begin{aligned} \delta_k &= p_{k+1} - p_k = \frac{g(p_k)}{1 + p_k} \\ g(p_k) &= -p_k^2 + \beta p_k + \alpha \\ \beta &= \alpha + a^2 - 1 \end{aligned}\right\} \tag{27.2.43}$$

Equilibrium points are obtained by setting $\delta_k = 0$, that is, by solving the quadratic equation $g(p_k) = 0$. The two equilibrium points are given by

$$p^* = \frac{\beta + \sqrt{\beta^2 + 4\alpha}}{2} \quad \text{and} \quad p_* = \frac{\beta - \sqrt{\beta^2 + 4\alpha}}{2}.$$

Evaluating the derivative $g'(p_k) = -2p_k + \beta$ at these two points, we get

$$g'(p^*) = -\sqrt{\beta^2 + 4\alpha} < 0 \quad \text{and} \quad g'(p_*) = \sqrt{\beta^2 + 4\alpha} > 0.$$

Hence, p^* is a **stable** (**attractor**) and p_* is an **unstable** (**repellor**) equilibrium point, from which we can readily conclude that

$$\lim_{k \to \infty} p_k = p^*.$$

A typical plot of $g(p_k)$ and δ_k as a function of p_k is given in Figure 27.2.3 and the variation of p^* as a function of α and a are given in Table 27.2.2. It follows from (27.2.42) and the definition of the equilibrium that (since $h = 1$)

$$p^* = \frac{a^2 p^*}{p^* + 1} + \alpha. \tag{27.2.44}$$

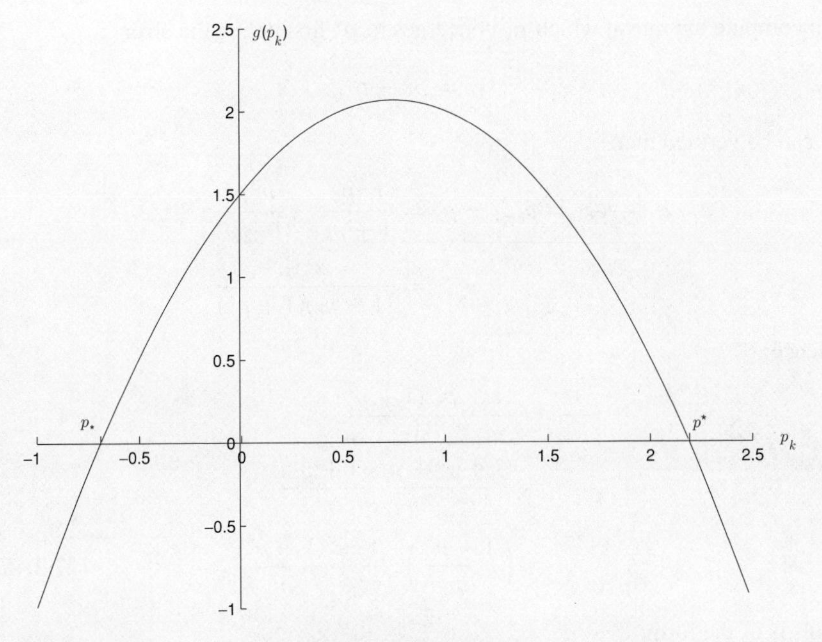

(a) Plot of $g(p_k)$ for $\alpha = 1.5$ and $a = 1.0$

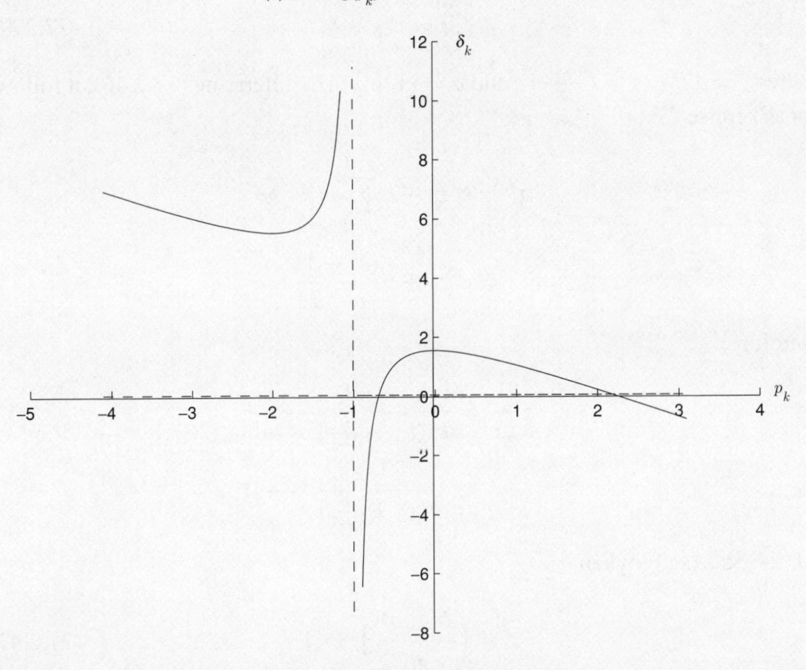

(b) Plot of δ_k for $\alpha = 1.5$ and $a = 1.0$

Fig. 27.2.3 An illustration of $g(p_k)$ and δ_k.

To compute the rate at which p_k converges to p^* first define the error

$$y_k = p_k - p^*.$$

It can be verified that

$$\begin{aligned} y_{k+1} = p_{k+1} - p^* &= \frac{a^2 p_k}{1+p_k} - \frac{a^2 p^*}{1+p^*} \\ &= \frac{a^2 y_k}{(1+p_k)(1+p^*)}. \end{aligned}$$

Hence

$$\begin{aligned} \frac{1}{y_{k+1}} &= \frac{(1+p_k)(1+p^*)}{a^2 y_k} \\ &= \frac{(1+y_k+p^*)(1+p^*)}{a^2 y_k} \\ &= \left(\frac{1+p^*}{a}\right)^2 \frac{1}{y_k} + \frac{(1+p^*)}{a^2}. \end{aligned} \tag{27.2.45}$$

This is of the form

$$z_{k+1} = c z_k + b \tag{27.2.46}$$

where $z_k = 1/y_k$, $c = (\frac{1+p^*}{a})^2$, and $b = (1+p^*)/a^2$. Iterating (27.2.46), it follows that (Exercise 27.1)

$$\begin{aligned} z_k &= c^k z_0 + b \sum_{j=0}^{k-1} c^j \\ &= c^k z_0 + b \frac{(c^k - 1)}{c-1}. \end{aligned}$$

Therefore,

$$\begin{aligned} y_k = \frac{1}{z_k} &= \frac{1}{c^k[z_0 + \frac{b}{c-1}] - \frac{b}{c-1}} \\ &< \frac{1}{c^k[z_0 + \frac{b}{c-1}]} \longrightarrow 0 \end{aligned}$$

as $k \to \infty$ exactly when

$$c = \left(\frac{1+p^*}{a}\right)^2 > 1 \tag{27.2.47}$$

and hence the error $y_k \to 0$ at an exponential rate. (Refer to Chapter 10 for the definition of rate of convergence). The special case when $a = 1$ and $\alpha = 0$ is covered in Example 27.2.4 .

Table 27.2.3 *Exponential Convergence of Variance*

Time step k	p_k $a = 0.01$ $\alpha = 0.5$	$a = 0.5$ $\alpha = 1.0$	$a = 1.0$ $\alpha = 1.0$	$a = 1.5$ $\alpha = 1.0$	$a = 2.0$ $\alpha = 1.0$
0	0.1000	0.1000	0.1000	0.1000	0.1000
1	0.5000	1.0208	1.0833	1.1875	1.3333
2	0.5000	1.0839	1.3421	1.7917	2.4545
3	0.5000	1.0855	1.3643	1.8795	2.6615
4	0.5000	1.0856	1.3659	1.8886	2.6837
5	0.5000	1.0856	1.3660	1.8895	2.6859
6	0.5000	1.0856	1.3660	1.8896	2.6861
7	0.5000	1.0856	1.3660	1.8896	2.6861

Table 27.2.3 illustrates the exponential convergence of the solution of the Riccati equation (27.2.42) for various combination of the values of α and a. Using (27.2.38) we can now characterize the limit of $\widehat{p}_k$ when $h = 1$. Rewriting the expression

$$\frac{\widehat{p}_k}{r} = \frac{(p^{\mathrm{f}}_k/r)}{(p^{\mathrm{f}}_k/r) + 1} = \frac{p_k}{p_k + 1}$$

with $p_k = p^{\mathrm{f}}_k/r$ as in (27.2.42). Since p_k converges to p^* in the limit, it immediately follows that

$$\widehat{p}^* = \lim_{k\to\infty} \widehat{p}_k = \frac{rp^*}{p^* + 1}.$$

Example 27.2.3 Stability of the Filter

Consider the homogeneous part of the forecast error with $h = 1$ in (27.2.40)

$$\bar{e}^{\mathrm{f}}_{k+1} = a(1 - K_k)\bar{e}^{\mathrm{f}}_k. \tag{27.2.48}$$

From (27.2.38) we get

$$K_k = \frac{p^{\mathrm{f}}_k}{p^{\mathrm{f}}_k + r} = \frac{p_k}{1 + p_k} \quad \text{and} \quad 1 - K_k = \frac{1}{p_k + 1} \tag{27.2.49}$$

where $p_k = p^{\mathrm{f}}_k/r$. Since $p_k \to p^*$ at an exponential rate (see Example 27.2.2 and also refer to Table 27.2.3), it follows that there exists $N > 0$ such that for all $k > N$, the recurrence (27.2.48) can be rewritten as

$$\bar{e}^{\mathrm{f}}_{k+1} = \left(\frac{a}{1 + p^*}\right) \bar{e}^{\mathrm{f}}_k = \left(\frac{1}{\sqrt{c}}\right) \bar{e}^{\mathrm{f}}_k \tag{27.2.50}$$

where it follows from (27.2.47) that $\sqrt{c} > 1$ for all a and α except when $a = 1$ and $\alpha = 0$ (The case when $a = 1$ and $\alpha = 0$ is covered in Example 27.2.4). Hence

$$\bar{e}^{\mathrm{f}}_k = \left(\frac{1}{\sqrt{c}}\right)^{k-N} \bar{e}^{\mathrm{f}}_N \to 0 \quad \text{as} \quad k \to \infty.$$

Since the homogeneous part of the recurrence for $\widehat{e}_{k+1}$ in (27.2.40) is also of the same form as (27.2.48), it immediately follows that for $k > N$ (with $h = 1$),

$$\widehat{e}_{k+1} = \left(\frac{1}{\sqrt{c}}\right)\widehat{e}_k + \left(\frac{1}{1+p^*}\right) w_{k+1} - \left(\frac{p^*}{1+p^*}\right) v_k. \tag{27.2.51}$$

Consequently, for large k, the estimate $\widehat{x}_k$ closely follows the state x_k except for the random perturbations given by the last two terms on the *r.h.s.* of (27.2.51).

Remark 27.2.2 From Example 27.2.1 it follows that the dynamics (27.2.37) is **stable** for $0 < a \leq 1$ and is **unstable** for $a > 1$ where stability implies that the state remains bounded. Examples 27.2.2 and 27.2.3 illustrate that, given a relevant set of observations though noisy, we can obtain the minimum variance estimate of the state of the system in both the stable and the unstable modes – a clear demonstration of the power of the Kalman filtering technique.

Example 27.2.4 Static Case
Let $a = 1$ and $h = 1$ and in addition $w_k = 0$ and hence $q = 0$. In this case we have a problem of estimating an unknown random constant (Refer to Chapters 16 and 17). Then

$$x_k \equiv x$$
$$z_k = x + v_k.$$

Then, the Kalman filter equations become

$$\begin{aligned}
x^{\mathrm{f}}_k &= \widehat{x}_{k-1}, \quad p^{\mathrm{f}}_k = \widehat{p}_{k-1}\\
\widehat{x}_k &= \widehat{x}_{k-1} + K_k\,[z_k - \widehat{x}_{k-1}]\\
K_k &= p^{\mathrm{f}}_k\,[p^{\mathrm{f}}_k + r]^{-1} = \widehat{p}_{k-1}[\widehat{p}_{k-1} + r]^{-1}\\
\widehat{p}_k &= p^{\mathrm{f}}_k\, r\,[p^{\mathrm{f}}_k + r]^{-1} = \widehat{p}_{k-1}\, r[\widehat{p}_{k-1} + r]^{-1}\\
\widehat{p}_k &= \frac{\widehat{p}_{k-1}\, r}{\widehat{p}_{k-1} + r}
\end{aligned} \tag{27.2.52}$$

which reduces to

$$p_k = \frac{p_{k-1}}{p_{k-1} + 1} \quad \text{where} \quad p_k = \frac{\widehat{p}_k}{r}. \tag{27.2.53}$$

Iterating this first-order nonlinear recurrence (Exercise 27.6), it can be shown that

$$p_k = \frac{p_0}{1 + kp_0} \longrightarrow 0 \text{ as } k \to \infty \tag{27.2.54}$$

at a rate $1/k$. Now using the definition of p_k and (27.2.54), we can rewrite

$$K_k = \frac{\widehat{p}_{k-1}}{\widehat{p}_{k-1} + r} = \frac{p_{k-1}}{p_{k-1} + 1} = \frac{p_0}{1 + kp_0}.$$

The equation for the estimate in (27.2.54) is given by

$$\widehat{x}_k = \widehat{x}_{k-1} + \frac{p_0}{1 + kp_0}[z_k - \widehat{x}_{k-1}].$$

As $k \to \infty$, the extra information provided by the innovation term $(z_k - \widehat{x}_{k-1})$ becomes vanishingly small and $\widehat{x}_k$ tends to a constant with $p_k \to 0$.

Exercises

27.1 Consider the scalar linear first-order recurrence

$$\mathbf{x}_k = \mathbf{a}_k \mathbf{x}_{k-1} + \mathbf{b}_k, \quad \mathbf{x}_0 = \mathbf{b}_0.$$

(a) By iterating verify that the solution is given by

$$\mathbf{x}_k = \sum_{j=0}^{k} \left(\prod_{s=j+1}^{k} \mathbf{a}_s \right) \mathbf{b}_j.$$

(b) When $\mathbf{a}_k \equiv \mathbf{a}$, verify that $(\mathbf{x}_0 = \mathbf{b}_0)$

$$\mathbf{x}_k = \mathbf{a}^k \mathbf{b}_0 + \sum_{j=1}^{k} \mathbf{a}^{k-j} \mathbf{b}_j.$$

27.2 Using the solution of the first-order linear recurrence given in part (*a*) of Exercise (27.1), solve the first-order linear recurrences for $\widehat{\mathbf{e}}_{k+1}$ and $\mathbf{e}^{\mathrm{f}}_{k+1}$ in (27.2.40).

27.3 By substituting for $\mathbf{P}^{\mathrm{f}}_k$ into $\widehat{\mathbf{P}}_{k+1}$ in (27.2.38), verify that $\widehat{\mathbf{P}}_{k+1}$ is given by the first order nonlinear recurrence of the form

$$\widehat{\mathbf{P}}_{k+1} = \frac{c_1 \widehat{\mathbf{P}}_k + c_2}{c_3 \widehat{\mathbf{P}}_k + c_4}$$

Find expressions for c_1, c_2, c_3, and c_4 in terms a, h, and α. Specialize to the case when $a = 1, h = 1$.

27.4 Using the method for solving the Riccati equation (27.2.42), solve the following related first-order nonlinear recurrence equation

$$\mathbf{P}_{k+1} = \frac{\mathbf{P}_k + \alpha}{\mathbf{P}_k + \beta}$$

for some constants α and β.

27.5 **Linearizing the Riccati equation** Consider the nonlinear first-order recurrence

$$\mathbf{P}_{k+1} = \frac{\mathbf{P}_k}{1 + \mathbf{P}_k} + \alpha.$$

(a) Substituting $\mathbf{P}_k = \mathbf{y}_k / \mathbf{x}_k$, verify that the above equation can be rewritten as

$$\frac{\mathbf{y}_{k+1}}{\mathbf{x}_{k+1}} = \frac{\alpha \mathbf{x}_k + (1 + \alpha)\mathbf{y}_k}{\mathbf{x}_k + \mathbf{y}_k}$$

or equating the numerators and denominators equivalently as a linear first order recurrence in matrix form as

$$\begin{pmatrix} \mathbf{x}_{k+1} \\ \mathbf{y}_{k+1} \end{pmatrix} = \begin{pmatrix} 1 & 1 \\ \alpha & 1 + \alpha \end{pmatrix} \begin{pmatrix} \mathbf{x}_k \\ \mathbf{y}_k \end{pmatrix}$$

that is,

$$\mathbf{z}_{k+1} = \mathbf{S}\mathbf{z}_k \qquad (*)$$

with

$$\mathbf{z}_k = \begin{pmatrix} \mathbf{x}_k \\ \mathbf{y}_k \end{pmatrix} \quad \text{and} \quad \mathbf{S} = \begin{pmatrix} 1 & 1 \\ \alpha & 1 + \alpha \end{pmatrix}.$$

(b) By iterating $(*)$ above verify that

$$\mathbf{z}_k = \mathbf{S}^k \mathbf{z}_0.$$

(c) Compute the determinant, eigenvalues and eigenvectors of the matrix $\mathbf{S}$. Quantify the limit of $\mathbf{S}^k$ and hence of $\mathbf{z}_k$ as $k \to \infty$. Verify that you get the same conclusion as in Example 27.2.1.

27.6 Verify that $\mathbf{P}_k$ in (27.2.54) is the solution of the recurrence in (27.2.53).

Notes and references

Section 27.1 The classification of estimation problems into filtering, smoothing and prediction is standard. Refer to Wiener (1949) Kolmogorov (1941).

Section 27.2 The derivation of the Kalman filter equation is now standard. The original papers by Kalman (1960) and Kalman and Bucy (1961) still continue to be a great source of inspiration and guidance. For a thorough discussion of various aspects of linear filtering refer to Kailath (1974). Books by Bryson and Ho (1975), Bucy and Joseph (1968), Gelb (1974), Jazwinski (1970), Maybeck (1979), Segers (2002) and Sorenson (1985) provide a comprehensive treatment of filtering and smoothing for linear dynamical systems. The recent elegant monograph by Bucy (1994) is notable for development of the discrete time formulation. Also refer to Swirling (1971) for an overview.

28

Linear filtering: part II

In Chapter 27 we derived the basic algorithm for linear, sequential filter due to Kalman. In this chapter we continue the analysis of the properties of this classical algorithm. In particular we cover the following topics: an interpretation of the Kalman filter from the point of view of **orthogonal projection**; **rederivation of the filter equations when the model noise $\mathbf{w}_k$ and the observation noise $\mathbf{v}_k$ are correlated**; inclusion and estimate of **bias terms** in both model and observation; **computational aspects** of the covariance matrices $\mathbf{P}_k^{\text{f}}$ and $\widehat{\mathbf{P}}_k$; **sensitivity** of the filter with respect to variations in the model and noise covariance matrices; and a discussion of the **stability** of the filter. We conclude this chapter with a derivation and a discussion of the so-called **square root filter**.

A good working knowledge of the contents of Chapter 27 is a prerequisite for this chapter.

28.1 Kalman filter and orthogonal projection

The principle of (both deterministic and statistical) least squares is intimately related to the notion of orthogonal (and oblique) projections; refer to Chapters 6 and 14 for details. The Kalman filter estimate being the minimum variance estimate, not surprisingly, is related to the **orthogonal projection**.

Recall that two random vectors are orthogonal if their correlation is zero. We now demonstrate that the minimum variance estimate $\widehat{\mathbf{x}}_k$ given in Figure 27.2.2 is indeed orthogonal to the error $\widehat{\mathbf{e}}_k$. This property is established by induction.

The initial condition $\mathbf{x}_0$ for the model is such that $E(\mathbf{x}_0) = \mathbf{m}_0$ and $\text{Var}(\mathbf{x}_0) = \mathbf{P}_0$. The obvious choice for the initial estimate $\mathbf{x}_0$ is

$$\widehat{\mathbf{x}}_0 = E(\mathbf{x}_0) = \mathbf{m}_0 \quad \text{and} \quad \widehat{\mathbf{e}}_0 = \mathbf{x}_0 - \widehat{\mathbf{x}}_0.$$

Hence

$$\begin{aligned} E[\widehat{\mathbf{x}}_0(\widehat{\mathbf{e}}_0)^{\text{T}}] &= E[\mathbf{m}_0(\mathbf{x}_0 - \mathbf{m}_0)^{\text{T}}] \\ &= \mathbf{m}_0 E[(\mathbf{x}_0 - \mathbf{m}_0)^{\text{T}}] = 0. \end{aligned}$$

Thus, the basis for induction, namely $\widehat{\mathbf{x}}_0$ is orthogonal to $\widehat{\mathbf{e}}_0$ is established. Now assume that $\widehat{\mathbf{x}}_{k-1}$ is orthogonal to $\widehat{\mathbf{e}}_{k-1}$. Then from (27.2.16) and (27.2.17) we have

$$\widehat{\mathbf{x}}_k = \widehat{\mathbf{M}}_{k-1}\widehat{\mathbf{x}}_{k-1} + \mathbf{K}_k[\mathbf{H}_k\mathbf{M}_{k-1}\widehat{\mathbf{e}}_{k-1} + \mathbf{H}_k\mathbf{w}_k + \mathbf{v}_k] \tag{28.1.1}$$

and

$$\widehat{\mathbf{e}}_k = (\mathbf{I} - \mathbf{K}_k\mathbf{H}_k)[\mathbf{M}_{k-1}\widehat{\mathbf{e}}_{k-1} + \mathbf{w}_k] - \mathbf{K}_k\mathbf{v}_k. \tag{28.1.2}$$

Now compute the outer product $\widehat{\mathbf{x}}_k(\widehat{\mathbf{e}}_k)^{\mathrm{T}}$ and take expectation. Since $\widehat{\mathbf{x}}_{k-1}$ is orthogonal to $\widehat{\mathbf{e}}_{k-1}$, after considerable algebra (Exercise 28.1), we obtain (using the definition of $\mathbf{P}^{\mathrm{f}}_k$ in (27.2.13))

$$\begin{aligned} E[\widehat{\mathbf{x}}_k(\widehat{\mathbf{e}}_k)^{\mathrm{T}}] &= \mathbf{K}_k\mathbf{H}_k[\mathbf{M}_{k-1}\widehat{\mathbf{P}}_{k-1}\mathbf{M}^{\mathrm{T}}_{k-1} + \mathbf{Q}_k][\mathbf{I} - \mathbf{K}_k\mathbf{H}_k]^{\mathrm{T}} - \mathbf{K}_k\mathbf{R}_k\mathbf{K}^{\mathrm{T}}_k \\ &= \mathbf{K}_k\mathbf{H}_k\mathbf{P}^{\mathrm{f}}_k[\mathbf{I} - \mathbf{K}_k\mathbf{H}_k]^{\mathrm{T}} - \mathbf{K}_k\mathbf{R}_k\mathbf{K}^{\mathrm{T}}_k. \end{aligned} \tag{28.1.3}$$

From (27.2.23) and (27.2.25) we have

$$\widehat{\mathbf{P}}_k = (\mathbf{I} - \mathbf{K}_k\mathbf{H}_k)\mathbf{P}^{\mathrm{f}}_k \quad \text{and} \quad \mathbf{K}_k = \widehat{\mathbf{P}}_k\mathbf{H}^{\mathrm{T}}_k\mathbf{R}^{-1}_k.$$

Substituting these on the r.h.s.of (28.1.3) it can be verified that $E[\widehat{\mathbf{x}}_k(\widehat{\mathbf{e}}_k)^{\mathrm{T}}] = 0$, that is $\widehat{\mathbf{x}}_k$ is orthogonal to $\widehat{\mathbf{e}}_k$.

It should be interesting to note that Kalman's original derivation (Kalman [1960]) is based on the principle of orthogonal projection. This result in fact brings out the thread of unity that underlies the method of least squares including both the **deterministic** and **statistical** as well as the **dynamic** and **static** framework.

28.2 Effect of correlation between the model noise $\mathbf{w}_k$ and the observation noise $\mathbf{v}_k$

The derivation of the linear filter equations was predicated on the assumption that the model noise $\mathbf{w}_k$ and the observation noise $\mathbf{v}_k$ are **uncorrelated**. In this section we examine the effect of relaxing this assumption and its impact on the filter equations.

Let

$$E[\mathbf{w}_k\mathbf{v}^{\mathrm{T}}_m] = \mathbf{C}_k\delta_{km} \tag{28.2.1}$$

where (δ_{km} is 1 if $m = k$ and 0 otherwise) $\mathbf{C}_k \in \mathbb{R}^{n\times m}$. Consider the error $\widehat{\mathbf{e}}_k$ in the estimate given by (27.2.17):

$$\widehat{\mathbf{e}}_k = (\mathbf{I} - \mathbf{K}_k\mathbf{H}_k)[\mathbf{M}_{k-1}\widehat{\mathbf{e}}_{k-1} + \mathbf{w}_k] - \mathbf{K}_k\mathbf{v}_k. \tag{28.2.2}$$

The computation of the error covariance matrix $\widehat{\mathbf{P}}_k = E[\widehat{\mathbf{e}}_k(\widehat{\mathbf{e}}_k)^{\mathrm{T}}]$ must now account for the correlation in (28.2.1).

Taking the outer product $\widehat{\mathbf{e}}_k(\widehat{\mathbf{e}}_k)^{\mathrm{T}}$ and taking expectation, after considerable simplification we obtain (Exercise 28.2)

$$\begin{aligned}\widehat{\mathbf{P}}_k = [\mathbf{I} - \mathbf{K}_k\mathbf{H}_k][\mathbf{M}_{k-1}\widehat{\mathbf{P}}_{k-1}\mathbf{M}_{k-1}^{\mathrm{T}} + \mathbf{Q}_k][\mathbf{I} - \mathbf{K}_k\mathbf{H}_k]^{\mathrm{T}} + \mathbf{K}_k\mathbf{R}_k\mathbf{K}_k^{\mathrm{T}}\\ - (\mathbf{I} - \mathbf{K}_k\mathbf{H}_k)\mathbf{C}_k\mathbf{K}_k^{\mathrm{T}} - \mathbf{K}_k\mathbf{C}_k^{\mathrm{T}}(\mathbf{I} - \mathbf{K}_k\mathbf{H}_k)^{\mathrm{T}}.\end{aligned} \tag{28.2.3}$$

Using the definition of $\mathbf{P}_k^{\mathrm{f}}$ in (27.2.13), after some non-trivial simplification, we obtain

$$\begin{aligned}\widehat{\mathbf{P}}_k = \mathbf{P}_k^{\mathrm{f}} - (\mathbf{P}_k^{\mathrm{f}}\mathbf{H}_k^{\mathrm{T}} + \mathbf{C}_k)\mathbf{K}_k^{\mathrm{T}} - \mathbf{K}_k(\mathbf{H}_k\mathbf{P}_k^{\mathrm{f}} + \mathbf{C}_k^{\mathrm{T}})\\ + \mathbf{K}_k[\mathbf{H}_k\mathbf{P}_k^{\mathrm{f}}\mathbf{H}_k^{\mathrm{T}} + \mathbf{R}_k + (\mathbf{H}_k\mathbf{C}_k + \mathbf{C}_k^{\mathrm{T}}\mathbf{H}_k^{\mathrm{T}})]\mathbf{K}_k^{\mathrm{T}}.\end{aligned} \tag{28.2.4}$$

It can be verified that if $\mathbf{C}_k \equiv 0$, then (28.2.4) reduces to (27.2.19) as it should. Notice that the $\widehat{\mathbf{P}}_k$ in (28.2.4) is a quadratic in $\mathbf{K}_k$ and our goal is to find the $\mathbf{K}_k$ that will minimize the trace of $\widehat{\mathbf{P}}_k$. Since this problem has been encountered at least twice – first in Chapter 17 and again in Chapter 27, we only indicate the major steps, leaving the verification as an exercise (Exercise 28.3).

By way of simplifying the notation, define

$$\left.\begin{aligned}\mathbf{A} &= \mathbf{H}_k + \mathbf{C}_k(\mathbf{P}_k^{\mathrm{f}})^{-1}\\ \mathbf{B} &= \mathbf{H}_k\mathbf{P}_k^{\mathrm{f}}\mathbf{H}_k^{\mathrm{T}} + \mathbf{R}_k + (\mathbf{H}_k\mathbf{C}_k + \mathbf{C}_k^{\mathrm{T}}\mathbf{H}_k^{\mathrm{T}})\end{aligned}\right\} \tag{28.2.5}$$

Using these, we can rewrite (28.2.4) as

$$\widehat{\mathbf{P}}_k = \mathbf{P}_k^{\mathrm{f}} - \mathbf{P}_k^{\mathrm{f}}\mathbf{A}^{\mathrm{T}}\mathbf{K}_k^{\mathrm{T}} - \mathbf{K}_k\mathbf{A}\mathbf{P}_k^{\mathrm{f}} + \mathbf{K}_k\mathbf{B}\mathbf{K}_k^{\mathrm{T}}. \tag{28.2.6}$$

Now add and subtract $\mathbf{P}_k^{\mathrm{f}}(\mathbf{A}^{\mathrm{T}}\mathbf{B}^{-1}\mathbf{A})\mathbf{P}_k^{\mathrm{f}}$ to the r.h.s. of (28.2.6) we get

$$\begin{aligned}\widehat{\mathbf{P}}_k = \mathbf{P}_k^{\mathrm{f}} - \mathbf{P}_k^{\mathrm{f}}(\mathbf{A}^{\mathrm{T}}\mathbf{B}^{-1}\mathbf{A})\mathbf{P}_k^{\mathrm{f}}\\ + [\mathbf{K}_k - \mathbf{P}_k^{\mathrm{f}}\mathbf{A}^{\mathrm{T}}\mathbf{B}^{-1}]\mathbf{B}[\mathbf{K}_k - \mathbf{P}_k^{\mathrm{f}}\mathbf{A}^{\mathrm{T}}\mathbf{B}^{-1}]^{\mathrm{T}}.\end{aligned} \tag{28.2.7}$$

The minimizing $\mathbf{K}_k$ is then given by

$$\begin{aligned}\mathbf{K}_k &= \mathbf{P}_k^{\mathrm{f}}\mathbf{A}^{\mathrm{T}}\mathbf{B}^{-1}\\ &= [\mathbf{P}_k^{\mathrm{f}}\mathbf{H}_k^{\mathrm{T}} + \mathbf{C}_k][\mathbf{H}_k\mathbf{P}_k^{\mathrm{f}}\mathbf{H}_k^{\mathrm{T}} + \mathbf{R}_k + (\mathbf{H}_k\mathbf{C}_k + \mathbf{C}_k^{\mathrm{T}}\mathbf{H}_k^{\mathrm{T}})]^{-1}\end{aligned} \tag{28.2.8}$$

and the minimum value of $\widehat{\mathbf{P}}_k$ is given by (using 28.2.8)

$$\begin{aligned}\widehat{\mathbf{P}}_k &= \mathbf{P}_k^{\mathrm{f}} - \mathbf{P}_k^{\mathrm{f}}[\mathbf{A}^{\mathrm{T}}\mathbf{B}^{-1}\mathbf{A}]\mathbf{P}_k^{\mathrm{f}}\\ &= \mathbf{P}_k^{\mathrm{f}} - [\mathbf{P}_k^{\mathrm{f}}\mathbf{H}_k^{\mathrm{T}} + \mathbf{C}_k^{\mathrm{T}}][\mathbf{H}_k\mathbf{P}_k^{\mathrm{f}}\mathbf{H}_k^{\mathrm{T}} + \mathbf{R}_k + (\mathbf{H}_k\mathbf{C}_k + \mathbf{C}_k^{\mathrm{T}}\mathbf{H}_k^{\mathrm{T}})]^{-1}[\mathbf{H}_k\mathbf{P}_k^{\mathrm{f}} + \mathbf{C}_k]\\ &= \mathbf{P}_k^{\mathrm{f}} - \mathbf{K}_k[\mathbf{H}_k\mathbf{P}_k^{\mathrm{f}} + \mathbf{C}_k].\end{aligned} \tag{28.2.9}$$

When $\mathbf{C}_k \equiv 0$, this expression reduces to (27.2.23).

Stated in other words, the effect of correlation between $\mathbf{w}_k$ and $\mathbf{v}_k$ is directly reflected in the addition of the term $\mathbf{C}_k$ in the expression for $\mathbf{K}_k$ and $\widehat{\mathbf{P}}_k$ in (28.2.8) and (28.2.9), respectively.

28.3 Model bias/parameter estimation

It is often the case that the chosen linear dynamic model is only an approximation to the reality. The difference between the model and the reality is called the **model error or bias**. Since the model error is unknown, it is often convenient to think of this error as a random variable and divide it in two mutually exclusive parts: one part is to account for the **high frequency** (small wavelength) which is usually captured by the white noise term $\mathbf{w}_k$ in (27.2.1) and the other is to account for the **low frequency** (larger wavelength) which can be modelled by an **unknown random constant vector** $\boldsymbol{\alpha} \in \mathbb{R}^p$. This second component is accommodated by expanding the standard model equation as follows:

$$\mathbf{x}_{k+1} = \mathbf{M}_k\mathbf{x}_k + \mathbf{A}_k\boldsymbol{\alpha}_k + \mathbf{w}_{k+1} \tag{28.3.1}$$

where $\boldsymbol{\alpha}_k \equiv \boldsymbol{\alpha}$ is the unknown random vector and $\mathbf{A}_k \in \mathbb{R}^{n\times p}$ is the known sequence of matrices. It is assumed that $\mathrm{Cov}(\boldsymbol{\alpha}) = \mathbf{Q}_\alpha \in \mathbb{R}^{p\times p}$ is known.

Similarly, consider the observations

$$\mathbf{z}_k = \mathbf{H}_k\mathbf{x}_k + \mathbf{B}_k\boldsymbol{\beta}_k + \mathbf{v}_k \tag{28.3.2}$$

where $\boldsymbol{\beta}_k \equiv \boldsymbol{\beta} \in \mathbb{R}^q$ denotes the unknown low-frequency errors in the observation; $\mathbf{B}_k \in \mathbb{R}^{m\times q}$ is the known sequence of matrices and $\mathbf{v}_k$ denotes the usual (high-frequency) white noise sequence. Again, it is assumed that $\mathrm{Cov}(\boldsymbol{\beta}) = \mathbf{R}_\beta$ is known.

There is another useful interpretation for $\boldsymbol{\alpha}$ and $\boldsymbol{\beta}$. These may denote the unknown (deterministic) constants in the model. Notwithstanding its origin, our goal is to estimate both $\boldsymbol{\alpha}$ and $\boldsymbol{\beta}$ along with $\mathbf{x}_k$.

To this end, we define a new (expanded) state vector consisting of all the unknowns – $\mathbf{x}_k$, $\boldsymbol{\alpha}_k$, and $\boldsymbol{\beta}_k$ as follows
Let

$$\boldsymbol{\xi}_k = \begin{pmatrix} \mathbf{x}_k \\ \boldsymbol{\alpha}_k \\ \boldsymbol{\beta}_k \end{pmatrix} \in \mathbb{R}^{n+p+q}. \tag{28.3.3}$$

Then (28.3.1) and (28.3.2) can be rewritten as

$$\begin{pmatrix} \mathbf{x}_{k+1} \\ \boldsymbol{\alpha}_{k+1} \\ \boldsymbol{\beta}_{k+1} \end{pmatrix} = \begin{bmatrix} \mathbf{M}_k & \mathbf{A}_k & 0 \\ 0 & \mathbf{I}_p & 0 \\ 0 & 0 & \mathbf{I}_q \end{bmatrix} \begin{bmatrix} \mathbf{x}_k \\ \boldsymbol{\alpha}_k \\ \boldsymbol{\beta}_k \end{bmatrix} + \begin{bmatrix} \mathbf{w}_{k+1} \\ 0 \\ 0 \end{bmatrix}.$$

That is, the new model equation is

$$\boldsymbol{\xi}_{k+1} = \overline{\mathbf{M}}_k\boldsymbol{\xi}_k + \overline{\mathbf{w}}_{k+1} \tag{28.3.4}$$

where $\mathbf{I}_r$ denotes an identity matrix of size r;

$$\overline{\mathbf{M}}_k = \begin{bmatrix} \mathbf{M}_k & \mathbf{A}_k & 0 \\ 0 & \mathbf{I}_p & 0 \\ 0 & 0 & \mathbf{I}_q \end{bmatrix} \in \mathbb{R}^{(n+p+q)\times(n+p+r)}$$

and

$$\overline{\mathbf{w}}_k = \begin{bmatrix} \mathbf{w}_k \\ 0 \\ 0 \end{bmatrix} \in \mathbb{R}^{n+p+q}$$

where

$$\boldsymbol{\xi}_0 = \begin{pmatrix} \mathbf{x}_0 \\ \boldsymbol{\alpha}_0 \\ \boldsymbol{\beta}_0 \end{pmatrix} \quad \text{and} \quad \text{Cov}(\boldsymbol{\xi}_0) = \begin{bmatrix} \mathbf{P}_0 & 0 & 0 \\ 0 & \mathbf{Q}_\alpha & 0 \\ 0 & 0 & \mathbf{R}_\beta \end{bmatrix}.$$

Similarly, rewriting (28.3.2) as

$$\mathbf{z}_k = \begin{bmatrix} \mathbf{H}_k, & 0, & \mathbf{B}_k \end{bmatrix} \begin{bmatrix} \mathbf{x}_k \\ \boldsymbol{\alpha}_k \\ \boldsymbol{\beta}_k \end{bmatrix} + \mathbf{v}_k$$

or

$$\overline{\mathbf{z}}_k = \overline{\mathbf{H}}_k \boldsymbol{\xi}_k + \mathbf{v}_k \tag{28.3.5}$$

where

$$\overline{\mathbf{H}}_k = \begin{bmatrix} \mathbf{H}_k, & 0, & \mathbf{B}_k \end{bmatrix} \in \mathbb{R}^{m\times(n+p+q)}.$$

This expanded set of equations (28.3.4) and (28.3.5) are in the standard form as in (27.2.1) and (27.2.2) respectively. Hence we can directly apply the standard Kalman filter equations in Figure 27.2.2 to estimate $\boldsymbol{\xi}_k$ (Exercise 28.4).

28.4 Divergence of Kalman filter

Despite its elegance and simplicity, implementations of the Kalman filter algorithm have exhibited **unstable** behavior in the sense that the error $\widehat{\mathbf{e}}_k$ in the estimate $\widehat{\mathbf{x}}_k$ grows without bound. This divergence may result from one or more of the following factors:

(1) **Model bias and/or errors** These include errors in $\mathbf{M}_k$ and $\mathbf{H}_k$ in (27.1.1) and (27.1.2) respectively.

(2) **Errors in prior statistics** These errors relate to the various assumptions on $\mathbf{x}_0$ and $\mathbf{w}_k$ in (27.1.1) and $\mathbf{v}_k$ in (27.1.2). For example, it is assumed that the system noise sequence $\{\mathbf{w}_k\}$ is such that it has mean zero and is serially uncorrelated

with $\mathrm{Cov}(\mathbf{w}_k) = \mathbf{Q}_k$. One or more of these assumptions may not hold. Similarly, assumptions on $\mathbf{x}_0$ and $\{\mathbf{v}_k\}$ may not hold. Further, it is assumed that $\{\mathbf{w}_k\}$, $\{\mathbf{v}_k\}$ and $\mathbf{x}_0$ are mutually uncorrelated which may not be true.

(3) **Round-off errors** Numerical inaccuracies resulting from **finite precision** arithmetic can cause havoc to the integrity of the computation, such as loss of symmetry and/or positive definiteness of the covariance matrices $\widehat{\mathbf{P}}_k$ and $\mathbf{P}^f_k$.

(4) **Nonlinearity in the System** Exact solution to the nonlinear filtering gives rise to an infinite dimensional problem of determining the evolution of the conditional density function over the state space as a function of time. Refer to Chapter 29 for details. Approximations are the only recourse to solving this problem. Any mismatch between the chosen approximation scheme and the type of nonlinearity can cause difficulty in the computation of covariance matrices.

Effect of model errors and errors in prior statistics are examined in Section 28.5. In fact, Example 28.5.1 illustrates the divergence resulting from model error. Various methods for computing the covariance matrices and useful suggestions to maintain symmetry and positive definiteness are discussed in Section 28.6.

A standard approach to taming the effect of round-off is to reduce the **condition number** of the matrices involved in the computation. Recall that if the condition number of a matrix is 10^d, then small errors (including the errors in the data and in the round-off) in the computation are magnified by the factor 10^d. Consequently, if we are dealing with finite precision arithmetic accurate up to d decimals, then small errors in time can wipe out the overall quality of the computations. One natural method that is best suited for dealing with symmetric and positive definite (such as the covariance) matrices is to reformulate the computations using their so-called **square root** matrices. For example, if $\mathbf{P}$ is a symmetric and positive definite matrix, then there exists a nonsingular matrix $\mathbf{S}$ called the square root of $\mathbf{P}$ such that $\mathbf{P} = \mathbf{S}\mathbf{S}^{\mathrm{T}}$ (refer to Chapter 9). Then, the **spectral condition** number $\kappa_2(\mathbf{P})$ is given by

$$\kappa_2(\mathbf{P}) = \frac{\lambda_1(\mathbf{P})}{\lambda_n(\mathbf{P})} \tag{28.4.1}$$

where

$$\lambda_1(\mathbf{P}) \geq \lambda_2(\mathbf{P}) \geq \cdots \geq \lambda_n(\mathbf{P}) > 0$$

are the eigenvalues of $\mathbf{P}$. It can be verified (Appendix B) that

$$\kappa_2(\mathbf{P}) = \kappa_2(\mathbf{S}\mathbf{S}^{\mathrm{T}}) = [\kappa_2(\mathbf{S})]^2. \tag{28.4.2}$$

That is, if 10^d is the condition number of $\mathbf{P}$, then $10^{d/2}$ is that of $\mathbf{S}$. In other words, by performing the filter computations using $\mathbf{S}$ instead $\mathbf{P}$ we can definitely shield the integrity of the computations against runaway round-off errors. A version of the **square root algorithms** for linear filtering is described in Section 28.7.

Theory of nonlinear filtering in discrete time is covered exclusively in Chapter 29.

Analysis of divergence of filters is intimately related to the **stability** properties of the filter dynamics which are covered in Section 28.8.

28.5 Sensitivity of the linear filter

In this section we analyze the impact of the difference between values of the parameters that describe the **real or actual** system and those used by the **model**, when the filter equations are derived based on the model and not on the actual system.

Let the actual system dynamics be given by

$$\left.\begin{aligned} \bar{\mathbf{x}}_{k+1} &= \overline{\mathbf{M}}_k \bar{\mathbf{x}}_k + \overline{\mathbf{w}}_{k+1} \\ \bar{\mathbf{z}}_k &= \overline{\mathbf{H}}_k \bar{\mathbf{x}}_k + \bar{\mathbf{v}}_k \end{aligned}\right\} \tag{28.5.1}$$

where

$$E(\overline{\mathbf{w}}_k) = 0, \qquad \text{Cov}(\overline{\mathbf{w}}_k) = \overline{\mathbf{Q}}_k$$

$$E(\bar{\mathbf{v}}_k) = 0, \qquad \text{Cov}(\bar{\mathbf{v}}_k) = \overline{\mathbf{R}}_k$$

and the initial condition $\bar{\mathbf{x}}_0$ is such that $E(\bar{\mathbf{x}}_0) = \overline{\mathbf{m}}_0$ and $\text{Cov}(\bar{\mathbf{x}}_0) = \bar{\mathbf{P}}_0$. Not knowing the actual system, let the filter computations be based on a model given by

$$\left.\begin{aligned} \mathbf{x}_{k+1} &= \mathbf{M}_k \mathbf{x}_k + \mathbf{w}_{k+1} \\ \mathbf{z}_k &= \mathbf{H}_k \mathbf{x}_k + \mathbf{v}_k \end{aligned}\right\} \tag{28.5.2}$$

with

$$E(\mathbf{w}_k) = 0, \qquad \text{Cov}(\mathbf{w}_k) = \mathbf{Q}_k$$

$$E(\mathbf{v}_k) = 0, \qquad \text{Cov}(\mathbf{v}_k) = \mathbf{R}_k.$$

The filter equations are derived based on the specifications of the model (28.5.2) but the filter operates using the **real data**, $\bar{\mathbf{z}}_k$. For easy reference the filter equations are reproduced below (Refer to Figure 27.2.2 and (27.2.22))

$$\left.\begin{aligned} \mathbf{x}^{\text{f}}_{k+1} &= \mathbf{M}_k \widehat{\mathbf{x}}_k \\ \mathbf{P}^{\text{f}}_{k+1} &= \mathbf{M}_k \widehat{\mathbf{P}}_k \mathbf{M}_k^{\text{T}} + \mathbf{Q}_{k+1} \\ \widehat{\mathbf{x}}_{k+1} &= (\mathbf{I} - \mathbf{K}_{k+1}\mathbf{H}_{k+1})\mathbf{x}^{\text{f}}_{k+1} + \mathbf{K}_{k+1}\mathbf{z}_{k+1} \\ \widehat{\mathbf{P}}_{k+1} &= (\mathbf{I} - \mathbf{K}_{k+1}\mathbf{H}_{k+1})\mathbf{P}^{\text{f}}_{k+1}(\mathbf{I} - \mathbf{K}_{k+1}\mathbf{H}_{k+1})^{\text{T}} + \mathbf{K}_{k+1}\mathbf{R}_{k+1}\mathbf{K}^{\text{T}}_{k+1} \\ \mathbf{K}_{k+1} &= \mathbf{P}^{\text{f}}_{k+1}\mathbf{H}^{\text{f}}_{k+1}[\mathbf{H}_{k+1}\mathbf{P}^{\text{f}}_{k+1}\mathbf{H}^{\text{T}}_{k+1} + \mathbf{R}_{k+1}]^{-1} \end{aligned}\right\} \tag{28.5.3}$$

Notice that the correct filter equations must use $\overline{\mathbf{M}}_k$, $\overline{\mathbf{H}}_k$, $\overline{\mathbf{Q}}_k$, $\overline{\mathbf{R}}_k$, and $\bar{\mathbf{P}}_0$ in place of $\mathbf{M}_k$, $\mathbf{H}_k$, $\mathbf{Q}_k$, $\mathbf{R}_k$, and $\mathbf{P}_0$, respectively. Hence the covariance matrices $\mathbf{P}^{\mathrm{f}}_{k+1}$ and $\widehat{\mathbf{P}}_{k+1}$ based on the **computed** error

$$\mathbf{e}^{\mathrm{f}}_{k+1} = \mathbf{x}_{k+1} - \mathbf{x}^{\mathrm{f}}_{k+1} \quad \text{and} \quad \widehat{\mathbf{e}}_{k+1} = \mathbf{x}_{k+1} - \widehat{\mathbf{x}}_{k+1} \tag{28.5.4}$$

respectively are indeed in error. The **actual forecast covariance**, $\mathbf{G}^{\mathrm{f}}_{k+1}$ and the **actual estimate covariance** $\widehat{\mathbf{G}}_{k+1}$ are to be computed based on the **actual forecast and estimation errors** given by

$$\mathbf{g}^{\mathrm{f}}_{k+1} = \bar{\mathbf{x}}_{k+1} - \mathbf{x}^{\mathrm{f}}_{k+1} \quad \text{and} \quad \widehat{\mathbf{g}}_{k+1} = \bar{\mathbf{x}}_{k+1} - \widehat{\mathbf{x}}_{k+1} \tag{28.5.5}$$

respectively. Our goal is to analyze the dependence of the errors in the covariance matrices given by

$$\left.\begin{aligned} \mathbf{E}^{\mathrm{f}}_{k+1} &= \mathbf{G}^{\mathrm{f}}_{k+1} - \mathbf{P}^{\mathrm{f}}_{k+1} \\ \widehat{\mathbf{E}}_{k+1} &= \widehat{\mathbf{G}}_{k+1} - \widehat{\mathbf{P}}_{k+1} \end{aligned}\right\} \tag{28.5.6}$$

on (1) the **model errors**

$$\left.\begin{aligned} \Delta\mathbf{M}_k &= \overline{\mathbf{M}}_k - \mathbf{M}_k \\ \Delta\mathbf{H}_k &= \overline{\mathbf{H}}_k - \mathbf{H}_k \end{aligned}\right\} \tag{28.5.7}$$

(2) **noise covariance errors**

$$\left.\begin{aligned} \Delta\mathbf{Q}_k &= \overline{\mathbf{Q}}_k - \mathbf{Q}_k \\ \Delta\mathbf{R}_k &= \overline{\mathbf{R}}_k - \mathbf{R}_k \end{aligned}\right\} \tag{28.5.8}$$

and (3) the **initial covariance error**

$$\Delta\mathbf{P}_0 = \bar{\mathbf{P}}_0 - \mathbf{P}_0. \tag{28.5.9}$$

For convenience, we divide the analysis into two parts.

(A) Analysis of the error $\mathbf{E}^{\mathrm{f}}_{k+1}$

The actual forecast error $\mathbf{g}^{\mathrm{f}}_{k+1}$ is given by

$$\begin{aligned} \mathbf{g}^{\mathrm{f}}_{k+1} &= -\mathbf{x}^{\mathrm{f}}_{k+1} + \bar{\mathbf{x}}_{k+1} \\ &= -\mathbf{M}_k\widehat{\mathbf{x}}_k + \overline{\mathbf{M}}_k\bar{\mathbf{x}}_k + \mathbf{w}_{k+1}. \end{aligned} \tag{28.5.10}$$

Adding and subtracting $\mathbf{M}_k\bar{\mathbf{x}}_k$ to the r.h.s. and simplifying we obtain

$$\mathbf{g}^{\mathrm{f}}_{k+1} = \mathbf{M}_k\widehat{\mathbf{g}}_k + \Delta\mathbf{M}_k\bar{\mathbf{x}}_k + \overline{\mathbf{w}}_{k+1}. \tag{28.5.11}$$

Hence, the actual forecast covariance is given by (Exercise 28.5)

$$\begin{aligned}\mathbf{G}^{\mathrm{f}}_{k+1} &= E[\mathbf{g}^{\mathrm{f}}_{k+1}(\mathbf{g}^{\mathrm{f}}_{k+1})^{\mathrm{T}}]\\ &= \mathbf{M}_k\widehat{\mathbf{G}}_k\mathbf{M}_k^{\mathrm{T}} + \mathbf{M}_k E[\widehat{\mathbf{g}}_k(\bar{\mathbf{x}}_k)^{\mathrm{T}}](\Delta\mathbf{M}_k)^{\mathrm{T}}\\ &\quad + \Delta\mathbf{M}_k E[\bar{\mathbf{x}}_k(\widehat{\mathbf{g}}_k)^{\mathrm{T}}]\mathbf{M}_k^{\mathrm{T}} + \Delta\mathbf{M}_k E[\bar{\mathbf{x}}_k\bar{\mathbf{x}}_k^{\mathrm{T}}](\Delta\mathbf{M}_k)^{\mathrm{T}} + \overline{\mathbf{Q}}_{k+1}.\end{aligned} \tag{28.5.12}$$

Let

$$\left.\begin{aligned}\widehat{\mathbf{G}}_k(\bar{\mathbf{x}}) &= E[\widehat{\mathbf{g}}_k\bar{\mathbf{x}}_k^{\mathrm{T}}]\\ \text{and}\qquad \overline{\mathbf{X}}_k &= E[\bar{\mathbf{x}}_k\bar{\mathbf{x}}_k^{\mathrm{T}}]\end{aligned}\right\} \tag{28.5.13}$$

denote the cross moment matrix between $\widehat{\mathbf{g}}_k$ and $\bar{\mathbf{x}}_k$ and that of $\bar{\mathbf{x}}_k$ with itself respectively. Using these definitions, we obtain

$$\begin{aligned}\mathbf{G}^{\mathrm{f}}_{k+1} &= \mathbf{M}_k\widehat{\mathbf{G}}_k\mathbf{M}_k^{\mathrm{T}} + \mathbf{M}_k\widehat{\mathbf{G}}_k(\bar{\mathbf{x}})(\Delta\mathbf{M}_k)^{\mathrm{T}}\\ &\quad + \Delta\mathbf{M}_k\widehat{\mathbf{G}}_k^{\mathrm{T}}(\bar{\mathbf{x}})\mathbf{M}_k^{\mathrm{T}} + \Delta\mathbf{M}_k\overline{\mathbf{X}}_k(\Delta\mathbf{M}_k)^{\mathrm{T}} + \overline{\mathbf{Q}}_{k+1}.\end{aligned} \tag{28.5.14}$$

Hence using (28.5.3) and (28.5.12) the error in the computed forecast covariance is given by (Exercise 28.6)

$$\begin{aligned}\mathbf{E}^{\mathrm{f}}_{k+1} &= -\mathbf{P}^{\mathrm{f}}_{k+1} + \mathbf{G}^{\mathrm{f}}_{k+1}\\ &= \mathbf{M}_k\widehat{\mathbf{E}}_k\mathbf{M}_k^{\mathrm{T}} + \Delta\mathbf{Q}_k + \mathbf{M}_k\widehat{\mathbf{G}}_k(\bar{\mathbf{x}})(\Delta\mathbf{M}_k)^{\mathrm{T}}\\ &\quad + (\Delta\mathbf{M}_k)\widehat{\mathbf{G}}_k(\bar{\mathbf{x}})\mathbf{M}_k^{\mathrm{T}} + (\Delta\mathbf{M}_k)\overline{\mathbf{X}}_k(\Delta\mathbf{M}_k)^{\mathrm{T}}.\end{aligned} \tag{28.5.15}$$

(B) Analysis of the error $\widehat{\mathbf{E}}_{k+1}$

The actual error in the estimate is given by

$$\begin{aligned}\widehat{\mathbf{g}}_{k+1} &= -\widehat{\mathbf{x}}_{k+1} + \bar{\mathbf{x}}_{k+1}\\ &= -(\mathbf{I} - \mathbf{K}_{k+1}\mathbf{H}_{k+1})\mathbf{x}^{\mathrm{f}}_{k+1} + \mathbf{K}_{k+1}\bar{\mathbf{z}}_{k+1} + \bar{\mathbf{x}}_{k+1}.\end{aligned}$$

Adding and subtracting $(\mathbf{I} - \mathbf{K}_{k+1}\mathbf{H}_{k+1})\bar{\mathbf{x}}_{k+1}$ to the r.h.s. and simplifying we obtain

$$\widehat{\mathbf{g}}_{k+1} = (\mathbf{I} - \mathbf{K}_{k+1}\mathbf{H}_{k+1})\mathbf{g}^{\mathrm{f}}_{k+1} - \mathbf{K}_{k+1}[\bar{\mathbf{z}}_{k+1} - \mathbf{H}_{k+1}\bar{\mathbf{x}}_{k+1}].$$

Now using the **actual** measurement equation for $\bar{\mathbf{z}}_{k+1}$ in (28.5.1), we get

$$\widehat{\mathbf{g}}_{k+1} = (\mathbf{I} - \mathbf{K}_{k+1}\mathbf{H}_{k+1})\mathbf{g}^{\mathrm{f}}_{k+1} - \mathbf{K}_{k+1}\Delta\mathbf{H}_{k+1}\bar{\mathbf{x}}_{k+1} + \mathbf{K}_{k+1}\bar{\mathbf{v}}_{k+1}. \tag{28.5.16}$$

The covariance of this actual error after considerable simplification (Exercise 28.7) is given by

$$\begin{aligned}\widehat{\mathbf{G}}_{k+1} = &(\mathbf{I} - \mathbf{K}_{k+1}\mathbf{H}_{k+1})\mathbf{G}^{\mathrm{f}}_{k+1}(\mathbf{I} - \mathbf{K}_{k+1}\mathbf{H}_{k+1})^{\mathrm{T}} \\ &- (\mathbf{I} - \mathbf{K}_{k+1}\mathbf{H}_{k+1})\mathbf{G}^{\mathrm{f}}_{k+1}(\bar{\mathbf{x}})(\Delta\mathbf{H}_{k+1})^{\mathrm{T}}\mathbf{K}^{\mathrm{T}}_{k+1} \\ &- \mathbf{K}_{k+1}\Delta\mathbf{H}_{k+1}[\mathbf{G}^{\mathrm{f}}_{k+1}(\bar{\mathbf{x}})]^{\mathrm{T}}(\mathbf{I} - \mathbf{K}_{k+1}\mathbf{H}_{k+1})^{\mathrm{T}} \\ &+ \mathbf{K}_{k+1}(\Delta\mathbf{H}_{k+1})\mathbf{X}_{k+1}(\Delta\mathbf{H}_{k+1})^{\mathrm{T}}\mathbf{K}^{\mathrm{T}}_{k+1} + \mathbf{K}_{k+1}\overline{\mathbf{R}}_{k+1}\mathbf{K}^{\mathrm{T}}_{k+1}\end{aligned} \tag{28.5.17}$$

where (analogous to (28.5.11))

$$\left.\begin{aligned}\mathbf{G}^{\mathrm{f}}_{k+1}(\bar{\mathbf{x}}) &= E[\mathbf{g}^{\mathrm{f}}_{k+1}\bar{\mathbf{x}}^{\mathrm{T}}_{k+1}] \\ \text{and} \qquad \overline{\mathbf{X}}_{k+1} &= E[\bar{\mathbf{x}}_{k+1}\bar{\mathbf{x}}^{\mathrm{T}}_{k+1}]\end{aligned}\right\} \tag{28.5.18}$$

Hence using (28.5.3) and (28.5.16) the actual error in $\widehat{\mathbf{P}}_{k+1}$ is then given by (Exercise 28.8)

$$\begin{aligned}\widehat{\mathbf{E}}_{k+1} &= -\widehat{\mathbf{P}}_{k+1} + \widehat{\mathbf{G}}_{k+1} \\ &= (\mathbf{I} - \mathbf{K}_{k+1}\mathbf{H}_{k+1})\mathbf{E}^{\mathrm{f}}_{k+1}(\mathbf{I} - \mathbf{K}_{k+1}\mathbf{H}_{k+1})^{\mathrm{T}} \\ &\quad + \mathbf{K}_{k+1}(\Delta\mathbf{R}_{k+1})\mathbf{K}^{\mathrm{T}}_{k+1} \\ &\quad - (\mathbf{I} - \mathbf{K}_{k+1}\mathbf{H}_{k+1})\mathbf{G}^{\mathrm{f}}_{k+1}(\bar{\mathbf{x}})(\Delta\mathbf{H}_{k+1})^{\mathrm{T}}\mathbf{K}^{\mathrm{T}}_{k+1} \\ &\quad - \mathbf{K}_{k+1}\Delta\mathbf{H}_{k+1}[\mathbf{G}^{\mathrm{f}}_{k+1}(\bar{\mathbf{x}})]^{\mathrm{T}}(\mathbf{I} - \mathbf{K}_{k+1}\mathbf{H}_{k+1})^{\mathrm{T}} \\ &\quad + \mathbf{K}_{k+1}(\Delta\mathbf{H}_{k+1})\overline{\mathbf{X}}_{k+1}(\Delta\mathbf{H}_{k+1})^{\mathrm{T}}\mathbf{K}^{\mathrm{T}}_{k+1}.\end{aligned} \tag{28.5.19}$$

A special case

Suppose that the model is perfect but there are errors in $\mathbf{Q}_k$, $\mathbf{R}_k$, and $\mathbf{P}_0$. This implies that $\Delta\mathbf{M}_k \equiv 0$ and $\Delta\mathbf{H}_k \equiv 0$. Substituting this into (28.5.13) and (28.5.17) we get

$$\left.\begin{aligned}&\mathbf{E}^{\mathrm{f}}_{k+1} = \mathbf{M}_k\widehat{\mathbf{E}}_k\mathbf{M}^{\mathrm{T}}_k + \Delta\mathbf{Q}_k \\ \text{and} \quad &\widehat{\mathbf{E}}_{k+1} = (\mathbf{I} - \mathbf{K}_{k+1}\mathbf{H}_{k+1})\mathbf{E}^{\mathrm{f}}_{k+1}(\mathbf{I} - \mathbf{K}_{k+1}\mathbf{H}_{k+1})^{\mathrm{T}} \\ &\qquad\quad + \mathbf{K}_{k+1}(\Delta\mathbf{R}_{k+1})\mathbf{K}^{\mathrm{T}}_{k+1}.\end{aligned}\right\}$$

Now combining these, we obtain the following recurrence:

$$\begin{aligned}\widehat{\mathbf{E}}_{k+1} = &(\mathbf{I} - \mathbf{K}_{k+1}\mathbf{H}_{k+1})\mathbf{M}_k\widehat{\mathbf{E}}_k\mathbf{M}^{\mathrm{T}}_k(\mathbf{I} - \mathbf{K}_{k+1}\mathbf{H}_{k+1})^{\mathrm{T}} \\ &+ (\mathbf{I} - \mathbf{K}_{k+1}\mathbf{H}_{k+1})\Delta\mathbf{Q}_k(\mathbf{I} - \mathbf{K}_{k+1}\mathbf{H}_{k+1})^{\mathrm{T}} \\ &+ \mathbf{K}_{k+1}(\Delta\mathbf{R}_{k+1})\mathbf{K}^{\mathrm{T}}_{k+1}\end{aligned}$$

where $\widehat{\mathbf{E}}_0 = \widehat{\mathbf{P}}_0 - \widehat{\mathbf{G}}_0 = \mathbf{P}_0 - \bar{\mathbf{P}}_0 = \Delta\mathbf{P}_0$.

It can be verified (Jazwinski [1970]) that if $\widehat{\mathbf{E}}_0 \le 0$ with $\Delta\mathbf{Q}_k \le 0$ and $\Delta\mathbf{R}_k \le 0$ for all k, then $\widehat{\mathbf{E}}_k \le 0$ for all k as well. Stated in other words, if the actual covariances are bounded by their computed values, that is, $\overline{\mathbf{Q}}_k \le \mathbf{Q}_k$, $\overline{\mathbf{R}}_k \le \mathbf{R}_k$ and $\bar{\mathbf{P}}_0 \le \mathbf{P}_0$, then

$$\mathbf{G}^{\mathrm{f}}_{k+1} \le \mathbf{P}^{\mathrm{f}}_{k+1} \quad \text{and} \quad \widehat{\mathbf{G}}_{k+1} \le \widehat{\mathbf{P}}_{k+1}.$$

In practice, since the actual covariances are **not** known, one might conservatively estimate them using $\mathbf{Q}_k$ and $\mathbf{R}_k$. If the computed values are not satisfactory, one could then revise the estimates $\mathbf{Q}_k$ and $\mathbf{R}_k$ and redo the analysis all over again.

We conclude this section on the sensitivity analysis with one example that illustrates the divergence of the filter arising from the model error.

Example 28.5.1 Let $x \in \mathbb{R}$ denote the altitude of a space vehicle that is actually climbing at a constant speed s. The **actual** (scalar) system dynamics that describes this motion is given by

$$\overline{x}_{k+1} = \overline{x}_k + s = \overline{x}_0 + (k+1)s$$

where it is assumed that there is no model error. That is, $w_k \equiv 0$ and $Q_k \equiv 0$. The initial state $\overline{x}_0$ is random with mean $\overline{m}_0$ and variance $\overline{P}_0$. Let $\overline{H}_k \equiv 1$ and the **actual** observations of the system state are given by

$$\overline{z}_k = \overline{x}_k + v_k$$

where v_k is the white noise sequence with $E(v_k) = 0$ and $\mathrm{Var}(v_k) \equiv r$.

Let the filter be designed on the (wrong) assumption that the altitude is a **fixed** constant. That is, the model is given by ($M_k \equiv 1$)

$$x_{k+1} = x_k = x_0$$

with x_0 as the initial condition with mean m_0 and variance P_0.

In other words, $\Delta M_k \ne 0$ and $\Delta P_0 \ne 0$ but $\Delta H_k = 0$, $\Delta Q_k = 0$, $\Delta R_k = 0$. Specializing (28.5.3), the filter equations are given by

$$x^{\mathrm{f}}_{k+1} = \widehat{x}_k, \qquad P^{\mathrm{f}}_{k+1} = \widehat{P}_k, \qquad K_{k+1} = \frac{\widehat{P}_k}{\widehat{P}_k + r}$$

and

$$\begin{aligned}\widehat{P}_{k+1} &= (1 - K_{k+1})^2 \widehat{P}_k + K^2_{k+1} r \\ &= \frac{\widehat{P}_k r}{\widehat{P}_k + r}.\end{aligned}$$

Solving this recurrence, we obtain

$$\widehat{P}_k = \frac{\widehat{P}_0 r}{k\widehat{P}_0 + r} \quad \text{and} \quad K_{k+1} = \frac{\widehat{P}_0}{(k+1)\widehat{P}_0 + r}. \tag{28.5.20}$$

Hence, the estimate is given by

$$\begin{aligned}\widehat{x}_{k+1} &= \widehat{x}_k + K_{k+1}[\overline{z}_{k+1} - \widehat{x}_k]\\ &= \widehat{x}_k + K_{k+1}[(\overline{x}_k - \widehat{x}_k) + s + v_{k+1}].\end{aligned}$$

The actual error in the forecast, using (28.5.10) is

$$\begin{aligned}g^{\mathrm{f}}_{k+1} &= \overline{x}_{k+1} - x^{\mathrm{f}}_{k+1}\\ &= \widehat{g}_k + s\end{aligned}$$

and the actual error in the estimate is

$$\begin{aligned}\widehat{g}_{k+1} &= \overline{x}_{k+1} - \widehat{x}_{k+1}\\ &= (\overline{x}_k + s) - \{\widehat{x}_k + K_{k+1}[(\overline{x}_k - \widehat{x}_k) + s + v_{k+1}]\}\\ &= (1 - K_{k+1})(\widehat{g}_k + s) - K_{k+1}v_{k+1}\\ &= \alpha_{k+1}\widehat{g}_k + \beta_{k+1} \qquad (28.5.21)\end{aligned}$$

where

$$\left.\begin{aligned}\alpha_{k+1} &= 1 - K_{k+1} = \frac{k\widehat{P}_0 + r}{(k+1)\widehat{P}_0 + r}\\ \text{and}\qquad \beta_0 &= \widehat{g}_0 \quad \text{and} \quad \beta_k = \alpha_k s - K_k v_k\end{aligned}\right\} \qquad (28.5.22)$$

(28.5.21) is a linear first-order recurrence whose solution (refer to Exercise 27.1) is given by (after substituting for β_i using (28.5.22) and simplifying)

$$\begin{aligned}\widehat{g}_k &= \sum_{i=0}^{k}\left(\prod_{j=i+1}^{k}\alpha_j\right)\beta_i\\ &= \left(\prod_{j=1}^{k}\alpha_j\right)\beta_0 + \sum_{i=1}^{k}\left(\prod_{j=i}^{k}\alpha_j\right)s - \sum_{i=1}^{k}\left(\prod_{j=i+1}^{k}\alpha_j\right)K_i v_i.\end{aligned} \qquad (28.5.23)$$

It can be verified that using (28.5.22) and (28.5.20)

$$\prod_{j=i}^{k}\alpha_j = \frac{(i-1)\widehat{P}_0 + r}{k\widehat{P}_0 + r}$$

and

$$\left(\prod_{j=i+1}^{k}\alpha_j\right)K_i = \frac{\widehat{P}_0}{k\widehat{P}_0 + r}.$$

Substituting these into (28.5.23) and after some algebra we get

$$\widehat{g}_k = \left(\frac{r}{k\widehat{P}_0 + r}\right)\widehat{g}_0 + \sum_{i=1}^{k}\frac{[(i-1)\widehat{P}_0 + r]}{[k\widehat{P}_0 + r]}s - \sum_{i=1}^{k}\frac{\widehat{P}_0}{(k\widehat{P}_0 + r)}v_i. \qquad (28.5.24)$$

Table 28.6.1 *Different forms of* $\widehat{\mathbf{P}}_k$

Different Forms	$\widehat{\mathbf{P}}_k$
Form A	$\mathbf{P}^{\mathrm{f}}_k - \mathbf{K}_k\mathbf{H}_k\mathbf{P}^{\mathrm{f}}_k$
Form B	$(\mathbf{I} - \mathbf{K}_k\mathbf{H}_k)\mathbf{P}^{\mathrm{f}}_k$
Form C	$\mathbf{P}^{\mathrm{f}}_k - \mathbf{P}^{\mathrm{f}}_k\mathbf{H}^{\mathrm{T}}_k[\mathbf{H}_k\mathbf{P}^{\mathrm{f}}_k\mathbf{H}^{\mathrm{T}}_k + \mathbf{R}_k]^{-1}\mathbf{H}_k\mathbf{P}^{\mathrm{f}}_k$
Form D	$[(\mathbf{P}^{\mathrm{f}}_k)^{-1} + \mathbf{H}^{\mathrm{T}}_k\mathbf{R}^{-1}_k\mathbf{H}_k]^{-1}$
Form E	$(\mathbf{I} - \mathbf{K}_k\mathbf{H}_k)\mathbf{P}^{\mathrm{f}}_k(\mathbf{I} - \mathbf{K}_k\mathbf{H}_k)^{\mathrm{T}} + \mathbf{K}_k\mathbf{R}_k\mathbf{K}^{\mathrm{T}}_k$

The first term on the r.h.s. of (28.5.24) $\to 0$ as $k \to \infty$ and by the law of large numbers, the last term also tends to zero in the mean square sense. It can be verified that the middle is given by

$$\frac{s}{(k\widehat{P}_0 + r)}\left[\widehat{P}_0\frac{k(k-1)}{2} + rk\right] \longrightarrow \infty \quad \text{as} \quad k \to \infty.$$

Thus, the actual error $\widehat{g}_k$ in the estimate $\widehat{x}_k$ grows **unbounded**.

28.6 Computation of covariance matrices

From the derivation of the filter equation it is clear that there are various ways of organizing the computation of $\widehat{\mathbf{P}}_k$. Refer to Table 28.6.1 where $\mathbf{P}^{\mathrm{f}}_k$ is computed using

$$\mathbf{P}^{\mathrm{f}}_k = \mathbf{M}_k\widehat{\mathbf{P}}_{k-1}\mathbf{M}^{\mathrm{T}}_k + \mathbf{Q}_k. \tag{28.6.1}$$

Two of the key properties of $\widehat{\mathbf{P}}_k$ that must be monitored during the computation are: symmetry and positive definiteness. We now examine these various forms from the point of preserving these two properties.

(a) In form A, $\widehat{\mathbf{P}}_k$ is expressed as a difference of two symmetric matrices. This form, while preserving symmetry, might lead to loss of positive definiteness resulting from the cancellation of large numbers.
(b) In form B, $\widehat{\mathbf{P}}_k$ is expressed as a product of symmetric and non-symmetric matrices which could lead to loss of both symmetry and positive definiteness.
(c) Form C expresses $\widehat{\mathbf{P}}_k$ as a difference of two symmetric matrices of which one involves the inverse of an $m \times m$ matrix. This form is preferable when $m < n$.
(d) Form D known as the **information form** (Chapter 26) is preferable when $n < m$.
(e) Form E, also known as the **Joseph's form**, gives $\widehat{\mathbf{P}}_k$ as the sum of positive definite and positive semidefinite matrices. This form while involving a lot more computation, has an important and a desirable property of being robust with respect to perturbation in $\mathbf{K}_k$.

To verify this claim, let $\mathbf{\Delta} \in \mathbb{R}^{n\times m}$ denote the perturbation in $\mathbf{K}_k \in \mathbb{R}^{n\times m}$. Let $\delta\widehat{\mathbf{P}}_k$ be the perturbation $\widehat{\mathbf{P}}_k$ induced by the perturbation in $\mathbf{K}_k$. Then, using the form E, we obtain (where we drop all the subscripts for convenience) that

$$\begin{aligned}\widehat{\mathbf{P}}_k + \delta\widehat{\mathbf{P}}_k &= [(\mathbf{I} - \mathbf{KH}) - \mathbf{\Delta H}]^{\mathrm{T}}\mathbf{P}^{\mathrm{f}}[(\mathbf{I} - \mathbf{KH}) - \mathbf{\Delta H}] \\ &\quad + (\mathbf{K} + \mathbf{\Delta})\mathbf{R}(\mathbf{K} + \mathbf{\Delta})^{\mathrm{T}}. \end{aligned} \tag{28.6.2}$$

Using $\widehat{\mathbf{P}}_k = (\mathbf{I} - \mathbf{K}_k\mathbf{H}_k)\mathbf{P}^{\mathrm{f}}_k$, after simplification we obtain

$$\begin{aligned}\delta\widehat{\mathbf{P}}_k &= -(\widehat{\mathbf{P}}_k\mathbf{H}_k^{\mathrm{T}} - \mathbf{K}_k\mathbf{R}_k)\mathbf{\Delta}^{\mathrm{T}} - \mathbf{\Delta}(\mathbf{H}_k\widehat{\mathbf{P}}_k - \mathbf{R}_k\mathbf{K}^{\mathrm{T}}) \\ &\quad + \mathbf{\Delta}(\mathbf{H}_k\mathbf{P}^{\mathrm{f}}_k\mathbf{H}_k^{\mathrm{T}} + \mathbf{R}_k)\mathbf{\Delta}^{\mathrm{T}}. \end{aligned} \tag{28.6.3}$$

Since $\mathbf{K}_k = \widehat{\mathbf{P}}_k\mathbf{H}_k^{\mathrm{T}}\mathbf{R}_k^{-1}$, the first-order terms in $\mathbf{\Delta}$ vanish leaving behind

$$\delta\widehat{\mathbf{P}}_k = \mathbf{\Delta}(\mathbf{H}_k\mathbf{P}^{\mathrm{f}}_k\mathbf{H}_k^{\mathrm{T}} + \mathbf{R}_k)\mathbf{\Delta}^{\mathrm{T}} \tag{28.6.4}$$

which is of second order in $\mathbf{\Delta}$, which verifies the claim. It can be verified that all the other forms A through D do not share this property (Exercise 28.11)

In all the forms A through E in Table 28.6.1 computation of $\widehat{\mathbf{P}}_k$ depends on the availability of $\mathbf{P}^{\mathrm{f}}_k$. It turns out that the computation of $\mathbf{P}^{\mathrm{f}}_k$ using (28.6.1) is indeed the most expensive part requiring an equivalent of $2n$ model runs. When n is very large, this could be a major bottleneck which can be alleviated in part by using parallel computation.

Another special case where numerical difficulties are known to arise is when the measurements are more accurate in the sense that the spectral radius of $\mathbf{R}_k$ is much smaller compared to that of $\widehat{\mathbf{P}}_k$. This case is discussed in Section 27.2 – refer especially to the comment on the **Impact of Perfect Observations**.

We conclude this discussion with the following useful guidelines: (1) compute only the upper triangular part and restore symmetry or compute the full matrix $\mathbf{P}$ and replace it with $\frac{1}{2}(\mathbf{P} + \mathbf{P}^{\mathrm{T}})$, the symmetric part of $\mathbf{P}$ and (2) use Joseph's form in computing $\widehat{\mathbf{P}}_k$.

28.7 Square root algorithm

In this section we describe a family of ideas leading to numerically stable implementations of the Kalman filter equations. This idea is rooted in the fundamental property of any symmetric positive definite (SPD) matrix, namely that it can be expressed as the product of its square root matrix.

Recall from Chapter 9 that any symmetric and positive definite matrix $\mathbf{A}$ can be expressed as a product of factors in two different ways:

$$\mathbf{A} = \mathbf{L}\mathbf{L}^{\mathrm{T}} \quad \text{and} \quad \mathbf{A} = \mathbf{S}^2 \tag{28.7.1}$$

where $\mathbf{L}$ is a lower triangular matrix called the Cholesky factor of $\mathbf{A}$ and $\mathbf{S}$ is a symmetric and positive definite matrix called the square root of $\mathbf{A}$.

There is another natural way to factorize $\mathbf{A}$ that is based on the eigen decomposition. Let $(\lambda_i, \mathbf{x}_i)$ be the eigenvalue-vector pair of $\mathbf{A}$, $i = 1$ to n. Let $\mathbf{X} = [\mathbf{x}_1, \mathbf{x}_2, \ldots, \mathbf{x}_n] \in \mathbb{R}^{n\times n}$ be the orthonormal matrix of eigenvectors of $\mathbf{A}$, that is $\mathbf{X}^{\mathrm{T}}\mathbf{X} = \mathbf{X}\mathbf{X}^{\mathrm{T}} = \mathbf{I}$ and $\mathbf{\Lambda} = \mathrm{Diag}(\lambda_1, \lambda_2, \ldots, \lambda_n)$ be the diagonal matrix of eigenvalues of $\mathbf{A}$ where without loss of generality

$$\lambda_1 \geq \lambda_2 \geq \cdots \geq \lambda_n > 0.$$

Then from $\mathbf{AX} = \mathbf{X\Lambda}$ (Appendix B), we get

$$\mathbf{A} = \mathbf{X\Lambda X}^{\mathrm{T}} = \mathbf{X\Lambda}^{1/2}\mathbf{\Lambda}^{1/2}\mathbf{X}^{\mathrm{T}} = \bar{\mathbf{X}}\bar{\mathbf{X}}^{\mathrm{T}} \tag{28.7.2}$$

where $\bar{\mathbf{X}} = [\bar{\mathbf{x}}_1, \bar{\mathbf{x}}_2, \ldots, \bar{\mathbf{x}}_n]$ and $\bar{\mathbf{x}}_i = \mathbf{x}_i\sqrt{\lambda_i}$ is a factor of $\mathbf{A}$. Notice that the ith column of $\bar{\mathbf{X}}$ is the ith eigenvector scaled by the square root of the ith eigenvalue λ_i of $\mathbf{A}$.

The following example, illustrates these three forms of factorization. It can be verified by direct computation that if

$$\mathbf{A} = \begin{bmatrix} 1 & 3/2 \\ 3/2 & 7/2 \end{bmatrix}$$

then $\mathbf{A} = \mathbf{LL}^{\mathrm{T}} = \mathbf{S}^2 = \bar{\mathbf{X}}\bar{\mathbf{X}}^{\mathrm{T}}$ where

$$\mathbf{L} = \begin{bmatrix} 1 & 0 \\ 3/2 & \sqrt{5}/2 \end{bmatrix}, \qquad \mathbf{S} = \begin{bmatrix} 0.8161 & 0.5779 \\ 0.5779 & 1.7793 \end{bmatrix}$$

and

$$\bar{\mathbf{X}} = \begin{bmatrix} -0.4939 & 0.8695 \\ 0.2313 & 1.8565 \end{bmatrix}.$$

Remark 28.7.1 Reduced rank factorization
Partition $\bar{\mathbf{X}}$ into two submatrices $\bar{\mathbf{X}}(1:r) \in \mathbb{R}^{n\times n}$ and $\bar{\mathbf{X}}(r+1:n) \in \mathbb{R}^{n\times(n-r)}$ consisting of the first r columns and the last $(n-r)$ columns respectively. That is,

$$\bar{\mathbf{X}}(1:r) = [\bar{\mathbf{x}}_1, \bar{\mathbf{x}}_2, \ldots, \bar{\mathbf{x}}_r] \quad \text{and} \quad \bar{\mathbf{X}}(r+1:n) = [\bar{\mathbf{x}}_{r+1}, \ldots, \bar{\mathbf{x}}_n]. \tag{28.7.3}$$

Then, it can be verified that

$$\mathbf{A} = \bar{\mathbf{X}}(1:r)\bar{\mathbf{X}}^{\mathrm{T}}(1:r) + \bar{\mathbf{X}}(r+1:n)\bar{\mathbf{X}}^{\mathrm{T}}(r+1:n). \tag{28.7.4}$$

Since $\lambda_1, \lambda_2, \ldots, \lambda_r$ denote the r largest or the dominant eigenvalues of $\mathbf{A}$, we can approximate

$$\mathbf{A} \approx \bar{\mathbf{X}}(1:r)\bar{\mathbf{X}}^{\mathrm{T}}(1:r)$$

where the $\mathrm{Rank}(\bar{\mathbf{X}}(1:r)) = r < n$. This class of reduced rank approximation is often used to reduce the heavy computational burden (refer to Chapter 27) involved in updating the covariance matrices in the Kalman filter. Refer to Chapter 30 for details.

Potter in 1963 developed the basic ideas of the square root algorithm by considering the special case when there was no dynamics noise ($w_k \equiv 0$) and the observations are scalars ($m = 1$ and $z_k \in \mathbb{R}$). The elegance and the simplicity of this idea combined with its core strength of inducing good numerical stability provided great impetus for extending this idea in several directions by numerous authors. The book by Bierman (1977) entitled **Factorization Methods for Discrete Sequential Estimation** is devoted in its entirety to the analysis of the square root algorithms and is a good source for implementable versions of this class of algorithms. In the following we present a succinct summary of the developments in this area.

Let

$$\mathbf{P}_k^{\mathrm{f}} = \mathbf{s}_k^{\mathrm{f}}(\mathbf{s}_k^{\mathrm{f}})^{\mathrm{T}}, \quad \widehat{\mathbf{P}}_k = \widehat{\mathbf{s}}_k(\widehat{\mathbf{s}}_k)^{\mathrm{T}} \quad \text{and} \quad \mathbf{Q}_k = \mathbf{s}_k^q(\mathbf{s}_k^q)^{\mathrm{T}} \tag{28.7.5}$$

be the given factorization. The goal is to rewrite the Kalman filter algorithm in Figure 27.2.1 where the covariance update is replaced by their corresponding square root update relations.

Forecast Step Consider the update of the forecast covariance matrix $\mathbf{P}_{k+1}^{\mathrm{f}}$ given by (refer to Figure 27.2.2)

$$\mathbf{P}_{k+1}^{\mathrm{f}} = \mathbf{M}_k \widehat{\mathbf{P}}_k \mathbf{M}_k^{\mathrm{T}} + \mathbf{Q}_{k+1}.$$

Using the factorizations given in (28.7.5), we can rewrite the above relation as

$$\begin{aligned}\mathbf{P}_{k+1}^{\mathrm{f}} &= \mathbf{M}_k \widehat{\mathbf{s}}_k (\widehat{\mathbf{s}}_k)^{\mathrm{T}} \mathbf{M}_k^{\mathrm{T}} + \mathbf{s}_{k+1}^q (\mathbf{s}_{k+1}^q)^{\mathrm{T}} \\ &= [\mathbf{M}_k \widehat{\mathbf{s}}_k, \mathbf{s}_{k+1}^q] \begin{bmatrix} (\mathbf{M}_k \widehat{\mathbf{s}}_k)^{\mathrm{T}} \\ (\mathbf{s}_{k+1}^q)^{\mathrm{T}} \end{bmatrix} \\ &= \bar{\mathbf{s}}_{k+1}^{\mathrm{f}} (\bar{\mathbf{s}}_{k+1}^{\mathrm{f}})^{\mathrm{T}} \end{aligned} \tag{28.7.6}$$

where

$$\bar{\mathbf{s}}_{k+1}^{\mathrm{f}} = [\mathbf{M}_k \widehat{\mathbf{s}}_k, \mathbf{s}_{k+1}^q] \in \mathbb{R}^{n \times 2n}. \tag{28.7.7}$$

Notice that while we have achieved our goal of rewriting the update equation for the forecast covariance matrix $\mathbf{P}_{k+1}^{\mathrm{f}}$ in terms of its square root matrix $\bar{\mathbf{s}}_{k+1}^{\mathrm{f}}$ in (28.7.7), this action has also created an undesirable side effect of requiring $\bar{\mathbf{s}}_{k+1}^{\mathrm{f}}$ to be an $n \times 2n$ instead of an $n \times n$ matrix. This doubling of the number of columns has a doubling effect on both storage and time. Our immediate task is therefore to transform $\bar{\mathbf{s}}_{k+1}^{\mathrm{f}} \in \mathbb{R}^{n \times 2n}$ to a new matrix $\mathbf{s}_{k+1}^{\mathrm{f}} \in \mathbb{R}^{n \times n}$ such that

$$\mathbf{P}_{k+1}^{\mathrm{f}} = \mathbf{s}_{k+1}^{\mathrm{f}} (\mathbf{s}_{k+1}^{\mathrm{f}})^{\mathrm{T}}. \tag{28.7.8}$$

This can be readily accomplished, thanks to the **QR-decomposition** method using the **Gram–Schmidt orthogonalization** procedure described in Chapter 9. According to Section 9.2, given $(\overline{\mathbf{s}}^{\mathrm{f}}_{k+1})^{\mathrm{T}} \in \mathbb{R}^{2n\times n}$, there exists a $\mathbf{Q} \in \mathbb{R}^{2n\times n}$ such that $\mathbf{Q}^{\mathrm{T}}\mathbf{Q} = \mathbf{I}$, the identity matrix and an upper triangular matrix $(\mathbf{s}^{\mathrm{f}}_{k+1})^{\mathrm{T}} \in \mathbb{R}^{n\times n}$ such that

$$(\overline{\mathbf{s}}^{\mathrm{f}}_{k+1})^{\mathrm{T}} = \mathbf{Q}(\mathbf{s}^{\mathrm{f}}_{k+1})^{\mathrm{T}}.$$

Substituting this into (28.7.6), we readily obtain

$$\begin{aligned}\mathbf{P}^{\mathrm{f}}_{k+1} &= \overline{\mathbf{s}}^{\mathrm{f}}_{k+1}(\overline{\mathbf{s}}^{\mathrm{f}}_{k+1})^{\mathrm{T}} \\ &= \mathbf{s}^{\mathrm{f}}_{k+1}\mathbf{Q}^{\mathrm{T}}\mathbf{Q}(\mathbf{s}^{\mathrm{f}}_{k+1})^{\mathrm{T}} \\ &= \mathbf{s}^{\mathrm{f}}_{k+1}(\mathbf{s}^{\mathrm{f}}_{k+1})^{\mathrm{T}}.\end{aligned}$$

Thus, given $\widehat{\mathbf{x}}_k$, $\widehat{\mathbf{s}}_k$, and $\mathbf{s}^{q}_{k+1}$, we can write the forecast step as follows:

$$\begin{aligned}\mathbf{x}^{\mathrm{f}}_{k+1} &= \mathbf{M}_k\widehat{\mathbf{x}}_k \\ \mathbf{s}^{\mathrm{f}}_{k+1} &= \overline{\mathbf{s}}^{\mathrm{f}}_{k+1}\mathbf{Q} = [\mathbf{M}_k\widehat{\mathbf{s}}_k, \mathbf{s}^{q}_{k+1}]\mathbf{Q}.\end{aligned}$$

Data Assimilation Step We begin by rewriting the expression for the Kalman gain. To this end, define

$$\mathbf{A} = (\mathbf{H}_{k+1}\mathbf{s}^{\mathrm{f}}_{k+1})^{\mathrm{T}} \in \mathbb{R}^{n\times m}. \tag{28.7.9}$$

Then using (28.7.5) and (28.7.9), we get (refer to Figure 27.2.2)

$$\begin{aligned}\mathbf{K}_{k+1} &= \mathbf{P}^{\mathrm{f}}_{k+1}\mathbf{H}^{\mathrm{T}}_{k+1}[\mathbf{H}_{k+1}\mathbf{P}^{\mathrm{f}}_{k+1}\mathbf{H}^{\mathrm{T}}_{k+1} + \mathbf{R}_{k+1}]^{-1} \\ &= \mathbf{s}^{\mathrm{f}}_{k+1}\mathbf{A}[\mathbf{A}^{\mathrm{T}}\mathbf{A} + \mathbf{R}_{k+1}]^{-1}.\end{aligned} \tag{28.7.10}$$

Using this in the covariance update relation for $\widehat{\mathbf{P}}_{k+1}$ (refer to Figure 27.2.2), we obtain

$$\begin{aligned}\widehat{\mathbf{P}}_{k+1} &= (\mathbf{I} - \mathbf{K}_{k+1}\mathbf{H}_{k+1})\mathbf{P}^{\mathrm{f}}_{k+1} \\ &= \mathbf{P}^{\mathrm{f}}_{k+1} - \mathbf{K}_{k+1}\mathbf{H}_{k+1}\mathbf{P}^{\mathrm{f}}_{k+1} \\ &= \mathbf{s}^{\mathrm{f}}_{k+1}[\mathbf{I} - \mathbf{A}(\mathbf{A}^{\mathrm{T}}\mathbf{A} + \mathbf{R}_{k+1})^{-1}\mathbf{A}^{\mathrm{T}}](\mathbf{s}^{\mathrm{f}}_{k+1})^{\mathrm{T}}.\end{aligned} \tag{28.7.11}$$

The goal of every square root algorithm is to factorize the $n \times n$ matrix inside the square bracket above in terms of its square root matrix. This is achieved in the following steps:

(1) Compute the matrix $\mathbf{B} \in \mathbb{R}^{m\times n}$ as the matrix solution of the $m \times m$ system

$$(\mathbf{A}^{\mathrm{T}}\mathbf{A} + \mathbf{R}_{k+1})\mathbf{B} = \mathbf{A}^{\mathrm{T}}. \tag{28.7.12}$$

Model $\mathbf{x}_{k+1} = \mathbf{M}_k\mathbf{x}_k + \mathbf{w}_{k+1}$

Observation $\mathbf{z}_k = \mathbf{H}_k\mathbf{x}_k + \mathbf{v}_k$

Forecast Step

$$\mathbf{x}^{\mathrm{f}}_{k+1} = \mathbf{M}_k\widehat{\mathbf{x}}_k$$

$$\mathbf{s}^{\mathrm{f}}_{k+1} = [\mathbf{M}_k\widehat{\mathbf{s}}_k, \mathbf{s}^{q}_{k+1}]\mathbf{Q}$$

where $\mathbf{Q} \in \mathbb{R}^{2n\times n}$ is such that $\mathbf{Q}^{\mathrm{T}}\mathbf{Q} = \mathbf{I}$

Data Assimilation Step

$$\widehat{\mathbf{x}}_{k+1} = \mathbf{x}^{\mathrm{f}}_{k+1} + \mathbf{K}_{k+1}[\mathbf{z}_{k+1} - \mathbf{H}_{k+1}\mathbf{x}^{\mathrm{f}}_{k+1}]$$

$$\mathbf{K}_{k+1} = \mathbf{s}^{\mathrm{f}}_{k+1}\mathbf{A}[\mathbf{A}^{\mathrm{T}}\mathbf{A} + \mathbf{R}_{k+1}]^{-1}$$

$$\mathbf{A} = (\mathbf{H}_{k+1}\mathbf{s}^{\mathrm{f}}_{k+1})^{\mathrm{T}}$$

$$\widehat{\mathbf{s}}_{k+1} = \mathbf{s}^{\mathrm{f}}_{k+1}\mathbf{C}$$

where

$$\mathbf{C}\mathbf{C}^{\mathrm{T}} = (\mathbf{I} - \mathbf{A}\mathbf{B}), \quad \mathbf{B} = (\mathbf{A}^{\mathrm{T}}\mathbf{A} + \mathbf{R}_{k+1})^{-1}\mathbf{A}^{\mathrm{T}}$$

Fig. 28.7.1 Covariance form of the square root algorithm.

(2) Find the square root $\mathbf{C} \in \mathbb{R}^{n\times n}$ satisfying

$$(\mathbf{I} - \mathbf{A}\mathbf{B}) = \mathbf{C}\mathbf{C}^{\mathrm{T}}. \tag{28.7.13}$$

(3) Substituting (28.7.12) and (28.7.13) into (28.7.11) we get

$$\begin{aligned}\mathbf{P}^{\mathrm{f}}_{k+1} &= \mathbf{s}^{\mathrm{f}}_{k+1}\mathbf{C}\mathbf{C}^{\mathrm{T}}(\mathbf{s}^{\mathrm{f}}_{k+1})^{\mathrm{T}} \\ &= \widehat{\mathbf{s}}_{k+1}(\widehat{\mathbf{s}}_{k+1})^{\mathrm{T}}\end{aligned} \tag{28.7.14}$$

where the required square root of $\widehat{\mathbf{P}}_{k+1}$ is given by

$$\widehat{\mathbf{s}}_{k+1} = \mathbf{s}^{\mathrm{f}}_{k+1}\mathbf{C}. \tag{28.7.15}$$

A summary of the square root algorithm is given in Figure 28.7.1.

We now describe an example of the square root algorithm for the special case of scalar observations.

Example 28.7.1 Potter's algorithm. Let $m = 1$, $\mathbf{H}_k = \mathbf{H} \in \mathbb{R}^{1\times n}$, a row vector of size n and $R_k \equiv r$, a positive scalar. Since the forecast step remains the same, we only need to consider the data assimilation step. In this case,

$$\mathbf{H}_{k+1}\mathbf{P}^{\mathrm{f}}_{k+1}\mathbf{H}^{\mathrm{T}}_{k+1} = \mathbf{H}\mathbf{P}^{\mathrm{f}}_{k+1}\mathbf{H} \text{ is a } \textbf{scalar} \text{ as is } R_k$$

and

$$\mathbf{A} = (\mathbf{H}_{k+1}\mathbf{s}^{\mathrm{f}}_{k+1})^{\mathrm{T}} = (\mathbf{H}\mathbf{s}^{\mathrm{f}}_{k+1})^{\mathrm{T}} \in \mathbb{R}^n, \text{ a } \textbf{column vector}.$$

Define a scalar

$$\alpha = (\mathbf{A}^{\mathrm{T}}\mathbf{A} + r)^{-1}.$$

Then, from (28.7.10), the Kalman gain is given by

$$\mathbf{K}_{k+1} = \alpha \mathbf{s}^{\mathrm{f}}_{k+1}\mathbf{A} \in \mathbb{R}^n, \text{ a column vector.}$$

and from (28.7.11) the covariance matrix is given by

$$\widehat{\mathbf{P}}_{k+1} = \mathbf{s}^{\mathrm{f}}_{k+1}[\mathbf{I} - \alpha\mathbf{A}\mathbf{A}^{\mathrm{T}}](\mathbf{s}^{\mathrm{f}}_{k+1})^{\mathrm{T}} \tag{28.7.16}$$

where the symmetric matrix $[\mathbf{I} - \alpha\mathbf{A}\mathbf{A}^{\mathrm{T}}]$ is called the **rank-one update** of $\mathbf{I}$ by the rank-one-symmetric matrix $\mathbf{A}\mathbf{A}^{\mathrm{T}}$. It turns out that this matrix can be easily expressed as a **square of a symmetric matrix** as

$$(\mathbf{I} - \alpha\mathbf{A}\mathbf{A}^{\mathrm{T}}) = (\mathbf{I} - \beta\mathbf{A}\mathbf{A}^{\mathrm{T}})^2. \tag{28.7.17}$$

Expanding the r.h.s. of (28.7.17) and equating the coefficients of the corresponding terms, β is then obtained as the solution of the quadratic equation

$$(\mathbf{A}^{\mathrm{T}}\mathbf{A})\beta^2 - 2\beta + \alpha = 0$$

that is,

$$\beta = \frac{1 \pm \sqrt{1 - \alpha\mathbf{A}^{\mathrm{T}}\mathbf{A}}}{\mathbf{A}^{\mathrm{T}}\mathbf{A}}.$$

Substituting $\mathbf{A}^{\mathrm{T}}\mathbf{A} = \alpha^{-1}(1 - \alpha r)$ and simplifying, it can be verified that

$$\beta = \alpha(1 \pm \sqrt{\alpha r})^{-1}.$$

Using $\beta = \alpha(1 + \sqrt{\alpha r})^{-1}$ in (28.7.17) and substituting in (28.7.16) we get

$$\begin{aligned}\widehat{\mathbf{P}}_{k+1} &= \mathbf{s}^{\mathrm{f}}_{k+1}[\mathbf{I} - \beta\mathbf{A}\mathbf{A}^{\mathrm{T}}]^2(\mathbf{s}^{\mathrm{f}}_{k+1})^{\mathrm{T}} \\ &= \widehat{\mathbf{S}}_{k+1}\widehat{\mathbf{S}}_{k+1}\end{aligned}$$

or

$$\widehat{\mathbf{S}}_{k+1} = \mathbf{s}^{\mathrm{f}}_{k+1}[\mathbf{I} - \beta\mathbf{A}\mathbf{A}^{\mathrm{T}}].$$

A number of observations are in order.

(1) **Whitening filter and scalar observations**

Let $\mathbf{z} \in \mathbb{R}^m$ be a vector of observations given by

$$\mathbf{z} = \mathbf{H}\mathbf{x} + \mathbf{v} \tag{28.7.18}$$

where $\mathbf{H} \in \mathbb{R}^{m\times n}$, $\mathbf{x} \in \mathbb{R}^n$, and $\mathbf{v} \in \mathbb{R}^m$, is such that $E(\mathbf{v}) = 0$ and $\mathrm{Cov}(\mathbf{v}) = \mathbf{R} \in \mathbb{R}^{m\times m}$ is a symmetric and positive definite matrix. Let $\mathbf{R} = \mathbf{L}\mathbf{L}^{\mathrm{T}}$ be the Cholesky decomposition of $\mathbf{R}$ (Refer to Chapter 9 for details). Multiplying both sides of (28.7.18) by $\mathbf{L}^{-1}$, we obtain a transformed set of observations given by

$$\overline{\mathbf{z}} = \overline{\mathbf{H}}\mathbf{x} + \overline{\mathbf{v}} \tag{28.7.19}$$

where $\overline{\mathbf{z}} = \mathbf{L}^{-1}\mathbf{z}$, $\overline{\mathbf{H}} = \mathbf{L}^{-1}\mathbf{H}$, and $\overline{\mathbf{v}} = \mathbf{L}^{-1}\mathbf{v}$. It follows that $E(\mathbf{v}) = 0$ and

$$\begin{aligned} \text{Cov}(\mathbf{v}) &= E(\overline{\mathbf{v}}\overline{\mathbf{v}}^{\mathrm{T}}) \\ &= \mathbf{L}^{-1}\text{Cov}(\mathbf{v})\mathbf{L} = \mathbf{I} \end{aligned}$$

that is, the components of $\overline{\mathbf{v}}$ are uncorrelated and have unit variance. Hence this process of creating $\overline{\mathbf{v}}$ from $\mathbf{v}$ is called the **whitening filter**. Consequently, we can treat the m components of $\overline{\mathbf{z}}$ as a sequence of m scalar observations where the ith observation is given by

$$\overline{\mathbf{z}}_i = \overline{\mathbf{H}}_{i*}\mathbf{x} + \overline{\mathbf{v}}_i \tag{28.7.20}$$

where $\overline{\mathbf{H}}_{i*}$ is the ith row of $\overline{\mathbf{H}}$. Thus, in the data assimilation phase of the square root algorithm, we can either use the one step of the matrix operations as in Figure 28.7.1 directly or convert $\mathbf{z}_k$ in $\overline{\mathbf{z}}_k$ using the whitening filter described above and use the m steps of the Potter's algorithm. We invite the reader to compare the computational complexity of these alternate implementations.

(2) **Duality in square root algorithm** Much like there are two equivalent or dual forms of Kalman filtering – the **covariance form** in Figure 27.2.2 and the **information form** in Figure 26.4.2, there are also two equivalent or dual forms for the square root version of this algorithm. The algorithm in Figure 28.7.1 is the covariance form. Refer to the interesting survey by Kaminski et al. (1971) for a detailed discussion of the duality of the square root algorithm.

28.8 Stability of the filter

Analysis of the stability of the filter relates to characterizing the **asymptotic** properties of the filter quantities – $\mathbf{x}^{\mathrm{f}}_{k+1}$, $\widehat{\mathbf{x}}_{k+1}$ and their covariances $\mathbf{P}^{\mathrm{f}}_{k+1}$ and $\widehat{\mathbf{P}}_{k+1}$. Analysis of the filter stability for the scalar linear case is covered in Examples 27.2.1 through 27.2.3. Extension of these results to the vector case is rather involved and is beyond our scope. For completeness, we provide an overview of the key results without proof. The details can be obtained from many sources listed in the notes and references.

We begin with a definition of stability of discrete time dynamical systems. Let $\mathbf{M} : \mathbb{R}^n \rightarrow \mathbb{R}^n$ and let

$$\mathbf{x}_{k+1} = \mathbf{M}(\mathbf{x}_k) \tag{28.8.1}$$

be the given dynamical system. The set

$$E = \{\mathbf{x} | \mathbf{M}(\mathbf{x}) = \mathbf{x}\} \subseteq \mathbb{R}^n \tag{28.8.2}$$

defines the **invariant set** or the **equilibrium** points of (28.8.1). Then the above dynamical system is said to be **uniformly asymptotically stable** if and only if for every $\varepsilon > 0$ there exists a $\delta > 0$ and an integer k_0 such that for all initial conditions $\mathbf{x}_0$, if $\| \mathbf{x}_0 - \mathbf{x}^e \| < \delta$ then $\| \mathbf{x}_k - \mathbf{x}^e \| < \varepsilon$ for all $k > k_0$ where $\mathbf{x}^e$ is an equilibrium point. Stated in words, the trajectories of (28.8.1) starting close to an equilibrium point will eventually be attracted towards it, if the system is uniformly asymptotically stable.

In analyzing the filter stability, first we rewrite the filter equations in Figure 27.2.2 as follows:

$$\begin{aligned}
\widehat{\mathbf{x}}_{k+1} &= \mathbf{x}^f_{k+1} + \mathbf{K}_{k+1}[\mathbf{z}_{k+1} - \mathbf{H}_{k+1}\mathbf{x}^f_{k+1}] \\
&= (\mathbf{I} - \mathbf{K}_{k+1}\mathbf{H}_{k+1})\mathbf{x}^f_{k+1} + \mathbf{K}_{k+1}\mathbf{z}_{k+1} \\
&= (\mathbf{I} - \mathbf{K}_{k+1}\mathbf{H}_{k+1})\mathbf{M}_k\widehat{\mathbf{x}}_k + \mathbf{K}_{k+1}\mathbf{z}_{k+1} \\
&= \widehat{\mathbf{P}}_{k+1}(\mathbf{P}^f_{k+1})^{-1}\mathbf{M}_k\widehat{\mathbf{x}}_k + \mathbf{K}_{k+1}\mathbf{z}_{k+1} \\
&= \mathbf{\Phi}_k\widehat{\mathbf{x}}_k + \mathbf{K}_{k+1}\mathbf{z}_{k+1}
\end{aligned} \tag{28.8.3}$$

where the new state transition matrix is

$$\mathbf{\Phi}_k = \widehat{\mathbf{P}}_{k+1}(\mathbf{P}^f_{k+1})^{-1}\mathbf{M}_k. \tag{28.8.4}$$

Also recall that the dynamics of the variance of $\widehat{\mathbf{x}}_{k+1}$ is

$$\widehat{\mathbf{P}}_{k+1} = \mathbf{P}^f_{k+1} - \mathbf{P}^f_{k+1}\mathbf{H}^T_{k+1}[\mathbf{H}_{k+1}\mathbf{P}^f_{k+1}\mathbf{H}^T_{k+1} + \mathbf{R}_{k+1}]^{-1}\mathbf{H}_{k+1}\mathbf{P}^f_{k+1}. \tag{28.8.5}$$

Let

$$\mathbf{y}_{k+1} = \mathbf{\Phi}_k\mathbf{y}_k \tag{28.8.6}$$

denote the **homogeneous** part of (28.8.3). Iterating, we obtain, for $k \geq N > 0$

$$\begin{aligned}
\mathbf{y}_k &= \mathbf{\Phi}_{k-1}\mathbf{\Phi}_{k-2}\cdots\mathbf{\Phi}_{k-N}\mathbf{y}_N \\
&= \mathbf{\Phi}(k-1 : N)\mathbf{y}_N
\end{aligned} \tag{28.8.7}$$

where

$$\mathbf{\Phi}(j:i) = \begin{cases} \mathbf{\Phi}_{j-1}\mathbf{\Phi}_{j-2}\cdots\mathbf{\Phi}_i & \text{if } j \geq i \\ \mathbf{I} & \text{if } j < i \end{cases} \tag{28.8.8}$$

We now state (without proof) a very fundamental result due to Deyst and Price (1968) on the asymptotic properties of $\widehat{\mathbf{P}}_k$ in (28.8.5) and $\mathbf{y}_k$ in (28.8.6).

Referring to the linear Kalman filter equations in Figure 27.2.2, let the system matrix $\mathbf{M}_k$, the system noise covariance matrix $\mathbf{Q}_k$ and the observation noise covariance matrix $\mathbf{R}_k$ satisfy the following:

Condition C Let a_1 and a_2 be two positive real constants such that

$$a_2\mathbf{I} \leq \sum_{i=k-N}^{k-1} \mathbf{M}(k-1:i)\mathbf{Q}_i\mathbf{M}^{\mathrm{T}}(k-1:i) \leq a_1\mathbf{I} \tag{28.8.9}$$

hold for all $k \geq N > 0$.

Condition O Let b_1 and b_2 be two positive real constants such that

$$b_1\mathbf{I} \leq \sum_{i=k-N}^{k-1} \mathbf{M}^{-\mathrm{T}}(k-1:i)\mathbf{H}_i^{\mathrm{T}}\mathbf{R}_i^{-1}\mathbf{H}_i\mathbf{M}^{-1}(k-1:i) \leq b_2\mathbf{I} \tag{28.8.10}$$

for all $k \geq N > 0$ where (recall from Chapter 26)

$$\mathbf{M}(k:i) = \begin{cases} \mathbf{M}_{k-1}\mathbf{M}_{k-2}\cdots\mathbf{M}_i & \text{if } k \geq i \\ \mathbf{I} & \text{if } k < i \end{cases}$$

Deyst and Price (1968) have proved that under conditions C and O, the following are true:

(1) The covariance matrix $\widehat{\mathbf{P}}_k$ in (28.8.5) is such that

$$\left(\frac{a_2}{1+a_2b_2}\right)\mathbf{I} \leq \widehat{\mathbf{P}}_k \leq \left(\frac{1+a_1b_1}{b_1}\right)\mathbf{I} \quad \text{for all } k \geq N \tag{28.8.11}$$

that is, $\widehat{\mathbf{P}}_k$ remain bounded for all $k \geq N$ and

(2) the solution $\mathbf{y}_k$ of the homogeneous equation (28.8.6) is uniformly asymptotically stable.

A number of observations are in order.

(1) The reader can verify that the matrix sum in the middle term of the two-sided inequality in (28.8.10) is closely related to the observability matrix defined in Section 26.3. Consequently, the condition is known as the **uniform, complete observability** condition. [Jazwinski (1970)]

(2) The condition C is known as the uniform complete controllability condition. [Jazwinski (1970)]

(3) We invite the reader to verify that the scalar linear dynamics covered in Examples 27.2.1–27.2.3 satisfy the conditions C and O.

Exercises

28.1 Using (28.1.1) and (28.1.2) verify the correctness of (28.1.3).

28.2 Verify the correctness of (28.2.3).

28.3 Verify the correctness of (28.2.7).

28.4 Rewrite the standard Kalman filter equations for the expanded model (28.3.4) and (28.3.5).

28.5 Using (28.5.9) derive the expression for $\mathbf{G}^{\mathrm{f}}_{k+1}$ in (28.5.12).

28.6 Verify the correctness of (28.5.15).

28.7 Verify the computation of $\widehat{\mathbf{G}}_{k+1}$ in (28.5.17).

28.8 Verify the correctness of the expression for $\widehat{\mathbf{E}}_{k+1}$ in (28.5.19).

28.9 Show that the matrix $\overline{\mathbf{X}}_k$ defined in (28.5.11) satisfies the following recurrence:

$$\overline{\mathbf{X}}_{k+1} = \mathbf{M}_k \overline{\mathbf{X}}_k \mathbf{M}_k^{\mathrm{T}} + \overline{\mathbf{Q}}_k.$$

28.10 Show that the matrix $\widehat{\mathbf{G}}_k(\bar{\mathbf{x}})$ defined in (28.5.11) satisfies the recurrence (use the relation 28.5.16)

$$\widehat{\mathbf{G}}_{k+1}(\bar{\mathbf{x}}) = (\mathbf{I} - \mathbf{K}_{k+1}\mathbf{H}_{k+1})\mathbf{G}^{\mathrm{f}}_{k+1}(\bar{\mathbf{x}}) + \mathbf{K}_{k+1}(\Delta\mathbf{H}_{k+1})\overline{\mathbf{X}}_{k+1}.$$

28.11 Compute $\delta\widehat{\mathbf{P}}_k$ for all the forms A through D of $\mathbf{P}_k$ resulting from the perturbation in $\mathbf{K}_k$.

28.12 Let $\mathbf{a} \in \mathbb{R}^n$ and $\mathbf{b} \in \mathbb{R}^n$ and α, β be two scalars. Define matrices $\mathbf{E}_a$ and $\mathbf{E}_b$ as

$$\mathbf{E}_\alpha = (\mathbf{I} - \alpha\mathbf{a}\mathbf{b}^{\mathrm{T}}) \quad \text{and} \quad \mathbf{E}_\beta = (\mathbf{I} - \beta\mathbf{a}\mathbf{b}^{\mathrm{T}})$$

called the **rank-one update** of **I** or **elementary matrices**.

(a) Compute $\mathbf{E}_\alpha \mathbf{E}_\beta$ and specialize when $\alpha = \beta$.

Notes and references

Section 28.1 For a derivation of the Kalman filter equations based on the orthogonal projections refer to the original paper by Kalman (1960).

Section 28.2 Again refer to the original paper by Kalman (1960). Our derivation is patterned after Jazwinski (1970).

Section 28.3 This section is adapted from Sorenson's survey chapter that appeared in the Advances in Control System (Sorenson (1966)).

Section 28.4 For a comprehensive discussion of the analysis of the divergence of Kalman filters refer to Schlee, Standish and Toda (1967), Price (1968) and Fitzgerald (1971). Also refer to Jazwinski (1970) and Maybeck (1982).

Section 28.5 There is a wide array of literature on the sensitivity of Kalman filters – Fagin (1964), Griffin and Sage (1968) are two representative papers in this category. Jazwinski (1970) contains a good summary of this literature.

Section 28.6 Maybeck (1982) contains a comprehensive discussion of the computational aspects of covariance matrices.

Section 28.7 Square root filters began with the work of Potter and Stern (1963). The book by Bierman (1977) in its entirety is devoted to the analysis of various types – covariance and information forms of square root filtering. Golub (1965) and Hansen and Lawson (1969) deal with the information form of this filters. Kaminski, Bryson and Schmidt (1971) and Maybeck (1982) provide an information survey of square root algorithms. Morf and Kailath (1975) provide a new way of analyzing the square root algorithms.

Section 28.8 Bucy and Joseph (1968) provide a comprehensive treatment of filter stability in continuous time. Deyst and Price (1968) and Bucy (1994) contains analysis of filter stability in discrete time.

29

Nonlinear filtering

This chapter provides an overview of the methods for recursively estimating the state of a nonlinear stochastic dynamical system based on a set of observations that (*a*) depend (nonlinearly) on the state being estimated and (*b*) are corrupted by additive white noise. The exact solution to this problem involves characterizing the evolution of the posterior probability density function over the state space, $\mathbb{R}^n$. This evolution equation can easily be derived from first principles. However, except in special cases (linear dynamics and linear observations) it is often difficult to explicitly characterize the form of the density as a function of space and time. Numerical methods are the only recourse to solving this class of infinite dimensional problems. Given this challenge and the difficulty, researchers have sought for alternate characterization, namely to compute the evolution of the moments of distribution of states being estimated. Ideally, one would require infinitely many moments to provide an equivalent characterization of the distribution. This infinite dimensional problem is further exacerbated by the fact that the rth moment often depends on the qth moment, for $q > r$. Computational feasibility demands that we find a "good" finite dimensional approximation to this infinite system of coupled moments.

One useful idea is to find the **closure property** among these moments, namely to find the least positive integer p such that the first p moments depend only among themselves and **not** on moments of order larger than p. If such a p can be found, then the first p moments would constitute a natural finite dimensional approximation to the density function that is being sought. It turns out that except for brute force enumerative method, there is no clever strategy for finding such a moment closure. One example of a nonlinear dynamics exhibiting the second moment ($p = 2$) closure is reported in Thompson (1985a). Against all these odds, dictated by computational feasibility one often unwittingly settles for computing the first few moments – mean, variance etc. (closed or not) of the state being estimated.

In Section 29.1 we first derive the exact equations for the evolution of the probability density. In Section 29.2 we develop several **ad hoc** but useful approximations known as **second-order filter**, **extended Kalman filters**, **linearized filter**, etc.

29.1 Nonlinear stochastic dynamics

Let $\mathbf{x}_k \in \mathbb{R}^n$ denote the **state** of a dynamical system at time k evolved according to

$$\mathbf{x}_{k+1} = \mathbf{M}(\mathbf{x}_k) + \sigma(\mathbf{x}_k)\mathbf{w}_{k+1} \tag{29.1.1}$$

where $\mathbf{M} : \mathbb{R}^n \rightarrow \mathbb{R}^n$ denotes the **field** that defines the **flow** of the system in the **state space**, $\mathbb{R}^n$. The term $\mathbf{w}_k \in \mathbb{R}^r$ denotes the sequence of noise vectors and $\sigma(\mathbf{x}_k) \in \mathbb{R}^{n\times r}$ is the matrix that transforms the noise vector from $\mathbb{R}^r$ to $\mathbb{R}^n$. It is assumed that $\sigma(\mathbf{x}_k)$ is of **full rank** for all $\mathbf{x}_k$. In general, the $\sigma(\mathbf{x}_k)\mathbf{w}_{k+1}$ term is meant to capture the **model errors**. In the special case, $\sigma(\mathbf{x}_k) \equiv \mathbf{I} \in \mathbb{R}^{n\times n}$, we are left with only the $\mathbf{w}_k$ term.

The initial condition It is assumed that the initial condition $\mathbf{x}_0$ for (29.1.1) is **random** and is drawn from a multivariate normal distribution, $N(\widehat{\mathbf{m}}_0, \widehat{\mathbf{P}}_0)$ where $\widehat{\mathbf{m}}_0 \in \mathbb{R}^n$ is the mean vector and $\widehat{\mathbf{P}}_0 \in \mathbb{R}^{n\times n}$ is the covariance matrix which is assumed to be positive definite. That is, the **prior information** about the initial state $\mathbf{x}_0$ is summarized by the probability density function $\mathbf{x}_0$ and is given by

$$P(\mathbf{x}_0) = \frac{1}{(2\pi)^{\frac{n}{2}}|\widehat{\mathbf{P}}_0|^{\frac{1}{2}}} \exp\left\{-\frac{1}{2}[\mathbf{x}_0 - \widehat{\mathbf{m}}_0](\widehat{\mathbf{P}}_0)^{-1}[\mathbf{x}_0 - \widehat{\mathbf{m}}_0]^{\mathrm{T}}\right\}. \tag{29.1.2}$$

The state noise vector $\mathbf{w}_k$ It is assumed that $\mathbf{w}_k$ are independent of $\mathbf{x}_0$ and that $\mathbf{w}_k$ are drawn from a common multivariate normal distribution with mean zero and the covariance matrix $\mathbf{Q}_k$. It is further assumed that $\mathbf{w}_k$ are serially uncorrelated and hence are independent (since $\mathbf{w}_k \sim N(0, \mathbf{Q}_k)$), that is

$$E[\mathbf{w}_k\mathbf{w}_j^{\mathrm{T}}] = \begin{cases} \mathbf{Q}_k, & \text{if } k = j \\ 0, & \text{otherwise} \end{cases} \tag{29.1.3}$$

Hence, the conditional density of $\sigma(\mathbf{x}_k)\mathbf{w}_{k+1}$ given $\mathbf{x}_k$ is given by

$$P(\sigma(\mathbf{x}_k)\mathbf{w}_{k+1}|\mathbf{x}_k) = N(0, \sigma(\mathbf{x}_k)\mathbf{Q}_{k+1}\sigma^{\mathrm{T}}(\mathbf{x}_k)). \tag{29.1.4}$$

Characterization of the probability density of $\mathbf{x}_k$
When $\mathbf{x}_k$ is defined by (29.1.1), it is clear that the conditional probability of $\mathbf{x}_{k+1} \in \mathbf{A}$ for some $\mathbf{A} \subseteq \mathbb{R}^n$ given the entire history of evolution $\{\mathbf{x}_k, \mathbf{x}_{k-1}, \ldots, \mathbf{x}_2, \mathbf{x}_1, \mathbf{x}_0\}$ is the same as the conditional probability of $\mathbf{x}_{k+1} \in \mathbf{A}$ given the present state $\mathbf{x}_k$. That is,

$$\mathrm{Prob}[\mathbf{x}_{k+1} \in \mathbf{A}|\mathbf{x}_k, \mathbf{x}_{k-1}, \ldots, \mathbf{x}_2, \mathbf{x}_1, \mathbf{x}_0] = \mathrm{Prob}[\mathbf{x}_{k+1} \in \mathbf{A}|\mathbf{x}_k].$$

Thus, given the present state $\mathbf{x}_k$, the future characterization of $\mathbf{x}_{k+1}$ is independent of the past history $\mathbf{x}_{k-1}, \mathbf{x}_{k-2}, \ldots, \mathbf{x}_1, \mathbf{x}_0$. This property is called the **Markov property** and stochastic sequences such as $\{\mathbf{x}_k\}$ generated by (29.1.1) are called **discrete time, continuous state space Markov processes.**

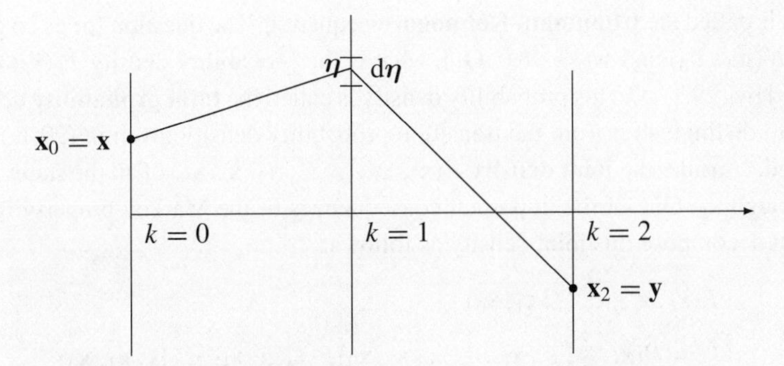

Fig. 29.1.1 Two-steps transition: an illustration.

An immediate import of this Markov property is that we can now characterize the conditional distribution of $\mathbf{x}_{k+1}$ given $\mathbf{x}_k$. Indeed, using (29.1.1) and (29.1.4),

$$\begin{aligned} P[\mathbf{x}_{k+1} - \mathbf{M}(\mathbf{x}_k)|\mathbf{x}_k] &= P[\sigma(\mathbf{x}_k)\mathbf{w}_{k+1}|\mathbf{x}_k] \\ &= N[0, \sigma(\mathbf{x}_k)\mathbf{Q}_{k+1}\sigma^{\mathrm{T}}(\mathbf{x}_k)] \end{aligned}$$

or

$$\begin{aligned} P[\mathbf{x}_{k+1}|\mathbf{x}_k] &= N[\mathbf{M}(\mathbf{x}_k), \sigma(\mathbf{x}_k)\mathbf{Q}_{k+1}\sigma^{\mathrm{T}}(\mathbf{x}_k)] \\ &= \frac{1}{(2\pi)^{\frac{n}{2}}|\sigma_k\mathbf{Q}_{k+1}\sigma_k^{\mathrm{T}}|^{\frac{1}{2}}} \exp\left\{-\frac{1}{2}[\mathbf{x}_{k+1} - \mathbf{M}(\mathbf{x}_k)]\right. \\ &\quad \left. \cdot (\sigma_k\mathbf{Q}_{k+1}\sigma_k^{\mathrm{T}})^{-1}[\mathbf{x}_{k+1} - \mathbf{M}(\mathbf{x}_k)]^{\mathrm{T}}\right\} \end{aligned} \tag{29.1.5}$$

where $\sigma_k = \sigma(\mathbf{x}_k)$ for simplicity in notation.

This conditional probability density function $P(\mathbf{x}_{k+1}|\mathbf{x}_k)$ is known as **the one-step transition probability density** of the Markov process $\{\mathbf{x}_k\}$. Using the straightforward probabilistic argument, we can extend the single step transition probability to multiple steps. For example, consider $P[\mathbf{x}_2|\mathbf{x}_0]$, the probability of going from a specified initial state $\mathbf{x}_0$ to a specified state $\mathbf{x}_2 = \mathbf{y}$. This is given by the product of (*a*) the one-step transition probability of starting at $\mathbf{x}_0 = \mathbf{x}$ and going into a small neighborhood $d\boldsymbol{\eta}$ around $\mathbf{x}_1 = \boldsymbol{\eta}$ which is given by $P[\mathbf{x}_1 = \boldsymbol{\eta}|\mathbf{x}_0 = \mathbf{x}]d\boldsymbol{\eta}$ and (*b*) the one-step transition probability of starting at $\boldsymbol{\eta}$ and going into $\mathbf{x}_2 = \mathbf{y}$. Since $\boldsymbol{\eta}$ is arbitrary, we have to sum this product over all $\boldsymbol{\eta} \in \mathbb{R}^n$. Hence

$$P[\mathbf{x}_2|\mathbf{x}_0] = \int_{\mathbf{x}_1} P[\mathbf{x}_2|\mathbf{x}_1]P[\mathbf{x}_1|\mathbf{x}_0]d\mathbf{x}_1.$$

Refer to Figure 29.1.1 for an illustration.

Generalizing this, for integers $k < p < q$, we have

$$P[\mathbf{x}_k|\mathbf{x}_q] = \int_{\mathbf{x}_p} P[\mathbf{x}_k|\mathbf{x}_p]P[\mathbf{x}_p|\mathbf{x}_q]d\mathbf{x}_p \tag{29.1.6}$$

which is called the **Chapman–Kolmogorov** equation. The question for us is: given $\mathbf{x}_0 \sim N(\widehat{\mathbf{m}}_0, \widehat{\mathbf{P}}_0)$ and $\mathbf{w}_k \sim N(0, \mathbf{Q}_k)$, what is the probability density $P_k(\mathbf{x}_k)$ of $\mathbf{x}_k$ defined by (29.1.1)? This probability density is called the **total probability** density of $\mathbf{x}_k$ to distinguish it from the transition probability density given in (29.1.5). To this end, consider the **joint density** $P(\mathbf{x}_k, \mathbf{x}_{k-1}, \ldots, \mathbf{x}_2, \mathbf{x}_1, \mathbf{x}_0)$ of all the states from $\mathbf{x}_0$ through $\mathbf{x}_k$. One of the important consequences of the Markov property is that we can decompose this joint density as follows:

$$
\begin{aligned}
&P(\mathbf{x}_k, \mathbf{x}_{k-1}, \ldots, \mathbf{x}_1, \mathbf{x}_0) \\
&\quad = P(\mathbf{x}_k, |\mathbf{x}_{k-1}, \mathbf{x}_{k-2}, \ldots, \mathbf{x}_1, \mathbf{x}_0) P(\mathbf{x}_{k-1}, \mathbf{x}_{k-2}, \ldots, \mathbf{x}_1, \mathbf{x}_0) \\
&\quad = P(\mathbf{x}_k|\mathbf{x}_{k-1}) P(\mathbf{x}_{k-1}, \mathbf{x}_{k-2}, \ldots, \mathbf{x}_1, \mathbf{x}_0)
\end{aligned} \tag{29.1.7}
$$

where the first equality follows from the properties of conditional probabilities and the second from the fact that $\{\mathbf{x}_k\}$ is a **Markov process**. That is, the joint density can be recursively characterized as in (29.1.7). By applying this repeatedly, we obtain a decomposition

$$
P(\mathbf{x}_k, \mathbf{x}_{k-1}, \ldots, \mathbf{x}_1, \mathbf{x}_0) = P(\mathbf{x}_k|\mathbf{x}_{k-1}) P(\mathbf{x}_{k-1}|\mathbf{x}_{k-2}) \cdots P(\mathbf{x}_1|\mathbf{x}_0) P_0(\mathbf{x}_0). \tag{29.1.8}
$$

Now, combining this with (29.1.5), we get an explicit expression for the joint density as

$$
\begin{aligned}
&P(\mathbf{x}_k, \mathbf{x}_{k-1}, \ldots, \mathbf{x}_1, \mathbf{x}_0) \\
&\quad = \left\{ \prod_{i=1}^{k} N\left(\mathbf{M}(\mathbf{x}_{i-1}), \sigma_{i-1}\mathbf{Q}_i\sigma_{i-1}^{\mathrm{T}}\right) \right\} N(\widehat{\mathbf{m}}_0, \widehat{\mathbf{P}}_0) \\
&\quad = C_k \exp\left[-\frac{1}{2} G_k \right] N(\widehat{\mathbf{m}}_0, \widehat{\mathbf{P}}_0)
\end{aligned} \tag{29.1.9}
$$

where

$$
G_k = \sum_{i=1}^{k} [\mathbf{x}_i - \mathbf{M}(\mathbf{x}_{i-1})]^{\mathrm{T}} \left(\sigma_{i-1}\mathbf{Q}_i\sigma_{i-1}^{\mathrm{T}}\right)^{-1} [\mathbf{x}_i - \mathbf{M}(\mathbf{x}_{i-1})]
$$

and

$$
C_k = \prod_{i=1}^{k} \frac{1}{(2\pi)^{\frac{n}{2}} |\sigma_{i-1}\mathbf{Q}_i\sigma_{i-1}^{\mathrm{T}}|^{\frac{1}{2}}}.
$$

The expression for the total probability density $\mathbf{P}_k(\mathbf{x}_k)$ is the **marginal density** function obtained by the k-fold integration of this joint density with respect to the states $\mathbf{x}_k, \mathbf{x}_{k-1}, \ldots, \mathbf{x}_1$ and $\mathbf{x}_0$. That is

$$
\mathbf{P}_k[\mathbf{x}_k] = \int_{\mathbf{x}_{k-1}} \cdots \int_{\mathbf{x}_1} \int_{\mathbf{x}_0} P(\mathbf{x}_k, \mathbf{x}_{k-1}, \ldots, \mathbf{x}_1, \mathbf{x}_0) \mathrm{d}\mathbf{x}_{k-1} \cdots \mathrm{d}\mathbf{x}_1 \mathrm{d}\mathbf{x}_0. \tag{29.1.10}
$$

Substituting (29.1.7) and using the definition of this total probability density, we obtain a recursive characterization as

$$\mathbf{P}_k(\mathbf{x}_k) = \int_{\mathbf{x}_{k-1}} P(\mathbf{x}_k|\mathbf{x}_{k-1})P_{k-1}(\mathbf{x}_{k-1})\mathrm{d}\mathbf{x}_{k-1}. \tag{29.1.11}$$

Recall that $\mathbf{P}_0(\mathbf{x}_0)$ is normal and from (29.1.5) it follows that the one-step transition density $P(\mathbf{x}_1|\mathbf{x}_0)$ is also normal. Yet, since $\mathbf{M}(\cdot)$ is a nonlinear function,

$$\begin{aligned} P_1(\mathbf{x}_1) &= \int_{\mathbf{x}_0} P(\mathbf{x}_1|\mathbf{x}_0)P_0(\mathbf{x}_0)\mathrm{d}\mathbf{x}_0 \\ &= C\int_{\mathbf{x}_0} \exp\left[-\frac{1}{2}\alpha(\mathbf{x}_1, \mathbf{x}_0)\right]\mathrm{d}\mathbf{x}_0 \end{aligned}$$

where

$$\begin{aligned} \alpha(\mathbf{x}_1, \mathbf{x}_0) = {} & [\mathbf{x}_1 - \mathbf{M}(\mathbf{x}_0)]^{\mathrm{T}}(\sigma_0\mathbf{Q}_1\sigma_0^{\mathrm{T}})^{-1}[\mathbf{x}_1 - \mathbf{M}(\mathbf{x}_0)] \\ & + [\mathbf{x}_0 - \widehat{\mathbf{m}}_0]^{\mathrm{T}}(\widehat{\mathbf{P}}_0)^{-1}[\mathbf{x}_0 - \widehat{\mathbf{m}}_0] \end{aligned}$$

and C, a normalizing constant, is **not** in general a normal density. From this and the recursive relation (29.1.10), it immediately follows that $\mathbf{P}_k(\mathbf{x}_k)$ is **not** in general a normal density.

Thus, while conceptually (29.1.11) provides a complete characterization of the evolution of the probability density function of the states of the Markov process, much of the challenge involved in stochastic dynamic system is largely due to the nonlinearity in the recursive multivariate integration in (29.1.10).

We conclude this discussion with the following:

Example 29.1.1 Consider the case of a scalar, linear dynamics (with $a \neq 0$ and $\sigma(x_k) \equiv 1$)

$$x_{k+1} = ax_k + w_{k+1}$$

where $x_0 \sim N(m, p_0)$ and $w_k \sim N(0, q_k)$. Since $x_1 = ax_0 + w_1$, we obtain that

$$P_1(x_1) = C\int_{x_0} \exp\left\{-\frac{1}{2}\left[\frac{(x_1 - ax_0)^2}{q_1} + \frac{(x_0 - m)^2}{p_0}\right]\right\}\mathrm{d}x_0 \tag{29.1.12}$$

where

$$C = (2\pi)^{-1}(p_0q_1)^{-\frac{1}{2}}.$$

Define

$$\alpha = \frac{p_0a^2 + q_1}{p_0q_1} \quad \text{and} \quad \beta = \frac{ap_0x_1 + mq_1}{p_0q_1}.$$

Now, expanding and simplifying the terms inside the square brackets on the r.h.s.of (29.1.12) we get

$$
\begin{aligned}
&\frac{(x_1 - ax_0)^2}{q_1} + \frac{(x_0 - m)^2}{p_0} \\
&= \alpha x_0^2 - 2\beta x_0 + \frac{x_1^2}{q_1} + \frac{m^2}{p_0} \\
&= \alpha\left(x_0 - \frac{\beta}{\alpha}\right)^2 + \frac{x_1^2}{q_1} + \frac{m^2}{p_0} - \frac{\beta^2}{\alpha} \quad \text{(completing a perfect square in } x_0\text{)} \\
&= \alpha\left(x_0 - \frac{\beta}{\alpha}\right)^2 + \frac{1}{(p_0 a^2 + q_1)}(x_1 - am)^2. \qquad (29.1.13)
\end{aligned}
$$

Substituting back into (29.1.12)

$$
\begin{aligned}
P_1(x_1) = &\frac{1}{\sqrt{2\pi}\, p_0^{\frac{1}{2}} q_1^{\frac{1}{2}}} \exp\left[-\frac{1}{2}\frac{(x_1 - am)^2}{(p_0 a^2 + q_1)}\right] \\
&\times \frac{1}{\sqrt{2\pi}} \int_{x_0} \exp\left[-\frac{1}{2}\alpha(x_0 - \beta/\alpha)^2\right] \mathrm{d}x_0.
\end{aligned}
$$

Change the variable using

$$
z = \sqrt{\alpha}(x_0 - \beta/\alpha)
$$

we get

$$
\begin{aligned}
P_1(x_1) &= \frac{1}{\sqrt{2\pi}(p_0 a^2 + q_1)^{\frac{1}{2}}} \exp\left[-\frac{1}{2}\frac{(x_1 - am)^2}{(p_0 a^2 + q_1)}\right] \frac{1}{\sqrt{2\pi}} \int_z \exp\left[-\frac{1}{2}z^2\right] \mathrm{d}z \\
&= N(am, p_0 a^2 + q_1)
\end{aligned}
$$

since the value of the last integral is unity. (Exercise 29.1)

Example 29.1.2 We now illustrate an alternate method for a vector case of the linear dynamics. Consider ($\sigma(\mathbf{x}_k) \equiv \mathbf{I} \in \mathbb{R}^{n\times n}$)

$$
\mathbf{x}_{k+1} = \mathbf{M}_k \mathbf{x}_k + \mathbf{w}_{k+1}
$$

where $\mathbf{M}_k \in \mathbb{R}^{n\times n}$, $\mathbf{x}_0 \sim N(\mathbf{m}_0, \mathbf{P}_0)$, $\mathbf{P}_0 \in \mathbb{R}^{n\times n}$ and $\mathbf{w}_k \sim N(0, \mathbf{Q}_k)$ with $\mathbf{Q}_k \in \mathbb{R}^{n\times n}$. Consider

$$
\mathbf{m}_1 = E(\mathbf{x}_1) = \mathbf{M}_0 \mathbf{m}_0
$$

and

$$
\begin{aligned}
\mathbf{P}_1 = \mathrm{Cov}(\mathbf{x}_1) &= E[(\mathbf{x}_1 - E(\mathbf{x}_1))(\mathbf{x}_1 - E(\mathbf{x}_1))^{\mathrm{T}}] \\
&= E[\mathbf{M}_0(\mathbf{x}_0 - \mathbf{m}_0)(\mathbf{x}_0 - \mathbf{m}_0)^{\mathrm{T}}\mathbf{M}_0^{\mathrm{T}}] + E[\mathbf{w}_1 \mathbf{w}_1^{\mathrm{T}}] \\
&= \mathbf{M}_0 \mathbf{P}_0 \mathbf{M}_0^{\mathrm{T}} + \mathbf{Q}_1.
\end{aligned}
$$

Furthermore, since the sum of two uncorrelated (hence independent) Gaussian variates is Gaussian (Exercise 26.3), it follows that

$$\mathbf{x}_1 \sim N(\mathbf{M}_0\mathbf{m}_0, \mathbf{M}_0\mathbf{P}_0\mathbf{M}_0^{\mathrm{T}} + \mathbf{Q}_1) = N(\mathbf{m}_1, \mathbf{P}_1).$$

Inductively, we readily obtain

$$\mathbf{x}_{k+1} \sim N(\mathbf{m}_{k+1}, \mathbf{P}_{k+1})$$

with

$$\mathbf{m}_{k+1} = \mathbf{M}_k\mathbf{m}_k$$

and

$$\mathbf{P}_{k+1} = \mathbf{M}_k\mathbf{P}_k\mathbf{M}_k^{\mathrm{T}} + \mathbf{Q}_k.$$

29.2 Nonlinear filtering

Let $\mathbf{x}_k$ evolve according to (29.1.1). It is often the case that the state $\mathbf{x}_k$ is **not** directly observable. It is assumed however, that a sequence of observations $\mathbf{z}_k$ where

$$\mathbf{z}_k = h(\mathbf{x}_k) + \mathbf{v}_k \tag{29.2.1}$$

are available where $\mathbf{v}_k \in \mathbb{R}^m$ is a **Gaussian white** noise sequence:

$$\begin{aligned} \mathbf{v}_k &\sim N(0, \mathbf{R}_k) \\ E(\mathbf{v}_k\mathbf{v}_p^{\mathrm{T}}) &= \begin{cases} \mathbf{R}_k, & \text{if } \ p = k \\ 0, & \text{otherwise} \end{cases} \end{aligned} \tag{29.2.2}$$

For simplicity in notation, let $\mathbf{z}[1:k] = \{\mathbf{z}_1, \mathbf{z}_2, \ldots, \mathbf{z}_k\}$. Our goal is to derive a recursive framework for the evolution of the conditional density of $\mathbf{x}_k$ given the set $\mathbf{z}[1:k]$ of observations.

To this end, first define the **filter** conditional density given the observations $\mathbf{z}[1:k]$ as

$$f_k(\mathbf{x}_k) = P[\mathbf{x}_k|\mathbf{z}[1:k]] \tag{29.2.3}$$

and **one-step predictor** conditional density given $\mathbf{z}[1:k]$ as

$$P_{k+1}(\mathbf{x}_{k+1}) = P[\mathbf{x}_{k+1}|\mathbf{z}[1:k]]. \tag{29.2.4}$$

Using the properties of conditional probabilities (Appendix F), we obtain

$$P[\mathbf{x}_k, \mathbf{x}_{k-1}, \ldots, \mathbf{x}_0|\mathbf{z}[1:k]] = \frac{P[\mathbf{x}_k, \mathbf{x}_{k-1}, \ldots, \mathbf{x}_0, \mathbf{z}[1:k]]}{P[\mathbf{z}[1:k]]} \tag{29.2.5}$$

where the numerator is the **joint density** of $\{\mathbf{x}_k, \mathbf{x}_{k-1}, \ldots, \mathbf{x}_0\}$ and $\mathbf{z}[1:k]$ and the denominator is the **marginal density** given by

$$P[\mathbf{z}[1:k]] = \int_{\mathbf{x}_k} \cdots \int_{\mathbf{x}_0} P[\mathbf{x}_k, \mathbf{x}_{k-1}, \ldots, \mathbf{x}_1, \mathbf{x}_0, \mathbf{z}[1:k]] \mathrm{d}\mathbf{x}_k \cdots \mathrm{d}\mathbf{x}_0. \quad (29.2.6)$$

The joint density on the numerator of (29.2.5) can also be written as

$$\begin{aligned} &P[\mathbf{x}_k, \mathbf{x}_{k-1}, \ldots, \mathbf{x}_0, \mathbf{z}[1:k]] \\ &= P[\mathbf{z}[1:k] | \mathbf{x}_k, \mathbf{x}_{k-1}, \ldots, \mathbf{x}_0] P[\mathbf{x}_k, \mathbf{x}_{k-1}, \ldots, \mathbf{x}_0]. \end{aligned} \quad (29.2.7)$$

Substituting (29.2.7) in (29.2.5), we obtain a version of the **Bayes' rule**

$$\begin{aligned} &P[\mathbf{x}_k, \mathbf{x}_{k-1}, \ldots, \mathbf{x}_0 | \mathbf{z}[1:k]] \\ &= \frac{P[\mathbf{z}[1:k] | \mathbf{x}_k, \mathbf{x}_{k-1}, \ldots, \mathbf{x}_0] P[\mathbf{x}_k, \mathbf{x}_{k-1}, \ldots, \mathbf{x}_0]}{P[\mathbf{z}[1:k]]}. \end{aligned} \quad (29.2.8)$$

The importance of this relation stems from the observation that the required **filter density** $f_k(\mathbf{x})$ can be obtained by integrating (29.2.8) w.r.t. $\mathbf{x}_{k-1}, \mathbf{x}_{k-2}, \ldots, \mathbf{x}_1$ and $\mathbf{x}_0$. That is,

$$f_k(\mathbf{x}_k) = \int_{\mathbf{x}_{k-1}} \cdots \int_{\mathbf{x}_0} P[\mathbf{x}_k, \mathbf{x}_{k-1}, \ldots, \mathbf{x}_1, \mathbf{x}_0 | \mathbf{z}[1:k]] \mathrm{d}\mathbf{x}_{k-1} \mathrm{d}\mathbf{x}_{k-2} \cdots \mathrm{d}\mathbf{x}_0. \quad (29.2.9)$$

Similarly, the **one-step predictor density** is given by

$$P_{k+1}(\mathbf{x}) = \int_{\mathbf{x}_k} \cdots \int_{\mathbf{x}_0} P[\mathbf{x}_{k+1}, \mathbf{x}_k, \mathbf{x}_{k-1}, \ldots, \mathbf{x}_1, \mathbf{x}_0 | \mathbf{z}[1:k]] \mathrm{d}\mathbf{x}_k \mathrm{d}\mathbf{x}_{k-1} \cdots \mathrm{d}\mathbf{x}_0 \quad (29.2.10)$$

where the integrand using the Bayes' rule is given by

$$\begin{aligned} &P[\mathbf{x}_{k+1}, \mathbf{x}_k, \ldots, \mathbf{x}_0 | \mathbf{z}[1:k]] \\ &= \frac{P[\mathbf{z}[1:k] | \mathbf{x}_{k+1}, \mathbf{x}_k, \ldots, \mathbf{x}_0] P[\mathbf{x}_{k+1}, \mathbf{x}_k, \ldots, \mathbf{x}_0]}{P[\mathbf{z}[1:k]]}. \end{aligned} \quad (29.2.11)$$

Several observations are in order.
(1) While (29.2.9) and (29.2.10) provide expressions for the required filter and the one-step predictor densities, these are **not** in the recursive form we are seeking.
(2) The key to obtaining this recursive form is to exploit the **Markov property** of the stochastic process $\{\mathbf{x}_k\}$ defined by (29.1.1).

Since $\{\mathbf{x}_k\}$ is a Markov process, from (29.1.8) we get

$$P[\mathbf{x}_k, \mathbf{x}_{k-1}, \ldots, \mathbf{x}_1, \mathbf{x}_0] = \left\{ \prod_{i=1}^{k} P[\mathbf{x}_i | \mathbf{x}_{i-1}] \right\} P_0(\mathbf{x}_0) \quad (29.2.12)$$

where recall $P_0(\mathbf{x}_0)$ is the **prior density**. Now by applying the basic identity

$$P[A, B|C] = P[A|B, C] P[B|C]$$

we get

$$\begin{aligned}
&P[\mathbf{z}_1, \mathbf{z}_2, \ldots, \mathbf{z}_k | \mathbf{x}_k, \mathbf{x}_{k-1}, \ldots, \mathbf{x}_0] \\
&\quad = P[\mathbf{z}_1 | \mathbf{z}_2, \mathbf{z}_3, \ldots, \mathbf{z}_k, \mathbf{x}_k, \mathbf{x}_{k-1}, \ldots, \mathbf{x}_0] \\
&\qquad \cdot P[\mathbf{z}_2, \mathbf{z}_3, \ldots, \mathbf{z}_k | \mathbf{x}_k, \mathbf{x}_{k-1}, \ldots, \mathbf{x}_0].
\end{aligned} \tag{29.2.13}$$

But (29.2.1) implies that $\mathbf{z}_1$ depends only on $\mathbf{x}_1$ and **not** on $\mathbf{z}_2, \mathbf{z}_3, \ldots, \mathbf{z}_k$ nor on $\mathbf{x}_2, \mathbf{x}_3, \ldots$ and $\mathbf{x}_k$. Thus, using the Markov property

$$P[\mathbf{z}_1 | \mathbf{z}_2, \mathbf{z}_3, \ldots, \mathbf{z}_k, \mathbf{x}_k, \mathbf{x}_{k-1}, \ldots, \mathbf{x}_0] = P[\mathbf{z}_1 | \mathbf{x}_1]. \tag{29.2.14}$$

Again, using the Markov property

$$\begin{aligned}
&P[\mathbf{z}_2, \mathbf{z}_3, \ldots, \mathbf{z}_k | \mathbf{x}_k, \mathbf{x}_{k-1}, \ldots, \mathbf{x}_1, \mathbf{x}_0] \\
&\quad = P[\mathbf{z}_2, \mathbf{z}_3, \ldots, \mathbf{z}_k | \mathbf{x}_k, \mathbf{x}_{k-1}, \ldots, \mathbf{x}_2].
\end{aligned} \tag{29.2.15}$$

Now combining (29.2.14) – (29.2.15) with (29.2.13), we get a recursive relation

$$\begin{aligned}
&P[\mathbf{z}_1, \mathbf{z}_2, \ldots, \mathbf{z}_k | \mathbf{x}_k, \mathbf{x}_{k-1}, \ldots, \mathbf{x}_1, \mathbf{x}_0] \\
&\quad = P[\mathbf{z}_1 | \mathbf{x}_1] P[\mathbf{z}_2, \mathbf{z}_3, \ldots, \mathbf{z}_k | \mathbf{x}_k, \mathbf{x}_{k-1}, \ldots, \mathbf{x}_2].
\end{aligned} \tag{29.2.16}$$

By applying the above argument repeatedly to the second factor on the r.h.s.of (29.2.16), we get

$$P[\mathbf{z}_1, \mathbf{z}_2, \ldots, \mathbf{z}_k | \mathbf{x}_k, \mathbf{x}_{k-1}, \ldots, \mathbf{x}_1, \mathbf{x}_0] = \prod_{i=1}^{k} P[\mathbf{z}_i | \mathbf{x}_i]. \tag{29.2.17}$$

Substituting (29.2.12) and (29.2.17) into (29.2.8), the integrand in the integral defining $f_k(\mathbf{x})$ in (29.2.9) becomes

$$\begin{aligned}
&P[\mathbf{x}_k, \mathbf{x}_{k-1}, \ldots, \mathbf{x}_0 | \mathbf{z}[1:k]] \\
&\quad = \frac{1}{P[\mathbf{z}[1:k]]} \left\{ \prod_{i=1}^{k} P[\mathbf{z}_i | \mathbf{x}_i] \right\} \left\{ \prod_{i=1}^{k} P[\mathbf{x}_i | \mathbf{x}_{i-1}] \right\} P_0(\mathbf{x}_0).
\end{aligned} \tag{29.2.18}$$

Similarly, the integrand in the integral defining $P_{k+1}(\mathbf{x})$ in (29.2.10) is given by

$$\begin{aligned}
&P[\mathbf{x}_{k+1}, \mathbf{x}_k, \ldots, \mathbf{x}_0 | \mathbf{z}[1:k]] \\
&\quad = P[\mathbf{x}_{k+1} | \mathbf{x}_k, \mathbf{x}_{k-1}, \ldots, \mathbf{x}_0, \mathbf{z}[1:k]] P[\mathbf{x}_k, \mathbf{x}_{k-1}, \ldots, \mathbf{x}_0 | \mathbf{z}[1:k]] \\
&\quad = P[\mathbf{x}_{k+1} | \mathbf{x}_k] P[\mathbf{x}_k, \mathbf{x}_{k-1}, \ldots, \mathbf{x}_0 | \mathbf{z}[1:k]] \\
&\quad = \frac{1}{P[\mathbf{z}[1:k]]} \left\{ \prod_{i=1}^{k} P[\mathbf{z}_i | \mathbf{x}_i] \right\} \left\{ \prod_{i=1}^{k+1} P[\mathbf{x}_i | \mathbf{x}_{i-1}] \right\} P_0(\mathbf{x}_0).
\end{aligned} \tag{29.2.19}$$

Hence, we obtain the first of the recursive relations by substituting (29.2.19) into (29.2.10) and using the definition of $f_k(\mathbf{x})$ in (29.2.9) namely

$$P_{k+1}(\mathbf{x}_{k+1}) = \int_{\mathbf{x}_k} P(\mathbf{x}_{k+1}|\mathbf{x}_k) f_k(\mathbf{x}_k) d\mathbf{x}_k. \tag{29.2.20}$$

Remark 29.2.1 This relation (29.2.20) is the **infinite-dimensional** analog of the finite-dimensional predictive equation

$$\mathbf{x}^{\mathrm{f}}_{k+1} = \mathbf{M}(\widehat{\mathbf{x}}_k)$$

for the case of linear Kalman filter equations (Chapter 27).

Again, from (29.2.18) and (29.2.19) we get

$$P[\mathbf{x}_k, \mathbf{x}_{k-1}, \ldots, \mathbf{x}_0|\mathbf{z}[1:k]] = \frac{P[\mathbf{z}[1:k-1]]}{P[\mathbf{z}[1:k]]} P[\mathbf{z}_k|\mathbf{x}_k] P[\mathbf{x}_k, \mathbf{x}_{k-1}, \ldots, \mathbf{x}_0|\mathbf{z}[1:k-1]] \tag{29.2.21}$$

where $P[\mathbf{z}[1:0]] = 1$ by definition. The second of the recursive relations is obtained by substituting (29.2.21) into (29.2.9) and using the definition of $P_{k-1}(\mathbf{x})$, namely

$$f_k(\mathbf{x}_k) = \frac{P[\mathbf{z}[1:k-1]]}{P[\mathbf{z}[1:k]]} P[\mathbf{z}_k|\mathbf{x}_k] P_k(\mathbf{x}_k). \tag{29.2.22}$$

Remark 29.2.2 This relation (29.2.22) is the infinite-dimensional analog of the estimation or the data assimilation step

$$\widehat{\mathbf{x}}_{k+1} = \mathbf{x}^{\mathrm{f}}_{k+1} + \mathbf{K}_{k+1}[\mathbf{z}_{k+1} - \mathbf{H}_k \mathbf{x}^{\mathrm{f}}_{k+1}]$$

for the case of linear Kalman filter equations (Chapter 27).

The resulting recursive framework is summarized in Figure 29.2.1.

From a computational point of view, the implementation of the nonlinear recursive filter equations requires the following three quantities:

(1) the **prior density**, $P_0(\mathbf{x}_0)$ of the initial state $\mathbf{x}_0$
(2) the **one-step state transition density** $P[\mathbf{x}_{k+1}|\mathbf{x}_k]$ of the Markov process $\{\mathbf{x}_k\}$ for $k = 0, 1, 2, \ldots$ which given the dynamics in (29.1.1) is uniquely determined by the properties of the **model noise** process $\{\mathbf{w}_k\}$ and
(3) the **conditional density** $P[\mathbf{z}_k|\mathbf{x}_k]$ of the observations which given (29.2.1) is uniquely determined by the properties of the **observation noise** process $\{\mathbf{v}_k\}$.

In the following example we derive the specific form of the nonlinear recursive filter for the special case when the above three densities are Gaussian.

Example 29.2.1 Following Section 29.1, let the prior density be given by: $\mathbf{x}_0 \sim N(\widehat{\mathbf{m}}_0, \widehat{\mathbf{P}}_0) = P_0(\mathbf{x}_0)$, and let $\mathbf{w}_k \sim N(0, \mathbf{Q}_k)$. Then the one-step state transition

Model $\mathbf{x}_{k+1} = \mathbf{M}(\mathbf{x}_k) + \sigma(\mathbf{x}_k)\mathbf{w}_{k+1}, \quad k = 0, 1, 2, \ldots$

Prior distribution for $\mathbf{x}_0 = P_0(\mathbf{x}_0) = f_0(\mathbf{x}_0)$ is given

Observation $\mathbf{z}_k = h(\mathbf{x}_k) + \mathbf{v}_k, \quad k = 1, 2, 3, \ldots$

Recursive nonlinear filter $k = 0, 1, 2, \ldots$

(a) **one-step predictor probability density**

$$P_{k+1}(\mathbf{x}_{k+1}) = \int_{\mathbf{x}_k} P[\mathbf{x}_{k+1}|\mathbf{x}_k] f_k(\mathbf{x}_k) d\mathbf{x}_k$$

(b) **Filter probability density**

$$f_{k+1}(\mathbf{x}_{k+1}) = \frac{P[\mathbf{z}[1:k]]}{P[\mathbf{z}[1:k+1]]} P[\mathbf{z}_{k+1}|\mathbf{x}_{k+1}] P_{k+1}(\mathbf{x}_{k+1})$$

Fig. 29.2.1 Nonlinear recursive filter equations.

density $P[\mathbf{x}_{k+1}|\mathbf{x}_k]$ is given by (29.1.5). From (29.2.1), it follows that

$$P[\mathbf{z}_k - h(\mathbf{x}_k)|\mathbf{x}_k] \sim N[0, \mathbf{R}_k]$$

or

$$P[\mathbf{z}_k|\mathbf{x}_k] \sim N[h(\mathbf{x}_k), \mathbf{R}_k]. \tag{29.2.23}$$

Combining these, we obtain the following filter equations:
(1)

$$f_0(\mathbf{x}_0) = P_0(\mathbf{x}_0) = \frac{1}{(2\pi)^{\frac{n}{2}} |\widehat{\mathbf{P}}_0|^{\frac{1}{2}}} \exp\left\{-\frac{1}{2}[\mathbf{x}_0 - \widehat{\mathbf{m}}_0]^{\mathrm{T}} (\widehat{\mathbf{P}}_0)^{-1} [\mathbf{x}_0 - \widehat{\mathbf{m}}_0]\right\}.$$

(2) Using (29.2.20) and (29.1.5) we get the one-step predictor density

$$P_{k+1}(\mathbf{x}_{k+1}) = \beta_1 \int_{\mathbf{x}_k} \exp\left\{-\frac{1}{2}[\mathbf{x}_{k+1} - \mathbf{M}(\mathbf{x}_k)]^{\mathrm{T}} [\sigma_k \mathbf{Q}_{k+1} \sigma_k^{\mathrm{T}}]^{-1} [\mathbf{x}_{k+1} - \mathbf{M}(\mathbf{x}_k)]\right\} d\mathbf{x}_k.$$

(3) Using (29.2.22) and (29.2.23) the filter density is given by

$$f_{k+1}(\mathbf{x}_{k+1}) = \beta_2 \exp\left\{-\frac{1}{2}[\mathbf{z}_{k+1} - h(\mathbf{x}_{k+1})]^{\mathrm{T}} \mathbf{R}_{k+1}^{-1} [\mathbf{z}_{k+1} - h(\mathbf{x}_{k+1})]\right\} P_{k+1}(\mathbf{x}_{k+1})$$

where β_1 and β_2 are normalizing constants (Exercise 29.1).

We now derive the Kalman filter as a special case of this nonlinear recursive filter.

Example 29.2.2 Consider the special case when $\mathbf{M}(\mathbf{x}_k) = \mathbf{M}_k \mathbf{x}_k$ for some non-singular matrix $\mathbf{M}_k \in \mathbb{R}^{n \times n}$ and $h(\mathbf{x}_k) = \mathbf{H}_k \mathbf{x}_k$ for some matrix $\mathbf{H}_k \in \mathbb{R}^{m \times n}$. Here again, the prior information on the initial condition is given by

$$\mathbf{x}_0 \sim N(\widehat{\mathbf{m}}, \widehat{\mathbf{P}}_0) = f(\mathbf{x}_0).$$

The one-step state transition density is given by

$$P(\mathbf{x}_{k+1}|\mathbf{x}_k) \sim N(\mathbf{M}_k \mathbf{x}_k, \mathbf{Q}_{k+1})$$

and

$$P(\mathbf{z}_k|\mathbf{x}_k) \sim N(\mathbf{H}_k \mathbf{x}_k, \mathbf{R}_k).$$

Then the one-step predictor density of $\mathbf{x}_1$ is given by

$$P_1(\mathbf{x}_1) = \beta_1 \int_{\mathbf{X}_k} \exp\left[-\frac{1}{2}(\mathbf{x}_1 - \mathbf{x}_0)\mathbf{Q}_1^{-1}(\mathbf{x}_1 - \mathbf{x}_0)^{\mathrm{T}}\right.$$
$$\left. - \frac{1}{2}(\mathbf{x}_0 - \widehat{\mathbf{m}})(\widehat{\mathbf{P}})^{-1}(\mathbf{x}_0 - \widehat{\mathbf{m}})^{\mathrm{T}}\right] \mathrm{d}\mathbf{x}_0$$

and the filter density

$$f_1(\mathbf{x}_1) = \beta_2 \exp\left[-\frac{1}{2}(\mathbf{z}_1 - \mathbf{H}\mathbf{x}_0)^{\mathrm{T}}\mathbf{R}_1^{-1}(\mathbf{z}_1 - \mathbf{H}\mathbf{x}_0)^{\mathrm{T}}\right] P_1(\mathbf{x}_1|\mathbf{x}_0).$$

Notice that the density function $P_k(\mathbf{x})$ and $f_k(\mathbf{x})$ are real-valued functions defined over the entire state space, $\mathbb{R}^n$. Since $\mathbf{M}(\mathbf{x})$ and $h(\mathbf{x})$ are nonlinear vector functions, obtaining a closed form expression for these densities even in the special case treated in Example 29.2.1 is often difficult. Finite-dimensional approximations are the only recourse available. There are basically two avenues for obtaining such approximations and the following is a summary of these ideas.

(a) **Approximating the density functions** Bucy (1969) and Bucy and Senne (1971) describe a method of approximating the required densities by first defining a discrete, floating or moving grid of a suitably large but a fixed size in $\mathbb{R}^n$ and then obtain a discrete representation of the density on this discrete grid. In a related development, Sorenson and Stubberud (1968) describe an approximation using the Edgeworth Series expansion and in Sorenson and Alspach (1971) develop a method of approximation using convex combination of Gaussian densities. We refer the reader to these papers for details.

(b) **Dynamics of evolution of moments** It is well known that corresponding to every probability density function there is a unique **moment-generating** function and vice versa. Accordingly, any density function can be equivalently represented by an **infinite** set of moments. By using only the first k moments, we can provide an approximation to the density we are seeking. This idea when combined with the Taylor Series expansion naturally leads to a system of recursive relations for the evolutions of the moments. This approach is described in detail in the rest of this chapter.

29.3 Nonlinear filter: moment dynamics

Let $\mathbf{x}_k \in \mathbb{R}^n$ denote the state of a nonlinear system evolving according to (29.1.1) and let $\mathbf{z}_k$ denote the observations specified in (29.2.1). Our goal in this section is to derive the exact dynamics of evolution of the (first two) conditional moments of the optimal (least squares) estimate of $\mathbf{x}_k$. The approach is a direct extension of the derivation of the Kalman filter equations in Chapter 27 and has two steps: (*a*) the forecast step and (*b*) the data assimilation step.

Forecast Step Let $\widehat{\mathbf{x}}_k$ be the **optimal** (in the sense of least squares) **unbiased** (hence minimum variance) estimate of $\mathbf{x}_k$ at time k and let $\widehat{\mathbf{P}}_k$ be the covariance of $\widehat{\mathbf{x}}_k$. Let $\mathbf{z}(1:k) = \{\mathbf{z}_1, \mathbf{z}_2, \ldots, \mathbf{z}_k\}$ denote the set of all observations at time k. The question of interest to us is: given $\widehat{\mathbf{x}}_k$, what is the best (in the sense of least squares) forecast $\mathbf{x}^{\mathrm{f}}_{k+1}$ of $\mathbf{x}_{k+1}$? Recall from Chapter 16 that the conditional expectation of $\mathbf{x}_{k+1}$ given $\mathbf{z}(1:k)$ is indeed the best estimate. Accordingly, we define

$$\begin{aligned}\mathbf{x}^{\mathrm{f}}_{k+1} &= E[\mathbf{x}_{k+1}|\mathbf{z}(1:k)]\\ &= E[\mathbf{M}(\mathbf{x}_k) + \mathbf{w}_{k+1}|\mathbf{z}(1:k)]\\ &= E[\mathbf{M}(\mathbf{x}_k)|\mathbf{z}(1:k)]\\ &= \widehat{\mathbf{M}}(\mathbf{x}_k)\end{aligned} \tag{29.3.1}$$

since $\mathbf{w}_{k+1}$ is uncorrelated with $\mathbf{z}(1:k)$. As the conditional probability density of $\mathbf{x}_k$ given $\mathbf{z}(1:k)$ is **not** known, it is not possible to explicitly compute the conditional mean, $\widehat{\mathbf{M}}(\mathbf{x}_k)$. Much of the challenge associated with obtaining the exact moment dynamics relates to this difficulty of computing $\widehat{\mathbf{M}}(\mathbf{x}_k)$. Also recall that $\widehat{\mathbf{M}}(\mathbf{x}_k) \neq \mathbf{M}(\mathbf{x}_k)$. Thus, any practical algorithm for forecasting must seek ways to approximate $\widehat{\mathbf{M}}(\mathbf{x}_k)$. Depending on the nature of the nonlinearity in $\mathbf{M}(\mathbf{x})$ and the closeness of $\widehat{\mathbf{x}}_k$ to $\mathbf{x}_k$, we can obtain a variety of useful approximations using the Taylor series expansion. While we pursue the computation of such approximations in Section 29.4, in the following our primary goal is to derive exact expressions for the moments of $\widehat{\mathbf{x}}_k$.

To this end define

$$f_k = \mathbf{M}(\mathbf{x}_k) - \widehat{\mathbf{M}}(\mathbf{x}_k). \tag{29.3.2}$$

Taking conditional expectations on both sides using (29.3.1), we obtain

$$\widehat{f}_k = E[f|\mathbf{z}(1:k)] = 0. \tag{29.3.3}$$

The error $\mathbf{e}^{\mathrm{f}}_{k+1}$ in the forecast $\mathbf{x}^{\mathrm{f}}_{k+1}$ is given by

$$\begin{aligned}\mathbf{e}^{\mathrm{f}}_{k+1} &= \mathbf{x}_{k+1} - \mathbf{x}^{\mathrm{f}}_{k+1}\\ &= \mathbf{M}(\mathbf{x}_k) + \mathbf{w}_{k+1} - \mathbf{x}^{\mathrm{f}}_{k+1}\\ &= \mathbf{f}_k + \mathbf{w}_{k+1}.\end{aligned} \tag{29.3.4}$$

Using (29.3.3) and from the properties of $\mathbf{w}_k$, it is immediate that

$$E[\mathbf{e}^{\mathrm{f}}_{k+1}|\mathbf{z}(1:k)] = 0. \tag{29.3.5}$$

That is, the forecast $\mathbf{x}^{\mathrm{f}}_{k+1}$ in (29.3.1) is an **unbiased** estimate. This when combined with the fact that it is also a least squares estimate, guarantees that it is also a **minimum variance** estimate.

The conditional variance $\mathbf{P}^{\mathrm{f}}_{k+1}$ of $\mathbf{x}^{\mathrm{f}}_{k+1}$ is then (using (29.3.4)) given by

$$\begin{aligned}
\mathbf{P}^{\mathrm{f}}_{k+1} &= E[(\mathbf{e}^{\mathrm{f}}_{k+1})(\mathbf{e}^{\mathrm{f}}_{k+1})^{\mathrm{T}}|\mathbf{z}(1:k)] \\
&= E[(\mathbf{f}_k + \mathbf{w}_{k+1})(\mathbf{f}_k + \mathbf{w}_{k+1})^{\mathrm{T}}|\mathbf{z}(1:k)] \\
&= E[\mathbf{f}_k\mathbf{f}^{\mathrm{T}}_k|\mathbf{z}(1:k)] + \mathbf{Q}_{k+1}
\end{aligned} \tag{29.3.6}$$

since

$$E[\mathbf{f}_k\mathbf{w}^{\mathrm{T}}_{k+1}|\mathbf{z}(1:k)] = E[\mathbf{f}_k|\mathbf{z}(1:k)]E[\mathbf{w}^{\mathrm{T}}_{k+1}] = 0.$$

Data Assimilation Step Let $\mathbf{z}_{k+1}$ be the new observation that is made available at time $(k+1)$. Now we have two pieces of information about $\mathbf{x}_{k+1}$: first is the **forecast** $\mathbf{x}^{\mathrm{f}}_{k+1}$ obtained from the previous stage and second is the **new observation** $\mathbf{z}_{k+1}$. The goal is to combine them to obtain the best (in the sense of least squares) unbiased estimate $\widehat{\mathbf{x}}_{k+1}$ of $\mathbf{x}_{k+1}$. While there are many ways of combining them, motivated by computational considerations, we are in particular seeking this estimate as a **linear combination** of $\mathbf{x}^{\mathrm{f}}_{k+1}$ and $\mathbf{z}_{k+1}$ (Chapter 17). That is, we are seeking the **best, linear, unbiased estimate** (BLUE) $\widehat{\mathbf{x}}_{k+1}$. Since we are dealing with a nonlinear problem, at the risk of a slight repetition, we start from the first principles.

Let $\mathbf{a} \in \mathbb{R}^n$ and $\mathbf{K} \in \mathbb{R}^{n\times m}$ and define

$$\widehat{\mathbf{x}}_{k+1} = \mathbf{a} + \mathbf{K}\mathbf{z}_{k+1} \tag{29.3.7}$$

where $\mathbf{a}$ and $\mathbf{K}$ are to be determined in such a way that $\widehat{\mathbf{x}}_{k+1}$ is a BLUE. The error $\widehat{\mathbf{e}}_{k+1}$ in this estimate is given by

$$\widehat{\mathbf{e}}_{k+1} = \mathbf{x}_{k+1} - \widehat{\mathbf{x}}_{k+1}. \tag{29.3.8}$$

Substituting (29.2.1) and (29.3.4) we get

$$\widehat{\mathbf{e}}_{k+1} = \mathbf{x}^{\mathrm{f}}_{k+1} + \mathbf{e}^{\mathrm{f}}_{k+1} - \mathbf{a} - \mathbf{K}h(\mathbf{x}_{k+1}) - \mathbf{K}\mathbf{v}_{k+1}. \tag{29.3.9}$$

Taking conditional expectations on both sides given $\mathbf{z}(1:k)$, (since $\mathbf{e}^{\mathrm{f}}_{k+1}$ is unbiased) we get the condition for the unbiasedness of $\widehat{\mathbf{e}}_{k+1}$ as

$$\begin{aligned}
0 &= E[\widehat{\mathbf{e}}_{k+1}|\mathbf{z}(1:k)] \\
&= \mathbf{x}^{\mathrm{f}}_{k+1} - \mathbf{a} - \mathbf{K}\widehat{h}(\mathbf{x}_{k+1})
\end{aligned} \tag{29.3.10}$$

where $\widehat{h}(\mathbf{x}_{k+1}) = E[h(\mathbf{x}_{k+1})|\mathbf{z}(1:k)]$ and $E[\mathbf{v}_{k+1}|\mathbf{z}(1:k)] = 0$. That is, for unbiasedness of $\widehat{\mathbf{x}}_{k+1}$, it is necessary that

$$\mathbf{a} = \mathbf{x}^{\mathrm{f}}_{k+1} - \mathbf{K}\widehat{h}(\mathbf{x}_{k+1}). \tag{29.3.11}$$

Substituting (29.3.11) into (29.3.7) results in the following linear structure for the new estimate

$$\widehat{\mathbf{x}}_{k+1} = \mathbf{x}^{\mathrm{f}}_{k+1} + \mathbf{K}[\mathbf{z}_{k+1} - \widehat{h}(\mathbf{x}_{k+1})]. \tag{29.3.12}$$

This expression is very similar to its linear counterpart in Chapter 27, except that $\widehat{h}(\mathbf{x}_{k+1})$ can not be computed explicitly since we do not have the conditional density of $\mathbf{x}_{k+1}$ given $\mathbf{z}(1:k)$. Defining

$$\mathbf{g}_k = h(\mathbf{x}_{k+1}) - \widehat{h}(\mathbf{x}_{k+1}) \tag{29.3.13}$$

it is immediate that $E[\mathbf{g}_k|\mathbf{z}(1:k)] = 0$. Substituting (29.3.13) into (29.3.12) and using (29.2.1), the expression for the error $\widehat{\mathbf{e}}_{k+1}$ in (29.3.8) becomes

$$\widehat{\mathbf{e}}_{k+1} = (\mathbf{e}^{\mathrm{f}}_{k+1} - \mathbf{K}\mathbf{g}_k) - \mathbf{K}\mathbf{v}_{k+1}. \tag{29.3.14}$$

Since the two terms $(\mathbf{e}^{\mathrm{f}}_{k+1} - \mathbf{K}\mathbf{g}_k)$ and $\mathbf{K}\mathbf{v}_{k+1}$ are uncorrelated, the covariance $\widehat{\mathbf{P}}_{k+1}$ of $\widehat{\mathbf{x}}_{k+1}$ is then given by

$$\begin{aligned}
\widehat{\mathbf{P}}_{k+1} &= E[(\widehat{\mathbf{e}}_{k+1})(\widehat{\mathbf{e}}_{k+1})^{\mathrm{T}}|\mathbf{z}(1:k)] \\
&= E[(\mathbf{e}^{\mathrm{f}}_{k+1} - \mathbf{K}\mathbf{g}_k)(\mathbf{e}^{\mathrm{f}}_{k+1} - \mathbf{K}\mathbf{g}_k)^{\mathrm{T}}|\mathbf{z}(1:k)] \\
&\quad + E[\mathbf{K}\mathbf{v}_{k+1}\mathbf{v}^{\mathrm{T}}_{k+1}\mathbf{K}^{\mathrm{T}}|\mathbf{z}(1:k)] \\
&= \mathbf{P}^{\mathrm{f}}_{k+1} - \mathbf{K}\mathbf{A}_k - \mathbf{A}^{\mathrm{T}}_k\mathbf{K}^{\mathrm{T}} + \mathbf{K}\mathbf{D}_k\mathbf{K}^{\mathrm{T}}
\end{aligned} \tag{29.3.15}$$

where

$$\left.\begin{aligned}
\mathbf{A}_k &= E[\mathbf{g}_k(\mathbf{e}^{\mathrm{f}}_{k+1})^{\mathrm{T}}|\mathbf{z}(1:k)] \\
\mathbf{D}_k &= \mathbf{C}_k + \mathbf{R}_{k+1} \\
\text{and} \qquad \mathbf{C}_k &= E[\mathbf{g}_k\mathbf{g}^{\mathrm{T}}_k|\mathbf{z}(1:k)]
\end{aligned}\right\} \tag{29.3.16}$$

It can be verified that $\mathbf{C}_k$ is symmetric and it is assumed that $\mathbf{R}_{k+1}$ and $\mathbf{D}_k$ are both positive definite.

We now turn to the important task of determining the matrix $\mathbf{K}$ that will force the estimate $\widehat{\mathbf{x}}_{k+1}$ in (29.3.12) to be of minimum variance. To this end (by invoking the technique described in Chapter 17 and used in Chapter 27) we add and subtract $\mathbf{A}^{\mathrm{T}}_k\mathbf{D}^{-1}_k\mathbf{A}_k$ to the r.h.s.of (29.3.15) and simplifying we get

$$\widehat{\mathbf{P}}_{k+1} = \mathbf{P}^{\mathrm{f}}_{k+1} - \mathbf{A}^{\mathrm{T}}_k\mathbf{D}^{-1}_k\mathbf{A}_k + (\mathbf{K} - \mathbf{A}^{\mathrm{T}}_k\mathbf{D}^{-1}_k)\mathbf{D}_k(\mathbf{K} - \mathbf{A}^{\mathrm{T}}_k\mathbf{D}^{-1}_k)^{\mathrm{T}}. \tag{29.3.17}$$

Hence, by setting

$$\mathbf{K} = \mathbf{A}^{\mathrm{T}}_k\mathbf{D}^{-1}_k \tag{29.3.18}$$

Model	$\mathbf{x}_{k+1} = \mathbf{M}(\mathbf{x}_k) + \mathbf{w}_{k+1}$
Observation	$\mathbf{z}_k = h(\mathbf{x}_k) + \mathbf{v}_k$
Forecast Step	$\mathbf{x}^{\mathrm{f}}_{k+1} = \widehat{\mathbf{M}}(\mathbf{x}_k)$
	$\mathbf{f}_k = \mathbf{M}(\mathbf{x}_k) - \widehat{\mathbf{M}}(\mathbf{x}_k)$
	$\mathbf{P}^{\mathrm{f}}_{k+1} = E[\mathbf{f}_k\mathbf{f}_k^{\mathrm{T}}\vert\mathbf{z}(1:k)] + \mathbf{Q}_{k+1}$
Data Assimilation Step	
	$\widehat{\mathbf{x}}_{k+1} = \mathbf{x}^{\mathrm{f}}_{k+1} + \mathbf{K}[\mathbf{z}_{k+1} - \widehat{h}(\mathbf{x}_{k+1})]$
	$\mathbf{g}_k = h(\mathbf{x}_{k+1}) - \widehat{h}(\mathbf{x}_{k+1})$
	$\mathbf{A}_k = E[\mathbf{g}_k(\mathbf{e}^{\mathrm{f}}_{k+1})^{\mathrm{T}}\vert\mathbf{z}(1:k)]$
	$\mathbf{C}_k = E[\mathbf{g}_k\mathbf{g}_k^{\mathrm{T}}\vert\mathbf{z}(1:k)]$
	$\mathbf{D}_k = (\mathbf{C}_k + \mathbf{R}_{k+1})$
	$\mathbf{K} = \mathbf{A}_k^{\mathrm{T}}\mathbf{D}_k^{-1}$
	$\widehat{\mathbf{P}}_{k+1} = \mathbf{P}^{\mathrm{f}}_{k+1} - \mathbf{A}_k^{\mathrm{T}}\mathbf{D}_k^{-1}\mathbf{A}_k$
	$\quad = \mathbf{P}^{\mathrm{f}}_{k+1} - \mathbf{K}\mathbf{A}_k$

Fig. 29.3.1 Exact dynamics of first two moments.

we eliminate the last term in (29.3.17), thereby forcing $\widehat{\mathbf{x}}_{k+1}$ to be of minimum variance with

$$\widehat{\mathbf{P}}_{k+1} = \mathbf{P}^{\mathrm{f}}_{k+1} - \mathbf{A}_k^{\mathrm{T}}\mathbf{D}_k^{-1}\mathbf{A}_k \qquad (29.3.19)$$

as its covariance.

A summary of the exact dynamics of evolution of the first two moments is given in Figure 29.3.1.

Example 29.3.1 Consider the special case when $\mathbf{M}(\mathbf{x}_k) = \mathbf{M}_k\mathbf{x}_k$ for some nonsingular matrix $\mathbf{M}_k \in \mathbb{R}^{n\times n}$ and $h(\mathbf{x}_k) = \mathbf{H}_k\mathbf{x}_k$ for some matrix $\mathbf{H}_k \in \mathbb{R}^{m\times n}$ of full rank. The forecast step becomes

$$\mathbf{x}^{\mathrm{f}}_{k+1} = \mathbf{M}_k\widehat{\mathbf{x}}_k$$

$$\mathbf{f}_k = \mathbf{M}_k(\mathbf{x}_k - \widehat{\mathbf{x}}_k) = \mathbf{M}_k\widehat{\mathbf{e}}_k$$

and

$$\mathbf{P}^{\mathrm{f}}_{k+1} = \mathbf{M}_k\widehat{\mathbf{P}}_k\mathbf{M}_k^{\mathrm{T}} + \mathbf{Q}_{k+1}.$$

Similarly, the data assimilation step becomes

$$\widehat{\mathbf{x}}_{k+1} = \mathbf{x}^{\mathrm{f}}_{k+1} + \mathbf{K}[\mathbf{z}_{k+1} - \mathbf{H}_k\mathbf{x}^{\mathrm{f}}_{k+1}]$$

since by definition

$$\widehat{h}(\mathbf{x}_{k+1}) = E[\mathbf{H}_k\mathbf{x}_{k+1}|\mathbf{z}(1:k)] = \mathbf{H}_k\mathbf{x}^{\mathrm{f}}_{k+1}.$$

Hence

$$\begin{aligned}
\mathbf{g}_k &= \mathbf{H}_k(\mathbf{x}_{k+1} - \mathbf{x}^{\mathrm{f}}_{k+1}) = \mathbf{H}_k\mathbf{e}^{\mathrm{f}}_{k+1}\\
\mathbf{A}_k &= E[\mathbf{H}_k\mathbf{e}^{\mathrm{f}}_{k+1}(\mathbf{e}^{\mathrm{f}}_{k+1})^{\mathrm{T}}|\mathbf{z}(1:k)] = \mathbf{H}_k\mathbf{P}^{\mathrm{f}}_{k+1}\\
\mathbf{C}_k &= \mathbf{H}_k\mathbf{P}^{\mathrm{f}}_{k+1}\mathbf{H}^{\mathrm{T}}_k\\
\mathbf{D}_k &= (\mathbf{H}_k\mathbf{P}^{\mathrm{f}}_{k+1}\mathbf{H}^{\mathrm{T}}_k + \mathbf{R}_{k+1})\\
\mathbf{K} &= \mathbf{P}^{\mathrm{f}}_{k+1}\mathbf{H}^{\mathrm{T}}_k[\mathbf{H}_k\mathbf{P}^{\mathrm{f}}_{k+1}\mathbf{H}^{\mathrm{T}}_k + \mathbf{R}_{k+1}]^{-1}
\end{aligned}$$

and

$$\begin{aligned}
\widehat{\mathbf{P}}_{k+1} &= \mathbf{P}^{\mathrm{f}}_{k+1} - \mathbf{P}^{\mathrm{f}}_{k+1}\mathbf{H}^{\mathrm{T}}_k[\mathbf{H}_k\mathbf{P}^{\mathrm{f}}_{k+1}\mathbf{H}^{\mathrm{T}}_k + \mathbf{R}_{k+1}]^{-1}\mathbf{H}_k\mathbf{P}^{\mathrm{f}}_{k+1}\\
&= (\mathbf{I} - \mathbf{K}\mathbf{H}_k)\mathbf{P}^{\mathrm{f}}_{k+1}.
\end{aligned}$$

That is, we obtain the Kalman filter equations in Figure 27.2.2.

29.4 Approximation to moment dynamics

The major impediments to using the exact dynamics of moments are due to the difficulty of computing $\widehat{\mathbf{M}}(\mathbf{x}_k)$ in the forecast step and $\widehat{h}(\mathbf{x}_{k+1})$ in the data assimilation step. In this section, we derive a family of approximations leading to practical algorithms.

The basic idea is to expand $\mathbf{M}(\mathbf{x}_k)$ in a rth-order Taylor Series expansion around the current estimate $\widehat{\mathbf{x}}_k$ and then compute an approximation to $\widehat{\mathbf{M}}(\mathbf{x}_k)$ using $\mathbf{M}(\widehat{\mathbf{x}}_k)$ and the value of the first r moments of $\widehat{\mathbf{x}}_k$. By varying the value of $r = 1, 2, 3\ldots$ we can obtain a family of approximations. A similar approach is again used to approximate $\widehat{h}(\mathbf{x}_{k+1})$ using $h(\mathbf{x}^{\mathrm{f}}_{k+1})$ and the moments of $\mathbf{x}^{\mathrm{f}}_{k+1}$.

We illustrate this approach by deriving the second-order filter using $r = 2$.

(A) Second-order filter

We begin with the forecast step first.

Forecast step Expanding $\mathbf{M}(\mathbf{x}_k)$ in a second-order Taylor Series (Appendix C), we obtain

$$\mathbf{M}(\mathbf{x}_k) \approx \mathbf{M}(\widehat{\mathbf{x}}_k) + \mathbf{D}_{\mathbf{M}}(\widehat{\mathbf{x}}_k)\widehat{\mathbf{e}}_k + \frac{1}{2}\mathbf{D}^2_{\mathbf{M}}(\widehat{\mathbf{x}}_k, \widehat{\mathbf{e}}_k) \tag{29.4.1}$$

where $\mathbf{D_M}(\widehat{\mathbf{x}}_k) \in \mathbb{R}^{n\times n}$ is the **Jacobian** of $\mathbf{M}(\mathbf{x})$ at $\widehat{\mathbf{x}}_k$,

$$\mathbf{D}^2_{\mathbf{M}}(\widehat{\mathbf{x}}_k, \widehat{\mathbf{e}}_k) = \begin{bmatrix} (\widehat{\mathbf{e}}_k)^{\mathrm{T}}\nabla^2\mathbf{M}_1\widehat{\mathbf{e}}_k \\ (\widehat{\mathbf{e}}_k)^{\mathrm{T}}\nabla^2\mathbf{M}_2\widehat{\mathbf{e}}_k \\ \vdots \\ (\widehat{\mathbf{e}}_k)^{\mathrm{T}}\nabla^2\mathbf{M}_n\widehat{\mathbf{e}}_k \end{bmatrix}. \tag{29.4.2}$$

$\mathbf{M}(\mathbf{x}) = (\mathbf{M}_1(\mathbf{x}), \mathbf{M}_2(\mathbf{x}), \ldots, \mathbf{M}_n(\mathbf{x}))^{\mathrm{T}}$ and $\nabla^2\mathbf{M}_i = \nabla^2\mathbf{M}_i(\mathbf{x}_k) \in \mathbb{R}^{n\times n}$ is the $n \times n$ **Hessian** of $\mathbf{M}_i(\mathbf{x}_k)$ at $\widehat{\mathbf{x}}_k$, $i = 1, 2, \ldots, n$.

Taking conditional expectation of both sides of (29.4.1), since $E[\widehat{\mathbf{e}}_k|\mathbf{z}(1:k)] = 0$, we get

$$\widehat{\mathbf{M}}(\mathbf{x}_k) \approx \mathbf{M}(\widehat{\mathbf{x}}_k) + \frac{1}{2}E[\mathbf{D}^2_{\mathbf{M}}(\widehat{\mathbf{x}}_k, \widehat{\mathbf{e}}_k)|\mathbf{z}(1:k)]. \tag{29.4.3}$$

The key to computing the value of the second term on the r.h.s.of (29.4.3) is contained in the following:

Example 29.4.1 Let $\mathbf{y} = (y_1, y_2)^{\mathrm{T}}$ be a random vector such that $E(\mathbf{y}) = 0$ and

$$\mathbf{P} = E[\mathbf{y}\mathbf{y}^{\mathrm{T}}] = \begin{bmatrix} \sigma_1^2 & \sigma_{12} \\ \sigma_{12} & \sigma_2^2 \end{bmatrix}.$$

Let

$$\mathbf{A} = \begin{bmatrix} a & b \\ b & c \end{bmatrix}$$

be a symmetric matrix. Then,

$$\mathbf{y}^{\mathrm{T}}\mathbf{A}\mathbf{y} = ay_1^2 + 2by_1y_2 + cy_2^2.$$

Taking expectations, we get

$$\begin{aligned} E[\mathbf{y}^{\mathrm{T}}\mathbf{A}\mathbf{y}] &= aE[y_1^2] + 2bE[y_1y_2] + cE[y_2^2] \\ &= a\sigma_1^2 + 2b\sigma_{12} + c\sigma_2^2 \\ &= \mathrm{tr}[\mathbf{AP}] \quad (\mathrm{tr}(\mathbf{A}) = \text{trace of } \mathbf{A}, \text{Appendix B}) \\ &= \mathrm{tr}[\mathbf{A}E(\mathbf{y}\mathbf{y}^{\mathrm{T}})] \\ &= \mathrm{tr}[E(\mathbf{A}\mathbf{y}\mathbf{y}^{\mathrm{T}})] \\ &= E[\mathrm{tr}(\mathbf{A}\mathbf{y}\mathbf{y}^{\mathrm{T}})] \\ &= E[\mathrm{tr}(\mathbf{y}^{\mathrm{T}}\mathbf{A}\mathbf{y})] \\ &= E[\mathbf{y}^{\mathrm{T}}\mathbf{A}\mathbf{y}]. \end{aligned}$$

Notice that this result easily carries over to any mean zero random vector $\mathbf{y} \in \mathbb{R}^n$ and $\mathbf{P}$ as its covariance matrix and any symmetric matrix $\mathbf{A} \in \mathbb{R}^{n\times n}$.

In view of this result, the ith component of the vector in the second term on the r.h.s. of (29.4.3) becomes

$$\begin{aligned} E[(\widehat{\mathbf{e}}_k)^{\mathrm{T}}\nabla^2\mathbf{M}_i\widehat{\mathbf{e}}_k] &= \mathrm{tr}\{\nabla^2\mathbf{M}_i E[\widehat{\mathbf{e}}_k(\widehat{\mathbf{e}}_k)^{\mathrm{T}}]\} \\ &= \mathrm{tr}\{\nabla^2\mathbf{M}_i\widehat{\mathbf{P}}_k\}. \end{aligned} \tag{29.4.4}$$

By way of simplifying the notation, define

$$\partial^2(\mathbf{M}, \widehat{\mathbf{P}}_k) = \begin{bmatrix} \mathrm{tr}\{\nabla^2\mathbf{M}_1\widehat{\mathbf{P}}_k\} \\ \mathrm{tr}\{\nabla^2\mathbf{M}_2\widehat{\mathbf{P}}_k\} \\ \vdots \\ \mathrm{tr}\{\nabla^2\mathbf{M}_n\widehat{\mathbf{P}}_k\} \end{bmatrix}. \tag{29.4.5}$$

Combining (29.4.2)–(29.4.5), we obtain the second-order accurate forecast

$$\mathbf{x}^{\mathrm{f}}_{k+1} = \widehat{\mathbf{M}}(\mathbf{x}_k) \approx \mathbf{M}(\widehat{\mathbf{x}}_k) + \frac{1}{2}\partial^2(\mathbf{M}, \widehat{\mathbf{P}}_k). \tag{29.4.6}$$

Hence, referring to (29.3.2), we get

$$\mathbf{f}_k = \mathbf{M}(\mathbf{x}_k) - \widehat{\mathbf{M}}(\mathbf{x}_k) = \mathbf{D_M}\widehat{\mathbf{e}}_k + \boldsymbol{\eta}_k \tag{29.4.7}$$

where $\mathbf{D_M} = \mathbf{D_M}(\widehat{\mathbf{x}}_k)$ and

$$\boldsymbol{\eta}_k = \frac{1}{2}[\mathbf{D}^2_{\mathbf{M}}(\widehat{\mathbf{x}}_k, \widehat{\mathbf{e}}_k) - \partial^2(\mathbf{M}, \widehat{\mathbf{P}}_k)]. \tag{29.4.8}$$

It can be verified that $E[\mathbf{f}_k|\mathbf{z}(1:k)] = 0$ and that the error in the forecast is given by (refer to 29.3.4)

$$\mathbf{e}^{\mathrm{f}}_{k+1} = \mathbf{f}_k + \mathbf{w}_{k+1} \tag{29.4.9}$$

where $E[\mathbf{e}^{\mathrm{f}}_{k+1}|\mathbf{z}(1:k)] = 0$. Now combining (29.4.7) and (29.4.9), since $\mathbf{w}_{k+1}$ is uncorrelated with $\mathbf{f}_k$, the forecast covariance $\mathbf{P}^{\mathrm{f}}_{k+1}$ is given by

$$\begin{aligned} \mathbf{P}^{\mathrm{f}}_{k+1} &= E[\mathbf{e}^{\mathrm{f}}_{k+1}(\mathbf{e}^{\mathrm{f}}_{k+1})^{\mathrm{T}}|\mathbf{z}(1:k)] \\ &= E[\mathbf{f}_k\mathbf{f}^{\mathrm{T}}_k|\mathbf{z}(1:k)] + \mathbf{Q}_{k+1} \\ &= \mathbf{D_M}\widehat{\mathbf{P}}_k\mathbf{D}^{\mathrm{T}}_{\mathbf{M}} + \mathbf{Q}_{k+1} + \mathbf{D_M}E[\widehat{\mathbf{e}}_k\boldsymbol{\eta}^{\mathrm{T}}_k|\mathbf{z}(1:k)] \\ &\quad + E[\boldsymbol{\eta}_k(\widehat{\mathbf{e}}_k)^{\mathrm{T}}]\mathbf{D}^{\mathrm{T}}_{\mathbf{M}} + E[\boldsymbol{\eta}_k\boldsymbol{\eta}^{\mathrm{T}}_k]. \end{aligned} \tag{29.4.10}$$

Since $\boldsymbol{\eta}_k$ is quadratic in $\widehat{\mathbf{e}}_k$, herein lies the evidence of the dependence of the second moment of $\mathbf{x}^{\mathrm{f}}_{k+1}$ on the second-, third-, and the fourth-order moments of $\mathbf{x}_k$ – **a lack of moment closure** alluded to at the end of Section 29.2. By dropping the third- and higher-order terms in (29.4.10), we obtain a second-order approximation to the forecast covariance given by

$$\mathbf{P}^{\mathrm{f}}_{k+1} \approx \mathbf{D_M}\widehat{\mathbf{P}}_k\mathbf{D}^{\mathrm{T}}_{\mathbf{M}} + \mathbf{Q}_{k+1}. \tag{29.4.11}$$

Remark 29.4.1 Impact of nonlinearity When $\mathbf{M}(\cdot)$ is nonlinear, since $\widehat{\mathbf{M}}(\mathbf{x}_k) \neq \mathbf{M}(\widehat{\mathbf{x}}_k)$, first we are forced to settle for an approximation, such as for example the second-order accurate forecast $\mathbf{x}^{\mathrm{f}}_{k+1}$ in (29.4.6). Computation of the forecast covariance is again riddled with its own set of problems related to the lack of the moment closure, as is evident from (29.4.10). Thus, once again we are forced to settle for a second-order approximation to the forecast covariance $\mathbf{P}^{\mathrm{f}}_{k+1}$ given in (29.4.11). The ultimate utility of this class of approximations will largely depend on the nature and type of nonlinearity in the dynamics.

Data Assimilation Step Expanding $h(\mathbf{x}_{k+1})$ in a second-order Taylor Series (Appendix C) around $\mathbf{x}^{\mathrm{f}}_{k+1}$, (analogous to (29.4.1))

$$h(\mathbf{x}_{k+1}) = h(\mathbf{x}^{\mathrm{f}}_{k+1}) + \mathbf{D_h}(\mathbf{x}^{\mathrm{f}}_{k+1})\mathbf{e}^{\mathrm{f}}_{k+1} + \frac{1}{2}\mathbf{D}^2_{\mathbf{h}}(\mathbf{x}^{\mathrm{f}}_{k+1}, \mathbf{e}^{\mathrm{f}}_{k+1}) \tag{29.4.12}$$

where $\mathbf{D_h}(\mathbf{x}^{\mathrm{f}}_{k+1}) \in \mathbb{R}^{m\times n}$ is the Jacobian of $h(\mathbf{x})$ evaluated at $\mathbf{x}^{\mathrm{f}}_{k+1}$; $\mathbf{D}^2_{\mathbf{h}}(\mathbf{x}^{\mathrm{f}}_{k+1}, \mathbf{e}^{\mathrm{f}}_{k+1}) \in \mathbb{R}^m$ whose ith component is given by $(\mathbf{e}^{\mathrm{f}}_{k+1})^{\mathrm{T}}\nabla^2 h_i(\mathbf{e}^{\mathrm{f}}_{k+1})(\mathbf{e}^{\mathrm{f}}_{k+1})$, and $\nabla^2 h_i$ is the Hessian of $h_i(\mathbf{x})$ evaluated at $\mathbf{x}^{\mathrm{f}}_{k+1}$. Taking the conditional expectations of both sides of (29.4.12) given $\mathbf{z}(1 : k)$, and by repeating the arguments leading to (29.4.5), we obtain

$$\widehat{h}(\mathbf{x}_{k+1}) = h(\mathbf{x}^{\mathrm{f}}_{k+1}) + \frac{1}{2}\partial^2(h, \mathbf{P}^{\mathrm{f}}_{k+1}) \tag{29.4.13}$$

where

$$\partial^2(h, \mathbf{P}^{\mathrm{f}}_{k+1}) = \begin{bmatrix} \mathrm{tr}\{\nabla^2 h_1 \mathbf{P}^{\mathrm{f}}_{k+1}\} \\ \mathrm{tr}\{\nabla^2 h_2 \mathbf{P}^{\mathrm{f}}_{k+1}\} \\ \vdots \\ \mathrm{tr}\{\nabla^2 h_m \mathbf{P}^{\mathrm{f}}_{k+1}\} \end{bmatrix}. \tag{29.4.14}$$

Referring to Figure 29.3.1, we now derive the computable version of the second-order approximation as follows:

$$\widehat{\mathbf{x}}_{k+1} = \mathbf{x}^{\mathrm{f}}_{k+1} + \mathbf{K}[\mathbf{z}_{k+1} - h(\mathbf{x}^{\mathrm{f}}_{k+1}) - \frac{1}{2}\partial^2(h, \mathbf{P}^{\mathrm{f}}_{k+1})]. \tag{29.4.15}$$

From (29.4.12) and (29.4.13), we get

$$\mathbf{g}_k = h(\mathbf{x}_{k+1}) - h(\mathbf{x}^{\mathrm{f}}_{k+1}) = \mathbf{D_h}\mathbf{e}^{\mathrm{f}}_{k+1} + \boldsymbol{\xi}_k \tag{29.4.16}$$

where

$$\boldsymbol{\xi}_k = \frac{1}{2}[\mathbf{D}^2_{\mathbf{h}}(\mathbf{x}^{\mathrm{f}}_{k+1}, \mathbf{e}^{\mathrm{f}}_{k+1}) - \partial^2(h, \mathbf{P}^{\mathrm{f}}_{k+1})]. \tag{29.4.17}$$

Model $\mathbf{x}_{k+1} = \mathbf{M}(\mathbf{x}_k) + \mathbf{w}_{k+1}$

Observation $\mathbf{z}_k = h(\mathbf{x}_k) + \mathbf{v}_k$

Forecast Step

$$\mathbf{x}^{\mathrm{f}}_{k+1} = \mathbf{M}(\widehat{\mathbf{x}}_k) + \frac{1}{2}\partial^2(\mathbf{M}, \widehat{\mathbf{P}}_k)$$
$$\mathbf{P}^{\mathrm{f}}_{k+1} = \mathbf{D_M}\widehat{\mathbf{P}}_k\mathbf{D}^{\mathrm{T}}_{\mathbf{M}} + \mathbf{Q}_{k+1}$$

Data Assimilation Step

$$\widehat{\mathbf{x}}_{k+1} = \mathbf{x}^{\mathrm{f}}_{k+1} + \mathbf{K}[\mathbf{z}_{k+1} - h(\mathbf{x}^{\mathrm{f}}_{k+1}) - \frac{1}{2}\partial^2(h, \mathbf{P}^{\mathrm{f}}_{k+1})]$$
$$\mathbf{K} = \mathbf{P}^{\mathrm{f}}_{k+1}\mathbf{D}^{\mathrm{T}}_{\mathbf{h}}[\mathbf{D_h}\mathbf{P}^{\mathrm{f}}_{k+1}\mathbf{D}^{\mathrm{T}}_{\mathbf{h}} + \mathbf{R}_{k+1}]^{-1}$$
$$\widehat{\mathbf{P}}_{k+1} = (\mathbf{I} - \mathbf{K}\mathbf{D_h})\mathbf{P}^{\mathrm{f}}_{k+1}$$

Fig. 29.4.1 The second-order filter.

is a quadratic in $\mathbf{e}^{\mathrm{f}}_{k+1}$. It can be verified that $E[\mathbf{g}_k|\mathbf{z}(1:k)] = 0$. Substituting (29.4.16)–(29.4.17) for $\mathbf{g}_k$ it follows that

$$\begin{aligned}
\mathbf{A}_k &= E[\mathbf{g}_k(\mathbf{e}^{\mathrm{f}}_{k+1})^{\mathrm{T}}|\mathbf{z}(1:k)] \\
&\approx \mathbf{D_h}\mathbf{P}^{\mathrm{f}}_{k+1} \\
\mathbf{C}_k &= E[\mathbf{g}_k\mathbf{g}^{\mathrm{T}}_k|\mathbf{z}(1:k)] \\
&\approx \mathbf{D_h}\mathbf{P}^{\mathrm{f}}_{k+1}\mathbf{D}^{\mathrm{T}}_{\mathbf{h}} \\
\mathbf{D}_k &= \mathbf{C}_k + \mathbf{R}_{k+1} = \mathbf{D_h}\mathbf{P}^{\mathrm{f}}_{k+1}\mathbf{D}^{\mathrm{T}}_{\mathbf{h}} + \mathbf{R}_{k+1} \\
\mathbf{K} &= \mathbf{P}^{\mathrm{f}}_{k+1}\mathbf{D}^{\mathrm{T}}_{\mathbf{h}}[\mathbf{D_h}\mathbf{P}^{\mathrm{f}}_{k+1}\mathbf{D}^{\mathrm{T}}_{\mathbf{h}} + \mathbf{R}_{k+1}]^{-1}
\end{aligned}$$

and

$$\widehat{\mathbf{P}}_{k+1} = (\mathbf{I} - \mathbf{K}\mathbf{D_h})\mathbf{P}^{\mathrm{f}}_{k+1}.$$

For easy reference, the second-order filter is summarized in Figure 29.4.1.

(B) First-order (extended Kalman) filter

As observed at the beginning of this section, by expanding $\mathbf{M}(\mathbf{x}_k)$ around $\widehat{\mathbf{x}}_k$ and $h(\mathbf{x}_{k+1})$ around $\mathbf{x}^{\mathrm{f}}_{k+1}$ in a **first-order** Taylor series expansion (Appendix C) and by repeating the computations described in Part A (for the second-order filter), we obtain the first-order filter, also known as the **extended Kalman** filter, since it reduces to the Kalman filter (Chapter 27) when $\mathbf{M}(\mathbf{x})$ and $h(\mathbf{x})$ are linear in $\mathbf{x}$. Since this is a special case of the second-order filter, we only indicate the major steps.

Forecast Step The forecast equation for the state and its covariance are obtained by merely dropping the second-order terms in (29.4.1), (29.4.6) and (29.4.7).

Model	$\mathbf{x}_{k+1} = \mathbf{M}(\mathbf{x}_k) + \mathbf{w}_{k+1}$
Observation	$\mathbf{z}_k = h(\mathbf{x}_k) + \mathbf{v}_k$
Forecast Step	$\mathbf{x}^f_{k+1} = \mathbf{M}(\widehat{\mathbf{x}}_k)$
	$\mathbf{P}^f_{k+1} = \mathbf{D_M}\widehat{\mathbf{P}}_k\mathbf{D}^T_\mathbf{M} + \mathbf{Q}_{k+1}$
Data Assimilation Step	
	$\widehat{\mathbf{x}}_{k+1} = \mathbf{x}^f_{k+1} + \mathbf{K}[\mathbf{z}_{k+1} - h(\mathbf{x}^f_{k+1})]$
	$\mathbf{K} = \mathbf{P}^f_{k+1}\mathbf{D}^T_\mathbf{h}[\mathbf{D_h}\mathbf{P}^f_{k+1}\mathbf{D}^T_\mathbf{h} + \mathbf{R}_{k+1}]^{-1}$
	$\widehat{\mathbf{P}}_{k+1} = (\mathbf{I} - \mathbf{K}\mathbf{D_h})\mathbf{P}^f_{k+1}$

Fig. 29.4.2 First-order/extended Kalman filter (EKF).

Data Assimilation Step The expression for the new estimate, its covariance and the gain matrix $\mathbf{K}$ are obtained by again dropping the second-order terms in (29.4.12), (29.4.13) and (29.4.16).

The resulting first-order or extended Kalman filter equations are given in Figure 29.4.2. Several observations are in order.

(1) Comparing the algorithms in Figures 29.4.1 and 29.4.2, it follows that the expression for the covariances $\mathbf{P}^f_{k+1}$ and $\widehat{\mathbf{P}}_k$ are identical. This similarity is only **skin deep** and their actual values for the first- and second-order filters must be different since their respective forecast and estimation equations are different.
(2) **Bias Correction** From Figures 29.4.1 and 29.4.2 it is immediate that the second-order filter has an extra term, $\frac{1}{2}\partial^2(\mathbf{M}, \widehat{\mathbf{P}}_k) \in \mathbb{R}^n$ in the forecast equation whose ith component is given by $tr[\nabla^2\mathbf{M}_i\widehat{\mathbf{P}}_k]$. This extra term is a direct result of the nonlinearity in $\mathbf{M}(\mathbf{x})$ and is known as the **forecast bias** correction term. Similarly, the extra term $\frac{1}{2}\partial^2(h, \mathbf{P}^f_{k+1}) \in \mathbb{R}^m$ in the data assimilation equation for the second-order filter is such that its ith component is given by $tr[\nabla^2 h_i \mathbf{P}^f_{k+1}]$. Again, this extra term is the direct consequence of the nonlinearity in $h(\mathbf{x})$ and is known as the **analysis bias** correction. It is the inclusion of these terms that makes the second-order filter more accurate than its first-order counterpart.
(3) **Extension** The basic principle that underlies the derivation of the second-order filter verbatim carries over to any rth-order filter. An example of the fourth-order filter for the scalar case is pursued in Exercise 29.6.
(4) **Linearized Kalman Filter** The distinguishing feature of the first-order filter is that given $(\widehat{\mathbf{x}}_k, \widehat{\mathbf{P}}_k)$ it **linearizes** the nonlinear dynamics $\mathbf{M}(\mathbf{x})$ **locally** around $\widehat{\mathbf{x}}_k$ and computes $\mathbf{x}^f_{k+1}$ and $\mathbf{P}^f_{k+1}$, based on which it again **linearizes** $h(\mathbf{x})$ **locally** around $\mathbf{x}^f_{k+1}$ to compute $\widehat{\mathbf{x}}_{k+1}$ and $\widehat{\mathbf{P}}_{k+1}$. Instead of using this repeated local linearization, one can also consider an alternate strategy of a **global**

linearization. Assume that we are given a prespecified **base** or **nominal** trajectory of the given nonlinear system. We can obtain a **global linearization** as a **first-order perturbation** along the given base trajectory leading to the so-called **tangent linear sytsem** (TLS) (refer to Chapter 24). Likewise, we can also linearize $h(\mathbf{x})$ along the base trajectory and obtain a sequence of linear increments to the observations about the same base trajectory. Now, using the TLS and the linearized observation increments, we can obtain a system of filter equations for the recursive estimation of the perturbation around the base trajectory called the **linearized Kalman filter** equations. See Exercise 29.5 for details.

Exercises

29.1 Let $P(z|x) \sim N(x, 1)$ and $P(x) \sim N(m_x, 1)$.
(a) Verify that $P(x|z) = \beta P(z|x)P(x)$ is a normal density exactly when

$$\beta = \frac{1}{\sqrt{2\pi}\sqrt{2}} \exp\left[-\frac{1}{2}\left(\frac{z - m_x}{\sqrt{2}}\right)^2\right]$$

and that

$$P(z|x) \sim N(\widehat{x}, \sigma^2)$$

where

$$\widehat{x} = \frac{1}{2}(z + m_x) \quad \text{and} \quad \sigma^2 = \frac{1}{2}.$$

(b) Using the Bayes' rule $P(x|z)P(z) = P(z|x)P(x)$ find $P(z)$ and compare it with β given in a.

29.2 Starting with $x_2 = ax_1 + w_2$ when $P(x_1) \sim N(am, P_0a^2 + q_1)$ and $w_2 \sim N(0, q_2)$, repeat the computations in Example 29.1.1 and compute $P(x_2)$. Generalize it to compute $P(x_k)$, when $x_k = ax_{k-1} + w_k$.

29.3 Repeat the derivation in Example 29.1.1 for the vector case where

$$\mathbf{x}_{k+1} = \mathbf{A}\mathbf{x}_k + \mathbf{w}_{k+1}$$

where $\mathbf{A} \in \mathbb{R}^{n\times n}$ is a nonsingular matrix, $\mathbf{x}_0 \sim N(m, \mathbf{P}_0)$ and $\mathbf{w}_k \sim N(0, \mathbf{Q}_k)$ where $\mathbf{P}_0 \in \mathbb{R}^{n\times n}$ and $\mathbf{Q}_k \in \mathbb{R}^{n\times n}$.

29.4 Let $Y_1 \sim N(\mu_1, \sigma_1^2)$ and $Y_2 \sim N(\mu_2, \sigma_2^2)$ be two independent Gaussian random variables. Then verify that $Y = Y_1 + Y_2 \sim N(\mu_1 + \mu_2, \sigma_1^2 + \sigma_2^2)$.
Hint: The density of Y is given by the **convolution** of the densities of Y_1 and Y_2. That is,

$$P_Y(y) = \int_{-\infty}^{\infty} P_{Y_1}(y)P_{Y_2}(t - y)\mathrm{d}t$$

and follow the method of the Example 29.1.1.

29.5 **Linearized Kalman filter** Consider the nonlinear model $\mathbf{x}_{k+1} = \mathbf{M}(\mathbf{x}_k)$ where $\mathbf{M} : \mathbb{R}^n \to \mathbb{R}^n$ and the observations $\mathbf{z}_k = h(\mathbf{x}_k)$ where $h : \mathbb{R}^n \to \mathbb{R}^m$. Let $\bar{\mathbf{x}}_0$ be the initial **base state** and let $\bar{\mathbf{x}}_k$ for $k = 1, 2, 3, \ldots$ be the base trajectory of the nonlinear model. Let $\mathbf{x}_0 = \bar{\mathbf{x}}_0 + \delta\mathbf{x}_0$ be an initial state "close" to $\bar{\mathbf{x}}_0$ where $\delta\mathbf{x}_0 \in \mathbb{R}^n$ is called the **initial perturbation**.

(a) Let $\delta\mathbf{x}_k$ be the perturbation at time k. Using the first-order Taylor Series expansion, verify that the dynamics of the first-order perturbations is given by the **tangent linear system** (TLS)

$$\delta\mathbf{x}_{k+1} = \mathbf{D}_{\mathbf{M}}(\bar{\mathbf{x}}_k)\delta\mathbf{x}_k$$

where $\delta\mathbf{x}_0$ is the initial condition and $\mathbf{D}_{\mathbf{M}}(\bar{\mathbf{x}}_k)$ is the Jacobian of $\mathbf{M}(\mathbf{x})$ at $\bar{\mathbf{x}}_k$.

(b) By linearizing $\mathbf{z}_k$ along the base trajectory $\{\bar{\mathbf{x}}_k\}$ verify that the first-order observation increments are given by (using first-order Taylor Series)

$$\delta\mathbf{z}_k = \mathbf{z}_k - h(\bar{\mathbf{x}}_k) = \mathbf{D}_{\mathbf{h}}(\bar{\mathbf{x}}_k)\delta\mathbf{x}_k$$

where $\mathbf{D}_{\mathbf{h}}(\bar{\mathbf{x}}_k)$ is the Jacobian of $h(\mathbf{x})$ at $\bar{\mathbf{x}}_k$.

(c) Consider the linear system

$$\delta\mathbf{x}_{k+1} = \mathbf{D}_{\mathbf{M}}(\bar{\mathbf{x}}_k)\delta\mathbf{x}_k + \mathbf{w}_{k+1} \tag{1}$$

and the linear observation increments

$$\delta\mathbf{z}_k = \mathbf{D}_{\mathbf{h}}(\bar{\mathbf{x}}_k) + \mathbf{v}_k \tag{2}$$

where the model noise $\mathbf{w}_{k+1} \in \mathbb{R}^n$ and the observation noise $\mathbf{v}_k \in \mathbb{R}^m$ satisfy the usual conditions set out in Chapter 27. Except for the notation, the equations (1) and (2) are exactly the same as those in (27.2.1) and (27.2.2) respectively. Rewrite the Kalman filter equations in Figure 27.2.2 using the new notation in (1) and (2). The resulting set of equations is called **linearized Kalman filter**.

Notes and references

Section 29.1 Randomness in a dynamical system can arise in one of three ways: random initial/boundary conditions, random forcing, or random coefficients. In this chapter we are largely concerned with randomness from the initial/boundary conditions and from forcing. There is a vast body of literature on stochastic dynamical systems. Satty (1967) and Snoog (1973) provide an elementary introduction. The modern theory of stochastic dynamic system relies on the theory of Markov process developed by A. N. Kolmogorov in the early 1930s and on the theory of stochastic differential equations based on the stochastic calculus developed by K. Ito (1944). In particular, the **Kolmogorov's forward** equation (also known as the **Fokker–Planck** equation) succinctly summarizes the evolution of the probability density of the states of a Markov process whose evolution is described by **Ito** type stochastic differential equation. This Kolmogorov's equation accounts for randomness due to both the initial conditions and forcing. In the special case when there is no random forcing the stochastic differential equation reduces to ordinary differential equations with random initial conditions. In this case, the Kolmogorov's forward equation reduces to the well-known **Liouville–Gibbs** equation. Refer to Arnold (1974), Friedman (1975), Gikhman and Skorokhod (1972), Grigoriu (2002), Oksendal (2003) for the theory of stochastic differential equations and the derivation of the Kolmogorov's forward equation. Refer to Satty (1967), Grigoriu (2002) and Snoog (1973) for the relation between Liouville–Gibbs equation and the Kolmogorov's forward equation.

Discrete time version of the theory of stochastic dynamical system is contained in Jazwinski (1970), Maybeck (1982) and Catlin (1989).

Section 29.2 Theory of nonlinear filtering is one of the well-understood aspects of stochastic dynamical systems and it began with the work of Kushner (1962). Also refer to Stratonovich (1962). For an alternate derivation refer to Zakai (1969). The Kushner–Stratonovich–Zakai equation defines the evolution of the probability density of the nonlinear filter estimate. This equation is a natural generalization of the Kolmogorov's forward equation. Since then the nonlinear filtering problem has received considerable attention and has been extended in several directions. For a systematic treatment of this topic refer to Bucy (1965), Bucy and Joseph (1968), Bucy (1970), Krishnan (1984), Kallianpur (1980), Liptser and Shiryaev (1977) and (1978), and Cohn (1997). Our treatment of nonlinear filtering follows Bucy and Joseph (1968) and Bucy (1994).

Section 29.3 This section is patterned after Wang and Lakshmivarahan (2004). Also refer to Henriksen (1980).

Section 29.4 Bucy (1965) and Kushner (1967) develop the theory of approximation to the nonlinear filters in continuous time. For discrete analogs refer to Jazwinski (1970), and Maybech (1979). Wishner et al. (1969) and Schwartz and Stear (1968) provide a comparison of various approximation schemes. Bucy and Senne (1971), Sorenson and Alspach (1971) and Sorenson and Stubberud (1968) develop methods for approximating the filter probability density. Refer to Bermaton (1985), Florchinger and LeGland (1984) and Kushner and Dupuis (1992) for details relating to other methods for approximating nonlinear filters.

30
Reduced-rank filters

While the basic principles of linear and nonlinear filtering are well understood – witness Chapters 27–29, they are **not** widely used in day-to-day operations at the national centers for weather prediction yet. This gap between the theory and its applications in Geophysical Sciences, especially in meteorology is largely a result of the excessive or prohibitively large computational requirements that render the implementation of this class of algorithms currently infeasible. To get a handle on this difficulty, recall from Table 27.2.1 that it requires $O(n^3)$ flops to update the covariance matrix $\mathbf{P}^{\mathrm{f}}_{k+1}$. When $n = 10^6$, this step alone requires of the order of 10^{18} flops. Assuming that we have access to the fastest computer that can deliver 1000 Giga flops - sec^{-1} $= 10^{12}$ flops - sec^{-1}, it would require 10^6 seconds to update $\mathbf{P}^{\mathrm{f}}_{k+1}$. Since there are only 31.536×10^6 seconds in a year, it would take nearly 12 days to compute $\mathbf{P}^{\mathrm{f}}_{k+1}$ from $\widehat{\mathbf{P}}_k$.

There are mainly two avenues to mitigate this **curse of dimensionality**. First is to resort to **parallel computation** which is getting increasing attention. Thanks to ever decreasing cost of computer hardware, today we can acquire powerful state of the art parallel processors at a fraction of the cost of yesteryears. For a given problem, the speedup achievable however, is largely dependent on (*i*) the algorithm, (*ii*) the number of processors and (*iii*) the topology of interconnection of the underlying network and (*iv*) how the tasks of the algorithm are mapped on to the processors. The second avenue which has become more popular is to compute a low- or **reduced-rank** approximation to the full-rank covariance matrix. All the low-rank filters differ only in the way in which the approximations are derived.

In this chapter we describe two types of reduced-rank approximations. First is a class of **explicit reduced-order** filters which are derived from the full-rank square root filters discussed in Section 28.7. Second is the class of **implicit reduced-order** filters for nonlinear problems where in the forecast $\mathbf{x}^{\mathrm{f}}_{k+1}$, the estimate $\widehat{\mathbf{x}}_{k+1}$ and their covariances $\mathbf{P}^{\mathrm{f}}_{k+1}$ and $\widehat{\mathbf{P}}_{k+1}$ respectively are computed using the standard **Monte Carlo** framework as the **sample** moments of an ensemble of size N much smaller compared to n, the dimension of the state space of the model.

Ensemble filters are described in Section 30.1. Section 30.2 contains a review of many of the known reduced-rank filters. The concluding section provides a summary of the known applications of the Kalman/sequential filtering methodology in the geophysical domain.

30.1 Ensemble filtering

It was shown in Chapter 28 that the exact method for nonlinear filtering involves recursively computing the **filter probability density** function $f_k(\mathbf{x}_k)$ and the **predictor probability density** function $P_{k+1}(\mathbf{x})$ over the state space $\mathbb{R}^n$. Except for the case when the model and the observations are linear and all the disturbances and the initial conditions are normally distributed, finding a closed form expression for these density functions is virtually impossible. Numerical methods are often the only avenue and when n is large these computations are practically infeasible. Ensemble filtering technique provides a feasible alternative by capturing (partial) information about the density functions by the distribution of the ensemble of states in $\mathbb{R}^n$ using the standard **Monte Carlo** framework. The Kalman filter (Chapter 26) and the approximate nonlinear filters (Chapter 28) operate by updating the mean $\widehat{\mathbf{x}}_k$ and its covariance $\widehat{\mathbf{P}}_k$. In contrast, in the ensemble approach, a filtering algorithm is applied to every strand of the ensemble from which the required mean and the variance are computed as the standard sample moments. In this section we describe the essence of the ensemble filtering methodology. We begin with two basic facts.

A result from point estimation theory

The basic premise of this ensemble analysis is centered around a very simple result from the theory of point estimation (Chapter 13). Let $f(x)$ be the probability density function of a random variable whose mean μ and variance σ^2 are **not** known. The standard method to estimate μ and σ^2 is to create an **ensemble** with a set of N independent samples, say $x_1, x_2, \ldots, x_N$, drawn from $f(x)$. Then the **sample mean**, $\overline{x}(N)$ and the **sample variance**, $s^2(N)$ are given by

$$\overline{x}(N) = \frac{1}{N}\sum_{i=1}^{N} x_i \quad \text{and} \quad s^2(N) = \frac{1}{N-1}\sum_{i=1}^{N}(x_i - \overline{x}(N))^2, \tag{30.1.1}$$

which are the **unbiased** estimates of μ and σ^2, respectively (Chapter 13). Further, it can be verified that

$$\mathrm{Var}[\overline{x}(N)] = \frac{\sigma^2}{N} \quad \text{and} \quad \mathrm{Var}[s^2(N)] = \frac{2\sigma^4}{N-1}. \tag{30.1.2}$$

Thus, the estimates in (30.1.1) are **asymptotically consistent**, that is,

$$\left.\begin{aligned} &\text{Prob}[|\overline{x}(N) - \mu| > \varepsilon] \longrightarrow 0 \\ \text{and} \quad & \\ &\text{Prob}[|s^2(N) - \sigma^2| > \varepsilon] \longrightarrow 0 \end{aligned}\right\} \tag{30.1.3}$$

as $N \to \infty$. That is, the sampling distribution of $\overline{x}(N)$ and $s^2(N)$ are centered and concentrated around μ and σ^2 respectively for large N. Expressions for the sampling variance in (30.1.2) imply that when the sample size is finite, we could observe a variation in the estimate of magnitude proportional to the standard deviation (also called **standard error**) which is of the order of $1/\sqrt{N}$.

These conclusions directly carry over to the problem of estimating the mean and the covariance of random vectors.

Generation of a random vector $\mathbf{x} \sim N(\mu, \Sigma)$

Ensemble filtering technique heavily relies on the ability to generate Gaussian random vectors with a prespecified mean μ and covariance Σ. We now describe an algorithm that will be used repeatedly in the following development.

(1) First factorize $\Sigma = \mathbf{L}\mathbf{L}^{\mathrm{T}}$ using the Cholesky method described in Chapter 9.
(2) Let $\mathbf{y} \sim N(0, \mathbf{I})$, the standard normal random vector with zero mean and unit variance.
(3) Then $\mathbf{x} = \mu + \mathbf{L}\mathbf{y}$ is the required random vector.

For,

$$E(\mathbf{x}) = \mu + \mathbf{L}E(\mathbf{y}) = \mu$$

and

$$\begin{aligned} \text{Cov}(\mathbf{x}) &= E[(\mathbf{x} - \mu)(\mathbf{x} - \mu)^{\mathrm{T}}] \\ &= LE(\mathbf{y}\mathbf{y}^{\mathrm{T}})\mathbf{L}^{\mathrm{T}} \\ &= \mathbf{L}\mathbf{L}^{\mathrm{T}} = \Sigma. \end{aligned}$$

Notice that this is the inverse of the **whitening** filter described in Chapter 28. Methods for generating the standard normal vector is pursued in Exercise 30.1.

We now turn our attention to developing the basic steps of the ensemble filtering. It is assumed that the model is nonlinear and the observations are linear functions of the state. Let the dynamical model be given by

$$\left.\begin{aligned} \mathbf{x}_{k+1} &= \mathbf{M}(\mathbf{x}_k) + \mathbf{w}_{k+1} \\ \text{and} \quad & \\ \mathbf{z}_k &= \mathbf{H}_k\mathbf{x}_k + \mathbf{v}_k \end{aligned}\right\} \tag{30.1.4}$$

It is assumed that

(A) the initial condition $\mathbf{x}_0 \sim N(\mathbf{m}_0, \mathbf{P}_0)$,
(B) the dynamic system noise $\mathbf{w}_k$ is a white Gaussian noise with $\mathbf{w}_k \sim N(0, \mathbf{Q}_k)$,

(C) the observation noise $\mathbf{v}_k$ is a white Gaussian noise with $\mathbf{v}_k \sim N(0, \mathbf{R}_k)$, and
(D) $\mathbf{x}_0$, $\{\mathbf{w}_k\}$ and $\{\mathbf{v}_k\}$ are mutually uncorrelated.

Creation of the initial ensemble

We begin by creating N initial ensemble members, say, $\widehat{\xi}_0(i)$, $i = 1, 2, \ldots, N$ drawn from the distribution $N(\mathbf{m}_0, \mathbf{P}_0)$. This is accomplished by first factoring $\mathbf{P}_0 = \mathbf{S}_0\mathbf{S}_0^{\mathrm{T}}$ and defining, for $i = 1, 2, \ldots, N$

$$\widehat{\xi}_0(i) = \mathbf{m}_0 + \mathbf{S}_0\mathbf{y}_0(i) \tag{30.1.5}$$

where $\mathbf{y}_0(i) \sim N(0, \mathbf{I})$. Clearly, the ensemble mean is given by

$$\left.\begin{aligned} \widehat{\mathbf{x}}_0(N) &= \tfrac{1}{N}\textstyle\sum_{i=1}^{N} \widehat{\xi}_0(i) \\ &= \mathbf{m}_0 + \mathbf{S}_0\widehat{\mathbf{y}}_0(N) \longrightarrow \mathbf{m}_0 \end{aligned}\right\} \tag{30.1.6}$$

since the sample mean

$$\widehat{\mathbf{y}}_0(N) = \frac{1}{N}\sum_{i=1}^{N} \mathbf{y}_0(i) \longrightarrow 0 \quad \text{as} \quad N \to \infty.$$

Similarly, the ensemble covariance is given by

$$\begin{aligned} \widehat{\mathbf{P}}_0(N) &= \frac{1}{N-1}\sum_{i=1}^{N}[\widehat{\xi}_0(i) - \widehat{\mathbf{x}}_0(N)][\widehat{\xi}_0(i) - \widehat{\mathbf{x}}_0(N)]^{\mathrm{T}} \\ &= \mathbf{S}_0\left[\frac{1}{N-1}\sum_{i=1}^{N}[\mathbf{y}_0(i) - \widehat{\mathbf{y}}_0(N)][\mathbf{y}_0(i) - \widehat{\mathbf{y}}_0(N)]^{\mathrm{T}}\right]\mathbf{S}_0^{\mathrm{T}} \\ &\to \mathbf{S}_0\mathbf{S}_0^{\mathrm{T}} = \mathbf{P}_0 \qquad \text{as} \qquad N \to \infty \end{aligned} \tag{30.1.7}$$

since the term inside the square bracket denotes the sample covariance of $\mathbf{y}(i)$ which tends to $\mathbf{I}$ as $N \to \infty$.

Ensemble forecast step

Inductively consider the time instant k. Given $(\widehat{\mathbf{x}}_k(N), \widehat{\mathbf{P}}_k(N))$, let $\widehat{\mathbf{P}}_k = \widehat{\mathbf{S}}_k\widehat{\mathbf{S}}_k^{\mathrm{T}}$. Create an ensemble

$$\widehat{\xi}_k(i) = \widehat{\mathbf{x}}_k(N) + \widehat{\mathbf{S}}_k\mathbf{y}_k(i)$$

where $\mathbf{y}_k(i) \sim N(0, \mathbf{I})$.

The N members of the ensemble forecast at time $(k+1)$ are generated

$$\xi_{k+1}^{\mathrm{f}}(i) = \mathbf{M}(\widehat{\xi}_k(i)) + \mathbf{w}_{k+1}(i) \tag{30.1.8}$$

where $\mathbf{w}_{k+1}(i) \sim N(0, \mathbf{Q}_{k+1})$ is generated using the method described above. Refer to Figure 30.1.1.

Herein lies one of the major differences between the ensemble filtering and the filtering techniques described in Chapter 27 through 29. Whereas the Kalman filtering and the approximate nonlinear filters rely on deterministic forecast, ensemble

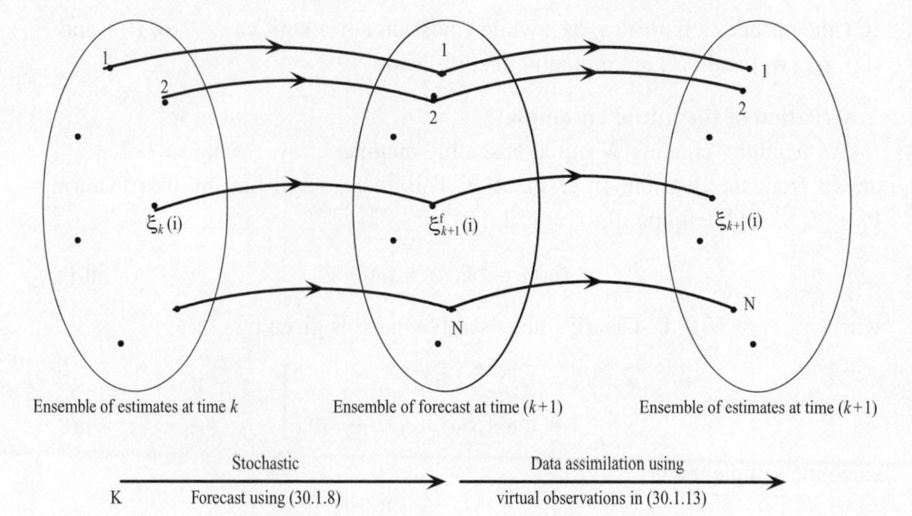

Fig. 30.1.1 A view of ensemble filtering.

filtering generates a set of random forecasts by adding the sample realizations of the system noise to the otherwise deterministic component of the forecast in (30.1.8).

The sample mean $\mathbf{x}^f_{k+1}(N)$ is then given by

$$\mathbf{x}^f_{k+1}(N) = \frac{1}{N}\sum_{i=1}^{N} \xi^f_{k+1}(i). \tag{30.1.9}$$

The forecast error $\mathbf{e}^f_{k+1}(i)$ is

$$\mathbf{e}^f_{k+1}(i) = \xi^f_{k+1}(i) - \mathbf{x}^f_{k+1}(N) \tag{30.1.10}$$

and the forecast covariance is given by

$$\mathbf{P}^f_{k+1}(N) = \frac{1}{N-1}\sum_{i=1}^{N} \mathbf{e}^f_{k+1}(i)\,[\mathbf{e}^f_{k+1}(i)]^{\mathrm{T}}. \tag{30.1.11}$$

Data assimilation step

Let $\mathbf{K} \in \mathbb{R}^{n\times m}$ be an arbitrary gain matrix. Given the actual observation $\mathbf{z}_{k+1}$, first define, the ith realization of the virtual observation, for $i = 1, 2, \ldots, N$

$$\mathbf{z}_{k+1}(i) = \mathbf{z}_{k+1} + \mathbf{v}_{k+1}(i) \tag{30.1.12}$$

where $\mathbf{v}_{k+1}(i) \sim N(0, \mathbf{R}_{k+1})$ is generated using the method described above. The estimate $\widehat{\xi}_{k+1}(i)$ is then given by

$$\widehat{\xi}_{k+1}(i) = \xi^f_{k+1}(i) + \mathbf{K}\,[\mathbf{z}_{k+1}(i) - \mathbf{H}_{k+1}\,\xi^f_{k+1}(i)]. \tag{30.1.13}$$

Herein lies another major difference. The estimate is obtained as a linear function of the forecast and the **virtual** observation $\mathbf{z}_{k+1}(i)$ created from the actual observations using (30.1.12). The reason for using the virtual instead of the actual observation will become apparent when we compute the sample covariance of the estimate in (30.1.13).

The sample mean of the estimate at time $(k+1)$ is then given by

$$\begin{aligned}\widehat{\mathbf{x}}_{k+1}(N) &= \frac{1}{N}\sum_{i=1}^{N}\widehat{\xi}_{k+1}(i)\\ &= \mathbf{x}^{\mathrm{f}}_{k+1}(N) + \mathbf{K}[\overline{\mathbf{z}}_{k+1}(N) - \mathbf{H}_{k+1}\mathbf{x}^{\mathrm{f}}_{k+1}(N)]\end{aligned} \tag{30.1.14}$$

where

$$\begin{aligned}\overline{\mathbf{z}}_{k+1}(N) = \frac{1}{N}\sum_{i=1}^{N}\mathbf{z}_{k+1}(i) &= \mathbf{z}_{k+1} + \frac{1}{N}\sum_{i=1}^{N}\mathbf{v}_{k}(i)\\ &= \mathbf{z}_{k+1} + \overline{\mathbf{v}}_{k+1}(N).\end{aligned} \tag{30.1.15}$$

Hence, the error in the ith strand of the estimate is

$$\begin{aligned}\widehat{\mathbf{e}}_{k+1}(i) &= \widehat{\xi}_{k+1}(i) - \widehat{\mathbf{x}}_{k+1}(N)\\ &= (\mathbf{I} - \mathbf{K}\mathbf{H}_{k+1})\mathbf{e}^{\mathrm{f}}_{k+1}(i) + \mathbf{K}[\mathbf{v}_{k+1}(i) - \overline{\mathbf{v}}_{k+1}(N)].\end{aligned} \tag{30.1.16}$$

Hence, the ensemble covariance of the estimate at time $(k+1)$ is

$$\begin{aligned}&\widehat{\mathbf{P}}_{k+1}(N)\\ &= \frac{1}{N-1}\sum_{i=1}^{N}\widehat{\mathbf{e}}_{k+1}(i)[\widehat{\mathbf{e}}_{k+1}(i)]^{\mathrm{T}}\\ &= (\mathbf{I} - \mathbf{K}\mathbf{H}_{k+1})\left[\frac{1}{N-1}\sum_{i=1}^{N}\mathbf{e}^{\mathrm{f}}_{k+1}(i)[\mathbf{e}^{\mathrm{f}}_{k+1}(i)]^{\mathrm{T}}\right](\mathbf{I} - \mathbf{K}\mathbf{H}_{k+1})^{\mathrm{T}}\\ &\quad + \mathbf{K}\left[\frac{1}{N-1}\sum_{i=1}^{N}[\mathbf{v}_{k+1}(i) - \overline{\mathbf{v}}_{k+1}(N)][\mathbf{v}_{k+1}(i) - \overline{\mathbf{v}}_{k+1}(N)]^{\mathrm{T}}\right]\mathbf{K}^{\mathrm{T}}\\ &\quad + (\mathbf{I} - \mathbf{K}\mathbf{H})\left\{\frac{1}{N-1}\sum_{i=1}^{N}\mathbf{e}^{\mathrm{f}}_{k+1}(i)[\mathbf{v}_{k+1}(i) - \overline{\mathbf{v}}_{k+1}(N)]^{\mathrm{T}}\right\}\mathbf{K}^{\mathrm{T}}\\ &\quad + \mathbf{K}\left\{\frac{1}{N-1}\sum_{i=1}^{N}[\mathbf{v}_{k+1}(i) - \overline{\mathbf{v}}_{k+1}(N)][\mathbf{e}^{\mathrm{f}}_{k+1}(i)]^{\mathrm{T}}\right\}(\mathbf{I} - \mathbf{K}\mathbf{H})^{\mathrm{T}}\\ &= (\mathbf{I} - \mathbf{K}\mathbf{H}_{k+1})\mathbf{P}^{\mathrm{f}}_{k+1}(N)(\mathbf{I} - \mathbf{K}\mathbf{H}_{k+1})^{\mathrm{T}} + \mathbf{K}\mathbf{R}_{k+1}(N)\mathbf{K}^{\mathrm{T}}\end{aligned} \tag{30.1.17}$$

Model $\mathbf{x}_{k+1} = \mathbf{M}(\mathbf{x}_k)\mathbf{x}_k + \mathbf{w}_{k+1}$

Observation $\mathbf{z}_k = \mathbf{H}_k\mathbf{x}_k + \mathbf{v}_k$

Initial ensemble

- Create the initial ensemble using (30.1.5)

Forecast step

- Create the ensemble of forecasts at time $(k+1)$ using (30.1.8)
- Compute $\mathbf{x}^{\mathrm{f}}_{k+1}(N)$ and $\mathbf{P}^{\mathrm{f}}_{k+1}(N)$ using (30.1.9) and (30.1.11) respectively

Data assimilation step

- Create the ensemble of estimates at time $(k+1)$ using (30.1.13) and (30.1.19)
- Compute $\widehat{\mathbf{x}}_{k+1}(N)$ and $\widehat{\mathbf{P}}_{k+1}(N)$ using (30.1.14) and (30.1.17) respectively

Fig. 30.1.2 Ensemble Kalman filter: a summary.

where for large N

$$\mathbf{R}_{k+1}(N) = \frac{1}{N-1}\sum_{i=1}^{N}[\mathbf{v}_{k+1}(i) - \overline{\mathbf{v}}_{k+1}(N)][\mathbf{v}_{k+1}(i) - \overline{\mathbf{v}}_{k+1}(N)]^{\mathrm{T}}$$
$$\rightarrow \mathbf{R}_{k+1} \tag{30.1.18}$$

and since the observation noise is uncorrelated with ensemble forecast we have

$$\frac{1}{N-1}\sum_{i=1}^{N}\mathbf{e}^{\mathrm{f}}_{k+1}(i)[\mathbf{v}_{k+1}(i) - \overline{\mathbf{v}}_{k+1}(N)]^{\mathrm{T}} \longrightarrow 0.$$

This expression for the covariance in (30.1.17) is quadratic in the gain matrix $\mathbf{K}$ and is exactly of the same form as in the linear Kalman filter – refer to equation (27.2.22). In view of this structural similarity, it follows that a natural choice for $\mathbf{K}$ is of the same for as the linear case in (27.2.25), namely

$$\mathbf{K} = \mathbf{P}^{\mathrm{f}}_{k+1}(N)\mathbf{H}^{\mathrm{T}}_{k+1}[\mathbf{H}_{k+1}\mathbf{P}^{\mathrm{f}}_{k+1}(N)\mathbf{H}^{\mathrm{T}}_{k+1} + \mathbf{R}_{k+1}]^{-1}. \tag{30.1.19}$$

A summary of the ensemble filter is given in Figure 30.1.2.

A number of observations are in order.

(a) **Errors due to finite sample size** Notice that expressions for $\widehat{\mathbf{P}}_{k+1}(N)$ in (30.1.17) and the Kalman gain $\mathbf{K}$ in (30.1.19) hold good only when the size N of the sample is really large. When N is small, errors due to the cross-product terms $\mathbf{e}^{\mathrm{f}}_{k+1}(i)[\mathbf{v}_{k+1}(i) - \bar{\mathbf{v}}(N)]^{\mathrm{T}}$ can cause serious errors in the estimate of $\widehat{\mathbf{P}}_{k+1}$ and hence the value of $\mathbf{K}$.

(b) **Covariance matrices** The approximate nonlinear filter described in Chapter 29 computes the forecast covariance using the recurrence

$$\mathbf{P}^{\mathrm{f}}_{k+1} = \mathbf{D}_{\mathbf{M}}(\widehat{\mathbf{x}}_k)\widehat{\mathbf{P}}_k\mathbf{D}^{\mathrm{T}}_{\mathbf{M}}(\widehat{\mathbf{x}}_k) + \mathbf{Q}_{k+1} \tag{30.1.20}$$

where $\mathbf{D}_{\mathbf{M}}(\widehat{\mathbf{x}}_k)$ is the Jacobian of $\mathbf{M}(\cdot)$ evaluated at $\widehat{\mathbf{x}}_k$. This form of the update is mainly due to the nature of the approximate forecast obtained using the Taylor series approximation. Provided $\mathbf{D}_{\mathbf{M}}(\widehat{\mathbf{x}}_k)$ is nonsingular (which is often the case) and $\mathbf{Q}_{k+1}$ is positive definite, $\mathbf{P}^{\mathrm{f}}_{k+1}$ computed using (30.1.20) is the most demanding part of the filter. In sharp contrast, in ensemble filtering, this step is replaced by N nonlinear model runs to generate $\xi^{\mathrm{f}}_{k+1}(i)$ followed by the computation of $\mathbf{P}^{\mathrm{f}}_{k+1}(N)$ as the sum of the N outer product matrices each of which is of size $n \times n$. In meteorological applications $n(\approx 10^6 \sim 10^8)$ is much larger than $N(\approx 10^2)$, the size of the ensemble. Hence $\mathbf{P}^{\mathrm{f}}_{k+1}(N)$ computed using (30.1.11) is such that

$$\mathrm{Rank}(\mathbf{P}^{\mathrm{f}}_{k+1}(N)) \le N < n.$$

By a similar argument, it also follows that

$$\mathrm{Rank}(\widehat{\mathbf{P}}_{k+1}(N)) \le N < n.$$

Hence ensemble filters belong to the family of **reduced-rank filters.**

(c) **Need for virtual observations $\mathbf{z}_{k+1}(i)$** If we used the actual observation $\mathbf{z}_{k+1}$ in place of the virtual observations $\mathbf{z}_{k+1}(i)$ in (30.1.13), then it can be verified that the expression for the error in (30.1.16) reduces to

$$\widehat{\mathbf{e}}_{k+1}(i) = (\mathbf{I} - \mathbf{K}\mathbf{H}_{k+1})\mathbf{e}^{\mathrm{f}}_{k+1}(i). \tag{30.1.21}$$

Hence,

$$\widehat{\mathbf{P}}_{k+1}(N) = (\mathbf{I} - \mathbf{K}\mathbf{H}_{k+1})\widehat{\mathbf{P}}_{k+1}(N)(\mathbf{I} - \mathbf{K}\mathbf{H}_{k+1}) \tag{30.1.22}$$

which is structurally different from (30.1.17) and results in an underestimation of this posterior covariance. Stated in other words, if we use the virtual observations and when the dynamics is linear, the ensemble filter for large N converges to the standard Kalman filter in Chapter 27. In view of this relation, ensemble filter is commonly known as the **ensemble Kalman** filter.

(d) **Modification of the gain instead of the observations** It is shown above that the use of virtual observations forces the ensemble posterior covariance to be the same as that of the standard Kalman filter. However, if the ensemble size is finite, use of this virtual observations often introduces sampling errors (Whitaker and Hamill (2002)). Another way to restore the equivalence between ensemble filtering and standard Kalman filtering is to seek a new gain matrix, say $\mathbf{W}$ (instead of the standard Kalman gain $\mathbf{K}$) while using no perturbation for the observations in the data assimilation step in (30.1.13). Using a new gain matrix $\mathbf{W} \in \mathbb{R}^{n \times m}$ and the actual observation, (30.1.13) becomes

$$\widehat{\xi}_{k+1}(i) = \xi^{\mathrm{f}}_{k+1}(i) + \mathbf{W}[\mathbf{z}_{k+1} - \mathbf{H}_{k+1}\xi^{\mathrm{f}}_{k+1}(i)]. \tag{30.1.23}$$

Then the ensemble mean in (30.1.14) becomes

$$\widehat{\mathbf{x}}_{k+1}(N) = \mathbf{x}^{\mathrm{f}}_{k+1}(N) + \mathbf{W}[\mathbf{z}_{k+1} - \mathbf{H}_{k+1}\widehat{\mathbf{x}}_{k+1}(N)]. \tag{30.1.24}$$

Subtracting (30.1.24) from (30.1.23) the error in the ith ensemble member is

$$\widehat{\mathbf{e}}_{k+1}(i) = (\mathbf{I} - \mathbf{W}\mathbf{H}_{k+1})\mathbf{e}^{\mathrm{f}}_{k+1}(i). \tag{30.1.25}$$

The new posterior ensemble covariance is then given by

$$\begin{aligned} \mathbf{P}_{k+1} &= (\mathbf{I} - \mathbf{W}\mathbf{H}_{k+1})\mathbf{P}^{\mathrm{f}}_{k+1}(\mathbf{I} - \mathbf{W}\mathbf{H}_{k+1})^{\mathrm{T}} \\ &= [(\mathbf{I} - \mathbf{W}\mathbf{H}_{k+1})\mathbf{S}^{\mathrm{f}}_{k+1}][(\mathbf{I} - \mathbf{W}\mathbf{H}_{k+1})\mathbf{S}^{\mathrm{f}}_{k+1}]^{\mathrm{T}} \end{aligned} \tag{30.1.26}$$

where $\mathbf{S}^{\mathrm{f}}_{k+1}(\mathbf{S}^{\mathrm{f}}_{k+1})^{\mathrm{T}} = \mathbf{P}^{\mathrm{f}}_{k+1}$ is the square root factorization of $\mathbf{P}^{\mathrm{f}}_{k+1}$ (refer to Section 28.7).

The question is: how to choose $\mathbf{W}$ such that this expression for the covariance matches that for the standard Kalman filter given by

$$\widehat{\mathbf{P}}_{k+1} = \mathbf{P}^{\mathrm{f}}_{k+1} - \mathbf{P}^{\mathrm{f}}_{k+1}\mathbf{H}_{k+1}[\mathbf{H}_{k+1}\mathbf{P}^{\mathrm{f}}_{k+1}\mathbf{H}_{k+1} + \mathbf{R}_{k+1}]^{-1}\mathbf{H}_{k+1}\mathbf{P}^{\mathrm{f}}_{k+1}. \tag{30.1.27}$$

Since (30.1.26) is already in the factored form, the key to the solution lies in factoring the r.h.s. of (30.1.27). It turns out that Andrews (1968) has already derived this factorization which is summarized below.

Dropping the time subscript $(k+1)$ for simplicity and substituting $\mathbf{P}^f = \mathbf{S}^{\mathrm{f}}(\mathbf{S}^{\mathrm{f}})^{\mathrm{T}}$ and $\mathbf{A} = (\mathbf{H}\mathbf{S}^{\mathrm{f}})^{\mathrm{T}}$, (30.1.27) becomes

$$\widehat{\mathbf{P}} = \mathbf{S}^{\mathrm{f}}[\mathbf{I} - \mathbf{A}(\mathbf{A}^{\mathrm{T}}\mathbf{A} + \mathbf{R})^{-1}\mathbf{A}^{\mathrm{T}}]\mathbf{S}^{\mathrm{f}}. \tag{30.1.28}$$

Let

$$(\mathbf{A}^{\mathrm{T}}\mathbf{A} + \mathbf{R}) = \mathbf{S}\mathbf{S}^{\mathrm{T}} \quad \text{and} \quad \mathbf{R} = \mathbf{F}\mathbf{F}^{\mathrm{T}}. \tag{30.1.29}$$

be the square root factorization. Then, using a sequence of mathematical manipulations, we get

$$\begin{aligned}
&\mathbf{A}(\mathbf{A}^{\mathrm{T}}\mathbf{A}+\mathbf{R})^{-1}\mathbf{A}^{\mathrm{T}} \\
&= \mathbf{A}\mathbf{S}^{-T}\mathbf{S}^{-1}\mathbf{A}^{\mathrm{T}} \\
&= \mathbf{A}\mathbf{S}^{-T}(\mathbf{S}+\mathbf{F})^{-1}(\mathbf{S}+\mathbf{F})(\mathbf{S}+\mathbf{F})^{\mathrm{T}}(\mathbf{S}+\mathbf{F})^{-T}\mathbf{S}^{-1}\mathbf{A}^{\mathrm{T}} \\
&= \mathbf{A}\mathbf{S}^{-T}(\mathbf{S}+\mathbf{F})^{-1}[(\mathbf{S}+\mathbf{F})(\mathbf{S}+\mathbf{F})^{\mathrm{T}}](\mathbf{S}+\mathbf{F})^{-T}\mathbf{S}^{-1}\mathbf{A}^{\mathrm{T}} \\
&= \mathbf{A}\mathbf{S}^{-T}(\mathbf{S}+\mathbf{F})^{-1}[\mathbf{S}(\mathbf{S}+\mathbf{F})^{\mathrm{T}}+(\mathbf{S}+\mathbf{F})\mathbf{S}^{\mathrm{T}}-\mathbf{A}^{\mathrm{T}}\mathbf{A}] \\
&\quad \cdot(\mathbf{S}+\mathbf{F})^{-T}\mathbf{S}^{-1}\mathbf{A}^{\mathrm{T}}.
\end{aligned}$$

Substituting this into (30.1.28), it can be verified that

$$\begin{aligned}
\widehat{\mathbf{P}} &= \mathbf{S}^{\mathrm{f}}[\mathbf{I}-\mathbf{A}(\mathbf{A}^{\mathrm{T}}\mathbf{A}+\mathbf{R})^{-1}\mathbf{A}^{\mathrm{T}}](\mathbf{S}^{\mathrm{f}})^{\mathrm{T}} \\
&= \mathbf{S}^{\mathrm{f}}[\mathbf{I}-\mathbf{A}\mathbf{S}^{-T}(\mathbf{S}+\mathbf{F})^{-1}\mathbf{A}^{\mathrm{T}}][\mathbf{I}-\mathbf{A}\mathbf{S}^{-T}(\mathbf{S}+\mathbf{F})^{-1}\mathbf{A}^{\mathrm{T}}]^{\mathrm{T}}\mathbf{S}_{\mathrm{f}}^{\mathrm{T}}.
\end{aligned} \tag{30.1.30}$$

Comparing this with (30.1.26), we get (suppressing the subscripts)

$$(\mathbf{I}-\mathbf{W}\mathbf{H})\mathbf{S}^{\mathrm{f}} = \mathbf{S}^{\mathrm{f}}[\mathbf{I}-\mathbf{A}\bar{\mathbf{S}}^{\mathrm{T}}(\mathbf{S}+\mathbf{F})^{-1}\mathbf{A}^{\mathrm{T}}]$$

which in turn suggests that $\mathbf{W}$ may be chosen as

$$\mathbf{W} = \mathbf{S}^{\mathrm{f}}\mathbf{A}\mathbf{S}^{-\mathrm{T}}(\mathbf{S}+\mathbf{F})^{-1} = \mathbf{P}^{f}\mathbf{H}^{\mathrm{T}}\mathbf{S}^{-\mathrm{T}}(\mathbf{S}+\mathbf{F})^{-1}. \tag{30.1.31}$$

Stated in other words, using this new gain matrix $\mathbf{W}$ in the data assimilation step we can restore the equivalence between the ensemble filter and the standard Kalman filter without using any perturbation for the observation. See Whitaker and Hamill (2002) and Tippett et al. (2003) for further details. Refer to Exercises 30.2 and 30.3 for two other related square root factorizations.

30.2 Reduced-rank square root (RRSQRT) filter

In this section we describe the basic ideas leading to the derivation of the reduced-rank filters from the full-rank filters described in Chapters 27–29. Recall from Section 9.1 and Section 28.7 that any real symmetric and positive definite matrix can be factored into the so-called **square root** form by using either the **Cholesky decomposition** (Chapter 9) or the **eigenvalue decomposition** (Section 28.7). The RRSQRT filter described in this section relies on using only the **dominant orthogonal** modes resulting from the eigenvalue decomposition of the covariance matrices.

As the name implies RRSQRT filter is the result of the modification of the standard full-rank square root filter described in Section 28.7. For simplicity in exposition, we assume that the system **model is linear** and the **observations are linear functions** of the state as in (27.2.1) and (27.2.2). Let p be a given fixed integer where $1 \leq p \leq n$. In the following we describe a method for obtaining the rank p square root filter.

(A) **Initialization** Let $\mathbf{x}_0$ be the initial condition with $E(\mathbf{x}_0) = \mathbf{m}_0$ and $\text{Cov}(\mathbf{x}_0) = \mathbf{P}_0$. Let $\mathbf{V}$ be the **orthonormal matrix** of eigenvectors and $\Lambda = \text{Diag}(\lambda_1, \lambda_2, \ldots, \lambda_n)$ be the matrix of the corresponding eigenvalues of $\mathbf{P}_0$ where it is assumed that

$$\lambda_1 \geq \lambda_2 \geq \cdots \geq \lambda_n > 0. \tag{30.2.1}$$

Then

$$\begin{aligned} \mathbf{P}_0 = \mathbf{V}\Lambda\mathbf{V}^{\mathrm{T}} &= (\mathbf{V}\Lambda^{1/2})(\mathbf{V}\Lambda^{1/2})^{\mathrm{T}} \\ &= \mathbf{S}_0\mathbf{S}_0^{\mathrm{T}} \end{aligned} \tag{30.2.2}$$

where $\mathbf{S}_0 = \mathbf{V}\Lambda^{1/2}$ is called the full-rank square root of $\mathbf{P}_0$, where the ith column of $\mathbf{S}_0$ is the ith eigenvector $\mathbf{v}_i$ scaled by $\sqrt{\lambda_i}$. Define $\mathbf{S}_0(1:p) \in \mathbb{R}^{n\times p}$ consisting of the first p columns called the dominant p modes of $\mathbf{P}_0$. This matrix $\mathbf{S}_0(1:p)$ is the rank p (approximate) square root of $\mathbf{P}_0$, that is, $\mathbf{P}_0 \sim \mathbf{S}_0(1:p)\mathbf{S}_0^{\mathrm{T}}(1:p)$ where $1 \leq p \leq n$. This process of obtaining $\mathbf{S}_0(1:p)$ from $\mathbf{S}_0$ is called **rank reduction** and is a very useful tool.

The RRSQRT filter is then initialized with $\widehat{\mathbf{x}}_0 = \mathbf{m}_0$ and $\widehat{\mathbf{S}}_0(1:p) = \mathbf{S}_0(1:p)$, where $\widehat{\mathbf{S}}_0(1:p)$ is the reduced-rank square root of $\mathbf{P}_0$, the covariance of $\mathbf{x}_0$.

(B) **Forecast Step** Assume inductively that we are given $(\widehat{\mathbf{x}}_k, \widehat{\mathbf{S}}_k(1:p))$ at time k, where $\widehat{\mathbf{S}}_k(1:p)$ is the rank p square root of $\widehat{\mathbf{P}}_k$, the covariance of $\widehat{\mathbf{x}}_k$. The forecast is then given by

$$\mathbf{x}^{\mathrm{f}}_{k+1} = \mathbf{M}_k\widehat{\mathbf{x}}_k. \tag{30.2.3}$$

We now describe the process of computing $\mathbf{S}^{\mathrm{f}}_{k+1}(1:p)$ the rank p square root of the covariance $\mathbf{P}^{\mathrm{f}}_{k+1}$ of $\mathbf{x}^{\mathrm{f}}_{k+1}$.

Recall from Section 27.2 that

$$\begin{aligned} \mathbf{P}^{\mathrm{f}}_{k+1} &= \mathbf{M}_k\widehat{\mathbf{P}}_k\mathbf{M}_k^{\mathrm{T}} + \mathbf{Q}_{k+1} \\ &= \mathbf{M}_k\widehat{\mathbf{S}}_k\widehat{\mathbf{S}}_k^{\mathrm{T}}\mathbf{M}_k^{\mathrm{T}} + \mathbf{S}^{Q}_{k+1}(\mathbf{S}^{Q}_{k+1})^{\mathrm{T}} \end{aligned} \tag{30.2.4}$$

where $\widehat{\mathbf{S}}_k$ and $\mathbf{S}^{Q}_{k+1}$ are the full-rank square root matrices of $\widehat{\mathbf{P}}_k$ and $\mathbf{Q}_{k+1}$ respectively. Then, it can be verified that

$$\mathbf{S}^{\mathrm{f}}_{k+1} = [\mathbf{M}\widehat{\mathbf{S}}_k, \mathbf{S}^{Q}_{k+1}] \in \mathbb{R}^{n\times 2n} \tag{30.2.5}$$

is the full-rank square root of $\mathbf{P}^{\mathrm{f}}_{k+1}$. In this framework, we do not know $\widehat{\mathbf{S}}_k$ but only its rank p approximation $\widehat{\mathbf{S}}_k(1:p)$. Similarly, consistent with the overall philosophy of reduced-rank approximation, it is assumed that we do not know $\mathbf{S}^{Q}_{k+1}$ but only its rank q approximation $\mathbf{S}^{Q}_{k+1}(1:q)$ consisting of the q dominant modes of $\mathbf{Q}_{k+1}$ for some integer $1 \leq q \leq n$.

Let

$$\mathbf{S}_{k+1} = [\mathbf{M}\widehat{\mathbf{S}}_k(1:p), \mathbf{S}^Q_{k+1}(1:q)] \in \mathbb{R}^{n\times(p+q)} \tag{30.2.6}$$

be the reduced-rank approximation to $\mathbf{S}^{\mathrm{f}}_{k+1}$ in (30.2.5). It can be verified that $\mathbf{S}_{k+1}$ is the rank $(p+q) \le n$ approximation to the square root of $\mathbf{P}^{\mathrm{f}}_{k+1}$. Notice that adding $\mathbf{S}^Q_{k+1}(1:q)$, the rank q approximation of $\mathbf{Q}_{k+1}$ in (30.2.6) has increased the number of columns and hence the rank of $\mathbf{S}_{k+1}$. Our immediate goal is to obtain $\mathbf{S}^{\mathrm{f}}_k(1:p)$ the rank p approximation to $\mathbf{S}_{k+1}$ in (30.2.6).

Rank reduction Two cases arise.

Case A: when $p+q<n$

In this case, the required rank p approximation to $\mathbf{S}_{k+1}$ is obtained by invoking the standard singular value decomposition(SVD). We state this process in the following algorithm. Refer to Chapter 9 for details.

Step 1 Let $\mathbf{V} \in \mathbb{R}^{(p+q)\times(p+q)}$ be the orthonormal matrix of eigenvectors and $\Lambda = \mathrm{Diag}(\lambda_1, \lambda_2, \ldots, \lambda_{p+q})$ be the corresponding matrix of eigenvalues of the Grammian $\mathbf{S}^{\mathrm{T}}_{k+1}\mathbf{S}_{k+1} \in \mathbb{R}^{(p+q)\times(p+q)}$, where

$$\lambda_1 \ge \lambda_2 \ge \cdots \ge \lambda_{p+q} > 0.$$

That is

$$\mathbf{S}^{\mathrm{T}}_{k+1}\mathbf{S}_{k+1} = \mathbf{V}\Lambda\mathbf{V}^{\mathrm{T}}. \tag{30.2.7}$$

It is well known (refer to Chapter 9) that the Grammian $\mathbf{S}^{\mathrm{T}}_{k+1}\mathbf{S}_{k+1} \in \mathbb{R}^{(p+q)\times(p+q)}$ and $\mathbf{S}_{k+1}\mathbf{S}^{\mathrm{T}}_{k+1} \in \mathbb{R}^{n\times n}$ share the same set of non-zero eigenvalues and that the matrix $\mathbf{U} = (\mathbf{S}_{k+1}\mathbf{V}\Lambda^{-1/2}) \in \mathbb{R}^{n\times(p+q)}$ is the matrix of eigenvectors of $\mathbf{S}_{k+1}\mathbf{S}^{\mathrm{T}}_{k+1}$. Thus, we have

$$\begin{aligned}\mathbf{S}_{k+1}\mathbf{S}^{\mathrm{T}}_{k+1} &= \mathbf{U}\Lambda\mathbf{U}^{\mathrm{T}}\\ &= (\mathbf{S}_{k+1}\mathbf{V}\Lambda^{-1/2})\Lambda(\Lambda^{-1/2}\mathbf{V}^{\mathrm{T}}\mathbf{S}^{\mathrm{T}}_{k+1})\\ &= (\mathbf{S}_{k+1}\mathbf{V})(\mathbf{S}_{k+1}\mathbf{V})^{\mathrm{T}}.\end{aligned} \tag{30.2.8}$$

Step 2 Let $\mathbf{V}(1:p) \in \mathbb{R}^{n\times p}$ consisting of the first p columns of $\mathbf{V}$ in (30.2.7). Then

$$\mathbf{S}^{\mathrm{f}}_{k+1}(1:p) = \mathbf{S}_{k+1}\mathbf{V}(1:p) \tag{30.2.9}$$

is the required rank p approximation to $\mathbf{S}_{k+1}$.

Case B: $(p+q)>n$

Let $\mathbf{V} \in \mathbb{R}^{n\times n}$ be the matrix of eigenvectors and $\Lambda = \mathrm{Diag}(\lambda_1, \lambda_2, \ldots, \lambda_n)$ be the corresponding eigenvalues of the Grammian $\mathbf{S}_{k+1}\mathbf{S}^{\mathrm{T}}_{k+1}$, where

$$\lambda_1 \ge \lambda_2 \ge \cdots \ge \lambda_n > 0.$$

Then

$$\begin{aligned} \mathbf{S}_{k+1}\mathbf{S}_{k+1}^{\mathrm{T}} &= \mathbf{V}\Lambda\mathbf{V}^{\mathrm{T}} \\ &= (\mathbf{V}\Lambda^{1/2})(\mathbf{V}\Lambda^{1/2})^{\mathrm{T}} \\ &= (\overline{\mathbf{V}})(\overline{\mathbf{V}})^{\mathrm{T}} \quad \text{with} \quad \overline{\mathbf{V}} = \mathbf{V}\Lambda^{1/2}. \end{aligned} \tag{30.2.10}$$

Then

$$\mathbf{S}_{k+1}^{\mathrm{f}}(1:p) = \overline{\mathbf{V}}(1:p) \tag{30.2.11}$$

is the rank p square root we are seeking.

This rank reduction process leading to (30.2.9) or (30.2.11) can be thought of as a projection process and can be succinctly denoted by

$$\mathbf{S}_{k+1}^{\mathrm{f}}(1:p) = \Pi_p(\mathbf{S}_{k+1}). \tag{30.2.12}$$

(C) **Data Assimilation Step** Given the forecast $\mathbf{x}_{k+1}^{\mathrm{f}}$ in (30.2.3) and the rank p square root $\mathbf{S}_{k+1}^{\mathrm{f}}(1:p)$ in (30.2.12), we now move on to the data assimilation step which is identical to its full-rank counterpart described in Figure 28.7.1.
(1) Compute $\mathbf{A} = (\mathbf{H}_{k+1}\mathbf{S}_{k+1}^{\mathrm{f}}(1:p))^{\mathrm{T}} \in \mathbb{R}^{p\times m}$.
(2) Compute $\mathbf{B} = (\mathbf{A}^{\mathrm{T}}\mathbf{A} + \mathbf{R}_{k+1})^{-1}\mathbf{A}^{\mathrm{T}} \in \mathbb{R}^{m\times p}$.
(3) Find the square root $\mathbf{C} \in \mathbb{R}^{p\times p}$ where

$$\mathbf{C}\mathbf{C}^{\mathrm{T}} = (\mathbf{I} - \mathbf{A}\mathbf{B}).$$

(4) The gain matrix is given by

$$\begin{aligned} \mathbf{K}_{k+1} &= \mathbf{S}_{k+1}^{\mathrm{f}}(1:p)\mathbf{A}[\mathbf{A}^{\mathrm{T}}\mathbf{A} + \mathbf{R}_{k+1}]^{-1} \\ &= \mathbf{S}_{k+1}^{\mathrm{f}}\mathbf{B}^{\mathrm{T}}. \end{aligned}$$

(5) The new estimate is

$$\widehat{\mathbf{x}}_{k+1} = \mathbf{x}_{k+1}^{\mathrm{f}} + \mathbf{K}_{k+1}[\mathbf{z}_{k+1} - \mathbf{H}_{k+1}\mathbf{x}_{k+1}^{\mathrm{f}}].$$

(6) The rank p square root $\widehat{\mathbf{S}}_{k+1}(1:p)$ of the covariance $\widehat{\mathbf{P}}_{k+1}$ of $\widehat{\mathbf{x}}_{k+1}$ is given by

$$\widehat{\mathbf{S}}_{k+1}(1:p) = \mathbf{S}_{k+1}^{\mathrm{f}}(1:p)\mathbf{C} \in \mathbb{R}^{n\times p}$$

Now, given the pair $(\widehat{\mathbf{x}}_{k+1}, \widehat{\mathbf{S}}_{k+1}(1:p))$ we can repeat the cycle of computation for the prespecified duration of interest.

Several observations are in order.

(a) **Potential for cost reduction** If p and q are such that $(p+q) \ll n$, then the net decrease in the covariance square root update could far exceed the cost increase resulting from the SVD portion in obtaining $\mathbf{S}_{k+1}^{\mathrm{f}}(1:p)$ in (30.2.7).

(b) **Lanczos algorithm** The standard algorithm for computing the leading or dominant eigenvalues and vectors is called the Lanczos algorithm (Golub and van Loan(1989)).

(c) **Non-negative definiteness** If $\mathbf{S} \in \mathbb{R}^{n\times p}$ is such that Rank($\mathbf{S}$) = p, then $\mathbf{S}\mathbf{S}^{\mathrm{T}}$ is also of rank p. Further, since $\mathbf{x}^{\mathrm{T}}\mathbf{S}\mathbf{S}^{\mathrm{T}}\mathbf{x} = (\mathbf{S}^{\mathrm{T}}\mathbf{x})(\mathbf{S}^{\mathrm{T}}\mathbf{x})^{\mathrm{T}} = \| \mathbf{S}^{\mathrm{T}}\mathbf{x} \|^2 \geq 0$ for all $\mathbf{x}$, it follows that $\mathbf{S}\mathbf{S}^{\mathrm{T}}$ is always non-negative definite. Hence the divergence problems associated with negative definite covariance matrix is completely avoided.

(d) **Better conditioning** Since the condition number of the square root $\mathbf{S}$ is the square root of the condition number of $\mathbf{S}\mathbf{S}^{\mathrm{T}}$, the round-off errors have much less impact in the square root version of the filter.

30.3 Hybrid filters

In this section we provide a summary of the basic ideas relating to the process of creating hybrid filters that combine several of the properties of the square root filter, reduced-rank filter, ensemble filter first- and second-order approximations to the nonlinear filter, to name a few. For definiteness, it is assumed that the **model is nonlinear** and the **observations are linear functions** of the state as given below:

$$\begin{aligned} \mathbf{x}_{k+1} &= \mathbf{M}(\mathbf{x}_k) + \mathbf{w}_{k+1} \\ \mathbf{z}_k &= \mathbf{H}_k\mathbf{x}_k + \mathbf{v}_k. \end{aligned} \qquad (30.3.1)$$

We first describe a general framework and then specify the details for specific versions.

At time k let $(\widehat{\mathbf{x}}_k, \widehat{\mathbf{S}}_k(1:p))$ be given where $\widehat{\mathbf{S}}_k(1:p)$ is the rank p square root of $\widehat{\mathbf{P}}_k$ where $1 \leq p \leq n$. When $p = n$, we get the full-rank square root as a specific case.

Step 1: Create an ensemble Let E be an operator that creates the first ensemble of states $\widehat{\xi}_k(i)$ for $i = 1, 2, \ldots, N_1$ from the given information at time k:

$$\{\widehat{\xi}_k(i) | 1 \leq i \leq N_1\} = E\{\widehat{\mathbf{x}}_k, \widehat{\mathbf{S}}_k(1:p)\}. \qquad (30.3.2)$$

Step 2: Propagate the ensemble Compute the second ensemble of size $(N_1 + N_2)$ as follows:

$$\left.\begin{aligned} \xi^{\mathrm{f}}_{k+1}(i) &= \mathbf{M}(\widehat{\xi}_k(i)) && \text{for} \quad 1 \leq i \leq N_1 \\ &\text{and} \\ \xi^{\mathrm{f}}_{k+1}(N_1 + i) &= \mathbf{M}(\widehat{\mathbf{x}}_k) + \mathbf{w}_{k+1}(i) && \text{for} \quad 1 \leq i \leq N_2 \end{aligned}\right\} \qquad (30.3.3)$$

Step 3: Compute the forecast Using the ensemble in step 2, compute the forecast $\mathbf{x}^{\mathrm{f}}_{k+1}$ and the reduced-rank square root $\mathbf{S}^{\mathrm{f}}_{k+1}(1:p)$ of the covariance $\mathbf{P}^{\mathrm{f}}_{k+1}$. This operation can be thought of as the inverse of (30.3.2) and is denoted by

$$(\mathbf{x}^{\mathrm{f}}_{k+1}, \mathbf{S}^{\mathrm{f}}_{k+1}(1:p)) = E^{-1}\{\xi^{\mathrm{f}}_{k+1}(i) | 1 \leq i \leq N_1 + N_2\}. \qquad (30.3.4)$$

Step 4: Data assimilation Given $\mathbf{x}^f_{k+1}$ and $\mathbf{S}^f_{k+1}(1:p)$, perform the data assimilation step to get $\widehat{\mathbf{x}}_{k+1}$ and $\widehat{\mathbf{S}}_{k+1}(1:p)$ and repeat the cycle.

Several comments are in order:

(1) This framework very naturally combines the ideas from the ensemble filter with those from the reduced-rank square root filter. It includes the full rank square root as a special case where $p = n$.

(2) The specific versions of this hybrid framework differ in the choice of the operator E and in the number N of ensemble members in (30.3.2).

(3) In Section 30.1, the ensemble was created only by using different realizations of the system noise vector $\mathbf{w}_k$. But in this hybrid framework the first ensemble in (30.3.2) is created by using the linear combinations of dominant modes in $\widehat{\mathbf{S}}_k(1:p)$. That is

$$\widehat{\xi}_k(i) = \widehat{\mathbf{x}}_k + \widehat{\mathbf{S}}_k(1:p)\mathbf{y}(i) \tag{30.3.5}$$

where $\mathbf{y}(i) \in \mathbb{R}^p$. Hence the spread in the second ensemble in (30.3.3) includes the effect of nonlinearity on the dominant modes as well as the realizations of the system noise.

(4) Given $\mathbf{x}^f_{k+1}$ and $\mathbf{S}^f_{k+1}(1:p)$, since the observations are linear functions of the states, the data assimilation step is exactly the same as described in Section 30.2. Hence in the following we concentrate only on the first three steps of the above framework.

(A) Hybrid filter 1 This first version has a flavor which is a combination of the **first-order**, **reduced-rank**, **square root** and **ensemble** filters. Let $\widehat{\mathbf{S}}_k(i) = \widehat{\mathbf{S}}_k(i:i)$ denote the ith column of $\widehat{\mathbf{S}}_k(1:p)$.

Step 1 Define the first ensemble of size $N_1 = p$ using the vector $\mathbf{y}(i)$ whose ith component is ε and the rest of all the components are zeros, for some $\varepsilon > 0$. Thus,

$$\begin{aligned}\widehat{\xi}_k(i) &= \widehat{\mathbf{x}}_k + \widehat{\mathbf{S}}_k(1:p)\mathbf{y}(i)\\ &= \widehat{\mathbf{x}}_k + \varepsilon\widehat{\mathbf{S}}_k(i).\end{aligned} \tag{30.3.6}$$

Step 2 The first $N_1 = p$ members of the second ensemble is given by $(1 \leq i \leq p)$

$$\xi^f_{k+1}(i) = \mathbf{M}(\widehat{\xi}_k(i)) = \mathbf{M}(\widehat{\mathbf{x}}_k + \varepsilon\widehat{\mathbf{S}}_k(i)). \tag{30.3.7}$$

To compute the $N_2 = q$ members of this second ensemble, let $\mathbf{S}^Q_{k+1}(1:q)$ be the rank q approximation to the full-rank square root $\mathbf{S}^Q_{k+1}$ of $\mathbf{Q}_{k+1}$. Then

$$\xi^f_{k+1}(p+j) = \mathbf{M}(\widehat{\mathbf{x}}_k) + \mathbf{S}^Q_{k+1}(j) \quad j = 1, 2, \ldots, q \tag{30.3.8}$$

where $\mathbf{S}^Q_{k+1}(j) = \mathbf{S}^Q_{k+1}(j:j)$ is the jth column of $\mathbf{S}^Q_{k+1}(1:q)$.

Step 3 Let

$$\mathbf{x}^f_{k+1} = \mathbf{M}(\widehat{\mathbf{x}}_k) \tag{30.3.9}$$

be the **deterministic** or the **central** forecast. Then using (30.3.7) and using the first-order Taylor series expansion, the forecast error is given by

$$\mathbf{e}^{\mathrm{f}}_{k+1}(i) = \frac{1}{\varepsilon}[\xi^{\mathrm{f}}_{k+1}(i) - \mathbf{x}^{\mathrm{f}}_{k+1}]$$

$$= \frac{1}{\varepsilon}[\mathbf{M}(\widehat{\mathbf{x}}_k + \varepsilon\widehat{\mathbf{S}}_k(i)) - \mathbf{M}(\widehat{\mathbf{x}}_k)] \tag{30.3.10}$$

$$= \mathbf{D_M}(\widehat{\mathbf{x}}_k)\widehat{\mathbf{S}}_k(i) \quad \text{for } 1 \le i \le p \tag{30.3.11}$$

which clearly brings out the relation of this filter to the **first-order** filter in Chapter 29. If the Jacobian is available, then (30.3.11) can be used or else the finite-difference approximation in (30.3.10) is used. Similarly,

$$\mathbf{e}^{\mathrm{f}}_{k+1}(p + j) = \mathbf{S}^{Q}_{k+1}(j) \quad \text{for } 1 \le j \le q. \tag{30.3.12}$$

Now assemble the $n \times (p + q)$ matrix $\mathbf{S}_{k+1}$ whose columns are the error vectors in (30.3.11) and (30.3.12):

$$\mathbf{S}_{k+1} = [\mathbf{e}^{\mathrm{f}}_{k+1}(1) \cdots \mathbf{e}^{\mathrm{f}}_{k+1}(p)\mathbf{e}^{\mathrm{f}}_{k+1}(p+1) \cdots \mathbf{e}^{\mathrm{f}}_{k+1}(p+q)]. \tag{30.3.13}$$

Using the rank-reduction procedure described in Section 30.2, compute

$$\mathbf{S}^{\mathrm{f}}_{k+1}(1 : p) = \Pi_p[\mathbf{S}_{k+1}]. \tag{30.3.14}$$

Given $(\mathbf{x}^{\mathrm{f}}_{k+1}, \mathbf{S}^{\mathrm{f}}_{k+1}(1 : p))$ perform the data assimilation step using the algorithm described in Section 30.2.

Using (30.3.11) and (30.3.12) in (30.3.13), it can be verified that

$$\mathbf{S}_{k+1}\mathbf{S}^{\mathrm{T}}_{k+1} = \mathbf{D_M}(\widehat{\mathbf{x}}_k)\left[\sum_{i=1}^{p} \widehat{\mathbf{S}}_k(i)[\widehat{\mathbf{S}}_k(i)]^{\mathrm{T}}\right]\mathbf{D}^{\mathrm{T}}_{\mathbf{M}}(\widehat{\mathbf{x}}_k)$$

$$+ \sum_{j=1}^{q} \mathbf{S}^{Q}_{k+1}(j)[\mathbf{S}^{Q}_{k+1}(j)]^{\mathrm{T}}$$

$$= \mathbf{D_M}(\widehat{\mathbf{x}}_k)\widehat{\mathbf{S}}_k(1 : p)[\widehat{\mathbf{S}}_k(1 : p)]^{\mathrm{T}}\mathbf{D}^{\mathrm{T}}_{\mathbf{M}}(\widehat{\mathbf{x}}_k)$$

$$+ \mathbf{S}^{Q}_{k+1}(1 : q)[\mathbf{S}^{Q}_{k+1}(1 : q)]^{\mathrm{T}} \tag{30.3.15}$$

which is an approximation to the exact dynamics of the covariance of the **first-order** filter (Chapter 29) given by

$$\mathbf{P}^{\mathrm{f}}_{k+1} = \mathbf{D_M}(\widehat{\mathbf{x}}_k)\widehat{\mathbf{P}}_k\mathbf{D}^{\mathrm{T}}_{\mathbf{M}}(\widehat{\mathbf{x}}_k) + \mathbf{Q}_{k+1}. \tag{30.3.16}$$

(B) Hybrid filter 2 This filter is very similar to filter 1 but uses a larger ensemble to attain the flavor of the **second-order** filter.

Step 1 Define the first ensemble of size $N_1 = 2p$ where

$$\widehat{\xi}_k(\pm i) = \widehat{\mathbf{x}}_k \pm \varepsilon\widehat{\mathbf{S}}_k(i) \quad \text{for } i = 1, 2, \ldots, p \tag{30.3.17}$$

whose sample average is given by

$$\frac{1}{2p}\sum_{i=1}^{p}\widehat{\xi}_k(\pm i) = \widehat{\mathbf{x}}_k. \tag{30.3.18}$$

Step 2 The first $N_1 = 2p$ members of the second ensemble are given by

$$\xi^{\mathrm{f}}_{k+1}(\pm i) = \mathbf{M}(\widehat{\xi}_k(\pm i)) \quad \text{for } i = 1, 2, \ldots, p \tag{30.3.19}$$

and the second $N_2 = q$ members are

$$\xi^{\mathrm{f}}_{k+1}(2p + j) = \mathbf{M}(\widehat{\mathbf{x}}_k) + \mathbf{S}^{Q}_{k+1}(j) \quad \text{for } j = 1, 2, \ldots, q. \tag{30.3.20}$$

Step 3 Define the forecast

$$\mathbf{x}^{\mathrm{f}}_{k+1} = \mathbf{M}(\widehat{\mathbf{x}}_k) + \frac{1}{2}\sum_{i=1}^{p}\frac{[\xi^{\mathrm{f}}_{k+1}(\pm i) - \mathbf{M}(\widehat{\mathbf{x}}_k)]}{\varepsilon^2}. \tag{30.3.21}$$

From (30.3.17) and (30.3.19) and using the second-order Taylor series (Appendix C), we get

$$\begin{aligned}\widehat{\xi}_{k+1}(\pm i) &= \mathbf{M}(\widehat{\xi}_k(\pm i)) = \mathbf{M}(\widehat{\mathbf{x}}_k \pm \varepsilon\widehat{\mathbf{S}}_k(i)) \\ &= \mathbf{M}(\widehat{\mathbf{x}}_k) \pm \varepsilon\mathbf{D}_{\mathbf{M}}(\widehat{\mathbf{x}}_k)\widehat{\mathbf{S}}_k(i) + \frac{\varepsilon^2}{2}\mathbf{D}^2_{\mathbf{M}}(\widehat{\mathbf{x}}_k, \widehat{\mathbf{S}}_k(i)).\end{aligned} \tag{30.3.22}$$

Substituting (30.3.22) into (30.3.21) and simplifying, we get

$$\mathbf{x}^{\mathrm{f}}_{k+1} = \mathbf{M}(\widehat{\mathbf{x}}_k) + \frac{1}{2}\sum_{i=1}^{p}\mathbf{D}^2_{\mathbf{M}}(\widehat{\mathbf{x}}_k, \widehat{\mathbf{S}}_k(i)) \tag{30.3.23}$$

where recall that $\mathbf{D}^2_{\mathbf{M}}(\widehat{\mathbf{x}}_k, \widehat{\mathbf{S}}_k(i))$ is a vector whose jth component is given by $[\widehat{\mathbf{S}}_k(i)]^{\mathrm{T}}\nabla^2\mathbf{M}_j[\widehat{\mathbf{S}}_k(j)]$ where $\nabla^2\mathbf{M}_j = \nabla^2\mathbf{M}_j(\widehat{\mathbf{x}}_k)$. Hence the jth component of the vector corresponding to the sum on the r.h.s. of (30.3.23) is given by

$$\begin{aligned}&\left[\sum_{i=1}^{p}\mathbf{D}^2_{\mathbf{M}}(\widehat{\mathbf{x}}_k, \widehat{\mathbf{S}}_k(i))\right]_j \\ &= \sum_{i=1}^{p}\left[\widehat{\mathbf{S}}_k(i)\right]^{\mathrm{T}}\nabla^2\mathbf{M}_j\widehat{\mathbf{S}}_k(i) \\ &= \sum_{i=1}^{p}\operatorname{tr}\left\{\widehat{\mathbf{S}}_k(i)\left[\widehat{\mathbf{S}}_k(i)\right]^{\mathrm{T}}\nabla^2\mathbf{M}_j\right\} \\ &= \operatorname{tr}\left[\left\{\sum_{i=1}^{p}\widehat{\mathbf{S}}_k(i)\left[\widehat{\mathbf{S}}_k(i)\right]^{\mathrm{T}}\right\}\nabla^2\mathbf{M}_j\right] \\ &= \operatorname{tr}\left\{\widehat{\mathbf{S}}_k(1:p)\left[\widehat{\mathbf{S}}_k(1:p)\right]^{\mathrm{T}}\nabla^2\mathbf{M}_j\right\} \\ &= \operatorname{tr}\left\{\nabla^2\mathbf{M}_j\widehat{\mathbf{S}}_k(1:p)\left[\widehat{\mathbf{S}}_k(1:p)\right]^{\mathrm{T}}\right\}\end{aligned} \tag{30.3.24}$$

which is a reduced-rank approximation to the correct value $\mathrm{tr}\left[\nabla^2 \mathbf{M}_j \widehat{\mathbf{P}}_k\right]$. Stated in other words, the forecast expression in (30.3.1) is an approximation to the second-order forecast equation in Figure 29.4.1. The finite difference term in (30.3.21) represents the **forecast bias** correction term (Chapter 29). Now define

$$\mathbf{e}^{\mathrm{f}}_{k+1}(\pm i) = \frac{1}{\varepsilon}\left[\xi^{\mathrm{f}}_{k+1}(\pm i) - \mathbf{x}^{\mathrm{f}}_{k+1}\right]$$

and

$$\mathbf{e}^{\mathrm{f}}_{k+1}(2p+j) = \xi^{\mathrm{f}}_{k+1}(2p+j) - \mathbf{M}(\widehat{\mathbf{x}}_k) = \mathbf{S}^{Q}_{k+1}(j). \tag{30.3.25}$$

Assemble the $n \times (2p+q)$ matrix

$$\begin{aligned}\mathbf{S}_{k+1} = \big[&\mathbf{e}^{\mathrm{f}}_{k+1}(1), \mathbf{e}^{\mathrm{f}}_{k+1}(-1), \ldots, \mathbf{e}^{\mathrm{f}}_{k+1}(+p), \mathbf{e}^{\mathrm{f}}_{k+1}(-p),\\ &\mathbf{e}^{\mathrm{f}}_{k+1}(2p+1), \ldots, \mathbf{e}^{\mathrm{f}}_{k+1}(2p+q)\big]\end{aligned} \tag{30.3.26}$$

and compute the $n \times p$ matrix by the rank reduction method to obtain

$$\mathbf{S}^{\mathrm{f}}_{k+1} = \Pi_p[\mathbf{S}_{k+1}]. \tag{30.3.27}$$

To further establish the connection between this filter and the second-order filter, expand the terms on the r.h.s. of the forecast error in the second order Taylor series to obtain

$$\begin{aligned}\mathbf{e}^{\mathrm{f}}_{k+1}(\pm i) &= \frac{1}{\varepsilon}\left[\xi^{\mathrm{f}}_{k+1}(\pm i) - \mathbf{x}^{\mathrm{f}}_{k+1}\right]\\ &= \frac{1}{\varepsilon}\left[\mathbf{M}\left(\widehat{\mathbf{x}}_k \pm \varepsilon \widehat{\mathbf{S}}_k(i)\right) - \mathbf{M}(\widehat{\mathbf{x}}_k) - \frac{1}{2}\mathbf{D}^2_{\mathbf{M}}\big(\widehat{\mathbf{x}}_k, \widehat{\mathbf{S}}_k(i)\big)\right]\\ &= \frac{1}{\varepsilon}\left[\pm\varepsilon \mathbf{D}_{\mathbf{M}}(\widehat{\mathbf{x}}_k)\widehat{\mathbf{S}}_k(i) + \frac{(\varepsilon^2-1)}{2}\mathbf{D}^2_{\mathbf{M}}\big(\widehat{\mathbf{x}}_k, \widehat{\mathbf{S}}_k(i)\big)\right].\end{aligned} \tag{30.3.28}$$

Hence, neglecting the third and higher terms in $\widehat{\mathbf{S}}_k(i)$, we get

$$\mathbf{e}^{\mathrm{f}}_{k+1}(+i)[\mathbf{e}^{\mathrm{f}}_{k+1}(+i)]^{\mathrm{T}} \approx \mathbf{D}_{\mathbf{M}}(\widehat{\mathbf{x}}_k)\widehat{\mathbf{S}}_k(i)[\widehat{\mathbf{S}}_k(i)]^{\mathrm{T}}\mathbf{D}^{\mathrm{T}}_{\mathbf{M}}(\widehat{\mathbf{x}}_k). \tag{30.3.29}$$

Similarly

$$\mathbf{e}^{\mathrm{f}}_{k+1}(2p+j)\left[\mathbf{e}^{\mathrm{f}}_{k+1}(2p+j)\right]^{\mathrm{T}} = \mathbf{S}^{Q}_{k+1}(j)\left[\mathbf{S}^{Q}_{k+1}(j)\right]^{\mathrm{T}}. \tag{30.3.30}$$

Hence, using (30.3.28) and (30.3.29) it can be verified that

$$\begin{aligned}\mathbf{S}_{k+1}\mathbf{S}^{\mathrm{T}}_{k+1} = &\,\mathbf{D}_{\mathbf{M}}(\widehat{\mathbf{x}}_k)\widehat{\mathbf{S}}_k(1:p)\left[\widehat{\mathbf{S}}_k(1:p)\right]^{\mathrm{T}}\mathbf{D}_{\mathbf{M}}(\widehat{\mathbf{x}}_k)\\ &+\mathbf{S}^{Q}_{k+1}(1:q)\left[\mathbf{S}^{Q}_{k+1}(1:q)\right]^{\mathrm{T}}.\end{aligned} \tag{30.3.31}$$

which is a reduced approximation to the second-order covariance (refer to Figure 29.4.1)

$$\mathbf{P}^{\mathrm{f}}_{k+1} = \mathbf{D}_{\mathbf{M}}(\widehat{\mathbf{x}}_k)\widehat{\mathbf{P}}_k\mathbf{D}_{\mathbf{M}}(\widehat{\mathbf{x}}_k) + \mathbf{Q}_{k+1}.$$

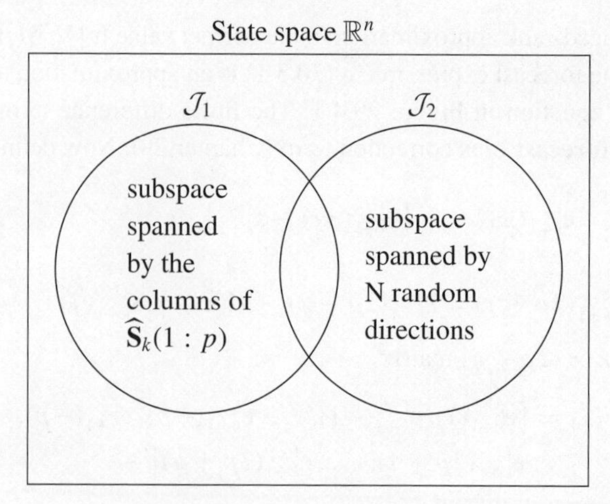

Fig. 30.3.1 Relative disposition of the ensembles.

(C) Hybrid filter 3: parallel filters

Recall that the deterministic ensemble generated by the Hybrid filter 1 at time k lies in the subspace $\mathcal{J}_1$ spanned by the leading p orthogonal modes of the covariance matrix $\widehat{\mathbf{P}}_k$. Hence this filter accounts for only that part of the covariance associated with this p-dimensional subspace. The ensemble in the ensemble filter in Section 30.1 on the other hand, covers the subspace $\mathcal{J}_2$ spanned by N randomly chosen directions. Thus, depending on the relative values of p and N (compared to n) and the **luck of the draw**, the subspace $\mathcal{J}_2$ may cover certain portion of the state space $\mathbb{R}^n$ not covered by $\mathcal{J}_1$. See Figure 30.3.1 for an illustration. Hence by running two filters in parallel, and by suitably combining their results, we could recover a larger fraction of the covariance than by running either filter alone. We now describe a general framework that exploits this principle.

(1) Initialization Let $\mathbf{x}_0$ be the initial condition such that $E(\mathbf{x}_0) = \mathbf{m}_0$ and $\mathrm{Cov}(\mathbf{x}_0) = \mathbf{P}_0$. Let $\widehat{\mathbf{S}}_0(1:p) \in \mathbb{R}^{n\times p}$ denote the p dominant modes of $\mathbf{P}_0$. Let $\widehat{\mathbf{E}}_0(1:N) \in \mathbb{R}^{n\times p}$ denote the N randomly chosen directions where the jth column is given by (refer to (30.1.5))

$$\widehat{\mathbf{E}}_0(j) = \widehat{\mathbf{E}}_0(j:j) = \frac{1}{\sqrt{N-1}}[\mathbf{m}_0 + \widehat{\mathbf{S}}_0\mathbf{y}_0(j)] \tag{30.3.32}$$

where $\mathbf{P}_0 = \widehat{\mathbf{S}}_0(\widehat{\mathbf{S}}_0)^{\mathrm{T}}$ is the full-rank factorization of $\mathbf{P}_0$ and $\mathbf{y}_0(j) \sim N(0, \mathbf{I})$ for $j = 1, 2, \ldots, N$. Construct the $n \times (p+N)$ matrix

$$\widehat{\mathbf{L}}_0 = \left[\widehat{\mathbf{S}}_0(1:p), \widehat{\mathbf{E}}_0(1:N)\right]. \tag{30.3.33}$$

(2) Forecast Step Assume that at time k, we have $\widehat{\mathbf{x}}_k$ and $\widehat{\mathbf{L}}_k = \left[\widehat{\mathbf{S}}_k(1:p), \widehat{\mathbf{E}}_k(1:N)\right]$.

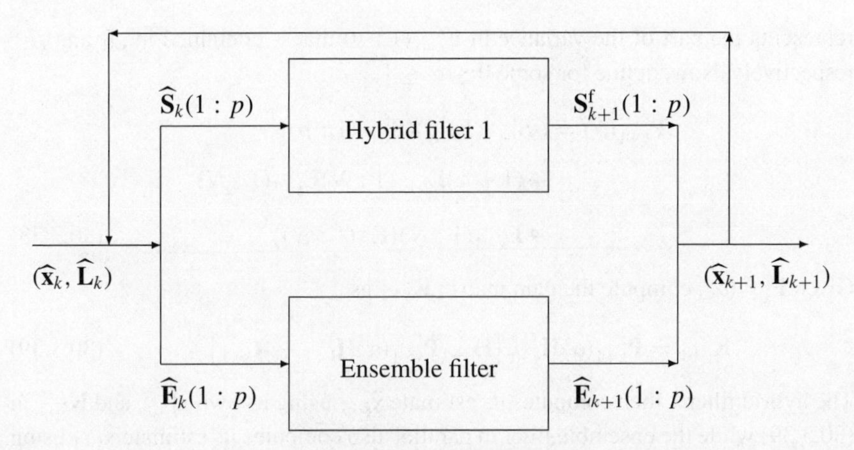

Fig. 30.3.2 Parallel hybrid filters.

(*a*) The hybrid filter 1, using $(\widehat{\mathbf{x}}_k, \widehat{\mathbf{S}}_k(1:p))$, generates the forecast $(\mathbf{x}^{\mathrm{f}}_{k+1}, \mathbf{S}^{\mathrm{f}}_{k+1}(1:p))$. Refer to Figure 30.3.2.
(*b*) The ensemble filter using $(\widehat{\mathbf{x}}_k, \widehat{\mathbf{E}}_k(1:N))$ generates the forecast $(\mathbf{x}^{\mathrm{f}}_{k+1}, \widehat{\mathbf{E}}_{k+1}(1:N))$.

(3) Data Assimilation Step Our goal in this step is to build $\mathbf{P}^{\mathrm{f}}_{k+1}$ that combines the information in $\mathbf{S}^{\mathrm{f}}_{k+1}(1:p)$ and $\mathbf{E}^{\mathrm{f}}_{k+1}(1:N)$. This is done by first splitting the covariance in $\mathbf{E}^{\mathrm{f}}_{k+1}(1:N)$ into two components – part of it that is contained in the subspace $\mathcal{J}_1$ and the rest of it that is contained in $\mathcal{J}_1^{\perp}$, the space orthogonal to $\mathcal{J}_1$. Algorithmically this splitting can be accomplished by using the orthogonal projection matrices (Chapter 6).

Recall if $\mathbf{S} \in \mathbb{R}^{n\times p}$ is a full-rank matrix of rank $p(< n)$, then the orthogonal projection on to range space of $\mathbf{S}$ is $n \times n$ matrix given by

$$\Gamma(\mathbf{S}) = \mathbf{S}(\mathbf{S}^{\mathrm{T}}\mathbf{S})^{-1}\mathbf{S}^{\mathrm{T}} \tag{30.3.34}$$

and the orthogonal projection on to the null space of $\mathbf{S}$ is given by

$$\Gamma^{\perp}(\mathbf{S}) = \mathbf{I} - \Gamma(\mathbf{S}). \tag{30.3.35}$$

Hence,

$$\mathbf{E}_{k+1}(1:p) = \Gamma(\mathbf{S}^{\mathrm{f}}_{k+1}(1:p))\mathbf{E}^{\mathrm{f}}_{k+1}(1:p) \tag{30.3.36}$$

and

$$\mathbf{E}^{\perp}_{k+1}(1:p) = \mathbf{E}^{\mathrm{f}}_{k+1}(1:p) - \mathbf{E}_{k+1}(1:p) \tag{30.3.37}$$

represents the part of the variance in $\mathbf{E}^{\mathrm{f}}_{k+1}(1:p)$ that is contained in $\mathcal{J}_1$ and $\mathcal{J}_1^{\perp}$ respectively. Now, define for some $0 \leq \alpha \leq 1$,

$$\begin{aligned}\mathbf{P}^{\mathrm{f}}_{k+1}(\alpha) = \alpha \mathbf{S}^{\mathrm{f}}_{k+1}(1:p)\big(\mathbf{S}^{\mathrm{f}}_{k+1}(1:p)\big)^{\mathrm{T}} \\ + (1-\alpha)\mathbf{E}_{k+1}(1:N)\mathbf{E}^{\mathrm{T}}_{k+1}(1:N) \\ + \mathbf{E}^{\perp}_{k+1}(1:N)\big(\mathbf{E}^{\perp}(1:N)\big)^{\mathrm{T}}.\end{aligned} \qquad (30.3.38)$$

Given $\mathbf{P}^{\mathrm{f}}_{k+1}(\alpha)$, compute the gain matrix $\mathbf{K}_{k+1}$ as

$$\mathbf{K}_{k+1} = \mathbf{P}^{\mathrm{f}}_{k+1}(\alpha)\mathbf{H}^{\mathrm{T}}_{k+1}\left[\mathbf{H}_{k+1}\mathbf{P}^{\mathrm{f}}_{k+1}(\alpha)\mathbf{H}^{\mathrm{T}}_{k+1} + \mathbf{R}_{k+1}\right]^{-1}. \qquad (30.3.39)$$

The hybrid filter 1 then computes its estimate $\widehat{\mathbf{x}}_{k+1}$ using its own $\mathbf{x}^{\mathrm{f}}_{k+1}$, and $\mathbf{K}_{k+1}$ in (30.3.39) while the ensemble filter in parallel also computes its estimate $\widehat{\mathbf{x}}_{k+1}$ using its own $\mathbf{x}^{\mathrm{f}}_{k+1}$ and $\mathbf{K}_{k+1}$ using the method in Section 30.1 and the cycle repeats.

A number of observations are in order.

(1) When $\alpha = 1$ in (30.3.38), the resulting filter is called the **partially orthogonal ensemble Kalman filter** (PoEnKF). Refer to Heemink, Verlann and Segers (2001) for more details.
(2) The ensemble filter instead of creating an ensemble over the whole space, could restrict its ensemble to cover only the part of the state space not covered by $\mathcal{J}_1$. Given $\mathbf{P}_0$, compute $\mathbf{P}_0^{\perp} = \mathbf{P}_0 - \widehat{\mathbf{S}}_0(1:p)\big[\widehat{\mathbf{S}}_0(1:p)\big]^{\mathrm{T}}$, the portion of the initial covariance that lies outside of $\mathcal{J}_1$. Then generate an ensemble $\widehat{\mathbf{E}}_0(1:N)$ using $\mathbf{P}_0^{\perp}$. This modification more efficiently captures the covariance outside of $\mathcal{J}_1$ and hence is called the **complementary orthogonal subspace filter for efficient ensemble** (COFFEE) filter. Refer to Heermink, Verlaan and Segers (2001) for details.
(3) Since the gain $\mathbf{K}_{k+1}$ in (30.3.39) is not obtained by directly minimizing the total variance in $\mathbf{P}^{\mathrm{f}}_{k+1}(\alpha)$, in computing $\widehat{\mathbf{P}}_{k+1}$ we should use the general formula (Chapter 27) given by

$$\widehat{\mathbf{P}}_{k+1} = (\mathbf{I} - \mathbf{K}_{k+1}\mathbf{H}_{k+1})\mathbf{P}^{\mathrm{f}}_{k+1}(\alpha)(\mathbf{I} - \mathbf{K}_{k+1}\mathbf{H}_{k+1})^{\mathrm{T}} + \mathbf{K}_{k+1}\mathbf{R}_{k+1}\mathbf{K}^{\mathrm{T}}_{k+1}$$

or in its square root form

$$\widehat{\mathbf{S}}_{k+1} = \left[(\mathbf{I} - \mathbf{K}_{k+1}\mathbf{H}_{k+1})\mathbf{S}^{\mathrm{f}}_{k+1}(\alpha), \mathbf{K}_{k+1}\mathbf{S}^{R}_{k+1}\right].$$

30.4 Applications of Kalman filtering: an overview

In this section we provide an overview of many of the applications of the Kalman filtering methodology to problems in meteorology, oceanography, hydrology and atmospheric chemistry. In fact many of the approximation methods including the ensemble filters, reduced-rank and hybrid filters described in this chapter were developed in the context of applications in oceanography. Early applications of Kalman filters were exclusively in aerospace and systems engineering problems and these are very well covered in many textbooks including Jazwinski (1970), Gelb (1974), Maybeck (1982) to name a few.

Meteorology Applications of the sequential statistical estimation based on Kalman filtering theory began in the early 1980s with the work of Cohn, Ghil and Isaacson (1981) and Cohn (1982) in which they analyzed the estimation problem related to the one-dimensional linearized shallow water model. Soon Parrish and Cohn (1985) extended this idea to linearized shallow water equations in two dimensions. A comprehensive and an authoritative survey entitled ***Data Assimilation in Meteorology and Oceanography*** by Ghil and Malanotte-Rizzoli (1991) contains a detailed account of the state of the art of various approaches to assimilation including Kalman filtering techniques.

Excessive computational requirements essentially hindered the application of Kalman filtering to large-scale operational forecasting problems. This situation provided the much-needed impetus for developing feasible **suboptimal filters**, a trend that continues to this day. In the following we provide a summary of the ideas leading to several successful suboptimal implementations. Following Todling and Cohn (1994) these ventures can be grouped into several categories: **covariance modelling**, **simplification of model dynamics**, **local approximation**, and the use of **steady-state approximations**. The idea of covariance modelling was central to the 3DVAR formulation wherein the form of the background or the forecast covariance is assumed to have a prespecified and fixed spatial variation. Refer to Gandin (1963), Lorenc (1981). Recall from Chapter 29 that the dynamics of the forecast error covariance which is computationally the most demanding part of the filter is given by

$$\mathbf{P}^{\mathrm{f}}_{k+1} = \mathbf{D}_{\mathbf{M}}(\widehat{\mathbf{x}}_k)\widehat{\mathbf{P}}_k\mathbf{D}^{\mathrm{T}}_{\mathbf{M}}(\widehat{\mathbf{x}}_k) + \mathbf{Q}_{k+1} \tag{30.4.1}$$

where $\mathbf{D}_{\mathbf{M}}(\widehat{\mathbf{x}}_k)$ is the Jacobian of vector field $\mathbf{M}(\mathbf{x})$ that defines the model. Several ideas have been proposed to simplify this dynamics. For example, Dee (1991) includes only the advective part in $\mathbf{D}_{\mathbf{M}}(\widehat{\mathbf{x}}_k)$. Another useful idea is to develop two systems – fine and coarse grid where the finer (larger) grid is used for obtaining the model forecast $\mathbf{x}^{\mathrm{f}}_{k+1}$ but a coarse (smaller) grid is used for updating the forecast covariance dynamics in (30.4.1). The computed forecast covariance is then lifted to the finer grid using a suitable interpolation. Refer to Fukumori and Malanotte-Rizzoli (1994) for specific details. Parrish and Cohn (1985) used a local approximation technique in which it was assumed that grid points separated by a distance larger than a threshold do not have any significant correlation. While this idea resulted in saving of storage and time, it has the potential to destroy the key property of positive definiteness of these matrices which could result in undesirable divergence. If the system model and the observations are linear and time invariant, one could precompute the limiting value for the forecast covariance and the corresponding gain and use this constant gain in sequentially updating the estimates. In this limiting case the Kalman filter becomes equivalent to the classical Wiener filter. This idea of using the steady state gain to achieve feasibility is demonstrated by Heemink (1988) and Heemink and Kloosterhuis (1990).

Also refer to Cohn and Parrish (1991) and Cohn (1993) for a detailed characterization of the behavior of the forecast error covariance. Daley and Ménard (1993) explore the spectral properties of Kalman filters. Verlaan and Heemink (2001) analyze data assimilation of the stochastic version of the Burgers' equation with advection and diffusion and present a comparison of the performance of various hybrid filters. Also refer to Ménard (1994) for an application of Kalman filters to Burgers' equation.

Oceanography Until recently observations from the ocean were too few and far between both in space and time. This paucity of data forced oceanographers to rely exclusively on the mathematical models that characterize and predict **ocean circulation**, **wind-driven ocean waves**, **storm surge** and **tidal flow**. The introduction of the special ERS – earth-observing satellites has helped to close this void and global ocean wave observations are now available with good resolution in near real time. This has provided a major impetus to the application of the data assimilation methods that suitably combine the information in observation and the predictive power of the models.

Ocean circulation model Earliest application of Kalman filtering techniques to oceanography is due to Barbieri and Schopf (1982), Miller (1986) and Budgell (1986), (1987), wherein sequential statistical estimation method was applied to the **ocean circulation model**. Bennett and Budgell (1987) using a truncated spectral expansion of the special form of the vorticity equation that represents the stratified synaptic-scale ocean circulation model analyze the conditions for the convergence of the Kalman filter. It is shown that the Kalman gain converges by suitably restricting the properties of the model noise. Refer to Miller (1986) for details. Evensen (1992) by using a multilayer quasi-geostrophic ocean circulation model demonstrate the difficulties of using the first-order filter. To overcome these numerical difficulties Evensen (1994) developed a Monte Carlo based approach to nonlinear filtering which has now come to be known as the **ensemble Kalman filtering**. Since then ensemble filtering has taken the center stage and is now widely applied in meteorology, oceanography and atmospheric chemistry. Evensen (1994) did not use the virtual observations. The need for introducing the virtual observations in ensemble filtering was later identified by Burgers et al. (1998). For other related work on the application of nonlinear filtering to problems in oceanography refer to Miller, Ghil and Gauthiez (1994), Miller, Carter and Blue (1999). The books by Bennett (1992)(2002) and Wunsch (1996) are exclusively devoted to data assimilation problems of interest in oceanography.

Prediction of wind-driven ocean waves is critical to the safety of both commercial shipping, leisure time cruise liners, off-shore drilling and exploration operations, in sediment transport and in ocean–atmosphere interaction. Heemink and Kloosterhuis (1990) describe an early application of Kalman filtering to nonlinear tidal models. Heemink, Bolding and Verlaan (1997) using a shallow-water model and a reduced-rank square root algorithm successfully predict the storm surge in the portion of the North Sea around the Netherlands. Voorrips, Heemink and Komen (1999) consider the wave data assimilation using hybrid filters.

Atmospheric Chemistry Documenting the space-time evolution of the concentration patterns of the air-pollutants is of central importance to the understanding of the quality of the air we breathe and the environment we live in especially in and around big, industrialized cities like Mexico City, Sao Paulo, Los Angeles, Beijing, to name a few. Kalman filtering techniques have been successfully applied to estimating the evolution of various chemical species in the atmosphere. Segers, Heemink, Verlaan and van Loan (2000) discuss the application of the RRSQRT filters to atmospheric chemical species estimation problem over western Europe. The recent monograph by Segers (2002) provides a comprehensive and self-contained account of the theory and application of Kalman filtering methodology to atmospheric chemistry problems. The recent book by Enting (2002) is a very good reference on data assimilation for atmospheric chemistry problems.

Hydrology Cahill et al. (1999) contains an application of Kalman filtering to the problem of estimating the parameters that determine hydraulic conductivity and soil moisture. Recently Zheng, Qiu and Xu (2004) describe an adaptive Kalman filtering approach to estimating the soil moisture content based on the covariance matching technique developed by Mehra (1972). For more details on adaptive Kalman filtering refer to Mehra (1972) and the references therein.

Exercises

30.1 **Random number generation** (Knuth (1980) Vol.2) In this exercise we provide a summary of the basic methods leading to the generation of normal random numbers.

(a) **Uniformly distributed random numbers in [0, 1]**

The standard method for generating random integers is based on the **mixed congruential generator** where the $(n+1)^{th}$ random integer x_{n+1} is given by

$$x_{n+1} \equiv (ax_n + c) \bmod m$$

where a is relatively prime to m. Then $u_n = x_n/m$ is the random number between [0, 1).

(b) If x is a uniformly distributed random number in [0, 1), then $y = ax + b$ is uniformly distributed in $[a, b)$.

(c) **Standard normal random variable** $y \sim N(0, 1)$

Let x_1 and x_2 be two independent random numbers uniformly distributed in [0, 1). Then

$$y_1 = \cos(2\pi x_1)\sqrt{-2\log x_1}$$
$$y_2 = \sin(2\pi x_2)\sqrt{-2\log x_2}$$

are two independent standard normal variables. This is called the **Box–Muller** method.

(d) If x is a standard normal variable, then $r = ax + b$ is a normal random, with mean b and variance a^2.

Generate a set $n = 10000$ of standard normal numbers and compute the mean and variance. Also plot the histogram.

30.2 **Bellantoni and Dodge (1967) factorization**

We can rewrite the covariance of the standard Kalman filter in (30.1.27) as

$$\widehat{\mathbf{P}}_{k+1} = \mathbf{S}^{\mathrm{f}}_{k+1}[\mathbf{I} - \mathbf{A}\mathbf{D}^{-1}\mathbf{A}^{\mathrm{T}}](\mathbf{S}^{\mathrm{f}}_{k+1})^{\mathrm{T}} \quad \text{(a)}$$

where $\mathbf{P}^{\mathrm{f}}_{k+1} = \mathbf{S}^{\mathrm{f}}_{k+1}(\mathbf{S}^{\mathrm{f}}_{k+1})^{\mathrm{T}}$, $\mathbf{A} = (\mathbf{H}_{k+1}\mathbf{S}^{\mathrm{f}}_{k+1})^{\mathrm{T}}$ and $\mathbf{D} = (\mathbf{A}^{\mathrm{T}}\mathbf{A} + \mathbf{R}_{k+1})$.

(a) Using the Sherman – Morrison – Woodbury formula (Appendix B) verify that

$$\begin{aligned}[\mathbf{I} - \mathbf{A}\mathbf{D}^{-1}\mathbf{A}^{\mathrm{T}}] &= [\mathbf{I} + \mathbf{A}[\mathbf{D} - \mathbf{A}^{\mathrm{T}}\mathbf{A}]^{-1}\mathbf{A}^{\mathrm{T}}]^{-1} \\ &= [\mathbf{I} + \mathbf{A}\mathbf{R}^{-1}_{k+1}\mathbf{A}^{\mathrm{T}}]^{-1} \quad \text{(b)}\end{aligned}$$

(b) Verify that $\mathbf{A}\mathbf{R}^{-1}_{k+1}\mathbf{A}^{\mathrm{T}} \in \mathbb{R}^{n\times n}$ is a symmetric positive semi-definite matrix where $\mathrm{Rank}(\mathbf{A}\mathbf{R}^{-1}_{k+1}\mathbf{A}^{\mathrm{T}}) = m < n$.

(c) Let $\mathbf{A}\mathbf{R}^{-1}_{k+1}\mathbf{A}^{\mathrm{T}} = \mathbf{V}\mathbf{\Lambda}\mathbf{V}^{\mathrm{T}}$ be an eigendecomposition where $\mathbf{V} \in \mathbb{R}^{n\times n}$ is the orthogonal matrix of eigenvectors and $\mathbf{\Lambda} = \mathrm{Diag}(\lambda_1, \lambda_2, \ldots, \lambda_m, 0, \ldots, 0)$ be the matrix of eigenvalues. Verify that

$$\begin{aligned}[\mathbf{I} - \mathbf{A}\mathbf{D}^{-1}\mathbf{A}^{\mathrm{T}}] &= [\mathbf{I} + \mathbf{V}\mathbf{\Lambda}\mathbf{V}^{\mathrm{T}}]^{-1} \\ &= (\mathbf{V}\mathbf{V}^{\mathrm{T}} + \mathbf{V}\mathbf{\Lambda}\mathbf{V}^{\mathrm{T}})^{-1} \\ &= \mathbf{V}(\mathbf{I} + \mathbf{V})^{-1}\mathbf{V}^{\mathrm{T}} = \bar{\mathbf{V}}\bar{\mathbf{V}}^{\mathrm{T}} \quad \text{(c)}\end{aligned}$$

is a square root factorization where $\bar{\mathbf{V}} = \mathbf{V}(\mathbf{I} + \mathbf{\Lambda})^{-1/2}$.

(d) Substituting (c) into (a) verify that

$$\widehat{\mathbf{P}}_{k+1} = (\mathbf{S}^{\mathrm{f}}_{k+1}\bar{\mathbf{V}})(\mathbf{S}^{\mathrm{f}}_{k+1}\bar{\mathbf{V}})^{\mathrm{T}} \quad \text{(d)}$$

Note: This factorization due to Bellantoni and Dodge (1967) when used with the ensemble filtering has come to be known as **ensemble transform Kalman filtering**. Refer to Bishop et al. (2001) and Tippett et al. (2003) for details.

30.3 Let $\mathbf{P}^{\mathrm{f}}_{k+1} = \mathbf{F}\Sigma\mathbf{F}^{\mathrm{T}}$ be an eigendecomposition where $\mathbf{F}$ is the orthogonal matrix of eigenvectors and Σ is the diagonal matrix of eigenvalues. Let $\mathbf{P}^{\mathrm{f}}_{k+1} = \bar{\mathbf{F}}\bar{\mathbf{F}}^{\mathrm{T}}$ be a square root factorization of $\mathbf{P}^{\mathrm{f}}_{k+1}$ where $\bar{\mathbf{F}} = \mathbf{F}\Sigma^{1/2}$.

(a) Using the above factorization rewrite

$$\widehat{\mathbf{P}}_{k+1} = \bar{\mathbf{F}}[\mathbf{I} - \mathbf{A}\mathbf{D}^{-1}\mathbf{A}^{\mathrm{T}}]\bar{\mathbf{F}}^{\mathrm{T}}$$

where $\mathbf{A}$ and $\mathbf{D}$ are as defined in Exercise 30.2.

(b) By following the arguments in Exercise 30.2 verify that

$$\widehat{\mathbf{P}}_{k+1} = \bar{\mathbf{F}}\bar{\mathbf{V}}\bar{\mathbf{V}}^{\mathrm{T}}\bar{\mathbf{F}}^{\mathrm{T}}$$

where $\bar{\mathbf{V}}$ is defined in Exercise 30.2.

(c) Rewrite (using $(\bar{\mathbf{F}})^{-1}\mathbf{P}^{\mathrm{f}}_{k+1}(\bar{\mathbf{F}})^{-T} = \mathbf{I}$) and verify that

$$\widehat{\mathbf{P}}_{k+1} = \bar{\mathbf{F}}\bar{\mathbf{V}}[(\bar{\mathbf{F}})^{-1}\mathbf{P}^{\mathrm{f}}_{k+1}(\bar{\mathbf{F}})^{-T}]\bar{\mathbf{V}}^{\mathrm{T}}\bar{\mathbf{F}}^{\mathrm{T}} = \widehat{\mathbf{S}}_{k+1}(\widehat{\mathbf{S}}_{k+1})^{\mathrm{T}}$$

where

$$\widehat{\mathbf{S}}_{k+1} = \bar{\mathbf{F}}\bar{\mathbf{V}}(\bar{\mathbf{F}})^{-1}\bar{\mathbf{F}} = \mathbf{L}\bar{\mathbf{F}}.$$

That is, the square root $\widehat{\mathbf{S}}_{k+1}$ of $\widehat{\mathbf{P}}_{k+1}$ is obtained as a linear transformation $\mathbf{L}$ of the square root $\bar{\mathbf{F}}$ of $\mathbf{P}^{\mathrm{f}}_{k+1}$ and $\mathbf{L} = \bar{\mathbf{F}}\bar{\mathbf{V}}(\bar{\mathbf{F}})^{-1}$.

Note: An implementation of the ensemble filtering using this factorization is known as **ensemble adjustment Kalman filtering**. Refer to Anderson (2001) and Tippett et al. (2003) for details.

Notes and references

Section 30.1 Ensemble Kalman filter was first introduced by Evensen (1994) and Evensen and van Leeuwen (1996). The need for using the virtual observations in ensemble filtering was first recognized by Burgers et al. (1998). Since then there has been a virtual explosion of literature, refer to Segers (2002) for a succinct description of many of the key ideas in this area.

Section 30.2 reduced-rank square root filters are systematically developed by Verlaan (1998), Verlaan and Heemink (1997), Voorrips et al. (1999) and Cañizares (1999).

Section 30.3 Various forms of hybrid filters are developed by Pham et al. (1998), Verron et al. (1999), Houtekamer and Mitchell (1998), Lermusiaux and Robinson (1999a), (1999b), Heemink et al. (2001). Segers (2002) provides a comprehensive coverage of reduced-rank filters and their implementation. Also refer to Houtekamer (1995) and Houtekamer and Derome (1995).

PART VIII

Predictability

31

Predictability: a stochastic view

In this and the following chapter we provide an overview of the basic methods for assessing predictability in dynamical systems. A stochastic approach to quantifying predictability is described in this chapter and the deterministic method which heavily relies on the ensemble approach is covered in Chapter 32. A classification of the predictability methods and various measures for assessing predictability are described in Section 31.1. Three basic methods – an analytical approach, approximate moment dynamics, and the Monte Carlo methods are described in Sections 31.2 through 31.4.

31.1 Predictability: an overview

Predictability has several dimensions. First, it relates to the ability to predict both the **normal** course of events as well as **extreme** or **catastrophic** events. Secondly, it also calls for assessing the goodness of the prediction where the goodness is often measured by the variance of the prediction.

Events to be predicted may be classified into three groups. Some events are perfectly predictable. Examples include lunar/solar eclipses, phases of the moon and their attendant impact on ocean tides, etc. While many events are not perfectly predictable, they can be predicted with relatively high accuracy in the sense that the variance of the prediction can be made small. Embedded in this idea is the notion of the classical **signal to noise** ratio. If this ratio is large, then the prediction is good. Examples include the prediction of maximum/minimum temperature in various cities of the world for tomorrow, prediction of tomorrow's interest rate for the 30 year home mortgage loan, prediction of the tax revenue by a state budget office for the last quarter of the current budget year, prediction of foreign exchange rate between U.S. dollar and Euro for tomorrow, etc. The third class of events include the prediction of the extreme or catastrophic events. Examples include the prediction of the probability of occurrence of an 8.0 (in RS) magnitude earthquake in the Los Angeles basin before the end of the year; probability that the Dow Jones

Industrial average will drop to 50% of its current value within the next six months, the probability of having 25" snow fall on New Year's Eve in Washington DC, to name a few.

There is a fundamental difference between predicting the normal events vs. extreme events. In predicting the normal events, our goal is to obtain a prediction with the least variance or maximum signal to noise ratio. But in predicting extreme events, we often want to quantify the probability of occurrence of least probable but high impact events. In this latter case we want to understand the conditions under which the future behavior of the system will exhibit maximum possible variance and/or variations. Analysis of extreme events lies at the heart of Risk Analysis in Actuarial Sciences and is routinely used by the Insurance industry.

Every prediction and its goodness is a direct function of the amount and quality of the information used in generating the prediction. This information set may often contain a mathematical model of the event, and/or actual observations of the evolving process. The topic of prediction when both the model and observations are available lies at the heart of filtering and prediction which is treated in Chapters 27–30 in Part VII. Refer to Figure 31.1.1. In this Part VIII we are primarily concerned with the prediction using only the model.

Recall that a model can be deterministic or stochastic and the initial conditions can again be deterministically or stochastically specified. In this chapter we examine the predictability from a stochastic view when the model is deterministic or stochastic but the initial condition is stochastic. A deterministic view of predictability when the model and the initial condition are deterministic is taken up in Chapter 32.

There are basically three approaches to assessing predictability in the stochastic context. The first method is called the analytical method that captures the dynamics of the evolution of the probability density of the states of the system. Given the probability density functions of the initial state, the one-step transition probability density of the Markov process is defined by the model. In the continuous time domain, this dynamics is given by the Kolmogorov forward equation or the Fokker–Planck equation. For the case when the model is deterministic but the initial condition is random, this equation reduces to the Liouville's equation. By solving these equations analytically or numerically, we can at least in principle compute the probability density of the state of the system for all future time. Using this density function, we can quantify answers to questions such as the following: given a subset S of the model space $\mathbb{R}^n$, what is the probability that the trajectory of the model enter the set S, which is given by

$$\text{Prob}\,[\mathbf{x}_k \in S] = \int_S P_k(\mathbf{x}_k)\,d\mathbf{x}_k \tag{31.1.1}$$

where $P_k(\mathbf{x}_k)$ is the probability density function of $\mathbf{x}_k$.

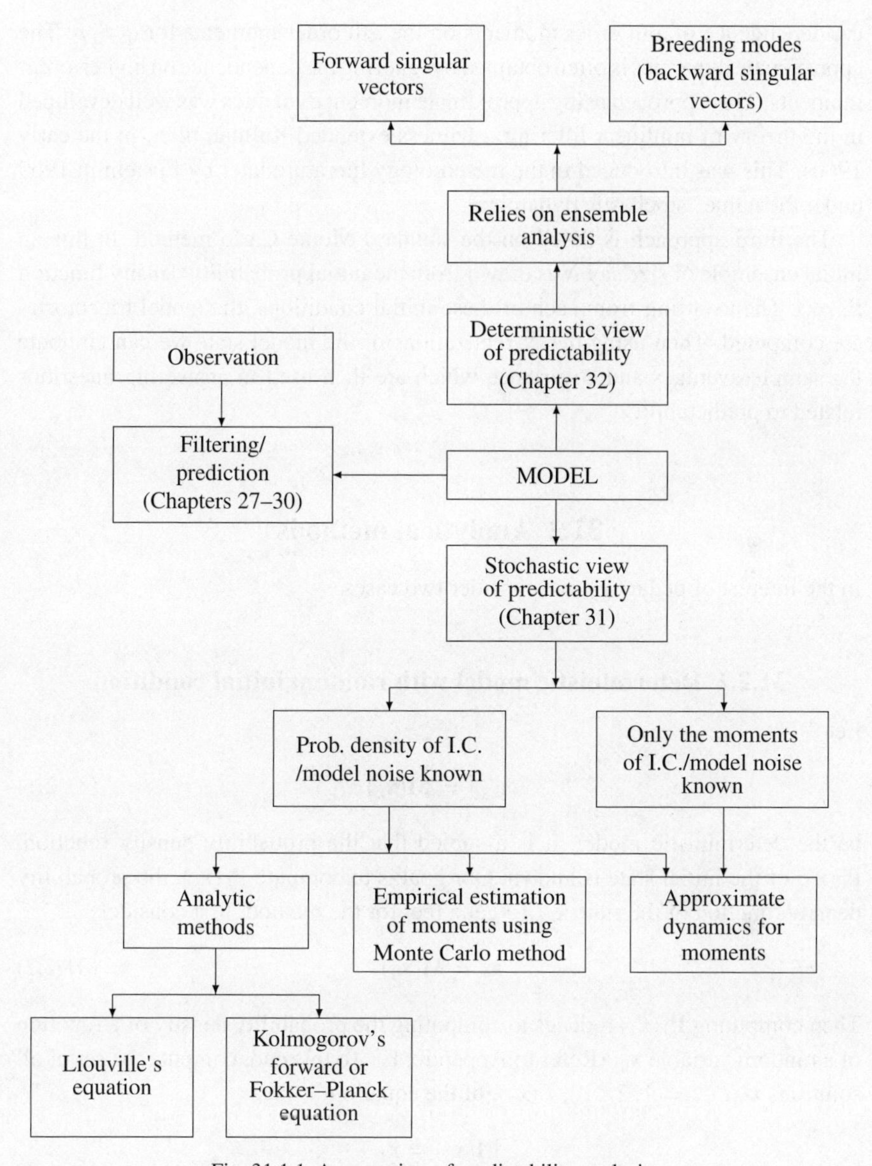

Fig. 31.1.1 An overview of predictability analysis.

While the Kolmogorov or Fokker–Planck and Liouville's approach provides the complete solution to the predictability problem, solving these equations is easier said than done. In an attempt to make the computations feasible, the interest shifts to quantifying the approximate dynamics of the first few moments – such as the mean and covariance of the state of the system. The second approach is based on solving this class of approximate moment dynamics. This approach is also riddled with its own challenges arising from the moment closure problem which relates to the dependence of pth order moments on the qth order moments for $q > p$. The approximate dynamics is often obtained by ignoring the dependence on higher-order moments. This approach using approximate moment dynamics was well developed in the theory of nonlinear filtering – witness extended Kalman filter, in the early 1960s. This was introduced in the meteorology literature later by Epstein in 1969 under the name "stochastic dynamics".

The third approach is based on the standard Monte Carlo method. In this an initial ensemble of size say N is drawn from the initial probability density function $\mathbf{P}_0(\mathbf{x}_0)$. Then starting from each of these initial conditions, the model trajectories are computed. Then using the N realizations of the model state we can compute the sample averages and covariance which are then used in answering questions related to predictability.

31.2 Analytical methods

In the interest of pedagogy, we consider two cases.

31.2.1 Deterministic model with random initial condition

Let

$$\mathbf{x}_{k+1} = \mathbf{M}(\mathbf{x}_k) \tag{31.2.1}$$

be the deterministic model. It is assumed that the probability density function, $\mathbf{P}_0(\mathbf{x}_0)$ of the initial state is known. Our goal is to compute $\mathbf{P}_k(\mathbf{x}_k)$, the probability density function of the state $\mathbf{x}_k$. To get a feel for the method, first consider

$$\mathbf{x}_1 = \mathbf{M}(\mathbf{x}_0). \tag{31.2.2}$$

Then computing $\mathbf{P}_1(\mathbf{x}_1)$ reduces to computing the probability density of a function of a random variable $\mathbf{x}_0$. (Refer to Appendix F). To this end, compute the set of all solutions $\mathbf{x}_0(i)$, $i = 1, 2, \ldots, L(\mathbf{x}_1)$, of the equation

$$\mathbf{M}(\mathbf{x}_0) = \mathbf{x}_1$$

that is,

$$S_M(\mathbf{x}_1) = \{\mathbf{x}_0(i) \mid M(\mathbf{x}_0(i)) = \mathbf{x}_1,\ i = 1, 2, \ldots, L(\mathbf{x}_1)\} \tag{31.2.3}$$

where it is tacitly assumed that $\mathbf{x}_1$ is in the range of the function $\mathbf{M}(\cdot)$. Then from Appendix F we obtain

$$\mathbf{P}_1(\mathbf{x}_1) = \sum_{\mathbf{x}_1(i) \in S_M(\mathbf{x}_1)} \frac{1}{\text{Det}\,[D_M(\mathbf{x}_0(i))]} \mathbf{P}_0(\mathbf{x}_0(i)). \tag{31.2.4}$$

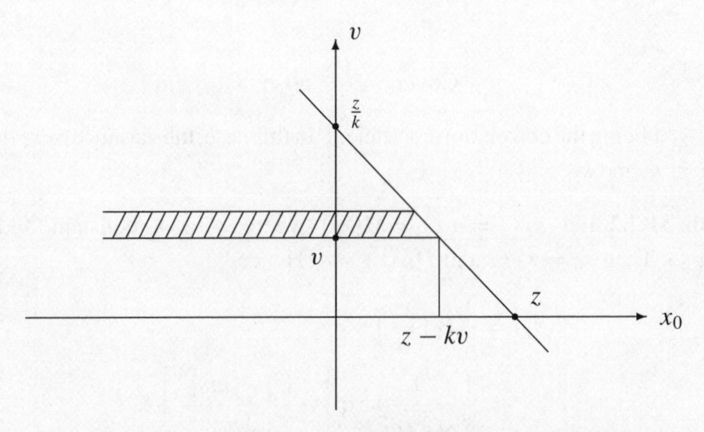

Fig. 31.2.1 An illustration of the computation in (31.2.6).

Once $\mathbf{P}_1(\mathbf{x}_1)$ is known, then $\mathbf{P}_2(\mathbf{x}_2)$ can be computed from $\mathbf{P}_1(\mathbf{x}_1)$ in the same way $\mathbf{P}_1(\mathbf{x}_1)$ was obtained from $\mathbf{P}_0(\mathbf{x}_0)$. By repeating this process, we obtain $\mathbf{P}_k(\mathbf{x}_k)$, $k = 1, 2, 3, \ldots$

Example 31.2.1 Let $x_{k+1} = x_k + v = x_0 + kv = M_k(x_0)$, we consider two cases.

Case 1 v is a known constant but x_0 is random and x_0 is uniformly distributed in $[a, b]$. Then applying the above procedure, $x_0 = x_k - kv$ and $D_{M_k}(x_0) = 1$. Hence

$$P_k(x_k) = P_0(x_k - kv) \tag{31.2.5}$$

i.e., x_k is uniformly distributed in the interval $[a + kv, b + kv]$. Thus, the distribution of x_k is obtained by translating that of x_0 by kv units to the right. Clearly the variance of x_k is constant but its mean increases linearly with k.

Case 2 Both v and x_0 are random with $P_0(x_0, v)$ as their joint probability to density function. Then, referring the Figure 31.2.1 we get

$$\begin{aligned}\text{Prob}[x_k \le z] &= \text{Prob}[x_0 + kv \le z] \\ &= \int_{-\infty}^{\infty}\int_{-\infty}^{z-k\,v} P_0(x_0, v)\,\mathrm{d}x_0\,\mathrm{d}v.\end{aligned} \tag{31.2.6}$$

By differentiating both sides of (31.2.6) we obtain the probability density of x_k.

Let m_0 and σ_0^2 be the mean and the variance of x_0, and m_v and σ_v^2 be those of v. Then

$$E(x_k) = E[x_0 + kv] = m_0 + km_v$$

and

$$\begin{aligned}\text{Var}(x_k) &= E[(x_0 - m_0) + k\,(v - m_v)]^2 \\ &= E[(x_0 - m_0)^2] + 2\,k\,E[(x_0 - m_0)(v - m_v)] + k^2\,E\,[(v - m_v)^2] \\ &= \sigma_0^2 + 2\,k\,\text{Cov}(x_0, v) + k^2\sigma_v^2\end{aligned} \tag{31.2.7}$$

where

$$\text{Cov}(x_0, v) = \rho\sigma_0\sigma_v \tag{31.2.8}$$

with $|\rho| \le 1$ being the correlation coefficient. In this case, the variance of x_k depends on $\sigma_0^2, \sigma_v^2, \rho$, and k.

Example 31.2.2 Let $x_{k+1} = a\,x_k = M(x_k)$ where a is a constant and $x_0 \sim N(m_0, \sigma_0^2)$. Then $x_0 = x_1/a$ and $D_M(x) = a$. Hence

$$\begin{aligned} P_1(x_1) &= \frac{1}{a} P_0\Big(\frac{x_1}{a}\Big) \\ &= \frac{1}{a}\frac{1}{\sqrt{2\pi}\sigma_0}\exp\left[-\frac{\left(\frac{x_1}{a} - m_0\right)^2}{2\sigma_0^2}\right] \\ &= \frac{1}{\sqrt{2\pi}\,(a\sigma_0)}\exp\left[-\frac{(x_1 - am_0)^2}{2a^2\sigma_0^2}\right] \\ &= N(am_0, a^2\sigma_0^2). \end{aligned} \tag{31.2.9}$$

By repeating this process, we get

$$P_k(x_k) = N(a^k\, m_0, (a^2)^k\, \sigma_0^2) \tag{31.2.10}$$

Thus, if $a < 1$, both the mean and the variance converge to zero and if $a > 1$ then both the mean and variance diverge to infinity.

Now consider the case when x_0 and a are random but independent. Then

$$E(x_k) = E[a^k\, x_0] = E(x_0)E(a^k) \tag{31.2.11}$$

and

$$\begin{aligned} \text{Var}(x_k) &= E(x_k^2) - [E(x_k)]^2 \\ &= E[x_0^2 a^{2k}] - [E(x_0)E(a^k)]^2 \\ &= E(x_0^2)E(a^{2k}) - [E(x_0)]^2[E(a^k)]^2. \end{aligned} \tag{31.2.12}$$

Thus, $\text{Var}(x_k)$ depends on the higher order-moments of a.

Remark 31.2.1 Liouville's equation In the continuous time case, the deterministic model dynamics in the state space form is given by

$$\dot{\mathbf{x}} = \mathbf{f}(t, \mathbf{x}(t)) \tag{31.2.13}$$

where $\mathbf{x} \in \mathbb{R}^n$ and $\mathbf{f}(t, \mathbf{x}) = (f_1(t, \mathbf{x}), f_2(t, \mathbf{x}), \ldots, f_n(t, \mathbf{x}))^{\mathrm{T}}$. Let $g(\mathbf{x})$ be the density of $\mathbf{x}_0$, and let $P(t, \mathbf{x}(t))$ be the density of $\mathbf{x}_t$. Then the dynamics of $P(t, \mathbf{x}(t))$ is given by Liouville's equation

$$\frac{\partial P}{\partial t} + \sum_{i=1}^{n} \frac{\partial}{\partial x_i(t)}[f_i(t, \mathbf{x}(t))\, P(t, \mathbf{x}(t))] = 0 \tag{31.2.14}$$

or

$$\frac{\partial P}{\partial t} + \sum_{i=1}^{n} f_i(t, \mathbf{x}(t)) \frac{\partial P}{\partial x_i(t)} + P(t, \mathbf{x}(t))[\nabla \cdot f(t, \mathbf{x}(t))] = 0 \quad (31.2.14)$$

with

$$P(0, \mathbf{x}(0)) = g(\mathbf{x}). \quad (31.2.15)$$

This is the continuity equation for the probability mass over $\mathbb{R}^n$. Except in simple cases, it is very difficult to solve (31.2.14) analytically. Numerical solution is the only avenue.

31.2.2 Stochastic model and random initial conditions

Let

$$\mathbf{x}_{k+1} = \mathbf{M}(\mathbf{x}_k) + \mathbf{w}_{k+1} \quad (31.2.16)$$

with $\mathbf{x}_0$ being the random initial condition satisfying the conditions that $\{\mathbf{w}_k\}$ is a white noise sequence and $\mathbf{x}_0$ and $\{\mathbf{w}_k\}$ are uncorrelated. In this case $\{\mathbf{x}_k\}$ is a Markov process and its one step-transition probability density is closely related to the probability density of $\{\mathbf{w}_k\}$. The expression for the probability density $P_k(\mathbf{x}_k)$ for this case is derived in Section 29.1 entitled Nonlinear Stochastic Dynamics.

Remark 31.2.2 Kolmogorov Forward or Fokker–Planck Equation In the continuous time case the model dynamics is given by

$$\mathrm{d}\mathbf{x}_t = \mathbf{f}(t, \mathbf{x})\, \mathrm{d}t + \sigma(t, \mathbf{x}_t) \mathrm{d}w_t \quad (31.2.17)$$

where $\mathbf{x}_t \in \mathbb{R}^n$, $\mathbf{f}(t, \mathbf{x}) = [f_1(t, \mathbf{x}), f_2(t, \mathbf{x}), \ldots, f_n(t, \mathbf{x})]^{\mathrm{T}}, \sigma(t, \mathbf{x}_t) = [\sigma_{ij}(t, \mathbf{x}_t)] \in \mathbb{R}^{n\times m}$ and $\mathrm{d}w_t = (\mathrm{d}w_{1t}, \mathrm{d}w_{2t}, \ldots, \mathrm{d}w_{m,t})^{\mathrm{T}}$ is an m-dimensional Brownian increment process, and $\mathbf{x}_0$ is a random vector with $g(\mathbf{x})$ as its probability density function. Let $P(t, \mathbf{x}_t)$ be the probability density function of $\mathbf{x}_t$. Then its dynamics of evolution is given by

$$\frac{\partial P[t, x]}{\partial t} = -\sum_{i=1}^{n} \frac{\partial}{\partial x_i}[f_i(t, \mathbf{x}) P(t, \mathbf{x})] + \frac{1}{2} \sum_{i,j=1}^{n} \frac{\partial^2}{\partial x_i \partial x_j} \{[\sigma(t, \mathbf{x})\sigma^{\mathrm{T}}(t, \mathbf{x})]_{ij} P(t, \mathbf{x})\} \quad (31.2.18)$$

with

$$P(0, \mathbf{x}(0)) = g(\mathbf{x}). \quad (31.2.19)$$

Numerical methods are often the only avenue for solving this equation.

In the special case when $\sigma(t, \mathbf{x}) \equiv 0$, the model equation (31.2.17) reduces to (31.2.13). In this case, the Kolmogorov forward equation (31.2.18) reduces to Liouville's equation (31.2.14).

31.3 Approximate moment dynamics

In the light of the difficulty involved in solving Liouville's and Kolmogorov's forward equations or their discrete counterparts, attention shifts to finding the (approximate) dynamics of evolution of the moments of the state of the system. In this section we derive the dynamics of the first two moments. We consider two cases.

31.3.1 Scalar case

Let $M : \mathbb{R} \to \mathbb{R}$ and x_k denote the state of a system that evolves according to a deterministic dynamic model

$$x_{k+1} = M(x_k). \tag{31.3.1}$$

It is assumed that the initial condition x_0 is random and hence x_k given by (31.3.1) is also random. Let

$$\begin{aligned} \mu_k &= E(x_k) & P_k &= E[x_k - \mu_k]^2 \\ \theta_k &= E[x_k - \mu_k]^3 & \Gamma_k &= E[x_k - \mu_k]^4 \end{aligned} \tag{31.3.2}$$

denote the mean, variance, the third, and the fourth central moments of x_k. Given μ_0 and P_0, our goal is to derive the dynamics of evolution of μ_k and P_k. From (31.3.1), it follows

$$\mu_{k+1} = E(x_{k+1}) = E[M(x_k)] = \widehat{M}(x_k). \tag{31.3.3}$$

Let

$$e_{k+1} = x_{k+1} - \mu_{k+1} = M(x_k) - \widehat{M}(x_k). \tag{31.3.4}$$

Since $E[e_{k+1}] = 0$, it follows that the variance P_{k+1} of x_{k+1} is given by

$$\begin{aligned} P_{k+1} &= E[e_{k+1}^2] \\ &= E[M^2(x_k)] - [\widehat{M}(x_k)]^2. \end{aligned} \tag{31.3.5}$$

Since the probability density $P_k(x_k)$ is **not** known, we cannot explicitly compute $\widehat{M}(x_k)$ and hence μ_{k+1} and P_{k+1}. This difficulty is circumvented by approximating $\widehat{M}(x_k)$ around μ_k using a second-order Taylor series expansion. Let

$$M(x_k) \approx M(\mu_k) + M_1 e_k + M_2\,(e_k)^2 \tag{31.3.6}$$

where $e_k = x_k - \mu_k$ and

$$M_k = \frac{1}{k!} \frac{\mathrm{d}^k M(x)}{\mathrm{d}x^k}\bigg|_{x=\mu_k}. \tag{31.3.7}$$

Taking expectations on both sides of (31.3.6) we obtain

$$\begin{aligned} \widehat{M}(x_k) &= M(\mu_k) + M_1\, E(e_k) + M_2\, E(e_k^2) \\ &= M(\mu_k) + M_2\, P_k \end{aligned} \tag{31.3.8}$$

from which the (approximate) dynamics of the mean is given by

$$\mu_{k+1} = M(\mu_k) + M_2 P_k\,. \tag{31.3.9}$$

Notice that the dynamics of the first moments depends on the variance which is the second central moment.

From

$$\begin{aligned} e_{k+1} &= M(x_k) - \widehat{M}(x_k) \\ &= M_1 e_k + M_2\, [e_k^2 - P_k] \end{aligned} \tag{31.3.10}$$

we obtain

$$\begin{aligned} P_{k+1} &= E[e_{k+1}^2] \\ &= M_1^2\, E[e_{k+1}^2] + 2M_1 M_2 E[e_k(e_k^2 - P_k)] \\ &\quad + M_2^2 E[e_k^2 - P_k]^2 \qquad (31.3.11) \\ &= M_1^2 P_k + 2M_1 M_2 \theta_k + M_2^2\, [\Gamma_k - P_k^2]. \qquad (31.3.12) \end{aligned}$$

Again notice that the second central moment depends on the third and the fourth central moments. This dependence of the kth moment on moments larger than k is called the moment closure problem. A practical way to deal with this difficulty is to further approximate by dropping all the moments beyond a prespecified moment. Keeping only the first two moments, we drop the terms containing θ_k and Γ_k from the r.h.s.of (31.3.12). By Jensen's inequality[†] since $P_k^2 < \Gamma_k$, for consistency we also drop the entire term $[\Gamma_k - P_k^2]$ leading to the following approximation:

$$P_{k+1} = M_1^2\, P_k. \tag{31.3.13}$$

Several observations are in order

(1) **Linear Case** When the dynamics is linear, that is, $M(x) = ax$, it follows that $\widehat{M}(x) = M(\widehat{x})$, $M_1 = a$ and $M_k \equiv 0$ for all $k \geq 2$. Hence the exact dynamics of the

[†] **Jensen's inequality** Let ϕ be a convex function. Then Jensen's inequality states that $\phi[E(y)] \leq E[\phi(y)]$. Let x be a random variable with $E(x) = 0$, $E(x^2) = P$, and $E(x^4) = \Gamma$. Let $y = x^2$. Then $E(y) = E(x^2) = P$. Consider $\phi(y) = y^2$. Then by Jensen's inequality $P^2 = \phi(P) = \phi(E(y)) \leq E[\phi(y)] = E(y^2) = E(x^4) = \Gamma$. When $x \sim N(0, \sigma^2)$, then $E(x^4) = \Gamma = 3\sigma^4$ and $P^2 = \sigma^4$.

mean and the variance are given by

$$\left.\begin{aligned} \mu_{k+1} &= a\mu_k \\ P_{k+1} &= a^2 P_k \end{aligned}\right\} \tag{31.3.14}$$

(2) **Model Error** If the nonlinear model in (31.3.1) has an error term modeled by white noise, then we get

$$x_{k+1} = M(x_k) + w_{k+1} \tag{31.3.15}$$

where $E(w_k) \equiv 0$ and $\text{Var}(w_k) = q_k$. Then by repeating the above derivation, it can be verified that the dynamics of the mean and variance are given by (Also refer to Sections 29.3 and 29.4)

$$\left.\begin{aligned} \mu_{k+1} &= M(\mu_k) + M_2 P_k \\ P_{k+1} &= M_1^2 P_k + Q_{k+1} \end{aligned}\right\} \tag{31.3.16}$$

The term $M_2 P_k$ is called the bias correction term. Refer to Section 29.4 for more details.

(3) **Quality of the approximation** When the model is nonlinear, there are two ways in which errors enter the approximation. First is from the Taylor series approximation which depends on $| e_k |$ and the second from the moment closure problem. For a given dynamics, we can in fact evaluate the usefulness of this approximation by using Monte Carlo simulation, a topic which is pursued in Section 31.4.

Example 31.3.1 Let

$$x_{k+1} = M(x_k) = ax_k^2 + bx_k + c \tag{31.3.17}$$

and let μ_0 and P_0 be the mean and the variance of the initial condition x_0. Then (31.3.9) gives

$$\mu_{k+1} = a\mu_k^2 + b\mu_k + c + 2aP_k \tag{31.3.18}$$

and from (31.3.13) we get

$$P_{k+1} = [2a\mu_k + b]^2 P_k\,. \tag{31.3.19}$$

In the special case when $a = 0$ and $c = 0$ we obtain

$$\mu_{k+1} = b\,\mu_k \quad \text{and} \quad P_{k+1} = b^2 P_k\,. \tag{31.3.20}$$

31.3.2 Vector case

For completeness, in the following we provide a short derivation of the dynamics of the mean and the variance for the vector case. Let $\mathbf{M} : \mathbb{R}^n \to \mathbb{R}^n$ and the model dynamics be given by

$$\mathbf{x}_{k+1} = \mathbf{M}(\mathbf{x}_k) \tag{31.3.21}$$

where the initial condition $\mathbf{x}_0$ is random. Let

$$\left.\begin{aligned} \boldsymbol{\mu}_{k+1} &= \mathbf{E}(\mathbf{x}_k) \\ \mathbf{P}_k &= \mathbf{E}\left[(\mathbf{x}_k - \boldsymbol{\mu}_k)(\mathbf{x}_k - \boldsymbol{\mu}_k)^{\mathrm{T}}\right] \end{aligned}\right\} \tag{31.3.22}$$

Taking expectations on both sides of (31.3.21), we get

$$\boldsymbol{\mu}_{k+1} = \mathbf{E}\,[\mathbf{M}(\mathbf{x}_k)] = \widehat{\mathbf{M}}(\mathbf{x}_k). \tag{31.3.23}$$

Expanding $\mathbf{M}(\mathbf{x}_k)$ in the second-order Taylor series around $\boldsymbol{\mu}_k$, we get

$$\mathbf{M}(\mathbf{x}_k) = \mathbf{M}(\boldsymbol{\mu}_k) + \mathbf{D}_{\mathbf{M}}\mathbf{e}_k + \frac{1}{2}\,\mathbf{D}^2_{\mathbf{M}}(\mathbf{e}_k, \mathbf{M}, \mathbf{e}_k) \tag{31.3.24}$$

where

$$\mathbf{D}^2_{\mathbf{M}}(\mathbf{e}_k, \mathbf{M}, \mathbf{e}_k) = \begin{bmatrix} \mathbf{e}_k^{\mathrm{T}}\nabla^2 M_1 \mathbf{e}_k \\ \mathbf{e}_k^{\mathrm{T}}\nabla^2 M_2 \mathbf{e}_k \\ \vdots \\ \mathbf{e}_k^{\mathrm{T}} t\nabla^2 M_n \mathbf{e}_k \end{bmatrix} \tag{31.3.25}$$

and the Jacobian $\mathbf{D}_{\mathbf{M}}$ and the Hessian $\nabla^2 M_i$ are evaluated at $\boldsymbol{\mu}_k$. Taking expectations on both sides of (31.3.24) and invoking the result from Example 29.4.1, we obtain

$$\begin{aligned} \widehat{\mathbf{M}}(\mathbf{x}_k) &= \mathbf{M}(\boldsymbol{\mu}_k) + \frac{1}{2}\,\mathbf{E}[\mathbf{D}^2_{\mathbf{M}}(\mathbf{e}_k, \mathbf{M}, \mathbf{e}_k)] \\ &= \mathbf{M}(\boldsymbol{\mu}_k) + \frac{1}{2}\,\partial^2(\mathbf{M}, \mathbf{P}_k) \end{aligned} \tag{31.3.26}$$

where

$$\partial^2(\mathbf{M}, \mathbf{P}_k) = \begin{bmatrix} \mathrm{tr}(\nabla^2 M_1\, \mathbf{P}_k) \\ \mathrm{tr}(\nabla^2 M_2\, \mathbf{P}_k) \\ \vdots \\ \mathrm{tr}(\nabla^2 M_n\, \mathbf{P}_k) \end{bmatrix}. \tag{31.3.27}$$

Hence, the dynamics of the mean is given by

$$\boldsymbol{\mu}_{k+1} = \mathbf{M}(\boldsymbol{\mu}_k) + \frac{1}{2}\partial^2(\mathbf{M}, \mathbf{P}_k)\,. \tag{31.3.28}$$

From

$$\begin{aligned} \mathbf{e}_{k+1} &= \mathbf{M}(\mathbf{x}_k) - \widehat{\mathbf{M}}(\mathbf{x}_k) \\ &= \mathbf{D}_{\mathbf{M}}\mathbf{e}_k + \boldsymbol{\eta}_k \end{aligned} \tag{31.3.29}$$

where

$$\boldsymbol{\eta}_k = \frac{1}{2}\,[\mathbf{D}^2_{\mathbf{M}}(\mathbf{e}_k, \mathbf{M}, \mathbf{e}_k) - \partial^2(\mathbf{M}, \mathbf{P}_k)] \tag{31.3.30}$$

we get

$$\begin{aligned}\mathbf{P}_{k+1} &= \mathbf{E}[\mathbf{e}_{k+1}\mathbf{e}_{k+1}^{\mathrm{T}}] \\ &= \mathbf{D_M}\,\mathbf{E}[\mathbf{e}_k\,\mathbf{e}_k^{\mathrm{T}}]\,\mathbf{D_M}^{\mathrm{T}} + \mathbf{D_M}\mathbf{E}[\mathbf{e}_k\boldsymbol{\eta}_k^{\mathrm{T}}] \\ &\quad + \mathbf{E}[\boldsymbol{\eta}_k\,\mathbf{e}_k^{\mathrm{T}}]\,\mathbf{D_M}^{\mathrm{T}} + \mathbf{E}[\boldsymbol{\eta}_k\,\boldsymbol{\eta}_k^{\mathrm{T}}].\end{aligned} \tag{31.3.31}$$

Since the components of $\boldsymbol{\eta}_k$ are quadratic in $\mathbf{e}_k$, dropping all the moments of $\mathbf{e}_k$ higher than the second moment, we obtain

$$\mathbf{P}_{k+1} = \mathbf{D_M}\mathbf{P}_k\mathbf{D_M}^{\mathrm{T}}. \tag{31.3.32}$$

When the model is linear, say

$$\mathbf{x}_{k+1} = \mathbf{M}_k\mathbf{x}_k \tag{31.3.33}$$

then $\mathbf{D_M} = \mathbf{M}_k$ and $\partial^2(\mathbf{M}, \mathbf{P}_k) \equiv 0$. Hence we get

$$\left.\begin{aligned}\boldsymbol{\mu}_{k+1} &= \mathbf{M}_k\boldsymbol{\mu}_k \\ \mathbf{P}_{k+1} &= \mathbf{M}_k\mathbf{P}_k\mathbf{M}_k^{\mathrm{T}}\end{aligned}\right\} \tag{31.3.34}$$

as the dynamics of the mean and the variance. Similarly, if there is model error modeled by a white noise sequence $\{\mathbf{w}_k\}$ with $\mathbf{E}[\mathbf{w}_k] = 0$ and $\mathbf{E}[\mathbf{w}_k\,\mathbf{w}_k^{\mathrm{T}}] = \mathbf{Q}_k$, then we get

$$\mathbf{x}_{k+1} = \mathbf{M}(\mathbf{x}_k) + \mathbf{w}_{k+1}.$$

In this case, the dynamics of the first two moments becomes (Refer to Section 29.4)

$$\left.\begin{aligned}\boldsymbol{\mu}_{k+1} &= \mathbf{M}(\boldsymbol{\mu}_k) + \tfrac{1}{2}\partial^2(\mathbf{M}, \mathbf{P}_k) \\ \mathbf{P}_{k+1} &= \mathbf{D_M}\mathbf{P}_k\mathbf{D_M}^{\mathrm{T}} + \mathbf{Q}_{k+1}\end{aligned}\right\} \tag{31.3.35}$$

We conclude this discussion with the following:

Example 31.3.2 Consider the following model used by Lorenz (1960), Thompson (1957) and Lakshmivarahan et al. (2003).

$$\left.\begin{aligned}\frac{d\mathbf{x}_1}{dt} &= \alpha_1\mathbf{x}_2\mathbf{x}_3 \\ \frac{d\mathbf{x}_2}{dt} &= \alpha_2\mathbf{x}_1\mathbf{x}_3 \\ \frac{d\mathbf{x}_3}{dt} &= \alpha_3\mathbf{x}_1\mathbf{x}_2\end{aligned}\right\} \tag{31.3.36}$$

Discretizing these using the standard forward Euler scheme, we obtain

$$\mathbf{x}_{k+1} = \mathbf{M}(\mathbf{x}_k)$$

where $\mathbf{x}_k = (x_{1k}, x_{2k}, x_{3k})^{\mathrm{T}}$, $\mathbf{M}(\mathbf{x}) = (M_1(\mathbf{x}), M_2(\mathbf{x}), M_3(\mathbf{x}))^{\mathrm{T}}$ where

$$\left.\begin{aligned} M_1(\mathbf{x}) &= x_{1k} + (\alpha_1 \Delta t) x_{2k} x_{3k} \\ M_2(\mathbf{x}) &= x_{2k} + (\alpha_2 \Delta t) x_{1k} x_{3k} \\ M_3(\mathbf{x}) &= x_{3k} + (\alpha_3 \Delta t) x_{1k} x_{2k} \end{aligned}\right\} \quad (31.3.37)$$

It can be verified that the Jacobian of $\mathbf{M}$ is given by

$$\mathbf{D_M}(\mathbf{x}) = \begin{bmatrix} 1 & (\alpha_1 \Delta t)\,\mathbf{x}_3 & (\alpha_1 \Delta t)\mathbf{x}_2 \\ (\alpha_2 \Delta t)\mathbf{x}_3 & 1 & (\alpha_2 \Delta t)\mathbf{x}_1 \\ (\alpha_3 \Delta t)\mathbf{x}_2 & (\alpha_3 \Delta t)\mathbf{x}_1 & 1 \end{bmatrix}. \quad (31.3.38)$$

Similarly, it can be verified that the Hessians are given by

$$\nabla^2 M_1(\mathbf{x}) = (\alpha_1 \Delta t) \begin{bmatrix} 0 & 0 & 0 \\ 0 & 0 & 1 \\ 0 & 1 & 0 \end{bmatrix}.$$

$$\nabla^2 M_2(\mathbf{x}) = (\alpha_2 \Delta t) \begin{bmatrix} 0 & 0 & 1 \\ 0 & 0 & 0 \\ 1 & 0 & 0 \end{bmatrix}$$

and

$$\nabla^2 M_3(\mathbf{x}) = (\alpha_3 \Delta t) \begin{bmatrix} 0 & 1 & 0 \\ 1 & 0 & 0 \\ 0 & 0 & 0 \end{bmatrix}.$$

If

$$\mathbf{P}_k = \begin{bmatrix} p_{11}(k) & p_{12}(k) & p_{13}(k) \\ p_{12}(k) & p_{22}(k) & p_{23}(k) \\ p_{13}(k) & p_{23}(k) & p_{33}(k) \end{bmatrix}$$

then the components of the vector $\partial^2(\mathbf{MP}_k) = (\mathrm{tr}[\nabla^2 M_1\, \mathbf{P}_k], \mathrm{tr}[\nabla^2 M_2\, \mathbf{P}_k], \mathrm{tr}[\nabla^2 M_3\, \mathbf{P}_k])^{\mathrm{T}}$ are given by

$$\left.\begin{aligned} \mathrm{tr}[\nabla^2 M_1\, \mathbf{P}_k] &= 2(\alpha_1 \Delta t) p_{23}(k) \\ \mathrm{tr}[\nabla^2 M_2\, \mathbf{P}_k] &= 2(\alpha_2 \Delta t) p_{13}(k) \\ \mathrm{tr}[\nabla^2 M_3\, \mathbf{P}_k] &= 2(\alpha_3 \Delta t) p_{12}(k) \end{aligned}\right\} \quad (31.3.39)$$

Hence, the dynamics of the mean is

$$\boldsymbol{\mu}_{k+1} = \mathbf{M}(\boldsymbol{\mu}_k) + \frac{1}{2}\partial^2(\mathbf{M}, \mathbf{P}_k) \quad (31.3.40)$$

which in component form can be written as

$$\left.\begin{aligned}\mu_{1,k+1} &= M_1(\boldsymbol{\mu}_k) + (\alpha_1 \Delta t) p_{23}(k)\\ \mu_{2,k+1} &= M_2(\boldsymbol{\mu}_k) + (\alpha_2 \Delta t) p_{13}(k)\\ \mu_{3,k+1} &= M_3(\boldsymbol{\mu}_k) + (\alpha_3 \Delta t) p_{12}(k)\end{aligned}\right\} \tag{31.3.41}$$

The dynamics of the variance is given by

$$\mathbf{P}_{k+1} = \mathbf{D}_{\mathbf{M}}(\boldsymbol{\mu}_k)\mathbf{P}_k\mathbf{D}_{\mathbf{M}}^{\mathrm{T}}(\boldsymbol{\mu}_k). \tag{31.3.42}$$

By choosing the parameters $\alpha_1 = -0.553$, $\alpha_2 = 0.451$, $\alpha_3 = 0.051$, and $\Delta t = 0.1$ and starting from the initial condition $\mathbf{x}_0 = (1.0, 0.1, 0.0)^{\mathrm{T}}$, a plot of the theoretical mean $\boldsymbol{\mu}_k$ is given in Figure 31.4.1. Since $\mathbf{P}_k$'s are 3×3 matrices, a plot of the trace, determinant and the Frobenius norm of the theoretical covariance matrix $\mathbf{P}_k$ are given in Figure 31.4.1.

Remark 31.3.1 Using the continuous time version of the model in (31.3.36) Thompson (1957) proved that the moment dynamics for this model has a natural closure property namely that the first two moment dynamics do not involve dependence on kth moment for $k > 2$. Thompson derived this closure property by exploiting the conservation property.

31.4 The Monte Carlo method

The third alternative is to compute the evolution of the moments as their sample counterparts using an ensemble of model states generated by invoking the standard Monte Carlo method. This method rests on the knowledge about the distribution of the initial state.

Let $\mathbf{P}_0(\mathbf{x}_0)$ be the given probability density function of the initial state $\mathbf{x}_0$. Then the Monte Carlo method consists in performing the following steps.

(1) **Generate the initial ensemble** Let $\mathbf{x}_0(j)$, $j = 1, 2, \ldots, N$ be the set of N random vectors generated as the sample realization from $\mathbf{P}_0(\mathbf{x}_0)$. This set defines the initial ensemble.

(2) **Compute the N strands of the model trajectory** Starting from each $\mathbf{x}_0(j)$, compute the trajectory $\mathbf{x}_k(j)$ for $k = 1, 2, 3, \ldots$ using the model:

$$\mathbf{x}_{k+1}(j) = \mathbf{M}(\mathbf{x}_k(j)) \tag{31.4.1}$$

for $j = 1, 2, \ldots, N$.

(3) **Compute the sample moments** Let

$$\bar{\mathbf{x}}_k = \frac{1}{N}\sum_{j=1}^{N} \mathbf{x}_k(j). \tag{31.4.2}$$

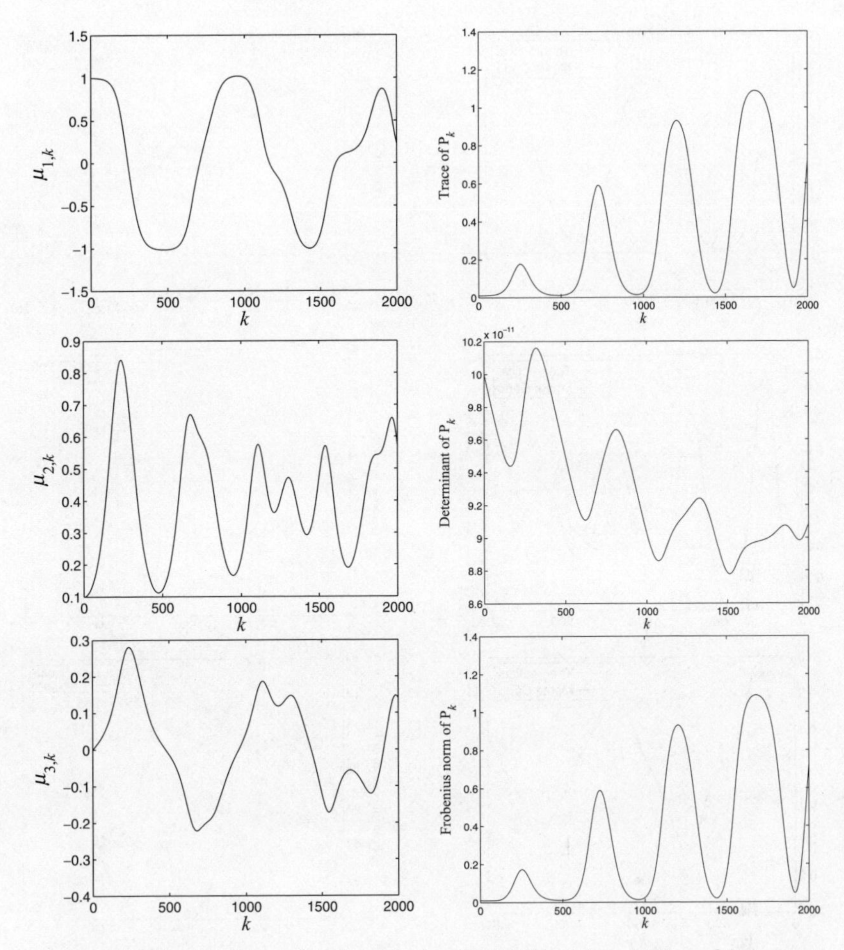

Fig. 31.4.1 A plot of $\boldsymbol{\mu}_k$ vs k computed using (31.3.40) and the evolution of the trace, determinant and the Frobenius norm of $\mathbf{P}_k$ computed using (31.3.42).

Then $\{\bar{\mathbf{x}}_k\}_{k\geq 0}$ defines the evolution of the sample mean state. Define

$$\mathbf{e}_k(j) = \mathbf{x}_k(j) - \bar{\mathbf{x}}_k \tag{31.4.3}$$

Compute the sample covariance matrix

$$\bar{\mathbf{P}}_{k+1} = \frac{1}{N-1}\sum_{j=1}^{N} \mathbf{e}_k(j)[\mathbf{e}_k(j)]^{\mathrm{T}} \tag{31.4.4}$$

where $\{\bar{\mathbf{P}}_k\}_{k\geq 0}$ defines the evolution of the sample covariance.

A number of remarks are in order here.

(1) **Accuracy** While the accuracy of the moment computation using the approximate moment dynamics in Section 31.3 was dictated by the order of the Taylor series expansion and the moment closure problem, in this Monte Carlo method,

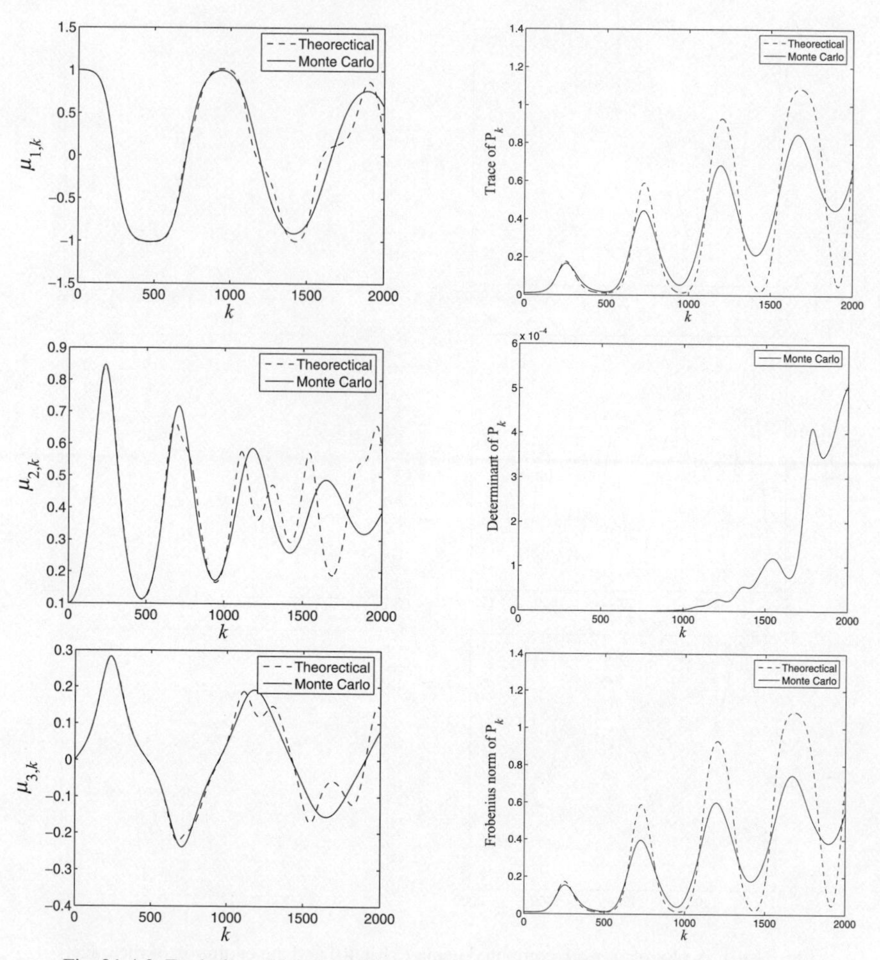

Fig. 31.4.2 Evolution of the ensemble mean, the trace, determinant and Frobenius norm of the ensemble covariance matrix.

the accuracy is only a function of the number of samples in the ensemble. As N increases, the sample mean and sample variance converge to the true mean and the variance. If $N < n$, the dimensional of the model space, the rank of $\bar{\mathbf{P}}_k$ is bounded by N. Thus, the Monte Carlo method gives a reduced-rank approximation to the actual variance.

2. **Testbed for comparison** In view of the above property, the Monte Carlo method is often used as a testbed for comparing the results obtained from using approximate moment dynamics.

We conclude this discussion with the following.

Example 31.4.1 In this example, we compare the results of Example 31.3.2 with the results obtained by using the Monte Carlo method on the model in (31.3.37). It is assumed that $\mathbf{x}_0 \sim N(\boldsymbol{\mu}_0, \mathbf{P}_0)$ where $\boldsymbol{\mu}_0$ and $\mathbf{P}_0$ are given below.

$$\boldsymbol{\mu}_0 = (1.0, 0.1, 0.0)^{\mathrm{T}} \quad \text{and} \quad \mathbf{P}_0 = \begin{bmatrix} .1^2 & 0 & 0 \\ 0 & .01^2 & 0 \\ 0 & 0 & .01^2 \end{bmatrix}.$$

We used the same values for the parameters α and Δt as in Example 31.3.2.

A plot of the evolution of the ensemble mean using a set of 1000 ensemble members is given in Figure 31.4.2. For purposes of comparison we have superimposed the theoretical mean in this same figure. Similarly, Figure 31.4.2 gives the evolution of the trace, determinant, and Frobenius norm of the covariance matrix $\mathbf{P}_k$ computed using the same set of 1000 ensemble members.

Exercises

31.1 **Epstein** (1969a) Let $T_k = T_0 - A\,k^{1/2}$ be dynamics of variation of the temperature T_k at time k where T_0 is the initial temperature and A is the parameter that decides the cooling rate. Let T_0 and A be random with m_0 and σ_0^2 being the mean and variance of T_0 and m_A and σ_A^2 those for A. Let ρ be the correlation coefficient between T_0 and A.
(i) Derive an expression for the mean and variance of T_k as a function of k.
(ii) Compute the rate of change of the variance of T_k w.r.t. to k.

31.2 **Epstein** (1969a) Let $q_k = Q + B\cos(\omega k - \beta)$ where Q is time invariance component, B is the amplitude of the time- varying part, ω is the angular frequency, and β is the phase. Assuming Q, B, ω, and β are independent random variables with m_Q, m_B, m_ω, and m_β as the mean and $\sigma_Q^2, \sigma_B^2, \sigma_\omega^2$, and σ_β^2 as their variance respectively. Compute the mean and variance of q_k.

Notes and references

Section 31.1 For an early account of the discussion on predictability refer to Thompson (1957) and Novikov (1959). For a review of the predictability problem refer to Thompson (1985b), and Houghton (1991). Epstein's (1969a,b) work was motivated by the earlier contribution by Gleeson (1967) and Freiberger and Grenander (1965). For a more recent review of predictability refer to Ehrendorfer (1997)(2002). Refer to Chu (1999) for an interesting view on the sources of predictability problem with respect to the famous Lorenz's model. Lorenz's 1993 book on *The Essence of Chaos* contains a readable account of the predictability question. Also refer to Leith (1971), (1974) and Leith and Kraichnan (1972). A stimulating essay on the reduction of variance with time in complicated systems (such as biological, social, and athletic systems) is found in "Losing the Edge" by paleontologist Stephen Jay Gould (Gould (1985)).

Section 31.2 Satty (1967) and Snoog (1973) contain derivations of Liouville's equation. For derivation of Kolmogorov's forward or the Fokker–Planck equation refer to Jazwinski (1970). Ehrendorfer (1994a) and (1994b) provide an excellent review of the problems and challenges in solving Liouville's equation. Various solution methods for solving the Fokker–Planck equation are discussed in Fuller(1969), Risken(1984) and Grasman(1999).

Section 31.3 In the context of nonlinear filtering, derivation of approximate moments dynamics was vigorously pursued in the early part of 1960. Refer to Jazwinski (1970) for details. For a discussion of the derivation and the use of approximate moment dynamics within the context of meteorology refer to Freiberger and Grenander (1965), Gleeson (1967), Epstein (1969a) and (1969b), Fleming (1971), Pitcher (1977), to name a few. Thompson (1985) presents an interesting example of an exact moment dynamics which is closed in the second-order moments.

Section 31.4 Monte Carlo methods for assessing predictability in meteorology was put forth by Leith (1971) and (1974). For a general discussion of the Monte Carlo methods refer to Hamersley and Handscomb (1964). Also refer to Metropolis and Ulam (1949).

32

Predictability: a deterministic view

In this chapter we describe a deterministic approach to stability and predictability of dynamic systems. While the classical stability theory deals with characterizing the growth and behavior of perturbations around an equilibrium state of a system, the goal of the predictability theory is to quantify the growth and behavior of infinitesimally small perturbations superimposed on an evolving trajectory – be it stable, unstable or chaotic – of the given dynamical system. Any two states that are infinitesimally close to each other are called **analogs**. Thus, predictability theory seeks to characterize the future evolution of analogous states. Since every trajectory starts from an initial state, predictability analysis is often recast as one of analyzing the sensitive dependence on initial state. Despite this difference in goals, both stability and predictability theories depend heavily on the same set of mathematical ideas and tools drawn from the spectral (eigenvalue) theory of finite dimensional (matrix) operators.

The goals and problems related to deterministic predictability theory are reviewed in the opening Section 32.1. Section 32.2 through 32.5 provide a succinct review of stability theory of dynamical systems. Predictability analysis using singular vectors is developed in Section 32.6. A summary of a fundamental theorem of Osledec leading to the definition of Lyapunov vectors and Lyapunov indices is given in Section 32.7. This section also contains two related algorithms for computing Lyapunov indices. The concluding Section 32.8 describes two methods for generating deterministic ensembles using which one can assess evolution of the spread among the trajectories measured among other things through the sample covariance.

32.1 Deterministic predictability: statement of problems

The quality of prediction of a geophysical phenomenon using a deterministic model depends on various factors: **model errors**, **errors in the initial/boundary conditions** and the **stability** of the given model dynamics. Since the choice of the model depends on the phenomenon being analyzed, any discussion of the impact of model errors can only be made within the context of a specific problem domain. Since our goal is to provide a general/generic discussion of the deterministic predictability, in the following we tacitly assume that the chosen deterministic model is perfect. This assumption enables us to concentrate on analyzing the effect of the other two factors.

Notwithstanding the methods used in estimating the unknown, it stands to reason to expect that any estimate based on finite and noisy sample of observations will always have an error component. These errors can be thought of as perturbations on the optimal state. Thus, if $\widehat{\mathbf{x}}_0$ is the initial estimate arising out of a data assimilation method, then

$$\widehat{\mathbf{x}}_0 = \mathbf{x}_0^* + \varepsilon_0 \tag{32.1.1}$$

where $\mathbf{x}_0^*$ is the unknown optimal state and ε_0 is the perturbation. The usefulness of the prediction obtained from this estimated initial condition $\widehat{\mathbf{x}}_0$, depends critically on the way in which the given deterministic model dynamics treats the perturbation ε_0; that is, whether the model amplifies/attenuates this initial error. This amplification/attenuation property of the model is directly related to the stability of the model in question. It will become evident from the discussion in the following Sections 32.2 through 32.5 that, if the model is stable, then the errors may grow but will remain bounded and if the model is asymptotically stable, then the initial perturbations will eventually die out. If the model is unstable, the initial perturbation will eventually grow without bound. On the other hand, if the model is chaotic, then while infinitesimally small errors grow at an exponential rate, the maximum value of the error is limited by the diameter of the invariant set or the attractor.

Against this backdrop, we now state two types of questions of interest in deterministic predictability. First is the **analysis** problem of quantifying the predictability limit which is directly related to the rate of amplification of initial errors. Refer to Figure 32.1.1. Recall that if the model is perfect and asymptotically stable – witness the dynamics of our Solar System – then there is virtually no limit to predictability. Thus, predictability limit is intimately associated with unstable models. Using the analogy of time constant, one can define the predictability limit as the time required for the (infinitesimally small) initial error to grow to e(= 2.7182) times its original value.

The second question related to predicting the high-impact, low-probability events. This converse problem calls for **synthesizing** an ensemble of initial perturbations using which we can gain an understanding of the possible modes of behavior of the model. Let $\mathbf{x}_T$ be the predicted state at time T, and let $\mathbf{h}(\mathbf{x}_T)$ be the model counterpart of the predicted observation (such as rainfall, flash flood etc.) If $\mathbf{z}_T$ is the actual observation at time T, then

$$\mathbf{e}_T = \| \mathbf{z}_T - \mathbf{h}(\mathbf{x}_T) \| \tag{32.1.2}$$

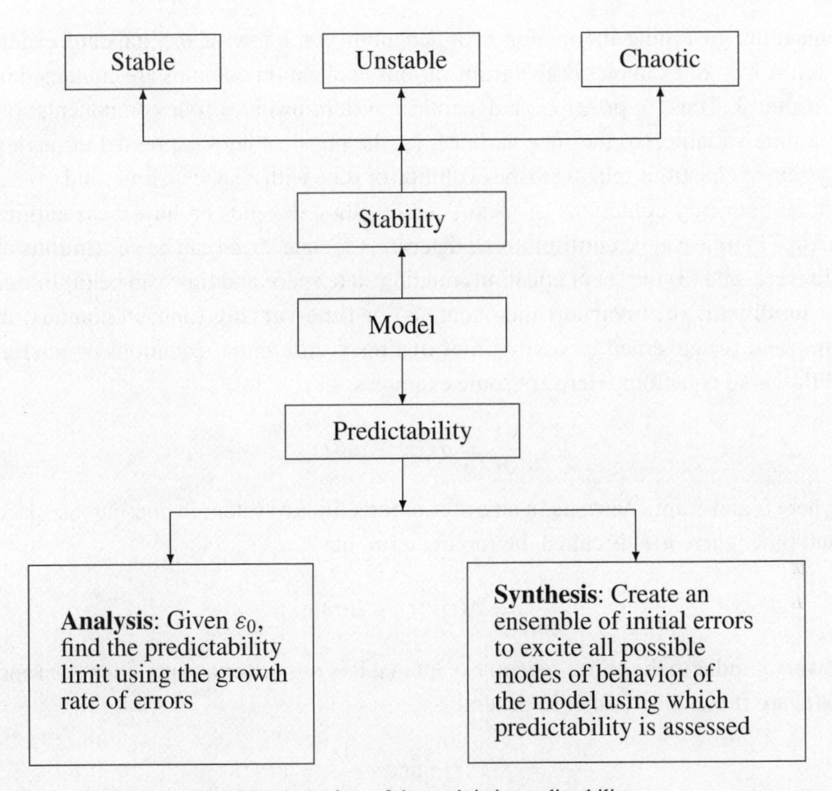

Fig. 32.1.1 A view of deterministic predictability.

is a measure of the error in prediction. Low-probability, high-impact events are characterized by very high value for $\mathbf{e}_T$. Let $\mathbf{x}_T^*$ be the state such that $\mathbf{z}_T = \mathbf{h}(\mathbf{x}_T^*)$. Since the model is perfect and unstable, large errors in prediction are essentially due to inappropriate initial conditions which in turn implies that the initial perturbation ε_0 is inadequate to generate a state that is closer to the derived state $\mathbf{x}_T^*$. So, the problem reduces to one of synthesizing an initial set or ensemble of initial conditions that will force the model to exhibit all possible modes of behavior at time T. Once an ensemble of predicted states at time T is available, we then can generate a wide variety of products which will help explain the observations better than the single forecast obtained from the single initial state $\widehat{\mathbf{x}}_0$.

32.2 Examples and classification of dynamical systems

Informally, any differential or difference equation relating to the evolution of a physical quantity (often representing the **state** of the system) in time represents a dynamical system. The differential equation governing the motion of a planet in elliptic orbits around the sun, the equations governing radioactive decay, and the equations governing the motion of a pendulum are a few of the standard examples. A host of examples drawn from various application domains are contained in Chapter 3. Thus, in principle, a dynamical system involves four components: (a) the time variable, (b) the state variable, (c) the physical laws expressed through a system of equations relating to the evolution of state with respect to time, and (d) the initial/boundary conditions. A useful classification depends on how these entities arise: (1) time can be **continuous** or **discrete**, (2) state space can be **continuous** or **discrete**, and (3) the set of equations relating state space and time can be (a) **linear** or **nonlinear**, (b) **invariant** (autonomous) or **time-varying** (nonautonomous) in time and (c) governed by a system of **ordinary** differential equations or **partial** differential equations. Here are some examples.

$$\frac{\mathrm{d}x}{\mathrm{d}t} = ax(t) + bu(t)$$

where a and b are constants in an **autonomous linear** system in continuous space and time where $u(t)$ is called the forcing term, but

$$\frac{\mathrm{d}x}{\mathrm{d}t} = A(t)x(t) + B(t)u(t)$$

is very similar to the above system except that it is **nonautonomous** since $A(t)$ and $B(t)$ are functions of time. In general

$$\frac{\mathrm{d}x}{\mathrm{d}t} = f(x, t) \quad \text{and} \quad \frac{\mathrm{d}x}{\mathrm{d}t} = g(x)$$

for general functions f and g are examples of **nonlinear** systems with the first one being nonautonomous but the second one is autonomous.

Most of the examples in Chapter 3 are governed by partial differential equations and we invite the reader to classify each of those examples. The following recurrence

$$x_{n+1} = ax_n(1 - x_n)$$

is an example of a discrete time, continuous state space, nonlinear, autonomous difference equation. Often, discretization of a system with continuous state and time leads to analogous systems with discrete space and time.

Example 32.2.1 Consider the standard first-order, constant coefficient (hence autonomous) ordinary differential equation

$$\dot{x} = \frac{\mathrm{d}x}{\mathrm{d}t} = \alpha x \tag{32.2.1}$$

where $x(0) = c$, given.

It is well known that

$$x(t) = x(0)\mathrm{e}^{\alpha t} = c\mathrm{e}^{\alpha t}. \tag{32.2.2}$$

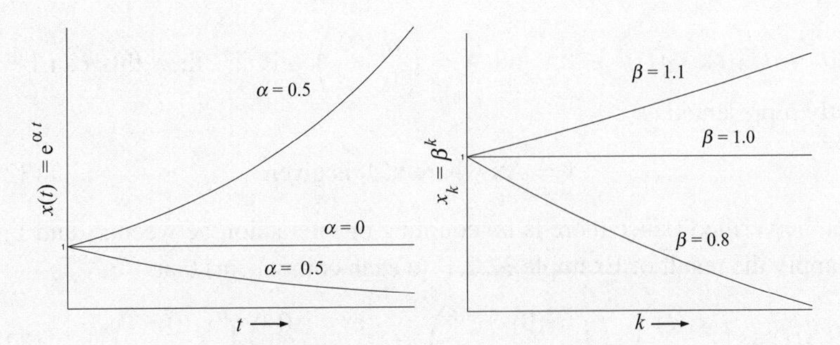

Fig. 32.2.1 Three possible modes of evolution.

The possible modes of behavior of $x(t)$ critically depend on α and are depicted in Figure 32.2.1. Notice that as α changes from being positive to negative, the qualitative behavior of $x(t)$ changes from **diverging** to **converging**. When α is away from zero, say $\alpha = 0.7$ or -0.5, even small changes in the values of α do **not** change the overall behavior of $x(t)$. In other words, qualitative behavior of $x(t)$ is relatively **"stable"** with regard to small perturbations in α, when it is bounded away from zero. However, when $\alpha = 0$, the story is quite different. Even small perturbations depending on their sign could lead to drastic changes in the behavior of $x(t)$. Such points in the parameter space of a dynamical system are known as the **bifurcation points**.

From this discussion it must be clear that estimation of parameters in a dynamical system, when it is operating at or near the bifurcation point, poses one of the greatest challenges in data assimilation.

We now consider the discrete time analog obtained through the standard Euler method using the forward approximation for the time derivative. If Δt is the time increment, denoting $x(n\Delta t) = x_n$, we obtain

$$x_{n+1} = (1 + \alpha\Delta t)x_n = \beta x_n \tag{32.2.3}$$

where $x_0 = c$, given. Then,

$$x_n = \beta^n x_0 \tag{32.2.4}$$

whose behavior is depicted in Figure 32.2.1. Since $\beta = (1 + \alpha\Delta t)$, $\beta = 1$ is the bifurcation point for this discrete time system.

Example 32.2.2 Consider a system of two uncoupled ordinary differential equations written in the matrix notation

$$\begin{pmatrix} \dot{x}_1 \\ \dot{x}_2 \end{pmatrix} = \begin{pmatrix} \alpha & 0 \\ 0 & \beta \end{pmatrix} \begin{pmatrix} x_1 \\ x_2 \end{pmatrix}. \tag{32.2.5}$$

If $\mathbf{x}(t) = (x_1(t), x_2(t))^{\mathrm{T}} \in \mathbb{R}^2$ and $\mathbf{A} = \begin{pmatrix} \alpha & 0 \\ 0 & \beta \end{pmatrix} \in \mathbb{R}^{2\times 2}$, then this can be succinctly represented as

$$\dot{\mathbf{x}} = \mathbf{A}\mathbf{x} \text{ where } \mathbf{x}(0) \text{ is given.} \tag{32.2.6}$$

It can be verified (since there is no coupling or interaction between x_1 and x_2, we can apply the result of Example 32.2.1 to each component) that

$$\mathbf{x}(t) = \begin{pmatrix} x_1(t) \\ x_2(t) \end{pmatrix} = \begin{pmatrix} x_1(0) & \mathrm{e}^{\alpha t} \\ x_2(0) & \mathrm{e}^{\beta t} \end{pmatrix} = \begin{pmatrix} \mathrm{e}^{\alpha t} & 0 \\ 0 & \mathrm{e}^{\beta t} \end{pmatrix} \begin{pmatrix} x_1(0) \\ x_2(0) \end{pmatrix} \tag{32.2.7}$$

is the solution. By expanding the exponentials in a series and collecting the like powers of t, it can be verified that

$$\begin{aligned} \begin{pmatrix} \mathrm{c}^{\alpha t} & 0 \\ 0 & \mathrm{e}^{\beta t} \end{pmatrix} &= \begin{pmatrix} 1 & 0 \\ 0 & 1 \end{pmatrix} + \begin{pmatrix} \alpha & 0 \\ 0 & \beta \end{pmatrix} t + \frac{1}{2!} \begin{pmatrix} \alpha^2 & 0 \\ 0 & \beta^2 \end{pmatrix} t^2 + \cdots \\ &= \mathbf{I} + \mathbf{A}t + \frac{1}{2!}\mathbf{A}^2 t^2 + \cdots \\ &= \mathrm{e}^{\mathbf{A}t} \end{aligned} \tag{32.2.8}$$

using the analogy of the exponential series for a scalar. Hence $\mathbf{x}(t) = \mathrm{e}^{\mathbf{A}t}\mathbf{x}(0)$ is the solution. The matrix $\mathrm{e}^{\mathbf{A}t}$ is often called the **state transition** matrix as it relates the state at time t to that at time 0.

A plot of this solution in the (x_1, x_2) plane as a function of time is called the **phase portrait**. In a sense, this phase portrait indicates how each point in $\mathbb{R}^2$ moves or flows along the solution as a function of time. The right-hand side in (32.2.5) is called the **field** and it denotes the direction of the tangent to the phase portrait. Figure 32.2.2 represents the phase portrait of the system

$$\begin{pmatrix} \dot{x}_1 \\ \dot{x}_2 \end{pmatrix} = \begin{pmatrix} -\frac{1}{2} & 0 \\ 0 & 2 \end{pmatrix} \begin{pmatrix} x_1 \\ x_2 \end{pmatrix}. \tag{32.2.9}$$

It can be seen that points on the x_1 axis move towards the origin. This is so because $x_1(t) = \mathrm{e}^{-\frac{1}{2}t}\, x_1(0)$. Likewise, since $x_2(t) = \mathrm{e}^{2t} x_2(0)$, the points along x_2–axis move to infinity. For points outside of x_1 and x_2 axis, the general flow pattern is such that x_1-component decreases to zero and the x_2-component goes to infinity. Thus the flow is toward x_2 axis and towards infinity along the x_2 axis.

Example 32.2.3 Consider a general (coupled) linear system

$$\dot{\mathbf{x}} = \mathbf{A}\mathbf{x} \tag{32.2.10}$$

where $\mathbf{x} \in \mathbb{R}^2$ and $\mathbf{A} \in \mathbb{R}^{2\times 2}$. By changing the variable in (32.2.10) using $\mathbf{x} = \mathbf{P}\mathbf{y}$, we obtain

$$\dot{\mathbf{y}} = (\mathbf{P}^{-1}\mathbf{A}\mathbf{P})\mathbf{y} = \mathbf{B}\mathbf{y}. \tag{32.2.11}$$

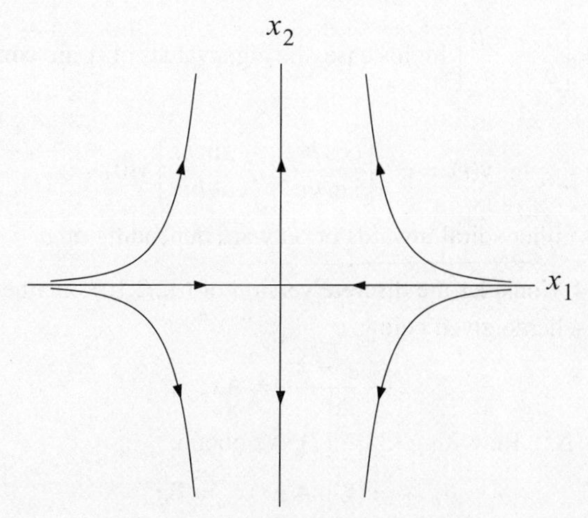

Fig. 32.2.2 Phase portrait of (32.2.9).

By choosing **P** to be the matrix of eigenvectors of **A**, it can be shown (Hirsch and Smale (1974)) that **B** takes one of the following three forms:

$$\begin{bmatrix} \alpha_1 & 0 \\ 0 & \alpha_2 \end{bmatrix}, \quad \begin{bmatrix} \alpha & 1 \\ 0 & \alpha \end{bmatrix}, \quad \text{or} \quad \begin{bmatrix} a & -b \\ b & a \end{bmatrix}. \tag{32.2.12}$$

It can be verified that the solution of (32.2.11) is given by

$$\mathbf{y}(t) = e^{\mathbf{B}t}\mathbf{y}(0). \tag{32.2.13}$$

We now specialize to each of the three forms of **B**.

Case 1: $\mathbf{B} = \begin{bmatrix} \alpha_1 & 0 \\ 0 & \alpha_2 \end{bmatrix}$ In this case, the two eigenvalues α_1 and α_2 of **A** are **real and distinct** and the solution is given by

$$\mathbf{y}(t) = \begin{bmatrix} e^{\alpha_1 t} & 0 \\ 0 & e^{\alpha_2 t} \end{bmatrix} \mathbf{y}(0). \tag{32.2.14}$$

Thus, the solution components $y_i(t)$ tend to zero or infinity depending on $\alpha_i < 0$ and $\alpha_i > 0$, for $i = 1, 2$.

Case 2: $\mathbf{B} = \begin{bmatrix} \alpha & 1 \\ 0 & \alpha \end{bmatrix}$ In this case, the eigenvalues of **A** are **real and equal** to α and

$$\mathbf{y}(t) = \begin{bmatrix} e^{\alpha t} & te^{\alpha t} \\ 0 & e^{\alpha t} \end{bmatrix} \mathbf{y}(0). \tag{32.2.15}$$

Here $y_i(t)$ tends to zero or infinity depending on $\alpha < 0$ or $\alpha > 0$.

Case 3: $\mathbf{B} = \begin{bmatrix} a & -b \\ b & a \end{bmatrix}$ In this case, the eigenvalues of **A** are **complex**, namely $a \pm ib$ and

$$\mathbf{y}(t) = e^{at} \begin{bmatrix} \cos bt & -\sin bt \\ \sin bt & \cos bt \end{bmatrix} \mathbf{y}(0) \tag{32.2.16}$$

The trajectories either spiral inwards or outward depending on $a < 0$ or $a > 0$.

Example 32.2.4 Consider the discrete version of (32.2.10) obtained by using the standard Euler scheme given below:

$$\frac{x_{n+1} - x_n}{\Delta t} = \mathbf{A} x_n \tag{32.2.17}$$

where $x_n = x(n\Delta t)$. Rewriting (32.2.17), we obtain

$$x_{n+1} = (\mathbf{I} + \mathbf{A}\Delta t) x_n = \mathbf{B} x_n \tag{32.2.18}$$

where $\mathbf{B} = (\mathbf{I} + \mathbf{A}\Delta t)$. It can be verified that if λ is an eigenvalue of **A**, then $(1 + \lambda \Delta t)$ is that of **B**. Now changing the variables using $\mathbf{x} = \mathbf{P}\mathbf{y}$, (32.2.18) becomes

$$y_{n+1} = (\mathbf{P}^{-1}\mathbf{B}\mathbf{P}) y_n = \mathbf{D} y_n$$

where $\mathbf{D} = \mathbf{P}^{-1}\mathbf{B}\mathbf{P}$ is one of the three forms:

$$\begin{bmatrix} \alpha & 0 \\ 0 & \beta \end{bmatrix}, \quad \begin{bmatrix} \alpha & 1 \\ 0 & \alpha \end{bmatrix}, \quad \text{or} \quad \begin{bmatrix} a & -b \\ b & a \end{bmatrix}.$$

It can be verified that $\mathbf{y}_n = (\mathbf{P}^{-1}\mathbf{B}^n\mathbf{P})\mathbf{y}_0 = \mathbf{D}^n \mathbf{y}_0$.

Case 1: $\mathbf{D} = \begin{bmatrix} \alpha & 0 \\ 0 & \beta \end{bmatrix}$ Then

$$\mathbf{y}_n = \begin{pmatrix} \alpha^n & 0 \\ 0 & \beta^n \end{pmatrix} \mathbf{y}_0. \tag{32.2.19}$$

Recall from Appendix B that the absolute value of the largest eigenvalue is called the **spectral radius** of **D** and is denoted by $\rho(\mathbf{D})$. Clearly $\mathbf{y}_n \to 0$ or ∞, depending on $\rho(\mathbf{D}) < 1$ or > 1.

Case 2: $\mathbf{D} = \begin{bmatrix} \alpha & 1 \\ 0 & \alpha \end{bmatrix}$ Then

$$\mathbf{y}_n = \begin{pmatrix} \alpha^n & n\alpha^{n-1} \\ 0 & \alpha^n \end{pmatrix} \mathbf{y}_0. \tag{32.2.20}$$

Again $\mathbf{y}_n \to 0$ or ∞ depending on $\rho(\mathbf{D}) < 1$ or > 1.

Case 3: $\mathbf{D} = \begin{bmatrix} a & -b \\ b & a \end{bmatrix}$

Let $r = \left(a^2 + b^2\right)^{1/2}$; $\cos\theta = a/r$ and $\sin\theta = b/r$. Then, from

$$\mathbf{D} = r \begin{bmatrix} \cos\theta & -\sin\theta \\ \sin\theta & \cos\theta \end{bmatrix} = \begin{bmatrix} r & 0 \\ 0 & r \end{bmatrix} \begin{bmatrix} \cos\theta & -\sin\theta \\ \sin\theta & \cos\theta \end{bmatrix}$$

it can be verified that the action of **D** on a vector **y** can be realized by a **rotation** of **y** by an angle θ in the anti-clockwise direction followed by a uniform **stretching** by a factor r. Thus

$$\mathbf{D}^n = r^n \begin{bmatrix} \cos n\theta & -\sin n\theta \\ \sin n\theta & \cos n\theta \end{bmatrix}.$$

From

$$\mathbf{y}_n = \begin{bmatrix} r^n & 0 \\ 0 & r^n \end{bmatrix} \begin{bmatrix} \cos n\theta & -\sin n\theta \\ \sin n\theta & \cos n\theta \end{bmatrix} \mathbf{y}_0 \tag{32.2.21}$$

it follows that $\mathbf{y}_n$ spirals inward to the origin if $r < 1$ and spirals outward to infinity if $r > 1$.

These examples clearly illustrate the basic fact that there is a one-to-one correspondence between the tools of analysis as well as the behavior of linear dynamics in continuous and in discrete time. Henceforth we shall switch between these formulations depending on convenience.

Against the backdrop of these examples, we now define a dynamical system more formally. Consider a system of linear, autonomous differential equations

$$\dot{\mathbf{x}} = \mathbf{A}\mathbf{x}, \;\; \text{with } \mathbf{x}(0) \text{ given} \tag{32.2.22}$$

where $\mathbf{x} \in \mathbb{R}^2$ and $\mathbf{A} \in \mathbb{R}^{2\times 2}$. Recall that the solution is given by $\mathbf{x}(t) = \mathrm{e}^{\mathbf{A}t}\mathbf{x}(0)$. Given any point $\mathbf{x} \in \mathbb{R}^2$, the vector $\mathbf{A}\mathbf{x}$ on the right-hand side of (32.2.22) defines the **vector field** at $\mathbf{x}$. By differentiating the solution $\mathbf{x}(t)$ with regard to t, it can be seen that

$$\frac{\mathrm{d}\mathbf{x}(t)}{\mathrm{d}t} = \mathbf{A}\mathrm{e}^{\mathbf{A}t}\mathbf{x}(0) = \mathbf{A}\mathbf{x}(t).$$

That is, the vector field defines the **tangent vector** to the solution curve. The linear operator represented by the matrix **A** mapping **x** to **Ax** thus **creates** the vector field in $\mathbb{R}^2$. For a given **A**, since $\mathbf{x}(t) = \mathrm{e}^{\mathbf{A}t}\mathbf{x}(0)$ is defined for all $t \in \mathbb{R}$ and $\mathbf{x}(0) \in \mathbb{R}^2$, we can think of the solution as a mapping $\phi : \mathbb{R} \times \mathbb{R}^2 \to \mathbb{R}^2$ where $\phi(t, \mathbf{y}) = \mathbf{x}(t)$ represents the (unique) state at time t starting from an initial state $\mathbf{y}$ at time zero. The term $\phi(t, \mathbf{y})$ is often written as $\phi_t(\mathbf{y})$ and $\mathbf{x}(t)$ as $\mathbf{x}_t$ for convenience. Now, for t fixed, $\phi_t : \mathbb{R}^2 \to \mathbb{R}^2$ is given by $\phi_t(\mathbf{y}) = \mathrm{e}^{\mathbf{A}t}\mathbf{y}$. The **infinite** collection $\{\phi_t\}_{t\in\mathbb{R}}$ of maps on $\mathbb{R}^2$ is called a **flow** or **dynamical system** corresponding to the differential equation (32.2.22).

In general, a dynamical system $\phi \in \mathbb{R}^n$ is a map $\phi : \mathbb{R} \times \mathbb{R}^n \to \mathbb{R}^n$ where $\phi(t, \mathbf{x}) = \phi_t(\mathbf{x})$ has continuous first derivatives in both t and $\mathbf{x}$ such that

(C1) $\phi_0 : \mathbb{R}^n \to \mathbb{R}^n$ is an **identity** map and

(C2) $\phi_{t+s}(\mathbf{x}) = \phi_t \cdot \phi_s(\mathbf{x}) = \phi_t\left(\phi_s(\mathbf{x})\right)$.

That is, (C2) denotes the **composition** rule for all t and s. The dynamical system is **linear** or **nonlinear** if the map $\phi_t : \mathbb{R}^n \to \mathbb{R}^n$ is linear or non-linear. The differential equation (32.2.10) defines a dynamical system in $\mathbb{R}^2$ since $\phi_t(\mathbf{y}) = e^{\mathbf{A}t}\mathbf{y}$ satisfies the conditions C1-C2 and is also differentiable in t and $\mathbf{y}$. Conversely, given a dynamical system $\{\phi_t\}_{t\in\mathbb{R}}$,

$$\mathbf{f}(\mathbf{x}) = \left.\frac{d\phi_t(\mathbf{x})}{dt}\right|_{t=0}$$

defines the vector field at $\mathbf{x}$ and defines a differential equation $d\mathbf{x}/dt = \mathbf{f}(\mathbf{x})$. Thus, every differential equation gives rise to a dynamical system and vice versa.

In closing this section we briefly consider dynamical systems induced by difference equations. To this end, let $Z = \{\ldots, -3, -2, -1, 0, 1, 2, 3, \ldots\}$ denote the set of all integers. Let $\mathbf{f} : \mathbb{R}^n \to \mathbb{R}^n$ be a continuously differentiable function such that its inverse is also continuously differentiable. Such a function is often called **diffeomorphism**. Given such an f, consider

$$\mathbf{x}_{k+1} = \mathbf{f}(\mathbf{x}_k) \tag{32.2.23}$$

where $\mathbf{x}_0$, the initial condition is given. Notice that $\mathbf{f}$ describes the one-step transition of the states of the system. Thus, $\mathbf{x}_1 = \mathbf{f}(\mathbf{x}_0)$ and $\mathbf{x}_2 = \mathbf{f}(\mathbf{x}_1) = \mathbf{f}(\mathbf{f}(\mathbf{x}_0)) = \mathbf{f}^{(2)}(\mathbf{x}_0)$ where $\mathbf{f}^{(2)}$ is called the two-fold iterate of $\mathbf{f}$. Clearly, $\mathbf{x}_k = \mathbf{f}^{(k)}(\mathbf{x}_0)$. The infinite family $\left\{\mathbf{f}^{(k)}\right\}_{k\in Z}$ of the iterates† of $\mathbf{f}$ defines the **flow** or **dynamical system** corresponding to (32.2.23).

The system (32.2.23) is linear or nonlinear depending on whether $\mathbf{f}$ is linear or nonlinear. If $\mathbf{f}$ is linear then, $\mathbf{x}_{k+1} = \mathbf{A}\mathbf{x}_k$ for some matrix $\mathbf{A}$ and $\mathbf{x}_k = \mathbf{A}^k\mathbf{x}_0$.

In conclusion, the long-term behavior of the solution of a linear system critically depends on the properties of the matrix $\mathbf{A}$ since $\mathbf{x}(t) = \left(e^{\mathbf{A}}\right)^t \mathbf{x}(0)$ or $\mathbf{x}_k = \mathbf{A}^k\mathbf{x}_0$ in continuous time and in discrete time respectively.

32.3 Characterization of stability of equilibria

Consider the differential equation

$$\dot{\mathbf{x}} = \mathbf{f}(\mathbf{x}, \boldsymbol{\alpha}) \tag{32.3.1}$$

where $\mathbf{f} : \mathbb{R}^n \times \mathbb{R}^p \to \mathbb{R}^n$ for some integers $n \geq 1$ and $p \geq 0$ with $\mathbf{f} = (f_1, f_2, \ldots, f_n)$, $f_i = f_i(\mathbf{x}, \boldsymbol{\alpha})$, $\mathbf{x} = (x_1, x_2, \ldots, x_n)^{\mathrm{T}}$ and $\boldsymbol{\alpha} = (\alpha_1, \alpha_2, \ldots, \alpha_p)^{\mathrm{T}}$. $\boldsymbol{\alpha}$ is called the set of **parameters**. The set

$$E = \{\mathbf{x} \mid \mathbf{f}(\mathbf{x}, \boldsymbol{\alpha}) = 0\} \tag{32.3.2}$$

is called the set of **equilibrium points** or **stationary points** of (32.3.1). Since the vector field is zero on this set, the flow, once it reaches this set, stays there forever.

† $\mathbf{f}^{(-3)}$ is to be interpreted as $\left(\mathbf{f}^{-1}\right)^{(3)}$, namely the three-fold iterate of the inverse of $\mathbf{f}$.

When (32.3.1) is a linear system (that is, $\dot{\mathbf{x}} = \mathbf{A}\mathbf{x}$), then there is only one equilibrium point, namely, the origin, $\mathbf{x} = 0$. When $\mathbf{f}$ is a nonlinear map, then there could be more than one equilibrium point. For example, consider the following system called Lorenz's system:

$$\left.\begin{aligned} \dot{x} &= -ax + ay \\ \dot{y} &= -xz + rx - y \\ \dot{z} &= xy - bz. \end{aligned}\right\} \tag{32.3.3}$$

With $\mathbf{x} = (x, y, z)^{\mathrm{T}}$, $\boldsymbol{\alpha} = (a, b, r)^{\mathrm{T}}$, $\mathbf{f} = (f_1, f_2, f_3)^{\mathrm{T}}$, $f_1(\mathbf{x}, \boldsymbol{\alpha}) = -ax + ay$, $f_2(\mathbf{x}, \boldsymbol{\alpha}) = -xz + rx - y$ and $f_3(\mathbf{x}, \boldsymbol{\alpha}) = xy - bz$, (32.3.3) can be expressed as $\dot{\mathbf{x}} = \mathbf{f}(\mathbf{x}, \boldsymbol{\alpha})$. The equilibria for (32.3.3) is obtained by setting the right hand side of (32.3.3) to zero and solving the resulting system of algebraic equations, namely

$$\left.\begin{aligned} x &= y \\ xz &= x(r-1) \\ xy &= bz. \end{aligned}\right\} \tag{32.3.4}$$

Clearly, $x = y = z = 0$ or the origin $(0, 0, 0)^{\mathrm{T}}$ is an equilibrium. It can be verified that there are also two other stationary states of (32.3.3) given by $(c, c, r-1)^{\mathrm{T}}$ and $(-c, -c, r-1)^{\mathrm{T}}$ where $c = \sqrt{b(r-1)}$. These two states come into play only when $r \geq 1$. Notice that the parameter a, while it does **not** affect the location of equilibria, nevertheless controls the evolution of the solution.

One of the fundamental concerns in dynamical systems is to describe the evolution of the flow starting from a state that is close to an equilibrium point. Clearly, the nature of the vector field in a region surrounding an equilibrium point must dictate the approach of the solution curve near that equilibrium point. In the remainder of this section we introduce various notions of stability that are useful in describing the qualitative behavior of the solution curve near equilibria.

Let $\mathbf{x}^{\mathrm{E}}$ denote an equilibrium point for the system in (32.3.1). The equilibrium $\mathbf{x}^{\mathrm{E}}$ is said to be **stable** if given any $\varepsilon > 0$, there exists a $\delta > 0$ such that if

$$\left\|\mathbf{x}(0) - \mathbf{x}^{\mathrm{E}}\right\| < \delta \quad \text{then} \quad \left\|\mathbf{x}(t) - \mathbf{x}^{\mathrm{E}}\right\| < \varepsilon \quad \text{for all} \quad t > 0. \tag{32.3.5}$$

Stated in words, $\mathbf{x}^{\mathrm{E}}$ is stable if for every sphere S_ε of radius ε centered at $\mathbf{x}^{\mathrm{E}}$, there exists a concentric sphere S_δ of radius δ such that the solution $\mathbf{x}(t)$ starting at an initial condition $\mathbf{x}(0)$ in S_δ remains in S_ε for all $t > 0$. Refer to Figure 32.3.1 for an illustration. Thus, stability of $\mathbf{x}^{\mathrm{E}}$ relates to the **boundedness** of solutions starting at initial points close to $\mathbf{x}^{\mathrm{E}}$.

An equilibrium point $\mathbf{x}^{\mathrm{E}}$ is said to be **asymptotically stable** if it is stable and in addition

$$\left\|\mathbf{x}(t) - \mathbf{x}^{\mathrm{E}}\right\| \to 0 \quad \text{as} \quad t \to \infty, \tag{32.3.6}$$

that is, the solution, in addition to remaining bounded, also asymptotically converges to $\mathbf{x}^{\mathrm{E}}$. Refer to Figure 32.3.2 for an illustration.

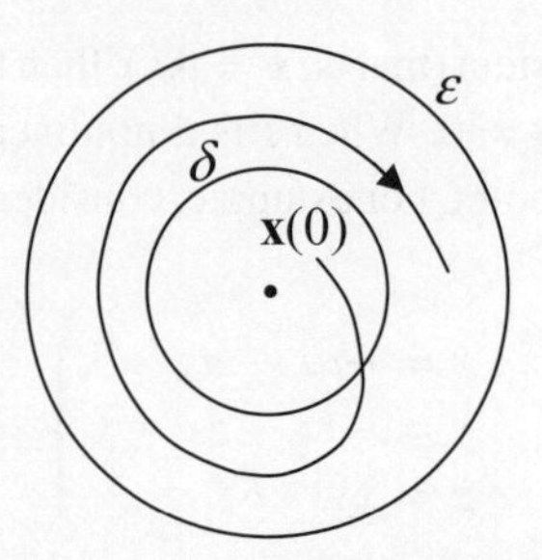

Fig. 32.3.1 $\mathbf{x}^{\mathrm{E}}$ is a stable equilibrium.

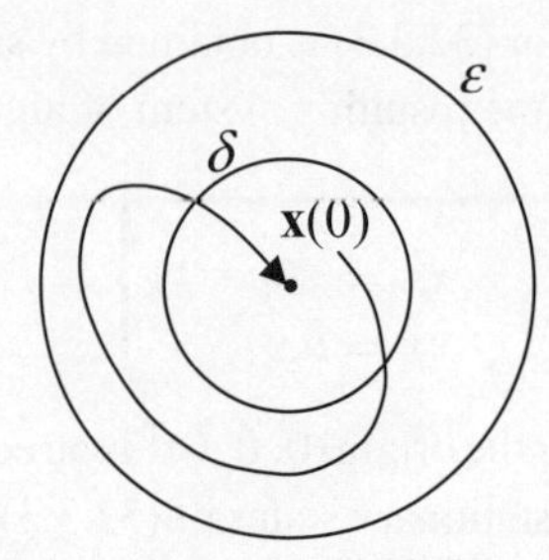

Fig. 32.3.2 Asymptotic stability of $\mathbf{x}^{\mathrm{E}+}$.

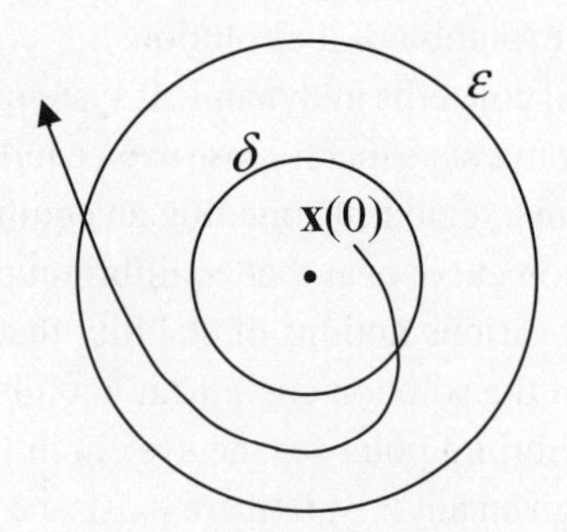

Fig. 32.3.3 $\mathbf{x}^{\mathrm{E}}$ being unstable.

An equilibrium point $\mathbf{x}^{\mathrm{E}}$ is **unstable** if it is **not** stable. That is, $\mathbf{x}^{\mathrm{E}}$ is unstable, if for every sphere S_ε of radius ε centered at $\mathbf{x}^{\mathrm{E}}$, there is a concentric sphere S_δ of radius δ such that there is at least one solution $\mathbf{x}(t)$, starting at $\mathbf{x}(0)$ in S_δ , that does **not** remain in S_ε for all $t > 0$. Refer to Figure 32.3.3 for an illustration.

The following example illustrates these definitions.

Example 32.3.1 Let $\dot{x}(t) = -a(t)x(t)$ be the linear, scalar, time varying system. The origin is the only equilibrium point and the solution is given by

$$\begin{aligned} x(t) &= x(0) \exp\left[-\int_{t_0}^{t} a(\tau)\mathrm{d}\tau\right] \\ &= x(0) \exp\left[-\alpha(t_0, t)\right] \end{aligned}$$

where

$$\alpha(t_0, t) = \int_{t_0}^{t} a(\tau)\mathrm{d}\tau.$$

The following conclusions readily follow:

(1) $x(t)$ is bounded and stable if $|\alpha(t_0, t)| < m$ for all $t \geq t_0$ where m is a finite real positive constant.
(2) $x(t) \to 0$ as $t \to \infty$ and hence asymptotically stable if $\alpha(t_0, t) \to \infty$ as $t \to \infty$.

Thus, a linear time-varying system can only be stable without being asymptotically stable depending on the properties of $a(t)$. On the other hand, if $a(t) \equiv a$, a constant, then

$$\begin{aligned} x(t) &= x(0)\mathrm{e}^{-at} \\ &\to 0 \qquad a > 0 \Rightarrow \text{asymptotically stable} \\ &\to \infty \qquad a < 0 \Rightarrow \text{unstable.} \end{aligned}$$

Remark 32.3.1 From the above definitions it is clear that the stability of equilibria is tested by perturbing the state in the phase space. There is at least one other type of stability of interest in dynamical systems, called the **structural stability**. This latter type is often tested by perturbing the field $\mathbf{f}(\mathbf{x}, \alpha)$ of the dynamical system $\dot{\mathbf{x}} = \mathbf{f}(\mathbf{x}, \alpha)$. A system is said to be structurally stable if small perturbations of the field result in a flow that is "topologically equivalent" to the flow defined by the unperturbed field. As observed in Example 32.2.1 , the solution curves for

$$\dot{x} = 0.7x \text{ and } \dot{y} = 0.71y \tag{32.3.7}$$

assuming the same initial conditions, while different, are qualitatively similar in the sense that they both diverge to infinity at slightly different rates. Thus, the system $\dot{x} = \alpha x$ is structurally stable when α is bounded away from zero. However, the linear system $\ddot{x} + w^2 x = 0$ describing the motion of a pendulum and represented equivalently by

$$\begin{pmatrix} \dot{x}_1 \\ x_2 \end{pmatrix} = \begin{pmatrix} 0 & 1 \\ -w^2 & 0 \end{pmatrix} \begin{pmatrix} x_1 \\ x_2 \end{pmatrix} \tag{32.3.8}$$

where $x_1 = x$ and $x_2 = \dot{x}_1$, is **structurally unstable**. For, addition of a term $a\dot{x}$ leads to $\ddot{x} + a\dot{x} + w^2 x = 0$ which now becomes

$$\begin{pmatrix} \dot{x}_1 \\ \dot{x}_2 \end{pmatrix} = \begin{pmatrix} 0 & 1 \\ -w^2 & -a \end{pmatrix} \begin{pmatrix} x_1 \\ x_2 \end{pmatrix}. \tag{32.3.9}$$

The eigenvalues of the matrix in (32.3.8) are purely imaginary and are given by $\pm iw$ and its solution is periodic. But the eigenvalues of the matrix in (32.3.9) are given by

$$\frac{-a \pm \sqrt{a^2 - 4w^2}}{2}.$$

Hence the solution to (32.3.9) either spirals inwards or outwards depending on the sign of a. Evidently, a periodic solution is **not** equivalent to a spiraling solution. Also notice that the origin, while it is the equilibrium point for both (32.3.8) and (32.3.9), it is stable but **not** asymptotically stable for (32.3.8) but with respect to (32.3.9) the origin is either asymptotically stable or unstable depending on the sign of the perturbation $a\dot{x}$.

As another example, consider the Burgers' equation

$$\frac{\partial u}{\partial t} + u\frac{\partial u}{\partial x} = \mu\frac{\partial^2 u}{\partial x^2}. \tag{32.3.10}$$

It is well known that the properties of the solution of (32.3.10) with $\mu = 0$ and $\mu > 0$ are qualitatively very different.

In a meteorological context, discussions relating to the structural stability must be brought to bear when one changes the field of the model by either adding or dropping a term or when changing the parameters in a highly parameterized model.

Remark 32.3.2 Analysis of stability of equilibria by changing the parameters of an otherwise fixed model has led to several startling discoveries over the years. Discovery of bifurcations of various types was one of the outcomes of this analysis. In the mid-sixties, Lorenz, using (32.3.3) and by changing the parameters, discovered the new phenomenon called **deterministic chaos**. What is even more interesting is the fact that Lorenz accidentally discovered the existence of deterministic chaos while analyzing the problems and challenges relating to meteorological prediction using simplified nonlinear models.

Remark 32.3.3 The notion of chaos is closely related to the fundamental notion of an **attractor** for a dynamical system. Simply put, an asymptotically stable equilibrium point is an attractor. For, by definition, if $\mathbf{x}^{\mathrm{E}}$ is asymptotically stable, then the solution $\mathbf{x}(t)$ gets attracted to $\mathbf{x}^{\mathrm{E}}$ as $t \to \infty$ so long as the initial condition $\mathbf{x}(0)$ is in some close neighborhood of $\mathbf{x}^{\mathrm{E}}$. Let $\mathbf{x}^{\mathrm{E}}$ be an asymptotically stable equilibrium point. Then there exists a largest subset B of $\mathbb{R}^n$ such that $\|\mathbf{x}(t) - \mathbf{x}^{\mathrm{E}}\| \to 0$ as $t \to \infty$ so long as $\mathbf{x}(0)$ is in B. Such a set is called the **basin of attraction** for $\mathbf{x}^{\mathrm{E}}$. While it is easy to visualize an asymptotically stable equilibrium as an attractor, not all attractors are equilibria. In fact, an attractor as a collection of points comes in various shapes, sizes, and geometry. A set A is said to be an attractor for the system (32.3.1) if $\mathbf{x}(t) \to A$ as $t \to \infty$ so long as the initial condition is in some close proximity of A. The basin B of attraction for an attractor A is the largest subset B in $\mathbb{R}^n$ such that the solution curve $\mathbf{x}(t)$ tends to the set A when $\mathbf{x}(0)$ is in B. Since points in an attractor are not necessarily equilibria, the field **does not** vanish in the set A. Thus, the solution curve after reaching the attractor A, will still evolve in time but always stays within A.

32.4 Classification of stability of equilibria

For ease in presentation, the discussion is divided into two parts.

32.4.1 Linear dynamics

Consider a linear autonomous system in $\mathbb{R}^2$ given by

$$\dot{\mathbf{x}} = \mathbf{A}\mathbf{x} \tag{32.4.1}$$

where

$$A = \begin{bmatrix} a_{11} & a_{12} \\ a_{21} & a_{22} \end{bmatrix}.$$

In this case the origin is the only equilibrium point. Changing the variable from $\mathbf{x}$ to $\mathbf{y}$ using $\mathbf{x} = \mathbf{P}\mathbf{y}$, we obtain an equivalent system $\dot{\mathbf{y}} = \mathbf{B}\mathbf{y}$ where $\mathbf{B} = \mathbf{P}^{-1}\mathbf{A}\mathbf{P}$ is one of the following three types (refer to Section 32.2 depending on the nature of the eigenvalues of $\mathbf{A}$:

$$\begin{bmatrix} \alpha_1 & 0 \\ 0 & \alpha_2 \end{bmatrix} \quad \begin{bmatrix} \alpha & 1 \\ 0 & \alpha \end{bmatrix} \text{ or } \begin{bmatrix} a & -b \\ b & a \end{bmatrix}.$$

Since $\mathbf{y}(t) = \mathrm{e}^{\mathbf{B}t}\mathbf{y}(0)$, we can obtain a very useful characterization of the long-term behavior of $\mathbf{y}(t)$ (and hence of $\mathbf{x}(t)$) by knowing the eigenvalues of $\mathbf{A}$.

From the analysis of Section 32.2, it follows that the solution curve $\mathbf{y}(t)$ gets attracted to the origin (that is, $\mathbf{y}(t) \to 0$ as $t \to \infty$) when the eigenvalues are either real and negative or complex with a negative real part. In this case the origin, as an equilibrium or stationary point is called a **sink**. Combining this with the definition in Section 32.3 it follows that a sink is asymptotically stable. On the other hand, if the eigenvalues of $\mathbf{A}$ are either real and positive or complex with a positive real part, then $\mathbf{y}(t) \to \infty$ as $t \to \infty$. In this case, the origin as an equilibrium is known as the **source**. Clearly, a source is an unstable equilibrium. Flows $\mathbf{x}(t) = \mathrm{e}^{\mathbf{A}t}\mathbf{x}(0)$ when the matrix $\mathbf{A}$ has non-zero real eigenvalues or complex eigenvalues with non-zero real parts are called **hyperbolic** flows. A further refinement of this classification is pursued in the following development.

Case 1: $\mathbf{B} = \begin{bmatrix} \alpha_1 & 0 \\ 0 & \alpha_2 \end{bmatrix}$ and α_1, α_2 are of **same** sign

In this case $y_1(t) = \mathrm{e}^{\alpha_1 t} y_1(0)$ and $y_2(t) = \mathrm{e}^{\alpha_2 t} y_2(0)$. The phase portraits for this case are given in Figures 32.4.1 and 32.4.2. When $\alpha_1 = \alpha_2$, the equilibrium is called a **focus**, and when $\alpha_1 < \alpha_2 < 0$, the equilibrium is called a **node**.

We invite the reader to sketch the phase portraits when $\alpha_1 = \alpha_2 > 0$ and $\alpha_1 > \alpha_2 > 0$ and verify that the origin is a source and hence unstable.

Case 2: $\mathbf{B} = \begin{bmatrix} \alpha_1 & 0 \\ 0 & \alpha_2 \end{bmatrix}$ and α_1, α_2 are of **opposite** sign

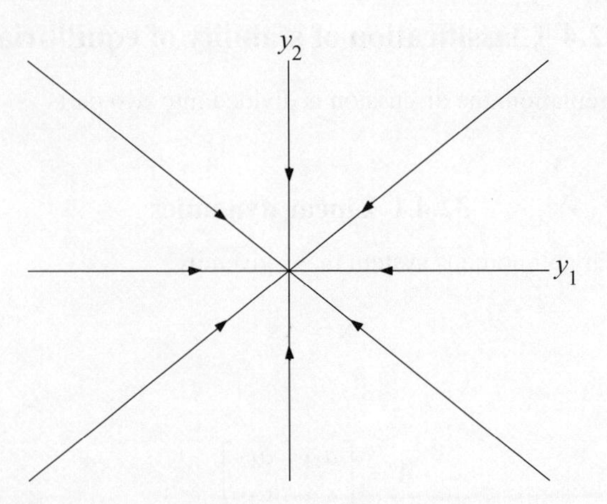

Fig. 32.4.1 $\alpha_1 = \alpha_2 < 0$. This equilibrium is called a focus.

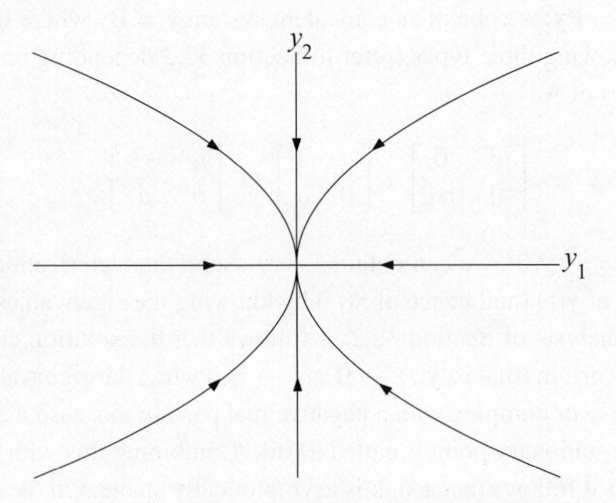

Fig. 32.4.2 $\alpha_1 < \alpha_2 < 0$. y_1 decreases faster than y_2. This equilibrium is called a node.

Let $\alpha_1 < 0 < \alpha_2$. In this case $y_1(t) \to 0$ and $y_2(t) \to \infty$ as $t \to \infty$. The equilibrium in this case is called a **saddle** point, and the phase portraits are given in Figure 32.4.3.

Notice that the origin attracts solutions along the y_1-axis but repels along the y_2-axis. Thus, the origin is simultaneously a sink in one direction and a source in another direction. For points outside of the y_1–y_2 axes, the phase trajectories move towards the y_2-axis (since $y_1(t) \to 0$) and away to infinity (since $y_2(t) \to \infty$) rather simultaneously. Hence a saddle point is an **unstable** equilibrium.

Case 3: $\mathbf{B} = \begin{bmatrix} \alpha & 1 \\ 0 & \alpha \end{bmatrix}$

In this case (referring to Example 32.2.3), we get $y_1(t) = [y_1(0) + y_2(0)t]\,\mathrm{e}^{\alpha t}$ and $y_2(t) = y_2(0)\mathrm{e}^{\alpha t}$. The phase portraits for this case (when $\alpha < 0$) are shown in

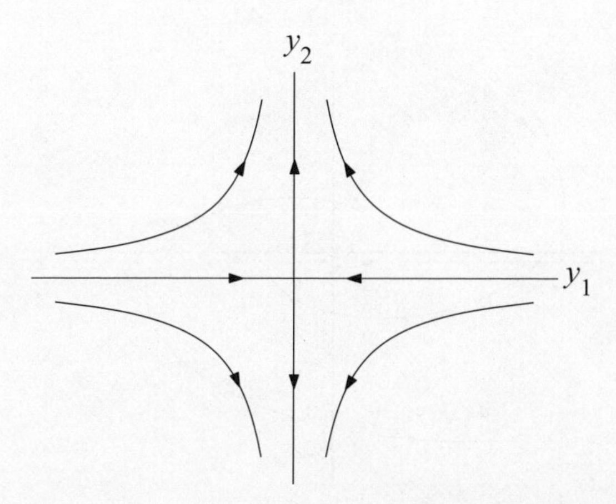

Fig. 32.4.3 $\alpha_1 < 0 < \alpha_2$. This equilibrium is a saddle point.

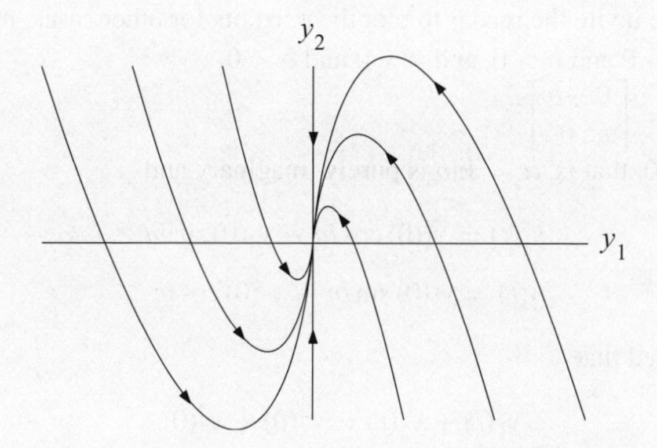

Fig. 32.4.4 $\alpha < 0$. This equilibrium is called an improper node.

Figure 32.4.4. The equilibrium is called an **improper node** and is **asymptotically stable**.

We invite the reader to plot the portraits for the case when $\alpha > 0$.

Case 4: $\mathbf{B} = \begin{bmatrix} a & b \\ b & a \end{bmatrix}$

In this case, we get

$$y_1(t) = [y_1(0) \cos bt - y_2(0) \sin bt]\, \mathrm{e}^{at},$$

$$y_2(t) = [y_1(0) \sin bt + y_2(0) \cos bt]\mathrm{e}^{at}.$$

In this case, when $a < 0$, the solution spirals inwards to the origin in the counter-clockwise direction if $b > 0$ and in the clockwise direction if $b < 0$. Consequently, the origin is asymptotically stable. When $a > 0$ the opposite effect is observed, and the origin is unstable. Figure 32.4.5 is an example of the phase portrait when

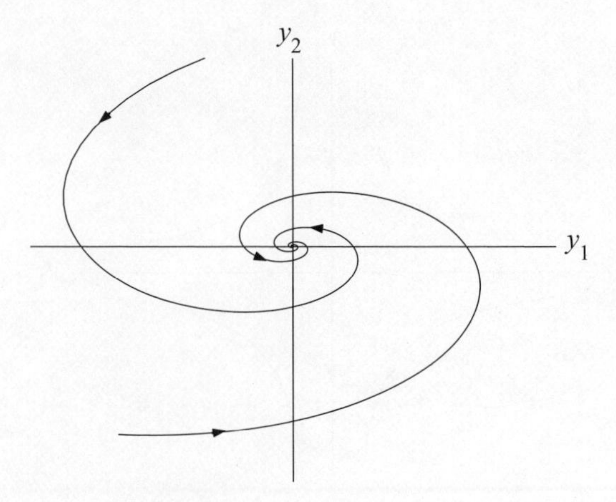

Fig. 32.4.5 $\alpha = a + ib$, $a < 0 < b$.

$a < 0 < b$. We invite the reader to plot the portraits for other cases, namely $a < 0$ and $b < 0$; $a > 0$ and $b > 0$, and $a > 0$ and $b < 0$.

Case 5: $\mathbf{B} = \begin{bmatrix} 0 & -b \\ b & 0 \end{bmatrix}$

When $a = 0$, that is, $\alpha = \pm ib$ is purely imaginary and

$$y_1(t) = y_1(0) \cos bt - y_2(0) \sin bt$$

$$y_2(t) = y_1(0) \sin bt + y_2(0) \cos bt.$$

It can be verified that

$$y_1^2(t) + y_2^2(t) = y_1^2(0) + y_2^2(0).$$

That is, the phase portraits are circles of radius $r = \left(y_1^2(0) + y_2^2(0)\right)^{1/2}$ depending only on the initial conditions. In this case the origin is stable but **not** asymptotically stable. Also refer to Remark 32.3.1 .

32.4.2 Nonlinear dynamics

Analysis of phase portraits of nonlinear systems is considerably more complex and involved. For, in general, a nonlinear system can have more than one equilibrium state and the overall behavior depends on the relative disposition of the initial condition and the equilibria, whether an equilibrium is a source or a sink. Consequently, often we may have to be content with the characterization of the phase portrait in a small neighborhood around an equilibrium state. This is often done by linearizing the given nonlinear dynamical equations.

Let $\mathbf{x}^*$ be an equilibrium state. Then the local properties of the phase portraits in a neighborhood are governed by the eigenvalues of the Jacobian of $\mathbf{f}(\mathbf{x}, \alpha)$ at

$\mathbf{x}^*$. Let $\mathbf{x} = \mathbf{x}^* + \mathbf{y}$ where $\mathbf{y}$ is very small (Refer to Appendix C for a definition of Jacobian). Then, expanding $\mathbf{f}(\mathbf{x}, \boldsymbol{\alpha})$ in a Taylor series and keeping only the first order terms in y we obtain

$$\begin{aligned}\dot{\mathbf{y}} = \dot{\mathbf{x}} &= \mathbf{f}(\mathbf{x}^* + \mathbf{y}, \boldsymbol{\alpha}) \\ &= \mathbf{f}(\mathbf{x}^*, \boldsymbol{\alpha}) + \mathbf{D_f}(\mathbf{x}^*, \boldsymbol{\alpha})\mathbf{y} \\ &= \mathbf{D_f}(\mathbf{x}^*, \boldsymbol{\alpha})\mathbf{y}\end{aligned} \tag{32.4.2}$$

where $\mathbf{f}(\mathbf{x}^*, \boldsymbol{\alpha}) = 0$ since $\mathbf{x}^*$ is an equilibrium point.

Since $\mathbf{f} : \mathbb{R}^n \times \mathbb{R}^p \to \mathbb{R}^n$, it follows that the Jacobian is given by

$$\mathbf{D_f}(\mathbf{x}, \boldsymbol{\alpha}) = \begin{bmatrix} \frac{\partial f_1}{\partial x_1} & \frac{\partial f_1}{\partial x_2} & \cdots & \frac{\partial f_1}{\partial x_n} \\ \frac{\partial f_2}{\partial x_1} & \frac{\partial f_2}{\partial x_2} & \cdots & \frac{\partial f_2}{\partial x_n} \\ \vdots & \vdots & & \vdots \\ \frac{\partial f_n}{\partial x_1} & \frac{\partial f_n}{\partial x_2} & \cdots & \frac{\partial f_n}{\partial x_n} \end{bmatrix}. \tag{32.4.3}$$

The equation (32.4.2) is called the tangent linear approximation to (32.3.1) at $\mathbf{x} = \mathbf{x}^*$. Since (32.4.2) is a linear system, all the preceding classifications of equilibria also apply to $\mathbf{x}^*$, however in a local sense. We now illustrate this using two typical examples.

Example 32.4.1 Consider a system of coupled nonlinear differential equations

$$\begin{aligned}\dot{x} &= \ \ x + y - x(x^2 + y^2) = f_1(x, y) \\ \dot{y} &= -x + y - y(x^2 + y^2) = f_2(x, y).\end{aligned} \tag{32.4.4}$$

It can be verified that $(0, 0)$ is the only equilibrium point. The Jacobian of $\mathbf{f}(x) = (f_1(x), f_2(x))^{\mathrm{T}}$ is given by

$$\mathbf{D_f}(x) = \begin{bmatrix} 1 - 3x^2 - y^2 & 1 - 2xy \\ -1 - 2xy & 1 - x^2 - 3y^2 \end{bmatrix}.$$

The $\mathbf{D_f}$ at the origin is given by

$$\mathbf{D_f}(0) = \begin{bmatrix} 1 & 1 \\ -1 & 1 \end{bmatrix}$$

and its eigenvalues are $1 + \mathrm{i}$ and $1 - \mathrm{i}$. Hence the trajectories starting close to the origin must spiral out and the origin as an equilibrium is a source. The question is: what happens to the trajectories as $t \to \infty$? If it were a linear system, we can readily conclude that the flow tends to infinity as $t \to \infty$. In this nonlinear system, something else happens. To understand this new phenomenon, let $x^2 + y^2 = 1$ in (32.4.4). The latter then reduces to $\dot{x} = y$ and $\dot{y} = -x$ which is equivalent to

$$\frac{\mathrm{d}y}{\mathrm{d}x} = -\frac{x}{y}. \tag{32.4.5}$$

Integrating (32.4.5), since $x^2 + y^2 = 1$, we readily obtain

$$x^2(t) + y^2(t) = 1.$$

That is, if the initial condition $x(0) = (x(0), y(0))^T$ is such that $x^2(0) + y^2(0) = 1$, then the trajectory is a circle of radius 1. Trajectories starting from initial conditions inside the circle spiral outwards and asymptotically merge with the circle of radius 1. Similarly, it can be verified that trajectories with initial conditions starting from outside of this unit circle, spiral inwards and again asymptotically merge with the unit circle. In other words, for all initial conditions $x(0) \neq (0, 0)^T$, the trajectory always merges with the cycle, called the **limit cycle**. Hence the limit cycle defined by $x^2(t) + y^2(t)$ is an **attractor** and no point on it is an equilibrium (refer to Remark 32.3.3). We hasten to add that while limit cycle represents an asymptotically periodic behavior, not every periodic behavior corresponds to limit cycle, witness the linear system in Case 5 above.

Example 32.4.2 It can be verified that the Jacobian of the nonlinear vector valued function $\mathbf{f}(\mathbf{x}, \alpha)$ corresponding to the Lorenz model in (32.2.3) is given by

$$\mathbf{D_f}(x) = \begin{bmatrix} -a & a & 0 \\ (r - z) & -1 & -x \\ y & x & -b \end{bmatrix}.$$

The three equilibria are given by $\mathbf{E}_1 = (0, 0, 0)^T$, $\mathbf{E}_2 = (c, c, r - 1)^T$ and $\mathbf{E}_3 = (-c, -c, r - 1)^T$ where $c = \sqrt{b(r - 1)}$. Notice that the equilibria $\mathbf{E}_2$ and $\mathbf{E}_3$ exist only when $r > 1$. The linear approximation to the Lorenz model at the equilibrium E_1 is given by

$$\dot{y} = \mathbf{A}y$$

where

$$\mathbf{A} = \begin{bmatrix} -a & a & 0 \\ r & -1 & 0 \\ 0 & 0 & -b \end{bmatrix}.$$

The three eigenvalues are given by

$$\lambda_1 = -b$$

$$\lambda_2 = \frac{-(a + 1) + \sqrt{(a - 1)^2 + 4ar}}{2}$$

$$\lambda_3 = \frac{-(a + 1) - \sqrt{(a - 1)^2 + 4ar}}{2}.$$

Since $\alpha > 0$, it follows that $\lambda_1 < 0$. Again λ_2 and λ_3 are negative if $(a + 1) > \sqrt{(a - 1)^2 + 4ar}$ which is true when $r < 1$. Refer to Table 32.4.1 for eigenvalues

Table 32.4.1

r ($a = 10, b = 8/3$)	Three eigenvalues of D_f at the equilibrium		
	E_1	E_2	E_3
0.0	-10.0		
	-2.67		
	-1.00		
0.5	-10.52		
	-2.67		
	-0.48		
1.0	-11.0		
	-2.67		
	0.00		
1.1	-11.09	-11.03	-11.03
	-2.67	-2.44	-2.44
	0.09	-0.20	-0.20
1.5	-11.44	-11.13	-11.13
	-2.67	$-1.27 + i0.88$	$-1.27 - i0.88$
	0.44	$-1.27 + i0.88$	$-1.27 - i0.88$
10	-16.47	-12.48	-12.48
	5.47	$-0.6 + i6.17$	$-0.6 + i6.17$
	-2.67	$-0.6 + i6.17$	$-0.6 + i6.17$
24.74	-21.86	-13.67	-13.67
	10.86	$0.0 + i9.63$	$0.0 + i9.63$
	-2.67	$0.0 + i9.63$	$0.0 + i9.63$
28	-22.83	-13.85	-13.85
	11.83	$0.09 + i10.19$	$0.09 + i10.19$
	-2.67	$0.09 + i10.19$	$0.09 + i10.19$

of the Jacobian matrix at the three equilibria for various values of r. Thus, $\mathbf{E}_1$ is a **stable node** when $r < 1$. Notice that $\lambda_3 = 0$ when $r = 1$ and $\lambda_3 > 0$ for $r > 1$. Indeed, $r = 1$ is a bifurcation point. When $r > 1$, since $\lambda_3 > 0$ and λ_1 and λ_2 are both negative $\mathbf{E}_1$ becomes a **saddle point**. Consequently, the solution vector along the eigenvector corresponding to λ_3 goes to infinity. However, all the trajectories lying solely in the plane defined by the eigenvectors corresponding to λ_1 and λ_2 converge to the origin.

Referring to Table 32.4.1, as r is increased from 0 through 28, $\mathbf{E}_1$ changes its character from being a stable node for $0 \leq r \leq 1$ to an unstable saddle point for $r > 1$. When $r = 1$ while $\mathbf{E}_1$ changes its character, two new equilibria $\mathbf{E}_2$ and $\mathbf{E}_3$ are introduced. Since the eigenvalues of the $\mathbf{D_f}$ at $\mathbf{E}_2$ and $\mathbf{E}_3$ are identical, we just comment about $\mathbf{E}_2$. As r increases from 1, two of the eigenvalues of $\mathbf{D_f}$ at $\mathbf{E}_2$ change their character from real and negative to complex with negative real part with the transition happening for r nearly equal to

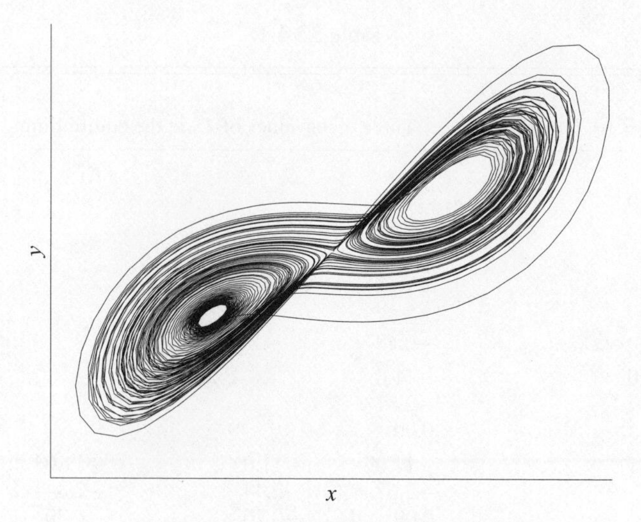

Fig. 32.4.6 Trajectories of Lorenz's system.

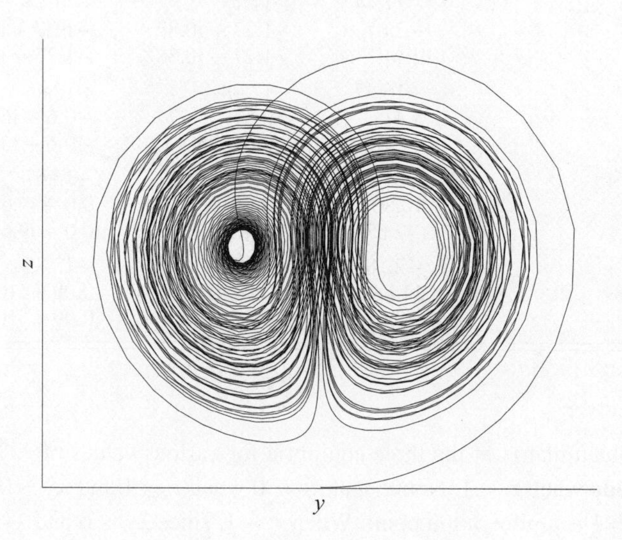

Fig. 32.4.7 Trajectories of Lorenz's system.

1.3459. As r is increased further another change occurs at $r = 24.74$ at which the real part of the complex eigenvalues changes from negative to positive. Thus, at $r = 24.74$, $\mathbf{E}_2$ (and hence $\mathbf{E}_3$) changes its character from one of asymptotically stable equilibrium to an unstable equilibrium. Lorenz extensively analyzed the case $r = 28$. In this case, all the three equilibria are unstable, yet the solution remains bounded for all t – a strange phenomenon indeed. The system exhibits an attractor A whose geometry is very complex. For completeness, in Figures 32.4.6 through 32.4.8, we have given the plots of the trajectories of (32.2.3) obtained using simple Euler discretization.

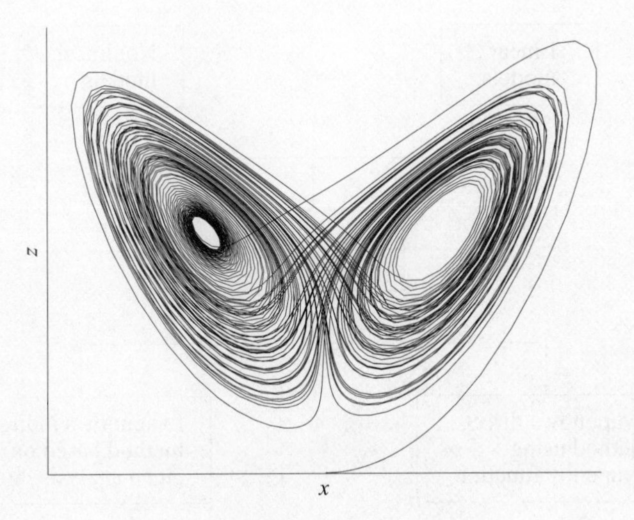

Fig. 32.4.8 Trajectories of Lorenz's System.

32.5 Lyapunov stability

In this section we provide an introduction to the Lyapunov stability theory. For convenience we divide the presentation into two parts – linear and nonlinear cases. Again, the stability analysis can be done in two ways – using Lyapunov's indirect method that relies on the eigen analysis and using Lyapunov's direct method that relies on finding a suitable function representing the "energy" in the system called the Lyapunov function. In the following we illustrate both the methods. Refer to Figure 32.5.1.

32.5.1 Lyapunov's indirect method

Consider an autonomous linear system in $\mathbb{R}^n$ given by

$$\mathbf{x}_{k+1} = \mathbf{M}\mathbf{x}_k \tag{32.5.1}$$

where $\mathbf{M} \in \mathbb{R}^n \times \mathbb{R}^n$. Consider two initial conditions $\bar{\mathbf{x}}_0$ and $\mathbf{x}_0 = \bar{\mathbf{x}}_0 + \varepsilon_0$ that are close to each other in the phase space where ε_0 denotes the perturbation. Define $\varepsilon_k = \mathbf{x}_k - \bar{\mathbf{x}}_k$ where

$$\mathbf{x}_{k+1} = \mathbf{M}\mathbf{x}_k \quad \text{and} \quad \bar{\mathbf{x}}_{k+1} = \mathbf{M}\bar{\mathbf{x}}_k.$$

Then

$$\varepsilon_{k+1} = \mathbf{x}_{k+1} - \bar{\mathbf{x}}_{k+1} = \mathbf{M}(\mathbf{x}_k - \bar{\mathbf{x}}_k) = \mathbf{M}\varepsilon_k. \tag{32.5.2}$$

That is, the dynamics of the error ε_k (which is the difference between $\mathbf{x}_k$ and $\bar{\mathbf{x}}_k$) is the same as the original system (32.5.1).

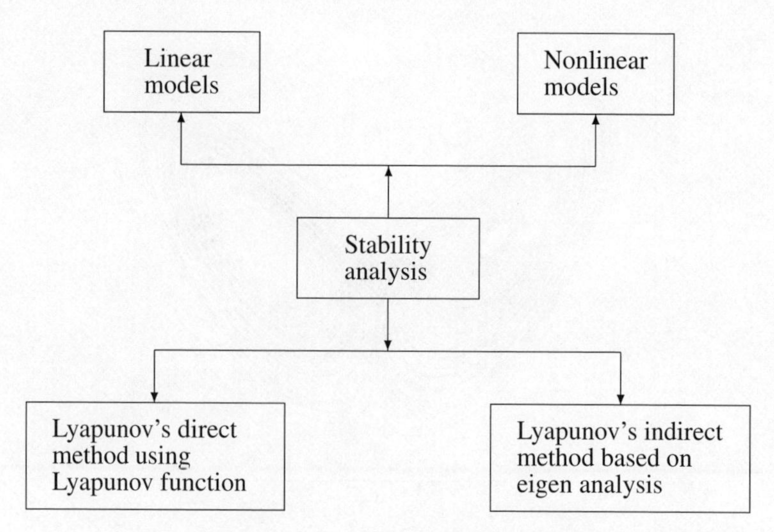

Fig. 32.5.1 A classification of stability analysis.

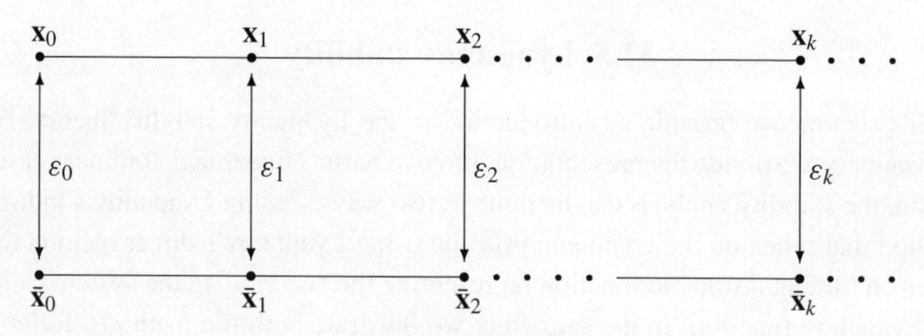

Fig. 32.5.2 An illustration of the error dynamics.

By iterating (32.5.2), it follows that

$$\varepsilon_k = \mathbf{M}^k \varepsilon_0. \tag{32.5.3}$$

Figure 32.5.2 provides an illustration of the error dynamics. In meteorological parlance the trajectory $\{\bar{\mathbf{x}}_k\}_{k=0}^{\infty}$ is called the **base state** and $\{\mathbf{x}_k\}_{k=0}^{\infty}$ is called the **perturbed state**.

Let $\mathbf{p}_i$ be the eigenvector corresponding to the eigenvalue λ_i of $\mathbf{M}$. That is,

$$\mathbf{M}\mathbf{p}_i = \lambda_i \mathbf{p}_i \quad \text{for } 1 \leq i \leq n. \tag{32.5.4}$$

Recall that the eigenvectors of $\mathbf{M}$ represent the **characteristic modes** of the linear system in question. The n relations in (32.5.4) can be succinctly rewritten as

$$\mathbf{M}[\mathbf{p}_1, \mathbf{p}_2, \ldots, \mathbf{p}_n] = [\lambda_1 \mathbf{p}_1, \lambda_2 \mathbf{p}_2, \ldots, \lambda_n \mathbf{p}_n]. \tag{32.5.5}$$

Denoting

$$\mathbf{P} = [\mathbf{p}_1, \mathbf{p}_2, \ldots, \mathbf{p}_n]$$

to be the matrix of eigenvectors of $\mathbf{M}$, (32.5.5) can be written as

$$\mathbf{MP} = \mathbf{P}\boldsymbol{\Lambda} \text{ or } \mathbf{P}^{-1}\mathbf{MP} = \boldsymbol{\Lambda} \tag{32.5.6}$$

where $\boldsymbol{\Lambda}$ is the diagonal matrix of eigenvalues of $\mathbf{M}$. Assuming that $\{\mathbf{p}_1, \mathbf{p}_2, \ldots, \mathbf{p}_n\}$ are linearly independent, we can express

$$\varepsilon_0 = a_1\mathbf{p}_1 + a_2\mathbf{p}_2 + \cdots + a_n\mathbf{p}_n. \tag{32.5.7}$$

Then

$$\begin{aligned}\varepsilon_1 &= \mathbf{M}\varepsilon_0 \\ &= \mathbf{M}(a_1\mathbf{p}_1 + a_2\mathbf{p}_2 + \cdots + a_n\mathbf{p}_n) \\ &= a_1\lambda_1\mathbf{p}_1 + a_2\lambda_2\mathbf{p}_2 + \cdots + a_n\lambda_n\mathbf{p}_n.\end{aligned} \tag{32.5.8}$$

Likewise using the recurrence (32.5.3) it can be verified that

$$\varepsilon_k = a_1\lambda_1^k\mathbf{p}_1 + a_2\lambda_2^k\mathbf{p}_2 + \cdots + a_n\lambda_n^k\mathbf{p}_n. \tag{32.5.9}$$

Several cases arise.

Case 1 The spectral radius $\rho(\mathbf{M}) < 1$

Then $|\lambda_i| < 1$ for all $1 \le i \le n$ and $\lambda_i^k \to 0$ as $k \to \infty$. In this case the linear system (32.5.3) and hence (32.5.1) is **asymptotically stable** and the error ε_k decreases in time and eventually vanishes to zero. Thus, $\varepsilon_k \to 0$ as $k \to \infty$, and the system is very **robust** in the sense that the system is insensitive to the initial errors.

Case 2 The spectral radius $\rho(\mathbf{M}) > 1$

Then there exists at least one eigenvalue whose absolute value is larger than 1. Let r be an integer such that $i \le r \le n$ and

$$\begin{aligned}|\lambda_i| &> 1, \quad \text{for} \quad 1 \le i \le r \\ |\lambda_i| &< 1, \quad \text{for} \quad r+1 \le i \le n\end{aligned} \tag{32.5.10}$$

Then the linear system (32.5.3) is **unstable** and $\varepsilon_k \to \infty$ as $k \to \infty$. In fact, we can quantify the rate of growth as follows: Combining (32.5.10) with (32.5.9), we see that

$$\varepsilon_k \approx a_1\lambda_1^k\mathbf{p}_1 + a_2\lambda_2^k\mathbf{p}_2 + \cdots + a_r\lambda_r^k\mathbf{p}_r. \tag{32.5.11}$$

The subspace spanned by the eigenvectors $\{\mathbf{p}_1, \mathbf{p}_2, \ldots, \mathbf{p}_r\}$ is called the **unstable manifold** and the subspace spanned by the eigenvectors $\{\mathbf{p}_{r+1}, \ldots, \mathbf{p}_n\}$ is called the **stable manifold**. The predictability limit k^* is the first time instant at which the ratio of the norm of ε_k to that of ε_0 exceeds a prespecified threshold, say α.

Table 32.5.1 *Stability properties of linear models*

Mode of behavior	Continuous time model $\dot{\mathbf{x}} = \mathbf{A}\mathbf{x}$	Discrete time model $\mathbf{x}_{k+1} = \mathbf{M}(\mathbf{x}_k)$
Stable oscillatory/ periodic behavior	Eigenvalues of **A** lie on the imaginary axis	Eigenvalues of **M** lie on the unit circle
Asymptotically stable behavior	Eigenvalues of **A** have negative real part, (i.e. lie on the left half of the complex plane)	Eigenvalues of **M** have absolute value less than one, (i.e. they lie inside the unit circle)
Unstable behavior	Eigenvalues of **A** have positive real part, (i.e. lie on the right half of the complex plane)	Eigenvalues of **M** have absolute value larger than one, (i.e. they lie outside the unit circle)

That is,

$$k^* = \min_k \left\{ \frac{\| \varepsilon_k \|}{\| \varepsilon_0 \|} \geq \alpha \right\}. \tag{32.5.12}$$

If $\alpha = 2$, then k^* is called the error-doubling time. Thus, a finite predictability limit exists exactly when the initial error has non-zero components that lie in the unstable manifold. If it does, then the error grows to infinity along the first r nodes. Thus, $\varepsilon_k \to \infty$ as $k \to \infty$. In this case, the system (32.5.1) is extremely sensitive to initial errors. Conditions for stability for the linear discrete and continuous time systems are given in Table 32.5.1.

32.5.2 Lyapunov's direct method

Let

$$\dot{\mathbf{x}} = \mathbf{f}(\mathbf{x}) \tag{32.5.13}$$

be the given nonlinear, autonomous system. Let $\mathbf{x}^{\mathrm{E}} = 0$, the origin, be an equilibrium point of (32.5.13). Then $\mathbf{x}^{\mathrm{E}}$ is an asymptotically stable equilibrium if there exists a function $V : \mathbb{R}^n \to \mathbb{R}$, called the **Lyapunov function** satisfying the following conditions:

(1) $V(\mathbf{x})$ has continuous partial derivatives.
(2) $V(\mathbf{x})$ is positive definite, that is, $V(\mathbf{x}) > 0$ for all $\mathbf{x} \neq 0$ and $V(\mathbf{0}) = 0$.
(3) $V(\mathbf{x}) \to \infty$ as $\| \mathbf{x} \| \to \infty$.
(4) $\dot{V}(\mathbf{x}) = [\nabla V(\mathbf{x})]^{\mathrm{T}}\dot{\mathbf{x}}$ is negative definite, that is, $\dot{V}(\mathbf{x}) < 0$ for all $\mathbf{x} \neq 0$ and $\dot{V}(\mathbf{0}) = 0$.

Notice that this so-called direct method does not require computation of the eigenvalues of the Jacobian of $\mathbf{f}(\mathbf{x})$ at the equilibrium point. We illustrate the power of this idea using two examples.

Example 32.5.1 Let $\dot{\mathbf{x}} = \mathbf{A}\mathbf{x}$ be the given linear, time invariant system with the origin as the only equilibrium point. This system is asymptotically stable if and only if given a symmetric positive definite matrix $\mathbf{Q}$, there exists a symmetric positive definite matrix $\mathbf{P}$ which is the unique solution of

$$\mathbf{A}^{\mathrm{T}}\mathbf{P} + \mathbf{P}\mathbf{A} = -\mathbf{Q}. \tag{32.5.14}$$

Thus, $V(\mathbf{x}) = \mathbf{x}^{\mathrm{T}}\mathbf{P}\mathbf{x}$ is a Lyapunov function for the given dynamics.

To verify this claim, Let $V(\mathbf{x}) = \mathbf{x}^{\mathrm{T}}\mathbf{P}\mathbf{x}$. Then†

$$\begin{aligned}\dot{V}(\mathbf{x}) &= 2(\mathbf{P}\mathbf{x})^{\mathrm{T}}\mathbf{A}\mathbf{x} \\ &= 2(\mathbf{x}^{\mathrm{T}}\mathbf{P}^{\mathrm{T}}\mathbf{A}\mathbf{x}) \\ &= \mathbf{x}^{\mathrm{T}}(\mathbf{A}^{\mathrm{T}}\mathbf{P} + \mathbf{P}\mathbf{A})\mathbf{x} \\ &= -\mathbf{x}^{\mathrm{T}}\mathbf{Q}\mathbf{x}.\end{aligned}$$

Since $\mathbf{Q}$ is positive definite, the claim follows. Notice that instead of solving for the eigenvalues of $\mathbf{A}$, this approach requires solving (32.5.14) for $\mathbf{P}$.

Example 32.5.2 Consider the nonlinear dynamics

$$\begin{aligned}\dot{x}_1(t) &= -x_1(t) + x_2(t)(x_1(t) + a) \\ \dot{x}_2(t) &= -x_1(t)(x_1(t) + a)\end{aligned}$$

for some constant a. It can be verified that origin is the only equilibrium point. Let $V(\mathbf{x}) = \frac{1}{2}\mathbf{x}^{\mathrm{T}}\mathbf{x}$. Then

$$\begin{aligned}\dot{V}(\mathbf{x}) = \mathbf{x}^{\mathrm{T}}\dot{\mathbf{x}} &= (x_1, x_2)\begin{pmatrix} -x_1 + x_2(x_1 + a) \\ -x_1(x_1 + a) \end{pmatrix} \\ &= -x_1^2.\end{aligned}$$

Hence, the origin is asymptotically stable.

A number of observations are in order.

(1) The idea behind this direct approach is that $V(\mathbf{x})$ plays the role of an "energy" function. Thus, for dissipative systems, this energy will diminish along the trajectory.

(2) This is not a necessary but only a sufficient condition. Thus, if we can not find the suitable function $V(\mathbf{x})$, it does not imply that the system is not stable.

† For any general matrix $\mathbf{B}$, $\mathbf{x}^{\mathrm{T}}\mathbf{B}\mathbf{x} = \frac{1}{2}\mathbf{x}^{\mathrm{T}}(\mathbf{B} + \mathbf{B}^{\mathrm{T}})\mathbf{x}$.

(3) Despite its elegance and simplicity, there is no guideline or prescription for obtaining a suitable function $V(\mathbf{x})$.
(4) This approach is very useful in obtaining qualitative behavior of systems.

32.6 Role of singular vectors in predictability

As mentioned in the opening paragraph of this chapter the goal of a deterministic approach to predictability is to quantify the growth of infinitesimally small errors superimposed on the trajectory of a given dynamical system. This is achieved by extending the Lyapunov (indirect) method (see Section 32.5) that relies on linear stability analysis.

Let $\mathbf{M} : \mathbb{R}^n \times \mathbb{R}^p \to \mathbb{R}^n$ and let

$$\mathbf{x}_{k+1} = \mathbf{M}(\mathbf{x}_k, \boldsymbol{\alpha}) \tag{32.6.1}$$

where $\boldsymbol{\alpha} \in \mathbb{R}^p$ is a set of parameters. Let $\{\bar{\mathbf{x}}_k\}_{k\geq 0}$ and $\{\mathbf{x}_k\}_{k\geq 0}$ be the two trajectories starting from the base initial state $\bar{\mathbf{x}}_0$ and the perturbed initial state $\mathbf{x}_0$ where the size of the initial perturbation or error $\varepsilon_0 = \mathbf{x}_0 - \bar{\mathbf{x}}_0$ is assumed to be infinitesimally small. Such a pair of states is called analogs. Let $\varepsilon_k = \mathbf{x}_k - \bar{\mathbf{x}}_k$ be the error at time k. Then using the first-order Taylor series (where it is tacitly assumed that ε_k is small) we obtain

$$\begin{aligned} \mathbf{x}_{k+1} &= \mathbf{M}(\mathbf{x}_k, \boldsymbol{\alpha}) = \mathbf{M}(\bar{\mathbf{x}}_k + \varepsilon_k, \boldsymbol{\alpha}) \\ &= \mathbf{M}(\bar{\mathbf{x}}_k, \boldsymbol{\alpha}) + \mathbf{D}_\mathbf{M}(\bar{\mathbf{x}}_k)\varepsilon_k = \bar{\mathbf{x}}_{k+1} + \mathbf{D}_\mathbf{M}(\bar{\mathbf{x}}_k)\varepsilon_k \end{aligned}$$

or

$$\varepsilon_{k+1} = \mathbf{D}_\mathbf{M}(\bar{\mathbf{x}}_k)\varepsilon_k \tag{32.6.2}$$

where $\mathbf{D}_\mathbf{M}(\bar{\mathbf{x}}_k)$ is the Jacobian of $\mathbf{M}$ at $\bar{\mathbf{x}}_k$. This nonautonomous linear system is also known as the tangent linear system which is a local approximation to the autonomous nonlinear system in (32.6.1). Iterating (32.6.2) we get

$$\varepsilon_{t+1} = \mathbf{D}_\mathbf{M}(t : s)\varepsilon_s \tag{32.6.3}$$

where for any two integers $t \geq s \geq 0$

$$\mathbf{D}_\mathbf{M}(t : s) = \mathbf{D}_\mathbf{M}(\bar{\mathbf{x}}_t)\mathbf{D}_\mathbf{M}(\bar{\mathbf{x}}_{t-1}) \cdots \mathbf{D}_\mathbf{M}(\bar{\mathbf{x}}_s) \tag{32.6.4}$$

is the product of the non-commuting Jacobian matrices evaluated along the base trajectory from time s to t. $\mathbf{D}_\mathbf{M}(t : s)$ is called the **resolvant** or the **propagator** which is essentially the **state transition matrix** from time s to $t + 1$. Define the ratio of the energy in the perturbation at time $t+1$ to that at time s as

$$\begin{aligned} r_{t+1}(\varepsilon_s) &= \frac{\| \varepsilon_{t+1} \|_{\mathbf{A}}^2}{\| \varepsilon_s \|_{\mathbf{B}}^2} = \frac{\| \mathbf{D_M}(t:s)\varepsilon_s \|_{\mathbf{A}}^2}{\| \varepsilon_s \|_{\mathbf{B}}^2} \\ &= \frac{\varepsilon_s^{\mathrm{T}} \mathbf{D}_{\mathbf{M}}^{\mathrm{T}}(t:s)\mathbf{A}\mathbf{D_M}(t:s)\varepsilon_s}{\varepsilon_s^{\mathrm{T}} \mathbf{B} \varepsilon_s} \end{aligned} \tag{32.6.5}$$

where $\mathbf{A}$ and $\mathbf{B}$ are two symmetric positive definite matrices denoting the choice of the energy measure at time $(t+1)$ and s, respectively. This ratio $r_{t+1}(\varepsilon_s)$ is called the **Rayleigh** coefficient and many known results in deterministic predictability theory are related to the properties of this ratio. Refer to Appendix B for a listing of these properties.

For purposes of simplifying the algebra it is assumed that the matrices $\mathbf{A}$ and $\mathbf{B}$ denoting the choice of energy in (32.6.5) are both identity matrices. Further, let us denote $\mathbf{D_M}(t:s)$ simply as $\mathbf{D_M}$ by suppressing the time indices. Then

$$r_{t+1}(\varepsilon_s) = \frac{\varepsilon_s^{\mathrm{T}} \mathbf{D}_{\mathbf{M}}^{\mathrm{T}} \mathbf{D_M} \varepsilon_s}{\varepsilon_s^{\mathrm{T}} \varepsilon_s}. \tag{32.6.6}$$

It is tacitly assumed that the Jacobians $\mathbf{D_M}(\bar{\mathbf{x}}_k)$ are nonsingular for all k. This in turn implies that the symmetric matrices $\mathbf{D}_{\mathbf{M}}^{\mathrm{T}}\mathbf{D_M}$ and $\mathbf{D_M}\mathbf{D}_{\mathbf{M}}^{\mathrm{T}}$ are both positive definite as well. Let

$$\mathbf{V}(t:s) = \mathbf{V} = [\mathbf{v}_1, \mathbf{v}_2, \ldots, \mathbf{v}_n] \tag{32.6.7}$$

and

$$\mathbf{\Lambda}(t:s) = \mathbf{\Lambda} = \mathrm{Diag}(\lambda_1, \lambda_2, \ldots, \lambda_n) \tag{32.6.8}$$

be the matrix of eigenvectors and the corresponding eigenvalues of the matrix $\mathbf{D}_{\mathbf{M}}^{\mathrm{T}}\mathbf{D_M}$ where it is assumed (without loss of generality) that

$$\lambda_1 > \lambda_2 > \cdots > \lambda_n. \tag{32.6.9}$$

Since $\mathbf{V}$ is also orthonormal, that is $\mathbf{V}^{\mathrm{T}}\mathbf{V} = \mathbf{V}\mathbf{V}^{\mathrm{T}} = \mathbf{I}$ we get

$$(\mathbf{D}_{\mathbf{M}}^{\mathrm{T}}\mathbf{D_M})\mathbf{v}_i = \mathbf{v}_i \lambda_i$$

and

$$(\mathbf{D}_{\mathbf{M}}^{\mathrm{T}}\mathbf{D_M})\mathbf{V} = \mathbf{V}\mathbf{\Lambda} \quad \text{or} \quad \mathbf{V}^{\mathrm{T}}(\mathbf{D}_{\mathbf{M}}^{\mathrm{T}}\mathbf{D_M})\mathbf{V} = \mathbf{\Lambda}. \tag{32.6.10}$$

Define

$$\mathbf{U}(t:s) = \mathbf{U} = [\mathbf{u}_1, \mathbf{u}_2, \ldots, \mathbf{u}_n]$$

where

$$\mathbf{u}_i = \frac{1}{\sqrt{\lambda_i}} \mathbf{D_M} \mathbf{v}_i. \tag{32.6.11}$$

Then, from

$$
\begin{aligned}
(\mathbf{D_M D_M^T})\mathbf{u}_i &= \frac{1}{\sqrt{\lambda_i}}(\mathbf{D_M D_M^T})\mathbf{D_M v}_i \\
&= \frac{1}{\sqrt{\lambda_i}}\mathbf{D_M}(\mathbf{D_M^T D_M v}_i) \\
&= \frac{1}{\sqrt{\lambda_i}}\mathbf{D_M v}_i\lambda_i \qquad \text{(using 32.6.10)} \\
&= \mathbf{u}_i\lambda_i
\end{aligned}
$$

or

$$(\mathbf{D_M D_M^T})\mathbf{U} = \mathbf{U\Lambda}. \tag{32.6.12}$$

That is, $\mathbf{D_M^T D_M}$ and $\mathbf{D_M D_M^T}$ share the same set of eigenvalues $\lambda_i, i = 1$ to n and their eigenvectors $\mathbf{v}_i$ and $\mathbf{u}_i$ are related through (32.6.11). Recall (Chapter 9) that $\sqrt{\lambda_i}$, $i = 1$ to n are called the **singular values** of $\mathbf{D_M}$ and the eigenvectors $\mathbf{v}_i$ are called the **right** or **forward** singular vectors and $\mathbf{u}_i$ are known as the **left** or **backward** singular vectors of $\mathbf{D_M}$. Rewriting (32.6.11) (and inserting the time dependence) we get the singular value decomposition of $\mathbf{D_M} = \mathbf{D_M}(t : s)$ as

$$\mathbf{D_M}(t : s) = \mathbf{U}(t : s)\mathbf{\Lambda}^{\frac{1}{2}}(t : s)\mathbf{V}^{\mathrm{T}}(t : s). \tag{32.6.13}$$

To understand the nature of the growth of errors, let us change the basis for $\mathbb{R}^n$ from the conventional coordinate system to those corresponding to the orthonormal columns of the right or the forward singular vectors in $\mathbf{V}$. Define

$$\varepsilon_s = \mathbf{V}\boldsymbol{\alpha} \tag{32.6.14}$$

where the elements of $\boldsymbol{\alpha} \in \mathbb{R}^n$ are the coordinates of ε_s in the new basis $\mathbf{V}$. Substituting (32.6.14) into (32.6.6) we get (since $\mathbf{VV}^{\mathrm{T}} = \mathbf{V}^{\mathrm{T}}\mathbf{V} = \mathbf{I}$)

$$
\begin{aligned}
r_{t+1}(\varepsilon_s) &= \frac{\boldsymbol{\alpha}^{\mathrm{T}}\mathbf{V}^{\mathrm{T}}\mathbf{D_M^T D_M V}\boldsymbol{\alpha}}{\boldsymbol{\alpha}^{\mathrm{T}}\boldsymbol{\alpha}} \\
&= \frac{\boldsymbol{\alpha}^{\mathrm{T}}\mathbf{\Lambda}\boldsymbol{\alpha}}{\boldsymbol{\alpha}^{\mathrm{T}}\boldsymbol{\alpha}} \qquad \text{(using 32.6.10).}
\end{aligned} \tag{32.6.15}
$$

Now, if $\boldsymbol{\alpha}$ is such that $\| \boldsymbol{\alpha} \| = 1$, then

$$
\begin{aligned}
r_{t+1}(\varepsilon_s) = \boldsymbol{\alpha}^{\mathrm{T}}\mathbf{\Lambda}\boldsymbol{\alpha} &= \sum \alpha_i^2\lambda_i \\
&= \sum_{i=1}^{n} \frac{\alpha_i^2}{\left(\sqrt{\frac{1}{\lambda_i}}\right)^2}.
\end{aligned} \tag{32.6.16}
$$

Thus, the action of the dynamical system is such that the errors ε_s at time s that lie on a unit sphere are mapped on to the surface of an ellipsoid at time $t + 1$, whose axes are the right or the forward singular vectors $\mathbf{v}_i$ and the length of their semi-axes are given by $\lambda_i^{-1/2}$ for $i = 1, 2, \ldots, n$. Further, since $\| \boldsymbol{\alpha} \| = 1$, it follows that

$$\lambda_n \leq r_{t+1}(\varepsilon_s) \leq \lambda_1 \tag{32.6.17}$$

and $r_{t+1}(\varepsilon_s)$ attains its maximum value of λ_1 exactly when $\alpha_1 = 1$ and $\alpha_j = 0$ for $j \neq 1$.

This discussion naturally leads to the question of quantifying the average rate of growth of errors during the time interval from time s to $t + 1$. This average growth/decay rate is embodied in the concept of the **Lyapunov index** which is the asymptotic average growth/decay rate which is achieved by keeping s fixed and letting $t \to \infty$. Before taking up the problem of computing the Lyapunov index (in Section 32.7), we conclude this section with a discussion of several properties that are germane to the definition of the corresponding Lyapunov vectors also defined in Section 32.7.

(1) Effect of forward dynamics Let $\varepsilon_s = \mathbf{v}_i$ which is a forward singular vector of $\mathbf{D_M}$. Under the action of the dynamics this ε_s then evolves into

$$\varepsilon_{t+1} = \mathbf{D_M}\varepsilon_s = \mathbf{D_M}\mathbf{v}_i. \tag{32.6.18}$$

Multiplying both sides on the left with $(\mathbf{D_M}\mathbf{D_M^T})$ and using (32.6.10) we obtain

$$(\mathbf{D_M}\mathbf{D_M^T})\varepsilon_{t+1} = \mathbf{D_M}(\mathbf{D_M^T}\mathbf{D_M})\mathbf{v}_i = \mathbf{D_M}\mathbf{v}_i\lambda_i = \varepsilon_{t+1}\lambda_i. \tag{32.6.19}$$

That is, ε_{t+1} is indeed an eigenvector of $\mathbf{D_M}\mathbf{D_M^T}$ corresponding to the eigenvalue λ_i. Since eigenvectors are unique (up to the ordering), it follows that $\varepsilon_{t+1} = \mathbf{u}_i$ which is a left or backward singular vector of $\mathbf{D_M}$.

Stated in the other words, if we start with an error in the direction of the forward singular vector $\mathbf{v}_i$ at time s, it grows into the corresponding backward singular vector $\mathbf{u}_i$ at time $t + 1$.

(2) Effect of inverse dynamics Recall from Appendix B that

$$\text{if} \quad \mathbf{Ax} = \lambda\mathbf{x} \quad \text{then} \quad \mathbf{A}^{-1}\mathbf{x} = \lambda^{-1}\mathbf{x}. \tag{32.6.20}$$

That is, if $(\lambda, \mathbf{x})$ are the eigenvalue-vector pair of $\mathbf{A}$, then $(\lambda^{-1}, \mathbf{x})$ are the eigenvalue-vector pair for $\mathbf{A}^{-1}$. This fact when combined with (32.6.10) and (32.6.12) leads to the following relations:

$$\left.\begin{aligned} (\mathbf{D_M^T}\mathbf{D_M})^{-1}\mathbf{V} &= (\mathbf{D_M^{-1}}\mathbf{D_M^{-T}})\mathbf{V} = \mathbf{V}\boldsymbol{\Lambda}^{-1} \\ (\mathbf{D_M}\mathbf{D_M^T})^{-1}\mathbf{U} &= (\mathbf{D_M^{-T}}\mathbf{D_M^{-1}})\mathbf{U} = \mathbf{U}\boldsymbol{\Lambda}^{-1} \end{aligned}\right\} \tag{32.6.21}$$

Multiplying both sides of (32.6.2) on the left with $\mathbf{D_M^{-1}}(\bar{\mathbf{x}}_k)$ we obtain the inverse dynamics

$$\varepsilon_k = \mathbf{D_M^{-1}}(\bar{\mathbf{x}}_k)\varepsilon_{k+1} \tag{32.6.22}$$

Iterating it backward from time $(t+1)$ to s we get

$$\varepsilon_s = \mathbf{D}_{\mathbf{M}}^{-1}(t:s)\varepsilon_{t+1}. \tag{32.6.23}$$

where

$$\begin{aligned}\mathbf{D}_{\mathbf{M}}^{-1}(t:s) &= \{\mathbf{D}_{\mathbf{M}}(\bar{\mathbf{x}}_t)\mathbf{D}_{\mathbf{M}}(\bar{\mathbf{x}}_{t-1})\cdots\mathbf{D}_{\mathbf{M}}(\bar{\mathbf{x}}_s)\}^{-1}\\ &= \mathbf{D}_{\mathbf{M}}^{-1}(\bar{\mathbf{x}}_s)\cdots\mathbf{D}_{\mathbf{M}}^{-1}(\bar{\mathbf{x}}_{t-1})\mathbf{D}_{\mathbf{M}}^{-1}(\bar{\mathbf{x}}_t).\end{aligned}$$

Now, let $\varepsilon_{t+1} = \mathbf{u}_i$, the ith left or backward singular vector. Then, (denoting $\mathbf{D}_{\mathbf{M}}^{-1}(t:s)$ simply as $\mathbf{D}_{\mathbf{M}}^{-1}$)

$$\varepsilon_s = \mathbf{D}_{\mathbf{M}}^{-1}\mathbf{u}_i. \tag{32.6.24}$$

Multiplying both sides on the left by $(\mathbf{D}_{\mathbf{M}}^{\mathrm{T}}\mathbf{D}_{\mathbf{M}})^{-1}$, we get

$$\begin{aligned}(\mathbf{D}_{\mathbf{M}}^{\mathrm{T}}\mathbf{D}_{\mathbf{M}})^{-1}\varepsilon_s &= \mathbf{D}_{\mathbf{M}}^{-1}\mathbf{D}_{\mathbf{M}}^{-\mathrm{T}}\varepsilon_s = \mathbf{D}_{\mathbf{M}}^{-1}\mathbf{D}_{\mathbf{M}}^{-\mathrm{T}}\mathbf{D}_{\mathbf{M}}^{-1}\mathbf{u}_i\\ &= \mathbf{D}_{\mathbf{M}}^{-1}(\mathbf{D}_{\mathbf{M}}\mathbf{D}_{\mathbf{M}}^{\mathrm{T}})^{-1}\mathbf{u}_i\\ &= \mathbf{D}_{\mathbf{M}}^{-1}\mathbf{u}_i\lambda_i^{-1} \quad \text{(use 32.6.21)}\\ &= \varepsilon_s\lambda_i^{-1}.\end{aligned} \tag{32.6.25}$$

Comparing this with the first relation in (32.6.21) it follows that ε_s is indeed equal to $\mathbf{v}_i$, the corresponding forward singular vector. Stated in other words, under the action of the inverse dynamics, the backward singular vector $\mathbf{u}_i$ grows into the forward singular vector, $\mathbf{v}_i$.

Using (32.6.23), the Rayleigh coefficient for the inverse dynamics becomes

$$\frac{\|\varepsilon_s\|^2}{\|\varepsilon_{t+1}\|^2} = \frac{\varepsilon_{t+1}^{\mathrm{T}}\mathbf{D}_{\mathbf{M}}^{-\mathrm{T}}\mathbf{D}_{\mathbf{M}}^{-1}\varepsilon_{t+1}}{\varepsilon_{t+1}^{\mathrm{T}}\varepsilon_{t+1}} = \frac{\varepsilon_{t+1}^{\mathrm{T}}(\mathbf{D}_{\mathbf{M}}\mathbf{D}_{\mathbf{M}}^{\mathrm{T}})^{-1}\varepsilon_{t+1}}{\varepsilon_{t+1}^{\mathrm{T}}\varepsilon_{t+1}}. \tag{32.6.26}$$

Once again, by way of changing the coordinate, define

$$\varepsilon_{t+1} = \mathbf{U}\boldsymbol{\alpha} \quad \text{with} \quad \|\boldsymbol{\alpha}\| = 1. \tag{32.6.27}$$

Substituting this into (32.6.26), the latter becomes

$$\begin{aligned}\frac{\|\varepsilon_s\|^2}{\|\varepsilon_{t+1}\|^2} &= \frac{\boldsymbol{\alpha}^{\mathrm{T}}\mathbf{U}^{\mathrm{T}}(\mathbf{D}_{\mathbf{M}}\mathbf{D}_{\mathbf{M}}^{\mathrm{T}})^{-1}\mathbf{U}\boldsymbol{\alpha}}{\boldsymbol{\alpha}^{\mathrm{T}}\boldsymbol{\alpha}}\\ &= \boldsymbol{\alpha}^{\mathrm{T}}\boldsymbol{\Lambda}^{-1}\boldsymbol{\alpha} \quad \text{(using 32.6.10)}\\ &= \sum_{i=1}^{n}\frac{\alpha_i^2}{\lambda_i}.\end{aligned} \tag{32.6.28}$$

Thus, under the action of the inverse dynamics, errors ε_{t+1} that lie on a unit sphere are mapped onto the surface of an ellipsoid at time s whose axes are left singular vector $\mathbf{u}_i$ and the length of the semi-axes are given by $\sqrt{\lambda_i}$ for $i = 1$ to n.

(3) Effect of adjoint dynamics

The dynamics that is adjoint to (32.6.2) is given by

$$\mathbf{y}_k = \mathbf{D}_\mathbf{M}^\mathrm{T}(\bar{\mathbf{x}}_k)\mathbf{y}_{k+1}. \tag{32.6.29}$$

Taking the inner product of both sides of (32.6.2) with respect to $\mathbf{y}_k$ and that of (32.6.29) with respect to ε_{k+1}, we get

$$\begin{aligned}\mathbf{y}_k^\mathrm{T}\varepsilon_{k+1} &= \mathbf{y}_k^\mathrm{T}\mathbf{D}_\mathbf{M}(\bar{\mathbf{x}}_k)\varepsilon_k = \langle \mathbf{y}_k, \mathbf{D}_\mathbf{M}(\bar{\mathbf{x}}_k)\varepsilon_k\rangle \\ &= \varepsilon_{k+1}^\mathrm{T}\mathbf{y}_k = \varepsilon_{k+1}^\mathrm{T}\mathbf{D}_\mathbf{M}^\mathrm{T}(\bar{\mathbf{x}}_k)\mathbf{y}_{k+1} = \langle \varepsilon_{k+1}, \mathbf{D}_\mathbf{M}^\mathrm{T}(\bar{\mathbf{x}}_k)\mathbf{y}_{k+1}\rangle.\end{aligned}$$

Using the adjoint property, we can rewrite the above relation as

$$\langle \mathbf{y}_k, \mathbf{D}_\mathbf{M}(\bar{\mathbf{x}}_k)\varepsilon_k\rangle = \langle \mathbf{y}_{k+1}, \mathbf{D}_\mathbf{M}(\bar{\mathbf{x}}_k)\varepsilon_{k+1}\rangle \tag{32.6.30}$$

which is a fundamental property that relates the adjoint and the forward variables.

Iterating (32.6.29), we get

$$\mathbf{y}_s = \mathbf{D}_\mathbf{M}^\mathrm{T}\mathbf{y}_{t+1} \tag{32.6.31}$$

where

$$\mathbf{D}_\mathbf{M}^\mathrm{T} = \mathbf{D}_\mathbf{M}^\mathrm{T}(\bar{\mathbf{x}}_s)\mathbf{D}_\mathbf{M}^\mathrm{T}(\bar{\mathbf{x}}_{s+1})\cdots\mathbf{D}_\mathbf{M}^\mathrm{T}(\bar{\mathbf{x}}_t).$$

Let $\mathbf{y}_{t+1} = \mathbf{u}_i$. Then

$$\mathbf{y}_s = \mathbf{D}_\mathbf{M}^\mathrm{T}\mathbf{y}_{t+1} = \mathbf{D}_\mathbf{M}^\mathrm{T}\mathbf{u}_i.$$

Multiplying both sides on the left with $(\mathbf{D}_\mathbf{M}^\mathrm{T}\mathbf{D}_\mathbf{M})$ we get

$$(\mathbf{D}_\mathbf{M}^\mathrm{T}\mathbf{D}_\mathbf{M})\mathbf{y}_s = \mathbf{D}_\mathbf{M}^\mathrm{T}(\mathbf{D}_\mathbf{M}\mathbf{D}_\mathbf{M}^\mathrm{T})\mathbf{u}_i = \mathbf{D}_\mathbf{M}^\mathrm{T}\mathbf{u}_i\lambda_i = \mathbf{y}_s\lambda_i.$$

That is, under the action of the adjoint dynamics, the backward singular vectors grow (in reverse time) into the forward singular vectors.

It is interesting to note that while in general adjoint dynamics is different from the inverse dynamics, with respect to the backward singular vectors their actions lead to identical results namely backward singular vectors grow into forward singular vectors in reverse time.

(4) Singular vector as eigenvector of covariance matrix

Let $\mathbf{P}_k$ be the covariance of the perturbation ε_k that is superimposed on the base state $\bar{\mathbf{x}}_k$. Then, to a first-order approximation, $\mathbf{P}_{k+1}$ is related to $\mathbf{P}_k$ via the recurrence (refer to Chapter 31)

$$\mathbf{P}_{k+1} = \mathbf{D}_\mathbf{M}(\bar{\mathbf{x}}_k)\mathbf{P}_k\mathbf{D}_\mathbf{M}^\mathrm{T}(\bar{\mathbf{x}}_k). \tag{32.6.32}$$

Iterating this from time s to $t+1$, we get

$$\mathbf{P}_{t+1} = \mathbf{D}_\mathbf{M}\mathbf{P}_s\mathbf{D}_\mathbf{M}^\mathrm{T}. \tag{32.6.33}$$

The question is how to choose ε_s that maximizes

$$J(\varepsilon_s) = \varepsilon_{t+1}^{\mathrm{T}}\varepsilon_{t+1} = \varepsilon_s^{\mathrm{T}}\mathbf{D}_{\mathbf{M}}^{\mathrm{T}}\mathbf{D}_{\mathbf{M}}\varepsilon_s \tag{32.6.34}$$

when

$$\varepsilon_s^{\mathrm{T}}\mathbf{P}_s^{-1}\varepsilon_s = 1. \tag{32.6.35}$$

Solving this constrained minimization problem using the Lagrangian multiplier method, it can be verified that the maximizing ε_s is given by the solution of the following generalized eigenvalue problem

$$\mathbf{D}_{\mathbf{M}}^{\mathrm{T}}\mathbf{D}_{\mathbf{M}}\varepsilon_s = \lambda\mathbf{P}_s^{-1}\varepsilon_s. \tag{32.6.36}$$

Let $\mathbf{P}_s = \mathbf{S}\mathbf{S}^{\mathrm{T}}$ be the Cholesky factorization of $\mathbf{P}_s$ (Chapter 9). Substituting this into (32.6.36) and multiplying both sides by $\mathbf{S}^{\mathrm{T}}$ we obtain

$$(\mathbf{D}_{\mathbf{M}}\mathbf{S})^{\mathrm{T}}(\mathbf{D}_{\mathbf{M}}\mathbf{S})\boldsymbol{\eta}_s = \lambda\boldsymbol{\eta}_s \tag{32.6.37}$$

where $\boldsymbol{\eta}_s = \mathbf{S}^{-1}\varepsilon_s$.

Define

$$\boldsymbol{\xi}_t = (\mathbf{D}_{\mathbf{M}}\mathbf{S})\boldsymbol{\eta}_s = \mathbf{D}_{\mathbf{M}}\varepsilon_s. \tag{32.6.38}$$

Now consider the action of $\mathbf{P}_{t+1}$ on ε_t:

$$\begin{aligned}
\mathbf{P}_{t+1}\boldsymbol{\xi}_t &= (\mathbf{D}_{\mathbf{M}}\mathbf{P}_s\mathbf{D}_{\mathbf{M}}^{\mathrm{T}})\boldsymbol{\xi}_t \quad \text{(use 32.6.33)}\\
&= (\mathbf{D}_{\mathbf{M}}\mathbf{S})(\mathbf{D}_{\mathbf{M}}\mathbf{S})^{\mathrm{T}}\boldsymbol{\xi}_t\\
&= (\mathbf{D}_{\mathbf{M}}\mathbf{S})(\mathbf{D}_{\mathbf{M}}\mathbf{S})^{\mathrm{T}}\mathbf{D}_{\mathbf{M}}\mathbf{S}\boldsymbol{\eta}_s \quad \text{(use 32.6.38)}\\
&= (\mathbf{D}_{\mathbf{M}}\mathbf{S})\lambda\boldsymbol{\eta}_s \quad \text{(use 32.6.37)}\\
&= \lambda\boldsymbol{\xi}_t \quad \text{(use 32.6.38)}.
\end{aligned} \tag{32.6.39}$$

That is, $\boldsymbol{\xi}_t$ is the eigenvector of the covariance matrix $\mathbf{P}_{t+1}$.

The moral of this story is as follows: from (32.6.18) – (32.6.19) it follows that if $\varepsilon_s = \mathbf{v}_i$, a forward singular vector, then $\boldsymbol{\xi}_t$ in (32.6.38) must be the backward singular vector which by (32.6.39) is also an eigenvector of the covariance matrix $\mathbf{P}_{t+1}$. Stated in other words under the action of the dynamics in (32.6.38), a forward singular vector grows into an eigenvector of the covariance matrix at time $(t+1)$.

We conclude this section with the following example.

Example 32.6.1 Let $n = 2$ and let $\mathbf{D}_{\mathbf{M}}(1:0) = \mathbf{A}$ where

$$\mathbf{A} = \begin{bmatrix} 0.5 & 1.0 \\ 2 & 1.5 \end{bmatrix}.$$

The eigenvectors and eigenvalues of $\mathbf{A}$ are given by

$$\mathbf{W} = \begin{bmatrix} -0.7071 & -0.4472 \\ 0.7071 & 0.8944 \end{bmatrix} \quad \text{and} \quad \mathbf{D} = \begin{bmatrix} -0.5 & 0 \\ 0 & 2.5 \end{bmatrix}.$$

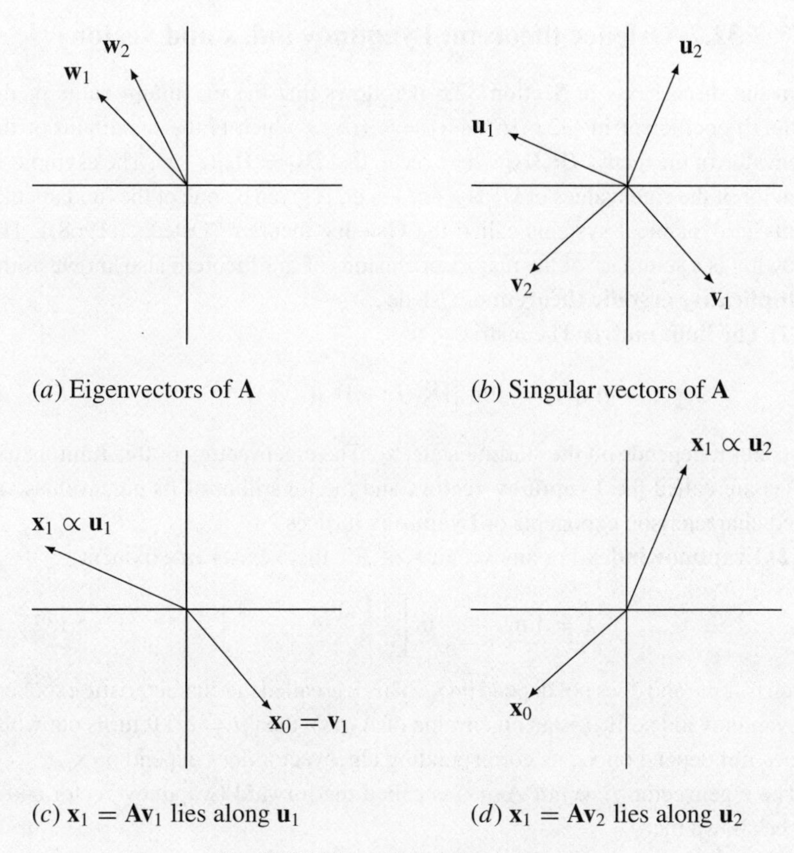

(a) Eigenvectors of $\mathbf{A}$ (b) Singular vectors of $\mathbf{A}$

(c) $\mathbf{x}_1 = \mathbf{A}\mathbf{v}_1$ lies along $\mathbf{u}_1$ (d) $\mathbf{x}_1 = \mathbf{A}\mathbf{v}_2$ lies along $\mathbf{u}_2$

Fig. 32.6.1 Illustration of singular vectors.

Then $(\mathbf{A}^{\mathrm{T}}\mathbf{A})\mathbf{V} = \mathbf{V}\mathbf{\Lambda}$ and $(\mathbf{A}\mathbf{A}^{\mathrm{T}})\mathbf{U} = \mathbf{U}\mathbf{\Lambda}$ are given by

$$\mathbf{A}^{\mathrm{T}}\mathbf{A} = \begin{bmatrix} 4.25 & 3.5 \\ 3.5 & 3.25 \end{bmatrix} \quad \mathbf{V} = \begin{bmatrix} 0.6552 & -0.7558 \\ -0.7555 & -0.6552 \end{bmatrix}$$

$$\mathbf{\Lambda} = \begin{bmatrix} 0.2142 & 0 \\ 0 & 7.2885 \end{bmatrix}$$

$$\mathbf{A}\mathbf{A}^{\mathrm{T}} = \begin{bmatrix} 1.25 & 2.5 \\ 2.5 & 6.25 \end{bmatrix} \quad \mathbf{U} = \begin{bmatrix} -0.9239 & 0.3827 \\ 0.3827 & 0.9239 \end{bmatrix}.$$

If $\mathbf{x}_0 = \mathbf{v}_1$, then it can be verified that $\mathbf{x}_1 = \mathbf{A}\mathbf{x}_0 = 0.463\mathbf{u}_1$. Likewise, when $\mathbf{x}_0 = \mathbf{v}_2$, we get $\mathbf{x}_1 = \mathbf{A}\mathbf{x}_0 = -2.6992\mathbf{u}_2$. Refer to Figure 32.6.1 for a graphical illustration.

32.7 Osledec theorem: Lyapunov index and vector

From the discussions in Section 32.6 it follows that the maximum value of the Rayleigh coefficient in (32.6.16) is $\lambda_1 = \lambda_1(t : s)$ which is the maximum of the eigenvalue of the matrix $\mathbf{D}_\mathbf{M}^\mathrm{T}\mathbf{D}_\mathbf{M}$ where recall that $\mathbf{D}_\mathbf{M} = \mathbf{D}_\mathbf{M}(t : s)$. The asymptotic behavior of the eigenvalues of $\mathbf{D}_\mathbf{M}^\mathrm{T}\mathbf{D}_\mathbf{M}$ as $t \to \infty$ is given by one of the fundamental results in dynamical systems called the **Osledec** theorem (Osledec (1968)). The following is a summary of the major conclusions of this theorem also known as the **multiplicative ergodic theorem** of Osledec.

(1) The limit matrix The matrix

$$\mathbf{\Lambda}_\mathbf{M}(s) = \lim_{t\to\infty} \left[\mathbf{D}_\mathbf{M}^\mathrm{T}(t : s)\mathbf{D}_\mathbf{M}(t : s)\right]^{\frac{1}{2(t-s)}} \tag{32.7.1}$$

exists but it depends on the starting state, $\mathbf{x}_s$. The eigenvectors of this limit matrix $\mathbf{\Lambda}_\mathbf{M}(s)$ are called the **Lyapunov vectors** and the logarithm of its eigenvalues are called characteristic exponents or **Lyapunov indices.**

(2) Lyapunov index For any vector $\varepsilon_s \in \mathbb{R}^n$, there exists an exponent

$$\lambda = \lim_{t\to\infty} \frac{1}{t-s} \ln \left[\frac{\| \mathbf{D}_\mathbf{M}(t : s)\varepsilon_s \|}{\| \varepsilon_s \|}\right] \tag{32.7.2}$$

which is finite and does not depend on $\mathbf{x}_s$. This λ is called the characteristic exponent or Lyapunov index. If μ is an eigenvalue of $\mathbf{\Lambda}_\mathbf{M}(s)$, then $\mu = \mathrm{e}^\lambda$. It turns out while λ does **not** depend on $\mathbf{x}_s$, its corresponding eigenvector does depend on $\mathbf{x}_s$.

The eigenvector $f_i^+(s)$ of $\mathbf{\Lambda}_\mathbf{M}(s)$ is called the forward Lyapunov vector and it can be shown that

$$f_i^+(s) = \lim_{t\to\infty} \mathbf{v}_i(t : s) \tag{32.7.3}$$

where $\mathbf{v}_i(t : s)$ is the eigenvector of $\mathbf{D}_\mathbf{M}^\mathrm{T}\mathbf{D}_\mathbf{M}$ which is also known as the right or the forward singular vector of $\mathbf{D}_\mathbf{M}(t : s)$.

(3) Embedded subspaces There exists a sequence of embedded subspaces

$$F_n^+(s) \subset F_{n-1}^+(s) \subset \cdots \subset F_1^+(s) = \mathbb{R}^n \tag{32.7.4}$$

with the following properties. Refer to Figure 32.7.1.

(a) Each $F_i^+(s)$ is invariant under the tangent flow operator $\mathbf{D}_\mathbf{M}(t : s)$, that is,

$$\mathbf{D}_\mathbf{M}(t : s)F_i^+(s) = F_i^+(t) \tag{32.7.5}$$

for all $t \geq s$.

(b) Perturbations in $F_i^+(s) \setminus F_{i+1}^+(s)$ (which is the set of all vectors in $F_i^+(s)$ but not in $F_{i+1}^+(s)$) grow at a rate λ_i, which is the ith Lyapunov index.

Before providing an algorithm for computing the Lyapunov indices, we first illustrate its meaning using the example of a scalar dynamics.

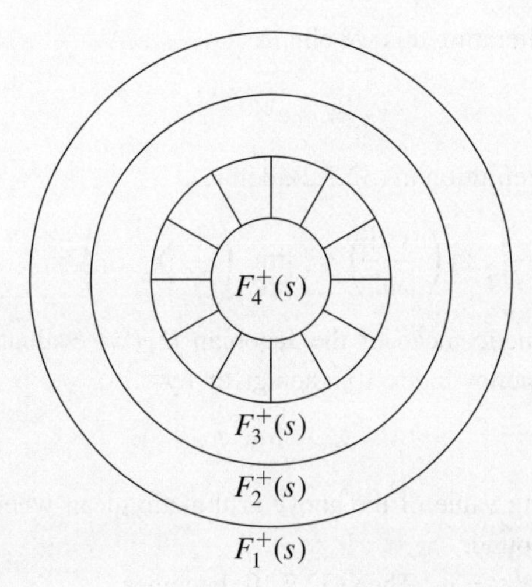

Fig. 32.7.1 An illustration of embedded subspaces. The hatched region denotes $F_3^+(s) \setminus F_4^+(s)$.

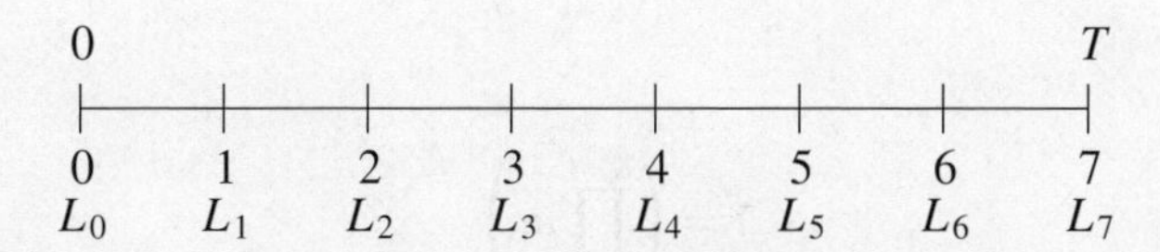

Fig. 32.7.2 An illustration of the approximation with $N = 7$, $L_{k-1} = \mathbf{D}_f(\bar{x}_t)|_{t=(k-1)\Delta}$.

Example 32.7.1 Let $\dot{x} = f(x)$. Let $\bar{x}_t$ be the base trajectory starting from $\bar{x}_0$. Let y_0 be the perturbation superimposed on $\bar{x}_0$. Then y_t, the perturbation at time t is given by the linear nonautonomous dynamics

$$\dot{y}_t = \mathbf{D}_f(\bar{x}_t)y \tag{32.7.6}$$

where $\mathbf{D}_f(\bar{x}_t) = \mathrm{d}f(\bar{x}_t)/\mathrm{d}x$. Since $\mathbf{D}_f(\bar{x}_t)$ varies along the base state, let us replace (32.7.6) using a cascaded system of equations as follows: Let $[0, T]$ denote the period of interest. Discretize this interval into N equal subintervals each of length, say ι, where $N\tau = T$. Refer to Figure 32.7.2.

The value of N is such that $\mathbf{D}_f(\bar{x}_t)$ is nearly constant in each subinterval. Define

$$L_{k-1} = \mathbf{D}_f(\bar{x}_t)|_{t=(k-1)\Delta}. \tag{32.7.7}$$

Then on the kth subinterval $(k-1)\tau \leq t \leq k\tau$ (32.7.6) is replaced by

$$\dot{y} = L_{k-1}y \tag{32.7.8}$$

for $k = 1, 2, \ldots, N$. Solving this we get

$$y_k = y_{k-1}\mathrm{e}^{L_{k-1}\tau} \tag{32.7.9}$$

where $y_k = y_{t=k\tau}$. Iterating this we obtain

$$y_N = y_0 e^{\left(\sum_{k=0}^{N-1} L_k\right)\tau}. \tag{32.7.10}$$

Now invoking the definition in (32.7.2), define

$$\lambda_T = \lim_{\tau \to 0} \frac{1}{N\tau} \ln \left(\frac{|y_N|}{|y_0|} \right) = \lim_{N \to \infty} \left(\frac{1}{N} \sum_{i=0}^{N-1} L_i \right) = \bar{L}_T \tag{32.7.11}$$

which is the arithmetic mean of the Jacobian $\mathbf{D}_f(\bar{x}_t)$ evaluated along the base trajectory. The Lyapunov index λ is then given by

$$\lambda = \lim_{T \to \infty} \bar{L}_T \tag{32.7.12}$$

which is the limiting value of the above arithmetic mean when the time horizon increases without bound.

Alternately, let $e^{L_k} = a_k$. Then (32.7.10) becomes

$$y_N = \left(\prod_{k=0}^{N-1} a_k \right)^{\mathrm{T}} y_0. \tag{32.7.13}$$

Let

$$\bar{a} = \left(\prod_{k=0}^{N-1} a_k \right)^{1/N} \tag{32.7.14}$$

be the geometric mean of the a_k's. Then

$$\lambda_T = \ln \bar{a} = \frac{1}{N} \sum_{k=0}^{N-1} \log a_k = \frac{1}{N} \sum_{k=0}^{N-1} L_k = \bar{L}_T$$

which is the same as in (32.7.11).

Since λ denotes the long-term average rate of growth of errors, we can see that y_t grows (assuming λ is positive) according to

$$y_t \approx e^{\lambda t} y_0. \tag{32.7.15}$$

Thus, consistent with the usual notion of **time constant** of a system, it follows that when $t = t_p = 1/\lambda$

$$\frac{|y_t|}{|y_0|} = e$$

that is, the value of $t_p = 1/\lambda$ can be used as a measure of the **predictability limit** for the system $\dot{x} = f(x)$.

We now describe two algorithms for computing the Lyapunov index.

Algorithm 1. Using the maximum eigenvalue of the propagator or resolvant Let $\lambda_i(t : s)$ for $i = 1$ to n denote the eigenvalues of the propagator or the resolvant

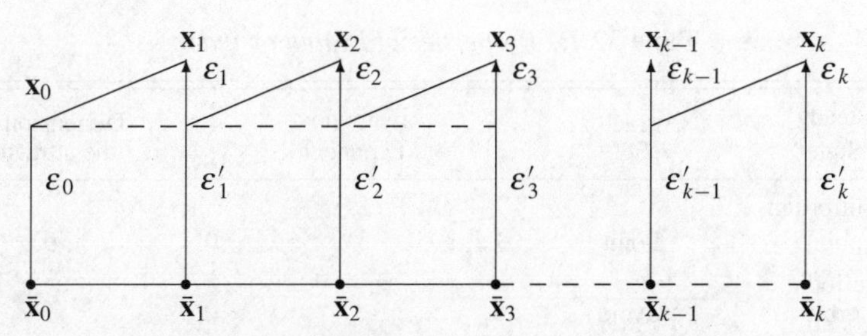

Fig. 32.7.3 Renormalization strategy: an illustration.

$\mathbf{D_M}(t:s)$. Let

$$|\lambda_1(t:s)| = \max_i\{|\lambda_i(t:s)|\}.$$

Then, the Lyapunov number is given by

$$L = \lim_{t\to\infty} |\lambda_1(t:s)|^{\frac{1}{(t-s)}} \tag{32.7.16}$$

which is the $(t-s)$th root of the absolute value of the maximum eigenvalue of $\mathbf{D_M}(t:s)$ as $t \to \infty$. The Lyapunov index is then given by

$$\lambda = \ln L. \tag{32.7.17}$$

While the above method is conceptually elegant, computationally it is often demanding since it requires repeated matrix–matrix multiplication and solution of an eigenvalue problem. There is an alternative and a practical way to compute Lyapunov index by a renormalization strategy that directly uses the nonlinear system instead of the linear approximation given by the tangent linear system.

Algorithm 2. A renormalization strategy

Let $\varepsilon_0 = \mathbf{x}_0 - \bar{\mathbf{x}}_0$ be the initial perturbation and let

$$\varepsilon_k = \mathbf{x}_k - \bar{\mathbf{x}}_k = \mathbf{M}(\mathbf{x}_{k-1}) - \mathbf{M}(\bar{\mathbf{x}}_{k-1}) \tag{32.7.18}$$

be the actual error in the nonlinear trajectory at time k. Refer to Figure 32.7.3. Then, from

$$\frac{\|\varepsilon_k\|}{\|\varepsilon_0\|} = \frac{\|\varepsilon_1\|}{\|\varepsilon_0\|}\frac{\|\varepsilon_2\|}{\|\varepsilon_1\|}\cdots\frac{\|\varepsilon_k\|}{\|\varepsilon_{k-1}\|}$$

we get

$$\log\frac{\|\varepsilon_k\|}{\|\varepsilon_0\|} = \sum_{k=0}^{k-1}\log\frac{\|\varepsilon_{k+1}\|}{\|\varepsilon_k\|}. \tag{32.7.19}$$

Then the first Lyapunov index is given by

$$\lambda = \lim_{N\to\infty}\lim_{\|\varepsilon_0\|\to 0}\frac{1}{N}\sum_{k=0}^{N-1}\log\frac{\|\varepsilon_{k+1}\|}{\|\varepsilon_k\|} \tag{32.7.20}$$

Table 32.7.1 *Properties of Lyapunov index*

Steady State	Attractor Set	Lyapunov exponent	Dimension of the attractor
Equilibrium point	Point	$\lambda_n < \lambda_{n-1} < \cdots < \lambda_1 < 0$	0
Periodic orbit	Cycle	$\lambda_1 = 0$ $\lambda_n < \lambda_{n-1} < \cdots < \lambda_2 < 0$	1
Two periodic orbits	Torus	$\lambda_1 = \lambda_2 = 0$ $\lambda_n < \lambda_{n-1} < \cdots < \lambda_3 < 0$	2
Chaotic	Fractal structure	$\lambda_1 > 0 \sum_{i=1}^{n} \lambda_i < 0$	non-integer

Step 1 Choose $\bar{\mathbf{x}}_0$ and compute the nonlinear trajectory $\bar{\mathbf{x}}_1, \bar{\mathbf{x}}_2, \bar{\mathbf{x}}_3, \ldots, \bar{\mathbf{x}}_N$ where N is a large fixed integer.

Step 2 Let ε_0 be a random vector such that $\| \varepsilon_0 \|$ is small and fixed. $\mathbf{x}_0 = \bar{\mathbf{x}}_0 + \varepsilon_0$. Set the accumulator $L = 0$.

Step 3 For $k = 0, 1, 2, \ldots, N-1$ do the following:

(a) **Renormalize:** $\varepsilon'_k = \frac{\varepsilon_k}{\|\varepsilon_k\|} \| \varepsilon_0 \|$

(b) **Compute the error:** $\mathbf{x}'_k = \bar{\mathbf{x}}_k + \varepsilon_k$

$$\mathbf{x}_{k+1} = \mathbf{M}(\mathbf{x}'_k)$$

$$\varepsilon_{k+1} = \mathbf{x}_{k+1} - \bar{\mathbf{x}}_{k+1}$$

(c) **Local amplification factor:** $a_k = \ln\left(\frac{\|\varepsilon_{k+1}\|}{\|\varepsilon_k\|}\right)$

(d) **Accumulate:** $L \leftarrow L + a_k$

Step 4 The first Lyapunov index $\lambda = L/N$.

Fig. 32.7.4 Computation of Lyapunov index using renormalization strategy.

which is the average of the logarithm of the amplification along the trajectory starting at $\bar{\mathbf{x}}_0$ as the initial perturbation shrinks in size and as the time horizon increases without bound. The algorithm is given in Figure 32.7.4.

A listing of the range of values for Lyapunov indices for various types of equilibrium or invariant sets of dynamical system is given in Table 32.7.1. Actual values of these indices for three typical systems are given in Table 32.7.2.

We conclude this section with the following remarks.

(1) **Sensitive dependence on initial conditions** It turns out for forced chaotic systems, the first Lyapunov index is strictly positive which is responsible for the sensitive dependence on initial conditions.
(2) The sum of all the Lyapunov indices is strictly negative for dissipative systems. Hence the last Lyapunov index is strictly negative. For this class of systems, one of the intermediate indices is zero.

Table 32.7.2 *Values of Lyapunov indices*

Model	Values of Lyapunov indices
Logistic $x_{k+1} = 4x_k(1 - x_k)$	$\lambda = 0.6931$
Henon (Exercise 32.11)	$\lambda_1 = 0.42, \lambda_2 = -1.62$
Lorenz(1963) (Example 32.4.2)	$\lambda_1 = 0.9, \lambda_2 = 0, \lambda_3 = -12.8$

(3) The growth rate of line segments is given by λ_1 and the growth rate of surface elements is given by $\lambda_1 + \lambda_2$. Similarly, $\sum_{i=1}^{k} \lambda_i$ denotes the growth rate of k-dimensional volumes. Thus for Lorenz's attractor (see Example 32.4.2), errors amplify at a rate $e^{\lambda_1} = e^{0.9} = 2.4596$ but volumes contract at a rate $e^{\lambda_1+\lambda_2+\lambda_3} = e^{-11.9} = 6.79 \times 10^{-6}$.

(4) **Predictability limit** Let the first r of the n Lyapunov indices be positive and let $\bar{\lambda}_p = \sum_{i=1}^{r} \lambda_i$. Then, it follows that two states that are infinitesimally close will diverge at a rate $\bar{\lambda}_p$ and their separation will grow as $e^{\bar{\lambda}_p t}$. Thus in time $t_p = 1/\bar{\lambda}_p$, their separation will grow by the factor e. Hence, we can use $(\bar{\lambda}_p)^{-1}$ as a useful measure of the **predictability limit**.

32.8 Deterministic ensemble approach to predictability

Let

$$\mathbf{x}_{k+1} = \mathbf{M}(\mathbf{x}_k) \tag{32.8.1}$$

denote the given deterministic model. As noted in Section 32.1 a useful way to capture the different modes of behavior of this deterministic model is to create an ensemble of initial states $\mathbf{x}_0(i)$, $i = 1, 2, \ldots, N$ centered around the given initial state $\bar{\mathbf{x}}_0$ by adding a perturbation $\varepsilon_0(i)$ to $\bar{\mathbf{x}}_0$ such that

$$\mathbf{x}_0(i) = \bar{\mathbf{x}}_0 + \varepsilon_0(i) \tag{32.8.2}$$

where the subscript denotes the discrete time index and i denotes the ith member of the initial ensemble. The basic idea is to compute the N strands of the model trajectories where

$$\mathbf{x}_{k+1}(i) = \mathbf{M}(\mathbf{x}_k(i)) \tag{32.8.3}$$

for $i = 1, 2, \ldots, N$ and $k = 0, 1, 2, \ldots$. Using this information, we can extract quite a variety of useful information about the forecast errors including sample statistics and histograms provided N is large. The **sample mean** and **covariance** are computed using

$$\bar{\mathbf{x}}_k = \frac{1}{N}\sum_{i=1}^{N}\mathbf{x}_k(i) \tag{32.8.4}$$

and

$$\bar{\mathbf{P}}_k = \frac{1}{N}\sum_{i=1}^{N}(\mathbf{x}_k(i) - \bar{\mathbf{x}}_k)(\mathbf{x}_k(i) - \bar{\mathbf{x}}_k)^{\mathrm{T}}. \tag{32.8.5}$$

This deterministic ensemble approach differs from the Monte Carlo method in that in the latter method the initial ensemble is created by using sample realizations from the given initial probability distribution. In the deterministic approach of interest in this chapter, since no such distribution is available, we have to turn to other ways to create such an ensemble.

In the following we describe two methods of generating the initial deterministic ensemble.

32.8.1 Ensemble generation using forward singular vectors

Referring to Section 32.6, let $\mathbf{D}_{\mathbf{M}}(T:0)$ be the propagator or the resolvant of the tangent linear dynamics over the period $[0, T]$. Let $\mathbf{V}(T:0)$ and $\boldsymbol{\Lambda}(T:0)$ be the matrices of eigenvectors and eigenvalues of $\mathbf{D}_{\mathbf{M}}^{\mathrm{T}}(T:0)\mathbf{D}_{\mathbf{M}}(T:0)$ where

$$\boldsymbol{\Lambda}(T:0) = \mathrm{Diag}(\lambda_1(T:0), \lambda_2(T:0), \ldots, \lambda_n(T:0))$$

where it is assumed that

$$\lambda_1(T:0) > \lambda_2(T:0) > \cdots > \lambda_r(T:0) > 1 > \lambda_{r+1}(T:0) > \cdots > \lambda_N(T:0). \tag{32.8.6}$$

Recall that the columns of $\mathbf{V}(T:0)$ are also known as the right or the forward singular vectors of $\mathbf{D}_{\mathbf{M}}(T:0)$ and the first r columns of $\mathbf{V}(T:0)$ correspond to the growing modes.

In this method, based on the linear analysis the ith member $\varepsilon_0(i)$ of the initial ensemble is chosen as the linear combination of the first r columns of $\mathbf{V}(T:0)$ which are the first r forward singular vectors. That is,

$$\varepsilon_0(i) = \sum_{j=1}^{r}\alpha_j(i)\mathbf{v}_j(T:0) \tag{32.8.7}$$

for $i = 1, 2, \ldots, N$, where $\boldsymbol{\alpha}(i) = (\alpha_1(i), \alpha_2(i), \ldots, \alpha_r(i))^{\mathrm{T}} \in \mathbb{R}^r$ is chosen randomly.

32.8.2 Ensemble generation using breeding strategy

In this approach an initial perturbation ε_0 is chosen randomly and allowed to grow by breeding it using the given nonlinear system as is done in the renormalizaition strategy described in the context of computing the leading Lyapunov exponent in Figure 32.7.3. If the given nonlinear system is indeed unstable, then the initial error will eventually grow and align itself along the direction of maximum rate of growth. In this nonlinear breeding method, the members of the initial ensemble are chosen along the grown or bred directions.

The key to understanding this method lies in quantifying the properties of this grown or bred directions.

(1) First, recall from Figure 32.7.3 that $\varepsilon_k = \mathbf{x}_k - \bar{\mathbf{x}}_k$ is computed using the nonlinear system starting from ε'_{k-1}. However, since ε'_{k-1} is the renormalized version of ε_{k-1} and $\| \varepsilon'_{k-1} \| = \| \varepsilon_0 \|$ which is assumed to be small, we can approximate the actual nonlinear error ε_k using the tangent linear dynamics as

$$\varepsilon_k = \mathbf{D_M}(\bar{\mathbf{x}}_{k-1})\varepsilon'_{k-1}. \tag{32.8.8}$$

(2) Secondly, recall that the eigenvectors $\mathbf{V}(T:0)$ of $\mathbf{D}^{\mathrm{T}}_{\mathbf{M}}(T:0)\mathbf{D_M}(T:0)$ span the space $\mathbb{R}^n$. Hence any random vector $\varepsilon_0 \in \mathbb{R}^n$ can be expressed uniquely as a linear combination of the columns of $\mathbf{V}(T:0)$, that is, $\varepsilon_0 = \mathbf{V}(T:0)\boldsymbol{\beta}$ for some random vector $\boldsymbol{\beta} \in \mathbb{R}^n$.
(3) Thirdly, recall from Section 32.6 that under the action of the tangent linear dynamics forward singular vector $\mathbf{v}_i(T:0)$ grows into the corresponding backward singular vector $\mathbf{u}_i(T:0)$.
(4) Lastly, it follows from Section 32.6 that the right eigenvectors (same as forward singular vectors) in $\mathbf{V}(T:0)$ and left eigenvectors (same as the backward singular vectors) in $\mathbf{U}(T:0)$ share the same set of eigenvalues in $\boldsymbol{\Lambda}(T:0)$. Hence from (32.8.6) it follows that the first $k \le r$ columns in $\mathbf{V}(T:0)$ and $\mathbf{U}(T:0)$ correspond to the growing modes.

Combining the above line of reasoning it follows that any random initial perturbation which is a linear combination of the forward singular vectors, under the action of the tangent linear dynamics, grow into the backward singular vectors. With the renormalization strategy, since (32.8.8) is a very good approximation to the actual nonlinear error, it follows that the grown or the bred modes lie in the subspace spanned by the leading backward singular vectors $\mathbf{u}_1(T:0)$, $\mathbf{u}_2(T:0), \ldots, \mathbf{u}_k(T:0)$.

Stated in other words, the initial members of the ensemble $\varepsilon_0(i)$ are given by

$$\varepsilon_0(i) = \sum_{j=1}^{k} \beta_j(i)\mathbf{u}_j(T:0). \tag{32.8.9}$$

Thus, in contrast to the linear method that generates an ensemble based on the forward singular vectors, this nonlinear method generates an ensemble based on the backward singular vectors.

Exercises

32.1 Let $\dot{x} = Ax$ where $x \in \mathbb{R}^2$ and $A \in \mathbb{R}^{2\times 2}$. Plot the vector field when A is given by

$$\begin{bmatrix} 2 & 0 \\ 0 & 1/2 \end{bmatrix}, \quad \begin{bmatrix} -2 & 0 \\ 0 & 1/2 \end{bmatrix}, \quad \begin{bmatrix} 2 & 0 \\ 0 & -1/2 \end{bmatrix}, \quad \begin{bmatrix} -2 & 0 \\ 0 & -1/2 \end{bmatrix}$$

$$\text{d}\begin{bmatrix} -1 & -1 \\ 1 & -1 \end{bmatrix}, \quad \begin{bmatrix} 0 & 0 \\ -3 & 0 \end{bmatrix} \text{ and } \begin{bmatrix} 2 & 0 \\ 1 & 2 \end{bmatrix}.$$

32.2 Draw the phase portraits of the dynamical systems in Exercise 32.1.

32.3 Compute e^{Bt} when

$$B = \begin{bmatrix} a & 1 & 0 \\ 0 & a & 1 \\ 0 & 0 & a \end{bmatrix}.$$

Hint: $B = aI + C$ where

$$C = \begin{bmatrix} 0 & 1 & 0 \\ 0 & 0 & 1 \\ 0 & 0 & 0 \end{bmatrix}.$$

Generalize your result to the case of an $n\times n$ matrix

$$B = \begin{bmatrix} a & 1 & 0 & 0 & \cdot & \cdot & 0 & 0 \\ 0 & a & 1 & 0 & \cdot & \cdot & 0 & 0 \\ \cdot & \cdot & \cdot & \cdot & \cdot & \cdot & \cdot & \cdot \\ \cdot & \cdot & \cdot & \cdot & \cdot & \cdot & \cdot & \cdot \\ 0 & 0 & 0 & 0 & \cdot & \cdot & a & 1 \\ 0 & 0 & 0 & 0 & \cdot & \cdot & 0 & a \end{bmatrix}.$$

32.4 Verify the following

(a) If $AB = BA$, then $\mathrm{e}^{A+B} = \mathrm{e}^{A} \cdot \mathrm{e}^{B}$.

(b) $\left(\mathrm{e}^{A}\right)^{-1} = \mathrm{e}^{-A}$.

(c) If a is an eigenvalue of A, then e^{a} is an eigenvalue of e^{A}. How are the eigenvectors related?

32.5 Prove that $d/\mathrm{d}t\left(\mathrm{e}^{At}\right) = A\mathrm{e}^{At} = \mathrm{e}^{At}A$.

32.6 Compute e^{At} when A is given by

$$\begin{bmatrix} -6 & -4 \\ 5 & 3 \end{bmatrix}, \quad \begin{bmatrix} 0 & 2 \\ 1 & 0 \end{bmatrix}, \quad \begin{bmatrix} \mathrm{i} & 0 \\ 0 & -\mathrm{i} \end{bmatrix}$$

where $\mathrm{i} = \sqrt{-1}$.

32.7 Show that any 3×3 real matrix is similar to one of the four following matrices.

$$\begin{bmatrix} \alpha_1 & 0 & 0 \\ 0 & \alpha_2 & 0 \\ 0 & 0 & \alpha_3 \end{bmatrix}, \begin{bmatrix} \alpha_1 & 0 & 0 \\ 0 & \alpha_1 & 0 \\ 0 & 0 & \alpha_2 \end{bmatrix}, \begin{bmatrix} \alpha_1 & 1 & 0 \\ 0 & \alpha_1 & 1 \\ 0 & 0 & \alpha_1 \end{bmatrix}, \begin{bmatrix} a & -b & 0 \\ b & a & 0 \\ 0 & 0 & \alpha_3 \end{bmatrix}.$$

32.8 Let $\dot{x} = Ax$ and $\|x\|_2 = \left(x_1^2 + x_x^2 + \cdots x_n^2\right)^{1/2}$, the standard Euclidean norm. Prove that

$$\frac{\mathrm{d}}{\mathrm{d}t}(\|x\|_2) = \frac{x^t Ax}{\|x\|_2}.$$

32.9 Using the discrete version of the nonlinear equations in (32.4.4) and using the standard Euler scheme given below, draw the field and the phase portrait starting at various initial conditions and verify the presence of the limit cycle. What is the basin of attraction for this attractor?

$$x_{k+1} = x_k(1 + \Delta t) + \Delta t[y_k - x_k(x_k^2 + y_k^2)]$$
$$y_{k+1} = y_k(1 + \Delta t) - \Delta t[x_k + y_k(x_k^2 + y_k^2)].$$

32.10 Let $a = 10$, and $b = 8/3$. Compute the Lyapunov index for various values of r given in Table (32.2.3) by starting at initial conditions close to three equilibria E_1, E_2, and E_3.

32.11 Find the equilibria and analyze their stability properties for the following dynamical systems.

(a) **Logistic model**

$$x_{k+1} = ax_k(1 - x_k) \text{ where } a \geq 0.$$

(b) **Two species population model**

$$x_{k+1} = ax_k - bx_k y_k$$
$$y_{k+1} = cy_k + dx_k y_k.$$

(c) **Henon model** with $a > 0$, $b > 0$

$$x_{k+1} = y_k + 1 - ax_k^2$$
$$y_{k+1} = bx_k.$$

For $b = 0.3$ and $a = 1.4$, the behavior is chaotic.

(d) **Lozi's model** with $a > 0$, $b > 0$

$$x_{k+1} = y_k + 1 - a|x_n|$$
$$y_{k+1} = bx_k.$$

For $a = 0.5$ and $a = 1.7$ this model exhibits chaotic behavior.

(e) **Rössler system**

$$\dot{x} = -(y + z)$$
$$\dot{y} = x + ay$$
$$\dot{z} = b + xz - cz.$$

This system exhibits chaotic behavior for $a = 0.2, b = 0.2$ and $c = 5.7$.

(f) **Lorenz (1990)**

$$\dot{x} = -y^2 - z^2 - ax + aF$$
$$\dot{y} = xy - bxz - y + G$$
$$\dot{z} = bxy + xz - z$$

with $a = 0.25$, $b = 4.0$, $F = 8.0$ and $G = 1$.

32.12 Identify the basin of attraction for each of the systems in Exercise 32.11.

32.13 Compute the first Lyapunov index for each of the systems in Exercise 32.11.

Notes and references

Section 32.2 The coverage of topics in this section is rather standard in a first course in differential equations. For a more detailed treatment of flows in continuous time and their properties refer to Hirsch and Smale (1974). Also see Coddington and Levinson (1955). Holmgren (1994) and Martelli (1992) provide extensive coverage of dynamical systems in discrete time.

Sections 32.3–32.5 The classification of equilibria based on the eigen analysis of the Jacobian is rather standard in nonlinear system theory. Refer to Hirsch and Smale (1974) and Cunningham (1958) for further details. Analysis of the qualitative behavior of nonlinear systems has a long and cherished history and a systematic investigation began with Poincaré at the turn of the century. The now classical Poincaré–Bendixson theorem provides a complete characterization of dynamical system in a plane. In particular it provides a criterion for detecting the presence of limit cycles. (Refer to Hirsch and Smale (1974) for details.) While Poincaré has made references to chaos, the actual demonstration of it happened only in 1963 by Lorenz (1963). Also refer to Lorenz (1993). Today the theory of chaos is rather widely applied in many areas (refer to Kiel (1994), Devaney (1989), and Peitgen, Jürgens and Saupe (1992)).

Analysis of stability of an equilibrium based on the eigenvalues of the Jacobian of the system at the equilibrium has come to be known as the **indirect** method for analyzing stability. There is an alternate method called Lyapunov's **direct** method which has become a standard method for analyzing stability of nonlinear differential equations. For an introduction to Lyapunov's theory refer to LaSalle and Lefschetz (1961) and Hirsch and Smale (1974). Lorenz (1963) contains a fascinating account of the discovery of deterministic chaos in deterministic dynamical systems. For an elaboration of the notion of attractors, and the concept of **strange attractors** refer to Peitgen, Jürgens and Saupe (1992). Certain attractors are called strange because they are endowed with the so-called **fractal dimension**, such as sets of dimension 1.5 as opposed to a line of dimension one or a plane of dimension two, etc. A recipe for determining the dimensions of strange attractors is contained in Martelli (1992) and Peitgen, Jürgens and Saupe (1992).

Parker and Chua (1989), Martelli (1992), and Peitgen, Jürgens, and Saupe (1992) contain an extensive coverage of sensitive dependence on initial condition and the role of Lyapunov index in determining this sensitivity. A complete analysis of the models in Exercise (32.11) is contained in Peitgen, Jürgens and Saupe (1992). Also refer to Lorenz (2005).

Section 32.6 This section follows the developments in Molteni and Palmer (1993), Buizza and Palmer (1995), Ehrendorfer and Tribbia (1997), Legras and Vautard (1996) and Mureau, Molteni and Palmer (1993). Also refer to Barkmeijer et al. (1998).

Section 32.7 The review paper by Eckmann and Ruelle (1985) provides a very readable and a succinct summary of Osledec theorem and its consequences. Fraedrich (1987) contains an interesting discussion of the application of Lyapunov index and other related measures in estimating climate predictability. Algorithms for computing Lyapunov indices are given in Peitgen, Jürgens and Saupe (1992) and Parker and Chua (1989).

Section 32.8 Generation of the deterministic ensemble using the forward singular vectors is described in Molteni and Palmer (1993) and Mureau, Molteni and Palmer (1993). Also refer to Palmer (2000) and Palmer and Hagedorn (2006). The notion of the ensemble generation using the breeding mode is developed in Toth and Kalney (1997). For a discussion of the relation between these approaches refer to Legras and Vautard (1996). Also refer to Anderson (1997). For an interesting discussion on short-range ensemble forecasting refer to the workshop reports by Brooks et al. (1995) and Hamill, Mullen et al. (2000). For a comparison of the performance of various ensemble methods refer to Hamill et al. (2000) and (2003). A parallel implementation of the ensemble Kalman filter is given in Keppenne (2000). For a historical account of the development of ensemble forecasting in meteorology, including a review of current methodology, see Lewis (2005).

Epilogue

It is inspiring to view data assimilation from that epochal moment two hundred years ago when the youthful Carl Friedrich Gauss experienced an epiphany and developed the method of least squares under constraint. In light of the great difficulty that stalwarts such as Laplace and Euler experienced in orbit determination, Gauss certainly experienced the *joie de vivre* of this creative work. Nevertheless, we suspect that even Gauss could not have foreseen the pervasiveness of his discovery.

And, indeed, it is difficult to view data assimilation aside from dynamical systems – that mathematical exploration commenced by Henri Poincaré in the pre-computer age. Gauss had the luxury of performing least squares on a most stable and forgiving system, the two-body problem of celestial mechanics. Poincaré and his successors, notably G. D. Birkhoff and Edward Lorenz, made it clear that the three-body problem was not so forgiving of slight inaccuracies in initial state – evident through their attack on the special three-body problem discussed earlier. Further, the failure of deterministic laws to explain Brownian motion and the intricacies of thermodynamics led to a stochastic–dynamic approach where variables were considered to be random rather than deterministic. In this milieu, probability melded with dynamical law and data assimilation expanded beyond the Gaussian scope.

There is a sense of majesty when working in this most challenging field of dynamic data assimilation. The majesty stems from the coupling of great advances made by the pioneers, and yet the real-world applications continue to present their challenges. For example, in atmospheric chemistry, the evolution of the reactive chemicals, numbering in the hundreds and often poorly observed, must be combined with the governing laws of the atmospheric motion, again in the presence of limited observations, to assimilate data for air quality forecasting. Even more challenging, the living system with its uncertain dynamics and its biochemistry, including the complicated feedback mechanisms that were first studied by Norbert Wiener, and the availability of remote and *in situ* data, offers a problem of Herculean dimension.

In meteorology, we are actively investigating assimilation in the context of ensemble forecasting. This work proceeds at a breakneck pace and there is an excitement – an excitement driven by the realization that forecasts can be improved by insightful data assimilation strategy. Indeed, the forces that motivate the work are not unlike those that spurred Gauss to devise a new methodology to accommodate model and data, and thereby offered guidance to the astronomers who searched for Ceres.

References

Albert, A. E. (1972). *Regression and Moore–Penrose Pseudoinverse*, Academic Press.

Anderson, D. A., J. C. Tannehill, & R. H. Pletcher. (1984). *Computational Fluid Mechanics and Heat Transfer*, Hemisphere Publishing Corporation.

Andersson, E., et al. (1998). "The ECMWF implementation of three-dimensional variational assimilation (3DVAR). II: experimental results". *Quarterly Journal of the Royal Meteorological Society*, **124**, 1831–1860.

Andersson, E., & H. Järvinen. (1999). "Variational quality control". *Quarterly Journal of the Royal Meteorological Society*, **125**, 697–722.

Anderson, J. L. (1997). "The impact of dynamical constraints on the selection of initial conditions for ensemble predictions: low-order perfect model results". *Monthly Weather Review*, **125**, 2969–2983.

(2001). "An ensemble adjustment filter for data assimilation". *Monthly Weather Review*, **129**, 2884–2903.

Andrews, A. (1968). "A square root formulation of the Kalman covariance equations". *American Institute of Aeronautics and Astronautics Journal*, **6**, 1165–1166.

Apostol, T. M. (1957). *Mathematical Analysis*, Addison-Wesley.

Arakawa, A. (1966). "Computational design of long term numerical integration of equations of fluid motion : I two dimensional incompressible flow". *Journal of Computational Physics*, **1**, 119–143.

Armijo, L. (1966). "Minimization of functions having Lipschitz continuous first partial derivatives". *Pacific Journal of Mathematics*, **16**, 1–3.

Arnold, L. (1974). *Stochastic Differential Equations*. Wiley.

Asselin, R. (1972). "Frequency filter for time integrations". *Monthly Weather Review*, **100**, 487–490.

Barbieri, R. W., & P. S. Schopf. (1982). *Oceanic applications of the Kalman filter*, NASA/Goddard Technical Memorandum-TM83993.

Barkmeijer, J., M. van Gijzen, & F. Bouttier. (1998). "Singular vectors and estimate of analysis-error covariance metric". *Quarterly Journal of the Royal Meteorological Society*, **124**, 1695–1713.

Barnes, S. L. (1964). "A technique for maximizing details in numerical weather map analysis". *Journal of Applied Meteorology*, **3**, 396–409.

Barrett, R., M. Berry, T. F. Chan, J. Demmel, J. Donato, J. Dongarra, V. Eijkhout, R. Pozo, C. Romine, & H. van der Vorst. (1994). *Templates for the Solution of Linear Systems: Building Blocks for Iterative Methods*, SIAM.

Bartels, R. H. (1987). *An Introduction to Splines for Use in Computer Graphics and Geometric Modelling*, M. Kaufmann Publishers.
Basilevsky, A. (1983). *Applied Matrix Algebra in the Statistical Sciences*, North-Holland.
Bateman, H. (1932). *Partial Differential Equations*, Cambridge University Press.
Beers, Y. (1957). *Introduction to the Theory of Errors*, Addison-Wesley.
Bell, E. (1937). *Men of Mathematics*, Simon & Schuster.
Bellantoni, J. F. & K. W. Dodge. (1967). "A square root formulation of the Kalman–Schmidt Filter". *American Institute of Aeronautics and Astronautics Journal*, **5**, 1309–1314.
Bellman, R. (1960). *Introduction to Matrix Analysis*, McGraw-Hill.
Bengtsson, L., M. Ghil, & E. Kallen. (1981). *Dynamical Meteorology: Data Assimilation Methods*, Springer-Verlag.
Bennett, A. F. (1992). *Inverse Methods in Physical Oceanography*, Cambridge Monographs on mechanics and applied mathematics, Cambridge University Press.
(2002). *Inverse Modeling of the Ocean and Atmosphere*, Cambridge University Press.
Bennett, A. F., & W. P. Budgell. (1987). "Ocean data assimilation and Kalman filters: spatial regularity". *Journal of Physical Oceanography*, **17**, 1583–1601.
Benton, E., & G. Platzman. (1972). "A table of solutions of the one-dimensional Burgers' equation", *Quarterly of Applied Mathematics*, 195–212.
Bergamini, D., & Editors of Life. (1963). *Mathematics*, Life Science Lib.
Bergman, K. (1979). "Multivariate analysis of temperatures and wind using optimum interpolation". *Monthly Weather Review*, **107**, 1423–1444.
Bergthorsson, P., & B. Döös. (1955). "Numerical weather map analysis". *Tellus*, **7**, 329–340.
Bermaton, J. F. (1985). "Discrete time Galerkin approximations to the nonlinear filtering solution". *Journal of Mathematical Analysis and Applications*, **110**, 364–383.
Bierman, G. J. (1977). *Factorization Methods for Discrete Sequential Estimation*, Academic Press.
Bishop, C. H., B. Etherton, & S. J. Majumdar (2001). "Adaptive sampling with the ensemble Kalman filter part I: theoretical aspects". *Monthly Weather Review*, **129**, 420–236.
Blackwell, D. (1969). *Basic Statistics*, McGraw-Hill.
Bohm, D. (1957). *Causality and Chance in Modern Physics*, Harper and Bros.
Boor, C. de. (1978). *A Practical Guide to Splines*, *Applied Mathematical Sciences*, Vol. 27, Springer-Verlag.
Box, G. E. P., & G. M. Jenkins. (1970). *Time Series Analysis: Forecasting and Control*, Holden-Day.
Bracewell, R. (1965). *The Fourier Transform and Its Applications*, McGraw-Hill.
Brammer, K., & G. Siffling. (1989). *Kalman–Bucy Filters*, Artech House.
Bratseth, A. M. (1986). "Statistical interpolation by means of successive corrections". *Tellus*, **38A**, 439–447.
Brauset, A. M. (1995). *A Survey of Preconditioned Iterative Methods*, Longman Scientific and Technical.
British Meteorological Office. (1961). *Handbook of Meteorological Instruments for Upper Air Observation, Part II*. M. O. 577, Her Majesty's Stationary Office.
Brooks, H. E., M. S. Tracton, D. J. Stensrud, G. DiMego, & Z. Toth. (1995). "Short-range ensemble forecasting: report from a workshop". *Bulletin of the American Meteorological Society*, **76**, 1617–1624.
Broyden, C. G. (1965). "A class of methods of solving nonlinear simultaneous equations". *Mathematics of Computation*, **19**, 577–593.
Bryson, A. E., & Y. C. Ho. (1975). *Applied Optimal Control*, Wiley.
Bucy, R. S. (1965). "Nonlinear filtering". *IEEE Transactions on Automatic Control*, **10**, 198.

(1969). "Bayes theorem and digital realizations for non-linear filters". *The Journal of Astronautical Sciences*, **XVI**, 80–94.

(1970). "Linear and nonlinear filtering". *Proceedings of the IEEE*, **58**, 854–864.

(1994). *Lectures on discrete time filtering*, Springer-Verlag.

Bucy, R. S., & P. D. Joseph. (1968). *Filtering for Stochastic Processes with Applications to Guidance*. Interscience Publications.

Bucy, R. S., & K. D. Senne. (1971). "Digital synthesis of nonlinear filters". *Automatica*, **7**, 287–298.

Budgell, W. P. (1986). "Nonlinear data assimilation for shallow water equations in branched channels". *Journal of Geophysical Research*, **91**, 10,633–10,644.

(1987). "Stochastic filtering of linear shallow water wave process". *SIAM Journal of Scientific and Statistical Computing*, **8**, 152–170.

Buizza, R., & T. Palmer. (1995). "The singular vector structure of the atmosphere circulation". *Journal of Atmospheric Sciences*, **52**, 1434–1456.

Burgers, G., P. J. van Leeuwen, & G. Evensen. (1998). "Analysis scheme in the ensemble Kalman filter". *Monthly Weather Review*, **126**, 1719–1724.

Burgers, J. M. (1939). "Mathematical examples illustrating relations occurring in the theory of turbulent fluid motion". *Transactions of Royal Netherlands Academy of Science*, **17**, 1–53.

(1975). "Some memories of early work in fluid mechanics at the Technical University of Delft". *Annual Review of Fluid Mechanics*, **7**, 1–11.

Cacuci, D. G. (2003). *Sensitivity and Uncertainty Analysis – Theory*, I, Chapman and Hall.

Cahill, A. T., F. Ungaro, M. B. Parlange, M. Mata, & D. R. Nielson. (1999). "Combined spatial and Kalman filter estimation of optimal soil hydraulic properties". *Water Resources Research*, **35**, 1079–1088.

Cañizares, T. R. (1999). "On the application of data assimilation in regional coastal models". Ph.D. Thesis, Delft University.

Carrier, G., & C. Pearson. (1976). *Partial Differential Equations: Theory and Technique*, Academic Press.

Catlin, D. E. (1989). *Estimation, Control, and Discrete Kalman Filter*, Springer-Verlag.

Charney, J. G., R. Fjortoft, & J. von Neumann. (1950). "Numerical integration of barotropic vorticity equation". *Tellus*, **2**, 237–254.

Chilés, J.-P., & P. Delfiner. (1999). *Geostatistics: Modeling Spatial Uncertainty*, Wiley.

Chu, P. C. (1999). "Two kinds of predictability in the Lorenz system". *Journal of Atmospheric Sciences*, **56**, 1427–1432.

Clebsch, A. (1857). "Über eine algemeine Transformation der Hydrodynamischen Gleichungen Crelle". *Journal für Mathematik*, **54(4)**, 293–312.

Coddington, E. A., & N. Levinson. (1955). Theory of Ordinary Differential Equations. McGraw-Hill, New York.

Cohen, I. B. (1960). *The Birth of a New Physics*, Doubleday.

Cohn, S. E. (1982). "Methods of sequential estimation for determining initial data in numerical weather prediction". PhD Thesis, Courant Institute of Mathematical Sciences, New York University.

(1993). "Dynamics of short-term univariate forecast error covariances". *Monthly Weather Review*, **121**, 3123–3149.

(1997). "An introduction to estimation theory". *Journal of the Meteorological Society of Japan*, **75**, 257–288.

Cohn, S. E., M. Ghil, & E. Isaacson. (1981). "Optimal interpolation and Kalman filter". *Proceedings of the Fifth Conference on Numerical Weather Prediction*, AMS, 36–42.

Cohn, S. E., & D. F. Parrish. (1991). "The behavior of forecast error covariance for a Kalman filter in two dimensions". *Monthly Weather Review*, **119**, 1757– 1785.

Cohn, S. E., A. DA Silva, J. Guo, M. Sienkiewicz, & D. Lamich. (1998). "Assessing the effects of data selection with DAO physical-space statistical analysis system". *Monthly Weather Review*, **126**, 2913–2926.

Courtier, P. (1997). "Dual formulation of four-dimensional variational assimilation". *Quarterly Journal of the Royal Meteorological Society*, **123**, 2449–2461.

Courtier, P., E. Andersson, W. Heckley, J. Pailleux, D. Vasiljevic, M. Hamrud, A. Hollingsworth, F. Rabier, & M. Fischer. (1998). "The ECMWF implementation of three-dimensional variational assimilation (3DVAR). I: Formulation". *Quarterly Journal of the Royal Meteorological Society*, **124**, 1783–1807.

Courtier, P., & D. Talagrand. (1990). "Variational assimilation with direct and adjoint shallow water equations". *Tellus*, **42A**, 531–549.

Cressman, G. (1959). "An operational objective analysis system". *Monthly Weather Review*, **87**, 367–374.

Cunningham, W. J. (1958). *Introduction to Nonlinear Analysis*. McGraw-Hill.

Daley, R. (1991). *Atmospheric Data Analysis*, Cambridge University Press.

Daley, R., & E. Barker. (2001). *Monthly Weather Review*, **129**, 869–883.

Daley, R., & R. Ménard. (1993). "Spectral characteristics of Kalman filter systems for atmospheric data assimilation". *Monthly Weather Review*, **121**, 1554–1565.

Davidon, W. C. (1959). *Variable metric methods for minimization*, Argonne National Labs Report, ANL-5990.

(1991). "Variable metric methods for minimization". *SIAM Journal on Optimization*, **1**, 1–17.

Dee, D. P. (1991). "Simplification of the Kalman filter for meteorological data assimilation". *Quarterly Journal of the Royal Meteorological Society*, **117**, 365–384.

Demaria, M. (1996). "A history of hurricane forecasting for the Atlantic Basin". *Historical Essays on Meteorology 1919–1995*, ed. J. R. Fleming, American Meteorological Society, 263–305.

Dennis, J. E., Jr., & R. B. Schnabel. (1996). *Numerical methods for unconstrained optimization and non-linear equations. Classics in Applied Mathematics*, **16**, SIAM.

Derber, J. C. (1989). "A variational continuous assimilation technique". *Monthly Weather Review*, **117**, 2437–2446.

Derber, J. C., & F. Bouttier. (1999). "A reformulation of the background error covariance in the ECMWF global data assimilation system". *Tellus*, **51A**, 195–221.

Derber, J. C., & A. Rosati. (1989). "A global oceanic data assimilation system". *Journal of Physical Oceanography*, **19**, 1333–1339.

Dermanis, A., A. Grün, & F. Sansò. (2000). *Geomatic Methods for the Analysis of Data in the Earth Sciences, Lecture Notes in Earth Sciences*, Vol. 95, Springer-Verlag.

Deutsch, R. (1965). *Estimation Theory*, Prentice Hall.

Devaney, R. (1989). *An Introduction to Chaotic Dynamical Systems*, second edition, Addison-Wesley.

Devenyi, D., & S. G. Benjamin. (1998). "Application of three-dimensional variational analysis in RUC-2". *12th Conference on Numerical Weather Prediction*.

Deyst, J. J., & C. F. Price. (1968). "Conditions for asymptotic stability of the discrete minimum variance linear estimator". *IEEE Transactions on Automatic Control*, **13**, 702–705.

Draper, N., & H. Smith. (1966). *Applied Regression Analysis*, Wiley.

du Plessis, R. (1967). *Poor Man's Exploration of Kalman Filtering or How I Stopped Worrying and Learned to Love Matrix Inversions*, North American Aviation, Inc.

Dunnington, G. (1955). *Carl Friedrich Gauss: Titan of Science*, Hafner.

Eckart, C. (1960). *Hydrodynamics of Ocean and Atmosphere*. Pergamon Press.

Eckmann, J. P., & D. Ruelle. (1985). "Ergodic theory of chaos and strange attractors". *Reviews of Modern Physics*, **57**, 617–656.

Eddy, A. (1967). "The statistical objective analysis of scalar data fields". *Journal of Applied Meteorology*, **6**, 597–609.

Ehrendorfer, M. (1994a). "The Liouville equation and its potential usefulness for the prediction of forecast skill. Part I: Theory". *Monthly Weather Review*, **122**, 703–713.

(1994b). "The Liouville equation and its potential usefulness for the prediction of forecast skill. Part II: Applications". *Monthly Weather Review*, **122**, 714–728.

(1997). "Predicting the uncertainty of numerical weather forecasts: a review". *Meteorologische Zeitschrift, N. F.*, **6**, 147–183.

(2002). "Predictability of atmospheric motions: evidence and question". *Fifth Workshop on Adjoint Applications in Dynamic Meteorology*.

Ehrendorfer, M., & J. J. Tribbia. (1997). "Optimal prediction of forecast error covariance through singular vectors". *Journal of Atmospheric Sciences*, **53**, 286–313.

Einstein, A., & L. Infeld. (1938). *The Evolution of Physics*, Simon and Schuster.

Eliassen, A. (1954). *Provisional report on the calculation of spatial covariance and autocorrelation of the pressure field*, Institute of Weather and Climate Research Academy of Science, Oslo, Norway, Rept. 5, 11.

(1990). *Oral History Interview* (6 Nov. 1990) (Interviewer: J. Green). Royal Meteorological Society.

Enting, I. G. (2002). *Inverse Problems in Atmospheric Constituent Transport*, Cambridge University Press.

Epstein, E. S. (1969a). "The role of initial uncertainties in prediction". *Journal of Applied Meteorology*, **8**, 190–198.

(1969b). "Stochastic dynamics prediction". *Tellus*, **XXI**, 739–759.

Errico, R. M., T. Vukicevic, & K. Rader. (1993). "Examination of the accuracy of the tangent linear model". *Tellus*, **45A**, 462–497.

Errico, R. M. (1997). "What is an adjoint method?" *Bulletin of the American Meteorological Society*, **78**, 2577–2591.

Errico, R. M., & R. Langland. (1999). "Notes on the appropriateness of 'bred modes' for generating initial perturbations used in ensemble prediction". *Tellus*, **51A**, 431–441.

Errico, R. M., & R. Langland. (1999). Reply to Comments on "Notes on appropriateness of 'bred modes' for generating initial perturbations". *Tellus*, **51A**, 450–451.

Evensen, G. (1992). "Using extended Kalman filter with a multilayer quasi-geostrophic ocean model". *Journal of Geophysical Research*, **97**, 17,905–17,924.

(1994). "Sequential data assimilation with a nonlinear quasi-geostrophic model using Monte Carlo methods for forecast error statistics". *Journal of Geophysical Research*, **99**, 10,143–10162.

Evensen, G., & P. J. van Leeuwen. (1996). "Assimilation of Geosat altimeter data for the Agulhas current using the ensemble Kalman filter with a quasi-geostrophic model". *Monthly Weather Review*, **124**, 85–96.

Fagin, S. L. (1964). "Recursive linear regression theory, optimal filter theory and error analysis of optimal systems", *IEEE International Convention Record*, **12**, 216–240.

Feller, W. (1957). *An Introduction to Probability Theory and Its Applications*, Vol. I, Wiley.

Fisher, M., & P. Courtier. (1995). *Estimating the covariance matrices of analysis and forecast error in variational data assimilation, European Centre for Medium Range Weather Forecasts Research Department*, Tech. Memo, 200.

Fitzgerald, R. J. (1971). "Divergence of the Kalman filter". *IEEE Transactions on Automatic Control*, **16**, 736–747.

Fleming, R. J. (1971). "On stochastic dynamic prediction I. The energetics of uncertainty and the question of closure". *Monthly Weather Review*, **99**, 851–872.

Fletcher, R., & M. J. D. Powell. (1963). "A rapidly convergent descent method for minimization". *Computer Journal*, **6**, 163–168.

Fletcher, R., & C. M. Reeves. (1964). "Function minimization by conjugate gradients". *Computer Journal*, **6**, 149–154.

Florchinger, P., & F. LeGland. (1984). "Time discretization of the Zakai equation for diffusion process observed in colored noise". *Analysis and Optimization of Systems*, ed. A. Bensoussan and J. L. Lions. Springer Lecture Notes on Control and Information Sciences, 228–237.

Ford, K. (1963). *The World of Elementary Particles*, Blaisdall.

Fraedrich, K. (1987). "Estimating weather and climate predictability on attractors". *Journal of Atmospheric Sciences*, **44**, 722–728.

Franke, R. (1988). "Statistical interpolation by iteration". *Monthly Weather Review*, **116**, 961–963.

Franke, R., & W. J. Gordon. (1983). *The Structure of optimal interpolation functions*, Technical Report, NPS-53-83-0005, Naval Postgraduate School, Monterey.

Freiberger, W. F., & V. Grenander. (1965). "On the formulation of statistical meteorology". *Review of International Statistical Institute*, **33**, 59–86.

Friedman, A. (1975). *Stochastic Differential Equations and Applications*, Vol. I and II, Academic Press.

Friedman, B. (1956). *Principles and Techniques of Applied Mathematics*, Wiley.

Fukumori, I., & P. Malanotte-Rizzoli. (1994). "An approximate Kalman filter for ocean data assimilation: an example with an idealized Gulf stream model". *Journal of Geophysical Research – Oceans*, **100**, 6777–6793.

Fuller, A. T. (1969). "Analysis of nonlinear stochastic systems by means of the Fokker–Planck equation". *International Journal of Control*, **9**, 603–655.

Gandin, L. S. (1963). *Objective Analysis of Meteorological Fields*, Hydromet Press (Translated from Russian by Israel Program for Scientific Translations, 1965).

Gary, R. M., & J. W. Goodman. (1995). *Fourier Transforms*, Kluwer Academic Publishers.

Gauss, C. (1809). *Theoria Motus Corporum Coelestium in Sectionibus Conicus Solem Ambientium* (Theory of the Motion of Heavenly Bodies Moving about the Sun in Conic Section). An English translation (by C. Davis) published by Little Brown, and Co. in 1857. The publication has been reissued by Dover in 1963, 142 pages.

Gautheir, P., L. Fillion, P. Koclas, & C. Charelet. (1996). "Implementation of a 3-D variational analysis at the Canadian Meteorological Center". *Proceedings of XIAMS Conference on Numerical Weather Prediction*, 19–23.

Gear, C. (1971). *Numerical Initial Value Problems in Ordinary Differential Equations*, Prentice Hall.

Gelb, A. (ed). (1974). *Applied Optimal Estimation*, The MIT Press.

Ghil, M., S. E. Cohen, J. Tavantzis, K. Bube, & E. Isaacson. (1981). "Application of estimation theory to numerical weather prediction". *Dynamic Meteorology: Data Assimilation Methods*, ed. L. Bengtsson, M. Ghil and E. Källén, Springer-Verlag, 139–224.

Ghil, M., K. Ibe, A. Bennett, P. Courtier, M. Kimoto, M. Nagata, M. Saiki, & M. Sato, eds. (1997). *Data Assimilation in Meteorology and Oceanography*, Meteorological Society of Japan.

Ghil, M., & P. Malanotte-Rizzoli. (1991). "Data assimilation in meteorology and oceanography". *Advances in Geophysics*, Academic Press, **33**, 141–265.

Gikhman, I. I., & A. V. Skorokhod. (1972). *Stochastic Differential Equations*, Springer-Verlag.

Gilchrist, B., & G. Cressman. (1954). "An experiment in objective analysis". *Tellus*, **6**, 309–318.

Gill, A. (1982). Atmosphere-Ocean Dynamics, Academic Press.

Gillispie, Ed. C. (1981). "Pierre-Simon Laplace", *Dictionary of Scientific Biography*, Vol. XV, Supp. I, Chas. Scribner's Sons, 273–403.

Gleeson, T. A. (1967). "On theoretical limits of predictability". *Journal of Applied Meteorology*, **6**, 355–359.

Goldstein, A. A. (1967). *Constructive Real Analysis*, Harper & Row.

Golub, G. H. (1965). "Numerical methods for solving linear least squares problems". *Numericsche Mathematik*, **7**, 206–216.

Golub, G., & C. van Loan. (1989). *Matrix Computations*, John Hopkins University Press.

Golub, G., & J. M. Ortega. (1993). *Scientific Computing: An Introduction with Parallel Computing*, Academic Press.

Gould, S. (1985). *The Flamingo's Smile: Reflections in Natural History*, Norton.

Grasman, J. (1999). *Asymptotic Methods for Fokker–Planck Equation and the Exit Problem in Applications*, Springer-Verlag.

Green, C. K. (1946). "Seismic sea wave of April 1, 1946, as recorded on tide gages". *Transactions of the American Geophysical Union*, **27**, 490–500.

Greenbaum, A. (1997). *Iterative Methods for Solving Linear Systems*, SIAM.

Greene, W. H. (2000). *Econometric Analysis*, Prentice Hall.

Griffin, R. E., & A. P. Sage. (1968). "Sensitivity analysis of discrete filtering and smoothing algorithms". *American Institute of Aeronautics and Astronautics Guidance, Control and Flight Dynamics Conference*. Paper No. 68-824.

Grigoriu, M. (2002). *Stochastic Calculus*, Birkhäuser.

Gustafsson, N., P. Lönnberg, & J. Pailleux. (1997). "Data Assimilation for high resolution limited area models". *Journal of the Meteorological Society of Japan*, **75**, 367–382.

Hageman, L. A., & D. M. Young. (1981). *Applied Iterative Methods*, Academic Press.

Hall, T. (1970). *Carl Friedrich Gauss: A Biography*, The MIT Press.

Halmos, P. R. (1958). *Finite Dimensional Vector Spaces*, Van Nostrand.

Hamerseley, J. M., & D. C. Handscomb. (1964). *Monte Carlo Methods*, Methuen and Co.

Hamill, T. M., S. L. Mullen, C. Snyder, Z. Toth, & D. Baumhefner. (2000). "Ensemble forecasting in the short to medium range: report from a workshop". *Bulletin of the American Meteorological Society*, **81**, 2653–2664.

Hamill, T. M., C. Snyder, & R. E. Morss. (2000). "A comparison of probabilistic forecasts from bred, singular-vector and perturbed observation ensembles". *Monthly Weather Review*, **128**, 1835–1851.

Hamill, T. M., C. Snyder, & J. S. Whitaker. (2003). "Ensemble forecasts and the properties of flow-dependent analysis-error covariance singular vectors". *Monthly Weather Review*, **131**, 1741–1758.

Hamilton, J. D. (1994). *Time Series Analysis*, Princeton University Press.

Hamming, R. W. (1989). *Digital Filters*, third edition Dover.

Hanke, M. (1995). *Conjugate Gradient Type Methods for Ill-Posed Problems*, Longman Scientific and Technical.

Hanson, R. J., & C. L. Lawson. (1969). "Extensions and applications of the householder algorithm for solving linear least squares problem". *Mathematics of Computation*, **23**, 787–812.

Hardy, G. (1967). *A Mathematicians Apology*, Cambridge University Press.

Harvey, A. (1989). *Forecasting, Structural Time Series Models and the Kalman Filter*, Cambridge University Press.

Hayden, C. M., & R. J. Purser. (1988). "Three-dimensional recursive filter objective analysis of meteorological fields". *8th Conference on Numerical Weather Prediction*, 185–190.

(1995). "Recursive filter objective analysis of meteorological fields: applications to nesdis operational processing". *Journal of Applied Meteorology*, **34**, 3–16.

Heemink, A. W. (1988). "Two-dimensional shallow water flow identification". *Applied Mathematics and Modelling*, **12**, 109–118.

Heemink, A. W., K. Bolding, & M. Verlaan. (1997). "Storm surge forecasting using Kalman filtering". *Journal of the Meteorological Society of Japan*, **75**, 1B, 305–318.

Heemink, A. W., & H. Kloosterhuis. (1990). "Data assimilation for non-linear tidal models". *International Journal for Numerical Methods in Fluids*, **11**, 1097–1112.

Heemink, A. W., M. Verlaan, & A. J. Segers. (2001). "Variance reduced ensemble Kalman filtering". *Monthly Weather Review*, **129**, 1718–1728.

Henriksen, R. (1980). "A correction of a common error in truncated second-order non-linear filters". *Modelling, Identification and Control*, **1**, 187–193.

Hestenes, M. (1975). *Optimization Theory: The Finite-Dimensional Case*, Wiley.

Hestenes, M. (1980). *Conjugate Direction Methods in Optimization*, Springer-Verlag.

Hestenes, M., & E. Stiefel. (1952). "Methods of conjugate gradients for solving linear systems". *Journal of Reaserch of the National Bureau of Standards*, **29**, 409–439.

Higham, N. J. (1996). *Accuracy and Stability of Numerical Algorithms*, SIAM.

Hirsch, M. W., & S. Smale. (1974). *Differential Equations, Dynamical Systems and Linear Algebra*, Academic Press.

Hoffman, R. N., & E. Kalnay. (1983). "Lagged average forecasting, an alternative to Monte Carlo forecasting". *Tellus*, **35A**, 100–118.

Hollingsworth, A. (1987). In "Short and medium range numerical Weather Prediction", ed. T. Matsuno. *Journal of the Meteorological Society of Japan* (Special Issue 1987), 11–60.

Hollingsworth, A., & P. Lönnberg. (1986). "The statistical structure of short-range forecast errors as determined from radiosonde data. I: the wind field". *Tellus*, **38A**, 111–136.

Holmgren, R. A. (1994). *A First Course in Discrete Dynamical Systems*. Springer-Verlag.

Holton, J. (1972). *An Introduction to Dynamic Meteorology*, Academic Press.

Houghton, J. H. (2002). *The Physics of Atmospheres*, third edition, Cambridge University Press.

Houghton, J. (1991). "The Bakerian Lecture, 1991. The predictability of weather and climate". *Philosophical Transactions of the Royal Society, London*, **337**, 521–572.

Houtekamer, P. L. (1995). "The construction of optimal perturbations". *Monthly Weather Review*, **123**, 2888–2898.

Houtekamer, P. L., & J. Derome. (1995). "Methods for ensemble prediction". *Monthly Weather Review*, **123**, 2181–2196.

Houtekamer, P. L., & H. L. Mitchell. (1998). "Data assimilation using an ensemble Kalman filter technique". *Monthly Weather Review*, **126**, 796–811.

Hoyle, F. (1962). *Astronomy*, Doubleday.

Huang, X. Yu. (2000). "Variational analysis using spatial filters", *Monthly Weather Review*, **128**, 2588–2600.

Ide, K., P. Courtier, M. Ghil, & A. Lorenc. (1997). "Unified notation for data assimilation: operational, sequential, and variational". *Journal of the Meteorological Society of Japan*, **75**, 181–189.

Ingleby, N. B., & A. C. Lorenc. (1993). "Bayesian quality control using multivariate normal distributions". *Quarterly Journal of the Royal Meteorological Society*, **119**, 1195–1225.

Isaacson, E. & H. Keller. (1966). *Analysis of Numerical Methods*, Wiley.
Ito, K. (1944). "Stochastic integrals". *Proceedings of the Imperial Academy*, Tokyo, **Vol.20**, pp. 519–524.
Jacobson, M. (2005). *Fundamentals of Atmospheric Modeling*, Cambridge University Press.
Jazwinski, A. H. (1970). *Stochastic Process and Filtering Theory*, Academic Press.
Jeans, J. (1961). *The Growth of Physical Science*, Fawcett.
Johnston, J., & J. DiNardo. (1997). *Econometric Methods*, McGraw-Hill.
Jordan, C. (1893–1896). *Cours d'analyse de École Polytechnique*, 2nd ed., 1–3, Gauthier-Vallars.
Journel, A. G. (1977). "Kriging in terms of projections" *Mathematical Geology*, **9**, 563–586.
Kailath, T. (1974). "A view of three decades of linear filtering theory". *IEEE Transactions on Information Theory*, **20**, 146–181.
Kallianpur, G. (1980). *Stochastic Filtering Theory*, Springer-Verlag.
Kalman, R. E. (1960). "A new approach to linear filtering and prediction problems". *Transactions of the American Society of Mechanical Engineering, Journal of Basic Engineering Series D*, **82**, 35–45.
Kalman, R. E., & R. S. Bucy. (1961). "New results in linear filtering and prediction theory". *Transactions of the American Society of Mechanical Engineering, Journal of Basic Engineering Series D*, **83**, 95–108.
Kalnay, E. (2003). *Atmospheric Modeling, Data Assimilation, and Predictability*, Cambridge University Press.
Kaminski, P. G., A. E. Bryson, Jr., & S. F. Schmidt. (1971). "Discrete square root filtering: a survey of current techniques". *IEEE Transactions on Automatic Control*, **16**, 727–736.
Kaplan, L. D. (1959). "Influence of atmospheric structure from remote radiation measurements". *Journal of the Optical Society of America*, **49**, 1004–7.
Kayo, I. P., P. Courtier, M. Ghil, & A. Lorenc. (1997). "Unified notation for data assimilation operational, sequential, variational". *Journal of the Meteorological Society of Japan*, **75**, 181–189.
Keppenne, C. L. (2000). "Data assimilation into a primitive-equation model with a parallel ensemble Kalman filter". *Monthly Weather Review*, **128**, 1971–1981.
Kiel, L. Douglas. (1994). *Managing Chaos and Complexity in Government*, Jossey-Bass Publishers.
Knuth, D. E. (1980). *The Art of Computer Programming*, Addison-Wesley.
Kolmogorov, A. N. (1941). "Interpolation, extrapolation of stationary random sequences". *Bulletin of Academy of Sciences, USSR, Series on Mathematics*, Vol. **5**. [Translation by RAND Corporation, memorandum RM-3090-PR April 1962).
Kolmogorov, A. N. & S. V. Fomin. (1975). *Introductory Real Analysis*, Dover.
Krige, D. G. (1951). "A statistical approach to some mine valuations and allied problems on the Witwatersrand". Unpublished Master's thesis, University of Witwatersrand.
Krishnan, V. (1984). *Nonlinear Filtering and Smoothing*, Wiley.
Kushner, H. J. (1962). "On the differential equations satisfied by conditional probability densities of Markov processes with applications". *SIAM Journal on Control*, **2**, 106–119.
(1967). "Approximations to optimal nonlinear filter". *IEEE Transactions on Automatic Control*, **12**, 546–556.
Kushner, H. J., & P. G. Dupuis. (1992). *Numerical Methods for Stochastic Control Problems in Continuous Time*, Springer-Verlag.
Lacarra, J. F., & O. Talagrand. (1988). "Short range evolution of small perturbations in a barotropic model". *Tellus*, **40A**, 81–95.

Lakshmivarahan, S., Y. Honda, & J. M. Lewis. (2003). "Second-order approximation to the 3DVAR cost function: application to analysis/forecast". *Tellus*, **55A**, 371–384.

Lanczos, C. (1970). *Variational Principles of Mechanics*, University of Toronto Press.

Landau, L., & E. Lifshitz. (1959). *Fluid Mechanics*, Pergamon Press.

Larsen, R. J., & M. L. Marx. (1986). *An Introduction to Mathematical Statistics and Its Applications*, Prentice-Hall.

LaSalle, J. P., & S. Lefschetz. (1961). *Stability by Lyapunov's Direct Method*, Academic Press.

Lawson, C. L., & R. J. Hanson. (1995). *Solving Least Squares Problems*, SIAM.

LeDimet, F. X. (1982). *A General Formalism of Variational Analysis*. Cooperative Institute for Mesoscale Meteorological Systems (CIMMS). University of Oklahoma, Report No. 11.

LeDimet, F. X., I. M. Navon, & D. N. Descau. (2002). "Second-order information in data assimilation". *Monthly Weather Review*, **130**, 629–648.

LeDimet, F. X., & O. Talagrand. (1986). "Variational algorithms for analysis and assimilation of meteorological observations, Theoretical aspects". *Tellus*, **38A**, 97–110.

Legras, B., & R. Vautard. (1996). "A Guide to Lyapunov Vectors". *Proceedings of the 1995 ECMWF Seminar on Predictability*, I, 143–156.

Leith, C. (1971). "Atmospheric predictability and two-dimensional turbulence". *Journal of Atmospheric Sciences*, **28**, 145–161.

Leith, C. (1974). "Theoretical skill of Monte Carlo forecasts". *Monthly Weather Review*, **102**, 409–418.

Leith, C., & R. H. Kraichnan. (1972). "Predictability of turbulent flows". *Journal of Atmospheric Sciences*, **19**, 1041–1058.

Lermusiaux, P., & A. Robinson. (1999a). "Data assimilation via error subspace statistical estimation. Part I: theory and schemes". *Monthly Weather Review*, **127**, 1385–1407.

(1999b). "Data assimilation via error subspace statistical estimation. Part II: middle Atlantic bright shelfbreak front simulations and ESSE validation". *Monthly Weather Review*, **127**, 1408–1432.

Levinson, N. (1947a). "The Wiener rms (root mean square) error criterion in filter design and prediction". *Journal of Mathematics and Physics*, **XXV**, **4**, 261–278 [Also see Appendix B in Wiener (1949)].

(1947b). "A heuristic exposition of Wiener's mathematical theory of prediction and filtering". *Journal of Mathematical Physics*, **25**, 110–119. [Also reprinted as Appendix C to Wiener's (1949) book]

Lewis, J. M. (1972). "An upper air analysis using the variational method". *Tellus*, **24**, 514–530.

(1990). *Introduction to Adjoint Method*, lecture notes, NCAR.

(2005). "Roots of ensemble forecasting", *Monthly Weather Review* **133**, 1865–1885.

Lewis, J. M., & J. C. Derber. (1985). "The use of adjoint equations to solve a variational adjustment problem with advective constraints". *Tellus*, **37A**, 309–322.

Lewis, J., & T. Grayson. (1972). "The adjustment of surface wind and pressure by Sasaki's variational matching technique". *Journal of Applied Meteorology*, **11**, 586–597.

Lewis, J. M., K. D. Raeder, & R. M. Errico. (2001). "Vapor flux associated with return flow over the Gulf of Mexico: a sensitivity study using adjoint modeling". *Tellus*, **53A**, 74–93.

Lilly, D. (1968). "Models of cloud-topped mixed layers under a strong inversion", *Quarterly Journal of the Royal Meteorological Society*, **94**, 292–309.

Liptser, R., & A. N. Shiryaev. (1977). *Statistics of Random Processes*, Vol. 1, Springer-Verlag.

(1978). *Statistics of Random Processes*, Vol. 2, Springer-Verlag.

Lorenc, A. C. (1981). "A Global Three-Dimensional Multivariate Statistical Interpolation Scheme". *Monthly Weather Review*, **109**, 701–721.
(1986). "Analysis methods for numerical weather prediction". *Quarterly Journal of the Royal Meteorological Society*, **112**, 1177–1194.
(1988). "Optimal nonlinear objective analysis". *Quarterly Journal of the Royal Meteorological Society*, **114**, 205–240.
(1992). "Iterative analysis using covariance functions and filters". *Quarterly Journal of the Royal Meteorological Society*, **118**, 569–591.
(1995). "Development of an operational variational assimilation scheme". *Journal of the Meteorological Society of Japan*, **75**, 415–420.
(1997). "Development of an operational variational assimilation scheme". *Journal of the Meteorological Society of Japan*, **75**, 339–346.
Lorenc, A. C., & O. Hammon. (1998). "Objective quality control of observations using Bayesian Methods. Theory, and a practical implementation". *Quarterly Journal of the Royal Meteorological Society*, **114**, 515–543.
Lorenz, E. N. (1960). "Maximum simplification of the dynamical equations". *Tellus*, **12**, 243–254.
(1963). "Deterministic non-periodic flow". *Journal of Atmospheric Sciences*, **20**, 130–141.
(1965). "Study of the predictability of a 28-variable atmospheric model". *Tellus*, **17**, 321–333.
(1966). "Atmospheric Predictability". *Advances in Numerical Weather Prediction*, 1965–66 Seminar Series, Travelers Research Center, Inc., 34–39.
(1969). "The predictability of a flow which possesses many scales of motion". *Tellus*, **21**, 289–308.
(1982). "Atmospheric predictability experiments with a large numerical model". *Tellus*, **34**, 505–513.
(1993). *The Essence of Chaos*, University of Washington Press.
(2005). "A look at some details of the growth of initial uncertainties". *Tellus*, **57A**, 1–11.
Luenberger, D. G. (1969). *Optimization in Vector Spaces*, Wiley.
(1973). *Introduction to Linear and Nonlinear Programming*, Addison-Wesley.
Martelli, M. (1992). *Discrete dynamical systems and chaos*. Longman Scientific and Technical (Pitman Monographs).
Matérn, B. (1960). "Spatial variation – stochastic models and their application to some problems in forest surveys and other sampling investigations". *Meddelanden från Statnes Skogsforskningsinstitut*, Vol. 49, No. 5, Almaenna Foerlaget, Stockholm. Springer, Berlin, Heidelberg.
Matheron, G. (1963). *Traité de Geostatisque Appliquée* Vol. 1 and 2, Editions Technip.
Maybeck, P. S. (1979). *Stochastic Models: Estimation and Control*, Vol. 1, Academic Press.
(1982). *Stochastic Models: Estimation and Control*, Vols 2 and 3, Academic Press.
Mehra, P. K. (1972). "Approaches to adaptive filtering". *IEEE Transactions on Automatic Control*, **17**, 693–698.
Melsa, J. L., & D. L. Cohn. (1978). *Decision and Estimation Theory*, McGraw-Hill.
Ménard, R. (1994). "Kalman filtering of Burgers' equation and its application to atmospheric data assimilation". Ph.D. Thesis, McGill University.
Menke, W. (1984). *Geophysical Data Analysis: Discrete Inverse Theory*, Academic Press.
Metropolis, N., & S. Ulam. (1949). "The Monte Carlo method", *Journal of the American Statistical Association*, **44**, 335–341.
Meyer, C. D. (2000). *Matrix Analysis and Applied Linear Algebra*, SIAM.

Miller, R. N. (1986). "Towards the application of the Kalman filter to regional open ocean modelling". *Journal of Physical Oceanography*, **16**, 72–86.

Miller, R. N., E. F. Carter, Jr., & S. T. Blue. (1999). "Data assimilation into nonlinear stochastic models". *Tellus*, **51A**, 167–194.

Miller, R. N., M. Ghil, & F. Gauthiez. (1994). "Advanced data assimilation in strongly nonlinear dynamical systems". *Journal of Atmospheric Sciences*, **51**, 1037–1056.

Miyakoda, K., & O. Talagrand. (1971). "The assimilation of past data in dynamical analysis: I". *Tellus*, **XXIII**, 310–327.

Molteni, F., & T. N. Palmer. (1993). "Predictability and finite-time instability of northern winter circulation". *Quarterly Journal of the Royal Meteorological Society*, **119**, 269–298.

Morf, M., & T. Kailath. (1975). "Square-root algorithms for least-square estimation", *IEEE Transactions on Automatic Control*, **20**, 487–497.

Morf, M., G. S. Sidhu, & T. Kailath. (1974). "Some new algorithms for recursive estimation in constant, linear, discrete-time systems". *IEEE Transactions on Automatic Control*, **19**, 315–383.

Moulton, F. (1902). *An Introduction to Celestial Mechanics*, Macmillan.

Mureau, R., F. Molteni, & T. N. Palmer. (1993). "Ensemble prediction using dynamically conditioned perturbations". *Quarterly Journal of the Royal Meteorological Society*, **119**, 299–323.

Nash, S. G., & A. Sofer. (1996). *Linear and Nonlinear Programming*, McGraw-Hill.

Novikov, E. A. (1959). "Contributions to the problem of predictability of synaptic process". *Bulletin Academy of Sciences, USSR, Geophysics Series* (English ed., AGV), 1209–1211.

Oksendal, B. (2003). *Stochastic Differential Equations*, Springer-Verlag.

Oppenheim, A. V., & R. W. Schaffer. (1975). *Digital Signal Processing*, Prentice Hall.

Orszag, S. A. (1971). "On the elimination of aliasing in finite-difference schemes by filtering high-wavenumber components". *Journal of Atmospheric Sciences*, **28**, 1074.

Ortega, J. M. (1988). *Introduction to Parallel and Vector Solution of Linear Systems*, Plenum Press.

Ortega, J. M. & W. Rheinboldt. (1970). *Iterative Solution of Nonlinear Equations in Several Variables*, Academic Press.

Oseledec, V. I. (1968). "A multiplicative ergodic theorem: Lyapunov characteristic numbers for dynamical systems". *Transactions of the Moscow Mathematical Society*, **19**, 197–231.

Palmer, T. (2000). "Predicting uncertainty in forecasts of weather and climate". Reports on *Progress in Physics*, **63**, 71–116.

Palmer, T., & R. Hagedorn. (eds.) (2006). *Predictability of Weather and Climate*, Cambridge University Press (in press).

Panofsky, H. (1949). "Objective weather map analysis". *Journal of Meteorology*, **5**, 386–392.

Papoulis, A. (1962). *The Fourier Integral and Its Applications*, McGraw-Hill.

(1984). *Probability, Random Variables, and Stochastic Processes*, second edition McGraw-Hill.

Parker, R. L. (1994). *Geophysical Inverse Theory*, Princeton University Press.

Parker, R. L., & L. O. Chua. (1989). *Practical Numerical Algorithms for Chaotic Systems*, Springer-Verlag.

Parrish, D. F., & S. E. Cohn. (1985). "*A Kalman filter for a two-dimensional shallow water model*". Office Note 304, NOAA/NMC.

Parrish, D. F., & J. C. Derber. (1992). "The National Meteorological Center's spectral statistical-interpolation analysis system". *Monthly Weather Review*, **120**, 1747–1764.

Parrish, D., J. Derber, J. Purser, W.-S. Wu, & Z.-X. Pu. (1997). "The NCEP global analysis system: Recent improvements and future plans". *Journal of the Meteorological Society of Japan*, **75**, 359–365.

Pascal, B. (1932). *Pensées*. (Trans. W. Trotter), J. M. Dent & Sons.

Pedlosky, J. (1979). *Geophysical Fluid Dynamics*, Springer-Verlag.

Peitgen, H., H. Jürgens, & D. Saupe. (1992). *Chaos and Fractals: New Frontiers in Science*. Springer-Verlag, New York.

Persson, A. (1998). "How do we understand the Coriolis force?". *Bulletin of the American Meteorological Society*, **79**, 1373–1385.

Peterson, D. P. (1968). "On the concept and implementation of sequential analysis for linear random filters". *Tellus*, **20**, 673–686.

Pfaendtner, J., S. Bloom, D. Lamich, M. Seablom, M. Sienkiewicz, J. Stobie, & A. Da Silva. (1995). *Documentation of the Goddard Earth Observing System (GEOS) Data Assimilation System Version I*. Nasa Technical Memorandom 104606, Vol. 4.

Pham, D. T., J. Verron, & M. C. Roubau. (1998). "A singular evolutive extended Kalman filter for data assimilation in oceanography". *Journal of Marine Systems*, **16**(3–4), 323–340.

Phelps, R., & J. Stein (eds.). (1962). *The German Scientific Tradition*. Holt Reinhart and Winston.

Pierre, D. A., & M. J. Lowe. (1975). *Mathematical Programming via Augmented Lagrangians*, Addison-Wesley.

Pindyck, R. S., & D. L. Rubinfeld. (1998). *Econometric Models and Economic Forecasts*, McGraw-Hill.

Pitcher, E. J. (1977). "Application of stochastic dynamic prediction to real data". *Journal of Atmospheric Sciences*, **34**, 3–21.

Platzman, G. (1964). "An exact integral of complete spectral equations for unsteady one-dimensional flow", *Tellus*, **21**, 422–431.

Platzman, G. (1968). "The Rossby Wave", *Quarterly Journal of the Royal Meteorological Society*, **94**, 225–248.

Poincaré, H. (1952). *Science and Hypothesis*, Dover.

Potter, J. E. & R. G. Stern. (1963). "Statistical filtering of space navigation measurements", *Proceedings of the 1963 AIAA Guidance and Control Conference*.

Price, C. F. (1968). "An analysis of the divergence problem in the Kalman filter". *IEEE Transactions on Automatic Control*, **13**, 699–702.

Proudman, J. (1953). *Dynamical Oceanography*, *Methuen*.

Purser, R. J. (1984). "A new approach to optimal assimilation of meteorological data by iterative Bayesian analysis". *Preprints of the 10th Conference on Weather forecasting and analysis. American Meteorological Society*, 102–105.

(1987). "Filtering meteorological fields". *Journal of Climate and Applied Meteorology*, **26**, 1764–1769.

(2005). *A geometrical approach to the synthesis of smooth anisotropic covariance operators for data assimilation*, US Department of Commerce, NOAA, Maryland.

Purser, R. J., & R. McQuigg. (1982). *A successive correction analysis scheme using recursive numerical filters*, Meteorological Office 011 Technical Report No.154.

Purser, R. J., W-S. Wu, D. F. Parrish, & N. M. Roberts (2003a). "Numerical aspects of the application of recursive filters to variational statistical analysis: Part I, spatially homogeneous and isotropic Gaussian covariances". *Monthly Weather Review*, **131**, 1524–1535.

(2003b). "Numerical aspects of the application of recursive filters to variational statistical analysis: Part II, spatially inhomogeneous and anisotropic general covariances". *Monthly Weather Review*, **131**, 1536–1548.

Rabier, P., A. McNally, E. Andersson, P. Courtier, P. Undén, A. Hollingsworth, & F. Bouttier. (1998). "The ECMWF implementation of three-dimensional variational assimilation (3DVAR). II: Structure Functions". *Quarterly Journal of the Royal Meteorological Society*, **124**, 1809–1829.

Rao, C. R. (1945). "Information and Accuracy Attainable in the Estimation of Statistical Parameters". *Bulletin of the Calcutta Mathematical Society*, **37**, 81–91.

(1973). *Linear Statistical Inference and Its Applications*, Wiley.

Rao, C. R., & S. K. Mitra. (1971). *Generalized Inverses of Matrices and its Applications*, Wiley.

Raymond, W. H. (1988). "High-order low-pass implicit tangent filters for use in finite area calculations", *Monthly Weather Review*, **116**, 2132–2141.

Raymond, W. H., & A. Garder. (1988). "A spatial filter for use in finite area calculations", *Monthly Weather Review*, **116**, 209–222.

(1991). "A review of recursive and implicit filters". *Monthly Weather Review*, **119**, 477–495.

Reddy, J. N., & D. K. Gartling. (2001). *The Finite Element Method in Heat Transfer and Fluid Mechanics*, CRC Press.

Rees, C. S., S. M. Sha, & Č. V. Stanojeviċ. (1981). *Theory and Applications of Fourier Analysis*, Marcel Dekker.

Reich, K. (1985). *Carl Friedrich Gauss 1777–1855* (in German), Moss & Partner.

Richardson, L. F. (1922). *Weather Prediction by Numerical Process*, Cambridge University Press, reprinted Dover 1965.

Richardson, L., & H. Stommel. (1948). "Note on eddy diffusion in the sea". *Journal of Meteorology*, **5**, 238–240.

Richtmyer, R. (1957). *Difference Methods for Initial-Value Problems*, Interscience Publications.

(1963). *A Survey of Difference Methods for Non-Steady Fluid Dynamics*, Nat. Cent. Atmos. Res. (NCAR), Tech. Notes, 63–2.

Richtmyer, R., & K. W. Morton. (1957). *Difference Methods for Initial Value Problem*, Interscience Publications.

Risken, H. (1984). *The Fokker–Planck Equation: Methods of Solution and Applications*, Springer-Verlag.

Robert, A. J. (1966). "The integration of a low order spectral form of the primitive meteorological equations". *Journal of the Meteorological Society of Japan*, Ser.2, **44**, 237–245.

Rockafellar, R. T. (1970). *Convex Analysis*, Princeton University Press.

Rossby, C., & Staff Members. (1939). "Relation between variations in the intensity of the zonal circulation of the atmosphere and the displacement of the semi-permanent centers of action", *Journal of Marine Systems*, **2**, 38–55.

Rutherford, I. (1972). "Data assimilation by statistical interpolation of forecast error fields". *Journal of Atmospheric Sciences*, 809–815.

Sage, A. P., & J. L. Melsa. (1971). *Estimation Theory with Applications to Communications and Control*, McGraw-Hill.

Sanders, F., & R. Burpee. (1968). "Experiments in barotropic hurricane track forecasting". *Journal of Applied Meteorology*, **7**, 313–323.

Sasaki, Y. (1955). "The fundamental study of the numerical prediction based on the variational principle". *Journal of the Meteorological Society of Japan*, **33**, 262–275.

(1958). "An objective analysis based on the variational method". *Journal of the Meteorological Society of Japan*, **36**, 77–88.

(1969). "Proposed inclusion of time variation terms, observational and theoretical, in numerical variational objective analysis", *Journal of the Meteorological Society of Japan*, **47**, 115–124.

(1970). "Some Basic Formalisms in Numerical Variational Analysis". *Monthly Weather Review*, **98**, 875–883.

Saaty, T. L. (1967). *Modern Nonlinear Equations*, McGraw-Hill chapter 8.

Scales, J. A., & R. Snieder. (2000). "The Anatomy of Inverse Problems". *Geophysics*, **65**, 1708–1710.

Schlatter, T. W. (1975). "Some experiments with a multivariate statistical objective analysis scheme". *Monthly Weather Review*, **103**, 246–257.

Schlee, F. H., C. J. Standish, & N. F. Toda. (1967). "Divergence in the Kalman filter". *American Institute of Aeronautics and Astronautics Journal*, **5**, 1114–1122.

Schwartz, L., & E. B. Stear. (1968). "A computational comparison of several nonlinear filters". *IEEE Transactions on Automatic Control*, **13**, 83–86.

Schweppe, F. C. (1973). *Uncertain Dynamic Systems*, Prentice Hall.

Seaman, R. S. (1977). "Absolute and differential accuracy of analysis achievable with specified observational network characteristics". *Monthly Weather Review*, **105**, 1211–1222.

(1988). "Some real data tests of the interpolation accuracy of Bratseth's successive correction method". *Tellus*, **40A**, 173–176.

Segers, A. (2002). *Data Assimilation in Atmospheric Chemistry Models Using Kalman Filtering*, Delft University Press.

Segers, A. J., A. W. Heemink, M. Verlaan, & M. van Loan. (2000). "A modified RRSQRT-filter for assimilating data in atmospheric chemistry models". *Environmental Modelling and Software*, **15**, 663–671.

Shapiro, R. (1970). "Smoothing, filtering, and boundary effects". *Reviews in Geophysics and Space Physics*, **8**, 359–387.

(1975). "Linear filtering". *Mathematics of Computation*, 1094–97.

Shuman, F. G. (1957). "Numerical methods in weather prediction: II smoothing and filtering". *Monthly Weather Review*, 357–361.

Sikorski, R. (1969). *Advanced Calculus: Functions of Several Variables*, Polish Scientific Publishers.

Snoog, T. T. (1973). *Random Differential Equations in Science and Engineering*, Academic Press.

Sorenson, H. W. (1966). "Kalman filtering techniques" in *Advances in Control Systems*, ed. C. T. Leondes, Academic Press, 219–292.

(1970). "Least squares estimation: from Gauss to Kalman". *IEEE Spectrum*, July, 63–68.

(1980). *Parameter Estimation: Principles and Practice*, Marcel Dekker.

Sorenson, H. W. (ed). (1985). *Kalman Filtering: Theory and Applications*, IEEE Press.

Sorenson, H. W., & D. L. Alspach. (1971). "Recursive Bayesian estimation using Gaussian sums". *Automatica*, **7**, 465–479.

Sorenson, H. W., & A. R. Stubberud. (1968). "Non-linear filtering by approximation of the *a posteriori* density". *International Journal of Control*, **8**, 33–51.

Stewart, G. W. (1973). *Introduction to Matrix Computations*, Academic Press.

Stratonovich, R. L. (1962). "Conditional Markov process". *Theory of Probability and Applications*, **5**, 156–178.

Struik, D. (1967). *A Concise History of Mathematics*, third revised edition, Dover.

Sun, J., W. Ficker, & D. Lilly. (1991). "Recovering three-dimensional wind and temperature fields from simulated single-Dropper radar data". *Journal Atmospheric Sciences*, **48**, 876–890.

Sverdrup, H., M. Johnson, & R. Fleming. (1942). *The Oceans*, Prentice Hall.

Swirling, P. (1959). "First-order error propagation in a stagewise smoothing procedure for satellite observations". *Journal of Astronautical Sciences*, **6**, 46–52.

(1971). "Modern state estimation methods from the viewpoint of the method of least squares". *IEEE Transactions on Automatic Control*, **16**, 707–719.

Talagrand, O. (1991). "The use of adjoint equations in numerical modeling of the atmospheric circulation". *Automatic Differentiation of Algorithms: Theory, Implementation and Application*, ed. A. Griewank & G. Corleiss, SIAM, 169–180.

Talagrand, O., & P. Courtier. (1987). "Variational assimilation of meteorological observations with the adjoint vorticity equation. Part I: Theory", *Quarterly Journal of the Royal Meteorological Society*, **113**, 1311–1328.

(1993). "Variational assimilation of conventional meteorological observations with multi-level primitive equation model". *Quarterly Journal of the Royal Meteorological Society*, **119**, 153–186.

Tarantola, A. (1987). *Inverse Problems Theory*, Elsevier.

Thacker, W. C. (1989). "The role of the Hessian matrix in fitting models to measurements". *Journal of Geophysical Research*, **94**, 6177–6196.

Thacker, W. C., & R. B. Long. (1988). "Fitting dynamics to data". *Journal of Geophysical Research*, **93**, 1127–1240.

Thiébaux, H. J., & M. A. Pedder. (1987). *Spatial Objective Analysis*, Academic Press.

Thompson, P. D. (1957). "Uncertainty of initial state as a factor in the predictability of large scale atmospheric flow patterns". *Tellus*, **9**, 275–295.

(1969). "Reduction of analysis error through constraints of dynamical consistency", *Journal of Applied Meteorology*, **8**, 738–742.

(1985a). "Prediction of probable errors in prediction". *Monthly Weather Review*, **113**, 248–259.

(1985b). "A review of the predictability problem", in G. Hollway, & B. J. West, eds., *Predictability of Fluid Motions*, American Institute of Physics, 1–10.

Tikhonov, A. & V. Arsenin. (1977). *Solutions of Ill-Posed Problems*, Wiley.

Tippett, M. K., J. L. Anderson, T. M. Hamill, & J. S. Whitaker. (2003). "Ensemble square root filters". *Monthly Weather Review*, **131**, 1485–1490.

Todling, R., & S. E. Cohn. (1994). "Suboptimal schemes for atmospheric data assimilation based on Kalman filter". *Monthly Weather Review*, **122**, 2530–2557.

Toth, Z., & E. Kalnay. (1997). "Ensemble forecasting at NCEP and breeding method". *Monthly Weather Review*, **125**, 3297–3319.

Toth, Z. , I. Szunyogh, E. Kalnay, & G. Iyengar (1999). "Comments on: notes on appropriateness of 'bred modes' for generating initial perturbations". *Tellus*, **51A**, 442–449.

Trapp, R. J., & C. A. Doswell. (2000). "Radar data objective analysis". *Journal of Atmospheric and Oceanic Technology*, **17**, 105–120.

Trefethen, L. N., & D. Bau III. (1997). *Numerical Linear Algebra*, SIAM.

Turbull, H. (1993). *The Great Mathematicians*, Barnes & Noble.

Varga, R. (2000). *Matrix Iterative Analysis*, second edition, Springer-Verlag.

Verlann, M. (1998). Efficient Kalman Filtering Algorithms for Hydrodynamical Models, Ph.D. thesis, Delft University of Technology.

Verlaan, M., & A. W. Heemink. (1997). "Tidal flow forecasting using reduced rank square root filters". *Stochastic Hydrology and Hydraulics*, **11**, 349–368.

(2001). "Nonlinearity in data assimilation: a practical method for analysis". *Monthly Weather Review*, **129**, 1578–1589.

Verron, J., L. Gourdeau, D. Pham, R. Murtugudde, & A. Busalacchi. (1999). "An extended Kalman filter to assimilate satellite altimeter data into a nonlinear numerical model of the tropical pacific ocean: method and violation". *Journal of Geophysical Research*, **104**(c3), 5441–5458.

Voorrips, A. C., A. W. Heemink, & G. J. Komen. (1999). "Wave data assimilation with Kalman filter". *Journal of Marine Systems*, **19**, 267–291.

Vukicevic, T., T. Greenwald, M. Zupanski, D. Zupanski, T. VonderHarr, & A. Jones. (2004). "Mesoscale cloud state estimation from visible and infrared satellite radiances". *Monthly Weather Review*, **132**, 3066–3077.

Wahba, G., & J. Wendelberger. (1980). "Some New Mathematical Methods for Variational Objective Analysis using Splines and Cross Validation". *Monthly Weather Review*, **108**, 1122–1143.

Wald, A. (1947). *Sequential Analysis*, Wiley.

Walker, J. S. (1988). *Fourier Analysis*, Oxford University Press.

Wang, Yunheng, & S. Lakshmivarahan. (2004). *A fourth order approximation to nonlinear filters in discrete time*, Technical Report, School of Computer Science, University of Oklahoma.

Wang, Z., K. Droegemeier, L. White, & I. M. Navon. (1997). "Application of a new adjoint Newton algorithm to the 3D ARPS storm scale model using simulated data". *Monthly Weather Review*, **125**, 2460–2478.

Wang, Z., I. M. Navon, F. X. LeDimet, & X. Zhou. (1992). "The second-order adjoint analysis: theory and application". *Meteorological and Atmospheric Physics*, **50**, 3–20.

Wang, Z., I. M. Navon, X. Zhou, & F. X. LeDimet. (1995). "A truncated Newton optimization algorithm in meteorological application with analytic Hessian/vector products". *Computer in Optimization and Applications*, **4**, 241–262.

Weaver, A., & P. Courtier. (2001). "Correlation modelling on the sphere using a generalized diffusion equation". *Quarterly Journal of the Royal Meteorological Society*, **127**, 1815–1846.

Whitaker, J. S., & T. M. Hamill. (2002). "Ensemble Data Assimilation without Perturbed Observations". *Monthly Weather Review*, **130**, 1913–1924.

Whittlesey, J. R. B. (1964). "A rapid method for digital filtering". *Communications of ACM*, **7**, 552–556.

Wiener, N. (1949). *Extrapolation, Interpolation and Smoothing of Stationary Time Series with Engineering Applications*, Wiley. [This was originally published as a classified defense document in February 1942].

Wiin-Nielson, A. (1991). "The birth of numerical weather prediction". *Tellus*, **43A**, 36–52.

Wilkinson, J. H. (1965). *The Algebraic Eigenvalue Problem*, Clarendon Press.

Wishner, R. P., J. A. Tabaczynski, & M. Athans. (1969). "A comparison of three non-linear filters". *Automatica*, **5**, 487–496.

Wolfram, S. (2002). *A New Kind of Science*, Wolfram Media.

Woolard, E. W. (1940). "The calculation of planetary motions". *National Mathematics Magazine*, **14**, 179–189.

Wu, W. S, R. J. Purser, & D. F. Parrish. (2002). "Three dimensional variational analysis with spatially inhomogeneous covariances". *Monthly Weather Review*, **130**, 2905–2916.

Wunsch, C. (1996). *The Ocean Circulation Inverse Problem*, Cambridge University Press.

Yaglom, A. M. (1962). *The Theory of Stationary Random Functions*, (translated from Russian by R. A. Silverman), Prentice Hall.

Young, D. M. (1971). *Iterative Solution of Large Linear Systems*, Academic Press.

Zakai, M. (1969). "On the optimal filtering of diffusion processes". *Zeitschrift fur Wahrscheinlichkeitstheorie und Verwandte Gebiete*, **11**, 230–243.

Zheng, S. W., C. J. Qiu, & Q. Xu. (2004). "Estimating soil water contents from soil temperature measurements by using an adaptive Kalman filter". *Journal of Applied Meteorology*, **43**, 379–389.

Zienkiewicz, D. C., & R. L. Taylor. (2000). *The Finite Element Method*. Fifth edition, Vols 1, 2 and 3, Butterworth-Heinemann.

Zupanski, D., M. Zupanski, D. Parrish, E. Rogers, & G. DiMego. (2002). "Fine resolution 4D-Var data assimilation for the blizzard of 2000". *Monthly Weather Review*, **130**, 1967–1988.

Index